Lecture Notes in Artificial Intelligence 3097

Edited by J. G. Carbonell and J. Siekmann

Subseries of Lecture Notes in Computer Science

T0181286

Springer

Berlin
Heidelberg
New York
Hong Kong
London
Milan
Paris
Tokyo

David Basin   Michaël Rusinowitch (Eds.)

Automated
Reasoning

Second International Joint Conference, IJCAR 2004
Cork, Ireland, July 4-8, 2004
Proceedings

 Springer

Series Editors

Jaime G. Carbonell, Carnegie Mellon University, Pittsburgh, PA, USA
Jörg Siekmann, University of Saarland, Saarbrücken, Germany

Volume Editors

David Basin
ETH Zentrum, Department of Computer Science
8092 Zurich, Switzerland
E-mail: basin@inf.ethz.ch

Michaël Rusinowitch
LORIA and INRIA Lorraine
Nancy, France
E-mail: Michael.Rusinowitch@loria.fr

Library of Congress Control Number: 2004107782

CR Subject Classification (1998): I.2.3, F.4.1, F.3, F.4, D.2.4

ISSN 0302-9743
ISBN 3-540-22345-2 Springer-Verlag Berlin Heidelberg New York

Springer-Verlag is a part of Springer Science+Business Media

springeronline.com

© Springer-Verlag Berlin Heidelberg 2004
Printed in Germany

Typesetting: Camera-ready by author, data conversion by PTP-Berlin, Protago-TeX-Production GmbH
Printed on acid-free paper   SPIN: 11018285   06/3142   5 4 3 2 1 0

Preface

This volume constitutes the proceedings of the *2nd International Joint Conference on Automated Reasoning* (IJCAR 2004) held July 4–8, 2004 in Cork, Ireland. IJCAR 2004 continued the tradition established at the first IJCAR in Siena, Italy in 2001, which brought together different research communities working in automated reasoning. The current IJCAR is the fusion of the following conferences:

CADE: The International Conference on Automated Deduction,

CALCULEMUS: Symposium on the Integration of Symbolic Computation and Mechanized Reasoning,

FroCoS: Workshop on Frontiers of Combining Systems,

FTP: The International Workshop on First-Order Theorem Proving, and

TABLEAUX: The International Conference on Automated Reasoning with Analytic Tableaux and Related Methods.

There were 74 research papers submitted to IJCAR as well as 12 system descriptions. After extensive reviewing, 26 research papers and 6 system descriptions were accepted for presentation at the conference and publication in this volume. In addition, this volume also contains papers from the three invited speakers and a description of the CADE ATP system competition.

We would like to acknowledge the enormous amount of work put in by the members of the program committee, the various organizing and steering committees, the IJCAR officials, the invited speakers, and the additional referees named on the following pages. We would also like to thank Achim Brucker and Barbara Geiser for their help in producing this volume.

May 2004   David Basin, Michaël Rusinowitch

Conference Chair and Local Organization

Toby Walsh (UCC, Ireland)
Barry O'Sullivan (UCC, Ireland)

Program Committee Chairs

David Basin (ETH Zürich, Switzerland)
Michaël Rusinowitch (LORIA and INRIA Lorraine, France)

Program Committee

Alessandro Armando	University of Genoa, Italy
Franz Baader	TU Dresden, Germany
Christoph Benzmüller	Saarland University, Germany
Armin Biere	ETH Zürich, Switzerland
Maria Paola Bonacina	Università degli Studi di Verona, Italy
Ricardo Caferra	LEIBNIZ-IMAG, France
Marta Cialdea Mayer	University of Rome, Italy
Nachum Dershowitz	Tel Aviv University, Israel
David Dill	Stanford University, USA
Amy Felty	University of Ottawa, Canada
Rajeev Goré	ANU Canberra, Australia
Bernhard Gramlich	TU Wien, Austria
Philippe de Groote	INRIA, France
Reiner Hähnle	Chalmers University, Sweden
Andreas Herzig	IRIT, France
Ian Horrocks	University of Manchester, UK
Jieh Hsiang	National Taiwan University, Taiwan
Deepak Kapur	University of New Mexico, USA
Claude Kirchner	LORIA and INRIA Lorraine, France
Reinhold Letz	TU München, Germany
Chris Lynch	Clarkson University, USA
Aart Middeldorp	University of Innsbruck, Austria
Hans Jürgen Ohlbach	LMU München, Germany
Paliath Narendran	SUNY Albany, USA
Tobias Nipkow	TU München, Germany
Leszek Pacholski	Wroclaw University, Poland
Frank Pfenning	Carnegie Mellon University, USA
David Plaisted	University of North Carolina, USA
Roberto Sebastiani	University of Trento, Italy

John Slaney	ANU Canberra, Australia
Viorica Sofronie-Stokkermans	Max-Planck-Institut, Germany
Ashish Tiwari	SRI International, USA
Ralf Treinen	ENS Cachan, France
Andrei Voronkov	University of Manchester, UK
Wolfgang Windsteiger	RISC Linz, Austria

Invited Speakers

Georg Gottlob (TU Wien, Austria)
José Meseguer (University of Illinois at Urbana-Champaign, USA)
Volker Weispfenning (University of Passau, Germany)

IJCAR Officials

Conference Chair: Toby Walsh (UCC, Ireland)
Program Committee Chairs:
  David Basin (ETH Zürich, Switzerland)
  Michaël Rusinowitch (LORIA and INRIA Lorraine, France)
Workshop Chair: Peter Baumgartner (Max-Planck-Institut, Germany)
Tutorial Chair: William Farmer (McMaster University, Canada)
Publicity Chair:
  Maria Paola Bonacina (Università degli Studi di Verona, Italy)
IJCAR Steering Committee:
  Alessandro Armando (University of Genoa, Italy)
  David Basin (ETH Zürich, Switzerland)
  Christoph Benzmüller (Saarland University, Germany)
  Maria Paola Bonacina, Coordinator (Università degli Studi di Verona, Italy)
  Ulrich Furbach (University of Koblenz-Landau, Germany)
  Reiner Hähnle (Chalmers University, Sweden)
  Fabio Massacci (University of Trento, Italy)
  Michaël Rusinowitch (LORIA and INRIA Lorraine, France)
  Toby Walsh (UCC, Ireland)
Local Organization: Barry O'Sullivan (UCC, Ireland)
Web page: Brahim Hnich (UCC, Ireland)
Registration: Eleanor O'Hanlon (UCC, Ireland)

IJCAR Sponsors

IJCAR gratefully acknowledges the sponsorship of:
  Science Foundation Ireland
  Cork Constraint Computation Centre
  University College Cork

Additional Referees

Pietro Abate
Husam Abu-Haimed
Andrew A. Adams
Peter Andrews
Jürgen Avenhaus
Jeremy Avigad
Arnaud Bailly
Sebastian Bala
Gertrud Bauer
Peter Baumgartner
Bernhard Beckert
Arnold Beckmann
Ramon Bejar
Sergey Berezin
Thierry Boy de la Tour
Chad Brown
Bruno Buchberger
Claudio Castellini
Serenella Cerrito
Iliano Cervesato
Witold Charatonik
Szu-Pei Chen
Alessandro Cimatti
Luca Compagna
Evelyne Contejean
Karl Crary
Stéphane Demri
Michael Dierkes
Jürgen Dix
Roy Dyckhoff
Germain Faure
Christian Fermüller
Murdoch Gabbay
Olivier Gasquet
Thomas Genet
Rosella Gennari
Silvio Ghilardi
Giuseppe de Giacomo
Laura Giordano

Jean Goubault-Larrecq
Elsa L. Gunter
Volker Haarslev
Ziyad Hanna
John Harrison
Miki Hermann
Thomas Hillenbrand
Nao Hirokawa
Joe Hurd
Ullrich Hustadt
Predrag Janičić
Tudor Jebelean
Tommi Juntilla
Lukasz Kaiser
Jaap Kamps
Emanuel Kieroński
Michael Kohlhase
Tomasz Kowalski
Temur Kutsia
Francis Kwong
Lei Li
Denis Lugiez
Carsten Lutz
Jacopo Mantovani
Felip Manya
João Marcos
Mircea Marin
William McCune
Andreas Meier
Gopalan Nadathur
Robert Nieuwenhuis
Hans de Nivelle
Greg O'Keefe
Michio Oyamaguchi
Jeff Pan
Fabrice Parennes
Lawrence Paulson
Nicolas Peltier
Martin Pollet

François Puitg
Stefan Ratschan
Antoine Reilles
Alexandre Riazanov
Christophe Ringeissen
Riccardo Rosati
Markus Rosenkranz
Pawel Rychlikowski
Mooly Sagiv
Sriram Sankaranayanan
Ulrike Sattler
Francesco Savelli
Steffen Schlager
Manfred
  Schmidt-Schauss
Christian Schulte
Klaus U. Schulz
Johann Schumann
Maria Sorea
Gernot Stenz
Jürgen Stuber
Lidia Tendera
Sergio Tessaris
Tinko Tinchev
Cesare Tinelli
Stefano Tonetta
Tomasz Truderung
Xavier Urbain
Sandor Vagvolgyi
Miroslav Velev
Laurent Vigneron
Lida Wang
Freek Wiedijk
Claus-Peter Wirth
Richard Zach
Hantao Zhang
Yunshan Zhu

Table of Contents

Competition

Rewriting Logic Semantics: From Language Specifications to Formal Analysis Tools

José Meseguer and Grigore Roşu

University of Illinois at Urbana-Champaign, USA

Abstract. Formal semantic definitions of concurrent languages, when specified in a well-suited semantic framework and supported by generic and efficient formal tools, can be the basis of powerful software analysis tools. Such tools can be obtained *for free* from the semantic definitions; in our experience in just the few weeks required to define a language's semantics even for large languages like Java. By combining, yet distinguishing, both equations and rules, rewriting logic semantic definitions unify both the semantic equations of equational semantics (in their higher-order denotational version or their first-order algebraic counterpart) and the semantic rules of SOS. Several limitations of both SOS and equational semantics are thus overcome within this unified framework. By using a high-performance implementation of rewriting logic such as Maude, a language's formal specification can be automatically transformed into an efficient interpreter. Furthermore, by using Maude's breadth first search command, we also obtain for free a semi-decision procedure for finding failures of safety properties; and by using Maude's LTL model checker, we obtain, also for free, a decision procedure for LTL properties of finite-state programs. These possibilities, and the competitive performance of the analysis tools thus obtained, are illustrated by means of a concurrent Caml-like language; similar experience with Java (source and JVM) programs is also summarized.

1 Introduction

Without a precise *mathematical semantics* compiler writers will often produce incompatible language implementations; and it will be meaningless to even attempt to formally verify a program. Formal semantics is not only a necessary *prerequisite* to any meaningful talk of software correctness, but, as we try to show in this paper, it can be a key technology to develop powerful software analysis tools. However, for this to happen in practice we need to have:

- a well-suited semantic framework, and
- a high performance implementation of such a framework.

We argue that rewriting logic is indeed a well-suited and flexible framework to give formal semantics to programming languages, including concurrent ones.

D. Basin and M. Rusinowitch (Eds.): IJCAR 2004, LNAI 3097, pp. 1–44, 2004.
© Springer-Verlag Berlin Heidelberg 2004

In fact we show that it unifies two well-known frameworks, namely equational semantics and structural operational semantics, combining the advantages of both and overcoming several of their respective limitations.

High performance is crucial to scale up both the execution and the formal analysis. In this regard, the existence of the Maude 2.0 system [19] implementing rewriting logic and supporting efficient execution as well as breadth-first search and LTL model checking, allows us to automatically turn a language's rewriting logic semantic definition into a quite sophisticated software analysis tool for that language *for free*. In particular, we can efficiently interpret programs in that language, and we can formally analyze programs, including concurrent ones, to find safety violations and to verify temporal logic properties by model checking.

The fact that rewriting logic specifications provide in practice an easy way to develop executable formal definitions of languages, which can then be subjected to different tool-supported formal analyses, is by now well established [83, 8,84,78,74,45,80,16,65,81,27,26,38]. However, ascertaining that this approach can scale up to large conventional languages such as Java and the JVM [27,26], and that the generic formal analysis methods associated to semantic definitions can compete in performance with special-purpose analysis tools developed for individual languages, is a more recent development that we have been investigating with our students and for which we give evidence in this paper.

1.1 Semantics: Equational Versus SOS

Two well-known semantic frameworks for programming languages are: equational semantics and structural operational semantics (SOS).

In *equational semantics*, formal definitions take the form of *semantic equations*, typically satisfying the *Church-Rosser* property. Both higher-order (denotational semantics) and first-order (algebraic semantics) versions have been shown to be useful formalisms. There is a vast literature in these two areas that we do not attempt to survey. However, we can mention some early denotational semantics papers such as [75,67] and the survey [56]. Similarly, we can mention [89,31,12] for early algebraic semantics papers, and [30] for a recent textbook.

We use the more neutral term *equational semantics* to emphasize the fact that denotational and algebraic semantics have many common features and can both be viewed as instances of a common equational framework. In fact, there isn't a rigid boundary between both approaches, as illustrated, for example, by the conversion of higher-order semantic equations into first-order ones by means of explicit substitution calculi or combinators, the common use of initiality in both initial algebras and in solutions of domain equations, and a continuous version of algebraic semantics based on continuous algebras.

Strong points of equational semantics include:

– it has a *model-theoretic*, denotational semantics given by *domains* in the higher-order case, and by *initial algebras* in the first-order case;

- it has also a *proof-theoretic*, operational semantics given by *equational reduction* with the semantic equations;
- semantic definitions can be easily turned into efficient interpreters, thanks to efficient higher-order functional languages (ML, Haskell, etc.) and first-order equational languages (ACL2, OBJ, ASF+SDF, etc.);
- there is good higher-order and first-order theorem proving support.

However, equational semantics has the following drawbacks:

- it is well suited for *deterministic* languages such as conventional sequential languages or purely functional languages, but is quite poorly suited to define the semantics of *concurrent languages*, unless the concurrency is that of a purely deterministic computation;
- one can *indirectly model*[1] some concurrency aspects with devices such as a scheduler, or lazy data structures, but a direct comprehensive modeling of all concurrency aspects remains elusive within an equational framework;
- semantic equations are typically *unmodular*, i.e., adding new features to a language often requires *extensive redefinition* of earlier semantic equations.

In SOS formal definitions take the form of *semantic rules*. SOS is a proof-theoretic approach, focusing on giving a detailed step-by-step formal description of a program's execution. The semantic rules are used as inference rules to reason about what computation steps are possible. Typically, the rules follow the syntactic structure of programs, defining the semantics of a language construct in terms of that of its componenta. The "locus classicus" is Plotkin's Aarhus lectures [62]; there is again a vast literature on the topic that we do not attempt to survey; for a good textbook introduction see [35].

Strong points of SOS include:

- it is an abstract and general formalism, yet quite intuitive, allowing detailed *step-by-step* modeling of program execution;
- has a simple *proof-theoretic* semantics using semantic rules as inference rules;
- is fairly well suited to model *concurrent languages*, and can also deal well with the detailed execution of deterministic languages;
- allows *mathematical reasoning and proof*, by reasoning inductively or coinductively about the inference steps.

However, SOS has the following drawbacks:

- although specific proposals have been made for *categorical models* for certain SOS formats, such as, for example, Turi's functorial SOS [79] and Gadducci and Montanari's tile models [29], it seems however fair to say that, so far, SOS has not commonly agreed upon model-theoretic semantics.

---

[1] Two good examples of indirectly modeling concurrency within a purely functional framework are the ACL2 semantics of the JVM using a scheduler [53], and the use of lazy data structures in Haskell to analyze cryptographic protocols [3].

- in its standard formulation it imposes a centralized *interleaving semantics* of concurrent computations, which may be unnatural in some cases, for example for highly decentralized and asynchronous mobile computations; this problem is avoided in "reduction semantics," which is different from SOS and is in fact a special case of rewriting semantics (see Section 5.2).
- although some tools have been built to execute SOS definitions (see for example [21]) tool support is considerably less developed than for equational semantics.
- standard SOS definitions are notoriously *unmodular*, unless one adopts Mosses' MSOS framework (see Section 5.3).

1.2   Rewriting Logic Unifies SOS and Equational Semantics

For the most part, equational semantics and SOS have lived separate lives. Pragmatic considerations and differences in taste tend to dictate which framework is adopted in each particular case. For concurrent languages SOS seems clearly superior and tends to prevail as the formalism of choice, but for deterministic languages equational approaches are also widely used. Of course there are also practical considerations of tool support for both execution and formal reasoning.

This paper addresses three fundamental questions:

1. can the semantic frameworks of SOS and equational semantics be unified in a mathematically rigorous way?
2. can the advantages of each formalisms be preserved and can their respective drawbacks be overcome in a suitable unification?
3. is it possible to efficiently execute and analyze programs using semantic language definitions in such a unified framework with suitable formal tools?

We answer each of the above questions in the affirmative by proposing rewriting logic [41,14] as such a unifying semantic framework. Roughly speaking,[2] a rewrite theory is a triple (Σ, E, R), with (Σ, E) an equational theory with signature of operations and sorts Σ and set of (possibly conditional) equations E, and with R a set of (possibly conditional) rewrite rules. Therefore, rewriting logic introduces a *key distinction* between semantic *equations* E, and semantic *rules* R. Computationally, this is a distinction between *deterministic* computations, and *concurrent* and possibly nondeterministic ones. That is, if (Σ, E, R) axiomatizes the semantics of a programming language $\mathcal{L}$, then the deterministic computations in $\mathcal{L}$ will be axiomatized by the semantic equations E, whereas the concurrent computations will be axiomatized by the rewrite rules R. The semantic unification of SOS and equational semantics is then achieved very naturally, since, roughly speaking, we can obtain SOS and equational semantics as,

---

[2] We postpone the issue of "frozen" arguments, which is treated in Section 2.2.

respectively, the special cases in which $E = \varnothing$ and we have only semantic rules[3], and $R = \varnothing$ and we have only semantic equations, respectively.

This unification makes possible something not available in either formalism, namely mixing semantic equations and semantic rules, using each kind of axiom for the purposes for which it is best suited: equations for deterministic computations, and rules for concurrent ones. This distinction between equations and rules is of more than academic interest. The point is that, since rewriting with rules R takes place *modulo* the equations E [41], many states are abstracted together by the equations E, and only the rules R contribute to the size of the system's state space, which can be drastically smaller than if all axioms had been given as rules, a fact of crucial importance for formal analyses of concurrent programs based on search and model checking.

This brings us to efficient tool support for both execution and formal analysis. Rewriting logic has several high-performance implementations [6,28,19], of which the most comprehensive so far, in expressiveness and in range of features, is probably the Maude system [19]. Maude can both efficiently execute a rewriting logic axiomatization of a programming language $\mathcal{L}$, thus providing an interpreter for $\mathcal{L}$, and also perform breadth-first search to find safety violations in a concurrent program, and model checking of linear time temporal logic (LTL) properties for such programs when the set of reachable states is finite. We illustrate these execution and analysis capabilities in Sections 3–4.

The rest of the paper is organized as follows. Basic concepts on rewriting logic and membership equational logic are gathered in Section 2. We then illustrate our language specification methodology by means of a nontrivial example – a substantial Caml-like language including functions, assignments, loops, exceptions, and threads – and briefly discuss another case study on Java semantics in Section 3. The formal analysis of concurrent programs is illustrated for our example language in Section 4. We revisit SOS and equational semantics and discuss the advantages of their unification within rewriting logic in Section 5. The paper gives some concluding remarks in Section 6.

2   Rewriting Logic Semantics

We explain here the basic concepts of rewriting logic, and how it can be used to define the semantics of a programming language. Since each rewrite theory has an underlying equational theory, different variants of equational logic give rise to corresponding variants of rewriting logic. The more expressive the underlying equational sublanguage, the more expressive will the resulting rewrite theories be. For this reason, we include below a brief summary of membership equational logic (MEL) [44], an expressive Horn logic with both equations $t = t'$

---

[3] The case of structural axioms is a separate issue that we postpone until Section 2; also rewrite rules and SOS rules, though closely related, do not correspond identically to each other, as explained in Section 5.2.

and membership predicates $t : s$ which generalizes order-sorted equational logic and supports sorts, subsorts, partiality, and sorts defined by equational axioms. Maude 2.0 [19] supports all the logical features of MEL and its rewriting logic super-logic with a syntax almost identical to the mathematical notation.

2.1   Membership Equational Logic

A membership equational logic (MEL) [44] *signature* is a triple (K, Σ, S) (just Σ in the following), with K a set of *kinds*, $\Sigma = \{\Sigma_{w,k}\}_{(w,k) \in K^* \times K}$ a many-kinded signature and $S = \{S_k\}_{k \in K}$ a K-kinded family of disjoint sets of sorts. The kind of a sort s is denoted by $[s]$. A MEL Σ-algebra A contains a set A_k for each kind $k \in K$, a function $A_f : A_{k_1} \times \cdots \times A_{k_n} \to A_k$ for each operator $f \in \Sigma_{k_1 \cdots k_n, k}$ and a subset $A_s \subseteq A_k$ for each sort $s \in S_k$, with the meaning that the elements in sorts are well-defined, while elements without a sort are *errors*. We write $T_{\Sigma,k}$ and $T_\Sigma(X)_k$ to denote respectively the set of ground Σ-terms with kind k and of Σ-terms with kind k over variables in X, where $X = \{x_1 : k_1, \ldots, x_n : k_n\}$ is a set of kinded variables. Given a MEL signature Σ, *atomic formulae* have either the form $t = t'$ (Σ-equation) or $t : s$ (Σ-membership) with $t, t' \in T_\Sigma(X)_k$ and $s \in S_k$; and Σ-*sentences* are conditional formulae of the form $(\forall X)\ \varphi$ if $\bigwedge_i p_i = q_i \wedge \bigwedge_j w_j : s_j$, where φ is either a Σ-equation or a Σ-membership, and all the variables in φ, p_i, q_i, and w_j are in X. A MEL theory is a pair (Σ, E) with Σ a MEL signature and E a set of Σ-sentences. We refer to [44] for the detailed presentation of (Σ, E)-algebras, sound and complete deduction rules, and initial and free algebras. In particular, given an MEL theory (Σ, E), its initial algebra is denoted $T_{\Sigma/E}$; its elements are E-equivalence classes of ground terms in T_Σ. Order-sorted notation $s_1 < s_2$ can be used to abbreviate the conditional membership $(\forall x : k)\ x : s_2$ if $x : s_1$. Similarly, an operator declaration $f : s_1 \times \cdots \times s_n \to s$ corresponds to declaring f at the kind level and giving the membership axiom $(\forall x_1 : k_1, \ldots, x_n : k_n)\ f(x_1, \ldots, x_n) : s$ if $\bigwedge_{1 \leq i \leq n} x_i : s_i$. We write $(\forall x_1 : s_1, \ldots, x_n : s_n)\ t = t'$ in place of $(\forall x_1 : k_1, \ldots, x_n : k_n)\ t = t'$ if $\bigwedge_{1 \leq i \leq n} x_i : s_i$.

2.2   Rewrite Theories

A *rewriting logic specification or theory* is a tuple $\mathcal{R} = (\Sigma, E, \phi, R)$, with:

- (Σ, E) a membership equational theory
- $\phi : \Sigma \longrightarrow \mathbb{N}$ a mapping assigning to each function symbol $f \in \Sigma$ (with, say, n arguments) a set $\phi(f) = \{i_1, \ldots, i_k\}$, $1 \leq i_1 < \ldots < i_k \leq n$ of *frozen argument positions* under which it is forbidden to perform any rewrites; and
- R a set of *labeled conditional rewrite rules* of the general form

$$r : (\forall X)\ t \longrightarrow t' \quad \text{if} \quad (\bigwedge_i u_i = u_i') \wedge (\bigwedge_j v_j : s_j) \wedge (\bigwedge_l w_l \longrightarrow w_l') \quad (\flat).$$

where the variables appearing in all terms are among those in X, terms in each rewrite or equation have the same kind, and in each membership $v_j : s_j$ the term v_j has kind $[s_j]$. Intuitively, $\mathcal{R}$ specifies a *concurrent system*, whose states are elements of the initial algebra $T_{\Sigma/E}$ specified by (Σ, E), and whose *concurrent transitions* are specified by the rules R, subject to the frozenness imposed by ϕ.

We can illustrate both a simple rewrite theory and the usefulness of frozen arguments by means of the following Maude module for nondeterministic choice:

```
mod CHOICE is protecting NAT .
  sorts Elt MSet .  subsorts Elt < MSet .
  ops a b c d e f g : -> Elt .
  op __ : MSet MSet -> MSet [assoc comm] .
  op card : MSet -> Nat [frozen] .
  eq card(X:Elt) = 1 .
  eq card(X:Elt M:MSet) = 1 + card(M:MSet) .
  rl [choice] : X:MSet Y:MSet => Y:MSet .
endm
```

In a Maude module,[4] introduced with the keyword `mod` followed by its name, and ended with the keyword `endm`, kinds are not declared explicitly; instead, each connected component of sorts in the subsort inclusion ordering implicitly determines a kind, which is viewed as the equivalence class of its corresponding sorts. Here, since the only two sorts declared, namely `Elt` and `MSet`, are related by a subsort inclusion[5] we have implicitly declared a new kind, which we can refer to by enclosing either of the sorts in square brackets, that is, by either `[Elt]` or `[MSet]`. There are however two more kinds, namely the kind `[Nat]` determined by the sort `Nat` of natural numbers and its subsorts in the imported `NAT` module, and the kind `[Bool]` associated to the Booleans, which by default are implicitly imported by any Maude module.

The operators in Σ are declared with `op` (`ops` if several operators are declared simultaneously). Here we have just three such declarations: (i) the constants `a` through `g` of sort `Elt`, (ii) a multiset union operator declared with infix[6] empty syntax (juxtaposition), and (iii) a multiset cardinality function declared with prefix notation `card`. The set E contains those equations and memberships of the imported modules, the two equations defining the cardinality function, and the equations of associativity and commutativity for the multiset union operator, which are not spelled out as the other equations, but are instead specified with the `assoc` and `comm` keywords. Furthermore, as pointed out in Section 2.1, the

---

[4] A Maude module specifies a rewrite theory $\mathcal{R} = (\Sigma, E, \phi, R)$; however, when $R = \varnothing$, then $\mathcal{R}$ becomes a membership equational theory. Maude has an *equational sublanguage* in which a membership equational theory (Σ, E) can be specified as a *functional module* with beginning and ending keywords `fmod` and `endfm`.

[5] A subsort inclusion is shorthand notation for a corresponding membership axiom.

[6] In general, prefix, infix, postfix, and general "mixfix" user-definable syntax is supported. In all cases except for prefix operators, each argument position is declared with an underbar symbol; for example, the usual infix notation for addition would be declared `_+_`, but here, since we use juxtaposition, no symbol is given between the two underbars `__` of the multiset union operator.

subsort inclusion declaration and the operator declarations at the level of sorts are in fact conditional membership axioms in disguise. The only rule in the set R is the [choice] rule, which arbitrarily chooses a nonempty sub-multiset of the given multiset. Maude then uses the assoc and comm declarations to apply the other equations and the [choice] rule in a built-in way *modulo* the associativity and commutativity of multiset union, that is, parentheses are not needed, and the order of the elements in the multiset is immaterial. It is then intuitively clear that if we begin with a multiset such as a a b b b c and repeatedly rewrite it in all possible ways using the [choice] rule, the terminating (deadlock) states will be the singleton multisets a, b, and c.

The multiset union operator has no special declaration, meaning that none of its two arguments are frozen, but the cardinality function is declared with the frozen attribute, meaning that all its arguments (in this case the single argument of sort MSet) are frozen, that is, $\phi(\text{card}) = \{1\}$. This declaration captures the intuition that it does not make much sense to rewrite below the cardinality function card, because then the multiset whose cardinality we wish to determine would become a *moving target*. If card had not been declared frozen, then the rewrites a b c $\longrightarrow$ b c $\longrightarrow$ c would induce rewrites $3 \longrightarrow 2 \longrightarrow 1$, which seems bizarre. The point is that we think of the kind [MSet] as the *state kind* in this example, whereas [Nat] is the *data kind*. By declaring card's single argument as frozen, we restrict rewrites to the state kind, where they belong.

2.3   Rewriting Logic Deduction

Given $\mathcal{R} = (\Sigma, E, \phi, R)$, the sentences that it proves are universally quantified rewrites of the form, $(\forall X)\, t \longrightarrow t'$, with $t, t' \in T_{\Sigma,E}(X)_k$, for some kind k, which are obtained by finite application of the following *rules of deduction*:

- **Reflexivity.** For each $t \in T_\Sigma(X)$, $\dfrac{}{(\forall X)\, t \longrightarrow t}$

- **Equality.** $\dfrac{(\forall X)\, u \longrightarrow v \quad E \vdash (\forall X) u = u' \quad E \vdash (\forall X) v = v'}{(\forall X)\, u' \longrightarrow v'}$

- **Congruence.** For each $f : k_1 \ldots k_n \longrightarrow k$ in Σ, with $\{1, \ldots, n\} - \phi(f) = \{j_1, \ldots, j_m\}$, with $t_i \in T_\Sigma(X)_{k_i}$, $1 \leq i \leq n$, and with $t'_{j_l} \in T_\Sigma(X)_{k_{j_l}}$, $1 \leq l \leq m$, $\dfrac{(\forall X)\, t_{j_1} \longrightarrow t'_{j_1} \quad \ldots \quad (\forall X)\, t_{j_m} \longrightarrow t'_{j_m}}{(\forall X)\, f(t_1, \ldots, t_{j_1}, \ldots, t_{j_m}, \ldots, t_n) \longrightarrow f(t_1, \ldots, t'_{j_1}, \ldots, t'_{j_m}, \ldots, t_n)}$

- **Replacement.** For each $\theta : X \longrightarrow T_\Sigma(Y)$ with, say, $X = \{x_1, \ldots, x_n\}$, and $\theta(x_l) = p_l$, $1 \leq l \leq n$, and for each rule in R of the form,

$$q : (\forall X)\, t \longrightarrow t' \ \text{ if } \ (\bigwedge_i u_i = u'_i) \wedge (\bigwedge_j v_j : s_j) \wedge (\bigwedge_k w_k \longrightarrow w'_k)$$

with $Z = \{x_{j_1}, \ldots, x_{j_m}\}$ the set of unfrozen variables in t and t', then,

$$(\bigwedge_r (\forall Y)\, p_{j_r} \longrightarrow p'_{j_r})$$

$$\frac{(\bigwedge_i (\forall Y)\, \theta(u_i) = \theta(u_i')) \wedge (\bigwedge_j (\forall Y)\, \theta(v_j) : s_j) \wedge (\bigwedge_k (\forall Y)\, \theta(w_k) \longrightarrow \theta(w_k'))}{(\forall Y)\, \theta(t) \longrightarrow \theta'(t')}$$

where for $x \in X - Z$, $\theta'(x) = \theta(x)$, and for $x_{j_r} \in Z$, $\theta'(x_{j_r}) = p_{j_r}'$, $1 \leq r \leq m$.

– **Transitivity**

$$\frac{(\forall X)\, t_1 \longrightarrow t_2 \qquad (\forall X)\, t_2 \longrightarrow t_3}{(\forall X)\, t_1 \longrightarrow t_3}$$

We can visualize the above inference rules as follows:

Reflexivity

Equality

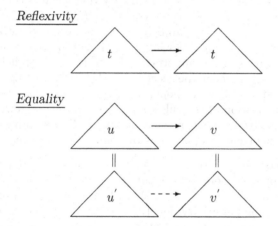

Congruence

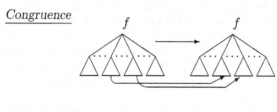

Replacement

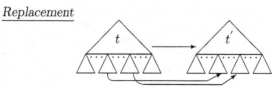

Transitivity

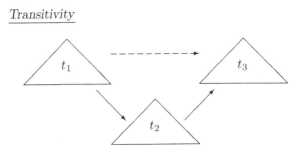

Intuitively, we should think of the above inference rules as different ways of *constructing* all the (finitary) *concurrent computations* of the concurrent system specified by $\mathcal{R}$. The **Reflexivity** rule says that for any state t there is an *idle transition* in which nothing changes. The **Equality** rule specifies that the states are in fact equivalence classes modulo the equations E. The **Congruence** rule is a very general form of "sideways parallelism," so that each operator f can be seen as a *parallel state constructor*, allowing its nonfrozen arguments to evolve in parallel. The **Replacement** rule supports a different form of parallelism, which could be called "parallelism under one's feet," since besides rewriting an instance of a rule's lefthand side to the corresponding righthand side instance, the state fragments in the substitution of the rule's variables can also be rewritten, provided the variables involved are not frozen. Finally, the **Transitivity** rule allows us to build longer concurrent computations by composing them sequentially.

For a rewrite theory to be *executable*, so that the above inference rules can be efficiently tool supported, some additional requirements should be met. First, E should decompose as a union $E = E_0 \cup A$, with A a set of equational axioms such as associativity, commutativity, identity, for which an effective matching algorithm modulo A exists, and E_0 a set of (ground) confluent and terminating[7] for each term t by applying the equations in E_0 modulo A to t until termination. Second, the rules R should be *coherent* with E_0 modulo A [87]; intuitively, this means that, to get the effect of rewriting in equivalence classes modulo E, we can always first simplify a term with the equations to its canonical form, and then rewrite with a rule in R. Finally, the rules in R should be *admissible* [18], meaning that in a rule of the form (♭), besides the variables appearing in t there can be extra variables in t', provided that they also appear in the condition and that they can all be *incrementally instantiated* by either matching a pattern in

---

[7] The termination condition may be dropped for programming language specifications in which some equationally defined language constructs may not terminate. Even the confluence modulo A may be relaxed, by restricting it to terms in some "observable kinds" of interest. The point is that there may be some "unobservable" kinds for which several different but semantically equivalent terms can be derived by equational simplification: all we need in practice is that the operations are confluent for terms in an observable kind, such as that of values, so that a unique canonical form is then reached for them if it exists.

a "matching equation" or performing breadth first search in a rewrite condition (see [18] for a detailed description of admissible equations and rules).

2.4   Rewriting Logic's Model Theory and Temporal Logic

Given a rewrite theory $\mathcal{R} = (\Sigma, E, \phi, R)$, its $\mathcal{R}$-reachability relation $\rightarrow_{\mathcal{R}}$ (also called $\mathcal{R}$-rewriting relation, or $\mathcal{R}$-provability relation) is defined proof-theoretically, for each kind k in Σ and each $[t], [t'] \in T_{\Sigma/E,k}$, by the equivalence,

$$[t] \rightarrow_{\mathcal{R}} [t'] \quad \Leftrightarrow \quad \mathcal{R} \vdash (\forall \varnothing)\, t \longrightarrow t',$$

which by the **Equality** rule is independent of the choice of t, t'. Model-theoretically, $\mathcal{R}$-reachability can be defined as the family of relations, indexed by the kinds k in Σ, interpreting the sorts $Arrow_k$ in the initial model of a membership equational theory $Reach(\mathcal{R})$ axiomatizing the reachability models of the rewrite theory $\mathcal{R}$ [14]. The initial reachability model is then the initial algebra $T_{Reach(\mathcal{R})}$. In particular, the one-step $\mathcal{R}$-rewrite relations for each kind k are the extensions in $T_{Reach(\mathcal{R})}$ of subsorts $Arrow_k^1 < Arrow_k$. We denote such a relation on E-equivalence classes of terms with the notation $[t] \rightarrow_{\mathcal{R},k}^1 [t']$. Thus, a rewrite theory $\mathcal{R}$ specifies for each kind k a transition system characterized by $\rightarrow_{\mathcal{R},k}^1$, which can be made total by adding idle transitions for deadlock states, denoted $(\rightarrow_{\mathcal{R},k}^1)^\bullet$. This is almost a Kripke structure: we still need to specify the state predicates in a set of predicates Π. This can be done equationally, by choosing a kind k as the kind of states, and giving equations defining when each predicate $p \in \Pi$ holds for a state $[t]$ of sort k, thus getting a labeling function L_Π.

This way, we can associate to a rewrite theory $\mathcal{R} = (\Sigma, E, \phi, R)$ with a designated kind k of states and state predicates Π, the Kripke structure $(T_{\Sigma/E,k}, (\rightarrow_{\mathcal{R},k}^1)^\bullet, L_\Pi)$. We can then define the semantics of any temporal logic formula over predicates Π in the usual way [17], for any desired temporal logic such as LTL, CTL*, the modal μ-calculus, and so on (see [45]). Furthermore, if the states reachable from an initial state form a finite set, then we can model check such formulas. Maude provides an explicit state LTL model checking for executable rewrite theories with a performance comparable to that of SPIN [24].

Reachability models for a rewrite theory $\mathcal{R}$ are a special case of more general true concurrency models, in which different concurrent computations from one state to another correspond to equivalence classes of proofs in rewriting logic. That is, concurrent computations are placed in bijective correspondence with proofs in a Curry-Howard like equivalence. The paper [14] shows that initial models exist for both reachability models and true concurrency models of a rewrite theory $\mathcal{R}$, and that both kinds of models make the rules of inference of rewriting logic sound and complete. We denote by $T_{Reach(\mathcal{R})}$, resp. $\mathcal{T}_{\mathcal{R}}$, the initial reachability, resp. true-concurrency, model of a rewrite theory $\mathcal{R}$.

2.5   Specifying Concurrency Models and Programming Languages

Because rewriting logic is *neutral* about concurrency constructs, it is a general *semantic framework* for concurrency that can express many concurrency models such as: equational programming, which is the special case of rewrite theories whose set of rules is empty and whose equations are Church-Rosser, possibly modulo some axioms A; lambda calculi and combinatory reduction systems [41, 39,72,69]; labeled transition systems [41]; grammars and string-rewriting systems [41]; Petri nets, including place/transition nets, contextual nets, algebraic nets, colored nets, and timed Petri nets [41,43,70,73,60,68]; Gamma and the Chemical Abstract Machine [41]; CCS and LOTOS [48,40,15,22,83,82,84,80]; the π calculus [85,69,78]; concurrent objects and actors [41,42,76,77]; the UNITY language [41]; concurrent graph rewriting [43]; dataflow [43]; neural networks [43]; real-time systems, including timed automata, timed transition systems, hybrid automata, and timed Petri nets [60,59]; and the tile logic [29] model of synchronized concurrent computation [49,13].

Since the above are typically executable, rewriting logic is a flexible *operational* semantic framework to specify such models. What is not immediately apparent is that it is also a flexible *mathematical* semantic framework for concurrency models. Well-known models of concurrency are isomorphic to the initial model $\mathcal{T}_{\mathcal{R}}$ of the rewrite theory $\mathcal{R}$ axiomatizing that particular model, or at least closely related to such an initial model: [39] shows that for rewrite theories $\mathcal{R} = (\Sigma, \varnothing, \phi, R)$ with the rules R left-linear, $\mathcal{T}_{\mathcal{R}}$ is isomorphic to a model based on residuals and permutation equivalence proposed by Boudol [7], and also that for $\mathcal{R}$ a rewrite theory of an orthogonal combinatory reduction system, including the λ-calculus, a quotient of $\mathcal{T}_{\mathcal{R}}$ is isomorphic to a well-known model of parallel reductions; [73] shows that for $\mathcal{R}$ a rewrite theory of a place/transition net, $\mathcal{T}_{\mathcal{R}}$ is isomorphic to the Best-Devillers net process model [5] and then generalizes this isomorphism to one between $\mathcal{T}_{\mathcal{R}}$ and a Best-Devillers-like model for the rewrite theory of an algebraic net; [15,22] show that for $\mathcal{R}$ axiomatizing CCS, a truly concurrent semantics causal model based on proved transition systems is isomorphic to a quotient of $\mathcal{T}_{\mathcal{R}}$; [50] shows that for $\mathcal{R}$ axiomatizing a concurrent object-oriented system satisfying reasonable requirements, a subcategory of $\mathcal{T}_{\mathcal{R}}$ is isomorphic to a partial order of events model which, for asynchronous object systems corresponding to actors, coincides with the finitary part of the Baker-Hewitt partial order of events model [2].

All the above remarks apply also to the specification of programming languages, which often implement specific concurrency models. In particular, both an operational semantics and a denotational semantics are provided for a language when it is specified as a rewrite theory. How is this generally done? We can define the semantics of a concurrent programming language, say $\mathcal{L}$, by specifying a rewrite theory, say $\mathcal{R}_{\mathcal{L}} = (\Sigma_{\mathcal{L}}, E_{\mathcal{L}}, \phi_{\mathcal{L}}, R_{\mathcal{L}})$, where $\Sigma_{\mathcal{L}}$ is $\mathcal{L}$'s *syntax* and the auxiliary operators (store, environment, etc.), $E_{\mathcal{L}}$ specifies the semantics of all the *deterministic features* of $\mathcal{L}$ and of the auxiliary semantic operations, the frozenness information $\phi_{\mathcal{L}}$ specifies what arguments can be rewritten with rules for each operator, and the rewrite rules $R_\mathcal{L}$ specify the semantics of all the *concurrent features* of $\mathcal{L}$. Section 3 does exactly this.

3   Specifying Deterministic and Concurrent Features

In this section we illustrate the rewriting logic semantics techniques advocated in this paper on a nontrivial Caml-like programming language. We show how several important programming language features, such as arithmetic and boolean expressions, conditional statements, high-order functions, lists, let bindings, recursion with let rec, side effects via variable assignments, blocks and loops, exceptions, and concurrency via threads and synchronization, can be succinctly, modularly and efficiently defined in rewriting logic. What we present in this section should be regarded as one possible way to define this language, a way which is by no means unique or optimal. The various features are shown in the following diagram:

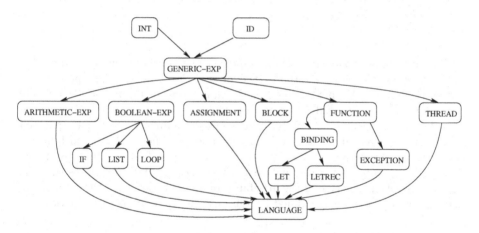

INT is a Maude builtin module defining arbitrary precision integers; ID defines identifiers as well as comma-separated lists of identifiers; GENERIC-EXP defines a special sort for expressions as well as comma-separated lists of such expressions; ARITHMETIC-EXP adds arithmetic operators, such as addition, multiplication, etc., and BOOLEAN-EXP adds boolean expressions; the latter are further used to define conditionals, loops and lists (lists contain an empty list check); BINDINGS defines bindings as special lists of pairs "identifier = expression", which are further needed to define both LET and LETREC; ASSIGNMENT defines variable assignments; BLOCK defines blocks enclosed with curly brackets "{" and "}" containing sequences of expressions separated by semicolon ";" (blocks are used for their side effects); FUNCTION defines high order functions, in a Caml style; EXCEPTION defines exceptions using try ... catch ... and throw ... keywords, where the "catch" part is supposed to evaluate to a function of one argument, to which a value can be "thrown" from the "try" part; THREAD defines new threads which can be created and destroyed dynamically; finally, LANGUAGE creates the desired programming language by putting all the features together. Each of the above has a syntactic and a semantic part, each specified as a Maude module. The entire Maude specification has less than 400 lines of code.

3.1   Defining the Syntax and Desugaring

Here we show how to define the syntax of a programming language in Maude, together with several simple desugaring translations that will later simplify the semantic definitions, such as translations of "for" loops into "while" loops. We first define identifiers, which add to Maude's builtin quoted identifiers (QID) several common (unquoted) one-character identifiers, together with comma-separated lists of identifiers that will be needed later:

```
fmod ID is protecting QID .
  sorts Id IdList .  subsorts Qid < Id < IdList .
  ops a b c d e f g h i j k l m n o p q r s t u v x y z w : -> Id .
  op nil : -> IdList .
  op _,_ : IdList IdList -> IdList [assoc id: nil prec 1] .
endfm
```

The attribute "prec 1" assigns a precedence to the comma operator, to avoid using unnecessary parentheses: the lower the precedence of an operator the tighter the binding. We next define generic expressions, including for now integers and white-space-separated sequences of "names", where a name is either an identifier or the special symbol "()". Sequences of names and comma-separated lists of expressions will be needed later to define lists and bindings and function declarations, respectively. The attribute "gather(E e)" states that the name sequencing operator is left associative and the "ditto" attribute states that the current operation inherits all the attributes of an operation with the same name and kind arrity previously defined (in our case the comma operator in ID):

```
fmod GENERIC-EXP-SYNTAX is protecting ID . protecting INT .
  sorts Unit Name NameSeq Exp ExpList .
  subsorts Unit Id < Name < NameSeq < Exp < ExpList .
  subsort Int < Exp .  subsort IdList < ExpList .
  op '(') : -> Unit .
  op __ : NameSeq NameSeq -> NameSeq [gather(E e) prec 1] .
  op _,_ : ExpList ExpList -> ExpList [ditto] .
endfm
```

The rest of the syntax adds new expressions to the language modularly. The next four modules add arithmetic, boolean, conditional and list expressions. Note that in the LIST module, the list constructor takes a comma-separated list of expressions and returns an expression:

```
fmod ARITHMETIC-EXP-SYNTAX is extending GENERIC-EXP-SYNTAX .
  ops _+_ _-_ _*_ : Exp Exp -> Exp [ditto] .
  ops _/_ _%_ : Exp Exp -> Exp [prec 31] .
endfm
```

```
fmod BOOLEAN-EXP-SYNTAX is extending GENERIC-EXP-SYNTAX .
  ops _==_ _<=_ _>=_ _<_ _>_ _and_ _or_ : Exp Exp -> Exp .
  op not_ : Exp -> Exp .
endfm

fmod IF-SYNTAX is extending GENERIC-EXP-SYNTAX .
  op if_then_else_ : Exp Exp Exp -> Exp .
endfm

fmod LIST-SYNTAX is extending GENERIC-EXP-SYNTAX .
  op list : ExpList -> Exp .
  ops car cdr null? : Exp -> Exp .
  op cons : Exp Exp -> Exp .
endfm
```

We next define functions. Like in Caml, we want to define functions using a syntax like "fun x y z -> x + y * z". However, "fun x y z -> ..." is syntactic sugar for "fun x -> fun y -> fun z -> ...", so to keep the semantics simple later we prefer to consider this uncurrying transformation as part of the syntax. Function application simply extends the name sequencing operator:

```
fmod FUNCTION-SYNTAX is extending GENERIC-EXP-SYNTAX .
  op fun_->_ : NameSeq Exp -> Exp .
  op __ : Exp Exp -> Exp [ditto] .
  var Zs : NameSeq . var Z : Name . var E : Exp .
  eq fun Zs Z -> E = fun Zs -> fun Z -> E .
endfm
```

Bindings of names to values are crucial in any functional programming language. Like in Caml, in our language bindings are "and"-separated pairs of equalities. However, note that "f x y z = ..." is just syntactic sugar for "f = fun x y z -> ...". Since the semantics of bindings will involve allocation of new memory locations for the bound identifiers, it will be very helpful to know upfront the number and the list of identifiers. Two equations take care of this:

```
fmod BINDING-SYNTAX is extending FUNCTION-SYNTAX .
  sorts Binding Bindings .   subsort Binding < Bindings .
  op _and_ : Bindings Bindings -> Bindings [assoc prec 100] .
  op '(_,_,_') : Nat IdList ExpList -> Bindings .
  op _=_ : NameSeq Exp -> Binding .
  var Zs : NameSeq . var Z : Name . var X : Id . var E : Exp .
  vars N N' : Nat .   vars Xl Xl' : IdList .   vars El El' : ExpList .
  eq (Zs Z = E) = (Zs = fun Z -> E) .
  eq (X = E) = (1, X, E) .
  eq (N, Xl, El) and (N', Xl', El') = (N + N', (Xl,Xl'), (El,El')) .
endfm
```

We can now define the two major binding language constructors, namely "let" and "let rec", the later typically being used to define recursive functions:

```
fmod LET-SYNTAX is extending BINDING-SYNTAX .
  op let_in_ : Bindings Exp -> Exp .
endfm

fmod LETREC-SYNTAX is extending BINDING-SYNTAX .
  op let rec_in_ : Bindings Exp -> Exp .
endfm
```

We next add several imperative features, such as variable assignment, blocks and loops. The variable assignment assumes the identifier already allocated at some location, and just changes the value stored at that location. Both "for" and "while" loops are allowed, but the former ones are immediately desugared:

```
fmod ASSIGNMENT-SYNTAX is extending GENERIC-EXP-SYNTAX .
  op _:=_ : Name Exp -> Exp .
endfm

fmod BLOCK-SYNTAX is extending GENERIC-EXP-SYNTAX .
  sort ExpBlock .  subsort Exp < ExpBlock .
  op _;_ : ExpBlock ExpBlock -> ExpBlock [assoc prec 100] .
  op {_} : ExpBlock -> Exp .
endfm

fmod LOOP-SYNTAX is extending BLOCK-SYNTAX .
  op while__ : Exp Exp -> Exp .
  op for(_;_;_)_ : Exp Exp Exp Exp -> Exp .
  vars Start Cond Step Body : Exp .
  eq for(Start ; Cond ; Step) Body = {Start ; while Cond {Body ; Step}} .
endfm
```

We next add syntax for two important features, exceptions and threads:

```
fmod EXCEPTION-SYNTAX is extending GENERIC-EXP-SYNTAX .
  op try_catch_ : Exp Exp -> Exp .
  op throw_ : Exp -> Exp .
endfm

fmod THREAD-SYNTAX is extending GENERIC-EXP-SYNTAX .
  ops spawn_ lock acquire_ release_ : Exp -> Exp .
endfm
```

We can now put all the syntax together, noticing that the syntax modules of most of the features above are independent from each other, so one can easily reuse them to build other languages, using Maude 2.01's renaming facility to adapt each module to the concrete syntax of the chosen language:

```
fmod LANGUAGE-SYNTAX is  extending ARITHMETIC-EXP-SYNTAX . extending BOOLEAN-EXP-SYNTAX .
  extending IF-SYNTAX .  extending FUNCTION-SYNTAX .      extending LIST-SYNTAX .
  extending LET-SYNTAX . extending LETREC-SYNTAX .        extending ASSIGNMENT-SYNTAX .
  extending LOOP-SYNTAX . extending EXCEPTION-SYNTAX .    extending THREAD-SYNTAX .
endfm
```

One can now parse programs using Maude's "parse" command. For example, the following program recursively calculating the product of elements in a list, will correctly parse as "Exp". Note that this program uses an exception to immediately return 0 whenever a 0 is encountered in the input list.

```
parse
  let p l = try let rec a l = if null?(l) then 1
                              else if car(l) == 0 then throw 0
                                   else car(l) * (a cdr(l))
                  in a l catch fun x -> x
   in p list(1,2,3,4,5,6,7,8,9,0,10,11,12,13,14,15,16,17,18,19,20)
  .
```

3.2   Defining the State Infrastructure

Before defining the semantics of a programming language, one needs to define the notion of programming language state. There are different possibilities to design the state needed to give semantics to a language, depending on its complexity and one's taste. However, any language worth its salt supports identifiers that are bound to values; since our language has side effects (due to variable assignments), we need to split the mapping of identifiers to values into a map of identifiers to locations, called an environment, and a map of locations to values, called a store.

Let us first define *locations*. A location is essentially an integer; to keep it distinct from other integers, we wrap it with the constructor "loc". An auxiliary operation creating a given number of locations starting with a given one will be very useful when defining bindings and functions, so we provide it here.

```
fmod LOCATION is protecting INT .
   sorts Location LocationList .   subsort Location < LocationList .
   op loc : Nat -> Location .
   op nil : -> LocationList .
   op _,_ : LocationList LocationList -> LocationList [assoc id: nil] .
   op locs : Nat Nat -> LocationList .
   eq locs(N:Nat,0) = nil .
   eq locs(N:Nat,#:Nat) = loc(N:Nat), locs(N:Nat + 1, #:Nat - 1) .
endfm
```

An elegant and efficient way to define a mapping in Maude is as a set of pairs formed with an associative (A) and commutative (C) union operator _||_ with identity (I) empty. Then *environments* can be defined as below. Note the use of ACI matching to evaluate or update an environment at an identifier, so that one does not need to traverse the entire set in order to find the desired element:

```
fmod ENVIRONMENT is protecting ID .   protecting LOCATION .
   sort Env .
   op empty : -> Env .
   op [_,_] : Id Location -> Env .
   op _||_ : Env Env -> Env [assoc comm id: empty] .
   op _[_<-_] : Env IdList LocationList -> Env .
   vars Env : Env . vars L L' : Location . var Xl : IdList . var Ll : LocationList .
   eq Env[nil <- nil] = Env .
   eq ([X:Id,L] || Env)[X:Id,Xl <- L',Ll] = ([X:Id,L'] || Env)[Xl <- Ll] .
   eq Env[X:Id,Xl <- L,Ll] = (Env || [X:Id,L])[Xl <- Ll] [owise] .
endfm
```

Values and *stores* can be defined in a similar way. Since we want our modules to be as independent as possible, to be reused for defining other languages, we prefer to *not* state at this moment the particular values that our language handles, such as integers, booleans, functions (i.e., closures), etc.. Instead, we define the values when they are first needed within the semantics. However, since lists of values are frequently used for various reasons, we believe that many languages need them so we introduce them as part of the VALUE module:

```
fmod VALUE is sorts Value ValueList .  subsort Value < ValueList .
  op noValue : -> Value .
  op nil : -> ValueList .
  op _,_ : ValueList ValueList -> ValueList [assoc id: nil] .
  op [_] : ValueList -> Value .
endfm

fmod STORE is protecting LOCATION .  protecting VALUE .
  sort Store .
  op empty : -> Store .
  op [_,_] : Location Value -> Store .
  op _||_ : Store Store -> Store [assoc comm id: empty] .
  op _[_<-_] : Store LocationList ValueList -> Store .
  var L : Location .  var M : Store .  vars V V' : Value .
  var Ll : LocationList .  var Vl : ValueList .
  eq M[nil <- nil] = M .
  eq ([L,V] || M)[L,Ll <- V',Vl] = ([L,V'] || M)[Ll <- Vl] .
  eq M[L,Ll <- V',Vl] = (M || [L,V'])[Ll <- Vl] [owise] .
endfm
```

Since our language has complex control-context constructs, such as exceptions and threads, we follow a *continuation passing style (CPS)* definitional methodology (see [63] for a discussion on several independent discoveries of continuations). The use of continuations seems to be novel in the context of semantic language definitions based on algebraic specification techniques. We have found continuations to be very useful in several of our programming language definitions, not only because they allow us to easily and naturally handle complex control-related constructs, but especially because they lead to an increased efficiency in simulations and formal analysis, sometimes more than an order of magnitude faster than using other techniques. Like for values, at this moment we prefer to avoid defining any particular continuation items; we only define the "stop" continuation, which will stop the computation, together with the essential operator stacking continuation items on top of an existing continuation:

```
fmod CONTINUATION is sorts Continuation ContinuationItem .
  op stop : -> Continuation .
  op _->_ : ContinuationItem Continuation -> Continuation .
endfm
```

We are now ready to put all the state infrastructure together and to define the *state* of a program in our language. A key decision in our definitional methodology is to consider states as sets of *state attributes*, which can be further nested at any degree required by the particular language definition. This way, the semantic equations and rules will be local, and will only have to mention those state attributes *that are needed* to define the semantics of a specific feature, thus increasing the clarity, modularity and efficiency of language definitions. The following specifies several state attributes, which are so common in modern languages that we define them as part of the generic state module. Other attributes will be defined later, as needed by specific language features.

```
fmod STATE is sorts StateAttribute LState . subsort StateAttribute < LState .
  extending ENVIRONMENT . extending STORE . extending CONTINUATION .
  op empty : -> LState .
  op _||_ : LState LState -> LState [assoc comm id: empty] .
  op k : Continuation -> StateAttribute .
```

```
op n : Nat -> StateAttribute .
op m : Store -> StateAttribute .
op t : LState -> StateAttribute .
op e : Env -> StateAttribute .
op x : Continuation -> StateAttribute .
endfm
```

"k" wraps a continuation, "n" keeps the current free memory location, "m" the store, or "memory", "t" the state of a thread, "e" the execution environment of a thread, so it will be part of the state of a thread, and "x" a continuation of exceptions that will also be part of a thread's state. A typical state of a program in our language will have the following structure,

```
t(k(...) || e(...) || x(...) || ...) || ... || t(k(...) || e(...) || x(...) || ...) ||
...
n(N) ||
m(...) ||
...
```

where the local states of one or more threads are wrapped as global state attributes using constructors t(...), where N is a number for the next free location, and m(...) keeps the store. Other state attributes can be added as needed, both inside threads and at the top level. Indeed, we will add top state attributes storing the locks that are taken by threads, and thread local attributes stating how many times each lock is taken by that thread. An important aspect of our semantic definitions, facilitated by Maude's ACI-matching capabilities, is that programming language features will be defined modularly, by referring to only those attributes that are needed. As we can see below, most of the semantic axioms refer to only one continuation! This way, one can add new features requiring new state attributes, *without having to change the already existing semantic definitions*. Moreover, equations and/or rules can be applied concurrently, thus increasing the efficiency of our interpreters and tools.

3.3   Defining the Semantics

The general intuition underlying CPS language definitions is that *control-contexts become data-contexts*, so they are manipulated like any other piece of data. Each continuation contains a sequence of execution obligations, which are stacked and processed accordingly. At each moment there is exactly one expression to be processed, namely the topmost expression in the stack. The following module gives CPS-semantics to generic expressions, that is, integers, identifiers, and lists of expressions. Integer values and several continuation items are needed:

```
mod GENERIC-EXP-SEMANTICS is protecting GENERIC-EXP-SYNTAX . extending STATE .
op int : Int -> Value .
op _->_ : ExpList   Continuation -> Continuation .
op _->_ : ValueList Continuation -> Continuation .
op _->_ : Env       Continuation -> Continuation .
var I : Int . var K : Continuation . var X : Id . vars Env Env' : Env .
var L : Location . var V : Value . var Vl : ValueList .
var M : Store . var E : Exp . var El : ExpList . var R : LState .
eq k(I -> K) = k(int(I) -> K) .
eq k(() -> K) = k((nil).ValueList -> K) .
```

```
  rl t(k(X -> K) || e([X,L] || Env) || R) || m([L,V] || M) =>
     t(k(V -> K) || e([X,L] || Env) || R) || m([L,V] || M) .
 ceq k((E,El) -> K) = k(E -> El -> K) if El =/= nil .
  eq k(V -> El -> K) = k(El -> V -> K) .
  eq k(Vl -> V -> K) = k(V,Vl -> K) .
  eq k(V -> Env -> K) || e(Env') = k(V -> K) || e(Env) .
endm
```

The definitions above deserve some discussion. A continuation of the form "E
-> K" should be read and thought of as one "containing E followed by the rest
of the computation/continuation K", and one of the form "V -> K" as one which
"calculated V as the result of the previous expression at the top, but still has to
process the computation/continuation K". Thus the first two equations above are
clear. Similarly, a continuation "Env -> K" states that the current environment
should be set to Env; this is needed in order to recover environments after processing bindings or function invocations. In fact, environments only need to be
restored after a value is calculated in the modified environment; the last equation does exactly that. The other three equations process a list of expressions
incrementally, returning a continuation containing a list of values at the top;
note that, again, exactly one expression is processed at each moment.

The trickiest axiom in the above module is the rewriting rule, fetching the
value associated to an identifier. It heavily exploits ACI matching and can intuitively be read as follows: if there is any thread whose continuation contains
an identifier X at the top, whose environment maps X to a location L, and whose
rest of resources R are not important, if V is the value associated to L in the
store (note that the store is not part of any thread, because it is shared by all of
them) then simply return V, the value associate to X, on top of the continuation;
the rest of the computation K will know what to do with V. It is very important
to note that this *must be a rule*! This is because the variable X may be shared by
several threads, some of them potentially writing it via a variable assignment, so
a variable read reflects a concurrent aspect of our programming language, whose
behavior may depend upon the behavior of other threads.

The CPS semantics of arithmetic and boolean operators is straightforward:
first place the operation to be performed in the continuation, then the expressions
involved as a list in the desired order of evaluation (they can have side effects);
after they are evaluated to a corresponding list of values, replace them by the
result of the corresponding arithmetic or boolean operator. Note that a new type
of value is needed for booleans:

```
fmod ARITHMETIC-EXP-SEMANTICS is extending ARITHMETIC-EXP-SYNTAX .
  extending GENERIC-EXP-SEMANTICS .
  ops + - * / % : -> ContinuationItem .
  vars E E' : Exp .  vars I I' : Int .  var K : Continuation .
  eq k(E + E' -> K) = k((E,E') -> + -> K) .
  eq k((int(I), int(I')) -> + -> K) = k(int(I + I') -> K) .
*** -, *, /, % are defined similarly
endfm
fmod BOOLEAN-EXP-SEMANTICS is protecting BOOLEAN-EXP-SYNTAX .
  extending ARITHMETIC-EXP-SEMANTICS .
  op bool : Bool -> Value .
  ops == >= <= > < and or not : -> ContinuationItem .
```

```
      vars E E' : Exp . var K : Continuation . vars I I' : Int . vars B B' : Bool .
      eq k((E > E') -> K) = k((E,E') -> > -> K) .
      eq k((int(I),int(I')) -> > -> K) = k(bool(I > I') -> K) .
  *** ==, >=, <=, < are defined similarly
      eq k((E and E') -> K) = k((E,E') -> and -> K) .
      eq k((bool(B),bool(B')) -> and -> K) = k(bool(B and B') -> K) .
  *** 'or' and 'not' are defined similarly
      endfm
```

The CPS semantics of conditionals "if BE then E else E'"is immediate:
freeze E and E' in the current continuation K and then place BE on top and
let it evaluate to a boolean value; then, depending upon the boolean value, un-
freeze either E or E' and continue the computation. Note that a Maude "runtime
error", i.e., a non-reducible term acting as a "core dump", will be obtained if
BE does not evaluate to a boolean value. In fact, our programming language can
be seen as a dynamically typed language; as shown in the lecture notes in [64],
it is actually not hard to define static type checkers, but we do not discuss this
aspect here:

```
  fmod IF-SEMANTICS is protecting IF-SYNTAX . extending BOOLEAN-EXP-SEMANTICS .
    op if : Exp Exp -> ContinuationItem .
    vars BE E E' : Exp .  var K : Continuation .  var B : Bool .
    eq k((if BE then E else E') -> K) = k(BE -> if(E,E') -> K) .
    eq k(bool(B) -> if(E,E') -> K) = k(if B then E else E' fi -> K) .
  endfm
```

Following the same CPS intuitions, the semantics of lists follows easily:

```
  fmod LIST-SEMANTICS is protecting LIST-SYNTAX .
    extending BOOLEAN-EXP-SEMANTICS .
    ops list car cdr cons null? : -> ContinuationItem .
    var E E' : Exp .  var El : ExpList .  var K : Continuation .
    var V : Value .  var Vl : ValueList .  var Env : Env .
    eq k(list(El) -> K) = k(El -> list -> K) .   eq k(Vl -> list -> K) = k([Vl] -> K) .
    eq k(car(E) -> K) = k(E -> car -> K) .        eq k([V,Vl] -> car -> K) = k(V -> K) .
  *** 'cdr', 'cons' and 'null' are defined similarly
  endfm
```

Due to the simplifying rule in FUNCTION-SYNTAX, we only need to worry about
giving semantics to functions of one argument, which can be either an identifier
or "()". Since our language is statically scoped, we need to introduce a new value
for *closures*, freezing the declaration environment of a function (1st equation).
Function applications are defined as usual, by first evaluating the two expressions
involved, the first being expected to evaluate to a closure, and then applying the
first value to the second (a new continuation item, "apply" is needed). Two
cases are distinguished here, when the argument of the function is the unit "()"
or when it is an identifier X. In both cases the current environment is stored in
the continuation, to be recovered later (after the evaluation of the body of the
function) via the last equation in GENERIC-EXP-SEMANTICS, and the body of the
function is evaluated in its declaration environment, that is frozen in the closure.
When the function has an argument, a new location also needs to be created.

```
fmod FUNCTION-SEMANTICS is protecting FUNCTION-SYNTAX . extending GENERIC-EXP-SEMANTICS .
  op cl : Name Exp Env -> Value .
  op apply : -> ContinuationItem .
  var A : Name .  vars F E : Exp .  var K : Continuation .  vars Env Env' : Env .
  var X : Id .  var N : Nat .  var R : LState .  var V : Value .  var M : Store .
  eq k((fun A -> E) -> K) || e(Env) = k(cl(A,E,Env) -> K) || e(Env) .
  eq k(F E -> K) = k(F,E -> apply -> K) .
  eq k((cl((),E,Env), nil) -> apply -> K) || e(Env') = k(E -> Env' -> K) || e(Env) .
  eq t(k((cl(X,E,Env), V) -> apply -> K) || e(Env') || R) || n(N) || m(M) =
     t(k(E -> Env' -> K) || e(Env[X <- loc(N)]) || R) || n(N + 1) || m(M[loc(N) <- V]) .
endfm
```

LET and LETREC create new memory locations and change the execution environment. With the provided infrastructure, they are however quite easy to define. Note first that the desugaring translation in the module BINDING-SYNTAX reduces any list of bindings to a triple (#,Xl,El), where # is the number of bindings, Xl is the list of identifiers to be bound, and El is the list of corresponding binding expressions. If let (#,Xl,El) in E is the current expression at the top of the continuation of a thread whose current environment is Env and whose rest of resources is R, and if N is the next available location (this is a global counter), then the CPS semantics works intuitively as follows: (1) freeze the current environment in the continuation, to be restored after the evaluation of E; (2) place E in the continuation; (3) generate # fresh locations and place in the continuation data-structure the information that these locations will be assigned to the identifiers in Xl at the appropriate moment, using an appropriate continuation item; (4) place the expressions El on top of the continuation; (5) once El is evaluated to a list of values Vl, they are stored at the new locations and the environment of the thread is modified accordingly, preparing for the evaluation of E; (6) after E is evaluated, the original environment will be restored, thanks to (1) and the last equation in GENERIC-EXP-SEMANTICS. All these technical steps can be compactly expressed with only two equations, again relying heavily on the ACI matching capabilities of Maude. Note that, despite their heavy use of memory, these equations do *not* need to be rules, because they can be executed deterministically regardless of the behavior of other threads. The fact that threads "compete" on the counter for the next available location N is immaterial, because there is no program whose behavior is influenced by which thread grabs N first.

```
fmod LET-SEMANTICS is protecting LET-SYNTAX . extending GENERIC-EXP-SEMANTICS .
  op '(_,_') : IdList LocationList -> ContinuationItem .
  vars # N : Nat .  var Xl : IdList .  var El : ExpList .  var E : Exp .
  var K : Continuation .  var Env : Env .  var R : LState .  var M : Store .
  var Ll : LocationList .  var Vl : ValueList .
  eq t(k(let (#,Xl,El) in E -> K) || e(Env) || R) || n(N) =
     t(k(El -> (Xl,locs(N,#)) -> E -> Env -> K) || e(Env) || R) || n(N + #) .
  eq t(k(Vl -> (Xl,Ll) -> K) || e(Env) || R) || m(M) =
     t(k(K) || e(Env[Xl <- Ll]) || R) || m(M[Ll <- Vl]) .
endfm
```

The let rec construct gives a statically scoped language an enormous power by allowing one to define recursive functions. Semantically, the crucial difference between let and let rec is that the latter evaluates the bindings expressions El in the modified environment rather than in the original environment. Therefore, one first creates the new environment by mapping Xl to # fresh locations, then evaluates El, then stores their values at the new locations, then evaluates E and then restores the environment. This way, functions declared with let rec see each other's names in their closures, so they can call each other:

```
fmod LETREC-SEMANTICS is protecting LETREC-SYNTAX . extending GENERIC-EXP-SEMANTICS .
  op _->_ : LocationList Continuation -> Continuation .
  vars # N : Nat .  var Xl : IdList .  var El : ExpList .  var E : Exp .
  var K : Continuation .  var Env : Env .  var R : LState .  var M : Store .
  var Ll : LocationList .  var Vl : ValueList .
  eq t(k(let rec (#,Xl,El) in E -> K) || e(Env) || R) || n(N) =
    t(k(El -> locs(N,#) -> E -> Env -> K) || e(Env[Xl <- locs(N,#)]) || R) || n(N + #) .
  eq t(k(Vl -> Ll -> K) || R) || m(M) = t(k(K) || R) || m(M[Ll <- Vl]) .
endfm
```

So far, none of the language constructs had side effects. Variable assignments, X := E, evaluate E and store its value at the existing location of X. Therefore, X is expected to have been previously bound, otherwise a "runtime error" will be reported. It is very important that the actual writing of the value is performed using a rewriting rule, not an equation! This is because variable writing is a concurrent action, potentially influencing the execution of other threads that may read that variable. To distinguish this concurrent value writing from the value writing that occurred as part of the semantics of let rec, defined using the last equation in the module LETREC-SEMANTICS, we use a different continuation constructor for placing a location on a continuation ("L => K"):

```
mod ASSIGNMENT-SEMANTICS is extending ASSIGNMENT-SYNTAX .
  extending GENERIC-EXP-SEMANTICS .
  var X : Name .  var E : Exp .  var Env : Env .  var K : Continuation .
  var L : Location .  var V : Value .  var M : Store .  var R : LState .
  op _=>_ : Location Continuation -> Continuation .
  eq k((X := E) -> K) || e([X,L] || Env) = k(E -> L => noValue -> K) || e([X,L] || Env) .
  rl t(k(V -> L => K) || R) || m(M) => t(k(K) || R) || m(M[L <- V]) .
endm
```

Blocks are quite straightforward: the semicolon-separated expressions are evaluated in order and the result of the evaluation of the block is the value of the last expression; the values of the other expressions except the last one are ignored. Therefore, the expressions in a block are used for their side effects:

```
fmod BLOCK-SEMANTICS is extending BLOCK-SYNTAX . extending GENERIC-EXP-SEMANTICS .
  op ignore : -> ContinuationItem .
  var E : Exp . var Eb : ExpBlock . var K : Continuation . var V : Value .
  eq k({E} -> K) = k(E -> K) .
  eq k({E ; Eb} -> K) = k(E -> ignore -> {Eb} -> K) .
  eq k(V -> ignore -> K) = k(K) .
endfm
```

There is nothing special in the CPS semantics of loops: the body of the loop followed by its condition are placed on top of the continuation at each iteration, and the loop is terminated when the condition becomes false. The evaluation of loops returns no value, so loops are also used just for their side effects:

```
fmod LOOP-SEMANTICS is extending LOOP-SYNTAX . extending BOOLEAN-EXP-SEMANTICS .
  op while(_,_) : Exp Exp -> ContinuationItem .
  vars BE E : Exp . var Vl : ValueList . var K : Continuation .
  eq k((while BE E) -> K) = k(BE -> while(BE,E) -> K) .
  eq k((Vl,bool(true)) -> while(BE,E) -> K) = k((E,BE) -> while(BE,E) -> K) .
  eq k((Vl,bool(false)) -> while(BE,E) -> K) = k(noValue -> K) .
endfm
```

We next define the semantics of exceptions. Whenever an expression of the form try E catch E' is encountered, E' is first evaluated. Then E is evaluated; if throw(E'') is encountered during the evaluation of E, then the entire control context within E is immediately discarded and the value of E'' is passed to E', which was previously supposed to evaluate to a function; if no exception is thrown during the evaluation of E, then E' is discarded and the value of E is returned as the value of try E catch E'. An interesting technicality here is that the above mechanism can be elegantly implemented by maintaining an additional continuation within each thread, wrapped within "x(...)", freezing and stacking the control contexts, i.e., the continuations, at the times when the try E catch E' expressions are encountered (note that these can be nested):

```
fmod EXCEPTION-SEMANTICS is protecting EXCEPTION-SYNTAX . extending FUNCTION-SEMANTICS .
  op try : Exp -> ContinuationItem .
  op popx : -> Continuation .
  op '(_,_') : Value Continuation -> ContinuationItem .
  op throw : -> ContinuationItem .
  vars E E' : Exp . vars K K' EX : Continuation . vars V V' : Value .
  eq k(try E catch E' -> K) = k(E' -> try(E) -> K) .
  eq k(V' -> try(E) -> K) || x(EX) = k(E -> popx) || x((V',K) -> EX) .
  eq k(V -> popx) || x((V',K) -> EX) = k(V -> K) || x(EX) .
  eq k(throw(E) -> K) = k(E -> throw -> K) .
  eq k(V -> throw -> K') || x((V',K) -> EX) = k((V',V) -> apply -> K) || x(EX) .
endfm
```

The only feature left to define is threads. In what follows we assume that one already has a definition for sets of integers with membership, INT-SET, and one for sets of pairs of integers, COUNTER-SET. Both of these are trivial to define, so we do not discuss them here. The former will be used to store all the synchronization objects, or *locks*, that are already acquired, and the latter to store how many times a thread has acquired a lock, so that it knows how many times it needs to release it. These are wrapped as program state attributes:

```
mod THREAD-SEMANTICS is protecting THREAD-SYNTAX . extending GENERIC-EXP-SEMANTICS .
  protecting INT-SET . protecting COUNTER-SET .
  op c : CounterSet -> StateAttribute .
  op b : IntSet -> StateAttribute .
```

A new type of value is needed, namely one for locks. A lock is just an integer, which is wrapped with lockv to keep it distinct from other integer values:

```
op lockv : Int -> Value .
```

Newly created threads are executed for their side effects. At the end of their execution, threads release their locks and kill themselves. Therefore, we introduce a new continuation, die, to distinguish the termination of a thread from the termination of the main program. The creation of a new thread is a no value operation. The new thread inherits the execution environment from its parent:

```
op die : -> Continuation .
ops lock acquire release : -> ContinuationItem .
var E : Exp . var K : Continuation . var Env : Env . var R : LState . var I : Int .
var V : Value . var Cs : CounterSet . var Is : IntSet . var N : Nat . var Nz : NzNat .
eq t(k(spawn(E) -> K) || e(Env) || R) = t(k(noValue -> K) || e(Env) || R) ||
   t(k(E -> die) || e(Env) || x(stop) || c(empty)) .
eq t(k(V -> die) || c([I,N] || Cs) || R) || b(I || Is) =
   t(k(V -> die) || c(Cs) || R) || b(Is) .
eq t(k(V -> die) || c(empty) || R) = empty .
```

Locks are values which can be handled like any other values in the language. In particular, they can be passed to and returned as results of functions; lock(E) evaluates E to an integer value I and then generates the value lockv(I):

```
eq k(lock(E) -> K) = k(E -> lock -> K) .
eq k(int(I) -> lock -> K) = k(lockv(I) -> K) .
```

Acquiring a lock needs to distinguish two cases. If the current thread already has the lock, reflected in the fact that it has a counter associated to that lock, then it just needs to increment that counter. This operation is not influenced by, and does not influence, the execution of other threads, so it can be defined using an ordinary equation. The other case, when a thread wants to acquire a lock which it does not hold already, needs to be a rewriting rule for obvious reasons: the execution of other threads may be influenced, so the global behavior of the program may be influenced. Once the new lock is taken, a thread local counter is created and initialized to 0, and the lock is declared "busy" in b(...). This rule is conditional, in that the lock can be acquired only if it is not busy:

```
eq k(acquire(E) -> K) = k(E -> acquire -> K) .
eq k(lockv(I) -> acquire -> K) || c([I, N] || Cs) =
   k(noValue -> K) || c([I, N + 1] || Cs) .
crl t(k(lockv(I) -> acquire -> K) || c(Cs) || R) || b(Is) =>
   t(k(noValue -> K) || c([I, 0] || Cs) || R) || b(I || Is)  if not(I in Is) .
```

Dually, releasing a lock also involves two cases. However, both of these can be safely defined with equations, because threads do not need to compete on releasing locks:

```
eq k(release(E) -> K) = k(E -> release -> K) .
eq k(lockv(I) -> release -> K) || c([I, Nz] || Cs) =
   k(noValue -> K) || c([I, Nz - 1] || Cs) .
eq t(k(lockv(I) -> release -> K) || c([I, 0] || Cs) || R) || b(I || Is) =
   t(k(noValue -> K) || c(Cs) || R) || b(Is) .
endm
```

All the features of our programming language have been given CPS rewriting logic semantics, so we can now put all the semantic specifications together and complete the definition of our language:

```
fmod LANGUAGE-SEMANTICS is extending ARITHMETIC-EXP-SEMANTICS .
  extending BOOLEAN-EXP-SEMANTICS .   extending IF-SEMANTICS .
  extending LET-SEMANTICS .           extending LETREC-SEMANTICS .
  extending FUNCTION-SEMANTICS .      extending LIST-SEMANTICS .
  extending ASSIGNMENT-SEMANTICS .
  extending BLOCK-SEMANTICS .         extending LOOP-SEMANTICS .
  extending EXCEPTION-SEMANTICS .     extending THREAD-SEMANTICS .
  op eval : Exp -> [Value] .
  op [_] : LState -> [Value] [strat(1 0)] .
  var E : Exp .   vars R S : LState .   var V : Value .
  eq eval(E) = [t(k(E -> stop) || e(empty) || x(stop) || c(empty)) ||
                n(0) || m(empty) || b(empty)] .
  eq [t(k(V -> stop) || R) || S] = V .
endfm
```

The main operator that enables all the semantic definitions above is `eval`, which may or may not return a proper value. As the definition of the auxiliary operator `[_]` shows, `eval` returns a proper value if and only if the original thread (the only one whose continuation is built on top of `stop`) evaluates to a proper value V. The definition of `eval` above also shows the various state attributes involved, as well as their nesting and grouping: the thread state attribute, `t(...)`, includes a continuation (`k`), an environment (`e`), an exception continuation (`x`), and a lock counter set (`c`); the other global state attributes, laying at the same top level as the thread attributes, are a counter for the next free location (`n`), a "memory" wrapping a mapping from locations to values (`m`), and a set of "busy" locks (`b`).

3.4   Getting an Interpreter for Free

Since Maude can efficiently execute rewriting logic specifications, an immediate benefit of defining the semantics of a programming language in rewriting logic is that we obtain an *interpreter* for that language with no extra effort. All what we have to do is to "rewrite" terms of the form `eval(E)`, which should reduce to values. For example, the evaluation of the following factorial program

```
rew eval(
  let rec f n =
    if n == 0 then 1 else n * f(n - 1)
  in f 100
) .
```

takes 5151 rewrites and terminates in 64ms[8]. with the following result:

```
result Value: int(93326215443944152681699238856266700490715968264381621468592963895217599
9932299156089414639761565182862536979208272237582511852109168640000000000000000000000000000)
```

The following recursive program calculating the product of elements in a list is indeed evaluated to `int(0)` (in 716 rewrites). This program is "inefficient" because the product function returns normally from its recursive calls when a 0 is encountered, which can be quite time consuming in many situations:

---

[8] All the performance results in this paper were measured on a 2.4GHz PC.

```
rew eval(
  let rec p l =
    if null?(l) then 1 else if car(l) == 0 then 0 else car(l) * (p cdr(l))
  in p list(1,2,3,4,5,6,7,8,9,0,10,11,12,13,14,15,16,17,18,19,20)
) .
```

Since our language has exceptions, a better version of the same program (reducing in 675 rewrites) is one which throws an exception when a 0 is encountered, thus exiting all the recursive calls at once:

```
rew eval(
  let p l = try let rec a l =
                  if null?(l) then 1
                  else if car(l) == 0 then throw 0 else car(l) * (a cdr(l))
                in a l
            catch fun x -> x
  in p list(1,2,3,4,5,6,7,8,9,0,10,11,12,13,14,15,16,17,18,19,20)
) .
```

To illustrate the imperative features of our language, let us consider Collatz' conjecture, stating that the procedure, dividing n by 2 if it is even and multiplying it by 3 and adding 1 if it is **odd**, eventually reaches 1 for any n). For our particular n below, it takes 73284 rewrites in 0.3s to evaluate to int(813):

```
rew eval(
  let n = 21342423543653426527423676545 and c = 0
  in {while n > 1 {
         if 2 * (n / 2) == n then n := n / 2 else n := 3 * n + 1 ;
         c := c + 1
       } ;
       c }
) .
```

Let us next illustrate some concurrent aspects of our language. The following program spawns a thread that assigns 1 to a and then recursively increments a counter c until a becomes indeed 1. Any possible value for the counter can be obtained, depending upon when the spawned thread is scheduled for execution:

```
rew eval(
  let a = 0 and c = 0 in {
    spawn(a := 1) ;
    let rec f() = if a == 1 then c else {c := c + 1 ; f()} in f()
  }
) .
```

We are currently letting Maude schedule the execution of threads based on its internal default scheduler for applications of rewrite rules, which in the example above leads to an answer int(0). Note, however, that one can also use Maude's fair rewrite command **frew** instead of **rew**, or one can even define one's own scheduler using Maude's meta-level capabilities. Even though we will not discuss thread scheduling aspects in this paper, in the next section we will show how one can use Maude's **search** capability to find executions leading to other possible results for the program above, for example int(10).

An inherent problem in multithreaded languages is that several threads may access the same location at the same time, and if at least one of these accesses is a write this can lead to *dataraces*. The following program contains a datarace:

```
rew eval(
  let x = 0 in {
    spawn(x := x + 1) ;
    spawn(x := x + 1) ;
    x
  }
) .
```

Maude's default scheduler happens to schedule the two spawned threads above such that no datarace occurs, the reported answer being int(2). However, under different thread interleavings the reported value of x can also be 0 or even 1. The latter reflects the datarace phenomenon: both threads read the value of x before any of them writes it, and then they both write the incremented value. Using search, we show in the next section that both int(0) and int(1) can be valid results of the program above. Thread synchronization mechanisms are necessary in order to avoid dataraces. We use locks for synchronization in our language. For example, the following program is datarace free, because each thread acquires the lock lock(1) before accessing its critical region. Note, however, that the final result of this program is still non-deterministic (can be either 2 or -1):

```
rew eval(
  let a = 1 and b = 1 and x = 0 and l = lock(1) in {
    spawn {acquire l ; x := x + 1 ; release l ; a := 0} ;
    spawn {acquire l ; x := x + 1 ; release l ; b := 0} ;
    if (a == 0) and (b == 0) then x else -1
  }
) .
```

3.5 Specifying Java and the JVM

The language presented above was selected and designed to be as simple as possible, yet including a substantial range of features, such as high-order and imperative features, static scoping, recursion, exceptions and concurrency. However, we are actively using the rewriting logic semantics approach to formally define different programming paradigms and large fragments of several languages, including Scheme, OCaml, ML, Pascal, Java, and JVM, several of them covered in a programming language design course at UIUC [64].

Java has been recently defined at UIUC by Feng Chen in about three weeks, using a CPS semantics as above, with 600 equations and 15 rewrite rules. Azadeh Farzan has developed a more direct rewriting logic specification for the JVM, not based on continuations, specifying about 150 out of 250 bytecode instructions with around 300 equations and 40 rewrite rules. The continuations-based style used in this paper should be regarded as just a definitional methodology, which may not be appropriate for some languages, especially for lower level ones. Both the Java and the JVM specifications include multithreading, inheritance, poly-morphism, object references, and dynamic object allocation. We do not support native methods nor many of the Java built-in libraries at present. The definition of Java follows closely the style used to define our sample language above, with states consisting of multisets of potentially nested state attributes. A new type of value is introduced for objects, wrapping also the name of the class that the object is an instance of, which is necessary in order to have access to that object's methods. The essential difference in the definitional styles of Java and the JVM is that the latter follows the object paradigm of Maude [42], considering objects also as part of the state multiset structure; and method calls are translated into messages that objects can send to each other, by placing them into the multiset state as well. Rewrites (with rewrite rules and equations) in this multiset model the changes in the state of the JVM.

4   Formal Analysis of Concurrent Programs

Specifying formally the rewriting logic semantics of a programming language in Maude, besides providing an increased understanding of all the details underlying a language design, also yields a prototype interpreter for free. Furthermore, a solid foundational framework for program analysis is obtained. It is conceptually meaningless to speak about rigorous verification of programs without a formal definition of the semantics of that language. Once a definition of a language is given in Maude, thanks to generic analysis tools for rewriting logic specifications that are efficiently implemented and currently provided as part of the Maude system, we additionally get the following important analysis tools also *for free*:

1. a *semi-decision procedure* to find failures of safety properties in a (possibly infinite-state) concurrent program using Maude's `search` command;
2. an LTL *model checker* for finite-state programs or program abstractions;
3. a *theorem prover* (Maude's ITP [20]) that can be used to semi-automatically prove programs correct.

We only focus on the first two items in this paper, because they are entirely auto-matic (except of course for the need to equationally define the atomic predicates of interest in temporal logic formulas).

4.1   Search

We have seen several examples where concurrent programs can have quite non-deterministic behaviors due to many possible thread interleavings, some of them leading to undesired behaviors, e.g., dataraces, due to lack of synchronization. Using Maude's `search` command, one can search a potentially infinite state space for behaviors of interest. Since such a search is performed in a breadth-first manner, if any safety violation exists then it will eventually be found, i.e., this is a semi-decision procedure for finding such errors. For example, the following two-threaded program which evaluates to 0 in Maude under its default scheduling, can be shown to evaluate to any possible integer value. It takes 11ms to find an interleaving that leads to int(10), after exploring 108 states:

```
search [1] eval( let a = 0 and c = 0 in {
                    spawn(a := 1) ;
                    let rec f() = if a == 1 then c else {c := c + 1 ; f()} in f()
                 } ) =>* int(10) .
```

One can show that the poorly synchronized program in Section 3.4 has a datarace,

```
search [1] eval( let x = 0 in {
                    spawn(x := x + 1) ;
                    spawn(x := x + 1) ;
                    x
                 } ) =>+ int(1) .
```

and also that the properly synchronized version of it is datarace free:

```
search [1] eval( let a = 1 and b = 1 and x = 0 and l = lock(1) in {
                    spawn {acquire l ; x := x + 1 ; release l ; a := 0} ;
                    spawn {acquire l ; x := x + 1 ; release l ; b := 0} ;
                    if (a == 0) and (b == 0) then x else -1
                 } ) =>+ int(1) .
```

The above returns "No solution", after exploring all 90 possible states in 23ms. If one wants to see the state space generated by the previous search command, one can type the command "show search graph". An interesting example showing that dataraces can be arbitrarily dangerous was proposed by J. Moore [54], where two threads performing the assignment "c := c + c" for some shared variable c, can lead to any possible integer value for c. The following shows how one can test whether the value 25 can be reached. It takes Maude about 1s to explore 4696 states and find a possible interleaving leading to the final result 25:

```
search [1] eval( let rec c = 1 and f() = {c := c + c ; f()} in {
                    spawn f() ;
                    spawn f() ;
                    c
                 } ) =>! int(25) .
```

4.2 Model Checking

When the state space of a concurrent program is finite, one can exhaustively analyze all its possible executions and check them against temporal logic properties. Currently, Maude provides a builtin explicit-state model checker for linear temporal logic (LTL) comparable in speed to SPIN [24], which can be easily used to model check programs once a programming language semantics is defined as a rewriting logic specification. The module MODEL-CHECKER, part of Maude's distribution, exports sorts State, Prop, and a binary "satisfaction" operator _|=_ : State Prop -> Bool. In order to define temporal properties to model-check, the user has to first define state predicates using the satisfaction operation.

To exemplify this analysis technique, we next consider the classical *dining philosophers* problem. The property of interest in this example is that the program terminates, so we only need one state predicate, terminates, which holds whenever a proper value is obtained as a result of the execution of the program (note that eval may not always return a proper value; its result was the kind [Value]):

```
fmod CHECK is extending MODEL-CHECKER .   extending LANGUAGE-SEMANTICS .
  subsort Value < State .
  op terminates : -> Prop .
  eq V:Value |= terminates = true .
endfm
```

We can model check a five dining philosophers program as follows:

```
red modelCheck(eval( let n = 5 and i = 1 and
                     f x = { acquire lock(x) ; acquire lock(x + 1) ;
                             --- eat
                             release lock(x + 1) ; release lock(x) }
              in { while i < n
                       { spawn(f i) ; i := i + 1 } ;
                     acquire lock(n) ; acquire lock(1) ;
                     --- eat
                     release lock(1) ; release lock(n) } ), <> terminates) .
```

Maude's model checker detects the deadlock and returns a counterexample trace in about 0.5s. If one fixes this program to avoid deadlocks, for example as follows:

```
red modelCheck(eval( let n = 5 and i = 1 and
                     f x = if x % 2 == 1
                           then { acquire lock(x); acquire lock(x + 1);
                                  --- eat
                                  release lock(x + 1); release lock(x) }
                           else { acquire lock(x + 1); acquire lock(x);
                                  --- eat
                                  release lock(x); release lock(x + 1) }
              in { while i < n
                       { spawn(f i) ; i := i + 1 } ;
                     if n % 2 == 1
                     then { acquire lock(n); acquire lock(1);
                            --- eat
                            release lock(1); release lock(n) }
                     else { acquire lock(1); acquire lock(n);
                            --- eat
                            release lock(n); release lock(1) } } ), <> terminates) .
```

then the model-checker analyzes the entire state space and returns true, meaning that the program will terminate for any possible thread interleaving.

4.3   Formal Analysis of Java Multithreaded Programs

In joint work with Azadeh Farzan and Feng Cheng, we are using Maude to develop JavaFAN (Java Formal ANalyzer) [27,26], a tool in which Java and JVM code can be executed and formally analyzed. JavaFAN is based on Maude rewriting logic specifications of Java and JVM (see Section 3.5). Since JavaFAN is intended to be a Java analysis tool rather than a programming language design platform, we have put a special emphasis on its *efficiency*. When several ways to give semantics to a feature were possible, we have selected the one which performed better on our benchmarks, instead of the mathematically simplest one. In this section we discuss JavaFAN and some of the experiments that we performed with it. They support the claim that the rewriting logic approach to formal semantics of programming languages presented in this paper is not only a clean theoretical model unifying SOS and equational semantics, but also a potentially powerful practical framework for developing software analysis tools.

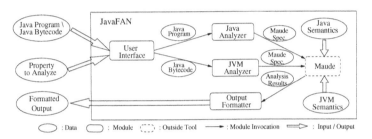

Fig. 1. Architecture of JavaFAN.

Figure 1 presents the architecture of JavaFAN. The *user interface* module hides Maude behind a user-friendly environment. It also plays the role of a dispatcher, sending the Java source or the bytecode to Java or JVM analyzers, respectively. The analyzers wrap the input programs into Maude modules and invoke Maude, which analyzes the code based on the formal specifications of the Java language and of the JVM. The output formatter collects the output of Maude, transforms it into a user-readable format, and sends it to the user.

We next discuss some of the examples analyzed with JavaFAN and comparisons to other similar tools. The *Remote Agent* (RA) is a spacecraft controller, part NASA's Deep Space 1 shuttle, that deadlocked 96 million miles from Earth due to a datarace. This example has been extensively studied in [32, 33]. JavaFAN's search found the deadlock in 0.1 seconds at the source code level and in 0.3 seconds at the bytecode level, while the tool in [61] finds it in more than 2 seconds. Another comparison with [61] was done on a 2 stage pipeline code, each stage executing as a separate thread, against a property taken from [61]. JavaFAN model checks the property in 17 minutes, while the tool in [61], without partial order reduction optimizations[9], does it in more than 100 minutes. JavaFAN can detect the deadlock for up to 9 philosophers. Other Java

---

[9] JavaFAN is currently just a brute force, unoptimized explicit state model checker.

model checkers, with support for heuristics and abstraction techniques such as Java PathFinder (JPF) [88,11,34], can do larger numbers. If the deadlock potential is removed, like in Section 4.2, thus diminishing the role of heuristics, then JavaFAN can prove the program deadlock-free for up to 7 philosophers, while JPF cannot deal with 4 philosophers (on the same program). All these examples as well as the JavaFAN system are available on the web [25].

4.4   Performance of the Formal Analysis Tools

There are two reasons for the efficiency of the formal analysis tools for languages whose rewriting logic semantics is given in Maude, and in particular for which JavaFAN compares favorably with more conventional Java analysis tools:

1. The high performance of Maude for execution, search, and model checking;
2. The optimized equational and rule definitions.

Maude's rewriting engine is highly optimized and can perform millions of rewrite steps per second, while its model checker is comparable in speed with SPIN [24]. In addition to these, we have used performance-enhancing specification techniques, including: expressing as equations the semantics of all deterministic computations, and as rules only concurrent computations (since rewriting happens *modulo* equations, only rules contribute to state space size); favoring *unconditional* equations and rules over less efficient conditional versions; and using a continuation passing style in semantic equations.

5   SOS and Equational Semantics Revisited

Now that rewriting logic semantics has been explained and has been illustrated in detail, we take a second look at how equational semantics and SOS are unified within rewriting logic semantics. We also explain how their respective limitations are overcome within this broader semantic framework.

5.1   Unification of Equational Semantics

If R is empty in a rewrite theory $\mathcal{R} = (\Sigma, E, \phi, R)$, then ϕ is irrelevant and $\mathcal{R}$ becomes an *equational theory*, and the initial model $T_{Reach(\mathcal{R})}$ becomes in essence the *initial algebra* $T_{\Sigma/E}$. Therefore, equational logic is a *sublogic* of rewriting logic, and *initial algebra semantics* is a special case of rewriting logic's initial model semantics. That is, equational semantics is a *special case* of rewriting logic semantics, namely the case when $R = \varnothing$. Higher-order semantic equations can be integrated in two alternative ways. On the one hand, we can make everything first-order by means of an *explicit substitution* calculus or the use of combinators. On the other hand, we can embed higher-order semantic equations within a higher-order version of rewriting logic such as Stehr's open calculus of constructions (OCC) [71]. Either way, since OCC and many other higher-order calculi can be faithfully represented in first-order rewriting logic [72,71], it is possible to execute such definitions in a rewriting logic language such as Maude.

Integrating equational semantics within rewriting logic makes the limitations in handling concurrency mentioned in Section 1.1 disappear, since all deterministic computations of a language can still be specified by equations, but the means missing in equational semantics to properly handle concurrency are now provided by rewrite rules. Furthermore, the extension from equational logic to rewriting logic is conservative and all the good proof- and model-theoretic properties are preserved in the extension. This leaves us with the pending issue of modularity, which is discussed in Section 5.3.

5.2   Unification of SOS and Reduction Semantics

SOS can also be integrated within rewriting logic. This has been understood from the early stages of rewriting logic [41,48,40], and has led to several implementations of SOS definitions [9,81]. Intuitively, an SOS rule of the form,

$$\frac{P_1 \longrightarrow P_1' \quad \dots \quad P_n \longrightarrow P_n'}{Q \longrightarrow Q'}$$

corresponds to a rewrite rule with *rewrites in its condition*. There is however an important difference between the meaning of a transition $P \longrightarrow Q$ in SOS and a sequent $P \longrightarrow Q$ in rewriting logic. In SOS a transition $P \longrightarrow Q$ is always a *one-step* transition. Instead, because of **Reflexivity** and **Transitivity**, a rewriting logic sequent $P \longrightarrow Q$ may involve many rewrite steps; furthermore, because of the **Congruence**, such steps may correspond to rewriting subterms.

Since the conditions in a conditional rewrite rule may involve many rewrite steps, whereas the transitions in the condition of an SOS rule are one-step transitions, in order to faithfully represent an SOS rule we have somehow to "dumb down" the rewriting logic inference system. Of course we do not want to actually change rewriting logic's inference rules: we just want to *get the effect* of such a change, so that in fact only the **Replacement** rule is used. This can be achieved by representing an SOS specification as a suitable rewrite theory that, due to its construction, precludes the application of the other inference rules in the logic. We explain how this can be done for an SOS specification consisting of unlabeled SOS rules of the general form described above. We can think of such an SOS specification as a pair $\mathcal{S} = (\Sigma, R)$, where Σ is a many-sorted signature, and the rules R are of the general form described above, where the Ps and Qs are Σ-terms having the same sort whenever they appear in the same transition. The SOS rules are then applied to substitution instances of the patterns appearing in each rule in the usual SOS way. The corresponding rewrite theory representing $\mathcal{S}$ is denoted $\widehat{\mathcal{S}}$ and is $\widehat{\mathcal{S}} = (\widehat{\Sigma}, OP, \phi, \widehat{R})$, where:

– $\widehat{\Sigma}$ is the MEL signature obtained from Σ by:

- adding a kind $[s]$ for each sort s in Σ, so that the set of sorts for kind $[s]$ is the singleton set $\{s\}$
- adding for each $f : s_1 \ldots s_n \longrightarrow s$ in Σ an operator $f : [s_1] \ldots [s_n] \longrightarrow [s]$
- adding for each kind $[s]$ two operators $[\_] : [s] \longrightarrow [s]$ and $\langle \_ \rangle : [s] \longrightarrow [s]$.
  - OP is the set of axioms associating to each operator $f : s_1 \ldots s_n \longrightarrow s$ in Σ the membership $\forall (x_1 : s_1, \ldots , x_n : s_n) \, f(x_1, \ldots , x_n) : s$, so that terms of the old sorts in Σ remain well-sorted in $\widehat{\Sigma}$
  - ϕ declares all arguments in all operators in $\widehat{\Sigma}$ *frozen*, and
  - $\widehat{R}$ has for each SOS rule in R a corresponding rewrite rule

$$[Q] \longrightarrow \langle Q' \rangle \quad \text{if} \quad [P_1] \longrightarrow \langle P_1' \rangle \wedge \ldots \wedge [P_n] \longrightarrow \langle P_n' \rangle,$$

The key result is then the following lemma, that we state without proof:

Lemma 1. *For any ground Σ-terms t, t' of the same sort, we have*

$$\mathcal{S} \vdash_{SOS} t \longrightarrow t' \quad \Leftrightarrow \quad \widehat{\mathcal{S}} \vdash_{RWL} [t] \longrightarrow \langle t' \rangle$$

where $\vdash_{SOS}$ and $\vdash_{RWL}$ denote the SOS and rewriting logic inference systems.

In general, SOS rules may have *labels*, *decorations*, and *side conditions*. In fact, there are many SOS rule variants and formats. For example, additional semantic information about stores or environments can be used to *decorate* an SOS rule. Therefore, showing in detail how SOS rules in each particular variant or format can be faithfully represented by corresponding rewrite rules would be a tedious business. Fortunately, Peter Mosses, in his modular structural operational semantics (MSOS) [57,58,55], has managed to neatly pack all the various pieces of semantic information usually scattered throughout a standard SOS rule *inside rule labels*, where now labels have a record structure whose fields correspond to the different semantic components (the store, the environment, action traces for processes, and so on) *before and after* the transition thus labeled is taken. The paper [47] defines a faithful representation of an MSOS specification $\mathcal{S}$ as a corresponding rewrite theory $\tau(\mathcal{S})$, provided the MSOS rules in $\mathcal{S}$ are in a suitable normal form. Such MSOS rules do in fact have labels that include any desired semantic information, and can have equational side conditions. A semantic equivalence result similar to the above lemma holds between transitions in $\mathcal{S}$ and one-step rewrites in $\tau(\mathcal{S})$ [47]. This shows the MSOS specifications are faithfully represented by their rewriting logic translations.

A different approach also subsumed by rewriting logic semantics is sometimes described as *reduction semantics*. It goes back to Berry and Boudol's Chemical Abstract Machine (Cham) [4], and has been adopted to give semantics to different concurrent calculi and programming languages (see [4,51] for two early references). Since the 1990 San Miniato Workshop on Concurrency, where both the Cham and rewriting logic were presented [23], it has been clearly understood that these are two closely related formalisms, so that each Cham can be naturally seen as a rewrite theory (see [41] Section 5.3.3, and [4]). In essence, a reduction semantics, either of the Cham type or with a different choice of basic primitives, can be naturally seen as a special type of rewrite theory $\mathcal{R} = (\Sigma, A, \phi, R)$, where A consists of *structural axioms*, e.g., associativity and commutativity of multiset union for the Cham[10], and R is typically a set of *unconditional* rewrite rules. The frozenness information ϕ is specified by giving explicit inference rules, stating which kind of *congruence* is permitted for each operator for rewriting purposes.

Limitations of SOS similar to those pointed out in Section 1.1 were also clearly perceived by Berry and Boudol, so that the Cham is proposed not as a variant of SOS, but as an *alternative semantic framework* (see [4], Section 2.3). Indeed, an important theme is overcoming the *rigidity of syntax*, forcing traditional SOS to express communication in a centralized, interleaving way, whereas the use of associativity and commutativity and the *locality* of rewrite rules allows a more natural expression of local concurrent interactions. On this point rewriting logic semantics and reduction semantics are in full agreement. Four further advantages added by rewriting logic semantics to overcome other limitations of SOS mentioned in Section 1.1 are: (i) the existence of a model-theoretic semantics having initial models, that smoothly integrates the model theory of algebraic semantics as a special case and serves as a basis for inductive and temporal logic reasoning; (ii) the more general use of equations not only as *structural axioms* A (e.g., AC of multiset union for the Cham) but also as *semantic equations* E_0 that are Church-Rosser modulo A, so that in general we have $E = E_0 \cup A$; (iii) allowing *conditional rewrite rules* which permits a natural integration of SOS within rewriting logic; and (iv) the existence of high-performance implementations supporting both execution and formal analysis. This brings us to the last limitation mentioned in Section 1.1 for both equational semantics and SOS, namely modularity.

5.3   Modularity

Both equational semantics and SOS are notoriously *unmodular*. That is, when a new kind of feature is added to the existing formal specification of a language's semantics, it is often necessary to introduce *extensive redefinitions* in the earlier specification. One would of course like to be able to define the semantics of each feature in a modular way *once and for all*, but this is easier said than done.

Rewriting logic as such does not solve the modularity problem. After all, equational definitions remain untouched when embedded in rewriting logic, and SOS definitions, except for the technicality of restricting rewrites to one step in conditions, are represented by quite similar conditional rewrite rules. Therefore, if the specifications were unmodular beforehand, it is unreasonable to expect that they will magically become modular when viewed as rewrite theories. Something else is needed, namely a *modular specification methodology*.

---

[10] As pointed out in [41], the Cham's heating and cooling rules and the airlock rule could also be seen as equations and could be made part of the set A.

In this regard, the already mentioned work of Mosses on MSOS [57,58,55] is very relevant and important, because it has given a simple and elegant solution to the SOS modularity problem. Stimulated by Mosses' work, the first author, in joint work with Christiano Braga, has investigated a methodology to make rewriting logic definitions of programming languages modular. The results of this research are reported in [47], and case studies showing the usefulness of these modularity techniques for specific language extensions are presented in [10]. In particular, since equational logic is a sublogic of rewriting logic, the modular methodology proposed in [47] specializes in a straightforward way to a new modular specification methodology for algebraic semantics. The two key ideas in [47] are the systematic use of ACI matching to make semantic definitions impervious to the later addition of new semantic entities, and the systematic use of *abstract interfaces* to hide the internal representations of semantic entities (for example a store) so that such internal representations can be changed in a language extension without a need to redefine the earlier semantic rules.

This methodology has influenced the specification style used in Section 3, even though the methodology in [47] is not followed literally. One limitation mentioned in [47] is the somewhat rigid style imposed by assuming configurations consisting of a program text and a record of semantic entities, which forces an interleaving semantics alien in spirit to rewriting logic's true concurrency semantics. One can therefore regard the specification style illustrated in Section 3 as a snapshot of our current steps towards a truly concurrent modular specification methodology, a topic that we hope to develop fully in the near future.

6   Concluding Remarks

We have introduced rewriting logic, have explained its proof theory and its model theoretic semantics, and have shown how it unifies both equational semantics and SOS within a common semantic framework. We have also explained how reduction semantics can be regarded as a special case of rewriting logic semantics. Furthermore, we have shown how rewriting logic semantic definitions written in a language like Maude can be used to get efficient program analysis tools, and have illustrated this by means of a substantial Caml-like language specification. The unification of equational semantics and SOS achieved this way combines the best features of these approaches and has the following advantages:

- a rewrite theory $\mathcal{R}$ has an initial model semantics given by $\mathcal{T}_{\mathcal{R}}$, and a proof-theoretic operational semantics given by rewrite proofs; furthermore, by the Completeness Theorem both semantics *agree*
- the initial model $\mathcal{T}_{\mathcal{R}}$ provides the mathematical basis for formal reasoning and theorem proving in first- and higher-order inductive theorem proving and in temporal logic deduction
- rewriting logic provides a *crucial distinction* between semantic *equations* E and semantic *rules* R, that is, a distinction between deterministic and concurrent computation not available in either equational semantics or SOS

- such a distinction is key not only conceptually, but also for efficiency reasons of drastically collapsing the state space
- rewriting logic has a *true concurrency* semantics, more natural than an interleaving semantics when defining concurrent languages with features such as distribution, asynchrony, and mobility
- when specified in languages like Maude, semantic definitions can be turned into efficient interpreters and program analysis tools for free
- when developed according to appropriate methodological principles, rewriting logic semantic definitions become *modular* and are easily extensible without any need for changes in earlier semantic rules.

An important aspect of the rewriting logic semantics we propose is the flexibility of choosing the desired *level of abstraction* at will when giving semantic definitions. Such a level of abstraction may be different for different modeling and analysis purposes, and can be easily changed as explained below. The point is that in a rewrite theory (Σ, E, ϕ, R), rewriting with the rules R happens *modulo* the equations in E. Therefore, the more semantic definitions we express as equations the more *abstract* our semantics becomes. Abstraction has important advantages for making search and model checking efficient, but changes what is *observable* in the model. In this sense, the Caml-like language specification in Section 3 is quite abstract; in fact, it has only three rewrite rules, with all other axioms given as equations. It is indeed possible to observe all global memory changes, since these are all expressed with rules, but some other aspects of the computation may not be observable at this level of abstraction. For example, nonterminating local sequential computations, such as a nonterminating function call or while loop, will remain within the same equivalence class. This may even lead to starvation of other threads in an interpreter execution. Generally speaking, when observing a program's computation in a more fine-grained way becomes important, this can be easily done by *transforming some equations into rules*. For example, one may wish to specify all potentially nonterminating constructs with rules. The most fine-grained way possible is of course to transform *all equations* (except for structural axioms such as ACI) into rules. These transformations are easy to achieve, since they amount to very simple changes in the specification. In fact, one may wish to use different variants of a language's specification, with certain semantic definitions specified as equations in one variant and as rules in another, because each variant may provide the best level of abstraction for a different set of purposes. The moral of the story is precisely that rewriting logic's distinction between equations and rules provides a useful "abstraction knob" by which we can fine tune a language's specification to best handle specific formal analysis purposes. There are a number of open research directions suggested by these ideas:

- for model checking scalability purposes it will be important to add techniques such as partial order reduction and predicate abstraction;
- besides search and model checking, using rewriting logic semantic definitions as a basis for *theorem proving* of program properties is also a direction that should be vigorously pursued; this semantics-based method is well-

understood for equational semantics [30] and has been used quite successfully by other researchers in the analysis of Java programs using both PVS [37] and ACL2 language specifications [52]; in the context of Maude, its ITP tool [20] has been already used to certify state estimation programs automatically synthesized from formal specifications [66,65] and also to verify sequential programs based on a language's semantic definition [46].

- rewriting is a simple and general model of computation, and rewriting-based semantic definitions already run quite fast on a language like Maude which is itself a semi-compiled interpreter; this suggests that, given an appropriate compilation technology for rewriting, one could directly compile programming languages into a *rewriting abstract machine*; a key issue in this regard is compiling conditional equations into unconditional ones [36,86];
- more experience is also needed in specifying different programming languages as rewrite theories; besides the work in the JavaFAN project, other language specification projects are currently underway at UIUC and at UFF Brazil, including Scheme, ML, OCaml, Haskell, and Pascal;
- more research is also needed on modularity issues; a key question is how to generalize to a true concurrency setting the modular methodology developed in [47]; an important goal would the development of a modular library of rewriting logic definitions of programming language features that could be used to easily define the semantics of a language by putting together different modules in the library.

There is, finally, what we perceive as a promising new direction in teaching programming languages, namely the development of courses and teaching material that use executable rewriting logic specifications as a key way to explain the precise meaning of each programming language feature. This can allow students to experiment with programming language concepts by developing executable formal specifications for them. We have already taught several graduate course at UIUC along these lines with very encouraging results, including a programming language design course and a formal verification course [64,46].

Acknowledgments. This research has been supported by ONR Grant N00014-02-1-0715 and NSF Grant CCR-0234524. We thank the IJCAR 2004 organizers for giving us the opportunity of presenting these ideas in an ideal forum. Several of the ideas presented in this work have been developed in joint work with students and colleagues; in particular: (1) the work on Maude is joint work of the first author with all the members of the Maude team at SRI, UIUC, and the Universities of Madrid, Málaga, and Oslo; (2) the work on Java and the JVM is joint work of both authors with Azadeh Farzan and Feng Cheng at UIUC; and (3) ideas on modular rewriting logic definitions have been developed in joint work of the first author with Christiano Braga at UFF Brazil. We thank Feng Chen for help with some of the examples in this paper, to Marcelo D'Amorim for help with editing, and to Mark-Oliver Stehr, Salvador Lucas and Santiago Escobar for their helpful comments on a draft version of this paper.

References

1. *Proceedings of WRLA'96, September 3–6, 1996*, volume 4 of *ENTCS*. Elsevier, Sept. 1996. http://www.elsevier.nl/locate/entcs/volume4.html.
2. H. Baker and C. Hewitt. Laws for communicating parallel processes. In *Proceedings of the 1977 IFIP Congress*, pages 987–992. IFIP Press, 1977.
3. D. Basin and G. Denker. Maude versus Haskell: an experimental comparison in security protocol analysis. In *Proc. 3rd. WRLA*. ENTCS, Elsevier, 2000.
4. G. Berry and G. Boudol. The chemical abstract machine. *Theoretical Computer Science*, 96(1):217–248, 1992.
5. E. Best and R. Devillers. Sequential and concurrent behavior in Petri net theory. *Theoretical Computer Science*, 55:87–136, 1989.
6. P. Borovanský, C. Kirchner, H. Kirchner, and P.-E. Moreau. ELAN from a rewriting logic point of view. *Theoretical Computer Science*, 285:155–185, 2002.
7. G. Boudol. Computational semantics of term rewriting systems. In *Algebraic Methods in Semantics*, pages 169–236. Cambridge University Press, 1985.
8. C. Braga. *Rewriting Logic as a Semantic Framework for Modular Structural Operational Semantics*. PhD thesis, Departamento de Informática, Pontificia Universidade Católica de Rio de Janeiro, Brasil, 2001.
9. C. Braga, E. H. Haeusler, J. Meseguer, and P. D. Mosses. Mapping modular SOS to rewriting logic. In *12th International Workshop, LOPSTR 2002, Madrid, Spain*, volume 2664 of *LNCS*, pages 262–277, 2002.
10. C. Braga and J. Meseguer. Modular rewriting semantics in practice. in Proc. *WRLA'04*, ENTCS.
11. G. Brat, K. Havelund, S. Park, and W. Visser. Model checking programs. In *ASE'00*, pages 3 – 12, 2000.
12. M. Broy, M. Wirsing, and P. Pepper. On the algebraic definition of programming languages. *ACM Trans. on Prog. Lang. and Systems*, 9(1):54–99, Jan. 1987.
13. R. Bruni. *Tile Logic for Synchronized Rewriting of Concurrent Systems*. PhD thesis, Dipartimento di Informatica, Università di Pisa, 1999. Technical Report TD-1/99. http://www.di.unipi.it/phd/tesi/tesi_1999/TD-1-99.ps.gz.
14. R. Bruni and J. Meseguer. Generalized rewrite theories. In *Proceedings of ICALP 2003, 30th International Colloquium on Automata, Languages and Programming*, volume 2719 of *LNCS*, pages 252–266, 2003.
15. G. Carabetta, P. Degano, and F. Gadducci. CCS semantics via proved transition systems and rewriting logic. In *Proceedings of WRLA'98, September 1–4, 1998*, volume 15 of *ENTCS*, pages 253–272. Elsevier, 1998. http://www.elsevier.nl/locate/entcs/volume15.html.
16. F. Chen, G. Roşu, and R. P. Venkatesan. Rule-based analysis of dimensional safety. In *Rewriting Techniques and Applications (RTA'03)*, volume 2706 of *LNCS*, pages 197–207, 2003.
17. E. Clarke, O. Grumberg, and D. Peled. *Model Checking*. MIT Press, 2001.
18. M. Clavel, F. Durán, S. Eker, P. Lincoln, N. Martí-Oliet, J. Meseguer, and J. Quesada. Maude: specification and programming in rewriting logic. *Theoretical Computer Science*, 285:187–243, 2002.
19. M. Clavel, F. Durán, S. Eker, P. Lincoln, N. Martí-Oliet, J. Meseguer, and C. Talcott. Maude 2.0 Manual. June 2003, http://maude.cs.uiuc.edu.
20. M. Clavel, F. Durán, S. Eker, and J. Meseguer. Building equational proving tools by reflection in rewriting logic. In *CAFE: An Industrial-Strength Algebraic Formal Method*. Elsevier, 2000. http://maude.cs.uiuc.edu.

21. D. Clément, J. Despeyroux, L. Hascoet, and G. Kahn. Natural semantics on the computer. In *Proceedings, France-Japan AI and CS Symposium*, pages 49–89. ICOT, 1986. Also, Information Processing Society of Japan, Technical Memorandum PL-86-6.
22. P. Degano, F. Gadducci, and C. Priami. A causal semantics for CCS via rewriting logic. *Theoretical Computer Science*, 275(1-2):259–282, 2002.
23. R. De Nicola and U. Montanari (eds.). Selected papers of the 2nd workshop on concurrency and compositionality, March 1990. *Theoretical Computer Science*, 96(1), 1992.
24. S. Eker, J. Meseguer, and A. Sridharanarayanan. The Maude LTL model checker. In *Proc. 4th. WRLA*. ENTCS, Elsevier, 2002.
25. A. Farzan, F. Chen, J. Meseguer, and G. Roşu. JavaFAN. http://fsl.cs.uiuc.edu/javafan.
26. A. Farzan, F. Cheng, J. Meseguer, and G. Roşu. Formal analysis of Java programs in JavaFAN. To appear in Proc. CAV'04, Springer LNCS, 2004.
27. A. Farzan, J. Meseguer, and G. Roşu. Formal JVM code analysis in JavaFAN. To appear in Proc. AMAST'04, Springer LNCS, 2004.
28. K. Futatsugi and R. Diaconescu. *CafeOBJ Report*. World Scientific, AMAST Series, 1998.
29. F. Gadducci and U. Montanari. The tile model. In G. Plotkin, C. Stirling and M. Tofte, eds., *Proof, Language and Interaction: Essays in Honour of Robin Milner*, MIT Press, 133–166, 2000.
30. J. A. Goguen and G. Malcolm. *Algebraic Semantics of Imperative Programs*. MIT Press, 1996.
31. J. A. Goguen and K. Parsaye-Ghomi. Algebraic denotational semantics using parameterized abstract modules. In *Formalizing Programming Concepts*, pages 292–309. Springer-Verlag, 1981. LNCS, Volume 107.
32. K. Havelund, M. Lowry, S. Park, C. Pecheur, J. Penix, W. Visser, and J. White. Formal analysis of the remote agent before and after flight. In *the 5th NASA Langley Formal Methods Workshop*, 2000.
33. K. Havelund, M. Lowry, and J. Penix. Formal Analysis of a Space Craft Controller using SPIN. *IEEE Transactions on Software Engineering*, 27(8):749–765, Aug. 2001. Previous version appeared in Proceedings of the 4th SPIN workshop, 1998.
34. K. Havelund and T. Pressburger. Model checking Java programs using Java PathFinder. *Software Tools for Technology Transfer*, 2(4):366 – 381, Apr. 2000.
35. M. Hennessy. *The Semantics of Programming Languages: An Elementary Introduction Using Structural Operational Semantics*. John Willey & Sons, 1990.
36. C. Hintermeier. How to transform canonical decreasing ctrss into equivalent canonical trss. In *4th International CTRS Workshop*, volume 968 of *LNCS*, 1995.
37. B. Jacobs and E. Poll. Java program verification at Nijmegen: Developments and perspective. Technical Report NIII-R0318, Computing Science Institute, University of Nijmegen, 2000.
38. E. B. Johnsen, O. Owe, and E. W. Axelsen. A runtime environment for concurrent objects with asynchronous method calls. In *Proc. 5th. Intl. Workshop on Rewriting Logic and its Applications*. ENTCS, Elsevier, 2004.
39. C. Laneve and U. Montanari. Axiomatizing permutation equivalence. *Mathematical Structures in Computer Science*, 6:219–249, 1996.
40. N. Martí-Oliet and J. Meseguer. Rewriting logic as a logical and semantic framework. In *Handbook of Philosophical Logic, 2nd. Edition*, pages 1–87. Kluwer Academic Publishers, 2002. First published as SRI Tech. Report SRI-CSL-93-05, August 1993.

41. J. Meseguer. Conditional rewriting logic as a unified model of concurrency. *Theoretical Computer Science*, 96(1):73–155, 1992.

42. J. Meseguer. A logical theory of concurrent objects and its realization in the Maude language. In *Research Directions in Concurrent Object-Oriented Programming*, pages 314–390. The MIT Press, 1993.

43. J. Meseguer. Rewriting logic as a semantic framework for concurrency: A progress report. In *CONCUR'96: Concurrency Theory, 7th International Conference, Pisa, Italy, August 26–29, 1996, Proceedings*, volume 1119 of *LNCS*, pages 331–372. Springer-Verlag, 1996.

44. J. Meseguer. Membership algebra as a logical framework for equational specification. In *Proc. WADT'97*, pages 18–61. Springer LNCS 1376, 1998.

45. J. Meseguer. Software specification and verification in rewriting logic. In *Models, Algebras, and Logic of Engineering Software, NATO Advanced Study Institute, July 30 – August 11, 2002*, pages 133–193. IOS Press, 2003.

46. J. Meseguer. Lecture notes on program verification. CS 376, University of Illinois, http://http://www-courses.cs.uiuc.edu/~cs376/, Fall 2003.

47. J. Meseguer and C. Braga. Modular rewriting semantics of programming languages. To appear in Proc. AMAST'04, Springer LNCS, 2004.

48. J. Meseguer, K. Futatsugi, and T. Winkler. Using rewriting logic to specify, program, integrate, and reuse open concurrent systems of cooperating agents. In *Proceedings of the 1992 International Symposium on New Models for Software Architecture, November 1992*, pages 61–106, 1992.

49. J. Meseguer and U. Montanari. Mapping tile logic into rewriting logic. In *Recent Trends in Algebraic Development Techniques, WADT'97, June 3–7, 1997*, volume 1376 of *LNCS*, pages 62–91. Springer-Verlag, 1998.

50. J. Meseguer and C. L. Talcott. A partial order event model for concurrent objects. In *CONCUR'99, August 24–27, 1999*, volume 1664 of *LNCS*, pages 415–430. Springer-Verlag, 1999.

51. R. Milner. Functions as processes. *Mathematical Structures in Computer Science*, 2(2):119–141, 1992.

52. J. Moore. Inductive assertions and operational semantics. In *Proceedings CHARME 2003*, volume 2860, pages 289–303. Springer LNCS, 2003.

53. J. Moore, R. Krug, H. Liu, and G. Porter. Formal models of Java at the JVM level – a survey from the ACL2 perspective. In *Proc. Workshop on Formal Techniques for Java Programs, in association with ECOOP 2001*, 2002.

54. J. S. Moore. http://www.cs.utexas.edu/users/xli/prob/p4/p4.html.

55. P. D. Mosses. Modular structural operational semantics. Manuscript, September 2003, to appear in *J. Logic and Algebraic Programming*.

56. P. D. Mosses. Denotational semantics. In *Handbook of Theoretical Computer Science, Vol. B*. North-Holland, 1990.

57. P. D. Mosses. Foundations of modular SOS. In *Proceedings of MFCS'99, 24th International Symposium on Mathematical Foundations of Computer Science*, pages 70–80. Springer LNCS 1672, 1999.

58. P. D. Mosses. Pragmatics of modular SOS. In *Proceedings of AMAST'02 Intl. Conf*, pages 21–40. Springer LNCS 2422, 2002.

59. P. C. Ölveczky. *Specification and Analysis of Real-Time and Hybrid Systems in Rewriting Logic*. PhD thesis, University of Bergen, Norway, 2000. http://maude.csl.sri.com/papers.

60. P. C. Ölveczky and J. Meseguer. Specification of real-time and hybrid systems in rewriting logic. *Theoretical Computer Science*, 285:359–405, 2002.

61. D. Y. W. Park, U. Stern, J. U. Sakkebaek, and D. L. Dill. Java model checking. In *ASE'01*, pages 253 – 256, 2000.
62. G. D. Plotkin. A structural approach to operational semantics. Technical Report DAIMI FN-19, Computer Science Dept., Aarhus University, 1981.
63. J. C. Reynolds. The discoveries of continuations. *LISP and Symbolic Computation*, 6(3–4):233–247, 1993.
64. G. Roşu. Lecture notes on program language design. CS 322, University of Illinois at Urbana-Champaign, Fall 2003.
65. G. Roşu, R. P. Venkatesan, J. Whittle, and L. Leustean. Certifying optimality of state estimation programs. In *Computer Aided Verification (CAV'03)*, pages 301–314. Springer, 2003. LNCS 2725.
66. G. Roşu and J. Whittle. Towards certifying domain-specific properties of synthesized code. In *Proceedings, International Conference on Automated Software Engineering (ASE'02)*. IEEE, 2002. Edinburgh, Scotland.
67. D. Scott. Outline of a mathematical theory of computation. In *Proceedings, Fourth Annual Princeton Conference on Information Sciences and Systems*, pages 169–176. Princeton University, 1970.
68. L. J. Steggles. Rewriting logic and Elan: Prototyping tools for Petri nets with time. In *Applications and Theory of Petri Nets 2001, 22nd CATPN 2001, June 25–29, 2001*, volume 2075 of *LNCS*, pages 363–381. Springer-Verlag, 2001.
69. M.-O. Stehr. CINNI — A generic calculus of explicit substitutions and its application to λ-, ς- and π-calculi. In *Proc. 3rd. Intl. Workshop on Rewriting Logic and its Applications*. ENTCS, Elsevier, 2000.
70. M.-O. Stehr. A rewriting semantics for algebraic nets. In *Petri Nets for System Engineering — A Guide to Modeling, Verification, and Applications*. Springer-Verlag, 2001.
71. M.-O. Stehr. Programming, Specification, and Interactive Theorem Proving — Towards a Unified Language based on Equational Logic, Rewriting Logic, and Type Theory. Doctoral Thesis, Universität Hamburg, Fachbereich Informatik, Germany, 2002. http://www.sub.uni-hamburg.de/disse/810/.
72. M.-O. Stehr and J. Meseguer. Pure type systems in rewriting logic: Specifying typed higher-order languages in a first-order logical framework. To appear in Springer LNCS Vol. 2635, 2004.
73. M.-O. Stehr, J. Meseguer, and P. Ölveczky. Rewriting logic as a unifying framework for Petri nets. In *Unifying Petri Nets*, pages 250–303. Springer LNCS 2128, 2001.
74. M.-O. Stehr and C. Talcott. Plan in Maude: Specifying an active network programming language. In *Proc. 4th. WRLA*. ENTCS, Elsevier, 2002.
75. C. Strachey. Fundamental concepts in programming languages. *Higher-Order and Symbolic Computation*, 13:11–49, 2000.
76. C. L. Talcott. Interaction semantics for components of distributed systems. In *Proceedings of FMOODS'96*, pages 154–169. Chapman & Hall, 1997.
77. C. L. Talcott. Actor theories in rewriting logic. *Theoretical Computer Science*, 285, 2002.
78. P. Thati, K. Sen, and N. Martí-Oliet. An executable specification of asynchronous Pi-Calculus semantics and may testing in Maude 2.0. In *Proc. 4th. WRLA*. ENTCS, Elsevier, 2002.
79. D. Turi. *Functorial Operational Semantics and its Denotational Dual*. PhD thesis, Free University, Amsterdam, 1996.
80. A. Verdejo. *Maude como marco semántico ejecutable*. PhD thesis, Facultad de Informática, Universidad Complutense, Madrid, Spain, 2003.

81. A. Verdejo and N. Martí-Oliet. Executable structural operational semantics in Maude. Manuscript, Dto. Sistemas Informáticos y Programación, Universidad Complutense, Madrid, August 2003.

82. A. Verdejo and N. Martí-Oliet. Executing E-LOTOS processes in Maude. In *INT 2000, Extended Abstracts*, pages 49–53. Technical report 2000/04, Technische Universitat Berlin, March 2000.

83. A. Verdejo and N. Martí-Oliet. Implementing CCS in Maude. In *Formal Methods For Distributed System Development. FORTE/PSTV 2000 IFIP TC6 WG6. October 10–13, 2000*, volume 183 of *IFIP*, pages 351–366, 2000.

84. A. Verdejo and N. Martí-Oliet. Implementing CCS in Maude 2. In *Proc. 4th. WRLA*. ENTCS, Elsevier, 2002.

85. P. Viry. Input/output for ELAN. In *Proceedings of WRLA'96, September 3–6, 1996* [1], pages 51–64. `http://www.elsevier.nl/locate/entcs/volume4.html`.

86. P. Viry. Elimination of conditions. *Journal of Symbolic Computation*, 28(3):381–401, 1999.

87. P. Viry. Equational rules for rewriting logic. *Theoretical Computer Science*, 285:487–517, 2002.

88. W. Visser, K. Havelund, G. Brat, and S. Park. Java PathFinder - second generation of a Java model checker. In *Proceedings of Post-CAV Workshop on Advances in Verification*, 2000.

89. M. Wand. First-order identities as a defining language. *Acta Informatica*, 14:337–357, 1980.

A Redundancy Criterion Based on Ground Reducibility by Ordered Rewriting

Bernd Löchner

FB Informatik, Technische Universität Kaiserslautern, Kaiserslautern, Germany,
`loechner@informatik.uni-kl.de`

Abstract. Redundancy criteria are an important means to restrict the search space of a theorem prover. In the presence of associative and commutative (AC) operators saturating provers soon generate many similar equations, most of them are redundant. We present a new criterion that is specialized for the AC-case and leads to significant speed-ups. The criterion uses a new sufficient test for the unsatisfiability of ordering constraints. The test runs in polynomial time, is easy to implement, and covers reduction orderings in a generic way, with possible extensions for LPO and KBO.

1 Introduction

Redundancy criteria are a means to extend and refine existing logical calculi in a modular way. This allows one to specialize the calculi to domains where redundancy can be suitably characterized. Compared to working modulo some theory redundancy criteria have the advantage that extending an existing prover does not affect core algorithms or data structures.

In previous work [AHL03] we have extended the unfailing completion approach [HR87,BDP89] by criteria that are based on ground joinability. This is especially rewarding in the presence of associative and commutative (AC) operators, for nearly all equations with AC-equal sides can be deleted. We have also investigated the use of more generic ground joinability tests. The first one is based on [MN90], which considers all possible ordering relations between variables. The second test uses confluence trees, which were originally introduced to decide the confluence of ordered rewrite systems [CNNR03]. Experiments have shown that both tests take effect especially when AC-operators are present.

However, the use of the generic tests is hampered by three effects: They tend to perform many case distinctions, much work is duplicated between independent tests of different equations, and the tests check for complete joinability proofs only, which make them fail when some necessary equation is not yet available. This means that we will take only part of the possible benefit even if we repeat the tests. The reason is that we perform too much work at once (cf. Sect. 2).

The main idea of the present paper is to replace the test on ground joinability of $s = t$ by a test on *ground reducibility*. In order to guarantee the existence of a smaller proof for $s = t$ we add new equations, if necessary. Of course, this is only a win if their number is small enough. As ground reducibility is undecidable for

D. Basin and M. Rusinowitch (Eds.): IJCAR 2004, LNAI 3097, pp. 45–59, 2004.
© Springer-Verlag Berlin Heidelberg 2004

ordered rewriting [CNNR03], we have to concentrate on special cases and use approximations. For the AC-case ground reducibility can be nicely described by ordering constraints and the set of overlaps is small in size (cf. Sect. 3).

Testing ordering constraints for satisfiability is an NP-complete problem for the lexicographic path ordering (LPO) and the Knuth-Bendix ordering (KBO) [Nie93,KV01], the orderings that are used most frequently for equational theorem proving. As we aim for a practical redundancy criterion, we approximate again and use a sufficient test for their unsatisfiability. In Sect. 4 we present a test, which is easy to implement and runs in polynomial time. It covers many orderings in a generic way and can be extended for specific orderings, such as LPO and KBO. In Sect. 5 experiments demonstrate that the unsatisfiability test can cope well with the ordering constraints produced by the criterion. Overall, the new criterion leads to remarkable speed-ups, especially for challenging proof tasks.

This work is embedded in the rich tradition of completion based theorem proving. It has early been noticed that AC-operators deserve special treatment. Especially working modulo AC, or more generally a theory, has gotten much attention. For space reasons, we refer to the recent overviews [DP01,NR01]. Another work that uses constraints for the AC-case is [PZ97].

2   Preliminaries

We use standard concepts from term rewriting [DP01] and equational theorem proving [NR01]. The set $\mathcal{F}$ contains function symbols (or operators). With $\mathcal{F}_{AC}$ we denote the subset of AC-operators. Let $+$ be an AC-operator. We will frequently use $\mathcal{A}$, $\mathcal{C}$, and $\mathcal{C}'$ to refer to associativity $(x + y) + z \to x + (y + z)$, commutativity $x + y = y + x$ and extended commutativity $x + (y + z) = y + (x + z)$. With $\mathcal{F}^e$ we denote an arbitrary extension of $\mathcal{F}$. As we use an extended signature semantics, we use $\mathrm{Term}(\mathcal{F}^e)$ as set of ground terms. With $\succ$ we denote a reduction ordering that is total on ground terms. We write $\mathrm{GSub}(t_1, \dots, t_n)$ for the set of all ground substitutions mapping the variables of terms $t_1, \dots, t_n$ into $\mathrm{Term}(\mathcal{F}^e)$. We write $\mathcal{O}(t)$ for the set of positions in term t. An *overlap* of equation $u = v$ into equation $s = t$ is defined as follows: Let $p \in \mathcal{O}(s)$ be a nonvariable position in s and $\sigma = \mathrm{mgu}(u, s|_p)$ the most general unifier of u and $s|_p$. Then $\sigma(s)[\sigma(v)]_p = \sigma(t)$ is an overlap if $\sigma(v) \not\succ \sigma(u)$. It is a *critical pair* if additionally $\sigma(t) \not\succ \sigma(s)$. Let $\mathrm{OL}(s = t, E)$ be the set of overlaps of elements of E into $s = t$.

The ordering constraints we use are built of atomic constraints of the form $s = t$, $s > t$, or $s \geq t$, which are satisfied by a substitution σ iff $\sigma(s) \equiv \sigma(t)$, $\sigma(s) \succ \sigma(t)$, or $\sigma(s) \succcurlyeq \sigma(t)$, respectively. A conjunction $C_1 \wedge \dots \wedge C_n$ is satisfied by σ iff all C_i are satisfied by σ, a disjunction $C_1 \vee \dots \vee C_n$ is satisfied by σ iff at least one C_i is satisfied. A constraint C is *satisfiable* with respect to $\succ$, if there is at least one ground substitution that satisfies C with respect to $\succ$. Otherwise, C is *unsatisfiable*. We use $\top$ and $\bot$ to denote the trivially satisfiable (resp. unsatisfiable) constraint. There is no need to use negation, as $\succ$ is total on ground terms.

Unfailing completion [HR87,BDP89] is a well-known method for (unit) equational theorem proving. We use an inference system that is based on [BDP89], which allows us to keep redundant equations for simplification [AHL03]. This is beneficial in practice, as it strengthens the simplification relation. For the purpose of this work it suffices to consider two sets: $R \subseteq \succ$ stores the rules and E contains the unoriented equations. Additionally to rewriting with R, *ordered rewriting* uses the set $E^{\succ}$ of orientable instances of equations of E for rewrite steps. We denote by $R(E)$ the set $R \cup E^{\succ}$.

The completeness of the approach is shown with the technique of proof transformations, with a *proof ordering* $\succ_P$ as main ingredient. For all proofs P that are modified by some operation on (R, E) we have to ensure that there is some proof P' with $P \succ_P P'$. A *ground proof* for $s = t$ is of the form $P = (t_0, \varrho_1, t_1, \ldots, \varrho_n, t_n)$, where $t_i \in \mathrm{Term}(\mathcal{F}^{\mathrm{e}})$, $t_0 \equiv s$, and $t_n \equiv t$. For each proof step $t_{i-1} \varrho_i t_i$ the justification ϱ_i records the direction of the rewrite step, its position p, and the used equation $u = v$ and substitution σ. We define the complexity $c(s \varrho t)$ of a proof step by

$$c(s \varrho t) = \begin{cases} (\{s\}, s|_p, (l, r), t) & \text{if } s \to_R t \text{ with } l \to r \text{ at } p \in \mathcal{O}(s) \\ (\{t\}, t|_p, (l, r), s) & \text{if } s \leftarrow_R t \text{ with } l \to r \text{ at } p \in \mathcal{O}(t) \\ (\{s\}, s|_p, (u, v), t) & \text{if } s \to_{E^{\succ}} t \text{ with } \sigma(u) \to \sigma(v) \text{ at } p \in \mathcal{O}(s) \\ (\{t\}, t|_p, (u, v), s) & \text{if } s \leftarrow_{E^{\succ}} t \text{ with } \sigma(u) \to \sigma(v) \text{ at } p \in \mathcal{O}(t) \\ (\{s, t\}, -, -, -) & \text{if } s \leftrightarrow_E t \text{ is an unoriented step} \end{cases}$$

On these tuples we define the ordering $\succ_t$ as the lexicographic combination $(\twoheadrightarrow, \succ_{\mathrm{st}}, \rhd, \succ)$. Here $\twoheadrightarrow$ is the multiset extension of the fixed reduction ordering $\succ$, $\succ_{\mathrm{st}}$ is the subterm ordering, and $\rhd$ is an Noetherian ordering on pairs of terms that we will explain in the following paragraph. The complexity $c(P)$ of proof P is the multiset of complexities of the proof steps that P contains. This allows us to define the following ordering on proofs: $P_1 \succ_P P_2$ iff $c(P_1) \twoheadrightarrow_t c(P_2)$, where $\twoheadrightarrow_t$ is the multiset extension of $\succ_t$.

To define the ordering $\rhd$, we use the following function Ψ to map term pairs to tuples. We have $\Psi(s, t) = (|s|, M(s), s, n)$, where $M(s)$ gives the multiset of function symbols in s, and n indicates whether the term pair is a rule $s \to t$ in R (then $n = 1$), or an equation $s = t$ in E (then $n = 3$). For a fixed number of "distinguished" equations, such as $\mathcal{C}$ or $\mathcal{C}'$, we can choose $n = 2$. On these tuples we define the ordering $\rhd_\Psi$ as the lexicographic combination $\rhd_\Psi = (>_\mathbb{N}, \supset, \rhd, >_\mathbb{N})$, where $\rhd$ denotes the encompassment ordering. We then define $\rhd$ by $(s, t) \rhd (u, v)$ iff $\Psi(s, t) \rhd_\Psi \Psi(u, v)$.

The advantage of this admittedly complex and technical proof ordering is that it extends the proof ordering of [BDP89] and makes therefore more proofs comparable. This leads to more redundant equations. Ordering $\rhd$ is used to guard e. g. the interreduction of left-hand sides of rules. If $l = r$ reduces $s \to t$ at top-level, we require $(s, t) \rhd (l, r)$. In an implementation, program invariants often imply the $\rhd$ relation. We can then skip explicit tests.

Our notion of redundancy is based on the proof ordering: an equation is redundant if every ground instance has a smaller proof. Unlike critical pair criteria it is not limited to critical pairs.

Definition 1. *An equation $s = t$ is redundant with respect to $R(E)$, which we write as $s = t \succ_P R(E)$, if for any $\sigma \in \mathrm{GSub}(s, t)$ either $\sigma(s) \equiv \sigma(t)$ or there* *is a ground proof P for $\sigma(s) = \sigma(t)$ in (R, E) with $\sigma(s) \rightarrow_{\{s=t\}^{\succ}} \sigma(t) \succ_{\mathcal{P}} P$ if $\sigma(s) \succ \sigma(t)$, or $\sigma(s) \leftarrow_{\{s=t\}^{\succ}} \sigma(t) \succ_{\mathcal{P}} P$ if $\sigma(t) \succ \sigma(s)$.*

In previous work [AHL03], we used a strengthened form of ground joinability as redundancy criterion. If an equation $s = t$ is ground joinable in $R(E)$, then there is for each ground instance a joinability proof in $R(E)$. To meet the requirements for redundancy one has to be careful with the first rewrite step of the larger side of the ground instance. If it occurs on the top-level position, the used rule or equation has to be strictly $\triangleright$-smaller than $s = t$.

For ordered rewriting ground joinability is undecidable, as even joinability of two ground terms is undecidable [Löc04]. Therefore, sufficient tests are required, which may be specific to certain theories or generic in nature. A simple, yet powerful test is based on AC-equality. An equation $s = t$ that is not contained in $\mathcal{ACC}'$ is redundant with respect to $\mathcal{ACC}'$ if $s =_{AC} t$. The two generic tests that we implemented are stronger, but for the price of much higher computational costs. The first considers all ordering relations between variables [MN90], which leads to a number of cases that is exponential in the number of variables. Therefore, we restrict the test to equations with at most 5 variables. The second is based on confluence trees [CNNR03]. Here, case distinctions are introduced by considering possible ordered rewrite steps. As full ordering constraints describe the different cases, the method is more powerful than the first, but needs a constraint solver. Hence, it is even more expensive than the first test. Both tests show mainly success when AC-operators are present.

In both generic tests much work is duplicated between independent tests of different equations: Consider two equations that differ only in two subterms s and t that are exchanged by commutativity. As both tests have to consider the cases when s is greater than t and vice versa, it is likely that both tests exchange in some subcases s and t and so perform (nearly) identical work twice. Furthermore, the tests check for complete joinability proofs only. They will fail if an equation that is necessary for some ground instance is not yet available. However, the completion process does not enumerate the equations in the order that is most suitable for the tests and our attempts to develop corresponding strategies failed. This means that we will take only part of the possible benefit even if we repeat the tests. To sum up, we perform too much work at once for testing redundancy during completion. This is not surprising, as the generic tests were originally designed to check properties of static rewrite systems.

3   Ground Reducibility as Redundancy Criterion

The main idea to reduce the work of redundancy tests is to use ground reducibility, which is more localized than ground joinability. Hence, we no longer search for complete joinability proofs for each ground instance, but focus on the first step instead. To complete the proofs, we add the necessary overlaps, which in turn make the original equation redundant. This is only a win if the overlaps do not outnumber the critical pairs.

Definition 2. *An equation $s = t$ is* ground reducible *with respect to $R(E)$ if for each $\sigma \in \text{GSub}(s, t)$ at least one of $\sigma(s)$ or $\sigma(t)$ is reducible by $R(E)$.*

For ordered rewriting ground reducibility is undecidable in general [CNNR03], therefore we have to use sufficient tests. We can distinguish three kinds of rewrite steps, which differ in the position at which they occur. If the position is a nonvariable position in the uninstantiated term, the step happens either at the *skeleton*, that is, all function symbols of the left-hand side of the rewriting equation match function symbols in the term, or it happens at the *fringe*, such that a part of the function symbols matches functions symbols in the substitution part. Furthermore, the step may happen completely in the *substitution part*, which means that the substitution itself is reducible. It is not necessary for theorem proving to consider such substitutions. Regarding the two other kinds of steps, if $\sigma(s) = \sigma(t)$ is reduced with $l = r$ to $s_1 = t_1$, then there is some overlap $u = v$ from $l = r$ into $s = t$ such that $s_1 = t_1$ is an instance of $u = v$.

To get smaller proofs rewrite steps at top-level have to be performed with $\triangleright$-smaller equations. This explains the additional condition in the following theorem, which provides the foundation of our work.

Theorem 1. *Let $s = t$ be ground reducible with respect to $R(E)$ such that all top-level reductions of maximal sides of ground instances are performed with equations in $R \cup E$ that are strictly $\triangleright$-smaller than $s = t$. Let $S = \mathrm{OL}(s = t, R \cup E)$. Then $s = t \succ_P R(E \cup S)$.*

Proof. Let $\sigma \in \mathrm{GSub}(s, t)$ be $R(E)$-irreducible. The case $\sigma(s) \equiv \sigma(t)$ is trivial. Consider $\sigma(s) \succ \sigma(t)$. Let P_0 be the proof $\sigma(s) \to_{\{s = t\}} \succ \sigma(t)$. As $s = t$ is ground reducible, $\sigma(s)$ or $\sigma(t)$ are reducible by $R(E)$. First, assume $\sigma(s)$ reduces to s_1 at position p with $l = r$ in $R(E)$. Let P_1 be the proof $\sigma(s) \to_{R(E)} s_1 \leftrightarrow_S \sigma(t)$. The first step is smaller than the proof step in P_0: If it does not occur on top-level, its complexity is smaller in the second component, otherwise in the third component, as then $(s, t) \triangleright (l, r)$ by assumption. The second step of P_1 is smaller than the step in P_0 by the first component of the complexity as both $\sigma(t)$ and s_1 are smaller than $\sigma(s)$. Therefore, $P_0 \succ_P P_1$. Second, consider the case that $\sigma(t)$ reduces to t_1 by $R(E)$. Then, $\sigma(s) \succ \sigma(t) \succ t_1$. Let P_2 be the proof $\sigma(s) \to_S t_1 \leftarrow_{R(E)} \sigma(t)$. The first step of P_2 is smaller than the proof step in P_0 by the last component of its complexity, as $\sigma(t) \succ t_1$. The second step is smaller by the first component of the complexity, which implies $P_0 \succ_P P_2$.

The case $\sigma(t) \succ \sigma(s)$ is similar to the case $\sigma(s) \succ \sigma(t)$. □

The patterns of the proofs resemble their counterparts used for justifying interreduction. There, we can also distinguish between simplification of larger and smaller sides. Considering the smaller sides enables more redundancy proofs, at the cost of overlapping into smaller sides. This is unproblematic as long as the number of overlaps is rather small, which holds true for the special case we consider next.

3.1   The AC-Case

An analysis of the successful ground joinability tests with the method of [MN90] reveals that the constructed proofs often start with an ordered reduction step with $\mathcal{C}$ or $\mathcal{C}'$. Hence, we concentrate on ground reducibility with respect to $\mathcal{ACC}'$. The simplification of ground terms with $\mathcal{ACC}'$ resembles bubble-sort: First, rewrite steps with $\mathcal{A}$ bracket the $\mathcal{AC}$-subterms to the right. Then, rewrite steps with $\mathcal{CC}'$ exchange adjacent subterms such that smaller terms come to the left.

Possible skeleton steps differ for the cases $\mathcal{A}$ and $\mathcal{CC}'$. As $\mathcal{A}$ is a linear rule, skeleton steps with $\mathcal{A}$ imply that the equation is already $\mathcal{A}$-reducible, which is easy to decide. For rewrite steps with $\mathcal{CC}'$, however, the ordering relation between the (instantiated) subterms is decisive. Consider for example equation $x + y = g(y) + (z + x)$. An instance σ is $\mathcal{CC}'$-reducible, if $\sigma(x) \succ \sigma(y)$, $\sigma(g(y)) \succ \sigma(z + x)$, $\sigma(z) \succ \sigma(x)$, or $\sigma(g(y)) \succ \sigma(z)$. Therefore, a $\mathcal{CC}'$-irreducible ground instance must satisfy the constraint $y \geq x \wedge z + x \geq g(y) \wedge x \geq z \wedge z \geq g(y)$. As this constraint is unsatisfiable, we know that the equation is ground reducible. Note that we can optimize the constraint. For positions where both $\mathcal{C}$ and $\mathcal{C}'$ are applicable, it suffices to consider $\mathcal{C}'$ only, as it leads to a stronger constraint. In the example $z \geq g(y)$ implies $z + x \geq g(y)$, hence we can omit the latter.

The function Γ constructs an ordering constraint for a term t such that the satisfying substitutions describe ground instances that are $\mathcal{CC}'$-irreducible at skeleton positions. It is defined as follows:

$$
\begin{aligned}
\Gamma(x) &= \top \\
\Gamma(f(t_1,\dots,t_n)) &= \Gamma(t_1) \wedge \dots \wedge \Gamma(t_n) && \text{if } f \notin \mathcal{F}_{AC} \\
\Gamma(t_1 + (t_1 + t_2)) &= \Gamma(t_1 + t_2) && \text{if } + \in \mathcal{F}_{AC} \\
\Gamma(t_1 + (t_2 + t_3)) &= t_2 \geq t_1 \wedge \Gamma(t_1) \wedge \Gamma(t_2 + t_3) && \text{if } + \in \mathcal{F}_{AC} \text{ and } t_1 \not\equiv t_2 \\
\Gamma(t_1 + t_1) &= \Gamma(t_1) && \text{if } + \in \mathcal{F}_{AC} \text{ and } \text{top}(t_1) \neq + \\
\Gamma(t_1 + t_2) &= t_2 \geq t_1 \wedge \Gamma(t_1) \wedge \Gamma(t_2) && \text{if } + \in \mathcal{F}_{AC}, t_1 \not\equiv t_2, \text{ and} \\
&&& \text{top}(t_2) \neq +
\end{aligned}
$$

Lemma 1. *If $\Gamma(t)$ is unsatisfiable, then t is $\mathcal{CC}'$-ground reducible.*

Proof. Let $\sigma \in \text{GSub}(t)$ and $\Gamma(t) = t_1 \geq t_1' \wedge \dots \wedge t_n \geq t_n'$. As $\Gamma(t)$ is unsatisfiable, substitution σ satisfies the complement of $\Gamma(t)$, which is equivalent to the constraint $t_1' > t_1 \vee \dots \vee t_n' > t_n$. Therefore, there is at least one i such that $\sigma(t_i') \succ \sigma(t_i)$, which implies that $\mathcal{C}$ or $\mathcal{C}'$ are applicable at the corresponding position in $\sigma(t)$. □

Note that function Γ does not capture AC-ground reducibility in a complete way, as it considers only skeleton steps. First, there are two kinds of fringe steps. Situations such as $x + s$ and $\sigma(x) \equiv s_1 + s_2$ lead to reduction steps with $\mathcal{A}$. Reduction steps with $\mathcal{C}'$ are possible for subterms of the form $s + x$ if $\sigma(x) \equiv s_1 + s_2$ and $\sigma(s) \succ s_1$. Both situations can be described by ordering constraints. But these need (local) quantification, which means that we leave (for LPO) the decidable fragment [CT97]. Furthermore, there are examples, such that each ground instance that is not skeleton reducible has a reducible substitution. Here, irreducibility constraints are appropriate. Deciding AC-ground reducibility with this approach requires the extension of decision procedures for the satisfiability of ordering constraints to cope with the additionally needed constraints. This is a topic of future research.

In this work, our main focus is on a practical criterion. For that, we use a test for unsatisfiability of constraints that is only sufficient. Hence, we consider only skeleton reductions. In practice we can expect the input of the redundancy test in $\mathcal{A}$-normal form. We can therefore concentrate on the $\mathcal{CC}'$-skeleton case. We write $\Gamma(s,t)$ as shorthand notation for $\Gamma(s) \wedge \Gamma(t)$. Combining the previous two results and considering that $\mathcal{C}$ and $\mathcal{C}'$ always fulfill the $\triangleright$ condition we get:

Corollary 1. *Let $s = t$ be an equation different from $\mathcal{C}$ and $\mathcal{C}'$ such that $\Gamma(s,t)$ is unsatisfiable. Let $S = \mathrm{OL}(s = t, \mathcal{CC}')$. Then $s = t \succ_P \mathcal{CC}' \cup S$.* $\square$

The set S is very easy to compute as only subterms are exchanged and even unification is unnecessary. The size of this set is linear in the number of AC-operators in $s = t$. This means a huge improvement in practice, as only say 5 or 10 equations are constructed. For challenging proof tasks the typical size of the set of critical pairs with an equation is in the order of several thousand equations.

3.2 A Refined Criterion for the AC-Case

Consider equation $x + (y + x) = a$. It can not be shown redundant by the method of the previous section, because there are irreducible ground instances. These are characterized by $\sigma(x) \equiv \sigma(y)$, that is, they unify nonidentical subterms.[1] Nevertheless, the equation seems to be redundant, as both $\mathcal{CC}'$-overlaps, which are $x + (x + y) = a$ and $y + (x + x) = a$, cover these instances, and both overlaps have irreducible ground instances that do not unify subterms. It is easy to exclude such unifying solutions by a modified constraint construction. Function $\hat{\Gamma}$ is identical to Γ, except that it uses $>$ constraints instead of $\geq$ constraints.

However, the proof ordering of Sect. 2 does not justify this approach, as none of the two overlaps is $\triangleright$-smaller than the original equation. This is not without reason, as the following example shows. Let $\succ$ be the LPO for $h >_{\mathcal{F}} g >_{\mathcal{F}} f$ and consider $f(h(x) + g(h(y)), h(y) + g(h(x))) = x + y$. For this equation the $\mathcal{CC}'$-irreducible ground instances unify x and y, as for $\succ$ the constraint $g(h(y)) \geq h(x) \wedge g(h(x)) \geq h(y) \wedge y \geq x$ is equivalent to $h(y) \geq h(x) \wedge h(x) \geq h(y) \wedge y \geq x$, which implies $x = y$. However, there is no $\mathcal{CC}'$-overlap that covers these instances and that has irreducible instances that do not unify x and y.

To retain the original idea to use $\hat{\Gamma}$ we have therefore to use an additional guard in the following predicate. Its definition is rather ad-hoc and technical, but the use of $\hat{\Gamma}$ is rewarding in practice. Let $S_{AC}(s = t)$ be given by $S_{AC}(s = t) = \{s' = t' \mid s =_{AC} s'$ and $t =_{AC} t'\}$.

Definition 3. *$U(s,t) = 1$ if $s = t$ not in $\mathcal{CC}'$, $\Gamma(s,t)$ is satisfiable, $\hat{\Gamma}(s,t)$ is unsatisfiable, and for all $\sigma \in \mathrm{GSub}(s,t)$ that satisfy $\Gamma(s,t)$ there is some $s' = t'$ in $S_{AC}(s = t)$ such that $\hat{\Gamma}(s',t')$ is satisfiable and $\sigma(s) = \sigma(t)$ is an instance of $s' = t'$. Otherwise, $U(s,t) = 0$.*

---

[1] It is tempting to replace the equation by $x + (x + x) = a$ in this situation. But this would endanger completeness as it would interact badly with interreductions. Formally, this is reflected by a violation of the proof ordering for such kind of steps.

The idea behind that definition is that proof steps using an equation $s = t$ with $U(s,t) = 1$ can be replaced by proof steps using an equation $s' = t'$ with $U(s',t') = 0$. Equations with $U(s,t) = 1$ are therefore redundant.

Lemma 2. *Let* $\sigma(s) = \sigma(t)$ *be a* $\mathcal{CC}'$*-irreducible ground instance of* $s = t$. *Then there is some* $s' = t'$ *in* $S_{\mathcal{AC}}(s = t)$ *such that* $\sigma(s) = \sigma(t)$ *is an instance of* $s' = t'$ *and* $U(s',t') = 0$.

Proof. The case $U(s,t) = 0$ is trivial. Consider $U(s,t) = 1$. Then there exists by definition of U some $s' = t'$ in $S_{\mathcal{AC}}(s = t)$ such that $\sigma(s) = \sigma(t)$ is an instance of $s' = t'$ and $\hat{\Gamma}(s',t')$ is satisfiable. Therefore, $U(s',t') = 0$.   □

For the example from the beginning of this section U evaluates to 1 for the original equation and to 0 for both overlaps, because their $\hat{\Gamma}$-constraints are satisfiable. With this predicate we modify the proof ordering by modifying ordering $\triangleright$ on term pairs. We extend function Ψ to Ψ' with $\Psi'(s,t) = (|s|, M(s), U(s,t), s, n)$ and $\triangleright_{\Psi'}$ is the lexicographic combination $\triangleright_{\Psi'} = (>_{\mathbb{N}}, \supset, >_{\mathbb{N}}, \triangleright, >_{\mathbb{N}})$.

Lemma 3. *Let* $s = t$ *be an equation with* $U(s,t) = 1$. *Then for each proof step* P_σ *with* $\sigma(s) \leftrightarrow_{\{s=t\}} \sigma(t)$ *there is some proof* P *in* $\mathcal{CC}' \cup S_{\mathcal{AC}}(s = t)$ *such that* $P_\sigma \succ_P P$ *and* P *uses only equations* $u = v$ *with* $U(u,v) = 0$. *Therefore,* $s = t \succ_P \mathcal{CC}' \cup S_{\mathcal{AC}}(s = t)$.

Proof. The case $\sigma(s) \equiv \sigma(t)$ is trivial. Consider $\sigma(s) \succ \sigma(t)$, which means that P_σ is $\sigma(s) \rightarrow_{\{s=t\}\succ} \sigma(t)$. Let $S = S_{\mathcal{AC}}(s = t)$. Let P be the proof of $\sigma(s) \xrightarrow{!}_{\mathcal{CC}'\succ} s_1 \leftrightarrow_S t_1 \xleftarrow{!}_{\mathcal{CC}'\succ} \sigma(t)$. Because $s_1 = t_1$ is $\mathcal{CC}'$-irreducible, there is some $s' = t'$ in S such that $U(s',t') = 0$ and $s_1 = t_1$ is a ground instance of $s' = t'$. The other steps use $\mathcal{C}$ and $\mathcal{C}'$, for which U evaluates to 0. To show $P_\sigma \succ_P P$ we have to consider two cases for the first step. If $\sigma(s) \succ s_1$ then $\sigma(s)$ gets reduced. The complexity of the first step is therefore smaller than the complexity of the step in P_σ, either by the second component (if the rewrite step is not top-level), or by the third component. If $\sigma(s) \equiv s_1$ the first step is performed top-level with $s' = t'$, so we have to consider the third component of the complexity. We have $(s,t) \triangleright (s',t')$ as $s \equiv s'$ and $U(s,t) > U(s',t')$. The remaining steps are smaller by the first component, as $\sigma(s)$ is the maximal term in P.
    The case $\sigma(t) \succ \sigma(s)$ is similar.   □

This means that we can refine the criterion to detect additionally equations $s = t$ with $U(s,t) = 1$. The conditions of $U(s,t)$ seem more restrictive than they are, as in practice we frequently encounter equations with "twisted" variables, such as the example from the beginning of this section, for which $U(s,t) = 1$ can be verified easily.

4   A Test for the Unsatisfiability of Ordering Constraints

Deciding the satisfiability of ordering constraints is an NP-complete problem for both LPO [Nie93] and KBO [KV01]. The decision procedure for LPO constraints devised in [NR02] performs in our experience well in practice [AL01], but its implementation is not an easy task. To our knowledge there exists no implementation of a decision procedure for KBO constraints[2]. Our aim is a sufficient test for unsatisfiability that is reasonably precise and efficient, easy to implement, and covers both LPO and KBO. Relying on an approximation is unproblematic for our purpose: Some redundant equations may remain undetected, but the correctness of the method is unaffected.

The input to our test is a conjunction C of atomic constraints. The main idea is to saturate C by using properties of the ordering to derive new constraints from existing ones. The properties are reflected by corresponding saturation rules. Recall that $\succ$ is a total reduction ordering on ground terms. Hence, it is e.g. irreflexive, transitive, and monotonic. For example, we can derive from $x > y \land g(y) > g(z) \land g(z) \geq g(x)$ by the monotonicity rule the new constraint $g(x) > g(y)$, then by the transitivity rule $g(z) > g(y)$ and $g(y) > g(y)$. This atomic constraint is clearly unsatisfiable, so we know that the original constraint is unsatisfiable as well.

Our method is inspired by the test of Johann and Socher-Ambrosius [JSA94], which decides for a constraint C whether there exists a simplification ordering $\succ$ and a ground substitution σ such that σ satisfies C with respect to $\succ$. However, our application is different, so we can strengthen the test and use the ordering $\succ$, which is known and fixed. Furthermore, it is possible for us to enrich the set of rules for the generic test by properties that are specific for LPO or KBO respectively. This includes the use of precedence $>_{\mathcal{F}}$ or weight function φ.

The constraints generated by the redundancy criteria have many subterms in common. So it is effective to identify them in a preprocessing phase, which furthermore simplifies the main saturation process. We describe the preprocessing in an abstract way by introducing a set $K = \{c_1, c_2, \dots\}$ of new constants to represent the different subterms in the original problem. This formulation is inspired by modern treatments of congruence closure algorithms [Kap97].

We call a term *flat*, if it is a variable x or of the form $f(c_1, \dots, c_n)$, that is, an operator applied to new constants only. Let $\mathfrak{C}$ be a conjunction of atomic constraints and $\mathfrak{D}$ a rewrite system, which consists of rules $t \to c$ where t is a flat term and $c \in K$. Initially, $\mathfrak{D}$ is empty and $\mathfrak{C}$ contains the original constraint C we want to test for unsatisfiability. The following two transformation rules assign new constants as names to subterms and propagate them through the problem.

1. NAMING: $\langle \mathfrak{C}[t], \mathfrak{D} \rangle \implies \langle \mathfrak{C}[c], \mathfrak{D} \cup \{t \to c\} \rangle$ if t is a flat term in $\mathfrak{D}$-normal form and c a new constant that is not present in $\mathfrak{C}$ or $\mathfrak{D}$.
2. PROPAGATION: $\langle \mathfrak{C}[t], \mathfrak{D} \rangle \implies \langle \mathfrak{C}[c], \mathfrak{D} \rangle$ if $t \to c$ is in $\mathfrak{D}$.

To apply the transformation rules, we perform a leftmost-innermost traversal of the terms. For example, consider the constraint $f(f(x)) > g(f(x))$. The first step is to give x the name c_1 and to record this by adding the rule $x \to c_1$ to $\mathfrak{D}$. Similar, we add the rules $f(c_1) \to c_2$ and $f(c_2) \to c_3$. Next, we traverse the second term. We first propagate already introduced names and replace x by c_1 and $f(c_1)$ by c_2. Then we add for the remaining term the rule $g(c_2) \to c_4$. The final form of the constraint is $c_3 > c_4$. It is easy to see that this transformation process performs

---

[2] This is confirmed by Konstantin Korovin (personal e-mail, December 2003).

a number of steps that is linear in the size of C. After this transformation each subterm t of C is uniquely represented by a new constant c. We use $T_{\mathfrak{D}}(c)$ to denote the subterm that c represents. In the example: $T_{\mathfrak{D}}(c_4) \equiv g(f(x))$. With this notation we adapt the definition that a ground substitution σ satisfies a constraint C with respect to an ordering $\succ$ to the new representation $\langle \mathfrak{C}, \mathfrak{D} \rangle$: σ satisfies the constraint $c > c'$ (resp. $c \ge c'$ or $c = c'$) if $\sigma(T_{\mathfrak{D}}(c)) \succ \sigma(T_{\mathfrak{D}}(c'))$ (resp. $\sigma(T_{\mathfrak{D}}(c)) \succcurlyeq \sigma(T_{\mathfrak{D}}(c'))$ or $\sigma(T_{\mathfrak{D}}(c)) \equiv \sigma(T_{\mathfrak{D}}(c'))$).

Lemma 4. *Let $\langle \mathfrak{C}, \mathfrak{D} \rangle$ be the result of preprocessing C. Then σ satisfies C with respect to $\succ$ if, and only if, σ satisfies $\langle \mathfrak{C}, \mathfrak{D} \rangle$ with respect to $\succ$. The transformation needs a number of steps that is linear in the size of C.* ☐

After this preprocessing we start the saturation process. This is based on the following saturation rules, which are suitable for all ground-total reduction orderings. We assume that a rule is only performed if the derived constraint is new to $\mathfrak{C}$. As equality constraints are symmetric, we do not distinguish $c = c'$ and $c' = c$: we assume $c = c'$ to be in $\mathfrak{C}$ iff $c' = c$ is in $\mathfrak{C}$. Furthermore, we assume that $\mathfrak{C}$ contains for each constant c the constraint $c = c$. This does not affect the set of solutions and allows us to simplify conditions such as "either $\mathfrak{C}$ contains $c_1 = c_2$, or $c_1 \equiv c_2$". Some rules use an ordering $\sqsupset$ on constraint symbols. We have $> \sqsupset \ge \sqsupset =$.

1. ORDERING: $\langle \mathfrak{C}, \mathfrak{D} \rangle \implies \langle \mathfrak{C} \cup \{c > c'\}, \mathfrak{D} \rangle$ if $T_{\mathfrak{D}}(c) \succ T_{\mathfrak{D}}(c')$.
2. TRANSITIVITY: $\langle \mathfrak{C}, \mathfrak{D} \rangle \implies \langle \mathfrak{C} \cup \{c_1 \varrho c_3\}, \mathfrak{D} \rangle$ if $\mathfrak{C}$ contains $c_1 \varrho_1 c_2$ and $c_2 \varrho_2 c_3$, and $\varrho = \max_{\sqsupset}\{\varrho_1, \varrho_2\}$.
3. MONOTONICITY: $\langle \mathfrak{C}, \mathfrak{D} \rangle \implies \langle \mathfrak{C} \cup \{c \varrho c'\}, \mathfrak{D} \rangle$ if $\mathfrak{D}$ contains the rules $f(c_1, \dots, c_n) \to c$ and $f(c'_1, \dots, c'_n) \to c'$, $\mathfrak{C}$ contains $c_i \varrho_i c'_i$ for all $i = 1, \dots, n$, and $\varrho = \max_{\sqsupset}\{\varrho_i \mid i = 1, \dots, n\}$.
4. DECOMPOSITION: $\langle \mathfrak{C}, \mathfrak{D} \rangle \implies \langle \mathfrak{C} \cup \{c_i = c'_i\}, \mathfrak{D} \rangle$ if $\mathfrak{D}$ contains the rules $f(c_1, \dots, c_n) \to c$ and $f(c'_1, \dots, c'_n) \to c'$, and $\mathfrak{C}$ contains $c = c'$.
5. CONTEXT: $\langle \mathfrak{C}, \mathfrak{D} \rangle \implies \langle \mathfrak{C} \cup \{c_i \varrho c'_i\}, \mathfrak{D} \rangle$ if $\mathfrak{D}$ contains the two rules $f(c_1, \dots, c_n) \to c$ and $f(c'_1, \dots, c'_n) \to c'$, $\mathfrak{C}$ contains $c \varrho c'$ with $\varrho \in \{>, \ge\}$, and $\mathfrak{C}$ contains $c_j = c'_j$ for $j = 1, \dots, n$ with $j \neq i$.
6. STRENGTHENING: $\langle \mathfrak{C}, \mathfrak{D} \rangle \implies \langle \mathfrak{C} \cup \{c > c'\}, \mathfrak{D} \rangle$ if $\mathfrak{C}$ contains $c \ge c'$ and $T_{\mathfrak{D}}(c)$ is not unifiable with $T_{\mathfrak{D}}(c')$.
7. CYCLE: $\langle \mathfrak{C}, \mathfrak{D} \rangle \implies \langle \mathfrak{C} \cup \{c = c'\}, \mathfrak{D} \rangle$ if $\mathfrak{C}$ contains $c \ge c'$ and $c' \ge c$.
8. ABSORPTION: $\langle \mathfrak{C} \cup \{c \ge c', c \varrho c'\}, \mathfrak{D} \rangle \implies \langle \mathfrak{C} \cup \{c \varrho c'\}, \mathfrak{D} \rangle$ if $\varrho \in \{>, =\}$.
9. CLASH: $\langle \mathfrak{C} \cup \{c = c'\}, \mathfrak{D} \rangle \implies \langle \bot, \mathfrak{D} \rangle$ if $\mathfrak{C}$ contains $c = c'$ and $T_{\mathfrak{D}}(c)$ is not unifiable with $T_{\mathfrak{D}}(c')$.
10. BOTTOM: $\langle \mathfrak{C} \cup \{c > c\}, \mathfrak{D} \rangle \implies \langle \bot, \mathfrak{D} \rangle$.

Note, that the saturation rules do not introduce new terms (i.e., constants), they merely add relations between existing ones. Applying substitutions as in [JSA94] can lead to an exponential growth of problem size.

Lemma 5. *Let $\succ$ be a reduction ordering that is total on ground terms. Let $\langle \mathfrak{C}, \mathfrak{D} \rangle \implies \langle \mathfrak{C}', \mathfrak{D} \rangle$ with one of the saturation rules. If σ satisfies $\langle \mathfrak{C}, \mathfrak{D} \rangle$ with respect to $\succ$, then σ satisfies $\langle \mathfrak{C}', \mathfrak{D} \rangle$ with respect to $\succ$.*

Proof. This is an easy consequence of properties concerning $\succ$ and the satisfiability of constraints. Note that CONTEXT relies on $\succ$ being total on ground terms.   □

The implementation of our method is based on three tables. The first one represents $\mathfrak{D}$ and offers for each c the access to rule $f(c_1, \ldots, c_n) \to c$ and directly to $T_{\mathfrak{D}}(c)$. Furthermore, it provides a function $parent(c, i)$, which gives for c all constants c' that have c as their i-th subterm. This is convenient for some of the more complicated rules. After the first phase of the algorithm, in which we build up $\mathfrak{D}$, the set of constants is fixed, say to $\{c_1, \ldots, c_N\}$. Then we can allocate a quadratic table for $\mathfrak{C}$ with $N \times N$ entries. The entries $\mathfrak{C}[c_i, c_i]$ are set to $=$, for the other entries we try to apply rule ORDERING. If this is done in a bottom-up way, the initialization of $\mathfrak{C}$ can be performed in $O(N^2)$ time, both for KBO and LPO. Then we insert the original constraints into $\mathfrak{C}$ and start the saturation with regard to the remaining rules.

In the third table, which we call $\mathfrak{L}$, we add for each modification of $\mathfrak{C}$ an entry with the old value, the new value, and the applied rule together with justifications. The size of $\mathfrak{L}$ is $2N^2$ in the worst case, as for each entry in $\mathfrak{C}$ at most two insertions can occur (first $\geq$, then either $>$ or $=$). Table $\mathfrak{L}$ provides several functionalities in a convenient way. First, with a simple index we can keep track for which constraints we already have applied all saturation rules, and for which this has to be done. Second, if one of the rules BOTTOM or CLASH applies, we can extract from $\mathfrak{L}$ a detailed justification why the original constraint is unsatisfiable. This is helpful not only for debugging purposes, but also for determining the subset of the atomic constraints that leads to unsatisfiability. Finally, $\mathfrak{L}$ facilitates an undo-mechanism. This is interesting, if we want to extend the constraint incrementally, and later retract such additions. With table $\mathfrak{L}$ we can do this in a stack-like manner, which is sufficient for example for the use in confluence trees.

Theorem 2. *The algorithm based on the transformation and saturation rules is correct: If it derives $\langle \perp, \mathfrak{D} \rangle$, then the original constraint is unsatisfiable. It needs polynomial space and runs in polynomial time in the size of the input.*

Proof. By Lemma 4 and Lemma 5 and induction on the number of saturation steps the algorithm preserves satisfiability. Considering time requirements note that for a given atomic constraint $c \varrho c'$ a fixed number of rules is tested for applicability. These need either constant time (CYCLE, ABSORPTION, BOTTOM), have a runtime that is bound by the maximal arity (DECOMPOSITION, CONTEXT), or have a worst-case complexity that either is linear in N (TRANSITIVITY, STRENGTHENING, CLASH) or is quadratic in N (MONOTONICITY). As there are considered maximally $2N^2$ atomic constraints, we achieve a polynomial complexity for the whole test.   □

4.1 Extensions for Particular Orderings

As mentioned in the introduction, two orderings are most frequently used in practice, namely LPO and KBO. For LPO we have the following additional rules, which reflect closely its definition (the third case of the definition is already covered by a combination of generic saturation rules):

1. BETA: $\langle \mathfrak{C}, \mathfrak{D} \rangle \implies \langle \mathfrak{C} \cup \{c > c'\}, \mathfrak{D} \rangle$ if $\mathfrak{D}$ contains the two rules $f(\ldots) \to c$ and $g(c_1', \ldots, c_n') \to c'$, $\mathfrak{C}$ contains $c > c_i'$ for $i = 1, \ldots, n$, and $f >_{\mathcal{F}} g$.
2. GAMMA: $\langle \mathfrak{C}, \mathfrak{D} \rangle \implies \langle \mathfrak{C} \cup \{c > c'\}, \mathfrak{D} \rangle$ if $\mathfrak{D}$ contains $f(c_1, \ldots, c_n) \to c$ and $f(c_1', \ldots, c_n') \to c'$, there is some i with $1 \le i < n$ such that $\mathfrak{C}$ contains $c_i > c_i'$ and contains $c_j \varrho_j c_j'$ with $\varrho_j \in \{\ge, =\}$ for all $j = 1, \ldots, i-1$ and contains $c > c_k'$ for all $k = i+1, \ldots, n$.

Lemma 6. *Let $\succ$ be a ground-total LPO for precedence $>_{\mathcal{F}}$. Let $\langle \mathfrak{C}, \mathfrak{D} \rangle \implies \langle \mathfrak{C}', \mathfrak{D} \rangle$ with BETA or GAMMA. If σ satisfies $\langle \mathfrak{C}, \mathfrak{D} \rangle$ with respect to $\succ$, then σ satisfies $\langle \mathfrak{C}', \mathfrak{D} \rangle$ with respect to $\succ$.* □

The KBO is parameterized by a weight function φ and a precedence $>_{\mathcal{F}}$. As φ establishes a homomorphism from terms to the naturals, we can analyze the effect of applying a substitution to the weights of the resulting terms with the help of the following function ϕ:

$$\phi(x) = x$$
$$\phi(f(t_1, \ldots, t_n)) = \varphi(f) + \phi(t_1) + \ldots + \phi(t_n)$$

For example, if $\phi(t) = 4 + 2 \cdot x$, then we know that $\varphi(\sigma(t)) = 4 + 2 \cdot \varphi(\sigma(x))$. The function ϕ establishes a natural mapping from constraints to linear Diophantine (in-)equations. Constraints $s > t$ and $s \ge t$ are mapped to $\phi(s) - \phi(t) \ge 0$, and constraints $s = t$ are mapped to $\phi(s) - \phi(t) = 0$. Hurd's sufficient test for KBO constraints uses basically this scheme [Hur03]. The decision procedure of Korovin and Voronkov [KV01] is based on a more elaborate translation.

One of our main goals is to keep the test polynomial, and solving linear Diophantine equations is well-known to be NP-complete. Therefore, we use ϕ only for local tests in the following rules. Let $\{x_1, \ldots, x_n\}$ be the variables of s and t, $\phi(s) = \alpha_0 + \sum \alpha_i x_i$ and $\phi(t) = \beta_0 + \sum \beta_i x_i$. We write $\phi(s) \ge \phi(t)$ if $\alpha_i \ge \beta_i$ for $i = 1, \ldots, n$ and $\alpha_0 + \sum \alpha_i \mu \ge \beta_0 + \sum \beta_i \mu$, where μ is the weight of the smallest term.

1. KBO-DOWN: $\langle \mathfrak{C}, \mathfrak{D} \rangle \implies \langle \mathfrak{C} \cup \{c_i \ge c_i'\}, \mathfrak{D} \rangle$ if $\mathfrak{D}$ contains the two rules $f(c_1, \ldots, c_n) \to c$ and $f(c_1', \ldots, c_n') \to c'$, $\mathfrak{C}$ contains $c \varrho c'$ with $\varrho \in \{>, \ge\}$, $1 \le i < n$, $\mathfrak{C}$ contains $c_j = c_j'$ for all $j < i$, and $\phi(T_{\mathfrak{D}}(c)) = \phi(T_{\mathfrak{D}}(c'))$.
2. KBO-UP: $\langle \mathfrak{C}, \mathfrak{D} \rangle \implies \langle \mathfrak{C} \cup \{c > c'\}, \mathfrak{D} \rangle$ if $\mathfrak{D}$ contains $f(c_1, \ldots, c_n) \to c$ and $f(c_1', \ldots, c_n') \to c'$, there is some i with $1 \le i < n$ such that $\mathfrak{C}$ contains $c_i > c_i'$ and $c_j = c_j'$ for all $j < i$, and $\phi(T_{\mathfrak{D}}(c)) \ge \phi(T_{\mathfrak{D}}(c'))$.

Lemma 7. *Let $\succ$ be a ground-total KBO for precedence $>_{\mathcal{F}}$ and weight function φ. Let $\langle \mathfrak{C}, \mathfrak{D} \rangle \implies \langle \mathfrak{C}', \mathfrak{D} \rangle$ with KBO-DOWN or KBO-UP. If σ satisfies $\langle \mathfrak{C}, \mathfrak{D} \rangle$ with respect to $\succ$, then σ satisfies $\langle \mathfrak{C}', \mathfrak{D} \rangle$ with respect to $\succ$.*

Proof. Let $s \equiv T_{\mathfrak{D}}(c)$ and $t \equiv T_{\mathfrak{D}}(c')$. Case KBO-DOWN: $\phi(s) = \phi(t)$ implies $\varphi(\sigma(s)) = \varphi(\sigma(t))$ for all ground substitutions σ. As s and t have the same top-symbol, $\sigma(s) \succ \sigma(t)$ implies $\sigma(T_{\mathfrak{D}}(c_i)) \succcurlyeq \sigma(T_{\mathfrak{D}}(c_i'))$. Case KBO-UP: $\phi(s) \geq \phi(t)$ implies $\varphi(\sigma(s)) \geq \varphi(\sigma(t))$ for all ground substitutions σ. Hence, $\sigma(s) \succ \sigma(t)$ either by the weight or by the lexicographic comparison of the subterms.   □

For the additional rules the test for applicability has a worst-case complexity quadratic in N. Therefore, the unsatisfiability test remains polynomial.

5 Experimental Evaluation

We implemented both variants of the AC-redundancy criterion in our theorem prover WALDMEISTER [LH02]. We consider six variants: STD is the standard system, GJ uses the redundancy test based on ground joinability [MN90], ACG uses the criterion of Sect. 3.1 and ACG+ the refined criterion of Sect. 3.2. The variants GJ/ACG and GJ/ACG+ combine the corresponding tests. Most experiments were performed on 1 GHz Pentium III machines with 4 GByte of memory. Exceptions are the RNG036-7 runs, which were performed on 2.6 GHz Xeon P4 with 2 GByte of memory.

As basis for our experiments we chose 218 unit equality problems of TPTP Ver. 2.6.0 [SS97] that contain AC operators. Within 5 minutes, 205 examples complete. They show, if at all, mostly modest improvements. The criterion based on AC-ground reducibility detects most of the equations that the method based on ground joinability detects and needs much less time to do so. It even detects additional equations, because it uses full ordering constraints. Space does not permit to go into further details. The real potential of the new criterion can be seen in Table 1. It contains the runtime data for four challenging proof tasks. Especially the refined criterion shows considerable improvements.

To analyze the accuracy of the sufficient unsatisfiability test, we took the four log files that version ACG+ produced for the challenging examples and constructed with functions Γ and $\hat{\Gamma}$ corresponding constraints. This makes 450,706 test cases, which we checked for unsatisfiability with respect to LPO and KBO. For LPO 112,146 cases can be shown trivially to be $\bot$ or $\top$ by simple ordering tests. For the remaining cases our implementation of the decision procedure of [NR02] shows 151,500 unsatisfiable, the sufficient test shows 149,365 unsatisfiable, i. e., it misses only 1.4 per cent. For KBO a comparison is more difficult, as we have no implementation of a decision procedure available. We use an implementation of the method of [Hur03] instead, which uses the Omega-library [Pug92] as decision procedure for the linear Diophantine inequations. Among all test cases 110,534 are trivial. Among the remaining cases 147,027 are shown unsatisfiable by at least one of the two methods. Whereas Hurd's method detects 104,936 cases, our test detects 141,157, which means that, compared to the union of the tests, it misses 4.0 per cent.

This accuracy is far better than originally expected. It seems that the generated constraints are rather easy. Although not yet optimized for speed, the unsatisfiability test needs less than one per cent of the prover's runtime.

Table 1. Running times in hours for challenging examples.

Problem	STD	GJ	ACG	ACG+	GJ/ACG	GJ/ACG+
ROB020-1	7.5	6.0	3.3	2.6	3.1	2.6
ROB007-1	61.9	39.4	20.7	13.3	17.7	13.4
LAT018-1	> 300	> 300	> 300	12.6	> 300	13.2
RNG036-7	> 500	341.6	> 500	151.6	243.1	112.0

6   Conclusions

We have presented a new redundancy criterion for equational theorem proving that is based on a restricted form of AC-ground reducibility. It captures in an efficient way a frequently occurring special case where other, more elaborate redundancy criteria show success. It fits well into the completion paradigm with its lazy approach, namely, to add overlaps and to treat them later. This simplifies the implementation, reduces the amount of work, and avoids the danger of multiple computations. Therefore, the new criterion shows a very good ratio between detection strength and computational cost. Its integration into the theorem prover WALDMEISTER demonstrates that methods based on constraint technology can significantly improve high-performance systems in practice. The prover can now cope with demanding AC specifications much better than before. Nevertheless, it has to be seen how this compares to an approach that integrates the theory into the calculus, i.e., works modulo AC. The unsatisfiability test for ordering constraints behaves better than originally expected. It will be interesting to be see, how it performs in other domains. For example, it can be used to guide the initial phase of the KBO solver of [KV01].

For generalizations of our redundancy criteria, two directions are immediate: The first one is to transfer the method to the full clausal case, i.e., as a redundancy criterion in the superposition calculus [BG94]. This seems straightforward for the basic case. For the refined criterion it is not so clear to us, because it relies on top-level steps. The other direction is to cover more general permutative theories or to use a more generic ground reducibility test.

References

[AHL03] J. Avenhaus, Th. Hillenbrand, and B. Löchner. On using ground joinable equations in equational theorem proving. *Journal of Symbolic Computation*, 36(1–2):217–233, 2003.

[AL01] J. Avenhaus and B. Löchner. CCE: Testing ground joinability. In R. Goré, A. Leitsch, and T. Nipkow, eds., *Proc. First International Joint Conference on Automated Reasoning*, vol. 2083 of *LNCS*, pp. 658–662. Springer, 2001.

[BDP89] L. Bachmair, N. Dershowitz, and D.A. Plaisted. Completion Without Failure. In *Resolution of Equations in Algebraic Structures*, vol. 2, Rewriting Techniques, pp. 1–30. Academic Press, 1989.

[BG94]   L. Bachmair and H. Ganzinger. Rewrite-Based Equational Theorem Prov-
            ing with Selection and Simplification. *Journal of Logic and Computation*,
            3(4):217–247, 1994.
[CNNR03]   H. Comon, P. Narendran, R. Nieuwenhuis, and M. Rusinowitch. Deciding
            the confluence of ordered term rewrite systems. *ACM Transactions on
            Computational Logic*, 4(1):33–55, 2003.
[CT97]   H. Comon and R. Treinen. The first-order theory of lexicographic path
            orderings is undecidable. *Theoretical Computer Science*, 176:67–87, 1997.
[DP01]   N. Dershowitz and D.A. Plaisted. Rewriting. In A. Robinson and
            A. Voronkov, eds., *Handbook of Automated Reasoning*, vol. I, chapter 9,
            pp. 535–610. Elsevier Science, 2001.
[HR87]   J. Hsiang and M. Rusinowitch. On word problems in equational theo-
            ries. In *Proc. 14th International Colloquium on Automata, Languages and
            Programming*, vol. 267 of *LNCS*. Springer-Verlag, 1987.
[Hur03]   J. Hurd. Using inequalities as term ordering constraints. Technical Report
            567, University of Cambridge Computer Laboratory, 2003.
[JSA94]   P. Johann and R. Socher-Ambrosius. Solving simplification ordering con-
            straints. In J.-P. Jouannaud, ed., *First Int. Conference on Constraints in
            Computational Logics*, vol. 845 of *LNCS*, pp. 352–367. Springer, 1994.
[Kap97]   D. Kapur. Shostak's congruence closure as completion. In H. Comon, ed.,
            *Proc. 8th International Conference on Rewriting Techniques and Applica-
            tions*, vol. 1232 of *LNCS*, pp. 23–37. Springer, 1997.
[KV01]   K. Korovin and A. Voronkov. Knuth-Bendix constraint solving is NP-
            complete. In *Proc. of 28th International Colloquium on Automata, Lan-
            guages and Programming*, vol. 2076 of *LNCS*, pp. 979–992. Springer, 2001.
[LH02]   B. Löchner and Th. Hillenbrand. A phytography of WALDMEISTER. *AI
            Communications*, 15(2–3):127–133, 2002. See www.waldmeister.org.
[Löc04]   B. Löchner. Joinability is undecidable for ordered rewriting. Technical
            Report, FB Informatik, Technische Universität Kaiserslautern, 2004.
[MN90]   U. Martin and T. Nipkow. Ordered rewriting and confluence. In M.E.
            Stickel, ed., *Proc. of the 10th International Conference on Automated De-
            duction*, vol. 449 of *LNCS*, pp. 366–380. Springer, 1990.
[Nie93]   R. Nieuwenhuis. Simple LPO constraint solving methods. *Information
            Processing Letters*, 47:65–69, 1993.
[NR01]   R. Nieuwenhuis and A. Rubio. Paramodulation-based theorem proving. In
            A. Robinson and A. Voronkov, eds., *Handbook of Automated Reasoning*,
            vol. I, chapter 7, pp. 371–443. Elsevier Science, 2001.
[NR02]   R. Nieuwenhuis and J. Rivero. Practical algorithms for deciding path or-
            dering constraint satisfaction. *Information & Computation*, 178:422–440,
            2002.
[Pug92]   W. Pugh. The Omega test: a fast and practical integer programming algo-
            rithm for dependence analysis. *Communications of the ACM*, 35(8):102–
            114, 1992. Library available at www.cs.umd.edu/projects/omega/.
[PZ97]   D.A. Plaisted and Y. Zhu. Equational reasoning using AC constraints. In
            *Proc. of the Fifteenth International Joint Conference on Artificial Intelli-
            gence*, vol. 1, pp. 108–113. Morgan Kaufmann, 1997.
[SS97]   C. B. Suttner and G. Sutcliffe. The TPTP problem library (TPTP v2.1.0).
            Technical Report 97/08, Department of Computer Science, James Cook
            University, Townsville, Australia, 1997. See www.tptp.org.

Efficient Checking of Term Ordering Constraints

Alexandre Riazanov and Andrei Voronkov

University of Manchester
{riazanov,voronkov}@cs.man.ac.uk

Abstract. Simplification orderings on terms play a crucial role in reducing the search space in paramodulation-based theorem proving. Such a use of orderings requires checking simple ordering constraints on substitutions as an essential part of many operations. Due to their frequency, such checks are costly and are a good target for optimisation. In this paper we present an efficient implementation technique for checking constraints in one of the most widely used simplification orderings, the Knuth-Bendix ordering. The technique is based on the idea of run-time algorithm specialisation, which is a close relative of partial evaluation.

1   Introduction

Many modern theorem provers for first-order logic (e. g., E[13], Gandalf[15], Waldmeister[1]) deal with equality by using inference systems based on the following *paramodulation* rule:

$$\frac{C \vee s \simeq t \quad D[u]}{(C \vee D[t])\theta} \text{ (unrestricted) paramodulation,}$$

where θ is a most general unifier of the terms s and u.

Unrestricted use of this rule is too prolific to be practical. To overcome this, a number of inference systems has been devised that use a *simplification ordering* $\succ$ on terms to restrict the application of paramodulation (see [7] for a comprehensive survey). Although *superposition* is probably the most restrictive instance of paramodulation, for the purposes of this paper it is sufficient to consider only inference systems based on a less restricted *ordered paramodulation* rule:

$$\frac{C \vee s \simeq t \quad D[u]}{(C \vee D[t])\theta} \text{ ordered paramodulation,}$$

where θ is a most general unifier of s and u, $t\theta \not\succeq s\theta$ and $\succ$ is the chosen simplification ordering. Checking the condition $t\theta \not\succeq s\theta$ can be viewed as checking whether the substitution θ satisfies the *constraint* $t \not\succeq s$.

Apart from restricting paramodulation, simplification orderings are used by some important simplification mechanisms, of which the most widely used one is *demodulation*: a positive unit equality $s \simeq t$ can be used to replace a clause $C[s\theta]$ by a "simpler" clause $C[t\theta]$, provided that the substitution θ satisfies the condition $s\theta \succ t\theta$. Again, this condition can be viewed as a constraint $s \succ t$ on substitutions.

D. Basin and M. Rusinowitch (Eds.): IJCAR 2004, LNAI 3097, pp. 60–74, 2004.

In a paramodulation-based theorem prover, checking ordering constraints on substi-tutions is a useful component of many operations, such as resolution, paramodulation and demodulation, and may be invoked frequently. It is not unusual when during a 10 minute run of our system Vampire [9] over 50,000,000 ordering constraint checks have to be performed. The simplification orderings used in Vampire and other systems are rather complex, so checking ordering constraints in a straightforward manner, that is by direct comparison of the terms $s\theta$ and $t\theta$, can be very costly, and in many cases even dominate the time taken by all other operations in a prover. As a consequence, checking ordering constraints must be implemented efficiently.

In this paper we show how checking ordering constraints is implemented in Vampire. We describe a *technique for efficient implementation of checking constraints* in the well-known *Knuth-Bendix ordering*. We illustrate our approach only for demodulation. Nevertheless, the proposed technique is directly applicable to, and works well in, all other operations in Vampire that require ordering constraint checking. The presented recipe is mainly intended for developers of systems implementing paramodulation with ordering-based restrictions. However, we hope that this paper will be of interest to a broader audience as a compact and self-contained illustration of the general method of *run-time algorithm specialisation*, on which our technique is based.

2 Preliminaries

2.1 Knuth-Bendix Ordering

Throughout the paper we will be dealing with a fixed signature $\mathcal{F}$ that contains *function symbols* denoted by f, g, h. If we want to emphasise that some function symbols are *constants*, i.e., have arity 0, they will be denoted by a, b, c. In addition to function symbols, we have an alphabet $\mathcal{V}$ of *variables*. The set of *terms* over $\mathcal{F}$ and $\mathcal{V}$, denoted by $\mathcal{T}(\mathcal{F}, \mathcal{V})$, is defined in the standard way: all variables in $\mathcal{V}$ are terms, and if $f \in \mathcal{F}$ is a function symbol of arity n and $t_1, \ldots, t_n$ are terms, then $f(t_1, \ldots, t_n)$ is also a term. Terms containing no variables are called *ground*.

An ordering $\succ$ on $\mathcal{T}(\mathcal{F}, \mathcal{V})$ is called a *simplification ordering* if:

1. $\succ$ is *stable under substitutions:* $s \succ t$ implies $s\theta \succ t\theta$ for all substitutions θ.
2. All function symbols of $\mathcal{F}$ are *monotonic w.r.t.* $\succ$: if $s \succ t$ then
   $f(s_1, \ldots, s_{i-1}, s, s_{i+1}, \ldots, s_n) \succ f(s_1, \ldots, s_{i-1}, t, s_{i+1}, \ldots, s_n)$ for all i.
3. $\succ$ satisfies the *subterm property*: $f(t_1, \ldots, t_n) \succ t_i$ for all $f \in \mathcal{F}$ and i.

This paper is mostly concerned with one specific simplification ordering: the *Knuth-Bendix ordering* [3]. From now on, $\succ$ will denote this ordering unless otherwise is explicitly stated.

The Knuth-Bendix ordering is parameterised by a *weight function* w from $\mathcal{F}$ to the set of non-negative integers and a *precedence relation*, which is a total order $\gg$ on the set $\mathcal{F}$ of function symbols. For simplicity, in this paper we assume that $w(f) > 0$ for all $f \in \mathcal{F}$. The weight function will be called *uniform* if $w(f) = 1$ for all $f \in \mathcal{F}$. In our examples we always assume that the weight function is uniform, unless otherwise is explicitly stated.

For every variable or a function symbol q and a term t, denote by $C_q(t)$ the number of occurrences of q in the term t. The *weight* of a term $t \in \mathcal{T}(\mathcal{F}, \mathcal{V})$, denoted by $|t|$, is a polynomial over $\mathcal{V}$ with integer coefficients defined as follows:

$$|t| = \sum_{f \in \mathcal{F}} C_f(t) \cdot w(f) + \sum_{x \in \mathcal{V}} C_x(t) \cdot x.$$

For example, if w is uniform, then $|f(f(a, x_1), f(x_1, x_2))| = 4 + 2 \cdot x_1 + x_2$. Note that if t is ground, then $|t|$ is an integer constant. Let μ denote the minimal weight of a ground term, which is in fact the minimal weight of a constant in $\mathcal{F}$. In our examples we will always assume $\mu = 1$.

The set of all linear polynomials over $\mathcal{V}$ with non-negative integer coefficients will be denoted by $\mathcal{P}(\mathcal{V})$. We will sometimes call such polynomials *weight expressions*. We call a *weight assignment* any function $\sigma : \mathcal{V} \to \mathcal{P}(\mathcal{V})$ such that for every $v \in \mathcal{V}$, if $\sigma(v)$ is constant, then $\sigma(v) \geq \mu$. If all polynomials $\sigma(x)$ are constant, σ will be called a *ground weight assignment*. We distinguish a weight assignment σ_μ, such that $\sigma_\mu(x) = \mu$ for all $x \in \mathcal{V}$. Weight assignments are extended to linear polynomials in the following way: $\sigma(\alpha_0 + \alpha_1 x_1 + \ldots + \alpha_n x_n)$ is the polynomial obtained by normalising $\alpha_0 + \alpha_1 \cdot \sigma(x_1) + \ldots + \alpha_n \cdot \sigma(x_n)$ using the standard arithmetic laws. For example, if $\sigma(x_1) = 2 + 3x_1 + 2x_2$ and $\sigma(x_2) = 3 + 2x_1 + x_2$, then $\sigma(1 + 2x_1 + 3x_2) = 1 + 2(2 + 3x_1 + 2x_2) + 3(3 + 2x_1 + x_2) = 14 + 12x_1 + 7x_2$. Note that if σ is a ground weight assignment, then $\sigma(p)$ is constant for all $p \in \mathcal{P}(\mathcal{V})$.

We define two binary relations $>$ and $\gg$ on $\mathcal{P}(\mathcal{V})$ as follows. Suppose $p_1 = \alpha_0 + \alpha_1 x_1 + \ldots + \alpha_n x_n$ and $p_2 = \beta_0 + \beta_1 x_1 + \ldots + \beta_n x_n$, where some of α_i, β_i may be zero. Then $p_1 > p_2$ if and only if $\sigma(p_1) > \sigma(p_2)$ for all ground weight assignments σ. Also, we let $p_1 \gg p_2$ if and only if $\sigma(p_1) \geq \sigma(p_2)$ for all ground weight assignments σ. Note that both $p_1 = p_2$ and $p_1 > p_2$ imply $p_1 \gg p_2$, but $p_1 \gg p_2$ is not equivalent to $p_1 > p_2 \vee p_1 = p_2$. For example, if a is a constant of the minimal weight μ, then $|f(x)| \gg |f(a)|$, but neither $|f(x)| > |f(a)|$ nor $|f(x)| = |f(a)|$.

Now the Knuth-Bendix ordering $\succ$ on $\mathcal{T}(\mathcal{F}, \mathcal{V})$ can be defined[1] inductively as follows:

$$s \succ t \Leftrightarrow \begin{cases} |s| > |t|, \\ \textbf{or else } |s| \gg |t|, \ s = f(\ldots), \ t = g(\ldots) \ and \ f \gg g, \\ \textbf{or else } |s| \gg |t|, \ s = f(s_1, \ldots, s_n), \ t = f(t_1, \ldots, t_n) \\ \qquad and \ for \ some \ i \ we \ have \ s_1 = t_1, \ldots, s_{i-1} = t_{i-1}, \ s_i \succ t_i. \end{cases}$$

In the sequel we will abbreviate "Knuth-Bendix ordering" to KBO. Note that a stronger variant $\succ'$ of KBO on $\mathcal{T}(\mathcal{F}, \mathcal{V})$ can be defined as follows: $s \succ' t$ if and only if the constraint $s \preceq t$ has no solutions, i.e. there is no grounding substitution θ, such that $s\theta \preceq t\theta$. Deciding $s \succ' t$ for given terms s and t can be done in polynomial time ([4]), but in practice approximations, such as ours, which allow simpler algorithms are usually used.

---

[1] Our definition is technically different from the classical one in [3]. The use of polynomials as term weights allows us to avoid considering numbers of variable occurences separately.

2.2 Demodulation and Ordering Constrains

Demodulation is the following simplification rule: in presence of a positive unit equality $s \simeq t$, a clause $C[s\theta]$ can be replaced by a "simpler" clause $C[t\theta]$ if the substitution θ satisfies the condition $s\theta \succ t\theta$. This condition can be viewed as an ordering constraint $s \succ t$ to be checked on different substitutions θ. This ordering constraint can be checked on a substitution θ by simply verifying $s\theta \succ t\theta$ using a general-purpose algorithm implementing $\succ$, but this may be inefficient since the same constraint can be checked on many different substitutions. Note that $s \succ t$ guarantees $s\theta \succ t\theta$, but during the proof search provers often generate non-orientable equations $s \simeq t$, so that neither $s \succ t$ nor $t \succ s$ hold and the condition $s\theta \succ t\theta$ actually depends on the substitution θ generated as a result of some unification or matching operation. In this paper we show how to do such checks $s\theta \succ t\theta$ for the KBO efficiently.

Demodulation is usually applied in either forward or backward mode. *Forward demodulation* is the following mechanism. Suppose we have just derived a new clause C. We would like to simplify it as much as possible by demodulation before adding to the current set of persistent clauses. For this purpose, we try to find among the current set of persistent clauses a positive unit equality $s \simeq t$ such that an instance $s\theta$ of the term s occurs in C. When such an equality is found and the constraint $s \succ t$ holds on θ, we replace $s\theta$ in C by $t\theta$. Demodulation inferences are performed on the clause exhaustively.

In *backward demodulation*, we try to use a newly derived positive unit equality $s \simeq t$ to rewrite some persistent clauses. To this end, we try to identify all persistent clauses that contain instances of s, i.e., terms of the form $s\theta$. As soon as such a clause $C[s\theta]$ is found, we have to check $s\theta \succ t\theta$ to ensure applicability of demodulation.

At this point we have to make an important observation. In backward demodulation, a constraint $s \succ t$ is tested against *potentially many substitutions θ in a row*. Such checks will be called *backward checks*. In forward demodulation, the constraints and substitutions occur in an unpredictable order, in this case we speak about *forward checks*.

In a typical paramodulation-based theorem prover, demodulation is only one of many operations that require checking ordering constraints. For example, in Vampire both forward and backward checks are also employed to implement superposition inferences. The techniques introduced in this paper are directly applicable to most, if not all, known instances of KBO constraint checking. The implementations used in Vampire for demodulation and superposition are nearly identical, so in our experiments we measure performance for both of them together.

2.3 Run-Time Algorithm Specialisation

In our system Vampire several powerful optimisations ([8,10]) use the idea of *runtime algorithm specialisation*. The idea is inspired by the use of *partial evaluation* in optimising program translation (see, e.g., [2]). Suppose that we need to execute some algorithm $alg(A, B)$ in a situation where a value of A is fixed for potentially many different values of B. We can try to find a specialisation of alg for every fixed A, i.e., such an algorithm alg_A, that executing $alg_A(B)$ is equivalent to executing $alg(A, B)$. The purpose of specialisation is to make $alg_A(B)$ work faster than $alg(A, B)$ by exploiting some particular properties of the fixed value A. Typically, $alg_A(B)$ can avoid some operations that $alg(A, B)$ would have to perform, if they are known to be redundant for this particular parameter A. In particular, we can often identify some tests that are true or false for A, unroll loops and recursion, etc. In general, we are free to use any algorithm as $alg_A(B)$ as long as it does the same job as $alg(A, B)$.

Unlike in partial evaluation, the values of A are not known statically, so the *specialisation takes place in run-time*. Moreover, we do not need any concrete representation of alg. We only have to *imagine alg* when we program the specialisation procedure. This implies that we cannot use any universal methods for specialising algorithms, which is usually the case with partial evaluation, and have to program a specialisation procedure for every particular algorithm alg. An important advantage of doing so is that we can use some powerful specific optimisations exploiting peculiarities of alg.

The specialised algorithm has to be represented in a form that can be interpreted. In many cases, usually when $alg_A(B)$ is to be computed on many values B in a row, we can compile alg_A into a code of a special abstract machine. Instructions of the abstract machine can be represented as records with one field storing an integer tag that identifies the instruction type, and other fields for instruction parameters. All instructions of a code can be stored in an array, or list, or tree. Interpretation is done by fetching instructions in some order, identifying their type and executing the actions associated with this type. In C or C++ we can use a **switch** statement to associate some actions with different instruction tags. Modern compilers usually compile a **switch** statement with integer labels from a narrow range rather efficiently by storing the address of the statement corresponding to a value i in the i-th cell of a special array. We exploit this by taking values for instruction tags from a small interval of integers.

There are situations when many instances of A are intended for long-term storage and the calls of $alg(A, B)$ occur with different B in an unpredictable order. For example, we may have to check $alg(A_1, B_1)$ first, then $alg(A_2, B_2)$, then $alg(A_1, B_3)$, and so on. In such circumstances, full-scale specialisation with compilation may not be appropriate due to excessive memory usage. However, we can sometimes find a compact specialised representation A' for every A that can be stored with, or instead of, A. We also define a variant alg' that works on this representation and any call to $alg(A, B)$ is replaced by $alg'(A', B)$, intended to do the same job faster.

3   General Framework

In this section we present a general framework for specialising KBO constraint checks. This framework will be used in the following sections to describe our implementations of forward and backward checks.

3.1   Specialising Weight Comparison

When implementing KBO constraint checking, it is convenient to have a function for comparing weight expressions. We introduce such function

$$compw : \mathcal{P}(\mathcal{V}) \times \mathcal{P}(\mathcal{V}) \to \{\text{=}, \text{>}, \text{<}, \text{≤}, \text{≥}, \text{?}\}$$

as follows.

$$compw(p_1, p_2) = \begin{cases} \text{\textcircled{=}} & \text{if } p_1 = p_2; \\ \text{\textcircled{>}} & \text{if } p_1 > p_2; \\ \text{\textcircled{<}} & \text{if } p_2 > p_1; \\ \text{\textcircled{$\geqslant$}} & \text{if } p_2 \gg p_1 \text{ but neither } p_2 = p_1 \text{ nor } p_2 > p_1; \\ \text{\textcircled{$\geqslant$}} & \text{if } p_1 \gg p_2 \text{ but neither } p_1 = p_2 \text{ nor } p_1 > p_2; \\ \text{\textcircled{?}} & \text{otherwise.} \end{cases}$$

Let us show how computation of $compw(|s\theta|, |t\theta|)$ can be specialised for fixed terms s and t. The specialisation process can be logically divided in two phases.

Phase 1. For every substitution θ define the weight assignment σ_θ as follows: $\sigma_\theta(x) = |x\theta|$ for all $x \in V$. Note that we can compute $compw(\sigma_\theta(|s|), \sigma_\theta(|t|))$ as the value for $compw(|s\theta|, |t\theta|)$. When s and t are fixed, the weight expressions $|s|$ and $|t|$ are precomputed, which allows us to avoid traversing the s and t parts in the terms $s\theta$ and $t\theta$ for various θ. Also, in the optimised comparison we compute $\sigma_\theta(x) = |x\theta|$ only once for each variable of s and t, while a general-purpose procedure would traverse the term $x\theta$ during computation of $|s\theta|$ and $|t\theta|$ as many times as there are occurrences of x in s and t.

For example, consider $s = f(x_0, f(x_0, f(x_1, f(x_2, x_3))))$. Instead of directly computing $|f(x_0\theta, f(x_0\theta, f(x_1\theta, f(x_2\theta, x_3\theta))))|$, we compute $4 + 2|x_0\theta| + |x_1\theta| + |x_2\theta| + |x_3\theta|$, thus avoiding traversal of the $f(., f(., f(., f(., .))))$ part and examining the term $x_0\theta$ twice.

Phase 2. Instead of computing $compw(\sigma_\theta(|s|), \sigma_\theta(|t|))$ we can compute $compw(\sigma_\theta(lft(|s|, |t|)), \sigma_\theta(rht(|s|, |t|)))$. The functions $lft, rht : \mathcal{P}(V) \times \mathcal{P}(V) \to \mathcal{P}(V)$ are defined as follows: $lft(p_1, p_2)$ is obtained by taking all members of $p_1 - p_2$ with positive coefficients, and $rht(p_1, p_2)$ is obtained by taking all members of $p_2 - p_1$ with positive coefficients [2]. The polynomials $lft(|s|, |t|)$ and $rht(|s|, |t|)$ may potentially have fewer different variables than $|s|$ and $|t|$, in which case we have to examine fewer terms $x\theta$.

For example, consider the following terms: $s = f(x_0, f(x_0, f(x_1, f(x_2, x_3))))$ and $t = f(x_1, f(x_2, f(x_3, x_3)))$. Comparison of $4 + 2|x_0\theta| + |x_1\theta| + |x_2\theta| + |x_3\theta|$ with $3 + |x_1\theta| + |x_2\theta| + 2|x_3\theta|$ is now reduced to comparing $1 + 2|x_0\theta|$ with $|x_3\theta|$, which does not require examining $x_1\theta$ and $x_2\theta$ at all. Moreover, we no longer need to multiply $|x_3\theta|$ by a constant.

Special cases. For fixed $p_1, p_2 \in \mathcal{P}(V)$, comparison of $\sigma_\theta(p_1)$ with $\sigma_\theta(p_2)$ for different θ can be specialised depending on the form of p_1 and p_2. Suppose that $p_1 = \alpha_1 x_1 + \ldots + \alpha_n x_n$, $\alpha_i > 0$, and $p_2 = \beta_0 > 0$. In this case $\sigma_\theta(p_1) \geqslant \sigma_\theta(p_2)$ is equivalent to $\alpha_1 \cdot mwgi(x_1\theta) + \ldots + \alpha_n \cdot mwgi(x_n\theta) \geq \beta_0$, where $mwgi(t)$ denotes the *minimal weight of a ground instance* of the term t, and can be computed as the weight of t with all variables replaced by constants of the minimal weight. Likewise, $\sigma_\theta(p_1) > \sigma_\theta(p_2)$ is equivalent to $\alpha_1 \cdot mwgi(x_1\theta) + \ldots + \alpha_n \cdot mwgi(x_n\theta) > \beta_0$ and $\sigma_\theta(p_1) = \sigma_\theta(p_2)$ holds

---

[2] Using $lft(p_1, p_2)$ and $rht(p_1, p_2)$ is conceptually the same as, but technically more convenient than considering positive and negative parts of $p_1 - p_2$.

if and only if $\alpha_1 \cdot mwgi(x_1\theta) + \ldots + \alpha_n \cdot mwgi(x_n\theta) = \beta_0$ and all $x_i\theta$ are ground. These observations give us two big advantages.

Firstly, $mwgi(x_i\theta)$ are *numbers* rather than polynomials, and thus are easier to compute, store and manipulate. For some representations of terms, $mwgi$ can even be computed in constant time. For example, if the used weight function is uniform, then $mwgi(t)$ is just the size of t, which can be computed in constant time if t is an *array-based flatterm* (see, e.g., [8]). In some situations it may also be useful to precompute and store $mwgi$ for some terms.

Secondly, for many substitutions θ we do not need to compute $\alpha_1 \cdot mwgi(x_1\theta) + \ldots + \alpha_n \cdot mwgi(x_n\theta)$ completely. Instead, we incrementally compute $\alpha_1 \cdot mwgi(x_1\theta)$, $\alpha_1 \cdot mwgi(x_1\theta) + \alpha_2 \cdot mwgi(x_2\theta), \alpha_1 \cdot mwgi(x_1\theta) + \alpha_2 \cdot mwgi(x_2\theta) + \alpha_3 \cdot mwgi(x_3\theta)$, and so on. If at the i-th step we notice that $\alpha_1 \cdot mwgi(x_1\theta) + \ldots + \alpha_i \cdot mwgi(x_i\theta) + \alpha_{i+1}\mu + \ldots + \alpha_n\mu > \beta_0$ we can immediately stop and claim $\sigma_\theta(p_1) > \sigma_\theta(p_2)$. Moreover, we can notice that the condition holds while collecting the weight $mwgi(x_i\theta)$.

Another special case corresponds to situations when $p_1 = \alpha_0$ and $p_2 = \beta_1 x_1 + \ldots + \beta_n x_n$. Note that $\sigma_\theta(p_1) > \sigma_\theta(p_2)$ if and only if all $x_i\theta$ are ground and $\alpha_0 \geq \beta_1 \cdot mwgi(x_1\theta) + \ldots + \beta_n \cdot mwgi(x_n\theta)$. Again, as soon as we notice that some $x_i\theta$ is non-ground or $\beta_1 \cdot mwgi(x_1\theta) + \ldots + \beta_i \cdot mwgi(x_i\theta) + \beta_{i+1}\mu + \ldots + \beta_n\mu > \alpha_0$, we can stop and claim that $\sigma_\theta(p_1) \not> \sigma_\theta(p_2)$.

The general case of $p_1 = \alpha_0 + \alpha_1 x_1 + \ldots + \alpha_n x_n$ and $p_2 = \beta_0 + \beta_1 x_1 + \ldots + \beta_n x_n$ degenerates into one of the above special cases when we discover that either $\sigma_\theta(p_1)$ or $\sigma_\theta(p_2)$ is a constant.

3.2   Unrolling Recursion and Loops

A straightforward algorithm for checking $s \succ t$ for arbitrary s and t is shown in Figure 1. It returns one of the three values $\{\boxed{=}, \boxed{>}, \boxed{\times}\}$, where $\boxed{\times}$ means failure. If we are to check $s\theta \succ t\theta$ for fixed terms s and t, we can specialise the general algorithm by unrolling recursion and loops, and detecting redundant operations. Let us show how to derive a specialised procedure $greater_{s,t}(\theta)$ which computes the same value as $greater(s\theta, t\theta)$.

We assume that none of $s \succ t$, $t \succ s$ and $s = t$ holds, otherwise the comparison does not depend on θ at all. If s or t is a variable, there is nothing we can do but to compute $greater(s\theta, t\theta)$, so in what follows we assume that $s = f(s_1, \ldots, s_m)$ and $t = g(t_1, \ldots, t_n)$ (f and g need not be different). The first specialisation step is to compute $compw(|s|, |t|)$. Our assumption implies $compw(|s|, |t|) \notin \{\boxed{>}, \boxed{<}\}$, so we are left with four cases.

1. *Case* $compw(|s|, |t|) = \boxed{=}$. In this case we can avoid computing $compw(|s\theta|, |t\theta|)$, since $|s\theta| = |t\theta|$. As a consequence, $greater_{s,t}(\theta) = greater'_{s,t}(\theta)$. Since the terms s and t are incomparable, we must have $f = g$ and $n \geq 1$, so $greater'_{s,t}(\theta) = lexgreater_{<s_1,\ldots,s_n>,<t_1,\ldots,t_n>}(\theta)$.
   The specialisation of lexicographic comparison $lexgreater_{<s_1,\ldots,s_n>,<t_1,\ldots,t_n>}$ is defined as follows.
   a) If $s_1 \succ t_1$, then $lexgreater_{<s_1,\ldots,s_n>,<t_1,\ldots,t_n>}(\theta)$ consists of a single line **return** $\boxed{>}$.

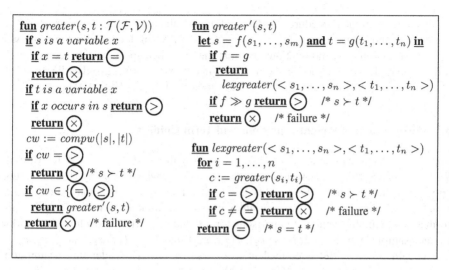

Fig. 1. Straightforward comparison of terms with KBO

b) If $t_1 \succ s_1$, then $lexgreater_{<s_1,\dots,s_n>,<t_1,\dots,t_n>}(\theta)$ consists of a single line **return** $\otimes$.

c) If $s_1 = t_1$, then we have $lexgreater_{<s_1,\dots,s_n>,<t_1,\dots,t_n>}(\theta)$ coincides with $lexgreater_{<s_2,\dots,s_n>,<t_2,\dots,t_n>}(\theta)$.

Note that in all of the above cases we skip a redundant test $greater(s_1\theta, t_1\theta)$ that would have to be performed by the straightforward procedure. In the remaining cases we assume that s_1 and t_1 are incomparable.

d) If $n = 1$, then $lexgreater_{<s_1,\dots,s_n>,<t_1,\dots,t_n>}$ is $greater_{s_1,t_1}(\theta)$.

e) If $n > 1$, then the code for $lexgreater_{<s_1,\dots,s_n>,<t_1,\dots,t_n>}$ is of the form

$$c := greater_{s_1,t_1}(\theta)$$
if $c \neq (=)$ **return** c
... code for $lexgreater_{<s_2,\dots,s_n>,<t_2,\dots,t_n>}(\theta)$

2. *Case* $compw(|s|,|t|) = (\leq)$. Here we know that $compw(|s\theta|,|t\theta|) \notin \{(>),(\geq)\}$. When $compw(|s\theta|,|t\theta|) = (=)$, $greater_{s,t}(\theta)$ returns $greater'_{s,t}(\theta)$, and fails otherwise.

3–4. *Cases* $compw(|s|,|t|) = (\geq)$ *and* $compw(|s|,|t|) = (?)$. In these cases we know little about $compw(|s\theta|,|t\theta|)$, but the lexicographic part can still be specialised:

fun $greater_{s,t}(\theta)$
  $cw := compw(|s\theta|,|t\theta|)$
  if $cw = (>)$ **return** $(>)$
  if $cw \in \{(=),(\geq)\}$ **return** $greater'_{s,t}(\theta)$
  return $\otimes$

In the body of the procedure $greater_{s,t}$ all inner calls to the procedures $greater_{s',t'}$, $greater'_{s',t'}$ and $lexgreater_{<s_1,...,s_n>,<t_1,...,t_n>}$ are fully unfolded, and in the result we have no recursion or iteration. Note also that the described specialisation technique is fully compatible with the specialisation of weight comparisons discussed earlier: instead of $compw(|s\theta|, |t\theta|)$ we can call $compw(\sigma_\theta(lft(|s|, |t|)), \sigma_\theta(rht(|s|, |t|)))$.

3.3 Propagation of Weight Equations and Term Unifiers

Suppose we have to compare $s = f(s_1, \ldots, s_n)\theta$ with $t = f(t_1, \ldots, t_n)\theta$ for various θ. When $|s\theta| = |t\theta|$, we have to compare $s_i\theta$ and $t_i\theta$, which in many cases requires comparing their weights. The equality $|s\theta| = |t\theta|$ may give us information about the weights of the instances of some variables in s and t. Sometimes we can use this information to simplify weight comparisons for $s_i\theta$ and $t_i\theta$. Let us show how this works by an example. Let $s = f(f(x_1, x_2), f(x_3, x_4))$ and $t = f(f(x_3, x_4), f(x_2, x_4))$. The equality $|s\theta| = |t\theta|$ is equivalent to $\sigma_\theta(x_1) = \sigma_\theta(x_4)$. Under this assumption we have $compw(|f(x_1, x_2)\theta|, |f(x_3, x_4)\theta|) = compw(\sigma_\theta(x_1 + x_2), \sigma_\theta(x_3 + x_4)) = compw(\sigma_\theta(x_2), \sigma_\theta(x_3))$, and thus we avoid comparing $\sigma_\theta(x_1)$ with $\sigma_\theta(x_4)$. If $|s\theta| \neq |t\theta|$ but $compw(|s\theta|, |t\theta|) = \text{(≥)}$, which indicates that $|s\theta| > |t\theta|$, we may still be able to simplify some further weight checks.

At run time this optimisation can hardly be useful, instead we built it into the specialisation procedure. In general, when we specialise comparison of some subterms in a constraint $s \succ t$, we use all relations on weights of variables, available from weight checks for the examined superterms. The relations are used to simplify the weight comparison in a lazy manner: we apply an equality only if it reduces the number of variables in the weight comparison in question and an inequality is applied only if it can reduce the weight comparison to a definite answer.

A similar optimisation which allows stronger specialisation is based on the following observation: if $greater(s_1\theta, t_1\theta)$ returns (=), it means that θ is a unifier of s_1 and t_1. In this situation, at the next step we can compare $(s_2\tau)\theta$ with $(t_2\tau)\theta$ instead of $s_1\theta$ and $t_1\theta$, where τ is an idempotent most general unifier of s_1 and t_1. This check can often be better specialised than the one for $s_2\theta$ and $t_2\theta$. For example, let $s = f(x_0, f(x_1, x_1))$ and $t = f(x_1, f(x_0, x_0))$. For any substitution θ making $x_0\theta = x_1\theta$, $greater(f(x_1, x_1)\theta, f(x_0, x_0)\theta) = greater(f(x_1, x_1)\theta, f(x_1, x_1)\theta) = \text{(=)}$. So, as soon as $x_0\theta = x_1\theta$, we can stop and claim $s\theta = t\theta$ without examining $f(x_1, x_1)\theta$ and $f(x_0, x_0)\theta$.

4 Forward Checks

4.1 Specialised Forward Checks

Forward checks are performed on constraints that are associated with persistent clauses. Any such constraint may potentially be checked many times on different substitutions, and thus is a good target for specialisation. There are usually many such constraints active at the same time, and a constraint may be persistent for a long time. Therefore, if we want to specialise the constraints, the specialised versions are subject to long-term storage. In these circumstances, full-scale specialisation would require too much memory for storing the specialised versions, and we are bound to look for a light-weight approximation with a very compact representation of specialised constraints.

Suppose $s \succ t$ is a constraint intended for forward checks. Whenever it is checked on a substitution θ, we have to compare the weights $|s\theta|$ and $|t\theta|$. We restrict ourselves to specialising only $compw(|s\theta|, |t\theta|)$. When this test is inconclusive, we proceed as in the straightforward check by computing $greater'(s\theta, t\theta)$. So, apart from the terms s and t themselves, the only thing to be stored is the specialisation of $compw(|s\theta|, |t\theta|)$, which is simply a pair of polynomials $lft(|s|, |t|)$ and $rht(|s|, |t|)$. In our implementation linear polynomials are stored as linked lists of pairs $< variable, coefficient >$, each corresponding to a monomial. The constant parts of polynomials are represented by $< \#, coefficient >$, where $\#$ is a special symbol. The main advantage of such representation is that some tails of such lists can be shared, thus reducing the strain on memory usage considerably.

4.2 Experimental Results

We compare experimentally the performance of straightforward constraint checking with the performance of our specialisation-based implementations. The comparison is done by running Vampire, version v5.40, with the OTTER algorithm (see [11]) on Pentium III machines[3], 1GHz, 256Kb cache, running Linux, giving it 10 minutes time limit and 350Mb memory limit. It is run on the 1132 pure equality problems (including 776 unit equality problems) from the TPTP library v2.6.0 ([14]). Experimental results are given for the whole benchmark suite as well as for the 10% of the problems with the largest times taken by straightforward checks. To make the assessment, we measure the time taken by the straightforward and optimised checks for every problem and compare the sums of these times for all problems in the test suite. The checks required by superposition are measured as well as the ones for demodulation. However, in our experiments superposition accounts only for less than 1% of the checks. When memory usage is mentioned, for each problem we measure the maximal amount of memory occupied by the data structures required by the optimisation in question. We experiment with both uniform and non-uniform weight functions. The non-uniform weight function is specified by the following simple formula: $w(f) = 1 + 10 \cdot arity(f)$.

Table 1 summarises the results of our experiments with the forward checks. The straightforward implementation is abbreviated as **str** and the specialised one as **spec**. The second column presents the improvement in time gained by specialisation. To show the reader a more thorough picture, we also present the percentage of time taken by straightforward checks in the overall time of Vampire runs. The overall time itself is presented in the next column to give the reader a feeling of complexity of the tests. The last two columns show the number of constraint checks performed and the number of invocations of the specialisation procedure.

---

[3] A smaller experiment has been also done on UltraSPARC II machines with Solaris. The results were consistent with the ones presented in this paper.

Table 1. Experimental results for forward checks

problem selection	weight func.	$\frac{time(\mathtt{str})}{time(\mathtt{spec})}$	$\frac{time(\mathtt{str})\cdot100\%}{total\ time}$	total time per prob., sec	checks per prob., 10^6	spec. calls per prob., 10^3
all 1132	uniform	2.20	42%	100	8.8	4.4
all 1132	non-un.	2.22	43%	101	8.6	4.4
113 hardest	uniform	2.08	78%	367	52.7	9.8
113 hardest	non-un.	2.07	80%	367	50.8	9.3

When memory overhead is concerned, average figures for heterogeneous problem sets are not sufficiently informative. So, we characterise the extreme cases here. In 44 tests with non-uniform weighting functions, the memory usage overhead exceeded 1 Megabyte. All these problems are relatively hard: on all of them Vampire worked for more than 320 seconds and used more than 270 Mb. The worst overhead of more than 6 Mb was incurred in a 510 second run on BOO066-1, in addition to 310 Mb used by other datastructures. Very similar results were obtained for the tests with the uniform weighting function.

In our early implementations we have experimented with a slightly deeper specialised representation of constraints. Namely, we stored the result of $greater'(s,t)$ in addition to $lft(|s|,|t|)$ and $rht(|s|,|t|)$. If $greater'(s,t) = \boxed{>}$, then $greater'(s\theta,t\theta) = \boxed{>}$ for any θ, and $greater'(s,t) = \boxed{<}$ also implies $greater'(s\theta,t\theta) = \boxed{<}$. To our surprise, this modification led to a noticeable slowdown, not to mention that it uses slightly more memory. This can be explained by the fact that in all our experiments the share of constraint checks that are affected is very low (always less than 3%, on average less than 1%) and the time spent for computing $greater'(s,t)$ does not pay off.

5   Compiled Backward Checks

In the backward mode a constraint is checked on potentially many substitutions in a row. In this situation we can afford a deeper specialisation with compilation. Indeed, it can be done in the same piece of memory for many constraints, the overhead introduced by the specialisation procedure is amortised well by the subsequent checks, and the benefits of specialisation increase with the growing number of constraint checks. Our current implementation makes use of all optimisations prescribed by the general framework described in Section 3. The specialised version of $greater(s\theta,t\theta)$ is now a rather complex algorithm and cannot be represented by a simple piece of data as in the case of forward checks. Instead, we compile it into a code of an abstract machine. A detailed description of the compilation process would be too long, so we only illustrate the idea by an example.

We are going to specialise $greater(s\theta,t\theta)$, for $s = f(x_0,g(h(x_1,x_2),h(x_2,x_1)),a)$ and $t = f(h(x_1,x_2),g(x_0,x_0),b)$. We assume $w(f) = w(g) = w(h) = w(b) = 1$ and $w(a) = 2$. Figure 2 shows the specialised version written as a pseudo-code. The

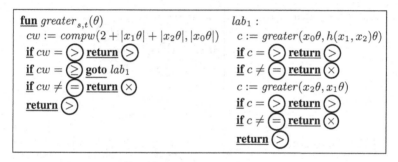

Fig. 2. Specialised check for $f(x_0, g(h(x_1, x_2), h(x_2, x_1)), a) \succ f(h(x_1, x_2), g(x_0, x_0), b)$

compiled version is shown in Figure 3 as a sequence of abstract machine instructions accompanied by their semantics. Note that the specialised version for the weight check $compw(2 + |x_1\theta| + |x_2\theta|, |x_0\theta|)$ is also compiled into a subroutine sub_1. In fact, since weight comparisons are still costly, we optimise them very thoroughly. We use a large set of highly specialised instructions in order to minimise the code size and the number of performed arithmetic operations, and to be able to use specialised versions of functions for weight collection and comparing polynomials. The abstract machine in our current implementation has about 70 instructions for weight comparison. The mnemonic names for these instructions are too long, so in Figure 3 we denote them by $instr$ with subscripts.

Analysing preliminary experiments we discovered that full-strength compilation of all constraints does not always pay off. During the first several seconds of a Vampire run the database of stored clauses is small. In these circumstances a constraint is usually checked on very few substitutions and the compilation effort is not properly amortised. To overcome this we came up with a simple adaptive scheme. We regulate *specialisation strength* by setting a numeric parameter which corresponds to the maximal number of possible weight comparison specialisations. When this limit is exceeded by the specialisation procedure, we stop unrolling recursion and compare the subterms that remain to be compared, by a call to the straightforward procedure. During a Vampire run we keep statistics on how many constraints have been used and how many substitutions checked. As the ratio of the number of checks over the number of constraints grows, we increase the specialisation strength linearly with an experimentally best coefficient.

Table 2 shows rather encouraging experimental results for the described implementation. The relatively small portion of the overall time taken by straightforward checks should not confuse us for two reasons. First, the table presents only the average figure, while on some problems backward checks take as much as 33% of the overall running time. Second, at the moment we mostly check constraints for demodulation. The constraints arising from orientation of equalities in superposition account only for less than 2% of the checks. The vector of development of Vampire points toward much more extensive use of ordering constraints. When we implement constraints arising from literal maximality conditions both in superposition and ordered resolution, and, possibly, some form of inherited constraints (see, e.g., [7]), the relative cost of constraint checking will most certainly increase significantly.

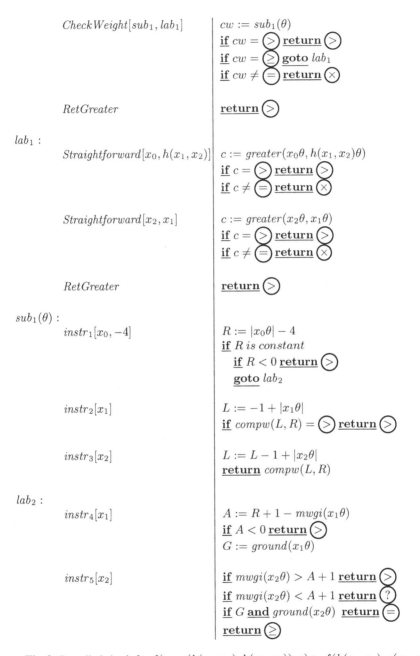

$CheckWeight[sub_1, lab_1]$	$cw := sub_1(\theta)$ **if** $cw =$ (>) **return** (>) **if** $cw =$ (≥) **goto** lab_1 **if** $cw \neq$ (=) **return** (×)
$RetGreater$	**return** (>)
lab_1 : $Straightforward[x_0, h(x_1, x_2)]$	$c := greater(x_0\theta, h(x_1, x_2)\theta)$ **if** $c =$ (>) **return** (>) **if** $c \neq$ (=) **return** (×)
$Straightforward[x_2, x_1]$	$c := greater(x_2\theta, x_1\theta)$ **if** $c =$ (>) **return** (>) **if** $c \neq$ (=) **return** (×)
$RetGreater$	**return** (>)

$sub_1(\theta)$:
 $instr_1[x_0, -4]$
	$R := |x_0\theta| - 4$
	 if R *is constant*
	    if $R < 0$ **return** (>)
	    goto lab_2

$instr_2[x_1]$	$L := -1 + |x_1\theta|$ **if** $compw(L, R) =$ (>) **return** (>)
$instr_3[x_2]$	$L := L - 1 + |x_2\theta|$ **return** $compw(L, R)$
lab_2 : $instr_4[x_1]$	$A := R + 1 - mwgi(x_1\theta)$ **if** $A < 0$ **return** (>) $G := ground(x_1\theta)$
$instr_5[x_2]$	**if** $mwgi(x_2\theta) > A + 1$ **return** (>) **if** $mwgi(x_2\theta) < A + 1$ **return** (?) **if** G **and** $ground(x_2\theta)$ **return** (=) **return** (≥)

Fig. 3. Compiled check for $f(x_0, g(h(x_1, x_2), h(x_2, x_1)), a) \succ f(h(x_1, x_2), g(x_0, x_0), b)$

Apart from adaptive compilation, we have experimented with a light-weight scheme where the specialisation strength is always 1. In this case the improvement is less impressive: the compiled checks are only about 2.5 times faster than the straightforward ones, which suggests that deep specialisation pays off very well.

Table 2. Experimental results for backward checks

problem selection	weight func.	$\frac{time(\text{str})}{time(\text{comp})}$	$\frac{time(\text{str})\cdot 100\%}{total\ time}$	total time per prob.	checks per prob., 10^6	comp. calls per prob., 10^3
all 1132	**uniform**	6.73	3.5%	135 sec	0.57	10.4
all 1132	**non-uniform**	6.79	3.5%	135 sec	0.57	10.3
113 hardest	**uniform**	8.98	6.8%	575 sec	4.74	37.7
113 hardest	**non-uniform**	9.18	6.9%	571 sec	4.52	37.2

Another question is how much the auxiliary optimisations based on weight equation and term unifier propagation (Section 3.3) add to the improvement obtained by weight comparison specialisation and loop/recursion unrolling. A large number of constraints is affected by these optimisations: more than 30% of constraints are simplified by weight equations and more than 50% are affected by unifier propagation. However, this does not translate into any noticeable speed-up in our experiments.

6 Related and Future Work

Developers of some other paramodulation-based systems recognise ordering constraint checking as one of the performance bottlenecks[4]. However, the problem of empirically efficient implementation of constraint checks has not been properly addressed in the literature so far. We have found only a couple of relevant references. In [5] the authors acknowledge the high frequency of constraint checks in problems with many unorientable equations and argue that *shared rewriting*, which is a form of caching for demodulation steps, helps to make the problem less prominent by reducing the number of demodulation invocations. In [12] the problem is addressed by optimising the comparison procedure for Recursive Path Orderings (RPO, for a definition see, e.g., [7]). The implementation caches the results of comparisons of subterms of the compared terms in order to avoid repeated tests.

Apart from the standard KBO, in Vampire we use a non-recursive variant of the KBO (see [9]), where the lexicographic part does not require comparing subterms with the ordering itself. Specialisation of constraint checking for this ordering roughly follows the framework presented here and is much simpler. It also generally gives better improvement over the straightforward checks: with the non-uniform weight function forward checks are nearly 7 times, and backward checks are 21 times faster than the corresponding straightforward ones. However, we refrain from a detailed discussion of this ordering in this paper since so far it has only been used in Vampire.

A natural continuation of our work would be to try optimising constraint checking for another popular simplification ordering, Lexicographic Path Ordering (for a definition see, e.g., [7]). We expect the idea of specialisation with compilation to work well for backward checks. A general framework can be formulated on the base of transformation of constraints into solved forms as in [6]. In the case of forward checks there seems

---

[4] Stephan Schulz, Bernd Löchner, private communications.

to be no obvious light-weight specialised representation of constraints, and solving this problem will probably require extensive experimentation with different variants in order to find a good balance between the speed of constraint checks and the overhead introduced by specialisation.

We also suspect that a simple adaptive scheme can be used to accelerate forward checks with KBO. Constraints checked more frequently should probably be more deeply specialised. In extreme cases, full specialisations, as used for backward checks, can be stored in a special cache of a limited size.

References

1. Th. Hillenbrand and B. Löchner. A Phytography of Waldmeister. *AI Communications*, 15(2-3):127–133, 2002.
2. N.D. Jones, C.K. Gomard, and P Sestoft. *Partial Evaluation and Automatic Program Generation*. Prentice Hall International, 1993.
3. D. Knuth and P. Bendix. Simple word problems in universal algebras. In J. Leech, editor, *Computational Problems in Abstract Algebra*, pages 263–297. Pergamon Press, Oxford, 1970.
4. K. Korovin and A. Voronkov. Orienting rewrite rules with the Knuth-Bendix order. *Information and Computation*, 183(2):165–186, 2003.
5. B. Löchner and S. Schulz. An Evaluation of Shared Rewriting. In H. de Nivelle and S. Schulz, editors, *Proc. of the 2nd International Workshop on the Implementation of Logics*, MPI Preprint, pages 33–48, Saarbrücken, 2001. Max-Planck-Institut für Informatik.
6. R. Nieuwenhuis and J. M. Rivero. Practical algorithms for deciding path ordering constraint satisfaction. *Information and Computation*, 178(2):422–440, 2002.
7. R. Nieuwenhuis and A. Rubio. Paramodulation-based theorem proving. In A. Robinson and A. Voronkov, editors, *Handbook of Automated Reasoning*, volume I, chapter 7, pages 371–443. Elsevier Science, 2001.
8. A. Riazanov and A. Voronkov. Partially adaptive code trees. In M. Ojeda-Aciego, I.P. de Guzmán, G. Brewka, and L.M. Pereira, editors, *JELIA 2000*, volume 1919 of *Lecture Notes in Artificial Intelligence*, pages 209–223, Málaga, Spain, 2000. Springer Verlag.
9. A. Riazanov and A. Voronkov. The design and implementation of Vampire. *AI Communications*, 15(2-3):91–110, 2002.
10. A. Riazanov and A. Voronkov. Efficient instance retrieval with standard and relational path indexing. In F. Baader, editor, *CADE-19*, volume 2741 of *Lecture Notes in Artificial Intelligence*, pages 380–396, Miami, Florida, USA, July 2003.
11. A. Riazanov and A. Voronkov. Limited resource strategy in resolution theorem proving. *Journal of Symbolic Computations*, 36(1-2):101–115, 2003.
12. J.M.A. Rivero. *Data Structures and Algorithms for Automated Deduction with Equality*. Phd thesis, Universitat Politècnica de Catalunya, Barcelona, May 2000.
13. S. Schulz. E - a braniac theorem prover. *AI Communications*, 15(2-3):111–126, 2002.
14. G. Sutcliffe and C. Suttner. The TPTP problem library. tptp v. 2.4.1. Technical report, University of Miami, 2001.
15. T. Tammet. Gandalf. *Journal of Automated Reasoning*, 18(2):199–204, 1997.

Improved Modular Termination Proofs Using Dependency Pairs

René Thiemann, Jürgen Giesl, and Peter Schneider-Kamp

LuFG Informatik II, RWTH Aachen, Ahornstr. 55, 52074 Aachen, Germany
{thiemann|giesl|psk}@informatik.rwth-aachen.de

Abstract. The dependency pair approach is one of the most powerful techniques for automated (innermost) termination proofs of term rewrite systems (TRSs). For any TRS, it generates inequality constraints that have to be satisfied by well-founded orders. However, proving *innermost* termination is considerably easier than termination, since the constraints for innermost termination are a subset of those for termination. We show that surprisingly, the dependency pair approach for termination can be improved by only generating the same constraints as for innermost termination. In other words, proving full termination becomes virtually as easy as proving innermost termination. Our results are based on splitting the termination proof into several modular independent subproofs. We implemented our contributions in the automated termination prover AProVE and evaluated them on large collections of examples. These experiments show that our improvements increase the power and efficiency of automated termination proving substantially.

1 Introduction

Most traditional methods for automated termination proofs of TRSs use *simplification orders* [7,26], where a term is greater than its proper subterms (*subterm property*). However, there are numerous important TRSs which are not *simply terminating*, i.e., termination cannot be shown by simplification orders. Therefore, the *dependency pair* approach [2,10,11] was developed which considerably increases the class of systems where termination is provable mechanically.

Example 1. The following variant of an example from [2] is not simply terminating, since $\mathsf{quot}(x, 0, \mathsf{s}(0))$ reduces to $\mathsf{s}(\mathsf{quot}(x, \mathsf{s}(0), \mathsf{s}(0)))$ in which it is embedded. Here, $\mathsf{div}(x, y)$ computes $\lfloor \frac{x}{y} \rfloor$ for $x, y \in \mathbb{N}$ if $y \neq 0$. The auxiliary function $\mathsf{quot}(x, y, z)$ computes $1 + \lfloor \frac{x-y}{z} \rfloor$ if $x \geq y$ and $z \neq 0$ and it computes 0 if $x < y$.

$$\mathsf{div}(0, y) \to 0 \tag{1}$$
$$\mathsf{div}(x, y) \to \mathsf{quot}(x, y, y) \tag{2}$$

$$\mathsf{quot}(0, \mathsf{s}(y), z) \to 0 \tag{3}$$
$$\mathsf{quot}(\mathsf{s}(x), \mathsf{s}(y), z) \to \mathsf{quot}(x, y, z) \tag{4}$$
$$\mathsf{quot}(x, 0, \mathsf{s}(z)) \to \mathsf{s}(\mathsf{div}(x, \mathsf{s}(z))) \tag{5}$$

In Sect. 2, we recapitulate dependency pairs. Sect. 3 proves that for termination, it suffices to require only the same constraints as for innermost termination.

D. Basin and M. Rusinowitch (Eds.): IJCAR 2004, LNAI 3097, pp. 75–90, 2004.
© Springer-Verlag Berlin Heidelberg 2004

This result is based on a refinement for termination proofs with dependency pairs by Urbain [29], but it improves upon this and related refinements [12,24] significantly. In Sect. 4 we show that the new technique of [12] to reduce the constraints for innermost termination by integrating the concepts of "argument filtering" and "usable rules" can also be adapted for termination proofs. Finally, based on the improvements presented before, Sect. 5 introduces a new method to remove rules of the TRS which reduces the set of constraints even further.

In each section, we demonstrate the power of the respective refinement by examples where termination can now be shown, while they could not be handled before. Our results are implemented in the automated termination prover AProVE [14]. The experiments in Sect. 6 show that our contributions increase power and efficiency on large collections of examples. Thus, our results are also helpful for other tools based on dependency pairs ([1], CiME [6], TTT [19]) and we conjecture that they can also be used in other recent approaches for termination of TRSs [5,9,27] which have several aspects in common with dependency pairs.

2   Modular Termination Proofs Using Dependency Pairs

We briefly present the *dependency pair* approach of Arts & Giesl and refer to [2,10,11,12] for refinements and motivations. We assume familiarity with term rewriting (see, e.g., [4]). For a TRS $\mathcal{R}$ over a signature $\mathcal{F}$, the *defined symbols* $\mathcal{D}$ are the roots of the left-hand sides of rules and the *constructors* are $\mathcal{C} = \mathcal{F} \setminus \mathcal{D}$. We restrict ourselves to finite signatures and TRSs. The infinite set of variables is denoted by $\mathcal{V}$ and $\mathcal{T}(\mathcal{F}, \mathcal{V})$ is the set of all terms over $\mathcal{F}$ and $\mathcal{V}$. Let $\mathcal{F}^\sharp = \{f^\sharp \mid f \in \mathcal{D}\}$ be a set of *tuple symbols*, where $f^\sharp$ has the same arity as f and we often write F for $f^\sharp$. If $t = g(t_1, \dots, t_m)$ with $g \in \mathcal{D}$, we write $t^\sharp$ for $g^\sharp(t_1, \dots, t_m)$.

Definition 2 (Dependency Pair). *The set of* dependency pairs *for a TRS* $\mathcal{R}$ *is* $DP(\mathcal{R}) = \{l^\sharp \to t^\sharp \mid l \to r \in \mathcal{R}, t \text{ is a subterm of } r \text{ with } \mathrm{root}(t) \in \mathcal{D}\}$.

So the dependency pairs of the TRS in Ex. 1 are

$$\mathsf{DIV}(x, y) \to \mathsf{QUOT}(x, y, y) \quad (6) \qquad \mathsf{QUOT}(\mathsf{s}(x), \mathsf{s}(y), z) \to \mathsf{QUOT}(x, y, z) \quad (7)$$
$$\mathsf{QUOT}(x, 0, \mathsf{s}(z)) \to \mathsf{DIV}(x, \mathsf{s}(z)) \quad (8)$$

For (innermost) termination, we need the notion of (innermost) *chains*. Intuitively, a dependency pair corresponds to a (possibly recursive) function call and a chain represents possible sequences of calls that can occur during a reduction. We always assume that different occurrences of dependency pairs are variable disjoint and consider substitutions whose domains may be infinite. Here, $\xrightarrow{i}_{\mathcal{R}}$ denotes innermost reductions where one only contracts innermost redexes.

Definition 3 (Chain). *Let* $\mathcal{P}$ *be a set of pairs of terms. A (possibly infinite) sequence of pairs* $s_1 \to t_1, s_2 \to t_2, \dots$ *from* $\mathcal{P}$ *is a* $\mathcal{P}$-chain *over the TRS* $\mathcal{R}$ *iff there is a substitution* σ *with* $t_i\sigma \to_{\mathcal{R}}^* s_{i+1}\sigma$ *for all* i. *The chain is an* innermost *chain iff* $t_i\sigma \xrightarrow{i}_{\mathcal{R}}^* s_{i+1}\sigma$ *and all* $s_i\sigma$ *are in normal form. An (innermost) chain is* minimal *iff all* $s_i\sigma$ *and* $t_i\sigma$ *are (innermost) terminating w.r.t.* $\mathcal{R}$.

To determine which pairs can follow each other in chains, one builds an *(innermost) dependency graph*. Its nodes are the dependency pairs and there is an arc from $s \to t$ to $u \to v$ iff $s \to t, u \to v$ is an (innermost) chain. Hence, every infinite chain corresponds to a cycle in the graph. In Ex. 1 we obtain the following graph with the cycles $\{(7)\}$ and $\{(6),(7),(8)\}$. Since it is undecidable whether two dependency pairs form an (innermost) chain, for automation one constructs *estimated* graphs containing the real dependency graph (see e.g., [2,18]).[1]

Theorem 4 (Termination Criterion [2]). *A TRS $\mathcal{R}$ is (innermost) terminating iff for every cycle $\mathcal{P}$ of the (innermost) dependency graph, there is no infinite minimal (innermost) $\mathcal{P}$-chain over $\mathcal{R}$.*

To automate Thm. 4, for each cycle one generates constraints which should be satisfied by a *reduction pair* $(\succsim, \succ)$ where $\succsim$ is reflexive, transitive, monotonic and stable (closed under contexts and substitutions) and $\succ$ is a stable well-founded order compatible with $\succsim$ (i.e., $\succsim \circ \succ \subseteq \succ$ and $\succ \circ \succsim \subseteq \succ$). But $\succ$ need not be monotonic. The constraints ensure that at least one dependency pair is strictly decreasing (w.r.t. $\succ$) and all remaining pairs and all rules are weakly decreasing (w.r.t. $\succsim$). Requiring $l \succsim r$ for all $l \to r \in \mathcal{R}$ ensures that in chains $s_1 \to t_1, s_2 \to t_2, \ldots$ with $t_i\sigma \to^*_{\mathcal{R}} s_{i+1}\sigma$, we have $t_i\sigma \succsim s_{i+1}\sigma$. For innermost termination, a weak decrease is not required for all rules but only for the *usable rules*. They are a superset of those rules that can reduce right-hand sides of dependency pairs if their variables are instantiated with normal forms.

Definition 5 (Usable Rules). *For $\mathcal{F}' \subseteq \mathcal{F} \cup \mathcal{F}^\sharp$, let $Rls(\mathcal{F}') = \{l \to r \in \mathcal{R} \mid root(l) \in \mathcal{F}'\}$. For any term t, the* usable rules *are the smallest set such that*

- $\mathcal{U}(x) = \varnothing$ *for $x \in \mathcal{V}$ and*
- $\mathcal{U}(f(t_1,\ldots,t_n)) = Rls(\{f\}) \cup \bigcup_{l \to r \in Rls(\{f\})} \mathcal{U}(r) \cup \bigcup_{j=1}^{n} \mathcal{U}(t_j)$.

For any set $\mathcal{P}$ of dependency pairs, we define $\mathcal{U}(\mathcal{P}) = \bigcup_{s \to t \in \mathcal{P}} \mathcal{U}(t)$.

For the automated generation of reduction pairs, one uses standard (monotonic) simplification orders. To build non-monotonic orders from simplification orders, one may drop function symbols and function arguments by an *argument filtering* [2] (we use the notation of [22]).

---

[1] Estimated dependency graphs may contain an additional arc from (6) to (8). However, if one uses the refinement of *instantiating* dependency pairs [10,12], then all existing estimation techniques would detect that this arc is unnecessary.

Definition 6 (Argument Filtering). *An argument filtering π for a signature $\mathcal{F}$ maps every n-ary function symbol to an argument position $i \in \{1, \dots, n\}$ or to a (possibly empty) list $[i_1, \dots, i_k]$ with $1 \leq i_1 < \dots < i_k \leq n$. The signature $\mathcal{F}_\pi$ consists of all symbols f with $\pi(f) = [i_1, \dots, i_k]$, where in $\mathcal{F}_\pi$, f has arity k. An argument filtering with $\pi(f) = i$ for some $f \in \mathcal{F}$ is* collapsing. *Every argument filtering π induces a mapping from $\mathcal{T}(\mathcal{F}, \mathcal{V})$ to $\mathcal{T}(\mathcal{F}_\pi, \mathcal{V})$, also denoted by π:*

$$\pi(t) = \begin{cases} t & \text{if } t \text{ is a variable} \\ \pi(t_i) & \text{if } t = f(t_1, ..., t_n) \text{ and } \pi(f) = i \\ f(\pi(t_{i_1}), ..., \pi(t_{i_k})) & \text{if } t = f(t_1, ..., t_n) \text{ and } \pi(f) = [i_1, ..., i_k] \end{cases}$$

For a TRS $\mathcal{R}$, $\pi(\mathcal{R})$ denotes $\{\pi(l) \to \pi(r) \mid l \to r \in \mathcal{R}\}$.

For an argument filtering π and reduction pair $(\succsim, \succ)$, $(\succsim_\pi, \succ_\pi)$ is the reduction pair with $s \succsim_\pi t$ iff $\pi(s) \succsim \pi(t)$ and $s \succ_\pi t$ iff $\pi(s) \succ \pi(t)$. Let $\succsim_{(\succsim)} = \succsim \cup \succ$ and $\succsim_{(\succsim)\pi} = \succsim_\pi \cup \succ_\pi$. In the following, we always regard filterings for $\mathcal{F} \cup \mathcal{F}^\sharp$.

Theorem 7 (Modular (Innermost) Termination Proofs [11]). *A TRS $\mathcal{R}$ is terminating iff for every cycle $\mathcal{P}$ of the dependency graph there is a reduction pair $(\succsim, \succ)$ and an argument filtering π such that both*

(a) $s \succsim_{(\succsim)\pi} t$ for all pairs $s \to t \in \mathcal{P}$ and $s \succ_\pi t$ for at least one $s \to t \in \mathcal{P}$
(b) $l \succsim_\pi r$ for all rules $l \to r \in \mathcal{R}$

$\mathcal{R}$ is innermost terminating if for every cycle $\mathcal{P}$ of the innermost dependency graph there is a reduction pair $(\succsim, \succ)$ and an argument filtering π satisfying both (a) and

(c) $l \succsim_\pi r$ for all rules $l \to r \in \mathcal{U}(\mathcal{P})$

Thm. 7 permits modular[2] proofs, since one can use different filterings and reduction pairs for different cycles. This is inevitable to handle large programs in practice. See [12,18] for techniques to automate Thm. 7 efficiently.

Innermost termination implies termination for locally confluent overlay systems and thus, for non-overlapping TRSs [17]. So for such TRSs one should only prove innermost termination, since the constraints for innermost termination are a subset of the constraints for termination. However, the TRS of Ex. 1 is not locally confluent: $\mathsf{div}(0, 0)$ reduces to the normal forms 0 and $\mathsf{quot}(0, 0, 0)$.

---

[2] In this paper, "modularity" means that one can split up the termination proof of a TRS $\mathcal{R}$ into several independent subproofs. However, "modularity" can also mean that one would like to split a TRS into subsystems and prove their termination more or less independently. For innermost termination, Thm. 7 also permits such forms of modularity. For example, if $\mathcal{R}$ is a hierarchical combination of $\mathcal{R}_1$ and $\mathcal{R}_2$, we have $\mathcal{U}(\mathcal{P}) \subseteq \mathcal{R}_1$ for every cycle $\mathcal{P}$ of $\mathcal{R}_1$-dependency pairs. Thus, one can prove innermost termination of $\mathcal{R}_1$ independently of $\mathcal{R}_2$. Thm. 11 and its improvements will show that similar modular proofs are also possible for termination instead of innermost termination. Then for hierarchical combinations, termination of $\mathcal{R}_1$ can be proved independently of $\mathcal{R}_2$, provided one uses an estimation of the dependency graph where no further cycles of $\mathcal{R}_1$-dependency pairs are introduced if $\mathcal{R}_1$ is extended by $\mathcal{R}_2$.

Example 8. An automated termination proof of Ex. 1 is virtually impossible with Thm. 7. We get the constraints $\mathsf{QUOT}(\mathsf{s}(x),\mathsf{s}(y),z) \succ_\pi \mathsf{QUOT}(x,y,z)$ and $l \succsim_\pi r$ for all $l \to r \in \mathcal{R}$ from the cycle $\{(7)\}$. However, they cannot be solved by a reduction pair $(\succsim, \succ)$ where $\succsim$ is a quasi-simplification order: For $t = \mathsf{quot}(x,0,\mathsf{s}(0))$ we have $t \succsim_\pi \mathsf{s}(\mathsf{quot}(x,\mathsf{s}(0),\mathsf{s}(0)))$ by rules (5) and (2). Moreover, $\mathsf{s}(\mathsf{quot}(x,\mathsf{s}(0),\mathsf{s}(0))) \succsim_\pi \mathsf{s}(t)$ by the subterm property, since $\mathsf{QUOT}(\mathsf{s}(x),\mathsf{s}(y),z) \succ_\pi \mathsf{QUOT}(x,y,z)$ implies $\pi(\mathsf{s}) = [1]$. But $t \succsim_\pi \mathsf{s}(t)$ implies $\mathsf{QUOT}(\mathsf{s}(t),\mathsf{s}(t),z) \succ_\pi \mathsf{QUOT}(t,t,z) \succsim_\pi \mathsf{QUOT}(\mathsf{s}(t),\mathsf{s}(t),z)$ which contradicts the well-foundedness of $\succ_\pi$.

In contrast, innermost termination of Ex. 1 can easily be proved. There are no usable rules because the dependency pairs have no defined symbols in their right-hand sides. Hence, with a filtering $\pi(\mathsf{QUOT}) = \pi(\mathsf{DIV}) = 1$, the constraints for innermost termination are satisfied by the embedding order.

Our goal is to modify the technique for termination such that its constraints become as simple as the ones for innermost termination. As observed in [29], the following definition is useful to weaken the constraint (b) for termination.

Definition 9 ($\mathcal{C}_\varepsilon$ [16]). *The TRS $\mathcal{C}_\varepsilon$ is defined as $\{\mathsf{c}(x,y) \to x, \mathsf{c}(x,y) \to y\}$ where c is a new function symbol. A TRS $\mathcal{R}$ is $\mathcal{C}_\varepsilon$-terminating iff $\mathcal{R} \cup \mathcal{C}_\varepsilon$ is terminating. A relation $\succsim$ is $\mathcal{C}_\varepsilon$-compatible[3] iff $\mathsf{c}(x,y) \succsim x$ and $\mathsf{c}(x,y) \succsim y$. A reduction pair $(\succsim, \succ)$ is $\mathcal{C}_\varepsilon$-compatible iff $\succsim$ is $\mathcal{C}_\varepsilon$-compatible.*

The TRS $\mathcal{R} = \{\mathsf{f}(0,1,x) \to \mathsf{f}(x,x,x)\}$ of Toyama [28] is terminating, but not $\mathcal{C}_\varepsilon$-terminating, since $\mathcal{R} \cup \mathcal{C}_\varepsilon$ admits the infinite reduction $\mathsf{f}(0,1,\mathsf{c}(0,1)) \to \mathsf{f}(\mathsf{c}(0,1),\mathsf{c}(0,1),\mathsf{c}(0,1)) \to^2 \mathsf{f}(0,1,\mathsf{c}(0,1)) \to \ldots$. This example shows that requiring $l \succsim_\pi r$ only for usable rules is not sufficient for termination: $\mathcal{R} \cup \mathcal{C}_\varepsilon$'s only cycle $\{\mathsf{F}(0,1,x) \to \mathsf{F}(x,x,x)\}$ has no usable rules and there is a reduction pair $(\succsim, \succ)$ satisfying the constraint (a).[4] So $\mathcal{R} \cup \mathcal{C}_\varepsilon$ is innermost terminating, but not terminating, since we cannot satisfy both (a) and $l \succsim r$ for the $\mathcal{C}_\varepsilon$-rules.

So a reduction of the constraints in (b) is impossible in general, but it is possible if we restrict ourselves to $\mathcal{C}_\varepsilon$-compatible reduction pairs. This restriction is not severe, since virtually all reduction pairs used in practice (based on LPO [20], RPOS [7], KBO [21], or polynomial orders[5] [23]) are $\mathcal{C}_\varepsilon$-compatible.

The first step in this direction was taken by Urbain [29]. He showed that in a hierarchy of $\mathcal{C}_\varepsilon$-terminating TRSs, one can disregard all rules occurring "later" in the hierarchy when proving termination. Hence, when showing the termination of functions which call div or quot, one has to require $l \succsim_\pi r$ for the div- and quot-rules. But if one regards functions which do not depend on div or quot, then one does not have to take the div- and quot-rules into account in constraint (b).

But due to the restriction to $\mathcal{C}_\varepsilon$-termination, [29] could not use the full power of dependency graphs. For example, recent dependency graph estimations [18] detect that the dependency graph for Toyama's TRS $\mathcal{R}$ has no cycle and thus, it is terminating. But since it is not $\mathcal{C}_\varepsilon$-terminating, it cannot be handled by [29].

---

[3] Instead of "$\mathcal{C}_\varepsilon$-compatibility", [29] uses the corresponding notion "π extendibility".

[4] For example, it is satisfied by the reduction pair $(\to^*_{\mathcal{R} \cup DP(\mathcal{R})}, \to^+_{\mathcal{R} \cup DP(\mathcal{R})})$.

[5] Any polynomial order can be extended to the symbol c such that it is $\mathcal{C}_\varepsilon$-compatible.

In [12], we integrated the approach of [29] with (arbitrary estimations of) dependency graphs, by restricting ourselves to $\mathcal{C}_\varepsilon$-compatible reduction pairs instead of $\mathcal{C}_\varepsilon$-terminating TRSs. This combines the advantages of both approaches, since now one only regards those rules in (b) that the current *cycle depends on*.

Definition 10 (Dependence). *Let $\mathcal{R}$ be a TRS. For two symbols f and g we say that f depends on g (denoted $f \sqsupset_0 g$) iff g occurs in an f-rule of $\mathcal{R}$ (i.e., in $Rls(\{f\})$). Moreover, every tuple symbol $f^\sharp$ depends on f. A cycle of dependency pairs $\mathcal{P}$ depends on all symbols occurring in its dependency pairs.[6] We write $\sqsupset_0^+$ for the transitive closure of $\sqsupset_0$. For every cycle $\mathcal{P}$ we define $\Delta_0(\mathcal{P}, \mathcal{R}) = \{f \mid \mathcal{P} \sqsupset_0^+ f\}$. If $\mathcal{P}$ and $\mathcal{R}$ are clear from the context we just write Δ_0 or $\Delta_0(\mathcal{P})$.*

In Ex. 1, we have div $\sqsupset_0$ quot, quot $\sqsupset_0$ div, and each defined symbol depends on itself. As QUOT $\sqsupset_0$ quot $\sqsupset_0$ div, Δ_0 contains quot and div for both cycles $\mathcal{P}$.

The next theorem shows that it suffices to require a weak decrease only for the rules that the cycle depends on. It improves upon Thm. 7 since the constraints of type (b) are reduced significantly. Thus, it becomes easier to find a reduction pair satisfying the resulting constraints. This increases both efficiency and power. For instance, termination of a well-known example of [25] to compute intervals of natural numbers cannot be shown with Thm. 7 and a reduction pair based on simplification orders, while a proof with Thm. 11 and LPO is easy [12].

Theorem 11 (Improved Modular Termination, Version 0 [12]). *A TRS $\mathcal{R}$ is terminating if for every cycle $\mathcal{P}$ of the dependency graph there is a $\mathcal{C}_\varepsilon$-compatible reduction pair $(\succsim, \succ)$ and an argument filtering π satisfying both constraint Thm. 7 (a) and*

(b) $l \succsim_\pi r$ for all rules $l \to r \in Rls(\Delta_0(\mathcal{P}, \mathcal{R}))$

Proof. The proof is based on the following key observation [29, Lemma 2]:

$$\text{Every minimal } \mathcal{P}\text{-chain over } \mathcal{R} \text{ is a } \mathcal{P}\text{-chain over } Rls(\Delta_0(\mathcal{P}, \mathcal{R})) \cup \mathcal{C}_\varepsilon. \quad (9)$$

For the proof of Thm. 11, by Thm. 4 we have to show absence of minimal infinite $\mathcal{P}$-chains $s_1 \to t_1, s_2 \to t_2, \ldots$ over $\mathcal{R}$. By (9), such a chain is also a chain over $Rls(\Delta_0(\mathcal{P}, \mathcal{R})) \cup \mathcal{C}_\varepsilon$. Hence, there is a substitution σ with $t_i\sigma \to^*_{Rls(\Delta_0(\mathcal{P}, \mathcal{R})) \cup \mathcal{C}_\varepsilon} s_{i+1}\sigma$ for all i. We extend π to c by $\pi(\mathsf{c}) = [1, 2]$. So $\mathcal{C}_\varepsilon$-compatibility of $\succsim$ implies $\mathcal{C}_\varepsilon$-compatibility of $\succsim_\pi$. By (b) we have $t_i\sigma \succsim_\pi s_{i+1}\sigma$ for all i as $\succsim_\pi$ is stable and monotonic. Using (a) and stability of $\succ_\pi$ leads to $s_i\sigma \succ_\pi t_i\sigma$ for infinitely many i and $s_i\sigma \succsim_\pi t_i\sigma$ for all remaining i contradicting $\succ_\pi$'s well-foundedness.   □

The proof shows that Thm. 11 only relies on observation (9). When refining the definition of Δ_0 in the next section, we only have to prove that (9) still holds.

---

[6] The symbol "$\sqsupset_0$" is overloaded to denote both the dependence between function symbols ($f \sqsupset_0 g$) and between cycles and function symbols ($\mathcal{P} \sqsupset_0 f$).

3 No Dependences for Tuple Symbols and Left-Hand Sides

Thm. 11 reduces the constraints for termination considerably. However for Ex. 1, the constraints according to Thm. 11 are the same as with Thm. 7. The reason is that both cycles $\mathcal{P}$ depend on quot and div and therefore, $Rls(\Delta_0(\mathcal{P})) = \mathcal{R}$. Hence, as shown in Ex. 8, an automated termination proof is virtually impossible.

To solve this problem, we improve the notion of "dependence" by dropping the condition that every tuple symbol $f^\sharp$ depends on f. Then the cycles in Ex. 1 do not depend on any defined function symbol anymore, since they contain no defined symbols. When modifying the definition of $\Delta_0(\mathcal{P})$ in this way in Thm. 11, we obtain no constraints of type (b) for Ex. 1, since $Rls(\Delta_0(\mathcal{P})) = \varnothing$. So now the constraints for termination of this example are the same as for innermost termination and the proof succeeds with the embedding order, cf. Ex. 8.[7]

Now the only difference between $\mathcal{U}(\mathcal{P})$ and $Rls(\Delta_0(\mathcal{P}))$ is that in $Rls(\Delta_0(\mathcal{P}))$, f also depends on g if g occurs in the left-hand side of an f-rule. Similarly, $\mathcal{P}$ also depends on g if g occurs in the left-hand side of a dependency pair from $\mathcal{P}$. The following example shows that disregarding dependences from left-hand sides (as in $\mathcal{U}(\mathcal{P})$) can be necessary for the success of the termination proof.

Example 12. We extend the TRS for division from Ex. 1 by the following rules.

$$\text{plus}(x, 0) \to x \qquad\qquad \text{times}(0, y) \to 0$$
$$\text{plus}(0, y) \to y \qquad\qquad \text{times}(\text{s}(0), y) \to y$$
$$\text{plus}(\text{s}(x), y) \to \text{s}(\text{plus}(x, y)) \qquad \text{div}(\text{div}(x, y), z) \to \text{div}(x, \text{times}(y, z))$$

Even when disregarding dependences $f^\sharp \sqsupset_0 f$, the constraints of Thm. 11 for this TRS are not satisfiable by reduction pairs based on RPOS, KBO, or polynomial orders: Any cycle containing the new dependency pair $\text{DIV}(\text{div}(x, y), z) \to \text{DIV}(x, \text{times}(y, z))$ would depend on both div and times and thus, all rules of the TRS would have to be weakly decreasing. Weak decrease of plus and times implies that one has to use an argument filtering with $\text{s}(x) \succ_\pi x$. But since $t \succsim_\pi \text{s}(t)$ for the term $t = \text{quot}(x, 0, \text{s}(0))$ as shown in Ex. 8, this gives a contradiction.

Cycles with $\text{DIV}(\text{div}(x, y), z) \to \text{DIV}(x, \text{times}(y, z))$ only depend on div because it occurs in the *left-hand side*. This motivates the following refinement of $\sqsupset_0$.

Definition 13 (Refined Dependence, Version 1). *For two function symbols f and g, the refined dependence relation $\sqsupset_1$ is defined as $f \sqsupset_1 g$ iff g occurs in the right-hand side of an f-rule and a cycle $\mathcal{P}$ depends on all symbols in the right-hand sides of its dependency pairs. Again, $\Delta_1(\mathcal{P}, \mathcal{R}) = \{f \mid \mathcal{P} \sqsupset_1^+ f\}$.*

With Def. 13, the constraints of Thm. 11 are the same as in the innermost case: $\mathcal{U}(\mathcal{P}) = Rls(\Delta_1(\mathcal{P}))$ and termination of Ex. 12 can be proved using LPO.

To show that one may indeed regard $\Delta_1(\mathcal{P})$ instead of $\Delta_0(\mathcal{P})$ in Thm. 11, we prove an adapted version of (9) with Δ_1 instead of Δ_0. As in the proofs for Δ_0

---

[7] If an estimated dependency graph has the additional cycle $\{(6), (8)\}$, here one may use an LPO with $\pi(\text{DIV}) = \pi(\text{QUOT}) = 2$, $\pi(\text{s}) = [\,]$, and the precedence $0 > \text{s}$.

in [24,29] and in the original proofs of Gramlich [16], we map any $\mathcal{R}$-reduction
to a reduction w.r.t. $Rls(\Delta_1) \cup \mathcal{C}_\varepsilon$. However, our mapping $\mathcal{I}_1$ is a modification
of these earlier mappings, since terms $g(t_1, \ldots, t_n)$ with $g \notin \Delta_1$ are treated
differently. Fig. 1 illustrates that by this mapping, every minimal chain over $\mathcal{R}$
corresponds to a chain over $Rls(\Delta_1) \cup \mathcal{C}_\varepsilon$, but instead of the substitution σ one
uses a different substitution $\mathcal{I}_1(\sigma)$. Thus, the observation (9) also holds for Δ_1
instead of Δ_0.

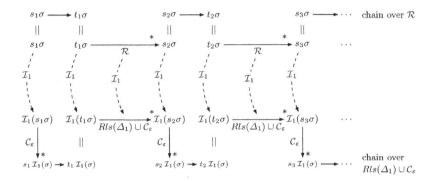

Fig. 1. Transformation of chains

Intuitively, $\mathcal{I}_1(t)$ "collects" all terms that t can be reduced to. However, we
only regard reductions on or below symbols that are not from Δ_1. Normal forms
whose roots are not from Δ_1 may be replaced by a fresh variable. To represent a
collection $t_1, \ldots, t_n$ of terms by just one term, one uses $\mathsf{c}(t_1, \mathsf{c}(t_2, \ldots \mathsf{c}(t_n, x)\ldots))$.

Definition 14. *Let $\Delta \subseteq \mathcal{F} \cup \mathcal{F}^\sharp$ and let $t \in \mathcal{T}(\mathcal{F} \cup \mathcal{F}^\sharp, \mathcal{V})$ be a terminating term.
We define $\mathcal{I}_1(t)$:*

$$\mathcal{I}_1(x) = x \qquad\qquad\qquad\qquad\qquad\qquad\qquad\qquad\qquad\qquad\quad \textit{for } x \in \mathcal{V}$$
$$\mathcal{I}_1(f(t_1, \ldots, t_n)) = f(\mathcal{I}_1(t_1), \ldots, \mathcal{I}_1(t_n)) \qquad\qquad\qquad\qquad\qquad\quad \textit{for } f \in \Delta$$
$$\mathcal{I}_1(g(t_1, \ldots, t_n)) = \mathcal{C}omp(\{g(\mathcal{I}_1(t_1), \ldots, \mathcal{I}_1(t_n))\} \cup \mathcal{R}ed_1(g(t_1, \ldots, t_n))) \quad \textit{for } g \notin \Delta$$

*where $\mathcal{R}ed_1(t) = \{\mathcal{I}_1(t') \mid t \to_\mathcal{R} t'\}$. Moreover, $\mathcal{C}omp(\{t\} \uplus M) = \mathsf{c}(t, \mathcal{C}omp(M))$
and $\mathcal{C}omp(\varnothing) = x_{new}$, where x_{new} is a fresh variable. To ensure that $\mathcal{C}omp$ is
well-defined we assume that in the recursive definition of $\mathcal{C}omp(\{t\} \uplus M)$, t is
smaller than all terms in M due to some total well-founded order $>_\mathcal{T}$ on terms.*
*For every terminating substitution σ (i.e., $\sigma(x)$ is terminating for all $x \in \mathcal{V}$),
we define the substitution $\mathcal{I}_1(\sigma)$ as $\mathcal{I}_1(\sigma)(x) = \mathcal{I}_1(\sigma(x))$ for all $x \in \mathcal{V}$.*

Note that Def. 14 is only possible for terminating terms t, since otherwise,
$\mathcal{I}_1(t)$ could be infinite. Before we can show that Thm. 11 can be adapted to the
refined definition Δ_1, we need some additional properties of $\mathcal{C}omp$ and $\mathcal{I}_1$. In
contrast to the corresponding lemmas in [24,29], they demonstrate that the rules

of $\Delta_0 \setminus \Delta_1$ are not needed and we show in Lemma 16 (ii) and (iii) how to handle dependency pairs and rules where the left-hand side is not from $\mathcal{T}(\Delta_1, \mathcal{V})$.[8]

Lemma 15 (Properties of $\mathcal{C}omp$). *If $t \in M$ then $\mathcal{C}omp(M) \to_{\mathcal{C}_\varepsilon}^+ t$.*

Proof. For $t_1 <_{\mathcal{T}} \cdots <_{\mathcal{T}} t_n$ and any $1 \le i \le n$ we have $\mathcal{C}omp(\{t_1, \ldots, t_n\}) = \mathsf{c}(t_1, \ldots \mathsf{c}(t_i, \ldots \mathsf{c}(t_n, x) \ldots) \ldots) \to_{\mathcal{C}_\varepsilon}^* \mathsf{c}(t_i, \ldots \mathsf{c}(t_n, x) \ldots) \to_{\mathcal{C}_\varepsilon} t_i.$ □

Lemma 16 (Properties of $\mathcal{I}_1$). *Let $\Delta \subseteq \mathcal{F} \cup \mathcal{F}^\sharp$ where $f \in \Delta$ and $f \sqsupseteq_1 g$ implies $g \in \Delta$. Let $t, s, t\sigma \in \mathcal{T}(\mathcal{F} \cup \mathcal{F}^\sharp, \mathcal{V})$ be terminating terms and let σ be a terminating substitution.*

(i)   *If $t \in \mathcal{T}(\Delta, \mathcal{V})$ then $\mathcal{I}_1(t\sigma) = t\,\mathcal{I}_1(\sigma)$.*
(ii)   *$\mathcal{I}_1(t\sigma) \to_{\mathcal{C}_\varepsilon}^* t\,\mathcal{I}_1(\sigma)$.*
(iii) *If $t \to_{\{l \to r\}} s$ by a root reduction step where $l \to r \in \mathcal{R}$ and $\mathrm{root}(l) \in \Delta$,*
       then $\mathcal{I}_1(t) \to_{\{l \to r\} \cup \mathcal{C}_\varepsilon}^+ \mathcal{I}_1(s)$.
(iv)   *If $t \to_{\mathcal{R}} s$ with $\mathrm{root}(t) \notin \Delta$, then $\mathcal{I}_1(t) \to_{\mathcal{C}_\varepsilon}^+ \mathcal{I}_1(s)$.*
(v)   *If $t \to_{\{l \to r\}} s$ where $l \to r \in \mathcal{R}$,*
       then $\mathcal{I}_1(t) \to_{\{l \to r\} \cup \mathcal{C}_\varepsilon}^+ \mathcal{I}_1(s)$ if $\mathrm{root}(l) \in \Delta$ and $\mathcal{I}_1(t) \to_{\mathcal{C}_\varepsilon}^+ \mathcal{I}_1(s)$ otherwise.

Proof.

(i)   The proof is a straightforward structural induction on t.
(ii) The proof is by structural induction on t. The only interesting case is $t = g(t_1, \ldots, t_n)$ where $g \notin \Delta$. Then we obtain

$$\begin{aligned}
\mathcal{I}_1(g(t_1, ..., t_n)\sigma) &= \mathcal{C}omp(\{g(\mathcal{I}_1(t_1\sigma), ..., \mathcal{I}_1(t_n\sigma))\} \cup \mathcal{R}ed_1(g(t_1\sigma, \ldots, t_n\sigma))) \\
&\to_{\mathcal{C}_\varepsilon}^+ g(\mathcal{I}_1(t_1\sigma), ..., \mathcal{I}_1(t_n\sigma)) \qquad \text{by Lemma 15} \\
&\to_{\mathcal{C}_\varepsilon}^* g(t_1\,\mathcal{I}_1(\sigma), \ldots, t_n\,\mathcal{I}_1(\sigma)) \quad \text{by induction hypothesis} \\
&= g(t_1, \ldots, t_n)\,\mathcal{I}_1(\sigma)
\end{aligned}$$

(iii) We have $t = l\sigma \to_{\mathcal{R}} r\sigma = s$. By the definition of $\sqsupseteq_1$, r is a term of $\mathcal{T}(\Delta, \mathcal{V})$. Using (ii) and (i) we get $\mathcal{I}_1(l\sigma) \to_{\mathcal{C}_\varepsilon}^* l\,\mathcal{I}_1(\sigma) \to_{\{l \to r\}} r\,\mathcal{I}_1(\sigma) = \mathcal{I}_1(r\sigma)$.
(iv) follows by $\mathcal{I}_1(t) = \mathcal{C}omp(\{\ldots\} \cup \mathcal{R}ed_1(t))$, $\mathcal{I}_1(s) \in \mathcal{R}ed_1(t)$, and Lemma 15.
(v)   We do induction on the position p of the redex. If $\mathrm{root}(t) \notin \Delta$, we use (iv). If $\mathrm{root}(t) \in \Delta$ and p is the root position, we apply (iii). Otherwise, p is below the root, $t = f(t_1, \ldots, t_i, \ldots, t_n)$, $s = f(t_1, \ldots, s_i, \ldots, t_n)$, $f \in \Delta$, and $t_i \to_{\{l \to r\}} s_i$. Then the claim follows from the induction hypothesis.   □

Now we show that in Thm. 11 one may replace Δ_0 by Δ_1.

Theorem 17 (Improved Modular Termination, Version 1). *A TRS $\mathcal{R}$ is terminating if for every cycle $\mathcal{P}$ of the dependency graph there is a $\mathcal{C}_\varepsilon$-compatible reduction pair $(\succsim, \succ)$ and an argument filtering π satisfying both constraint Thm. 7 (a) and*

(b)   *$l \succsim_\pi r$ for all rules $l \to r \in Rls(\Delta_1(\mathcal{P}, \mathcal{R}))$*

---

[8] Here, equalities in the lemmas of [24,29] are replaced by $\mathcal{C}_\varepsilon$-steps. This is possible by including the term $g(\mathcal{I}_1(t_1), \ldots, \mathcal{I}_1(t_n))$ in the definition of $\mathcal{I}_1(g(t_1, ..., t_n))$.

Proof. The proof is as for Thm. 11, but instead of (9) one uses this observation:

$$\text{Every minimal } \mathcal{P}\text{-chain over } \mathcal{R} \text{ is a } \mathcal{P}\text{-chain over } Rls(\Delta_1(\mathcal{P}, \mathcal{R})) \cup \mathcal{C}_\varepsilon. \quad (10)$$

To prove (10), let $s_1 \to t_1, s_2 \to t_2, \dots$ be a minimal $\mathcal{P}$-chain over $\mathcal{R}$. Hence, there is a substitution σ such that $t_i\sigma \to_{\mathcal{R}}^* s_{i+1}\sigma$ and all terms $s_i\sigma$ and $t_i\sigma$ are terminating. This enables us to apply $\mathcal{I}_1$ to both $t_i\sigma$ and $s_i\sigma$ (where we choose Δ to be $\Delta_1(\mathcal{P}, \mathcal{R})$). Using Lemma 16 (v) we obtain $\mathcal{I}_1(t_i\sigma) \to_{Rls(\Delta_1)\cup\mathcal{C}_\varepsilon}^* \mathcal{I}_1(s_{i+1}\sigma)$.

Moreover, by the definition of $\sqsupset_1$, all t_i are terms over the signature Δ_1. Thus, by Lemma 16 (i) and (ii) we get $t_i\,\mathcal{I}_1(\sigma) = \mathcal{I}_1(t_i\sigma) \to_{Rls(\Delta_1)\cup\mathcal{C}_\varepsilon}^* \mathcal{I}_1(s_{i+1}\sigma) \to_{\mathcal{C}_\varepsilon}^*$ $s_{i+1}\,\mathcal{I}_1(\sigma)$ stating that $s_1 \to t_1, s_2 \to t_2, \dots$ is also a chain over $Rls(\Delta_1)\cup\mathcal{C}_\varepsilon$. □

4 Dependences with Respect to Argument Filterings

For innermost termination, one may first apply the argument filtering π and determine the usable rules $\mathcal{U}(\mathcal{P}, \pi)$ afterwards, cf. [12]. The advantage is that the argument filtering may eliminate some symbols f from right-hand sides of dependency pairs and rules. Then, the f-rules do not have to be weakly decreasing anymore. We also presented an algorithm to determine suitable argument filterings, which is non-trivial since the filtering determines the resulting constraints.

We now introduce a corresponding improvement for termination by defining "dependence" w.r.t. an argument filtering. Then a cycle only depends on those symbols that are not dropped by the filtering. However, this approach is only sound for non-collapsing argument filterings. Consider the non-terminating TRS

$$\mathsf{f}(\mathsf{s}(x)) \to \mathsf{f}(\mathsf{double}(x)) \qquad \mathsf{double}(0) \to 0 \qquad \mathsf{double}(\mathsf{s}(x)) \to \mathsf{s}(\mathsf{s}(\mathsf{double}(x)))$$

In the cycle $\{\mathsf{F}(\mathsf{s}(x)) \to \mathsf{F}(\mathsf{double}(x))\}$, the filtering $\pi(\mathsf{double}) = 1$ results in $\{\mathsf{F}(\mathsf{s}(x)) \to \mathsf{F}(x)\}$. Since the filtered pair has no defined symbols, we would conclude that no rule must be weakly decreasing for this cycle. But then we can solve the cycle's only constraint $\mathsf{F}(\mathsf{s}(x)) \succ \mathsf{F}(x)$ and falsely prove termination.[9]

Example 18. We extend the TRS of Ex. 12 by rules for prime numbers.

$$\begin{aligned}
\mathsf{prime}(\mathsf{s}(\mathsf{s}(x))) &\to \mathsf{pr}(\mathsf{s}(\mathsf{s}(x)), \mathsf{s}(x)) & \mathsf{pr}(x, \mathsf{s}(0)) &\to \mathsf{true} \\
\mathsf{eq}(0, 0) &\to \mathsf{true} & \mathsf{pr}(x, \mathsf{s}(\mathsf{s}(y))) &\to \mathsf{if}(\mathsf{divides}(\mathsf{s}(\mathsf{s}(y)), x), x, \mathsf{s}(y)) \\
\mathsf{eq}(\mathsf{s}(x), 0) &\to \mathsf{false} & \mathsf{if}(\mathsf{true}, x, y) &\to \mathsf{false} \\
\mathsf{eq}(0, \mathsf{s}(y)) &\to \mathsf{false} & \mathsf{if}(\mathsf{false}, x, y) &\to \mathsf{pr}(x, y) \\
\mathsf{eq}(\mathsf{s}(x), \mathsf{s}(y)) &\to \mathsf{eq}(x, y) & \mathsf{divides}(y, x) &\to \mathsf{eq}(x, \mathsf{times}(\mathsf{div}(x, y), y))
\end{aligned}$$

The cycle $\{\mathsf{PR}(x, \mathsf{s}(\mathsf{s}(y))) \to \mathsf{IF}(\mathsf{divides}(\mathsf{s}(\mathsf{s}(y)), x), x, x, \mathsf{s}(y)), \mathsf{IF}(\mathsf{false}, x, y) \to \mathsf{PR}(x, y)\}$ depends on divides and hence, on div and times. So for this cycle, Thm. 17

---

[9] Essentially, we prove absence of infinite $\pi(\mathcal{P})$-chains over $\pi(\mathcal{R})$. But if π is collapsing, then the rules of $\pi(\mathcal{R})$ may have left-hand sides l with $root(l) \in \mathcal{C}$ or $l \in \mathcal{V}$. Thus, inspecting the defined symbols in a term $\pi(t)$ is not sufficient to estimate which rules may be used for the $\pi(\mathcal{R})$-reduction of $\pi(t)$.

requires the div- and times-rules to be weakly decreasing. This is impossible with reduction pairs based on RPOS, KBO, or polynomial orders, cf. Ex. 12.

But if we first use the filtering $\pi(\mathsf{IF}) = [2,3]$ and compute dependences afterwards, then the cycle no longer depends on divides, div, or times. If one modifies "dependence" in this way, then the constraints can again be solved by LPO.

Definition 19 (Refined Dependence, Version 2). *Let π be a non-collapsing argument filtering. For two function symbols f and g we define $f \sqsupset_2 g$ iff there is a rule $l \to r \in Rls(\{f\})$ where g occurs in $\pi(r)$. For a cycle of dependency pairs $\mathcal{P}$, we define $\mathcal{P} \sqsupset_2 g$ iff there is a pair $s \to t \in \mathcal{P}$ where g occurs in $\pi(t)$. We define $\Delta_2(\mathcal{P}, \mathcal{R}, \pi) = \{f \mid \mathcal{P} \sqsupset_2^+ f\}$ and omit $\mathcal{P}$, $\mathcal{R}$, π if they are clear from the context.*

To show that Δ_1 may be replaced by Δ_2 in Thm.17, we define a new mapping $\mathcal{I}_2$.

Definition 20. *Let π be a non-collapsing argument filtering, $\Delta \subseteq \mathcal{F} \cup \mathcal{F}^\sharp$, $t \in \mathcal{T}(\mathcal{F} \cup \mathcal{F}^\sharp, \mathcal{V})$ be terminating. We define $\mathcal{I}_2(t)$. Here, $Red_2(t) = \{\mathcal{I}_2(t') \mid t \to_\mathcal{R} t'\}$.*

$$
\begin{aligned}
\mathcal{I}_2(x) &= x &&\text{for } x \in \mathcal{V} \\
\mathcal{I}_2(f(t_1, \ldots, t_n)) &= f(\mathcal{I}_2(t_{i_1}), \ldots, \mathcal{I}_2(t_{i_k})) &&\text{for } f \in \Delta, \; \pi(f) = [i_1, \ldots, i_k] \\
\mathcal{I}_2(g(t_1, \ldots, t_n)) &= Comp(\{g(\mathcal{I}_2(t_{i_1}), \ldots, \mathcal{I}_2(t_{i_k}))\} \\
&\quad \cup Red_2(g(t_1, \ldots, t_n))\,) \;\text{for } g \notin \Delta, \; \pi(g) = [i_1, \ldots, i_k]
\end{aligned}
$$

Lemma 21 differs from the earlier Lemma 16, since $\mathcal{I}_2$ already applies the argument filtering π and in (v), we have "$*$" instead of "$+$", as a reduction on a position that is filtered away leads to the same transformed terms w.r.t. $\mathcal{I}_2$.

Lemma 21 (Properties of $\mathcal{I}_2$). *Let π be a non-collapsing argument filtering and let $\Delta \subseteq \mathcal{F} \cup \mathcal{F}^\sharp$ such that $f \in \Delta$ and $f \sqsupset_2 g$ implies $g \in \Delta$. Let $t, s, t\sigma \in \mathcal{T}(\mathcal{F} \cup \mathcal{F}^\sharp, \mathcal{V})$ be terminating and let σ be a terminating substitution.*

(i)   If $\pi(t) \in \mathcal{T}(\Delta_\pi, \mathcal{V})$ then $\mathcal{I}_2(t\sigma) = \pi(t)\mathcal{I}_2(\sigma)$.

(ii) $\mathcal{I}_2(t\sigma) \to_{\mathcal{C}_\varepsilon}^* \pi(t)\mathcal{I}_2(\sigma)$.

(iii) If $t \to_{\{l \to r\}} s$ by a root reduction step where $l \to r \in \mathcal{R}$ and $\mathrm{root}(l) \in \Delta$, then $\mathcal{I}_2(t) \to_{\{\pi(l) \to \pi(r)\} \cup \mathcal{C}_\varepsilon}^+ \mathcal{I}_2(s)$.

(iv) If $t \to_\mathcal{R} s$ with $\mathrm{root}(t) \notin \Delta$, then $\mathcal{I}_2(t) \to_{\mathcal{C}_\varepsilon}^+ \mathcal{I}_2(s)$.

(v)   If $t \to_{\{l \to r\}} s$ where $l \to r \in \mathcal{R}$, then
$\mathcal{I}_2(t) \to_{\{\pi(l) \to \pi(r)\} \cup \mathcal{C}_\varepsilon}^* \mathcal{I}_2(s)$ if $\mathrm{root}(l) \in \Delta$ and $\mathcal{I}_2(t) \to_{\mathcal{C}_\varepsilon}^* \mathcal{I}_2(s)$ otherwise.

Proof. The proof is analogous to the proof of Lemma 16. □

We are restricted to non-collapsing filterings when determining the rules that have to be weakly decreasing. But one can still use arbitrary (possibly collapsing) filterings in the dependency pair approach. For every filtering π we define its *non-collapsing variant* π' as $\pi'(f) = \pi(f)$ if $\pi(f) = [i_1, \ldots, i_k]$ and $\pi'(f) = [i]$ if $\pi(f) = i$. Now we show that in Thm. 17 one may replace Δ_1 by Δ_2.

Theorem 22 (Improved Modular Termination, Version 2). *A TRS $\mathcal{R}$ is terminating if for every cycle $\mathcal{P}$ of the dependency graph there is a $\mathcal{C}_\varepsilon$-compatible reduction pair $(\succsim, \succ)$ and an argument filtering π satisfying both constraint Thm. 7 (a) and*

(b) $l \succsim_{\pi} r$ for $l \to r \in Rls(\Delta_2(\mathcal{P}, \mathcal{R}, \pi'))$, where π' is π's non-collapsing variant

Proof. Instead of (10), now we need the following main observation for the proof.

If $s_1 \to t_1, s_2 \to t_2, \ldots$ is a minimal $\mathcal{P}$-chain over $\mathcal{R}$, then $\pi'(s_1) \to \pi'(t_1), \pi'(s_2) \to \pi'(t_2), \ldots$ is a $\pi'(\mathcal{P})$-chain over $\pi'(Rls(\Delta_2(\mathcal{P}, \mathcal{R}, \pi'))) \cup \mathcal{C}_\varepsilon$.   (11)

Similar to the proof of (10), $t_i \sigma \to_{\mathcal{R}}^* s_{i+1} \sigma$ implies that $\pi'(t_i)\mathcal{I}_2(\sigma) = \mathcal{I}_2(t_i \sigma)$ $\to_{\pi'(Rls(\Delta_2)) \cup \mathcal{C}_\varepsilon}^* \mathcal{I}_2(s_{i+1}\sigma) \to_{\mathcal{C}_\varepsilon}^* \pi'(s_{i+1})\mathcal{I}_2(\sigma)$ by Lemma 21 (i), (v), and (ii), which proves (11).

To show that (11) implies Thm. 22, assume that $s_1 \to t_1, s_2 \to t_2, \ldots$ is a minimal infinite $\mathcal{P}$-chain over $\mathcal{R}$. Then by (11) there is a substitution δ ($\mathcal{I}_2(\sigma)$ from above) with $\pi'(t_i)\delta \to_{\pi'(Rls(\Delta_2)) \cup \mathcal{C}_\varepsilon}^* \pi'(s_{i+1})\delta$ for all i. Let π'' be the argument filtering for the signature $\mathcal{F}_{\pi'} \cup \mathcal{F}_{\pi'}^\sharp$, which only performs the collapsing steps of π (i.e., if $\pi(f) = i$ and thus $\pi'(f) = [i]$, we have $\pi''(f) = 1$). All other symbols of $\mathcal{F}_{\pi'} \cup \mathcal{F}_{\pi'}^\sharp$ are not filtered by π''. Hence, $\pi = \pi'' \circ \pi'$. We extend π'' to the new symbol c by defining $\pi''(\mathsf{c}) = [1, 2]$. Hence, $\mathcal{C}_\varepsilon$-compatibility of $\succsim$ implies $\mathcal{C}_\varepsilon$-compatibility of $\succsim_{\pi''}$. Constraint (b) requires $\pi(l) \succsim \pi(r)$ for all rules of $Rls(\Delta_2)$. Therefore, we have $\pi'(l) \succsim_{\pi''} \pi'(r)$, and thus, all rules of $\pi'(Rls(\Delta_2)) \cup \mathcal{C}_\varepsilon$ are decreasing w.r.t. $\succsim_{\pi''}$. This implies $\pi'(t_i)\delta \succsim_{\pi''} \pi'(s_{i+1})\delta$ for all i. Moreover, (a) implies $\pi'(s_i)\delta \succ_{\pi''} \pi'(t_i)\delta$ for infinitely many i and $\pi'(s_i)\delta \succsim_{\pi''} \pi'(t_i)\delta$ for all remaining i. This contradicts the well-foundedness of $\succ_{\pi''}$.   □

Now we are nearly as powerful as for innermost termination. The only difference between $\Delta_2(\mathcal{P}, \mathcal{R}, \pi)$ and $\mathcal{U}(\mathcal{P}, \pi)$ is that $\mathcal{U}(\mathcal{P}, \pi)$ may disregard subterms of right-hand sides of dependency pairs if they also occur on the left-hand side [12], since they are instantiated to normal forms in innermost chains. But for the special case of constructor systems, the left-hand sides of dependency pairs are constructor terms and thus $\Delta_2(\mathcal{P}, \mathcal{R}, \pi) = \mathcal{U}(\mathcal{P}, \pi)$. The other differences between termination and innermost termination are that the innermost dependency graph is a subgraph of the dependency graph and may have fewer cycles. Moreover, the conditions for applying dependency pair transformations by narrowing, rewriting, or instantiation [2,10,12] are less restrictive for innermost termination. Finally for termination, we use $\mathcal{C}_\varepsilon$-compatible reduction pairs, which is not necessary for innermost termination. However, virtually all reduction pairs used in practice are $\mathcal{C}_\varepsilon$-compatible. So in general, innermost termination is still easier to prove than termination, but the difference has become much smaller.

5   Removing Rules

To reduce the constraints for termination proofs even further, in this section we present a technique to remove rules of the TRS that are not relevant for termination. To this end, the constraints for a cycle $\mathcal{P}$ may be pre-processed with a reduction pair $(\succsim, \succ)$. If all dependency pairs of $\mathcal{P}$ and all rules that $\mathcal{P}$ depends on are at least weakly decreasing (w.r.t. $\succsim$), then one may remove all those rules $\mathcal{R}_\succ$ that are strictly decreasing (w.r.t. $\succ$). So instead of proving absence of infinite $\mathcal{P}$-chains over $\mathcal{R}$ one only has to regard $\mathcal{P}$-chains over $\mathcal{R} \setminus \mathcal{R}_\succ$.

In contrast to related approaches to remove rules [15,23,30], we permit arbitrary reduction pairs and remove rules in the modular framework of dependency pairs instead of pre-processing a full TRS. So when removing rules for a cycle $\mathcal{P}$, we only have to regard the rules $\mathcal{P}$ depends on. Moreover, removing rules can be done repeatedly with different reduction pairs $(\succsim, \succ)$. Thm. 23 can also be adapted for innermost termination proofs with similar advantages as for termination.

Theorem 23 (Modular Removal of Rules). *Let $\mathcal{P}$ be a set of pairs, $\mathcal{R}$ be a TRS, and $(\succsim, \succ)$ be a reduction pair where $\succ$ is monotonic and $\mathcal{C}_\varepsilon$-compatible. If $l \underset{(\succsim)}{} r$ for all $l \to r \in Rls(\Delta_1(\mathcal{P}, \mathcal{R}))$ and $s \underset{(\succsim)}{} t$ for all $s \to t \in \mathcal{P}$ then the absence of minimal infinite $\mathcal{P}$-chains over $\mathcal{R} \backslash \mathcal{R}_\succ$ implies the absence of minimal infinite $\mathcal{P}$-chains over $\mathcal{R}$ where $\mathcal{R}_\succ = \{ l \to r \in Rls(\Delta_1(\mathcal{P}, \mathcal{R})) \mid l \succ r \}$.[10]*

Proof. Let $s_1 \to t_1, s_2 \to t_2, \ldots$ be an infinite minimal $\mathcal{P}$-chain over $\mathcal{R}$. Hence, $t_i \sigma \to_{\mathcal{R}}^* s_{i+1}\sigma$. We show that in these reductions, $\mathcal{R}_\succ$-rules are only applied for finitely many i. So $t_i \sigma \to_{\mathcal{R}\backslash\mathcal{R}_\succ}^* s_{i+1}\sigma$ for all $i \geq n$ for some $n \in \mathbb{N}$. Thus, $s_n \to t_n, s_{n+1} \to t_{n+1}, \ldots$ is a minimal infinite $\mathcal{P}$-chain over $\mathcal{R} \backslash \mathcal{R}_\succ$ which proves Thm. 23.

Assume that $\mathcal{R}_\succ$-rules are applied for infinitely many i. By Lemma 16 (v) we get $\mathcal{I}_1(t_i\sigma) \to_{Rls(\Delta_1)\cup\mathcal{C}_\varepsilon}^* \mathcal{I}_1(s_{i+1}\sigma)$. As $\succ$ is $\mathcal{C}_\varepsilon$-compatible and $\to_{Rls(\Delta_1)} \subseteq \underset{(\succsim)}{}$, we have $\mathcal{I}_1(t_i\sigma) \underset{(\succsim)}{} \mathcal{I}_1(s_{i+1}\sigma)$. Moreover, whenever an $\mathcal{R}_\succ$-rule is used in $t_i\sigma \to_{\mathcal{R}}^* s_{i+1}\sigma$, then by Lemma 16 (v), the same rule or at least one $\mathcal{C}_\varepsilon$-rule is used in the reduction from $\mathcal{I}_1(t_i\sigma)$ to $\mathcal{I}_1(s_{i+1}\sigma)$. (This would not hold for $\mathcal{I}_2$, cf. Lemma 21 (v).) Thus, then we have $\mathcal{I}_1(t_i\sigma) \succ \mathcal{I}_1(s_{i+1}\sigma)$ since $\succ$ is monotonic. As $\mathcal{R}_\succ$-reductions are used for infinitely many i, we have $\mathcal{I}_1(t_i\sigma) \succ \mathcal{I}_1(s_{i+1}\sigma)$ for infinitely many i. Using Lemma 16 (ii), (i), and $s \underset{(\succsim)}{} t$ for all pairs in $\mathcal{P}$, we obtain $\mathcal{I}_1(s_i\sigma) \to_{\mathcal{C}_\varepsilon}^* s_i\mathcal{I}_1(\sigma) \underset{(\succsim)}{} t_i\mathcal{I}_1(\sigma) = \mathcal{I}_1(t_i\sigma)$. By $\mathcal{C}_\varepsilon$-compatibility of $\succ$, we get $\mathcal{I}_1(s_i\sigma) \underset{(\succsim)}{} \mathcal{I}_1(t_i\sigma)$ for all i. This contradicts the well-foundedness of $\succ$. $\square$

Rule removal has three benefits. First, the rules $\mathcal{R}_\succ$ do not have to be weakly decreasing anymore after the removal. Second, the rules that $\mathcal{R}_\succ$ depends on do not necessarily have to be weakly decreasing anymore either. More precisely, since we only regard chains over $\mathcal{R} \backslash \mathcal{R}_\succ$, only the rules in $\Delta_1(\mathcal{P}, \mathcal{R} \backslash \mathcal{R}_\succ)$ or $\Delta_2(\mathcal{P}, \mathcal{R} \backslash \mathcal{R}_\succ, \ldots)$ must be weakly decreasing. And third, it can happen that $\mathcal{P}$ is not a cycle anymore. Then no constraints at all have to be built for $\mathcal{P}$. More precisely, we can delete all edges in the dependency graph between pairs $s \to t$ and $u \to v$ of $\mathcal{P}$ where $s \to t, u \to v$ is an $\mathcal{R}$-chain, but not an $\mathcal{R} \backslash \mathcal{R}_\succ$-chain.

Example 24. We extend the TRS of Ex. 18 by the following rules.

$$\mathsf{p}(\mathsf{s}(x)) \to x \qquad \mathsf{plus}(\mathsf{s}(x), y) \to \mathsf{s}(\mathsf{plus}(\mathsf{p}(\mathsf{s}(x)), y)) \qquad \mathsf{plus}(x, \mathsf{s}(y)) \to \mathsf{s}(\mathsf{plus}(x, \mathsf{p}(\mathsf{s}(y))))$$

---

[10] Using Δ_2 instead of Δ_1 makes Thm. 23 unsound. Consider $\{\mathsf{f}(\mathsf{a},\mathsf{b}) \to \mathsf{f}(\mathsf{a},\mathsf{a}), \mathsf{a} \to \mathsf{b}\}$. With $\pi(\mathsf{F}) = [1]$, an LPO-reduction pair makes the filtered dependency pair weakly decreasing and the rule strictly decreasing ($\mathsf{F}(\mathsf{a}) \succsim \mathsf{F}(\mathsf{a})$ and $\mathsf{a} \succ \mathsf{b}$). But then Thm. 23 would state that we can remove the rule and only prove absence of infinite chains of $\mathsf{F}(\mathsf{a},\mathsf{b}) \to \mathsf{F}(\mathsf{a},\mathsf{a})$ over the empty TRS. Then we could falsely prove termination.

For the cycle $\{\mathsf{PLUS}(\mathsf{s}(x), y) \to \mathsf{PLUS}(\mathsf{p}(\mathsf{s}(x)), y), \mathsf{PLUS}(x, \mathsf{s}(y)) \to \mathsf{PLUS}(x, \mathsf{p}(\mathsf{s}(y)))\}$ there is no argument filtering and reduction pair $(\succsim, \succ)$ with a quasi-simplification order $\succsim$ satisfying the constraints of Thm. 22. The reason is that due to p's rule, the filtering cannot drop the argument of p. So $\pi(\mathsf{PLUS}(\mathsf{p}(\mathsf{s}(x)), y)) \succsim \pi(\mathsf{PLUS}(\mathsf{s}(x), y))$ and $\pi(\mathsf{PLUS}(x, \mathsf{p}(\mathsf{s}(y)))) \succsim \pi(\mathsf{PLUS}(x, \mathsf{s}(y)))$ hold for any quasi-simplification order $\succsim$. Furthermore, the transformation technique of "narrowing dependency pairs" [2,10,12] is not applicable, since the right-hand side of each dependency pair above unifies with the left-hand side of the other dependency pair. Therefore, automated tools based on dependency pairs fail.

In contrast, by Thm. 23 and a reduction pair with the polynomial interpretation $\mathcal{P}ol(\mathsf{PLUS}(x, y)) = x + y$, $\mathcal{P}ol(\mathsf{s}(x)) = x + 1$, $\mathcal{P}ol(\mathsf{p}(x)) = x$, p's rule is strictly decreasing and can be removed. Then, p is a constructor. If one uses the technique of "instantiating dependency pairs" [10,12], for this cycle the second dependency pair can be replaced by $\mathsf{PLUS}(\mathsf{p}(\mathsf{s}(x)), \mathsf{s}(y)) \to \mathsf{PLUS}(\mathsf{p}(\mathsf{s}(x)), \mathsf{p}(\mathsf{s}(y)))$. Now the two pairs form no cycle anymore and thus, no constraints at all are generated.

If we also add the rule $\mathsf{p}(0) \to 0$, then again $\mathsf{p}(\mathsf{s}(x)) \to x$ can be removed by Thm. 23 but p does not become a constructor and we cannot delete the whole cycle. Still, the resulting constraints are satisfied by an argument filtering with $\pi(\mathsf{PLUS}) = [1, 2]$, $\pi(\mathsf{s}) = \pi(\mathsf{p}) = [\,]$ and an LPO with the precedence $\mathsf{s} > \mathsf{p} > 0$.

Note that here, it is essential that Thm. 23 only requires $l \succsim r$ for rules $l \to r$ that $\mathcal{P}$ depends on. In contrast, previous techniques [15,23,30] would demand that all rules including the ones for div and times would have to be at least weakly decreasing. As shown in Ex. 12, this is impossible with standard orders.

To automate Thm. 23, we use reduction pairs $(\succsim, \succ)$ based on linear polynomial interpretations with coefficients from $\{0, 1\}$. Since $\succ$ must be monotonic, n-ary function symbols can only be mapped to $\sum_{i=1}^{n} x_i$ or to $1 + \sum_{i=1}^{n} x_i$. Thus, there are only two possible interpretations resulting in a small search space. Moreover, polynomial orders can solve constraints where one inequality must be strictly decreasing and all others must be weakly decreasing in just one search attempt without backtracking [13]. In this way, Thm. 23 can be applied very efficiently. Since removing rules never complicates termination proofs, Thm. 23 should be applied repeatedly as long as some rule is deleted in each application.

Note that whenever a dependency pair (instead of a rule) is strictly decreasing, one has solved the constraints of Thm. 17 and can delete the cycle. Thus, one should not distinguish between rule- and dependency pair-constraints when applying Thm. 23 and just search for a strict decrease in any of the constraints.

6   Conclusion and Empirical Results

We presented new results to reduce the constraints for termination proofs with dependency pairs substantially. By Sect. 3 and 4, it suffices to require weak decrease of the dependent rules, which correspond to the usable rules regarded for innermost termination. So surprisingly, the constraints for termination and innermost termination are (almost) the same. Moreover, we showed in Sect. 5 that one may pre-process the constraints for each cycle and eliminate rules that are strictly decreasing. All our results can also be used together with dependency pair transformations [2,10,12] which often simplify (innermost) termination proofs.

We implemented our results in the system AProVE[11] [14] and tested it on the 130 terminating TRSs from [3,8,25]. The following table gives the percentage of the examples where termination could be proved within a timeout of 30 s and the time for running the system on all examples (including the ones where the proof failed). Our experiments were performed on a Pentium IV with 2.4 GHz and 1 GB memory. We used reduction pairs based on the embedding order, LPO, and linear polynomial interpretations with coefficients from $\{0, 1\}$ ("Polo"). The table shows that with every refinement from Thm. 7 to Thm. 22, termination proving becomes more powerful and for more complex orders than embedding, efficiency also increases considerably. Moreover, a pre-processing with Thm. 23 using "Polo" makes the approach even more powerful. Finally, if one also uses dependency pair transformations ("tr"), one can increase power further. To measure the effect of our contributions, in the first 3 rows we did not use the technique for innermost termination proofs, even if the TRS is non-overlapping. (If one applies the innermost termination technique in these examples, we can prove termination of 95 % of the examples in 23 s with "Polo".) Finally, in the last row ("Inn") we verified *innermost* termination with "Polo" and usable rules $\mathcal{U}(\mathcal{P})$ as in Thm. 17, with usable rules $\mathcal{U}(\mathcal{P}, \pi)$ as in Thm. 22, with a pre-processing as in Thm. 23, and with dependency pair transformations. This row demonstrates that termination is now almost as easy to prove as innermost termination. To summarize, our experiments show that the contributions of this paper are indeed relevant and successful in practice, since the reduction of constraints makes automated termination proving significantly more powerful and faster.

	Thm. 7	Thm. 11	Thm. 17	Thm. 22	Thm. 22, 23	Thm. 22, 23, tr
Emb	39 s, 28 %	7 s, 30 %	42 s, 38 %	50 s, 52 %	51 s, 65 %	82 s, 78 %
LPO	606 s, 51 %	569 s, 54 %	261 s, 59 %	229 s, 61 %	234 s, 75 %	256 s, 84 %
Polo	9 s, 61 %	8 s, 66 %	5 s, 73 %	5 s, 78 %	6 s, 85 %	9 s, 91 %
Inn			8 s, 78 %	8 s, 82 %	10 s, 88 %	31 s, 97 %

References

1. T. Arts. System description: The dependency pair method. In L. Bachmair, editor, *Proc. 11th RTA*, LNCS 1833, pages 261–264, Norwich, UK, 2000.
2. T. Arts and J. Giesl. Termination of term rewriting using dependency pairs. *Theoretical Computer Science*, 236:133–178, 2000.
3. T. Arts and J. Giesl. A collection of examples for termination of term rewriting using dependency pairs. Technical Report AIB-2001-09[12], RWTH Aachen, 2001.
4. F. Baader and T. Nipkow. *Term Rewriting and All That*. Cambridge, 1998.
5. C. Borralleras, M. Ferreira, and A. Rubio. Complete monotonic semantic path orderings. In D. McAllester, editor, *Proc. 17th CADE*, LNAI 1831, pages 346–364, Pittsburgh, PA, USA, 2000.

---

[11] http://www-i2.informatik.rwth-aachen.de/AProVE. Our contributions are integrated in AProVE 1.1-beta, which does not yet contain all options of AProVE 1.0.

6. E. Contejean, C. Marché, B. Monate, and X. Urbain. CiME. http://cime.lri.fr.
7. N. Dershowitz. Termination of rewriting. *J. Symb. Comp.*, 3:69–116, 1987.
8. N. Dershowitz. 33 examples of termination. In *Proc. French Spring School of Theoretical Computer Science*, LNCS 909, pages 16–26, Font Romeux, 1995.
9. O. Fissore, I. Gnaedig, and H. Kirchner. Cariboo: An induction based proof tool for termination with strategies. In C. Kirchner, editor, *Proc. 4th PPDP*, pages 62–73, Pittsburgh, PA, USA, 2002. ACM Press.
10. J. Giesl and T. Arts. Verification of Erlang processes by dependency pairs. *Appl. Algebra in Engineering, Communication and Computing*, 12(1,2):39–72, 2001.
11. J. Giesl, T. Arts, and E. Ohlebusch. Modular termination proofs for rewriting using dependency pairs. *Journal of Symbolic Computation*, 34(1):21–58, 2002.
12. J. Giesl, R. Thiemann, P. Schneider-Kamp, and S. Falke. Improving dependency pairs. In Vardi and Voronkov, editors, *Proc 10th LPAR*, LNAI 2850, 165–179, 2003.
13. J. Giesl, R. Thiemann, P. Schneider-Kamp, and S. Falke. Mechanizing dependency pairs. Technical Report AIB-2003-08[12], RWTH Aachen, Germany, 2003.
14. J. Giesl, R. Thiemann, P. Schneider-Kamp, and S. Falke. Automated termination proofs with AProVE. In v. Oostrom, editor, *Proc. 15th RTA*, LNCS, Aachen, 2004.
15. J. Giesl and H. Zantema. Liveness in rewriting. In R. Nieuwenhuis, editor, *Proc. 14th RTA*, LNCS 2706, pages 321–336, Valencia, Spain, 2003.
16. B. Gramlich. Generalized sufficient conditions for modular termination of rewriting. *Appl. Algebra in Engineering, Communication & Computing*, 5:131–158, 1994.
17. B. Gramlich. Abstract relations between restricted termination and confluence properties of rewrite systems. *Fundamenta Informaticae*, 24:3–23, 1995.
18. N. Hirokawa and A. Middeldorp. Automating the dependency pair method. In F. Baader, editor, *Proc. 19th CADE*, LNAI 2741, Miami Beach, FL, USA, 2003.
19. N. Hirokawa and A. Middeldorp. Tsukuba termination tool. In R. Nieuwenhuis, editor, *Proc. 14th RTA*, LNCS 2706, pages 311–320, Valencia, Spain, 2003.
20. S. Kamin and J. J. Lévy. Two generalizations of the recursive path ordering. Unpublished Manuscript, University of Illinois, IL, USA, 1980.
21. D. Knuth and P. Bendix. Simple word problems in universal algebras. In J. Leech, editor, *Computational Problems in Abstract Algebra*, pages 263–297. 1970.
22. K. Kusakari, M. Nakamura, and Y. Toyama. Argument filtering transformation. In G. Nadathur, editor, *Proc. 1st PPDP*, LNCS 1702, pages 48–62, Paris, 1999.
23. D. Lankford. On proving term rewriting systems are Noetherian. Technical Report MTP-3, Louisiana Technical University, Ruston, LA, USA, 1979.
24. E. Ohlebusch. *Advanced Topics in Term Rewriting*. Springer, 2002.
25. J. Steinbach. Automatic termination proofs with transformation orderings. In J. Hsiang, editor, *Proc. 6th RTA*, LNCS 914, pages 11–25, Kaiserslautern, Germany, 1995. Full version in Technical Report SR-92-23, Universität Kaiserslautern.
26. J. Steinbach. Simplification orderings: History of results. *Fund. I.*, 24:47–87, 1995.
27. R. Thiemann and J. Giesl. Size-change termination for term rewriting. In R. Nieuwenhuis, editor, *Proc. 14th RTA*, LNCS 2706, pages 264–278, Valencia, Spain, 2003.
28. Y. Toyama. Counterexamples to the termination for the direct sum of term rewriting systems. *Information Processing Letters*, 25:141–143, 1987.
29. X. Urbain. Automated incremental termination proofs for hierarchically defined term rewriting systems. In R. Goré, A. Leitsch, and T. Nipkow, editors, *Proc. IJCAR 2001*, LNAI 2083, pages 485–498, Siena, Italy, 2001.
30. H. Zantema. TORPA: Termination of rewriting proved automatically. In *Proc. 15th RTA*, LNCS, Aachen, 2004. Full version in TU/e CS-Report 03-14, TU Eindhoven.

---

[12] Available from http://aib.informatik.rwth-aachen.de

Deciding Fundamental Properties of Right-(Ground or Variable) Rewrite Systems by Rewrite Closure[*]

Guillem Godoy[1] and Ashish Tiwari[2]

[1] Technical University of Catalonia
Jordi Girona 1, Barcelona, Spain
ggodoy@lsi.upc.es
[2] SRI International, Menlo Park, CA 94025
tiwari@csl.sri.com

Abstract. Right-(ground or variable) rewrite systems (RGV systems for short) are term rewrite systems where all right hand sides of rules are restricted to be either ground or a variable. We define a minimal rewrite extension $\overline{R}$ of the rewrite relation induced by a RGV system R. This extension admits a rewrite closure presentation, which can be effectively constructed from R. The rewrite closure is used to obtain decidability of the reachability, joinability, termination, and confluence properties of the RGV system R. We also show that the word problem and the unification problem are decidable for confluent RGV systems. We analyze the time complexity of the obtained procedures; for shallow RGV systems, termination and confluence are exponential, which is the best possible result since all these problems are EXPTIME-hard for shallow RGV systems.

1 Introduction

It is being increasingly realized that theorem provers are most effective and useful in their incarnation as specialized decision procedures for restricted logics. This is true, in particular, for equational theories, where rewriting techniques have provided effective decision procedures. In the simplest case of equality between constants, rewrite rules, representing a Union-Find data structure, decide the word problem. At the next level, flat rewrite rules, in the form of abstract congruence closure, handle ground equational theories. This can be generalized to shallow-linear equational theories.

In going from special to more general equational theories, the complexity of deciding various fundamental problems, like the word problem, termination, confluence, reachability, and joinability, increases until all these problems become undecidable. It is, therefore, fruitful to know how far the approach towards "specialized theorem provers" can take us. In this context, we consider RGV term rewrite systems, where every rule $l \rightarrow r$ is such that r is either a ground term or a variable (and there are no restrictions on l). For example, the rules $0 + x \rightarrow x$, $x + 0 \rightarrow x$, $x + (-x) \rightarrow 0$, $(-x) + x \rightarrow 0$, and $-(-x) \rightarrow x$ are all of this kind. It is known that the word problem for this class of

---

[*] Research of the first author was supported in part by the Spanish CICYT project MAVERISH ref. TIC2001-2476-C03-01. Research of the second author was supported in part by NSF grant CCR-0326540.

term rewriting systems is undecidable [12]. In this paper, we show that the reachability, joinability, termination, and confluence properties are decidable. We also show that the word problem is decidable for confluent RGV systems. We note here that most of these properties are undecidable even for very restricted classes of term rewrite systems [9].

Takai, Kaji, and Seki [14] showed that right-linear finite-path-overlapping systems effectively preserve recognizability. Since RGV systems are right-linear and finite-path-overlapping, it follows that the reachability and joinability problems for RGV systems are decidable. We obtain the same results using a different approach. Takai, Kaji, and Seki [14] construct a tree-automaton that represents all terms reachable from a given term, whereas we construct a rewrite closure (which works for all initial choices of terms). The generality of rewrite closure allows us to obtain new decidability results for confluence and termination. We believe that, apart from the main decidability results, our approach to obtaining them is also significant. We build on the simple cases in a modular way so that an implementation can seamlessly use efficient implementation of, say, the union-find data structure for handling equality between constants and congruence closure for ground theories (which itself uses the union-find as a subroutine). This is achieved by generalizing the existing definitions minimally, for example, *flat rules* of the form $fc_1 \ldots c_m \to c$, are generalized to richer F-rules. Second, we make specialized use of more general concepts so that it is easy to see where further generalizations to larger classes of rewrite system fail. For instance, we use the canonical concept of rewrite closure, in parallel to congruence closure or convergent presentations, for nonsymmetric properties such as reachability. We show that RGV systems admit rewrite closure presentation, but only after we have introduced a new concept of minimal rewrite extensions $\overline{R}$.

The results described in this paper build upon our previous works [7,8]. Some of the techical proofs, which are easy extensions of previously published proofs, have been left out. The extension to RGV systems was motivated by the observation that several important axioms can be stated using collapsing rules (i.e. right variable rules). In our previous work, rewrite closure construction has been a crucial first step for deciding some properties of term rewrite systems. Intuitively, a rewrite closure for a rewrite system R is the union of two rewrite systems $F \cup B$ such that the relations induced by R and by $F \cup B$ coincide, and rewriting with $F \cup B$ can be always re-ordered to rewrite first with only F, and then with only B; moreover, F is (size-)decreasing whereas B is (size-)nondecreasing.

The following example introduces the key ideas in the construction of a rewrite closure for RGV rewrite systems. Consider the rewrite system $R = \{c_1 \to fc_1 \ , \ c_2 \to fc_2 \ , \ c_1 \to fc_1' \ , \ c_2 \to fc_2' \ , \ c_1' \to fc_3 \ , \ c_2' \to fc_3 \ , \ gxx \to x\}$ (we use the notation fc_1 to represent $f(c_1)$ and gxx to represent $g(x,x)$). Note that R contains a size-decreasing rule $gxx \to x$ and many size-increasing rules. The term gc_1c_2 rewrites into fc_3 by R: $gc_1c_2 \to_{c_1 \to fc_1'} gf(c_1')c_2 \to_{c_2 \to fc_2'} gf(c_1')f(c_2') \to_{c_1' \to fc_3} gf(fc_3)c_2' \to_{c_2' \to fc_3} gf(fc_3)f(fc_3) \to_{gxx \to x} f(fc_3)$. But the first four rewrite steps are increasing and the last one is decreasing, and hence, this derivation is not of the required form. In order to construct a rewrite closure for R, we replace the rule $gxx \to x$ by the constrained rule $gx_1x_2 \to x_3$ if $\{x_3; x_1, x_2\}$. A substitution σ is a solution to the constraint $\{x_3; x_1, x_2\}$ if $x_3\sigma$ is reachable from $x_1\sigma$ and $x_2\sigma$. Hence, this new rule represents all the rules $gs_1s_2 \to s_3$ such that s_3 is reachable from s_1 and s_2 by the increasing rules; and in particular, $gc_1c_2 \to f(fc_3)$. But these new rules are not necessarily decreasing, and consequently we cannot claim that we have a rewrite closure. The other ingredient we need is the introduction of new constants like $c_{\{1,2\}}$ that represent all the terms reachable from c_1 and c_2. Hence, we need to add new rules like $c_{\{1,2\}} \to fc_{\{1',2'\}}$ and $c_{\{1',2'\}} \to fc_3$, and now, we have the derivation $gc_1c_2 \to_{gx_1x_2 \to x_3 \text{ if } \{x_3;x_1,x_2\}}$ $c_{\{1,2\}} \to_{c_{\{1,2\}} \to fc_{\{1',2'\}}} fc_{\{1',2'\}} \to_{c_{\{1',2'\}} \to fc_3} f(fc_3)$ of the required form.

Outline of the paper. Section 2 introduces some basic notions and notations. In Section 3 we argue that we can make some simplifying assumptions on the initial RGV system R without loss of generality. These will simplify the arguments in the rest of the paper. Section 4 introduces the concept of minimum joining extension $\overline{R}$ of a rewrite system R, and shows that this extension preserves some properties of the original R. Section 5 defines the concept of constrained rewriting with joinability constraints. In Section 6 we construct a rewrite closure for $\overline{R}$. As a consequence, we obtain the decidability of reachability and joinability for RGV systems. In Section 7 we prove the decidability of termination for RGV systems. In Section 8 we give a characterization of the confluence of RGV systems, and use it to prove the decidability of confluence for RGV systems, and the decidability of the word problem for confluent RGV systems. In Section 9 we briefly analyse the complexity of all the presented algorithms.

2 Preliminaries

We use standard notation from the term rewriting literature. A signature Σ is a (finite) set of function symbols, which is partitioned as $\cup_i \Sigma_i$ such that $f \in \Sigma_n$ if arity of f is n. Symbols in Σ_0, called *constants*, are denoted by a, b, c, d, with possible subscripts. The elements of a set $\mathcal{X}$ of variable symbols are denoted by x, y with possible subscripts. The set $\mathcal{T}(\Sigma, \mathcal{X})$ of *terms* over Σ and $\mathcal{X}$, *position* p in a term, *subterm* $t|_p$ of term t at position p, and the term $t[s]_p$ obtained by replacing $t|_p$ by s are defined in the standard way. For example, if t is $f(a, g(b, h(c)), d)$, then $t|_{2.2.1} = c$, and $t[d]_{2.2} = f(a, g(b, d), d)$. By $t[s_1, s_2, \ldots, s_n]_{p_1, p_2, \ldots, p_n}$ we denote $t[s_1]_{p_1}[s_2]_{p_2} \ldots [s_n]_{p_n}$. By $Pos(t)$ we denote the set of all positions p such that $t|_p$ is defined. By $Vars(t)$ we denote the set of all variables occurring in t. The *height* of a term s is 0 if s is a variable or a constant, and $1 + max_i height(s_i)$ if $s = f(s_1, \ldots, s_m)$. Usually we will denote a term $f(t_1, \ldots, t_n)$ by the simplified form $ft_1 \ldots t_n$, and $t[s]_p$ by $t[s]$ when p is clear by the context or not important.

A substitution σ is sometimes presented explicitly as $\{x_1 \mapsto t_1, \ldots, x_n \mapsto t_n\}$. We assume standard definition for a *rewrite rule* $l \to r$, a *rewrite system* R, the *one step rewrite relation at position p induced by R* $\to_{R,p}$, and the *one step rewrite relation induced by R* (at any position) $\to_R$. If R is any rewrite system, then R^- denotes the rewrite system $\{r \to l \mid l \to r \in R\}$. If $p = \lambda$, then the rewrite step $\to_{R,p}$ is said to be applied *at the topmost position* (at the root) and is denoted by $s \to_R^r t$; it is denoted by $s \to_R^{nr} t$ otherwise. With $s \to_R^{||} t$ we denote a derivation $s \to_R^* t$ where all the rules are applied at disjoint positions, i.e. this derivation is not of the form $s \to_R^* \circ \to_{R,p} \circ \to_R^* \circ \to_{R,q} \circ \to_R^* t$ where p is a prefix of q or q is a prefix of p.

The notations $\leftrightarrow$, $\rightarrow^+$, and $\rightarrow^*$, are used respectively for the symmetric, transitive, and reflexive-transitive closure of a binary relation $\rightarrow$. A rewrite system R is *confluent* if the relation $\leftarrow_R^* \circ \rightarrow_R^*$ is contained in $\rightarrow^* \circ \leftarrow^*$, which is equivalent to the relation $\leftrightarrow_R^*$ being contained in $\rightarrow^* \circ \leftarrow^*$ (called the Church-Rosser property). A term t is *reachable* from s by R if $s \rightarrow_R^* t$. A term s is *R-irreducible* (or, in R-normal form) if there is no term t such that $s \rightarrow_R t$. We denote by $s \rightarrow_R^! t$, or $t = NF_R(s)$, the fact that an R-irreducible term t is reachable from s by R. Two terms s and t are said to be *equivalent* by R (or, R-equivalent) if $s \leftrightarrow_R^* t$. The terms s and t are *R-joinable*, denoted by $s \downarrow_R t$, if $s \rightarrow_R^* \circ \leftarrow_R^* t$. A *(rewrite) derivation or proof* (from s) is a sequence of rewrite steps (starting from s), that is, a sequence $s \rightarrow_R s_1 \rightarrow_R s_2 \rightarrow_R \cdots$.

A TRS R is *terminating* if there is no infinite derivation $s \rightarrow_R s_1 \rightarrow_R s_2 \rightarrow_R \cdots$. Termination is usually ensured by showing that the rewrite system R is contained in a *reduction ordering* [5]. A confluent and terminating term rewrite system is said to be *convergent*.

A term t is called *ground* if t contains no variables. A term rewrite system R is *right-(ground or variable)* (RGV) if for every $l \rightarrow r \in R$, the term r is either ground, or a variable.

3   Simplifying Assumptions on the Initial RGV System

Let R be a RGV term rewrite system over the signature Σ. We henceforth assume that a left hand side of a rule in R is not a variable. If this were the case, R would be trivially not terminating, and either it would be trivially confluent or this rule would be useless (of the form $x \rightarrow x$). For reachability and joinability, the procedures introduced here can be easily adapted to handle rules of this kind.

We assume, without loss of generality, that (a) the rewrite system R is over the signature $\Sigma = \Sigma_0 \cup \Sigma_m$, where $\Sigma_0 = \{c_1 \ldots c_n\}$ and $\Sigma_m = \{f\}$, and (b) R is partitioned into $F \cup B$ in such a way that F contains rules of the form $f s_1 \ldots s_m \rightarrow \alpha$ (where α is a constant or a variable, or, in short, a height 0 term); whereas B contains rules of the form $c \rightarrow f c_1 \ldots c_m$ and rules of the form $c \rightarrow d$. The first assumption is made to simplify the presentation of proofs and is not crucial for correctness.

These transformations are standard and have been formally presented before [7,8]. We will instead just give an example.

Example 1. Let $\Sigma_0 = \{a\}$, $\Sigma_1 = \{g\}$, and $\Sigma_2 = \{h\}$. Terms over the signature $\cup_i \Sigma_i$ can be mapped onto terms over a new signature $\Sigma_3' \cup \Sigma_0'$ where $\Sigma_3' = \{f\}$ and $\Sigma_0' = \{a, g, h\}$. The term a is mapped to $f(a, a, a)$, the term $g(x)$ to $f(x, g, g)$, and the term $h(a, g(x))$ to $f(f(a, a, a), f(x, g, g), h)$.

The second transformation is achieved by flattening right-hand side ground terms. For example, the rewrite rule $s \rightarrow f(f(a, a, a), f(a, g, g), h)$ is replaced by $s \rightarrow c_1$, while introducing the new B-rules $c_1 \rightarrow f c_2 c_3 h$, $c_2 \rightarrow f a a a$, and $c_3 \rightarrow f a g g$.

4   New Constants and the Minimum Joining Extension

Let R be a RGV system. Suppose $f x x \rightarrow x$ is a rewrite rule in R. We shall see in later sections that it is useful to interpret the rule $f x x \rightarrow x$ as representing all the instances

$fs_1s_2 \rightarrow s_3$ where s_3 is a term simultaneously reachable from s_1 and s_2 by R. In general, define

$$Reach_R(\{s_1, \ldots, s_k\}) = \{s : \forall i.s_i \rightarrow^*_R s\}.$$

We say that the set $\{s_1, \ldots, s_k\}$ is R-joinable if $Reach_R(\{s_1, \ldots, s_k\}) \neq \emptyset$.

The set $Reach_R(\{s_1, \ldots, s_k\})$ is especially important when every s_i is a constant. Let $\{c_1 \ldots c_n\}$ be the set of constants appearing in R. We introduce new constants of the form $c_{\{i_1 \ldots i_k\}}$, where $\{i_1 \ldots i_k\}$ is a subset of $\{1 \ldots n\}$. Our intention is to make a term s reachable from $c_{\{i_1 \ldots i_k\}}$ iff $s \in Reach_R(\{c_{i_1}, \ldots, c_{i_k}\})$. For uniformity, we first rename every constant c_i to $c_{\{i\}}$; i.e. we assume that the set of constants appearing in R is $\{c_{\{1\}} \ldots c_{\{n\}}\}$. The new constants are counterparts of the *packed states* in [14]. Let Σ' be $\Sigma \cup \{c_S \mid \emptyset \neq S \subseteq \{1 \ldots n\}$.

Definition 1. *The* minimum joining extension, $\overline{R}$, *of a rewrite system R is defined as the minimum rewrite relation[1] on $\mathcal{T}(\Sigma', \mathcal{X})$ such that (a) $\rightarrow^*_R \subseteq \overline{R}$, and (b) for all subsets $S = \{i_1, \ldots, i_k\} \subseteq \{1 \ldots n\}$,*

$$Reach_{\overline{R}}(\{c_S\}) = \begin{cases} \{c_S\} & \text{if } Reach_{\overline{R}}(\{c_{\{i_1\}}, \ldots, c_{\{i_k\}}\}) = \emptyset \\ Reach_{\overline{R}}(\{c_{\{i_1\}}, \ldots, c_{\{i_k\}}\}) & \text{otherwise} \end{cases}$$

Note that there exist rewrite relations that satisfy conditions (a) and (b); for example, the rewrite relation induced by $R \cup \{c_S \leftrightarrow c_{\{i\}} : \emptyset \neq S \subseteq \{1, \ldots, n\}, i \in S\}$ is one such relation. If $\{R_i\}_{i \in I}$ are rewrite relations that satisfy conditions (a) and (b), then $\bigcap_{i \in I} R_i$ also satisfies the two conditions. Therefore, the *minimum* is well defined.

Example 2. If $R = \{c_{\{1\}} \rightarrow fc_{\{4\}}, c_{\{2\}} \rightarrow fc_{\{4\}}, c_{\{3\}} \rightarrow fc_{\{4\}}\}$, then $\overline{R}$ is the rewrite relation induced by $\{c_S \rightarrow c_{S'} : \emptyset \neq S \subseteq S' \subseteq \{1, 2, 3\}\} \cup \{c_{\{1,2,3\}} \rightarrow fc_{\{4\}}\}$.

We will use $Const(\{i_1 \ldots i_k\})$ to denote the set $\{c_{\{i_1\}}, \ldots, c_{\{i_k\}}\}$ and infix notation $s\overline{R}t$ to denote $(s, t) \in \overline{R}$.

Lemma 1. *The following properties are true about the the minimum joining extension $\overline{R}$ of R:*

1. *If $\emptyset \neq S \subseteq S'$ and $Reach_{\overline{R}}(Const(S')) \neq \emptyset$ then $c_S \overline{R} c_{S'}$.*
2. *If $Reach_{\overline{R}}(Const(S_1)) \neq \emptyset$ and $Reach_{\overline{R}}(Const(S_2)) \neq \emptyset$ and $\{c_{S_1}, c_{S_2}\}$ is $\overline{R}$-joinable, then $Reach_{\overline{R}}(Const(S_1 \cup S_2)) \neq \emptyset$.*
3. *If $c_{S_1} \overline{R} c_{S'_1}, c_{S_2} \overline{R} c_{S'_2}$, and the sets $Reach_{\overline{R}}(Const(S_1))$, $Reach_{\overline{R}}(Const(S_2))$, $Reach_{\overline{R}}(Const(S'_1))$ and $Reach_{\overline{R}}(Const(S'_2))$ are not empty, and and $\{c_{S'_1}, c_{S'_2}\}$ is $\overline{R}$-joinable, then $c_{S_1 \cup S_2} \overline{R} c_{S'_1 \cup S'_2}$.*
4. *If $c_{S_1} \overline{R} f c_{S_{11}} \ldots c_{S_{1m}}$, and $c_{S_2} \overline{R} f c_{S_{21}} \ldots c_{S_{2m}}$, and the sets $Reach_{\overline{R}}(Const(S_{11} \cup S_{21})), \ldots, Reach_{\overline{R}}(Const(S_{1m} \cup S_{2m}))$ are not empty, then $c_{S_1 \cup S_2} \overline{R} f c_{S_{11} \cup S_{21}} \ldots c_{S_{1m} \cup S_{2m}}$.*

The following technical lemma provides an alternate characterization of $\overline{R}$.

---

[1] A rewrite relation is a reflexive, transitive, and monotonic relation that is also closed under substitution.

Lemma 2. *Let* R_J *be* $\{c_S \to t[c_{S_1}, \ldots, c_{S_l}]_P \mid \forall i \in S.\forall j \in \{1, \ldots, l\}.(c_{\{i\}} \to_R^*$
$t[c_{\{k_{i1}\}}, \ldots, c_{\{k_{il}\}}]_P \land k_{ij} \in S_j) \land Reach_R(Const(S_j)) \neq \emptyset)\}$. *Then* $\overline{R} =\to_{R_J}^*$
$\circ \to_R^* \circ \to_{R_J}^*$.

The intuition that the minimum joining extension $\overline{R}$ only conservatively extends the rewrite relation R is captured by the following two corollaries.

Corollary 1. *Let* $\overline{R}$ *be the minimum joining extension of* R. *Then,*

– $\overline{R}$ *and* $\to_R^*$ *coincide on pairs of terms over the original signature.*
– *The set* $Reach_R(Const(S))$ *is nonempty iff* $Reach_{\overline{R}}(Const(S))$ *is nonempty iff* $Reach_{\overline{R}}(\{c_S\}) \neq \{c_S\}$.

Corollary 2. *The following is true for the minimum joining extension* $\overline{R}$ *of* R:

– *If* s *is a term over the original signature and* $s\overline{R}t$, *then for all constant* c_S *appearing in* t, $Const(S)$ *is* R-*joinable.*
– *If* s *is a term containing only constants* c_S *such that* $Const(S)$ *is* R-*joinable, then there exists a term* t *over the original signature such that* $s\overline{R}t$.
– *If* s *is a term over the original signature and* $s\overline{R}t$, *then there exists* t' *over the original signature such that* $t\overline{R}t'$.
– *Joinability of* $\overline{R}$ *coincides with joinability of* $\to_R^*$ *for terms over the original signature.*

5   Constrained Rewriting with Joinability Constraints

In the following sections we will work with a particular form of constrained rewriting where the constraints are interpreted modulo joinability by a given rewrite system.

Definition 2. *A* constrained rule *is of the form* $l \to r$ *if* C, *where (i)* r *is ground or a variable, (ii) no variable appears twice in* $l \to r$, *and (iii)* C *is a set of subsets of* $Vars(l \to r) \cup \Sigma_0'$ *such that every variable appears in at most one subset; moreover, there is a distinguished element in one of these subsets if and only if this element is a variable and coincides with* r.

We will denote a subset with no distinguished element by $\{\alpha_1, \alpha_2, \ldots, \alpha_k\}$ and a subset with a distinguished element α by $\{\alpha; \alpha_1, \alpha_2, \ldots, \alpha_k\}$.

The interpretation of constraints is dependent upon (the joinability relation induced by some) rewrite system R.

Definition 3. *Let* R *be a rewrite system. A substitution* σ *is an* R-*solution of the constraint* C *if for each* $\{\alpha_1, \alpha_2, \ldots, \alpha_k\} \in C$, *the terms* $\alpha_1\sigma, \ldots, \alpha_k\sigma$ *are* R-*joinable, and for each* $\{\alpha; \alpha_1, \alpha_2, \ldots, \alpha_k\} \in C$, *the term* $\alpha\sigma$ *is in* $Reach_R(\{\alpha_1\sigma, \ldots, \alpha_k\sigma\})$.

Thus, for a given R, a constrained rewrite rule $l \to r$ **if** C denotes all instances $l\sigma \to r\sigma$, where σ is an R-solution of C.

Definition 4. *Given a set C of constrained rules, we write $s = u[l\sigma] \to_{C/R} u[r\sigma] = t$ if $(l \to r$ if $C)$ is a constrained rule in C and σ is an R-solution of C.*

Example 3. If $C = \{(x_1 + (-x_2)) + x_3 \to x_4$ if $\{\{x_4; x_1, x_2, x_3\}\}\}$ and R is as in Example 2, then $(c_{\{1\}} + -c_{\{2\}}) + c_{\{3\}} \to_{C/R} c_{\{1,2,3\}}$, and $(c_{\{1\}} + -c_{\{2\}}) + c_{\{3\}} \to_{C/R} fc_{\{4\}}$.

6 Rewrite Closure

A pair of term rewrite systems (F, B) is a rewrite closure for R (with respect to a reduction ordering $\succ$) if (i) $F \cup B^-$ is contained in $\succ \cup (\Sigma_0 \times \Sigma_0)$ and (ii) the relation $\to_R^*$ is equivalent to the relation $\to_F^* \circ \to_B^*$. We will now construct a rewrite closure for the minimal extension $\overline{R}$ of a RGV system R.

First, we transform the rules in F to *linear* constrained rules by "moving" the non-linearity information into constraints. For example, the rule $(x + -x) + x \to x$ is transformed into the constrained rule

$$(x_1 + -x_2) + x_3 \to x_4 \qquad \text{if} \qquad \{\{x_4; x_1, x_2, x_3\}\}.$$

Intuitively, this constrained rewrite rule represents all instances $((x_1 + -x_2) + x_3)\sigma \to x_4\sigma$, where σ is such that $x_4\sigma \in Reach_R(\{x_1\sigma, x_2\sigma, x_3\sigma\})$.

Definition 5. *A constrained form of an F-rule $l \to r$ is a constrained rule $l' \to r'$ if C where (i) C is a partition of $Vars(l') \cup Vars(r')$, and (ii) there is a mapping $\sigma : Vars(l') \cup Vars(r') \to Vars(l) \cup Vars(r)$ such that $l'\sigma = l$, $r'\sigma = r$ and $x\sigma = y\sigma$ iff x and y are in the same subset in C, and if r is a variable, then r' is a distinguished element in C.*

Transition Rules. Let $R = F_0 \cup B_0$ be a RGV system and $CR_0 = CF_0 \cup B_0$ be its constrained form, where CF_0 denotes the constrained versions of the F_0 rules. It is easy to see that the relations $\to_R^*$ and $\to_{CF_0/B_0 \cup B_0}^*$ are identical.

We saturate the set $CR = CR_0$ under the following transition rules. In the process, we could generate constrained rules of the form $c \to \alpha$ if C, where α is height 0, called CB-rules. Initially there are no CB-rules, and hence $CB_0 = \emptyset$. Starting with $CR_0 = CF_0 \cup B_0 \cup CB_0$, we apply the following inference rules exhaustively.

Ordered Chaining: $\dfrac{CR = CR_1 \cup \{c \to_B s, \; u[t] \to_{CF} d \text{ if } C\}}{CR \cup \{u[c] \to d \text{ if } C\sigma\}}$

if σ is the most-general unifier of s and t, and t is not a variable.

Extension 1: $\dfrac{CR}{CR \cup \{c_S \to_B c_{\{i_1 \ldots i_k\}}\}}$

if S is non-empty and $S \subseteq \{i_1 \ldots i_k\}$ and $\{c_{\{i_1\}} \ldots c_{\{i_k\}}\}$ is B-joinable.

Extension 2:
$$CR = CR_1 \cup \{c_{S_1} \to_B c_{S_1'}, \ c_{S_2} \to_B c_{S_2'}\}$$
$$\overline{CR \cup \{c_{S_1 \cup S_2} \to_B c_{S_1' \cup S_2'}\}}$$

if $\{c_{S_1'}, c_{S_2'}\}$ is B-joinable.

Extension 3:

$$CR = CR_1 \cup \{c_{S_1} \to_B f c_{S_{11}} \dots c_{S_{1m}}, \ c_{S_2} \to_B f c_{S_{21}} \dots c_{S_{2m}}\}$$
$$\overline{CR \cup \{c_{S_1 \cup S_2} \to_B f c_{S_{11} \cup S_{21}} \dots c_{S_{1m} \cup S_{2m}}\}}$$

if, for all i, the set $\{c_{S_{1i}}, c_{S_{2i}}\}$ is B-joinable.

Constraint Solving 1:
$$CR = CR_1 \cup \{c \to_{CB} d \text{ if } \mathcal{C}\}$$
$$\overline{CR \cup \{c \to_B d\}}$$

if $\mathcal{C}$ contains only B-joinable sets of constants.

Constraint Solving 2:
$$CR = CR_1 \cup \{c \to_{CB} x \text{ if } \mathcal{C}\}$$
$$\overline{CR \cup \{c \to_B c_{S_1 \cup \dots \cup S_k}\}}$$

if $\mathcal{C} = \mathcal{C}' \cup \{\{x; c_{S_1}, \dots, c_{S_k}\}\}$, where $\mathcal{C}'$ contains only B-joinable sets of constants, and $\{c_{S_1}, \dots, c_{S_k}\}$ is B-joinable.

Chaining B-rules 1:
$$CR = CR_1 \cup \{c \to_B d, \ d \to_B e\}$$
$$\overline{CR \cup \{c \to_B e\}}$$

Chaining B-rules 2:
$$CR = CR_1 \cup \{c \to_B d, \ d \to_B f d_1 \dots d_m\}$$
$$\overline{CR \cup \{c \to_B f d_1 \dots d_m\}}$$

The ordered chaining rule attempts to remove "peaks" from rewrite proofs, the extension rules add rules over the new constants that would be part of $\overline{R}$, the constraint solving rules simplify the constraints, and the last chaining rules add new B-rules so that all derivations using the final set of B-rules can be made increasing (defined below). Note that the rewrite rule $(u[c] \to d \text{ if } \mathcal{C}\sigma)$ introduced by ordered chaining can be a CF-, B-, or a CB-rule, but (i) the depth of $u[c]$ is bounded by the maximum depth of terms in $CF_0 \cup B_0$, (ii) the constraint $\mathcal{C}$ only contains depth zero terms, and (iii) every variable in $\mathcal{C}$ also occurs in either $u[c]$ or d. These observations together imply that saturation under the above inference rules can only generate finitely many different constrained rewrite rules (up to variable renaming). This implies that the inference rules can be implemented so as to guarantee termination.

Example 4. Let $R = \{0 + x \to x, \ -a \to 0 + a\}$. After flattening (where a new constant c_1 is introduced), moving nonlinearity into the constraints, and renaming constants we get $CF_0 = \{c_{\{0\}} + x_1 \to x_2 \text{ if } \{\{x_2; x_1\}\}, -c_{\{a\}} \to c_{\{1\}}\}$ and $B_0 = \{c_{\{1\}} \to c_{\{0\}} + c_{\{a\}}\}$. We have written $-c_{\{a\}}$ and $c_{\{0\}} + c_{\{a\}}$ instead of $f(c_{\{a\}}, -, -)$ and

$f(c_{\{0\}}, c_{\{a\}}, +)$ for readability. Saturating the set $CF_0 \cup B_0$ under the above transition rules, we add the following rewrite rules:

$c_{\{1\}} \to x_2$ if $\{\{x_2; c_{\{a\}}\}\}$	Ordered Chaining
$c_{\{1\}} \to c_{\{a\}}$	Constraint Solving 2
$-c_{\{1\}} \to c_{\{1\}}$	Ordered Chaining
$-c_{\{1\}} \to c_{\{a\}}$	Ordered Chaining
$c_{\{a\}} \to c_{\{a,1\}}$	Extension 1
$c_{\{1\}} \to c_{\{a,1\}}$	Extension 1
$c_{\{a,1\}} \to c_{\{a\}}$	Extension 2

We note that the transition rules do not add any more new rewrite rules, and the saturated set of rules is $CF_\infty = CF_0 \cup \{-c_{\{1\}} \to c_{\{1\}}, -c_{\{1\}} \to c_{\{a\}}\}$, $CB_\infty = \{c_{\{1\}} \to x_2$ if $\{\{x_2; c_{\{a\}}\}\}\}$ and $B_\infty = B_0 \cup \{c_{\{1\}} \to c_{\{a\}}, c_{\{a\}} \to c_{\{a,1\}}, c_{\{1\}} \to c_{\{a,1\}}, c_{\{a,1\}} \to c_{\{a\}}\}$.

Lemma 3. *Let $CR_\infty = CF_\infty \cup B_\infty \cup CB_\infty$ denote the result of saturating the initial set CR_0, obtained from the set $R = F_0 \cup B_0$ (as described before). Then, $\to^*_R \subseteq \to^*_{(CF_\infty \cup CB_\infty)/B_\infty \cup B_\infty} \subseteq \overline{R}$.*

Example 5. Let R be as defined in Example 4. Add the rewrite rule $-x + x \to x$ to R. We will notice that the set $CF'_\infty = CF_\infty \cup \{-x_1 + x_2 \to x_3$ if $\{\{x_3; x_1, x_2\}\}\}$ is a saturation of the resulting initial system. In the new initial system, the term $c_{\{a\}}$ is reachable from the term $- - c_{\{a\}} + c_{\{a\}}$ since

$$- - c_{\{a\}} + c_{\{a\}} \to -c_{\{1\}} + c_{\{a\}} \to -(c_{\{0\}} + c_{\{a\}}) + c_{\{a\}} \to -c_{\{a\}} + c_{\{a\}} \to c_{\{a\}}.$$

Using the saturated set,

$$- - c_{\{a\}} + c_{\{a\}} \to_{CF_\infty/B_\infty} -c_{\{1\}} + c_{\{a\}} \to_{B_\infty} -c_{\{a\}} + c_{\{a\}} \to_{CF_\infty/B_\infty} c_{\{a\}}.$$

A derivation $s \to^*_R t$ is said to be *increasing* [4] if whenever this derivation can be written as $s \to^*_R \circ \to_{R,p_1} \circ \to^*_R \circ \to_{R,p_2} \circ \to^*_R t$, then the position p_2 is not a prefix of p_1.

Lemma 4. *If $s \to_{CB_\infty/B_\infty} t$, then $s \to^*_{B_\infty} t$.*

Lemma 4 implies that the relation $\to^*_{(CF_\infty \cup CB_\infty)/B_\infty \cup B_\infty}$ coincides with the relation $\to^*_{CF_\infty/B_\infty \cup B_\infty}$. It would be possible to prove that this relation coincides with $\to^*_{CF_\infty/B_\infty} \circ \to^*_{B_\infty}$, but this does not imply that it is a rewrite closure presentation yet, since rewriting with $\to^*_{CF_\infty/B_\infty}$ can make terms bigger (if right-hand side is a variable, it can be instantiated by an arbitrarily large term). For this reason, we need to show that rewriting with $\to^*_{CF_\infty/B_\infty}$ can be restricted to size-decreasing rules while preserving the global relation $\to^*_{CF_\infty/B_\infty \cup B_\infty}$.

Definition 6. *Given a set C of constrained rules and a set D of rules, we write $s = u[l\sigma] \to_{C \sqrt{D}} u[r\sigma] = t$ if $(l \to r$ if $C)$ is a constrained rule in C and σ is a D-solution of C such that, if $\{x; \alpha_1, \ldots, \alpha_k\} \in C$, then $Pos(x\sigma) = Pos(\alpha_1\sigma) \cup \ldots \cup Pos(\alpha_k\sigma)$.*

The new constants introduced in Section 4 are crucial for the following technical result.

Lemma 5. *Let* $T = \{t_1 \ldots t_k\}$ *be a set of* B_∞-*joinable terms. Then, there exists an effectively computable set of terms* $S = \{s_1 \ldots s_l\}$ *such that* $Pos(s_1) = \ldots = Pos(s_l) = \bigcup_{i \in \{1,\ldots,k\}} (Pos(t_i))$ *and* $Reach_{B_\infty}(\{t_1, \ldots, t_k\}) = \bigcup_{i \in \{1,\ldots,l\}} Reach_{B_\infty}(s_i)$.

Lemma 5 allows us to restrict the rewrite relation $\to^*_{CF_\infty/B_\infty}$ to the smaller relation of Definition 6.

Corollary 3. $\to^*_{CF_\infty/B_\infty \cup B_\infty} = \to^*_{CF_\infty \sqrt{B_\infty} \cup B_\infty}$

Henceforth, we fix $\succ$ to denote some reduction ordering that is an extension of the lexicographic combination of the height and size ordering on terms. Any such ordering $\succ$ will possess the subterm property, that is, $s \succ t$ if t is a proper subterm of s. We note that $\succ$ orients all the rules in $CF\sqrt{B}$ and $B^- \subseteq (\succ \cup \Sigma_0 \times \Sigma_0)$. The following proposition is established using proof simplification arguments [1].

Theorem 1. $\to^*_{CF_\infty \sqrt{B_\infty} \cup B_\infty} = \to^*_{CF_\infty \sqrt{B_\infty}} \circ \to^*_{B_\infty} = \overline{R}$, *and hence, the pair* $(CF_\infty \sqrt{B_\infty}, B_\infty)$ *is a rewrite closure for* $\overline{R}$ *w.r.t. the ordering* $\succ$.

Example 6. Consider the following derivation from Example 5:

$$-- c_{\{a\}} + c_{\{a\}} \to_{CF_\infty/B_\infty} -c_{\{1\}} + c_{\{a\}} \to_{B_\infty} -c_{\{a\}} + c_{\{a\}} \to_{CF_\infty/B_\infty} c_{\{a\}}.$$

The peak in the above proof can be eliminated by including the B_∞ rewrite step in the constrained rewriting step to get

$$-- c_{\{a\}} + c_{\{a\}} \to_{CF_\infty/B_\infty} -c_{\{1\}} + c_{\{a\}} \to_{CF_\infty/B_\infty} c_{\{a\}},$$

since $c_{\{1\}}$ and $c_{\{a\}}$ are B_∞-joinable.

Theorem 2. *Reachability and joinability of RGV rewrite systems is decidable.*

Proof. Using item 1 of Corollary 1 and item 4 of Corollary 2, deciding reachability and joinability for a RGV system R reduces to deciding reachability and joinability for the corresponding $\overline{R}$.

Theorem 1 establishes the decidability of reachability and joinability for $\overline{R}$ as follows. Given two terms s and t, the reachability problem $s \to^*_{CF_\infty \sqrt{B_\infty} \cup B_\infty} t$ can be decided by simply obtaining all terms reachable from s by $CF_\infty \sqrt{B_\infty}$, and checking whether one of them reaches t by rewriting with B_∞. The joinability problem $s \downarrow^*_{CF_\infty \sqrt{B_\infty} \cup B_\infty} t$ can be decided by obtaining all terms $s_1, \ldots, s_k$ and $t_1, \ldots, t_l$ that are reachable, from s and t respectively, by $CF_\infty \sqrt{B_\infty}$, and checking whether some s_i and t_j are B_∞-joinable.

7   Deciding Termination

For this section and the following one we make an additional assumption on the initial R: for every rule of the form $l \rightarrow x$, the variable x occurs in l. Note that, otherwise, R is trivially not terminating; and moreover, it is not confluent, since by our initial assumptions there is no rule of the form $x \rightarrow t$, and hence the equivalent terms y and z can not be joined.

Definition 7. *Let G be a term rewrite system. A term t is* context-reachable *from s by G if $s \rightarrow_G^+ c[t]_p$ for some context $c[\_]_p$. If the position p is not λ, then t is said to be* strictly context-reachable *from s.*

Lemma 6 characterizes nontermination of RGV systems R as described above.

Lemma 6. *R is nonterminating iff there exists a constant context-reachable from itself.*

The decidability of termination for ground rewrite systems [13] can now be generalized.

Theorem 3. *Termination of a RGV system R is decidable.*

Proof. On the one side, we can use $\rightarrow_{B_\infty}^*$ to check if a constant of the form $c_{\{i\}}$ is strictly context-reachable from itself by $\bar{R}$: this property is easy to check for ground rewrite systems like B_∞, for example by using tree automata techniques (see [3]). On the other, we can check if $c_{\{i\}} \rightarrow_R^+ c_{\{i\}}$ by considering all the terms t such that $c_{\{i\}} \rightarrow_R t$ (only one step with R), and checking if $t \bar{R} c_{\{i\}}$.

8   Deciding Confluence

Our goal is to decide the confluence of a RGV system R, but we show that this is equivalent to checking the confluence of $\bar{R}$. This, in turn, means we can just focus on the rewrite closure presentation of $\bar{R}$.

Lemma 7. *$\bar{R}$ is confluent iff R is confluent.*

From now on, we denote the sets CF_∞ and B_∞ by CF and B respectively. Moreover, we view the rewrite closure construction procedure as a black box and, for example, refer to the set of constants appearing in $CF \cup B$ as $\{c_1, c_2, \ldots\}$.

We will use the following result from [8].

Theorem 4. *Let $F \cup B$ be a rewrite closure of R with respect to the reduction ordering $\succ$ and let B_{CC} be a convergent presentation of B with respect to the same ordering. The rewrite system R is confluent if and only if the following two conditions are true:*
(a) The rewrite system $F \cup B_{CC}$ is locally confluent.
(b) $\downarrow_{F \cup B_{CC}} \subseteq \downarrow_{F \cup B}$.

In fact, the proof of decidability of confluence follows the scheme of [8], and the following subsections contain an adequate adaptation (and in some cases a simplification) of the proofs given in [8] for proving the decidability of the confluence of right-ground systems.

8.1   Canonical Instances of Constrained Rules

Note that $CF\sqrt{B} \cup B$ is a rewrite closure. We wish to use Theorem 4 to decide confluence of $CF\sqrt{B} \cup B$. In particular, we wish to check the (local) confluence of the terminating system $CF\sqrt{B} \cup B_{CC}$. We note here that, in the present case, the set B_{CC} is an *abstract congruence closure* for B^-, as defined in [2], and contains flat rules of the form $fc_1 \ldots c_m \to c$ and $c \to d$. The problem, however, is that $CF\sqrt{B}$ possibly represents infinitely many unconstrained rewrite rules. We solve this problem by considering a finite unconstrained rewrite system CCF that contains only certain *canonical* instances of CF; and we will show that checking the (local) confluence of $CCF \cup B_{CC}$ is enough. For the correct construction of CCF, we remove the useless rules of CF, i.e. the rules with unsatisfiable constraints.

Deletion 1:
$$\frac{CR \cup \{s \to t \text{ if } \mathcal{C} \cup \{\{c_{S_1}, \ldots, c_{S_k}, x_1, \ldots, x_l\}\}\}}{CR}$$

if the constants $c_{S_1}, c_{S_2}, \ldots, c_{S_k}$ are not B-joinable.

Deletion 2:
$$\frac{CR \cup \{s \to t \text{ if } \mathcal{C} \cup \{\{x; c_{S_1}, \ldots, c_{S_k}, x_1, \ldots, x_l\}\}\}}{CR}$$

if the constants $c_{S_1}, c_{S_2}, \ldots, c_{S_k}$ are not B-joinable.

Note that the removal of rules with unsatisfiable constraints does not modify the relations $\to_{CF/B \cup B}$ and $\to_{CF\sqrt{B} \cup B}$.

We define the *canonical solution* σ of a constraint $\mathcal{C}$ as the result of the following computation: Initially, let σ be the identity substitution. First, each constant in $\mathcal{C}$ is normalized by B_{CC}. Thanks to the deletion rules, there is no more than one distinct constant in every set in $\mathcal{C}$. We apply the following two rules exhaustively:
1. If $\{x_1, \ldots, x_k, c\}$ or $\{x_1; x_2, \ldots, x_k, c\}$ is in $\mathcal{C}$, then set $x_i\sigma \mapsto c$ for all i, and remove this constraint from $\mathcal{C}$.
2. If $\{x_1, \ldots, x_k\}$ or $\{x_1; x_2, \ldots, x_k\}$ is in $\mathcal{C}$, then set $x_2\sigma \mapsto x_1, x_3\sigma \mapsto x_1, \ldots,$ $x_k\sigma \mapsto x_1$ and remove this constraint set from $\mathcal{C}$.
The substitution σ thus obtained is the *canonical solution* of the constraint $\mathcal{C}$.

We define the set

$$CCF = \{l\sigma \to r\sigma : l \to r \text{ if } \mathcal{C} \in CF \text{ and } \sigma \text{ is a canonical solution of } \mathcal{C}\}.$$

Proposition 1. *If $s \to_{CF\sqrt{B}} t$, then there exists t' such that $s \to^*_{CCF \cup B_{CC}} t' \leftarrow^*_{B_{CC}} t$. In fact, the relation $\leftrightarrow^*_{CF\sqrt{B} \cup B_{CC}}$ is identical to the relation $\leftrightarrow^*_{CCF \cup B_{CC}}$.*

Example 7. Consider $R = \{x + (-x) \to 0, x + 0 \to x, 0 \to -0\}$. The rewrite closure procedure of Section 6 starting from R gives $CF = \{x_1 + (-x_2) \to 0 \text{ if } \{\{x_1, x_2\}\},$ $x_1 + 0 \to x_2 \text{ if } \{\{x_2; x_1\}\}, x_1 + 0 \to 0 \text{ if } \{\{x_1, 0\}\}\}$ and $B = \{0 \to -0\}$, where $CF\sqrt{B} \cup B$ is a rewrite closure. Then $B_{CC} = \{-0 \to 0\}$ and $CCF = \{x + (-x) \to 0, x + 0 \to x, 0 + 0 \to 0\}$.

We now use Theorem 4 for characterizing the confluence of $CF\sqrt{B} \cup B$ in terms of CCF.

Corollary 4. *The rewrite system $CF\sqrt{B} \cup B$ is confluent if and only if the following two conditions are true:*
(i) $CCF \cup B_{CC}$ is (locally) confluent,[2]
(ii) $\downarrow_{CCF \cup B_{CC}} \subseteq \downarrow_{CF\sqrt{B} \cup B}$.

Condition (i) of Corollary 4 has crucial consequences: if $CF/B \cup B$ is confluent, then $CCF \cup B_{CC}$ is a finite and convergent presentation of the theory $CF/B \cup B$ and hence, it decides the word problem for $CF/B \cup B$.

This fact can be used to decide the unification problem for confluent RGV systems (this result has been recently obtained in [11]). Given two terms s and t and a substitution σ, $s\sigma \leftrightarrow^*_{CF\sqrt{B} \cup B} t\sigma$ iff $s\sigma \downarrow_{CCF \cup B_{CC}} t\sigma$, that is, the unification problem for a confluent right ground system reduces to the unification problem for a convergent system. Now, note that the rewrite system $CCF \cup B_{CC}$ is decreasing (its rules $l \to r$ are oriented by the ordering $\succ$ that cointains the proper subterm ordering), and hence unification modulo $CCF \cup B_{CC}$ is decidable [10].

Corollary 5. *The word problem and the unification problem are decidable for confluent RGV term rewrite systems.*

Now, deciding confluence reduces to checking conditions (i) and (ii). The rewrite system $CCF \cup B_{CC}$ is terminating; hence, checking condition (i) is decidable, and thus, the problem now reduces to checking condition (ii) of Corollary 4.

Example 8. Continuing the previous example, we notice that the rewrite system $CCF \cup B_{CC} = \{x + (-x) \to 0,\ x + 0 \to x,\ 0 + 0 \to 0,\ -0 \to 0\}$ is locally confluent. Hence to prove confluence of $CF\sqrt{B} \cup B$, we only need to test for the inclusion $\leftrightarrow^*_{CCF \cup B_{CC}} \subseteq \downarrow_{CF\sqrt{B} \cup B}$.

8.2   Top Stable Structures

Henceforth, we assume that $CCF \cup B_{CC}$ is confluent. We consider the problem of deciding if the equivalence relation $\leftrightarrow^*_{CCF \cup B_{CC}}$ is included in the joinability relation $\downarrow_{CF\sqrt{B} \cup B}$. We use an extension of the concept of top-stability [4] for checking this. For simplicity of proofs, we will use the interreduced (or fully-reduced) version $IRFB$ of $CCF \cup B_{CC}$. We define the interreduced version $IR(R)$ of R as follows:

$$IR(R) = \{l \to r : \exists l \to r' \in R . r' \to^!_R r \text{ and all proper subterms of } l \text{ are } R\text{-irreducible}\}$$

Proposition 2. *[8] If R is a convergent term rewriting system, then $R' = IR(R)$ is also a convergent presentation of R.*

Define $IRFB$ as $IR(CCF \cup B_{CC})$. Proposition 2 implies that $IRFB$ is a convergent presentation of $CCF \cup B_{CC}$.

---

[2] Note that $CCF \cup B_{CC}$ is terminating.

Example 9. In our running example, the set $CCF \cup B_{CC}$, as defined in Example 8, gives $IRFB = \{x + -x \to 0,\ x + 0 \to x,\ 0 + 0 \to 0,\ -0 \to 0\}$.

Let $IRFBG$ be the ground rules of $IRFB$. Formally, for every height 0 term α, we define $TopStable(\alpha)$ as

$$TopStable(\alpha) = \begin{cases} TopStable(NF_{IRFB}(\alpha)) & \text{if } \alpha \neq NF_{IRFB}(\alpha) \\ \{t = ft_1 \ldots t_m : t \to \alpha \in IRFB, \exists t^\uparrow : t^\uparrow \to^{*,nr}_{IRFBG} t \text{ and} \\ t^\uparrow \text{ is } CF/(CCF \cup B_{CC})\text{-irreducible}\} & \text{otherwise.} \end{cases}$$

Lemma 8. *The equivalence relation induced by $CCF \cup B_{CC}$ is included in the joinability relation induced by $CF\sqrt{B} \cup B$ iff*
(i) constants $\{c_1, \ldots, c_n\}$ that are equivalent modulo $CCF \cup B_{CC}$ are B-joinable and (ii) for every height zero term α, it is the case that (a) the set $TopStable(\alpha)$ has at most one term, and (b) if $t = ft_1 \ldots t_m \in TopStable(\alpha)$, then α is a constant, t is ground, and $\alpha \to^*_B fc_1 \ldots c_m$ for some constants $c_1, \ldots, c_m$ such that c_i and t_i are B_{CC}-equivalent.

8.3 Computing Top Stable Structures

The sets $TopStable(\alpha)$ are computed using a fixpoint computation that starts with $TS_0(c)$ and generates $TS_{i+1}(\alpha)$ by gradually adding more elements to $TS_i(\alpha)$. If $(t, t^\uparrow) \in TS_i(\alpha)$, it means that $t \in TopStable(\alpha)$ and $t^\uparrow$ is a witness for t. This construction is completely analogous to the one presented in [8].

Definition 8. *Two terms s and t are D-similar, denoted by $s =_D t$, if $Pos_{\leq D}(s) = Pos_{\leq D}(t)$ and for all $p \in Pos_{\leq D}(s)$, the root symbol of $s|_p$ is equal to the root symbol of $t|_p$.*

Formally, the fixpoint computation is defined as follows.

$$TS_0(\alpha) = \{ (l, l) :\ l \to \alpha \in IRFB, \text{ and } l \text{ is } CF/(CCF \cup B_{CC})\text{-irreducible}\}$$
$$TS_{j+1}(\alpha) = \{ (l, l') : l = l[c_1, \ldots, c_k] \to \alpha \in IRFB,$$
$$l' = l[l'_1, \ldots, l'_k] \text{ where } (\_, l'_i) \in TS_j(c_i),$$
$$l' \text{ is } CF/(CCF \cup B_{CC})\text{-irreducible and}$$
$$\text{there is no } (l, l'') \in TS_j \text{ with } l' =_D l''\}$$

The fixpoint computation is *prematurely terminated* if condition (ii) of Lemma 8 is ever violated for the sets $TS_i(\alpha)$. We denote by $TS_\infty(\alpha)$ the result of the fixpoint computation above in the case when it does not prematurely terminate. In this case, it is clear that for every pair $(t, t^\uparrow) \in TS_\infty(\alpha)$, $t \in TopStable(\alpha)$, that is, the fixpoint computation is sound. Lemma 9 shows that it is complete too.

Example 10. In our running example $IRFB = \{x + -x \to 0, x + 0 \to x, 0 + 0 \to 0, -0 \to 0\}$. The term $x + -x$ is reducible by $x_1 + -x_2 \to 0$ if $\{\{x_1, x_2\}\}$, the term $x + 0$ is reducible by $x_1 + 0 \to x_2$ if $\{\{x_2; x_1\}\}$, and the term $0 + 0$ is reducible by $x_1 + 0 \to 0$ if $\{\{x_1, 0\}\}$, but the term -0 is not. Hence, $TS_0(0) = \{(-0, -0)\}$. When computing $TS_1(x)$, we consider the rule $x + 0 \to x$ and the replacement of the 0 in its left-hand side by -0. This gives the term $x + (-0)$, which is $CF/CCF \cup B_{CC}$ irreducible. Hence, the procedure prematurely terminates and $CF\sqrt{B} \cup B$ is not confluent.

Lemma 9. *If the fixpoint computation above does not prematurely terminate, then, for all α and all $t \in TopStable(\alpha)$ and every witness $t^{\uparrow}$ for t, there exists a witness $t^{\uparrow'}$ for t such that $(t, t^{\uparrow'}) \in TS_{\infty}(\alpha)$ and $t^{\uparrow} \leftrightarrow_D t^{\uparrow'}$.*

8.4 Deciding Confluence

Theorem 5. *The confluence of RGV rewrite systems is decidable.*

Proof. Given a RGV rewrite system R, a rewrite closure $CR = CF/B \cup B$ for $\overline{R}$ is constructed using the transformations and transition rules introduced in Section 6 (Theorem 1). The rewrite system R is confluent iff CR is confluent. An abstract congruence closure B_{CC} of B is then computed [2].

We compute CCF from CF and B_{CC} (Section 8.1) and Corollary 4 reduces the confluence problem to checking (i) local confluence of $CCF \cup B_{CC}$, which is decidable, and (ii) the inclusion $\downarrow_{CCF \cup B_{CC}} \subseteq \downarrow_{CF \sqrt{B}} \cup B$. Lemma 8 reduces checking condition (ii) to verifying two conditions. The first of these two is easily checked since B is ground (for example, see [3]). For the second, we note that the relations $\to_B^*$ and $\leftrightarrow_{B_{CC}}^*$ are decidable since B and B_{CC} are ground rewrite systems. Hence, it is sufficient to show computability of the sets $TopStable(\alpha)$ and this is established by Lemma 9.

9 Complexity

Termination and confluence of *shallow* RGV systems can be decided in $2^{O(n)}$ time assuming that the maximum arity m of a function symbol is a constant. This upperbound is tight since these problems can be shown to be EXPTIME-hard by reductions similar to the ones presented in [6]. The time complexity is obtained by bounding the size of the rewrite closure. Note that the total number of possible rules is now $(2^n)^{2m} = 2^{2mn}$ and its computation has cost $(2^{2mn})^4 = 2^{8mn}$. For termination, a fixpoint computation algorithm can decide if constants are strictly context reachable from themselves by B with cost $(2^{2mn})^4 = 2^{8mn}$, and deciding reachability of some height 1 terms into constants has cost $(2^{2mn})^4 = 2^{8mn}$. For confluence, the number of $CCF \cup B_{CC}$ rules is 2^{2mn}, its computation has cost $(2^{2mn})^2 = 2^{4mn}$, and checking its confluence has cost $(2^{2mn})^3$. In the computation of topstable structures, the $TS_i(c)$ contain terms of height at most 2, but two introduced terms must be different in positions smaller than or equal to 1; hence, the $|TS_i(c)|$'s are bounded by 2^{mn}, and the total cost for computing them is $(2^{2mn})^4 = 2^{8mn}$. If we drop the *shallowness* assumption, then we get a $2^{O(n^2)}$ complexity for deciding these properties (assuming m is a constant).

10 Conclusion

We have presented new decidability results for several fundamental properties of RGV systems, such as reachability, joinability, termination, and confluence. These results are interesting from many different perspectives. First, since our approach is based on extending and generalizing existing concepts that have proven useful for other simpler classes of term rewriting systems, we clearly bring out what additional technical issues are presented by richer classes of term rewriting systems. Second, when specialized to ground systems, there is a correspondence between our approach based on rewrite closure and the tree transducer techniques. However, using our approach, we have been successful in lifting the decidability results to several subclasses of nonground systems. Third, these new decidability results lie at the fringes of the decidability space as several properties are undecidable for term rewrite systems with slightly different restrictions. For future work, it would be interesting to study if these techniques can be adapted to incorporate axioms that can not be expressed by RGV rules.

References

1. L. Bachmair. *Canonical Equational Proofs*. Birkhäuser, Boston, 1991.
2. L. Bachmair and A. Tiwari. Abstract congruence closure and specializations. In David McAllester, editor, *Conference on Automated Deduction, CADE '2000*, pages 64–78, Pittsburgh, PA, June 2000. Springer-Verlag. LNAI 1831.
3. H. Comon, M. Dauchet, R. Gilleron, F. Jacquemard, D. Lugiez, S. Tison, and M. Tommasi. Tree automata techniques and applications. Available at
   http://www.grappa.univ-lille3.fr/tata, 1997.
4. H. Comon, G. Godoy, and R. Nieuwenhuis. The confluence of ground term rewrite systems is decidable in polynomial time. In *42nd Annual IEEE Symposium on Foundations of Computer Science (FOCS)*, pages 298–307, Las Vegas, Nevada, USA, 2001.
5. N. Dershowitz and J.-P. Jouannaud. Rewrite systems. In J. van Leeuwen, editor, *Handbook of Theoretical Computer Science, Volume B: Formal Models and Sematics (B)*, pages 243–320. MIT press/Elsevier, 1990.
6. G. Godoy, A. Tiwari, and R. Verma. On the confluence of linear shallow term rewrite systems. In H. Alt, editor, *20th Intl. Symposium on Theoretical Aspects of Computer Science STACS 2003*, volume 2607 of *LNCS*, pages 85–96. Springer, February 2003.
7. G. Godoy, A. Tiwari, and R. Verma. Characterizing confluence by rewrite closure and right ground term rewrite systems. *Annals of Pure and Applied Logic*, 2004. To appear.
8. G. Godoy, A. Tiwari, and R. Verma. Characterizing confluence by rewrite closure and right ground term rewrite systems. *Applicable Algebra in Engineering, Communication and Computing*, 15(1), 2004. Tentative.
9. F. Jacquemard. Reachability and confluence are undecidable for flat term rewriting systems. *Inf. Process. Lett.*, 87(5):265–270, 2003.
10. S. Mitra. *Semantic unification for convergent systems*. PhD thesis, University of Illinois at Urbana-Champaign, 1994.
11. I. Mitsuhashi, M. Oyamaguchi, Y. Ohta, and T. Yamada. On the unification problem for confluent monadic term rewriting systems. *IPSJ Trans. Programming*, 44(4):54–66, 2003.
12. M. Oyamaguchi and Y. Ohta. The unification problem for confluent right-ground term rewriting systems. In Aart Middeldorp, editor, *Rewriting Techniques and Applications, 12th Intl, Conf,, RTA 2001*, volume 2051 of *LNCS*, pages 246–260. Springer, 2001.
13. D. A. Plaisted. Polynomial time termination and constraint satisfaction tests. pages 405–420, Montreal, Canada, 1993. Springer-Verlag. LNCS 690.
14. T. Takai, Y. Kaji, and H. Seki. Right-linear finite path overlapping term rewriting systems effectively preserve recognizability. In *Rewriting Techniques and Applications, RTA 2000*, LNCS 1833, pages 246–260, 2000.

Redundancy Notions for Paramodulation with Non-monotonic Orderings

Miquel Bofill[1] and Albert Rubio[2]*

[1] Universitat de Girona, Dept. IMA, Girona, Spain
mbofill@ima.udg.es
[2] Universitat Politècnica de Catalunya, Dept. LSI, Barcelona, Spain
rubio@lsi.upc.es

Abstract. Recently, ordered paramodulation and Knuth-Bendix completion were shown to remain complete when using non-monotonic orderings. However, these results only implied the compatibility with too weak redundancy notions and, in particular, demodulation could not be applied at all.

In this paper, we present a complete ordered paramodulation calculus compatible with powerful redundancy notions including demodulation, which strictly improves previous results.

Our results can be applied as well to obtain a Knuth-Bendix completion procedure compatible with simplification techniques, which can be used for finding, whenever it exists, a convergent TRS for a given set of equations and a (possibly non-totalizable) reduction ordering.

1 Introduction

Knuth-Bendix-like completion techniques and their extensions to ordered paramodulation for first-order clauses are among the most successful methods for automated deduction with equality [BG98,NR01]. For many years, all known completeness results for Knuth-Bendix completion and ordered paramodulation required the term ordering $\succ$ to be well-founded, monotonic and total (or extendible to a total ordering) on ground terms [HR91,BDH86,BD94,BG94]. In [BGNR99,BGNR03], the monotonicity requirement was dropped and well-foundedness and the subterm property were shown to be sufficient for ensuring refutation completeness of ordered paramodulation (note that any such ordering can be totalized without losing these two properties). And recently, in [BR02], it was shown that well-foundedness of the ordering suffices for completeness of ordered paramodulation for Horn clauses, i.e., the subterm property can be dropped as well. Apart from its theoretical value, these results have several potential applications in contexts where the usual requirements are too strong.

* Both authors partially supported by the Spanish CICYT project MAVERISH ref. TIC2001-2476-C03-01. M. Bofill is also supported by the Spanish CICYT project CADVIAL ref. TIC2001-2392-C03-01 and A. Rubio is also supported by the Spanish DURSI group 2001SGR 00254.

D. Basin and M. Rusinowitch (Eds.): IJCAR 2004, LNAI 3097, pp. 107–121, 2004.

For example, in deduction modulo built-in equational theories E, where E-*compatibility* of the ordering (i.e. $s =_E s' \succ t' =_E t$ implies $s \succ t$) is needed, finding E-*compatible* orderings fulfilling the required properties is extremely complex or even impossible. For instance, when E contains an idempotency axiom $f(x,x) \simeq x$, no total E-compatible reduction ordering exists: if $s \succ t$, by monotonicity one should have $f(s,s) \succ f(s,t)$, which by E-compatibility implies $s \succ f(s,t)$, and hence non-well-foundedness. Therefore, the techniques for dropping ordering requirements, among other applications, open the door to deduction modulo many more classes of equational theories.

However, ordered strategies are useful in practice only if compatibility with redundancy is shown. Simplification of formulae and elimination of redundant formulae are essential components of automated theorem provers. In fact, in most successful automated theorem provers, simplification is the primary mode of computation, whereas prolific deduction rules are used only sparingly.

Here, we present a calculus which strictly improves the one in [BGNR03] (due to lack of space, we restrict ourselves to the equational case). On the one hand, regarding the amount of inferences needed to be performed, our inference system is essentially the same as the one in [BGNR03], but, on the other hand, our calculus is compatible with powerful redundancy notions which include simplification by rewriting (i.e. demodulation). As a side effect, as in [BGNR03], we can apply our results to obtain a Knuth-Bendix completion procedure, but in this case compatible with simplification techniques. This procedure can be used for finding, whenever it exists, a convergent TRS for a given set of equations and a (possibly non-totalizable) reduction ordering.

In our calculus it is assumed that there is a known reduction ordering $\succ_r$, which is included in the (possibly non-monotonic) ordering $\succ$ w.r.t. which equations are oriented. Then we show that redundancy notions w.r.t. the reduction ordering $\succ_r$ can be applied while keeping refutation completeness. Roughly, a formula is redundant if it is a logical consequence of smaller (w.r.t. a well-founded ordering including $\succ_r$) formulae. Then demodulation w.r.t. $\succ_r$ fits in this abstract notion of redundancy since, in particular, a formula which has been simplified by rewriting follows from its smaller w.r.t. $\succ_r$ (due to monotonicity of $\succ_r$) "simplified version" and the (also smaller) equations used to simplify it. As a particular case, if all equations involved in the saturation process[1] turn out to be orientable with $\succ_r$, then demodulation can be fully applied. The following well-known simple challenge example illustrates this situation:

Example 1. Under our inference system,

$$f(f(x)) \simeq f(g(f(x)))$$

is saturated if $f(f(x)) \succ_r f(g(f(x)))$, and the ordering $\succ$ w.r.t. which ordered paramodulation is applied is an extension of $\succ_r$ (which cannot be totalized without losing monotonicity, since $g(f(x))$ and $f(x)$ should be incomparable in any monotonic extension). The only possible inferences are (i) an ordered

---

[1] The saturation of a set of equations (or clauses) S amounts to the closure of S under the inference system *up to redundancy*.

paramodulation with $f(f(x)) \simeq f(g(f(x)))$ on itself at the underlined position of $f(\underline{f(x)})$, which has $f(f(g(f(x)))) \simeq f(g(f(f(x))))$ as conclusion, and (ii) an ordered paramodulation with $f(f(x)) \simeq f(g(f(x)))$ on itself at the underlined position of $f(g(\underline{f(x)}))$, which has $f(f(f(x))) \simeq f(g(f(g(f(x)))))$ as conclusion. In either case, the conclusion can be simplified into a tautology by rewriting it with the equation $f(f(x)) \simeq f(g(f(x)))$ (oriented with the reduction ordering). Therefore, the system is saturated. It is worth noting that, without considering demodulation, the closure of $f(f(x)) \simeq f(g(f(x)))$ under ordered paramodulation w.r.t. $\succ$ diverges.   □

We must point out that, although demodulation w.r.t. $\succ_r$ is allowed in our inference system, not every equation can be demodulated: during our saturation process, some subterms of equations become "blocked" for demodulation. Roughly, we block for demodulation those subterms that occur as the small side (w.r.t. $\succ$) of an equation that cannot be handled with $\succ_r$. Later inferences will inherit these blockings. Note that, if paramodulation is applied w.r.t. a non-monotonic ordering $\succ$ and we use full demodulation w.r.t. $\succ_r$, we can lose completeness as shown in Example 4 of Sect. 5.

The reason for adding these blockings comes from our completeness proof, which is based on the model generation technique of [BG94] and its variant used in [BGNR03]. In the latter, in contrast to the former, the ordering used for orienting the equations in the inference system and the ordering used for induction in the completeness proof do not coincide. Concretely, equations are oriented w.r.t. a non-monotonic ordering $\succ$, whereas completeness is proved by induction w.r.t. $\xrightarrow{+}_R$, where R is the limit ground rewrite system that defines the model. This is the key for the completeness proof of ordered paramodulation with non-monotonic orderings given in [BGNR03]. Since completeness is proved by induction w.r.t. $\xrightarrow{+}_R$, the redundancy notions should be defined w.r.t. $\xrightarrow{+}_R$ as well, instead of w.r.t. $\succ$, which is what would be desirable in principle. Unfortunately, $\xrightarrow{+}_R$ is unknown during the saturation process, and it is not clear how it can be approximated sufficiently. This is why in [BGNR03] it was left open to what extent demodulation could be applied (although some practical redundancy notions like tautology deletion and subsumption were shown to preserve completeness).

Here our aim is to add for instance demodulation w.r.t. the reduction ordering $\succ_r$, which is included in the (possibly) non-monotonic extension $\succ$. The idea for doing that is, roughly, to find a well-founded ordering $\succ_i$, which combines as much as possible $\succ_r$ with $\xrightarrow{+}_R$, and then prove completeness by induction w.r.t. $\succ_i$. The problem is that, although $\succ_r$ is a well-founded and monotonic ordering and R is a terminating TRS whose rules are included in a well-founded extension $\succ$ of $\succ_r$, the relation $\rightarrow_R \cup \succ_r$ is not well-founded in general and, in fact, $\xrightarrow{+}_R$ can even contradict $\succ_r$. Take e.g. $f(b) \succ_r f(a)$ and a well-founded extension $\succ$ of $\succ_r$ such that $f(b) \succ f(a) \succ a \succ b$. Then possibly $a \rightarrow b \in R$ (since the rules in R are oriented w.r.t. $\succ$) and hence the infinite sequence $f(a) \rightarrow_R f(b) \succ_r f(a) \cdots$ can be built.

The idea to circumvent this non-well-foundedness problem is to *block* the "bad" right-hand sides, i.e., the ones that can be introduced by a rewriting step with R which is not included in $\succ_r$. Then we can combine $\succ_r$ with $\rightarrow_R$ if the comparisons with $\succ_r$ cannot take into account the blocked terms. For instance, in the previous example, b should be blocked, and then $f(b)$ with b blocked is no longer greater than $f(a)$.

A convenient way to represent terms with blocked positions is by means of superindexed subterms, also called *marked terms*. For example $f(b^x)$, where x is a variable, denotes the term $f(b)$ where the subterm b is blocked. Then, although for performing inferences $f(b^x)$ still corresponds to the term $f(b)$, for the redundancy notions it will be seen as $f(x)$ with a blocked term b in x. In this context $f(b^x)$ cannot be simplified with $f(b) \rightarrow f(a)$, but it could still be simplified with $f(x) \rightarrow g(x)$ into $g(b^x)$, if $f(x) \succ_r g(x)$. Marked terms resemble the term closures of [BGLS95]. However, their semantics is fairly different: here the blockings only have effect on the redundancy notions, i.e., the ordered paramodulation inferences are applied at both blocked and unblocked positions.

Detection of terms that have to be marked and introduction of marks in the equations will be part of our inference system. The terms to be marked come from the equations that cannot be oriented with $\succ_r$ (since we do not know in advance which equations will be part of R). These terms are kept in a set, which is used for introducing marks in the equations. Hence, if there are many equations that cannot be handled by the reduction ordering then many terms will become blocked and we will have almost no redundancy, but the same amount of inferences by paramodulation as in [BGNR03]. On the other hand, if many equations can be handled by the reduction ordering, then few terms will be blocked and we will have as much redundancy as in the paramodulation calculi for total reduction orderings.

The paper is structured as follows. Preliminaries are presented in Sect. 2. Marked terms and orderings on marked terms are defined in Sect. 3. In Sect. 4, our calculus, including the notions of redundancy for inferences, is presented for deduction with sets of equations. In Sect. 5 some practical notions of redundancy are presented. In Sect. 6 we briefly outline how, from our results, a Knuth-Bendix completion procedure for finding convergent TRSs can be obtained. Finally in Sect. 7 we give some conclusions. Due to lack of space, we have omitted many details and all proofs (see [BR04] for the long version of this paper).

2   Preliminaries

We use the standard definitions of [NR01]. $T(\mathcal{F}, \mathcal{X})$ $(T(\mathcal{F}))$ is the set of (ground) terms over a set of symbols $\mathcal{F}$ and a denumerable set of variables $\mathcal{X}$ (over $\mathcal{F}$). The subterm of t at *position* p is denoted by $t|_p$, the result of replacing $t|_p$ by s in t is denoted by $t[s]_p$, and syntactic equality of terms is denoted by $\equiv$. A *substitution* is a partial mapping from variables to terms. The application of a substitution σ to a term t is denoted by $t\sigma$. An *equation* is a multiset of terms $\{s, t\}$, denoted $s \simeq t$ or, equivalently, $t \simeq s$.

If $\rightarrow$ is a binary relation, then $\leftarrow$ is its inverse, $\leftrightarrow$ is its symmetric closure, $\xrightarrow{+}$ is its transitive closure and $\xrightarrow{*}$ is its reflexive-transitive closure. If $s \xrightarrow{*} t$ and there is no t' such that $t \rightarrow t'$ then t is called *irreducible* and a *normal form* of s (w.r.t. $\rightarrow$). The relation $\rightarrow$ is *well-founded* or *terminating* if there exists no infinite sequence $s_1 \rightarrow s_2 \rightarrow \ldots$ and it is *confluent* or *Church-Rosser* if the relation $\xleftarrow{*} \circ \xrightarrow{*}$ is contained in $\xrightarrow{*} \circ \xleftarrow{*}$. A relation $\rightarrow$ on terms is *monotonic* if $s \rightarrow t$ implies $u[s]_p \rightarrow u[t]_p$ for all terms s, t and u and positions p. An *equivalence relation* is a reflexive, symmetric and transitive binary relation. A *congruence* is a monotonic equivalence relation.

A *rewrite rule* is an ordered pair of terms (s,t), written $s \rightarrow t$, and a set of rewrite rules R is a *term rewrite system* (TRS). The *rewrite relation* with R on $T(\mathcal{F}, \mathcal{X})$, denoted by $\rightarrow_R$, is the smallest monotonic relation such that for all rules $l \rightarrow r$ in R, and substitutions σ, $l\sigma \rightarrow_R r\sigma$. If $s \rightarrow_R t$ then we say that s *rewrites into* t with R. R is called terminating, confluent, etc. if $\rightarrow_R$ is. It is called *convergent* if it is confluent and terminating. The congruence $\leftrightarrow_R$ defines an *equality* Herbrand interpretation where $\simeq$ is interpreted by $s \simeq t$ iff $s \leftrightarrow_R t$. Such an interpretation will be denoted by R^*.

A (strict partial) *ordering* $\succ$ on $T(\mathcal{F}, \mathcal{X})$ is an irreflexive and transitive binary relation. It is *stable under substitutions* if $s \succ t$ implies $s\sigma \succ t\sigma$ for all substitutions σ. Monotonic orderings that are stable under substitutions are called *rewrite orderings*. A *reduction ordering* is a well-founded rewrite ordering.

A rewrite system terminates if, and only if, its rules are contained in a reduction ordering. In fact, if R is a terminating TRS, then $\xrightarrow{+}_R$ is a reduction ordering. However in restricted cases checking the rules with a well-founded ordering is enough:

Lemma 1. *[BGNR03] Let $\succ$ be a well-founded ordering, and let R be a ground TRS such that, for all $l \rightarrow r$ in R, $l \succ r$ and r is irreducible by R at non-topmost positions. Then R is terminating.*

The multiset extension of an equivalence relation $\sim$ is denoted by $\sim^{mul}$. The multiset extension of $\succ$ with respect to $\sim$ is defined as the smallest ordering $\succ^{mul}$ on multisets of elements such that $M \cup \{s\} \succ^{mul} N \cup \{t_1, \ldots, t_n\}$ if $M \sim^{mul} N$ and $s \succ t_i$ for all i in $1 \ldots n$. If $\succ^{mul}$ is used without explicitly indicating which is the equivalence relation $\sim$ then $\sim$ is assumed to be the syntactic equality relation. If $\succ$ is well-founded on a set S, so is $\succ^{mul}$ on finite multisets over S.

There are some orderings that play a central role in our results (in particular, in the redundancy notions). The *subsumption* relation, denoted by $\geq$, is defined as $s \geq t$ if s is an instance of t, i.e., $s \equiv t\sigma$ for some σ. The equivalence relation associated with $\geq$ is denoted by $\doteq$. Note that if $s \doteq t$ then t is a variable renamed version (or a *variant*) of s. The *encompassment* relation, denoted by $\trianglerighteq$, is defined as $s \trianglerighteq t$ if a subterm of s is an instance of t, i.e., $s \trianglerighteq t\sigma$ for some σ; therefore, encompassment is the composition of the subterm and the subsumption relations. We have that, if $\succ_r$ is a reduction ordering, then $\triangleright \cup \succ_r$ is well-founded.

It is said that an ordering fulfills the *subterm property* if $\succ \supseteq \triangleright$, where $\triangleright$ denotes the strict subterm relation. A *west ordering* [BGNR03] is a well-founded ordering on $T(\mathcal{F})$ that fulfills the subterm property and that is total on $T(\mathcal{F})$ (it is called *west* after well-founded, subterm and total). Every well-founded ordering can be totalized [Wec91], and hence every well-founded ordering satisfying the subterm property can be extended to a west ordering. We also have that every reduction ordering can be extended to a west ordering [BGNR03].

3 Marked Terms

Marked terms resemble the closures of [BGLS95]. A *term closure*, following the definition of [BGLS95], is a pair $s \cdot \gamma$ consisting of a term s (the *skeleton*) and a substitution γ from variables to terms. For example, $f(x, g(y)) \cdot \{x \mapsto a, y \mapsto h(z)\}$ is a closure with skeleton $f(x, g(y))$ and substitution $\{x \mapsto a, y \mapsto h(z)\}$. We generalize closures in that the substitution can map a variable not only to a term but also to a closure. For example, $f(x, g(y)) \cdot \{x \mapsto a, y \mapsto h(z) \cdot \{z \mapsto b\}\}$ is a generalized term closure, called a *marked term*. Moreover, closures are used in [BGLS95] to express that certain positions are blocked for inferences, whereas in our case they are blocked only for redundancy purposes. In order to ease the reading, sometimes we will denote marked terms by superindexing the "marked" subterms with variables. For example,

$$f(x, g(y)) \cdot \{x \mapsto a, y \mapsto h(z) \cdot \{z \mapsto b\}\} \text{ will be written as } f(a^x, g(h(b^z)^y)).$$

An equation between marked terms is called a *marked equation*. Given a substitution γ of the form $\{x_1 \mapsto t_1 \cdot \gamma_1, \dots, x_n \mapsto t_n \cdot \gamma_n\}$, the domain of γ, denoted by $Dom(\gamma)$, is defined as the set of variables $\{x_1, \dots, x_n\}$, and the range of γ, denoted by $Ran(\gamma)$, is defined as the set of marked terms $\{t_1 \cdot \gamma_1, \dots, t_n \cdot \gamma_n\}$. The variables occurring in $Dom(\gamma)$ are called *marking variables*. We will restrict marked terms to those which fulfill that marking variables occur at most once in the skeleton. For example $f(x, y) \cdot \{x \mapsto a, y \mapsto a\}$ is a valid marked term, but $f(z, z) \cdot \{z \mapsto a\}$ is not. Therefore, although we generalize closures in one sense, in another sense we restrict them.[2]

The congruence $\equiv$ on marked terms is defined as the monotonic, reflexive, symmetric and transitive closure including (i) $t^{xy} \equiv t^x$, and (ii) $s[t^x]_p \equiv s[t^y]_p$ if neither x nor y occur in $s[t]_p$.[3] Note that if $s \cdot \gamma \equiv t \cdot \delta$, then $s \doteq t$, i.e., s is a variant of t.

Let $s \cdot \{x_1 \mapsto t_1 \cdot \gamma_1, \dots, x_n \mapsto t_n \cdot \gamma_n\}$ be a marked term and σ be a substitution of the form $\{y_1 \mapsto u_1 \cdot \delta_1, \dots, y_m \mapsto u_m \cdot \delta_m\}$, such that the variables in the domains of all substitutions are distinct. Then $(s \cdot \gamma)\sigma$ is defined as

$$s\{y_1 \mapsto u_1, \dots, y_m \mapsto u_m\} \cdot (\{x_1 \mapsto (t_1 \cdot \gamma_1)\sigma, \dots, x_n \mapsto (t_n \cdot \gamma_n)\sigma\} \cup \delta_1 \cup \dots \cup \delta_m),$$

---

[2] Both the generalization and the restriction are necessary for our orderings on marked terms being well-founded.

[3] There is no condition that prevents terms like t^{xy} from appearing. For example, a marked variable z in e.g. z^y may be instantiated with a marked term t^x, giving t^{xy}. However, we will assume that these variable-variable bindings are instantiated whenever they arise, i.e. that terms are kept in its simpler form (t^x in the example).

where the marking variables of the resulting skeleton are renamed if necessary to keep linearity. For example, $f(x, g(x)^z)\{x \mapsto a^y\} = f(a^y, g(a^y)^z)$, and $f(x, x)\{x \mapsto a^y\} = f(a^y, a^z)$.

By $Forget(s \cdot \gamma)$ we denote the term in $T(\mathcal{F}, \mathcal{X})$ obtained from $s \cdot \gamma$ by repeatedly applying all the substitutions. For example, $Forget(f(a^y, g(a^y)^z))$ is $f(a, g(a))$. A marked term $s \cdot \gamma$ is said to be ground if $Forget(s \cdot \gamma)$ is a ground term. If $(s \cdot \gamma)\sigma$ is a ground marked term, then σ is called a *ground substitution* for $s \cdot \gamma$, and $(s \cdot \gamma)\sigma$ is called a *ground instance* of $s \cdot \gamma$. This notation is extended to sets of terms and equations in the usual way.

The usual concepts of (and notation for) position, subterm and subterm replacement can be easily extended for marked terms. For example, $f(x, g(x)^z)|_2 = g(x)^z$, and $f(x, g(x)^z)[a^y]_{2 \cdot 1} = f(x, g(a^y)^z)$.

From these definitions, the notion of rewriting for terms can be extended to marked terms in the natural way, as well as the concept of unifier: a unifier of two marked terms $s \cdot \gamma$ and $t \cdot \delta$ is a substitution σ such that $(s \cdot \gamma)\sigma \equiv (t \cdot \delta)\sigma$. There exists a unique (up to renaming of variables) most general unifier between marked terms, which can be obtained by straightforwardly adapting any known unification algorithm for terms.

3.1 Orderings on Marked Terms

Let $\succ_r$ be a reduction ordering, $\succ$ be a west ordering including $\succ_r$, and R be a terminating ground TRS. In this section we define the two main orderings on marked terms, namely $\succ_{red}$, which will be used in the redundancy notions and will be closely related to $\succ_r$, and $\succ_i$, which will be used for induction in the completeness proof and will combine as much as possible $\succ_{red}$ with $\xrightarrow{+}_R$. For the reasons explained in the introduction, it is crucial for the completeness proof that $\succ_{red} \subseteq \succ_i$.

We first define the marking ordering $\succ_m$. It only applies to terms that just differ in the markings and it is similar to the subsumption ordering for terms, in the sense that a term is smaller if it has higher marks.

Definition 1. *Let $s \cdot \gamma$ and $t \cdot \delta$ be two marked terms. Then $s \cdot \gamma \succ_m t \cdot \delta$ iff $Forget(s \cdot \gamma) \equiv Forget(t \cdot \delta)$, $s \cdot \gamma \not\equiv t \cdot \delta$ and (i) $s > t$ or (ii) $s \doteq t$ and $Ran(\gamma) \succ_m^{mul} Ran(\delta)$.*

Since $\succ_{red}$ will be used in the redundancy notions, all its ingredients are known while performing paramodulation steps. Given two marked terms, $\succ_{red}$ roughly compares the skeletons w.r.t. the reduction ordering and the encompassment relation and, if the skeletons are the same and the terms only differ on the markings, then it compares the level of the marks in the substitution part.

Definition 2. *Let $s \cdot \gamma$ and $t \cdot \delta$ be two marked terms. Then $s \cdot \gamma \succ_{red} t \cdot \delta$ iff (i) $s \rhd \cup \succ_r t$ or (ii) $s \doteq t$ and $Ran(\gamma) \succ_m^{mul} Ran(\delta)$.*

Since $\succ_i$ will be used for induction in the completeness proof, it extends $\succ_{red}$ with an additional case which uses as ingredient a terminating rewrite system R that will be known only in the proof.

Definition 3. *Let $s \cdot \gamma$ and $t \cdot \delta$ be two marked terms. Then $s \cdot \gamma \succ_i t \cdot \delta$ iff (i) $s \rhd \cup \succ_r t$ or (ii) $s \doteq t$ and $Ran(\gamma) \succ_m^{mul} Ran(\delta)$ or (iii) $s \doteq t$ and $Forget(Ran(\gamma)) \xrightarrow{+}_R Forget(Ran(\delta))$.*

In what follows, we will consider $\succ_i$ and $\succ_{red}$ as the transitive closures of the orderings defined above.

Lemma 2. $\succ_i$ *is well-founded.*

Example 2. Here there are some examples of applications of $\succ_i$.

1. If $f(x) \succ_r g(x)$ for all x, then $f(a) \succ_i f(a^x) \succ_i g(a^x) \succ_i a^x$ by case (i) of the definition of $\succ_i$.
2. We have $h(g(f(a^x, a))^y) \succ_i h(g(f(a^x, a^z))^y) \succ_i h(g(f(a, a)^x)^y)$ by case (ii) of the definition of $\succ_i$.

4   Paramodulation with Equations

In this section we present an ordered paramodulation calculus for sets of equations. The calculus will work on both a set of marked equations E and a set of terms T, which contains the set of terms that are going to be marked.

In the following, let $\succ_r$ be a given reduction ordering and $\succ$ be a given west ordering including $\succ_r$. We assume that the non-marking variables of all equations are disjoint (if necessary, the variables can be renamed). On the other hand, to ease the reading, the marked terms of each equation are considered to have the same substitution (if necessary, the substitution can be extended and the marking variables of one of the two skeletons can be renamed).

Definition 4. *Let E be a set of marked equations and T be a set of terms, which satisfy the following condition: for every marked subterm $(s \cdot \gamma)^x$ occurring in E, we have that $Forget(s \cdot \gamma)$ is an instance of a term in T. The inference system $\mathcal{I}$ consists of the following two inference rules and a mark detection rule:*

Paramodulation:

$$\frac{l \cdot \delta \simeq r \cdot \delta \qquad s \cdot \gamma \simeq t \cdot \gamma}{(s \cdot \gamma[r \cdot \delta]_p \simeq t \cdot \gamma)\sigma} \quad if$$

1. $\sigma = mgu(l \cdot \delta, s \cdot \gamma|_p)$, the most general unifier of $l \cdot \delta$ and $s \cdot \gamma|_p$,
2. $Forget(s \cdot \gamma|_p)$ is not a variable,
3. $l \succ_r r$ or $r \cdot \delta$ is marked on top, and
4. for some ground substitution θ, we have $Forget((l \cdot \delta)\sigma\theta) \succ Forget((r \cdot \delta)\sigma\theta)$ and, if $p = \lambda$, then we also have $Forget((s \cdot \gamma)\sigma\theta) \succ Forget((t \cdot \gamma)\sigma\theta)$.

Marking:

$$\frac{s \cdot \gamma[u \cdot \delta]_p \simeq t \cdot \gamma}{(s \cdot \gamma[(u \cdot \delta)^x]_p \simeq t \cdot \gamma)\sigma} \quad if$$

1. $\sigma = mgu(Forget(u \cdot \delta), u')$ for some term u' in T,
2. u is not a variable, and
3. x is a fresh variable.

Mark Detection:

$$T \Longrightarrow T \cup \{Forget(r \cdot \delta)\} \qquad if$$

1. $l \cdot \delta \simeq r \cdot \delta \in E$,
2. $l \not\succ_r r$ and $r \cdot \delta$ is not marked on top, and
3. for some ground substitution θ, we have that $Forget((l \cdot \delta)\theta) \succ Forget((r \cdot \delta)\theta)$.

Some examples of applications of these rules can be found in Example 4 of Sect. 5.

4.1 Redundancy Notions: A Static View

Here we define some abstract redundancy notions for inferences, like the ones in [BG94] and [NR01]. Some practical redundancy notions fitting in these abstract notions are discussed afterwards, in Sect. 5.

Given a set of terms M, by $gnd_M(E)$ we denote the set of ground instances $e\theta$ of equations e in E such that θ is a substitution with marks in M. Given a well-founded ordering $>$ on marked equations, by $E^{<e}$ we denote the set of all d in E such that $e > d$. Now we define redundancy of inferences and mark detection rules. Some explanations are given below.

Definition 5. *Let π be an inference by $\mathcal{I}$ with premises $e_1, \ldots e_n$ and conclusion d. Then a* ground instance $\pi\theta$ *of the inference π is an inference by $\mathcal{I}$ with premises $e_1\theta, \ldots e_n\theta$ and conclusion $d\theta$ for some ground substitution θ.*

Let G be a set of ground marked equations. A ground inference by $\mathcal{I}$ with premises $e_1, \ldots e_n$ and conclusion d is redundant *in G if*

$$Forget(G^{\prec^{mul}_{red} e_n}) \cup Forget(G^{\preceq^{mul}_{red} d}) \models Forget(d).$$

Let E be a set of marked equations. An inference by $\mathcal{I}$ with premises $e_1, \ldots e_n$ and conclusion d is redundant *in E if, for all sets of terms M and ground substitutions θ with marks in M, $\pi\theta$ is redundant in $gnd_M(E)$.*

Let T be a set of terms. A mark detection rule $T \Longrightarrow T \cup \{t\}$ is redundant *in T if t is an instance of a term in T.*

Note that the set of terms M is universally quantified in the previous definition of redundancy for inferences. This is due to the fact that the set of terms T to be marked will change during the computation, and it is crucial to have a notion of redundancy which is preserved along the computation. Moreover, the set $G^{\preceq^{mul}_{red} d}$ may not be included in $G^{\prec^{mul}_{red} e_n}$ since, in general, the conclusion of an inference does not need to be smaller w.r.t. $\succ^{mul}_{red}$ than the rightmost premise.

Redundant inferences and mark detection rules are unnecessary, and therefore we are interested in computing the closure of the set of equations with respect to $\mathcal{I}$ up to redundancy.

Definition 6. *Let E be a set of equations and T be a set of terms. The pair $\langle E, T \rangle$ is saturated with respect to $\mathcal{I}$ if every inference by $\mathcal{I}$ is redundant in E and every mark detection rule is redundant in T.*

4.2   Model Generation

Given a set of marked equations E and a set of terms T, here we show how to build an equational model of E defined by means of a set of rules coming from ground instances of E and T. Since the model is defined inductively, we need to treat E and T in a uniform way. To this end, a new constant symbol $\bot$ is introduced and the west ordering is extended with $\bot$ as minimal element.

Let T_E denote the set of equations $t \simeq \bot$ for all t in T, and T_G denote the set of mark free ground instances of equations in T_E. The model will be defined inductively on the equations in T_G and $gnd_T(E)$ with respect to $\succ_{mod}$, a total well-founded ordering including: $(s \simeq t) \succ_{mod} (u \simeq v)$ if

1. $Forget(s \simeq t) \succ^{mul} Forget(u \simeq v)$.
2. $Forget(s \simeq t) \equiv^{mul} Forget(u \simeq v)$ and $(s \simeq t) \succ_m^{mul} (u \simeq v)$, which means that $u \simeq v$ has marks at higher positions than $s \simeq t$.

Now we define a rewrite system R_E from $gnd_T(E)$, which requires to define at the same time a rewrite system R_T from T_G.

Definition 7. *The rewrite system R_{ET} is the union of R_E and R_T. Let e be a ground (marked) equation in $gnd_T(E) \cup T_G$.*

1. *If e is an instance $(s \cdot \gamma)\theta \simeq (t \cdot \gamma)\theta$ in $gnd_T(E)$ of some equation $s \cdot \gamma \simeq t \cdot \gamma$ in E, then it generates the rule $(s \cdot \gamma)\theta \to (t \cdot \gamma)\theta$ in R_E if*
   a) $s \succ_r t$ *or* $t \cdot \gamma$ *is marked on top,*
   b) $Forget((s \cdot \gamma)\theta) \succ Forget((t \cdot \gamma)\theta)$,
   c) $(s \cdot \gamma)\theta$ *is irreducible by* $R_E^e \cup R_T^e$, *and*
   d) $(t \cdot \gamma)\theta$ *is irreducible by* $R_E^e \cup R_T^e$ *at non-topmost positions.*
2. *If e is an instance $t\theta \simeq \bot$ in T_G of an equation $t \simeq \bot$ in T_E, then it generates the rule $u \cdot \delta \to (u \cdot \delta)^x$ in R_T if $u \cdot \delta$ is the normal form of $t\theta$ w.r.t. R_T^e and $u \cdot \delta$ is irreducible by R_E^e,*

where R_E^e and R_T^e denote the set of rules in R_E and R_T respectively, generated by equations d such that $e \succ_{mod} d$.

The following lemma about irreducibility of left and right-hand sides of R_{ET} holds easily by construction.

Lemma 3. *For all rules $l \cdot \delta \to r \cdot \delta$ in R_{ET} we have that*

1. $r \cdot \delta$ *is irreducible by* R_{ET} *at non-topmost positions, and*
2. $l \cdot \delta$ *is irreducible by* $R_{ET} \setminus \{l \cdot \delta \to r \cdot \delta\}$.

Again by construction of R_{ET}, and by the fact that all marks correspond to instances of terms in T, we obtain the following.

Lemma 4. *Let $s \cdot \gamma$ be a marked subterm occurring, not as right-hand side, in some rule of R_E. Then $s \cdot \gamma$ is marked on top if, and only if, $Forget(s \cdot \gamma)$ is an instance of a term in T.*

Then, from Lemmas 3 and 4, we obtain the same irreducibility result for $Forget(R_E)$ as for R_{ET} since, roughly, forgetting the marks cannot cause reducibility (otherwise, we would already had reducibility in R_{ET}). From this irreducibility result, convergence of $Forget(R_E)$ follows.

Lemma 5. *$Forget(R_E)$ is convergent.*

Finally, we can prove that $Forget(R_E)^*$ is a model of $Forget(E)$.

Theorem 1. *If a pair $\langle E, T \rangle$ is saturated with respect to $\mathcal{I}$, then $Forget(R_E)^* \models Forget(E)$.*

The following is a sketch of the proof. Since we have that $Forget(R_E)$ is convergent[4], the proof is a straightforward adaptation of the one in [BG94, NR01], but where the induction is done w.r.t. $\succ_i^{mul}$, taking $R = Forget(R_E)$ (see Sect. 3.1 for the definition of $\succ_i$). A contradiction is derived from the existence of a minimal w.r.t. $\succ_i^{mul}$ ground instance $e\theta$ in $gnd_T(E)$ of an equation e in E, such that $R^* \not\models Forget(e\theta)$. As usual, we have that $e\theta$ has not generated a rule in R_E. If this is because $e\theta$ is reducible by R_E, then we have an inference by paramodulation and, if $e\theta$ is reducible by R_T, then we have an inference by marking and, in both cases, a contradiction is obtained from minimality of $e\theta$ and saturatedness of $\langle E, T \rangle$.

From $Forget(R_E)^* \models Forget(E)$ and convergence of $Forget(R_E)$, it follows that every (ground) equation following from $Forget(E)$ has a rewrite proof with $Forget(R_E)$. Then, roughly, every procedure that computes a saturated pair $\langle E, T \rangle$ with respect to $\mathcal{I}$ from an initial set of equations E_0, will eventually generate sufficient equations for a rewrite proof of any logical consequence of E_0.

Lemma 6. *Let $\langle E, T \rangle$ be a pair of marked equations and terms that is saturated with respect to $\mathcal{I}$, and $s \simeq t$ be an equation such that $Forget(E) \models s \simeq t$. Then for every ground instance $s\sigma \simeq t\sigma$ of $s \simeq t$ we have that $s\sigma \xrightarrow{*}_R u \xleftarrow{*}_R t\sigma$ for some term u, where R denotes $Forget(R_E)$.*

5   Practical Notions of Redundancy

In this section we provide sufficient conditions for proving redundancy of inferences, which includes simplification by rewriting for those rules included in the reduction ordering $\succ_r$.

Let E be a set of equations. By E_r we denote the set of rules $l \cdot \delta \to r \cdot \delta$ such that the equation $l \cdot \delta \simeq r \cdot \delta$ is in E and $l \succ_r r$. An inference with premises $e_1, \dots, e_n$ and conclusion d can be shown redundant in E in two steps:

---

[4] In fact, termination of the rewrite system suffices for the proof of this theorem, that is, confluence is not necessary here.

1. First, d is simplified by rewriting it into d' using marked rules $l \cdot \delta \to r \cdot \delta$ in E_r. In order to ensure that we are only using smaller (w.r.t. $\succ_{red}^{mul}$) instances of equations in E, we have the following sufficient conditions. If d is of the form $s \simeq t$, then (i) we can apply any rewriting step at a non-top position of the skeleton of s (resp. t) obtaining s' (resp. t'), and after that we can apply any rewriting step at any position of the skeleton of s' (resp. t'); (ii) we can apply a rewriting step at top position of s (resp. t) obtaining s' (resp. t') if the instance of the applied rule is smaller (w.r.t. $\succ_{red}^{mul}$) than e_n or smaller than or equal to d, and after that we can apply any rewriting step at any position of the skeleton of s' (resp. t'). Note that in all these cases we can show that for any ground instance of the inference, the ground instances of the applied rules are either smaller (w.r.t. $\succ_{red}^{mul}$) than the instance of e_n or smaller than or equal to the instance of d.

2. Second, we check whether $d_1, \ldots, d_n \models d'$ for some instances of equations in E s.t. $d' (\rhd_{sk} \cup \succ_m)^{mul} d_i$ for all i in $1 \ldots n$, where $s \cdot \gamma \rhd_{sk} t \cdot \delta$ if $t \cdot \delta$ occurs in $s \cdot \gamma$ at some position of its skeleton. This includes, among other cases, the case where d' is in E, the case where there is a version of d' with additional marks in E, and the case where d' is a tautology. If this situation does not apply, then we can add d' to E to make the inference redundant.

Example 3. In Example 1 of the introduction we showed how, by an ordered paramodulation with $\underline{f(f(x))} \simeq f(g(f(x)))$ on $f(f(y)) \simeq f(g(f(\underline{y})))$, we got the equation $f(f(f(x))) \simeq f(g(f(g(f(f(x))))))$, i.e. $f(f(f(z))) \simeq f(\overline{g(f(g(f(f(z)))))})$. In this case, $f(f(f(z)))$ could be rewritten in two steps with $f(f(x)) \to f(g(f(x)))$ into $f(g(f(g(f(z)))))$, and hence the conclusion of the inference could be simplified into a tautology. Observe that the first step is applied at a non-top position, obtaining $f(f(g(f(z))))$, and then the second step can be applied on top obtaining $f(g(f(g(f(z)))))$.

All previous results can be adapted in the usual way for showing *redundancy of clauses* (or equations, in this case), and not only *redundancy of inferences*. Roughly, an equation can be defined as redundant if it follows from smaller equations w.r.t. $\prec_{red}^{mul}$.

In the following example, we show how demodulation w.r.t. $\succ_r$ and redundancy with smaller equations w.r.t. the subterm relation can cause incompleteness if no blockings are introduced at all. This thus shows the need for some control or, in other words, that our blockings are not merely a technicality to prove completeness.

The example has been adapted to the equational case from an example on Horn clauses. In this case, to show that some existentially quantified equation e is a consequence of an equational theory E, a contradiction is derived after adding the negation of e to E. To this end, paramodulation and marking inferences also take place on the negation of e. These paramodulation inferences correspond, at the ground level, to the rewrite steps in the rewrite proof of an instance of e.

Example 4. Let $\succ_r$ be a reduction ordering such that $f(x,x) \succ_r f(a,b)$ and $f(x,x) \succ_r f(b,a)$. Then necessarily $a \not\succ_r b$ and $b \not\succ_r a$. Let E denote the following inconsistent set of equalities and disequalities:

1) $a \simeq b$

2) $f(x,x) \simeq f(a,b)$

3) $f(x,x) \simeq f(b,a)$

4) $f(x,x) \simeq f(b,b)$

5) $g(f(a,b)) \simeq g(g(f(b,a)))$

6) $x \not\simeq g(x)$

First of all, we will show how a contradiction can be derived under our inference system (which introduces the convenient blockings). Let $\succ$ be a west ordering which is an extension of $\succ_r$ and that includes

$$g(g(f(a,b))) \succ g(f(a,b)) \succ g(g(f(b,a))) \succ g(f(b,a))$$
$$\succ f(a,a) \succ f(b,b) \succ f(a,b) \succ f(b,a) \succ a \succ b.$$

Initially, the set T of terms that are going to be marked is empty. Then, since $a \succ b$ and $a \not\succ_r b$, the *mark detection* rule gives us $T = \{b\}$. After this, an inference by *marking* into 1 gives us

7) $a \simeq b^{x_0}$

Now, an inference by *paramodulation* with 7 into 5 is possible (recall that an inference by paramodulation with 1 into 5 is not possible since $a \simeq b$ does not fulfill $a \succ_r b$ and b is not marked on top), giving

8) $g(f(b^{x_1}, b)) \simeq g(g(f(b,a)))$

A new inference by *paramodulation* with 7 into 8 gives us

9) $g(f(b^{x_1}, b)) \simeq g(g(f(b, b^{x_4})))$

And, with two more inferences by *marking*, we get

10) $g(f(b^{x_1}, b^{x_2})) \simeq g(g(f(b, b^{x_4})))$

11) $g(f(b^{x_1}, b^{x_2})) \simeq g(g(f(b^{x_3}, b^{x_4})))$

It is worth noting that equation 1 follows from the smaller equation 7, equations 9 and 10 follow from the smaller equation 11, and equation 8 follows from the smaller equations 7 and 11. Hence, using *redundancy of clauses*, the equations 1, 8, 9 and 10 could be deleted.

Now recall that $\succ$ fulfills the subterm property. Therefore we have that $Forget(g(g(f(b^{x_3}, b^{x_4})))) \succ Forget(g(f(b^{x_1}, b^{x_2})))$, and hence an inference by *paramodulation* with 11 into 6 gives us a contradiction,

12) $g(f(b^{x_1}, b^{x_2})) \not\simeq g(f(b^{x_1}, b^{x_2}))$.

Now let us show that, if blockings were not introduced (in other words, if inferences by *paramodulation* with 1 instead of 7 were allowed), then refutation completeness would be lost. If an inference by *paramodulation* with 1 into 5 was possible, then it would give either $g(f(b,b)) \simeq g(g(f(b,a)))$ or $g(f(a,b)) \simeq g(g(f(b,b)))$, which could be simplified back into 5 respectively with 2 and 3, which are included in the reduction ordering. Moreover, inferences by *paramodulation* with 1 into 2 and 3 give us 4. And inferences by *paramodulation* between 2, 3 and 4 give us already existing equations or equations like $f(a,b) \simeq f(b,a)$, which follow from the smaller (w.r.t. the subterm relation) equation 1. Finally, since $g(f(a,b)) \succ g(g(f(b,a)))$, an inference by *paramodulation* with 5 into 6 gives us

$$7) \quad f(a,b) \not\simeq g(g(f(b,a)))$$

but, since $f(a,b) \simeq g(g(f(b,a)))$ does not follow from the positive equations, further inferences on 7 will not lead to a contradiction. Therefore the set E can be saturated without a contradiction being derived. □

6   Knuth-Bendix Completion

Let E be a set of equations and $\succ_r$ be a (possibly non-totalizable) reduction ordering. Then a *convergent TRS R for E and $\succ_r$* is a convergent TRS, logically equivalent to E, and such that $l \succ_r r$ for all its rules $l \to r$. The problem we deal with here is finding a convergent TRS for the given E and $\succ_r$ whenever it exists, and find it in finite time if it is finite.

The first procedure for this problem, not relying on the enumeration of all equational consequences, was presented in [BGNR99,BGNR03]. From our present results, that method can be improved by making it compatible with powerful redundancy notions, like simplification by rewriting. The proof (given in the long version of this paper [BR04]) is a straightforward adaptation of the one in [BGNR03] since, in the limit, an interreduced convergent TRS is generated and, hence, the same techniques can be applied.

7   Conclusions

We have proposed a complete ordered paramodulation calculus which works with non-monotonic orderings and is compatible with powerful redundancy notions. Our inference system works with a pair of orderings: a west ordering $\succ$ which is used for orienting the equations, and a reduction ordering $\succ_r$, included in $\succ$, which is used in the redundancy notions. Our method is based on adding some marks that block terms for redundancy (but not for inferences). These results extend easily to Knuth-Bendix completion and to Horn clauses, as shown in the long version of this paper [BR04]. We believe that these results extend to general clauses as well, but maybe some more mark detection rules need to be added.

As a final remark, we want to note that a way to obtain, in practice, the orderings $\succ$ and $\succ_r$, is to define the reduction ordering $\succ_r$ with the *monotonic semantic path orderings* [BFR00], since then it is directly included, by definition, in a *semantic path ordering* [KL80], which can be used as the west ordering $\succ$.

Acknowledgement. We would like to thank Robert Nieuwenhuis for his helpful comments on this work.

References

[BD94]   Leo Bachmair and Nachum Dershowitz. Equational inference, canonical proofs, and proof orderings. *Journal of the ACM*, 41(2):236–276, Feb. 1994.

[BDH86]   Leo Bachmair, Nachum Dershowitz, and Jieh Hsiang. Orderings for equational proofs. In *First IEEE Symposium on Logic in Computer Science (LICS)*, Cambridge, Massachusetts, USA, June 1986. IEEE Computer Society Press.

[BFR00]   C. Borralleras, M. Ferreira, and A. Rubio. Complete monotonic semantic path orderings. In David McAllester, editor, *Proceedings of the 17th Conference on Automated Deduction (CADE-17)*, volume 1831 of *LNAI*, Pittsburgh, USA, June 2000. Springer-Verlag.

[BG94]   Leo Bachmair and Harald Ganzinger. Rewrite-based equational theorem proving with selection and simplification. *Journal of Logic and Computation*, 4(3):217–247, 1994.

[BG98]   Leo Bachmair and Harald Ganzinger. Equational reasoning in saturation-based theorem proving. In W. Bibel and P. Schmitt, editors, *Automated Deduction: A Basis for Applications*. Kluwer, 1998.

[BGLS95]   L. Bachmair, H. Ganzinger, Chr. Lynch, and W. Snyder. Basic paramodulation. *Information and Computation*, 121(2):172–192, 1995.

[BGNR99]   Miquel Bofill, Guillem Godoy, Robert Nieuwenhuis, and Albert Rubio. Paramodulation with non-monotonic orderings. In *14th IEEE Symposium on Logic in Computer Science (LICS)*, Trento, Italy, July 1999.

[BGNR03]   Miquel Bofill, Guillem Godoy, Robert Nieuwenhuis, and Albert Rubio. Paramodulation and Knuth-Bendix Completion with Nontotal and Nonmonotonic Orderings. *Journal of Automated Reasoning*, 30(1):99–120, 2003.

[BR02]   Miquel Bofill and Albert Rubio. Well-foundedness is sufficient for completeness of ordered paramodulation. In Andrei Voronkov, editor, *Proceedings of the 18th Conference on Automated Deduction (CADE-18)*, volume 2392 of *LNAI*, Copenhagen, Denmark, July 2002. Springer-Verlag.

[BR04]   M. Bofill and A. Rubio. Redundancy notions for paramodulation with non-monotonic orderings. Available at www.lsi.upc.es/~albert/papers.html.

[HR91]   J. Hsiang and M Rusinowitch. Proving refutational completeness of theorem proving strategies: the transfinite semantic tree method. *Journal of the ACM*, 38(3):559–587, July 1991.

[KL80]   S. Kamin and J.-J. Levy. Two generalizations of the recursive path ordering. Unpublished note, Dept. of Computer Science, Univ. of Illinois, 1980.

[NR01]   Robert Nieuwenhuis and Albert Rubio. Paramodulation-based theorem proving. In J.A. Robinson and A. Voronkov, editors, *Handbook of Automated Reasoning*, volume 1, chapter 7, pages 372–444. Elsevier Science Publishers and MIT Press, 2001.

[Wec91]   W. Wechler. *Universal Algebra for Computer Scientists*, volume 25 of *EATCS Monographs on Theoretical Computer Science*. Springer-Verlag, Berlin, 1991.

A Resolution Decision Procedure for the Guarded Fragment with Transitive Guards

Yevgeny Kazakov and Hans de Nivelle

MPI für Informatik, Saarbrücken, Germany
{ykazakov|nivelle}@mpi-sb.mpg.de

Abstract. We show how well-known refinements of ordered resolution, in particular redundancy elimination and ordering constraints in combination with a selection function, can be used to obtain a decision procedure for the guarded fragment with transitive guards. Another contribution of the paper is a special scheme notation, that allows to describe saturation strategies and show their correctness in a concise form.

1 Introduction

The *guarded fragment* $\mathcal{GF}$ of first order logic has been introduced by Andréka, van Benthem & Németi (1998) to explain and generalize the good computational properties of modal and temporal logics. This is achieved essentially by restricting quantifications in first order formulae to the following "bounded" forms: $\forall \overline{x}.[G \to F]$ and $\exists \overline{y}.[G \wedge F]$, where G should be an atomic formula (so-called *guard*) containing all free variables of F. The guarded fragment is decidable in 2EXPTIME (Grädel 1999) and inherits many other nice computational properties from the modal logics like the finite model property, the interpolation property and invariance under an appropriate notion of bisimulation.

Many extensions of the guarded fragment have been found to capture the known formalisms: the *loosely guarded fragment* has been introduced by van Benthem (1997) to capture the until operator in temporal logics; Grädel & Walukiewicz (1999) have extended the guarded fragment by fixed-point constructors to capture the modal mu-calculus. All these extensions of the guarded fragment, however, cannot express the *transitivity axiom:* $\forall xyz.(xTy \wedge yTz \to xTz)$. Transitivity is important, since it is used to model discrete time (in temporal verification) and ordered structures (in program shape analysis). The question of whether transitivity can be safely integrated into the guarded fragment was answered negatively by Grädel (1999). He proved that the guarded fragment becomes undecidable as long as transitivity is allowed. This result was later sharpened by Ganzinger, Meyer & Veanes (1999) who showed that even the *two-variable* guarded fragment $\mathcal{GF}^2$ with transitivity is undecidable. The same paper, however, presents the first restriction of the guarded fragment, where transitivity can be allowed without loss of decidability. In this, so-called *monadic* $\mathcal{GF}^2$, binary relations are allowed to occur as guards only. The paper poses two natural questions: **(i)** Does $\mathcal{GF}$ remain decidable if *transitive* predicates are admitted

D. Basin and M. Rusinowitch (Eds.): IJCAR 2004, LNAI 3097, pp. 122–136, 2004.
© Springer-Verlag Berlin Heidelberg 2004

only as guards? (**ii**) What is the exact complexity of the monadic $\mathcal{GF}^2$? The first question was answered positively in (Szwast & Tendera 2001) where using a heavy model-theoretic construction, it was shown that the *guarded fragment with transitive guards* $\mathcal{GF}[\mathcal{TG}]$ is decidable in 2EXPTIME. Kieroński (2003) has proved the matching 2EXPTIME lower bound for the *monadic* $\mathcal{GF}^2$ with transitivity, answering hereby the second question.

A practical disadvantage of procedures based on enumeration of structures, like the one given for $\mathcal{GF}[\mathcal{TG}]$ in (Szwast & Tendera 2001), is that without further optimizations, those methods exhibit the full worst-case complexity. Resolution-based approach, is a reasonable alternative to model-theoretic procedures, as its goal-oriented nature and numerous refinements allow to scale well between "easy" and "hard" instances of problems. In this paper we demonstrate the practical power of resolution refinements, such as redundancy elimination and usage of ordering constraints in combination with selection function. We present a first resolution-based decision procedure for $\mathcal{GF}[\mathcal{TG}]$. Another aspect that is demonstrated in our paper is the usage of resolution as a specification language for decision procedures. We introduce a special scheme notation that allows to describe resolution strategies in a concise form. This may provide a formal foundation for using resolution for specifying decision procedures and proving their correctness.

2 Preliminaries

We shall use a standard notation for first-order logic clause logic. An *expression* is either a term or a literal. A *literal symbol* l is either a or $\neg a$, where a is a predicate symbol. An *expression symbol* e is either a functional symbol f or a literal symbol l. We write literals and expressions using literal symbols and expression symbols as follows: $L = l(t_1,...,t_n)$, $E = e(t_1,...,t_n)$. As usual, a clause is a disjunction of literals $C = L_1 \vee \cdots \vee L_n$. The empty clause is denoted by $\square$. We use the shortcuts ⋈ for conjunction or disjunction and $\overline{x}$ for some vector of variables.

The *depth of an expression* $dp(E)$ is recursively defined as follows: (**i**) $dp(x) := 0$; (**ii**)$dp(e(t_1,...,t_n)) := \max\{0, dp(t_1), ..., dp(t_n)\} + 1$. The *depth of the clause* $C = L_1 \vee \cdots \vee L_n$ is $dp(C) := \max\{0, dp(L_1), ..., dp(L_n)\}$. The *width of a formula* $wd(F)$ is the maximal number of free variables in subformulas of F.

2.1 The Framework of Resolution Theorem Proving

For describing the decision procedures we use the well-known ordered resolution calculus with selection $\mathcal{OR}^{\succ}_{Sel}$ enhanced with additional simplification rules. Our presentation of the calculus is very close to (Bachmair & Ganzinger 2001). The ordered resolution calculus $\mathcal{OR}^{\succ}_{Sel}$ is parametrized by an admissible ordering $\succ$ and a selection function Sel. A partial ordering $\succ$ on atoms is *admissible* (for $\mathcal{OR}^{\succ}_{Sel}$) if (**i**) $\succ$ is *liftable*: $A_1 \succ A_2$ implies $A_1\sigma \succ A_2\sigma$ for any substitution σ

and $(ii) \succ$ is a total reduction ordering on ground atoms. Although resolution remains complete for a much wider class of orderings, admissible orderings are better understood and widely used in existing theorem provers. Examples of admissible orderings are the *recursive path ordering with status RPOS* and the *Knuth-Bendix ordering KBO*.

The ordering $\succ$ is extended on literals by comparing $L = A$ as the multiset $\{A\}$ and $L = \neg A$ as the multiset $\{A, A\}$. The ordering on clauses is the multiset extension of the ordering on literals. Given a clause C, we say that a literal $L \in C$, is *maximal in C* if there is no L' in C, with $L' \succ L$. A *selection function Sel* assigns a set of negative literal to every clause, which we call *selected literals*. A literal L is *eligible* in a clause C if it is either selected: $L \in Sel(C)$, or otherwise nothing is selected and L is maximal in C.

The ordered resolution calculus $\mathcal{OR}^{\succ}_{Sel}$ consists of two inference rules below. We mark eligible literals with "star" and underline the expressions to be unified:

Ordered (Hyper-)Resolution

$$HR: \frac{C_1 \vee \underline{A_1}^* \ldots C_n \vee \underline{A_n}^* \quad D \vee \underline{\neg B_1}^* \vee \ldots \vee \underline{\neg B_n}^*}{C_1\sigma \vee \ldots \vee C_n\sigma \vee D\sigma} \left| \begin{array}{l} \text{where (i) } \sigma = mgu(A_i, B_i), \text{ (ii) } A_i \\ \text{and } \neg B_i \text{ are eligible } (1 \leq i \leq n). \end{array} \right.$$

Ordered Factoring

$$OF: \frac{C \vee \underline{A}^* \vee \underline{A'}}{C\sigma \vee A\sigma} \left| \begin{array}{l} \text{where (i) } \sigma = mgu(A, A'), \text{ (ii) } A \\ \text{is eligible.} \end{array} \right.$$

The conventional **Ordered Resolution** rule OR, is a partial case of the ordered (hyper-)resolution rule when $n = 1$. The calculus $\mathcal{OR}^{\succ}_{Sel}$ is refutationally complete for any choice of an admissible ordering $\succ$ and a selection function Sel. Moreover, the calculus is compatible with a general notion of redundancy which allows to make use of additional simplification rules.

A ground clause C is called *redundant* w.r.t. a set of the ground clauses N if C follows from the set $N_{\prec C}$ of the clauses from N that are smaller than C. A non-ground clause C is redundant w.r.t. N if every ground instance $C\sigma$ of C is redundant w.r.t. the set N^{gr} of all ground instances of N. A *ground inference $S \vdash C$* from the clause set S is called *redundant* w.r.t. a clause set N if its conclusion C follows from the set $N^{gr}_{\prec max(S)}$, where $max(S)$ is the maximal clause from S. A non-ground *inference $S \vdash C$* is redundant w.r.t. N if every ground instance $S\sigma \vdash C\sigma$ of the inference is redundant w.r.t. N. A clause set N is *saturated up to redundancy* if the conclusion of every non-redundant w.r.t. N inference from N is contained in N.

Theorem 1. (Bachmair & Ganzinger 2001) *Let N be a clause set that is saturated up to redundancy in $\mathcal{OR}^{\succ}_{Sel}$. Then N is satisfiable iff N does not contain the empty clause.*

For our decision procedures we do not need the full power of redundancy but rather additional simplification rules. A (non-deterministic) inference rule $S \vdash S_1 \parallel S_2 \cdots \parallel S_k$ producing one of the clause sets S_i from the clause set S is called *sound* if every model of S can be extended to a model for some S_i with $1 \leq i \leq k$. Additionally, if *every* set S_i makes some clause from S redundant, the rule is called a *simplification* rule.

Given a set of clauses N, a theorem prover based on ordered resolution non-deterministically computes a *saturation* of N by adding conclusions of inference rules to N and marking[1] redundant clauses as *deleted* so that they do not participate in further inferences. If the process terminates without deriving the empty clause $\Box$, then a set of the clauses $\mathcal{OR}^{\succ}_{Sel}(N)$ is computed that is saturated in $\mathcal{OR}^{\succ}_{Sel}$ up to redundancy. Theorem 1 then implies that the clause set N is satisfiable, since only satisfiability preserving transformations $N \Rightarrow \cdots \Rightarrow \mathcal{OR}^{\succ}_{Sel}(N)$ were applied to N. Note that termination of a saturation process is a key issue of using resolution as a decision procedure. If any application of inference rules is *a priori* guaranteed to terminate for a clause set N then satisfiability of N can be decided in finite time by enumerating all possible saturations.

In our paper we use the following simplification rule:

Elimination of Duplicate Literals $ED: \dfrac{[\![\, C \vee D \vee D \,]\!]}{C \vee D}$

An additional simplification rule will be introduced later, when a certain class of orderings is considered. We indicate redundant premises of rules by enclosing them in double brackets. The simplification rules are applied *eagerly*, that is before any resolution or factoring inference is made. In particular, in the sequel we assume that no clause contain several occurrences of the same literal.

Constraint clauses. The ordered resolution calculus on non-ground level is a directly lifted version of the calculus on the ground level: (*i*) each clause represents the set of its ground instances and (*ii*) whenever an inference is possible from the ground instances of some clauses, there should be a corresponding inference from the clauses themselves, that captures the result of the ground inference. This is due to the fact that $\mathcal{OR}^{\succ}_{Sel}$ is parametrized with a liftable ordering $\succ$ and does not use non-liftable conditions in inferences (like, say, in the paramodulation calculus, when paramodulation to a variable is not allowed). Therefore, in fact, any representation for sets of ground clauses can be admitted as long as the condition (*ii*) above holds. In our decision procedure we use *constraint clauses* of the form: $C \mid R$, where C is a (non-ground) clause and R is a set of *ordering constraints* of the form: $t \succ s$ or $t \succeq s$. Constraint clause $C \mid R$ represent the set of ground instances $C\sigma$ of C such that every constraint in $R\sigma$ is true. The ordered resolution calculus and all notions of redundancy can be straightforwardly adopted to be used with constraint clauses: instead of considering all substitutions (for determining a maximal literal, or showing redundancy) one should consider only substitutions satisfying the constraints. In particular, one could use different values for selection function for different constraint variants of the same clause.

---

[1] Clauses are not removed from the set to avoid repetition of generation/deletion of the same redundant clauses.

2.2   Schemes of Expressions and Clauses

To describe resolution-based decision procedures we have to reason about sets of clauses. We introduce a special notation that allows to represent sets of clauses in a compact form. We extend our vocabulary with additional symbols called *signature groups* that represent sets of functional symbols: *function groups*, predicate symbols: *predicate groups* or literal symbols: *literal groups*. We allow to use these symbols in expressions as usual functional and literal symbols and to distinguish them, we use small letters with a "hat" $\hat{g}$. For instance, if $\hat{f}_{all}$ denotes the set of all functional symbols, we write $\hat{f}_{all}(t)$ meaning a term of the form $f(t)$ where $f \in \hat{f}_{all}$ (the formal definition follows below). We adopt the following notation for referring to arguments of expressions. By writing $e\langle !t_1,...,!t_n, s_1,...,s_m \rangle$ we mean an expression starting with the expression symbol e, having all arguments $t_1,...,t_n$ and optional arguments $s_1,...,s_m$ (ordered in an arbitrary way). Formally, the set of *term schemes*, *literal schemes* and *clause schemes* are defined as follows:

$$\hat{T}m ::= \; x \mid \hat{f}(\hat{t}_1,...,\hat{t}_n) \mid \hat{f}\langle !\hat{t}_1,...,!\hat{t}_n, \hat{s}_1,...,\hat{s}_m \rangle, \; n \geq 0, m \geq 0.$$
$$\hat{L}t ::= \quad \hat{l}(\hat{t}_1,...,\hat{t}_n) \mid \hat{l}\langle !\hat{t}_1,...,!\hat{t}_n, \hat{s}_1,...,\hat{s}_m \rangle, \; n \geq 0, m \geq 0.$$
$$\hat{C}l ::= \; \hat{L} \mid !\hat{L} \mid \hat{C}_1 \vee \hat{C}_2.$$

where $\hat{f}$ is a functional group, $\hat{l}$ is a literal group, $\hat{t}_i, \hat{s}_j$ with $1 \leq i \leq n, 1 \leq j \leq m$ are term schemes, $\hat{L}$ is a literal scheme and $\hat{C}_1, \hat{C}_2$ are clause schemes. For convenience, we assume that every functional and literal symbol acts as a singleton group consisting of itself, so usual terms and clauses are term schemes and clause schemes as well.

Each term scheme $\hat{t}$, literal scheme $\hat{L}$ and clause scheme $\hat{C}$ represents a set $\langle \hat{t} \rangle$, $\langle \hat{L} \rangle$ and $\langle \hat{C} \rangle$ of terms, literals and clauses respectively, as defined below:

$$\langle \hat{T}m \rangle, \langle \hat{L}t \rangle : = \qquad \langle x \rangle : \{x\} \qquad\qquad\qquad\qquad\qquad\qquad\qquad |$$
$$\langle \hat{g}(\hat{t}_1,...,\hat{t}_n) \rangle : \{g(t_1,...,t_n) \mid g \in \hat{g}, \; t_i \in \langle \hat{t}_i \rangle, 1 \leq i \leq n\} \qquad |$$
$$\langle \hat{g}\langle !\hat{t}_1,...,!\hat{t}_n, \hat{s}_1,...,\hat{s}_m \rangle \rangle : \{g(h_1,...,h_k) \mid g \in \hat{g}, \; \{h_1,...,h_k\} \cap \langle \hat{t}_i \rangle \neq \emptyset, 1 \leq i \leq n,$$
$$\{h_1,...,h_k\} \subseteq \cup_{i=1}^{n} \langle \hat{t}_i \rangle \cup_{j=1}^{m} \langle \hat{s}_j \rangle\}.$$

$$\langle \hat{C}l \rangle = \qquad\qquad \langle \hat{L} \rangle : \{L_1 \vee \cdots \vee L_k \mid k \geq 0, \; L_i \in \langle \hat{L} \rangle, \; 1 \leq i \leq k\} \qquad |$$
$$\langle !\hat{L} \rangle : \{L_1 \vee \cdots \vee L_k \mid k \geq 1, \; L_i \in \langle \hat{L} \rangle, \; 1 \leq i \leq k\} \qquad |$$
$$\langle \hat{C}_1 \vee \hat{C}_2 \rangle : \{C_1 \vee C_2 \mid C_1 \in \langle \hat{C}_1 \rangle, \; C_2 \in \langle \hat{C}_2 \rangle\}.$$

We use the shortcuts $\hat{e}(., \overline{x}, .)$, $\hat{e}\langle ., \overline{x}, . \rangle$ and $\hat{e}\langle ., !\overline{x}, . \rangle$ where $\overline{x}$ is a vector $x_1,...,x_n$, to stand for $\hat{e}(., x_1,...,x_n, .)$, $\hat{e}\langle ., x_1,...,x_n, . \rangle$ and $\hat{e}\langle ., !x_1,...,!x_n, . \rangle$ respectively. We write $.. \vee \neg !\hat{A} \vee ..$ in clause schemes instead of $.. \vee !\neg \hat{A} \vee ..$, where $\hat{A}$ is either of the form $\hat{a}(...)$ or $\hat{a}\langle ... \rangle$. In fact we use variable vectors $\overline{x}$ and functional symbols without "hat" f as *parameters of clause schemes*. A clause scheme $\hat{C}(\overline{x}, f, ...)$ with parameters $\overline{x}, f, ...$ represents the union $\langle \hat{C} \rangle := \cup_\eta \langle C\eta \rangle$ for all substitutions η of vectors $x_1,...,x_n$ for $\overline{x}$, function symbols for f, etc.

Example 1. Suppose $\hat{a}$ is a predicate group consisting of all predicate symbols and $\hat{\alpha} := \{\hat{a}, \neg \hat{a}\}$ is a literal group consisting of all literal symbols.

Then the clause scheme $\hat{C} = \neg!\hat{a}\langle!\overline{x}\rangle \vee \hat{a}\langle!f(\overline{x}),\overline{x}\rangle$ has two parameters: $\overline{x}$ and f. Any clause $C \in \langle\hat{C}\rangle$ corresponds to some choice of these parameters $\overline{x} = x_1,...,x_n$, $f = f'$. The clause C should have a nonempty subset of negative literals containing all variables $x_1,...,x_n$ and no other arguments. Other literals of C should contain the subterm $f'(x_1,...,x_n)$ as an argument and possibly some variables from $x_1,...,x_n$. In particular, $\langle\hat{C}\rangle$ contains the clauses $\neg a(x,y,x) \vee b(y,f'(x,y))$, $\neg b(x,y) \vee \neg b(y,x)$ and $\neg p \vee \neg q(c,c)$, but not the clauses $\neg a(x,y,x) \vee b(f'(x,y),f'(y,x))$ or $\neg b(y,f'(x,y))$.

3 Deciding the Guarded Fragment by Resolution

In this section we demonstrate our technique by revisiting a resolution decision procedure for the guarded fragment without equality. The original procedure is due to (de Nivelle & de Rijke 2003). Resolution-based decision procedures (for an overview see Fermüller, Leitsch, Hustadt & Tammet 2001) usually consist of several main steps. First, a clause normal form transformation is applied to a formula of a fragment that produce *initial clauses*. Then a clause set containing the initial clauses is defined, that is shown later to be closed under inferences of the calculus. Decidability and complexity results follow from the fact that the defined clause class contains only finitely many different clauses over a fixed signature.

3.1 Clause Normal Form Translation

In order to describe the transformation to a *clause normal form* (CNF), it is convenient to use the *recursive definition* for the guarded fragment:

$$\mathcal{GF} ::= A \mid F_1 \vee F_2 \mid F_1 \wedge F_2 \mid \neg F_1 \mid \forall \overline{x}.(G \rightarrow F_1) \mid \exists \overline{x}.(G \wedge F_1).$$

where A is an atom, $F_i, i = 1,2$ are guarded formulas, and G is an atom called *the guard* containing all free variables of F_1. The translation of a guarded formula into CNF is done in two steps. First, the formula is transformed into *negation normal form* (NNF) in the standard way. Guarded formulas in NNF are defined by the following recursive definition:

$$[\mathcal{GF}]^{nnf} ::= (\neg)A \mid F_1 \vee F_2 \mid F_1 \wedge F_2 \mid \forall \overline{y}.(G \rightarrow F_1) \mid \exists \overline{y}.(G \wedge F_1).$$

Second, a so-called *structural transformation* is applied, that decomposes the formula by introducing *definitions* for all of its subformulae. We assume that to each subformula F' of F, a unique predicate $P_{F'} = p_{F'}(\overline{x})$ is assigned. Each predicate $P_{F'}$ has the arity equal to the number of free variables $\overline{x}$ of F'. Using the new predicates, the structural transformation can be defined as $\exists \overline{x}.P_F \vee [F]^{st}$, where $[F]^{st}$ is given below. In each row, $\overline{x}$ are the free variables of F.

$$[F]^{st}_g := [(\neg)A]^{st}_g : \forall \overline{x}.(P_F \rightarrow (\neg)A) \qquad\qquad\quad\; \mid\; \neg p_F(\overline{x}) \vee (\neg)a\langle\overline{x}\rangle$$

$$[F_1 \bowtie F_2]^{st}_g : \forall \overline{x}.(P_F \rightarrow [P_{F_1} \bowtie P_{F_2}]) \wedge [F_1]^{st}_g \wedge [F_2]^{st}_g \mid\; \neg p_F(\overline{x}) \vee p_{F_i}\langle\overline{x}\rangle\;[\vee\;p_{F_j}\langle\overline{x}\rangle]$$

$$[\forall \overline{y}.(G \rightarrow F_1)]^{st}_g : \forall \overline{x}.(P_F \rightarrow \forall \overline{y}.[G \rightarrow P_{F_1}]) \wedge [F_1]^{st}_g \mid\; \neg g\langle!\overline{x},!\overline{y}\rangle \vee \neg p_F(\overline{x}) \vee p_{F_1}\langle\overline{x},\overline{y}\rangle$$

$$[\exists y.F_1]^{st}_g : \forall \overline{x}.(P_F \rightarrow \exists y.P_{F_1}) \wedge [F_1]^{st}_g . \qquad\qquad \neg p_F(\overline{x}) \vee p_{F_1}\langle f(\overline{x}),!\overline{x}\rangle$$

The transformation unfolds a guarded formula according to its construction and introduces predicates and definitions for its guarded subformulae. A guarded formula F in negation normal form is satisfiable whenever $\exists \overline{x}. P_F \wedge [F]^{st}$ is: one can extend the model of F by interpreting the new predicate symbols according to their definitions. Every recursive call of the transformation contributes to a result with a conjunct describing a definition for an introduced predicate. Performing the usual skolemization and writing the result in a clause form, we obtain the clauses shown to the right of the definition for $[F]_g^{st}$. It is easy to see that the clauses for $P_F \wedge [F]_g^{st}$ fall into the set of clauses described by the following clause schemes:

$$1.\ \hat{a}\langle\hat{c}\rangle;$$
$$2.\ \neg!\hat{a}\langle!\overline{x}\rangle \vee \hat{\alpha}\langle f(\overline{x}),\overline{x}\rangle. \tag{G}$$

where the predicate group $\hat{a}$ consists of all (initial and introduced) predicate symbols and the literal group $\hat{\alpha}$ consists of all literal symbols.

3.2   Saturation of the Clause Set

The resolution calculus has two parameters that can be chosen: an admissible ordering and a selection function. These parameters should prevent clauses from growing during the inferences. We will set the ordering and selection function in such a way, that eligible literals would be (*i*) of maximal depth and (*ii*) contain all variables of the clause.

We assume that the ordering $\succ$ enjoys $L \succ K$ for $L \in \langle\hat{a}\langle!f(\overline{x}),\overline{x}\rangle\rangle$ and $K \in \langle\hat{a}\langle\overline{x}\rangle\rangle$, that is, any literal containing the functional symbol with all variables is greater then any other literal in the clause without functional subterms. This can be achieved by taking, say, any recursive path ordering $\succ_{rpos}$ on expressions with the precedence $>_P$ enjoying $f >_P p$ for any functional symbol f and predicate symbol p. We define the selection function Sel for the clauses without functional symbols to select a negative literal containing all variables of the clause if there is one.

We prove that the clause class from (G) is closed under the ordered resolution by making case analysis of possible inferences between clauses of this class. The complete case analysis is given below:

1 $\hat{\alpha}^*$		2 $\neg!\hat{g}\langle!\overline{x}\rangle \vee \hat{\alpha}\langle f(\overline{x}),\overline{x}\rangle$	
1.1 $\hat{a}\langle\hat{c}\rangle \vee \hat{a}\langle\hat{c}\rangle^*$	: OR.1	2.1 $\neg!\hat{g}\langle!\overline{x}\rangle \vee \hat{\alpha}\langle f(\overline{x}),\overline{x}\rangle \vee \hat{\alpha}\langle!f(\overline{x}),\overline{x}\rangle^*$	
1.2 $\hat{a}\langle\hat{c}\rangle \vee \neg\hat{a}\langle\hat{c}\rangle^*$	: OR.2	2.1.1 $\neg!\hat{g}\langle!\overline{x}\rangle \vee \hat{\alpha}\langle f(\overline{x}),\overline{x}\rangle \vee \hat{a}\langle!f(\overline{x}),\overline{x}\rangle^*$	: OR.1
1.3 $\hat{a}\langle\hat{c}\rangle \vee \hat{a}\langle\hat{c}\rangle^* \vee \hat{a}\langle\hat{c}\rangle$: OF		2.1.2 $\neg!\hat{g}\langle!\overline{x}\rangle \vee \hat{\alpha}\langle f(\overline{x}),\overline{x}\rangle \vee \neg\hat{a}\langle!f(\overline{x}),\overline{x}\rangle^*$	: OR.2
OR[1.1; 1.2]: $\hat{a}\langle\hat{c}\rangle$	: 1	2.1.3 $\neg!\hat{g}\langle!\overline{x}\rangle \vee \hat{\alpha}\langle f(\overline{x}),\overline{x}\rangle \vee \hat{a}\langle!f(\overline{x}),\overline{x}\rangle^* \vee \hat{a}\langle f(\overline{x}),\overline{x}\rangle$: OF	
OF[1.3] : $\hat{a}\langle\hat{c}\rangle \vee \hat{a}\langle\hat{c}\rangle$:1		OR[2.1.1; 2.1.2]: $\neg!\hat{g}\langle!\overline{x}\rangle \vee \hat{\alpha}\langle f(\overline{x}),\overline{x}\rangle$	: 2
		OF[2.1.3] : $\neg!\hat{g}\langle!\overline{x}\rangle \vee \hat{\alpha}\langle f(\overline{x}),\overline{x}\rangle \vee \hat{a}\langle!f(\overline{x}),\overline{x}\rangle$:2	
		2.2 $\neg\hat{g}\langle!\overline{x}\rangle^* \vee \neg!\hat{g}\langle!\overline{x}\rangle \vee \hat{\alpha}\langle\overline{x}\rangle$: OR.2	
		OR[1.1; 2.2] : $\hat{\alpha}$	: 1
		OR[2.1.1; 2.2]: $\neg!\hat{g}\langle!\overline{x}\rangle \vee \hat{\alpha}\langle f(\overline{x}),\overline{x}\rangle \vee \hat{\alpha}\langle f(\overline{x}),\overline{x}\rangle$:2	

The table is organized as follows. The clause schemes from (G) are spread in the table on different levels of precision. On the first level the schemes are given themselves. On the second level, different possibilities for eligible literals (marked by the asterisk) are considered. On the last level, possible inference rules that can be applied for a clause are identified and the expressions to be unified are underlined. For example, OR.1 marked to the right of the clause scheme 1.1 means that a clause represented by this scheme may act as a first premise of the ordered resolution rule. Below the last level, inferences between preceding clauses are drawn and their conclusions are identified as instances of clause schemes.

We have used the special form of literals in the clauses when the unifiers has been computed. For instance, the reason of why the resolution inference OR[2.1.1; 2.1.2] has produced the clause of the same depth is because the so-called *covering* expressions have been unified. An expression E is called *covering* if all functional subterms of E contain *all variables* of E. It is well known that the unifier for the two covering expressions is the renaming for the deepest of them:

Theorem 2. (Fermüller, Leitsch, Tammet & Zamov 1993) *Let E_1 and E_2 be two covering expressions with $dp(E_1) \geq dp(E_2)$ and let $\sigma = mgu(E_1, E_2)$. If $\overline{x} = free(E_1)$ then $\sigma : \overline{x} \to \overline{u}$ for some vector of variables $\overline{u}$. As a conclusion $dp(E_2\sigma) = dp(E_1\sigma) = dp(E_1)$.*

Theorem 3. (de Nivelle & de Rijke 2003) *Ordered resolution decides the guarded fragment in double exponential time.*

Proof. Given a formula $F \in \mathcal{GF}$ of size n, the structural transformation introduces at most linear number of new predicate symbols of arity not greater than n. Since every non-ground clause from (G) has a guard, the number of variables in such a clause does not exceed n. It can be shown that most $c = 2^{2^{O(n \log n)}}$ different clauses from (G) over the initial and introduced signature can be constructed. A saturation of the size c can be computed in time $O(c^2)$. So the resolution decision procedure for $\mathcal{GF}$ can be implemented in 2EXPTIME.   □

4   Deciding the Guarded Fragment with Transitivity

Some binary predicates of Σ, which we call *transitive predicates* have a *special* status. We usually denote them by the letters T, S and use the *infix* notation $(t_1 T t_2)$ rather than the *postfix* notation $a(t_1, t_2)$, as for the other predicates. For any group of transitive predicates $\hat{T} = \{T_1, \ldots, T_n\}$, the shortcuts $(x\hat{T}y)$ and $\neg(x\hat{T}y)$ represent respectively the disjunctions $(xT_1y) \vee \cdots \vee (xT_ny)$ and $\neg(xT_1y) \vee \cdots \vee \neg(xT_ny)$. We assume that every set of clauses N contains the *transitivity clause:* $\neg(xTy) \vee \neg(yTz) \vee xTz$ for every transitive predicate T.

The *guarded fragment with transitive guards* $\mathcal{GF}[\mathcal{TG}]$ is defined by:

$$\mathcal{GF}[\mathcal{TG}] ::= \mathsf{A} \mid \mathsf{F}_1 \vee \mathsf{F}_2 \mid \mathsf{F}_1 \wedge \mathsf{F}_2 \mid \neg\mathsf{F}_1 \mid \forall\overline{x}.(\mathsf{G} \to \mathsf{F}_1) \mid \exists\overline{x}.(\mathsf{G} \wedge \mathsf{F}_1).$$

where $\mathsf{F}_i, i = 1, 2$ are from $\mathcal{GF}[\mathcal{TG}]$, A is a non-transitive atom and G is a (possibly transitive) guard for F_1. Note that $\mathcal{GF}$ can be seen as a sub-fragment of $\mathcal{GF}[\mathcal{TG}]$, when there are no transitive predicates. It is easy to see from the CNF transformation for guarded formulas that transitive predicates can appear only in initial clauses of the form: $\neg xTy \vee \hat{a}\langle x, y \rangle$, $\neg xTx \vee \hat{a}\langle x \rangle$ or $\neg g(x) \vee T\langle x, f(x) \rangle$. We present a resolution decision procedure for $\mathcal{GF}[\mathcal{TG}]$ as an extension of the one for $\mathcal{GF}$ by carefully analyzing and blocking the cases when resolution with transitivity predicates can lead to unbounded generation of clauses.

4.1   Obstacles for Deciding the Guarded Fragment with Transitivity

The transitivity clauses do not behave well when they resolve with each other because the number of variables increases. The simple solution is to block the inferences between the transitivity axioms by setting the selection function Sel such that it selects one of the negative literals. However this is only a partial solution to the problem since saturation with other clauses, in which positive transitive literals "should" be maximal, generate arbitrary large clauses as shown on the example below (left part):

1. $\neg(\underline{xTy})^* \vee \neg(yTz) \vee xTz$;
2. $\alpha(x) \vee \underline{f(x)Tx}^*$;
OR[2; 1]: 3. $\underline{\alpha(x)} \vee \neg(xTz) \vee \underline{f(x)Tz}^*$;
OR[3; 1]: 4. $\alpha(x) \vee \neg(xTz) \vee \underline{\neg(zTz_1)} \vee \underline{f(x)Tz_1}^*$;

.... : ....

1. $\neg(\underline{xTy})^* \vee \neg(\underline{yTz})^* \vee xTz$;
2. $\alpha(x) \vee \underline{f(x)Tx}^*$;
HR[2, 2; 1]: 3. $\underline{\alpha(x)} \vee \underline{ff(x)Tx}^*$;
HR[3, 2; 1]: 4. $\alpha(x) \vee \underline{fff(x)Tx}^*$;

.... : ....

The reason for the growth of the clause size is that the atoms which were resolved in the inferences *do not contain all variables* of the clause. To keep the number of variables from growing it is possible to use the *hyperresolution*, namely to select both negative literals of the transitivity clause and resolve them simultaneously. However, this strategy may result in increase of the clause depth, as shown on the right part of the example. Note that the variable depth in hyperresolution inference with the transitivity clause grows only if for the terms h, t and s which where simultaneously unified with x, y and z respectively, either $h \succ max(t, s)$ or $s \succ max(t, h)$. In all other cases, say, when $h = t \succ s$ like in the inference below, neither variable depth nor the number of variables grows:

1. $\neg(\underline{xTy})^* \vee \neg(\underline{yTz})^* \vee xTz$;   2. $\alpha(x) \vee \underline{xTx}^*$;   3. $\beta(x) \vee \underline{f(x)Tx}^*$;
HR[2, 3; 1]: 4. $\alpha(f(x)) \vee \beta(x) \vee f(x)Tx$;

We are going to distinguish these cases of using the transitivity clauses by using *ordering constraints* in combination with a selection function. We split the transitivity clause into the *constraint* clauses of forms:

T	$TTxyz\neg(xTy) \vee \neg(yTz) \vee xTz$;	
T.1. $\neg(xTy)^* \vee \neg(yTz) \vee xTz$	$x \succ max(y, z)$;	
T.2. $\neg(xTy) \vee \neg(yTz)^* \vee xTz$	$z \succ max(y, x)$;	(T)
T.3. $\neg(xTy)^* \vee \neg(yTx)^* \vee xTx$	$x \succ y$;	
T.4. $\neg(xTy)^* \vee \neg(yTz)^* \vee xTz$	$y \succeq max(x, z)$;	

where selected literals are indicated with the asterisk. In the sequel, assume that every set of clauses contains transitivity clauses T.1 – T.4 from (T) for every transitive predicate T.

4.2   Redundancy of Inferences Involving Transitive Relations

In this section we prove the main technical lemmas that allow to gain a control over the saturation process in presence of transitivity clauses. We show that many inferences involving transitive predicates are redundant. It is not very convenient to show redundancy of inferences "by definition". We proof auxiliary lemmas using which redundancy of inferences can be shown in a much simpler way.

Lemma 1 (Four Clauses). *Let N be a clause set containing the ground clauses:*

C1. $C \vee C' \vee \underline{A}^*$;   C2. $D \vee D' \vee \neg \underline{A}^*$;   C3. $C \vee D \vee B$;   C4. $C' \vee D' \vee \neg B$;

Then the following ordered resolution inference:

$\mathsf{OR}[\mathsf{C1};\mathsf{C2}]\!:\!\mathsf{P}.$ $C \vee C' \vee D \vee D'$;   *is redundant provided that $A \succ B$.*

Proof. Obviously, the conclusion of the inference $\mathsf{OR}[\mathsf{C1};\mathsf{C2}]$ follows from the clauses C3 and C4. It remains to show that both C3 and C4 are smaller than the maximum of the clauses C1 and C2. We use the fact that the conclusion of the ordered resolution inference is always smaller than the premise with the negative eligible literal. Therefore, $C \vee D \prec \mathsf{P} \prec \mathsf{C2}$ and since $B \prec \neg A \prec \mathsf{C2}$, $\mathsf{C3} = C \vee D \vee B \prec \mathsf{C2}$. Similarly, $\mathsf{C4} \prec \mathsf{C2}$. We have shown that $max(\mathsf{C3},\mathsf{C4}) \prec max(\mathsf{C1},\mathsf{C2})$, thus the inference $\mathsf{OR}[\mathsf{C1};\mathsf{C2}]$ is redundant.   □

Lemma 1 can be generalized to show redundancy of hyperresolution inferences as follows:

Lemma 2. *Let N be a clause set containing the ground clauses:*

C1. $C_1 \vee \underline{A_1}^*$; $\cdots$ Cn. $C_n \vee \underline{A_n}^*$;   D1. $C'_1 \vee D'_1$; $\cdots$ Dm. $C'_m \vee D'_m$;
C. $C \vee \neg\underline{A_1}^* \vee \cdots \vee \neg\underline{A_n}^*$;

for $n, m > 1$ such that: (i) $C_1 \vee \cdots \vee C_n \vee C = C'_1 \vee \cdots \vee C'_m$, (ii) $D'_1 \wedge \cdots \wedge D'_m \models \bot$ and (iii) $max(A_1, \ldots, A_n) \succ max(D'_1, \ldots, D'_m)$. Then the (hyper-)resolution inference: $\mathsf{HR}[\mathsf{C1},\mathsf{C2},\ldots,\mathsf{Cn};\mathsf{C}]\!:\!\mathsf{P}.$ $C_1 \vee C_2 \vee \cdots \vee C_n \vee C$; *is redundant.*

Proof. The conclusion $C_1 \vee \cdots \vee C_n \vee C = C'_1 \vee \cdots \vee C'_n$ of the inference logically follows from the clauses D1,..., Dm because of the condition (ii). Moreover, $max(\mathsf{D1}, \ldots, \mathsf{Dm}) \prec max(\mathsf{C1}, \ldots, \mathsf{Cn}, \mathsf{C}) = \mathsf{C}$ since for any i with $1 \leq i \leq m$, $C'_i \prec \mathsf{P} \prec \mathsf{C}$ (condition (i)) and $D'_i \prec \neg A_1 \vee \cdots \vee \neg A_n$ (condition (iii)). Therefore, the inference $\mathsf{HR}[\mathsf{C1},\mathsf{C2},\ldots,\mathsf{Cn};\mathsf{C}]$ is redundant.   □

For proving redundancy of inferences involving transitive relations, we need to make additional assumption about the the ordering $\succ$ used in $\mathcal{OR}^{\succ}_{Sel}$. We say that the ordering $\succ$ is *T-argument monotone* if: (i) $\{t_1, t_2\} \succ_{mul} \{s_1, s_2\}$ implies $(t_1 T t_2) \succ (s_1 T s_2)$, and (ii) $b(t_1, t_2) \succ (t_1 T t_2) \succ u(t_1)$ for any non-transitive predicate b and unary predicate u. From now on we assume that the ordering $\succ$ is T-argument monotone. The intended ordering can be easily obtained from the ordering $\succ_{rpos}$, that has been used for deciding the guarded fragment, by requiring that all transitive predicates have the multiset status and $b >_P T >_P u$ for any non-transitive predicate b whose arity is greater than two, transitive predicate T and unary predicate u.

Lemma 3. *Let N be a clause set containing the clause:*

1. $C \vee \underline{t_1 T t_2}^*$;   *together with the result of the inference:*
(a) $\mathrm{OR}[\underline{1}; \mathrm{T.1}]$:2. $C \vee \neg(t_2 T z) \vee \underline{t_1 T z}^* \mid t_1 \succ max(t_2, z)$;   *or*
(b) $\mathrm{OR}[1; \mathrm{T.2}]$:2. $C \vee \neg(x T t_1) \vee \underline{x T t_2}^* \mid t_2 \succ max(t_1, x)$;

Then the following inferences are redundant respectively:

(a) $\mathrm{OR}[2; \mathrm{T.1}]$:$C \vee \neg(t_2 T z) \vee \neg(z T z_1) \vee t_1 T z_1 \mid t_1 \succ max(t_2, z, z_1)$;
(b) $\mathrm{OR}[2; \mathrm{T.2}]$:$C \vee \neg(x_1 T x) \vee \neg(x T t_1) \vee x_1 T t_2 \mid t_2 \succ max(t_1, x, x_1)$.

Proof. **(a)** The result of any instance of the inference $\mathrm{OR}[2; \mathrm{T.1}]$:

2a. $C \vee \neg(t_2 T s) \vee \underline{t_1 T s}^* \mid t_1 \succ max(t_2, s)$;
T.1a. $\neg(\underline{t_1 T s})^* \vee \neg(\underline{s T h}) \vee t_1 T h \mid t_1 \succ max(s, h)$;
$\mathrm{OR}[2a; \mathrm{T.1}a]$:$C \vee \neg(t_2 T s) \vee \neg(s T h) \vee t_1 T h \mid t_1 \succ max(t_2, s, h)$;

can be obtained from other instances of the constraint clauses 2 and T:

2b. $C \vee \neg(\underline{t_2 T h}) \vee t_1 T h \mid t_1 \succ max(t_2, h)$;
Tb. $\neg(t_2 T s) \vee \neg(s T h) \vee \underline{t_2 T h}$;

by resolving on the smaller atom: $t_2 T h \prec t_1 T s$. Therefore, by Lemma 1 the inference is redundant. The case (b) is proven symmetrically to (a).   □

Lemma 4. *Let N be a clause set containing the clauses:*

1. $C \vee \underline{t_1 T t_2}^*$;
2. $D \vee \underline{t_2 T t_3}^*$;
$\mathrm{OR}[1; \mathrm{T.2}]$:3. $C \vee \neg(x T t_1) \vee \underline{x T t_2}^* \mid t_2 \succ max(t_1, x)$;
$\mathrm{OR}[2; \mathrm{T.1}]$:4. $D \vee \neg(t_3 T z) \vee \underline{t_2 T z}^* \mid t_2 \succ max(t_3, z)$;
$\mathrm{HR}[1, 2; \mathrm{T.4}]$:5. $C \vee D \vee t_1 T t_3 \mid t_2 \succeq max(t_1, t_3)$;
$\mathrm{HR}[2, 3; \mathrm{T.3}]$:6. $D \vee C \vee \neg(t_3 T t_1) \vee t_2 T t_2 \mid t_2 \succ t_3$;

Then the following inferences are redundant:

(a) $\mathrm{HR}[1, 4; \mathrm{T.4}]$:$C \vee D \vee \neg(t_3 T z) \vee t_1 T z \mid t_2 \succ max(t_3, z); \; t_2 \succeq t_1$
(b) $\mathrm{HR}[3, 2; \mathrm{T.4}]$:$C \vee D \vee \neg(x T t_1) \vee x T t_3 \mid t_2 \succ max(t_1, x); \; t_2 \succeq t_3$
(c) $\mathrm{HR}[3, 4; \mathrm{T.4}]$:$C \vee D \vee \neg(x T t_1) \vee \neg(t_3 T z) \vee x T z \mid t_2 \succ max(t_1, t_3, x, z)$;
(d) $\mathrm{HR}[4, 3; \mathrm{T.3}]$:$D \vee C \vee \neg(t_3 T x) \vee \neg(x T t_1) \vee t_2 T t_2 \mid t_2 \succ max(t_1, t_3, x)$.

Proof. The proof is analogous to Lemma 4. The complete proof can be found in the extended version of the paper (de Nivelle & Kazakov 2004).   □

We have shown that redundancy and ordering constraints help to avoid many inferences involving transitivity. However, certain inferences may still result in increasing the number of variables in clauses as in the situation shown below:

1. $\alpha(x) \vee \underline{f(x) T x}^*$;
2. $\neg(\underline{x T y})^* \vee a(x) \vee \beta(y)$;
3. $\neg(\underline{x T y})^* \vee \neg a(x) \vee \beta'(y)$;
$\mathrm{OR}[1; \mathrm{T.1}]$:5. $\alpha(x) \vee \neg(x T z) \vee \underline{f(x) T z}^* \mid f(x) \succeq max(x, z)$;
$\mathrm{OR}[5; 2]$:6. $\alpha(x) \vee \neg(x T z) \vee \underline{a(f(x))}^* \vee \beta(z) \mid f(x) \succeq max(x, z)$;
$\mathrm{OR}[5; 3]$:7. $\alpha(x) \vee \neg(x T z_1) \vee \underline{\neg a(f(x))}^* \vee \beta'(z_1) \mid f(x) \succeq max(x, z_1)$;
$\mathrm{OR}[6; 7]$:8. $\alpha(x) \vee \alpha(x) \vee \neg(x T z) \vee \neg(x T z_1) \vee \beta(z) \vee \beta'(z_1) \mid f(x) \succeq max(x, z, z_1)$;

The problem here is that the functional term $f(x)$ which does not contain all variables of the clause appears as an argument of a non-transitive predicate. That has happened as result of resolution inferences OR[5; 2] and OR[5; 3]. To resolve this problem we introduce an additional inference rule:

Transitive Recursion

$$TR: \frac{\neg(x\hat{T}y)^* \vee \alpha(x) \vee \beta(y)}{\substack{\neg(x\hat{T}y) \vee \alpha(x) \vee u_{\alpha(\cdot)}^{\hat{T}}(y) \\ \neg(x\hat{T}y) \vee \neg u_{\alpha(\cdot)}^{\hat{T}}(x) \vee u_{\alpha(\cdot)}^{\hat{T}}(y) \\ \neg u_{\alpha(\cdot)}^{\hat{T}}(y) \vee \beta(y)}}$$

where (i) $\hat{T}$ is a not empty set of transitive predicates (ii) $u_\alpha^{\hat{T}}$ is a special unary predicate indexed by α and $\hat{T}$.

The inference rule extends the signature by introducing new unary predicate symbols $u_\alpha^{\hat{T}}$, whose intended interpretation is "the set of elements that are T-reachable from the ones where α is false".

Lemma 5. *The transitive recursion rule is a sound inference rule.*

Proof. Let $\mathcal{M}$ be a model for the premise of the rule, such that all predicates $T_1, \ldots, T_n$ from $\hat{T}$ are interpreted by transitive relations and let $xT'y := xT_1y \wedge \cdots \wedge xT_ny$. Obviously, T' is interpreted in $\mathcal{M}$ by a transitive relation. We extend $\mathcal{M}$ to a the model $\mathcal{M}'$ by interpreting the new predicate $u_\alpha^{\hat{T}}(x)$ as the formula $\exists x'.(\neg\alpha(x') \wedge x'T'x)$. In particular, $\mathcal{M}' \models \forall y.([\exists x.(\neg\alpha(x) \wedge xT'y)] \rightarrow u_\alpha^{\hat{T}}(y))$, so the first conclusion of the inference rule is true in $\mathcal{M}'$. The following sequence of implications: $u_\alpha^{\hat{T}}(x) \wedge xT'y \equiv \exists x'.[\neg\alpha(x') \wedge x'T'x \wedge xT'y] \Rightarrow$ (transitivity of T') $\Rightarrow \exists x'.[\neg\alpha(x') \wedge x'T'y] \equiv u_\alpha^{\hat{T}}(y)$ shows that the second conclusion is true in $\mathcal{M}'$. Finally, the last conclusion is a consequence of the premise of the rule: $u_\alpha^{\hat{T}}(y) \equiv \exists x'.(\neg\alpha(x') \wedge x'T'y) \Rightarrow \exists x'.\beta(y) \equiv \beta(y)$. $\square$

Note that the transitive recursion rule is not a reduction rule, although the premise logically follows from the conclusion (the premise may be smaller than a clause in the conclusion, say, when $\beta(y)$ is empty). However, the rule helps to avoid dangerous inferences involving transitive predicates, like in the example above, by making them redundant:

Lemma 6. *Let $\hat{T} = \{T_1, \ldots, T_n\}$ with $n \geq 1$ be the set of transitive predicates and N be a clause set containing the clauses:*

D. $\neg(x\hat{T}z)^* \vee \alpha(x) \vee \beta(z)$; 1^i. $C_i \vee tT_ih^*$ *with one of the following clauses:*
D_1. $\neg(x\hat{T}z)^* \vee \alpha(x) \vee u(z)$;   OR$[1^i; T.1]: 2_a^i.C_i \vee \neg(hT_iz) \vee tT_iz^* | t \succ max(h, z)$;
D_2. $\neg(x\hat{T}z)^* \vee \neg u(x) \vee u(z)$;   OR$[1^i; T.2]: 2_b^i.C_i \vee \neg(xT_it) \vee \overline{xT_ih^*} | h \succ max(t, x)$;
D_3. $\neg u(z) \vee \beta(z)$;   HR$[1^1, \ldots, 1^n; D_1]: 3.$ $C_1 \mathbb{W} C_n \vee \alpha(t) \vee u(h)$.

for $1 \leq i \leq n$. Then the following inferences are redundant respectively:

(a) HR$[2_a^1, \ldots, 2_a^n; D]$: $\mathbb{W}_{i=1}^n \{C_i \vee \neg hT_iz\} \vee \alpha(t) \vee \beta(z)$
(b) HR$[2_b^1, \ldots, 2_b^n; D]$: $\mathbb{W}_{i=1}^n \{C_i \vee \neg xT_it\} \vee \alpha(x) \vee \beta(h)$

Proof. Consider the case (a) (case (b) is proven symmetrically). For any instance of the inference HR$[2_a^1, \ldots, 2_a^n; D]$ satisfying the constraints:

2_a^i. $C_i \vee \neg(hT_is) \vee \underline{tT_is}^* \mid t \succ max(h,s)$;

D. $\neg(\underline{t\hat{T}s})^* \vee \alpha(t) \vee \beta(s)$;

$\text{HR}[2_a^1, \ldots, 2_a^n; \text{D}]$: $\bigvee_{i=1}^n \{C_i \vee \neg hT_is\} \vee \alpha(t) \vee \beta(s)$

the conclusion can be derived from the clause 3 and instances of D_2 and D_3:

3. $C_1 \vee C_n \vee \alpha(t) \vee \underline{u(h)}$.

D_2^a. $\neg(h\hat{T}s) \vee \neg\underline{u(h)} \vee \underline{u(s)}$;

D_3^a. $\neg\underline{u(s)} \vee \beta(s)$;

by resolving on $u(h)$ and $u(s)$, both of which are smaller than each tT_is used in the inference. Therefore the inference is redundant by Lemma 2.   □

Remark 1. Note that the inferences $\text{HR}[2_a^1, \ldots, 2_a^n; \text{D}_1]$ and $\text{HR}[2_a^1, \ldots, 2_a^n; \text{D}_2]$ ($\text{HR}[2_b^1, \ldots, 2_b^n; \text{D}_1]$ and $\text{HR}[2_b^1, \ldots, 2_b^n; \text{D}_2]$) are redundant as well, since we can apply Lemma 6 for $\beta(z) := u(z)$; and $\alpha(x) := \neg u(x)$, $\beta(z) := u(z)$ respectively.

4.3   Saturation of the Clause Set

We have prepared the ground for describing a resolution decision procedure for $\mathcal{GF}[\mathcal{TG}]$. However, to simplify the upcoming case analysis, we introduce an additional inference rule:

Literal Projection

$$LP: \frac{[\![\, C \vee L \,]\!]}{\begin{array}{l} p_L(x) \vee C \\ \neg p_L(x) \vee L \end{array}} \quad \begin{array}{l} \textit{where (i) } L \textit{ is non-unary literal with} \\ \textit{free}[L] = \{x\}; \textit{ (ii) } C \textit{ contains } x \textit{ in} \\ \textit{non-unary literal or in functional} \\ \textit{subterm and (iii) } p_L \textit{ is a unary} \\ \textit{predicate for } L. \end{array}$$

The literal projection rule is a variant of the general splitting rule, which allows to split a clause by introducing a *new* predicate over shared variables of its parts. The purpose of this rule is to avoid clauses with several positive transitive literals that can be produced, for instance, with the inference:

1. $\neg a(f(x))^* \vee xT_1x$;   2. $\neg b(x) \vee \underline{a(f(x))}^* \vee xT_2x$;

$\text{OR}[1;2]$: $\neg b(x) \vee xT_1x \vee xT_2x$;

Instead of producing the inference above, one can alternatively simplify the clauses 1 and 2 using the literal projection rule:

1a. $p_{T_1}(x) \vee \neg a(f(x))^*$ 2a. $p_{T_2}(x) \vee \neg b(x) \vee \underline{a(f(x))}^*$;

1b. $\neg p_{T_1}(x) \vee \underline{xT_1x}$;   2b. $\neg p_{T_2}(x) \vee xT_2x$;

$\text{OR}[1a;2a]$: $p_{T_1}(x) \vee p_{T_2}(x) \vee \neg b(x)$;

Note that the literal projection rule cannot be applied to literals containing a new predicate symbol since they are unary, therefore, only finitely many predicates p_L can be introduced.

We show the decidability of $\mathcal{GF}[\mathcal{TG}]$ in similar way as for $\mathcal{GF}$ by describing a clause class containing the input clauses for $\mathcal{GF}[\mathcal{TG}]$-formulae and closed under the ordered resolution inferences up to redundancy. This clause class is represented by the set of the clause schemes below:

$$
\begin{array}{ll}
1\colon & [\neg\hat{T},\hat{\gamma}]\langle\hat{c}\rangle; \\
2\colon & [\neg!\hat{d},\hat{\gamma}]\langle!\overline{x}\rangle \vee [\neg\hat{T},\hat{\gamma}]\langle!f(\overline{x}),\overline{x}\rangle \vee \hat{\beta}\langle f(\overline{x}),\overline{x}\rangle; \\
3\colon & \neg!\hat{p}^1(!x) \vee \neg\hat{p}^1(f(x)) \vee \hat{T}\langle f(x),x\rangle; \\
T\colon & \neg(xTy) \vee \neg(yTz) \vee xTz; \\
4\colon & \neg!\hat{p}^1(x) \vee \neg(xTz) \vee f(x)Tz \mid f(x) \succ z; \\
5\colon & \neg!\hat{p}^1(x) \vee \neg(zTx) \vee zTf(x) \mid f(x) \succ z; \\
R\colon & \neg(x!\hat{T}y) \vee \hat{\gamma}(x) \vee \hat{\gamma}(y);
\end{array}
$$

$$
\begin{aligned}
\hat{\alpha} &:= \{\hat{a},\neg\hat{a}\}; \\
\hat{T} &:= \{\hat{T},\neg\hat{T}\}; \\
\hat{b} &:= \{\hat{a},p_{\hat{a}},p_{\hat{T}(\cdot)}\}; \\
\hat{\beta} &:= \{\hat{b},\neg\hat{b}\}; \\
\hat{p} &:= \{\hat{b},u_{\hat{\beta}}^\tau\}; \hat{\gamma} := \{\hat{p},\neg\hat{p}\}; \\
\hat{d} &:= \{\hat{p},\hat{T}\}; \quad \hat{\delta} := \{\hat{d},\neg\hat{d}\}.
\end{aligned}
$$

(GT)

Where $\hat{a}$ consists of initial non-transitive predicate symbols and $\hat{T}$ consists of all transitive predicate symbols.

Theorem 4. *There is a strategy based on $\mathcal{OR}^{\succ}_{Sel}$ with ordering constraints and additional inference rules such that given a formula $F \in \mathcal{GF}[\mathcal{TG}]$ a finite clause set N containing the CNF transformation for F is produced such that: (i) N is closed under rules of $\mathcal{OR}^{\succ}_{Sel}$ up to redundancy and (ii) N is a subset of (GT).*

Proof. The limited space does not allow us to present the complete case analysis of possible inferences between the clauses of (GT). The proof can be found in (de Nivelle & Kazakov 2004), where the extended version of the paper is given. ☐

Corollary 1. (Szwast & Tendera 2001) $\mathcal{GF}[\mathcal{TG}]$ *is decidable in double exponential time.*

Proof. Given a formula $F \in \mathcal{GF}[\mathcal{TG}]$, it could be seen from construction of (GT) that clauses generated in the saturation for F contain at most linear number of initial predicate and functional symbols and at most exponential number of introduced (by inferences extending the signature) *unary* predicates. Simple calculations show that the number of clauses from (GT) that can be constructed from them is at most double exponential. Therefore the saturation can be computed in double exponential time. ☐

5   Conclusions and Future Work

The resolution decision procedure for $\mathcal{GF}[\mathcal{TG}]$ presented in the paper can shed light on the reasons why this fragment is so fragile with respect to decidability and which decidable extensions it may have. Note, that we in fact have already shown the decidability of a larger fragment: it is possible to admit non-empty *conjunctions* of transitive relations $x\hat{T}y$ as guards since the CNF-transformation maps them to the same decidable fragment. This might help to find a decidable counterpart for the *interval-based* temporal logics à-la Halpern Shoham (Halpern & Shoham 1986) because the relation between intervals can be expressed as a conjunction of (transitive) relations between their endpoints. As a future work, we try to extend our approach to the case with equality, as well as to other theories like theories of general compositional axioms: $\forall xyz.(xSy \wedge yTz \rightarrow xHz)$ and theories of linear, branching and dense partial orderings without endpoints.

References

Andréka, H., van Benthem, J. & Németi, I. (1998), 'Modal languages and bounded fragments of predicate logic', *Journal of Philosophical Logic* **27**, 217–274.

Bachmair, L. & Ganzinger, H. (2001), Resolution theorem proving, *in* A. Robinson & A. Voronkov, eds, 'Handbook of Automated Reasoning', Vol. I, Elsevier Science, chapter 2, pp. 19–99.

de Nivelle, H. & de Rijke, M. (2003), 'Deciding the guarded fragments by resolution', *Journal of Symbolic Computation* **35**, 21–58.

de Nivelle, H. & Kazakov, Y. (2004), Resolution decision procedures for the guarded fragment with transitive guards, Research Report MPI-I-2004-2-001, Max-Planck-Institut für Informatik, Stuhlsatzenhausweg 85, 66123 Saarbrücken, Germany.

Fermüller, C., Leitsch, A., Hustadt, U. & Tammet, T. (2001), Resolution decision procedures, *in* A. Robinson & A. Voronkov, eds, 'Handbook of Automated Reasoning', Vol. II, Elsevier Science, chapter 25, pp. 1791–1849.

Fermüller, C., Leitsch, A., Tammet, T. & Zamov, N. (1993), *Resolution Methods for the Decision Problem*, Vol. 679 of *LNAI*, Springer, Berlin, Heidelberg.

Ganzinger, H., Meyer, C. & Veanes, M. (1999), The two-variable guarded fragment with transitive relations, *in* 'Proc. 14th IEEE Symposium on Logic in Computer Science', IEEE Computer Society Press, pp. 24–34.

Grädel, E. (1999), 'On the restraining power of guards', *Journal of Symbolic Logic* **64(4)**, 1719–1742.

Grädel, E. & Walukiewicz, I. (1999), Guarded fixed point logic, *in* 'Proceedings of 14th IEEE Symposium on Logic in Computer Science LICS '99, Trento', pp. 45–54.

Halpern, J. Y. & Shoham, Y. (1986), A propositional modal logic of time intervals, *in* 'Proceedings 1st Annual IEEE Symp. on Logic in Computer Science, LICS'86, Cambridge, MA, USA, 16–18 June 1986', IEEE Computer Society Press, Washington, DC, pp. 279–292.

Kieroński, E. (2003), The two-variable guarded fragment with transitive guards is 2EXPTIME-hard, *in* A. D. Gordon, ed., 'FoSSaCS', Vol. 2620 of *Lecture Notes in Computer Science*, Springer, pp. 299–312.

Szwast, W. & Tendera, L. (2001), On the decision problem for the guarded fragment with transitivity, *in* 'Proc. 16th IEEE Symposium on Logic in Computer Science', pp. 147–156.

van Benthem, J. (1997), Dynamic bits and pieces, Technical Report LP-97-01, ILLC, University of Amsterdam.

Attacking a Protocol for Group Key Agreement by Refuting Incorrect Inductive Conjectures

Graham Steel, Alan Bundy, and Monika Maidl

School of Informatics,
University of Edinburgh,
Edinburgh, EH8 9LE, Scotland,
g.j.steel@ed.ac.uk, {bundy,monika}@inf.ed.ac.uk
http://dream.dai.ed.ac.uk/graham

Abstract. Automated tools for finding attacks on flawed security protocols often struggle to deal with protocols for group key agreement. Systems designed for fixed 2 or 3 party protocols may not be able to model a group protocol, or its intended security properties. Frequently, such tools require an abstraction to a group of fixed size to be made before the automated analysis takes place. This can prejudice chances of finding attacks on the protocol. In this paper, we describe CORAL, our system for finding security protocol attacks by refuting incorrect inductive conjectures. We have used CORAL to model a group key protocol in a general way. By posing inductive conjectures about the trace of messages exchanged, we can investigate novel properties of the protocol, such as tolerance to disruption, and whether it results in agreement on a single key. This has allowed us to find three distinct novel attacks on groups of size two and three.

1 Introduction

The aim of cryptographic security protocols is to prescribe a way in which users can communicate securely over an insecure network. A protocol describes an exchange of messages in which the principals involved establish shared secrets, in order perhaps to communicate privately or to protect themselves from impersonators. These protocols are designed to be secure even in the presence of an active attacker, who may intercept or delay messages and send faked messages in order to gain access to secrets. Unsurprisingly, given this hostile operating environment, they have proved very hard to get right. What's more, protocol flaws are often quite subtle. New attacks are often found on protocols many years after they were first proposed.

In the last five years or so, there has been an explosion of interest in the idea of applying formal methods to the analysis of these protocols. Researchers have used techniques from model checking, term rewriting, theorem proving and logic programming amongst others, [19,20,27]. However, very few of these are able to analyse protocols for group key agreement, where an unbounded number of parties may be involved in a single round, [21,27]. Significant attacks on such

D. Basin and M. Rusinowitch (Eds.): IJCAR 2004, LNAI 3097, pp. 137–151, 2004.

protocols have appeared in the literature, but these have been discovered by hand, [28]. A problem for many automated approaches is that they can only attack concrete models of protocols, and so require the size of the group to be chosen in advance. This can prejudice the chances of discovering an attack. In this paper, we model such a protocol in a general way, without predetermining group size. The protocol in question is the Asokan–Ginzboorg protocol for key establishment in an ad-hoc Bluetooth network, [2]. Our formalism is a first-order version of the inductive model proposed by Paulson, [27]. The use of a first-order model allows us to search for counterexamples using automatic methods, which is not supported in Paulson's approach. We show how our counterexample finder for inductive conjectures, CORAL, has been used to automatically discover three new attacks on the protocol. One requires the group to be of size two, and the other two require a group of size three or more. CORAL refutes incorrect inductive conjectures using the 'proof by consistency' technique. Proof by consistency was originally developed as a method for automating inductive proofs in first-order logic, but has the property of being refutation complete, i.e. it is able to refute in finite time conjectures which are inconsistent with the set of hypotheses. Recently, Comon and Nieuwenhuis have drawn together and extended previous research to show how it may be more generally applied, [12]. CORAL is the first full implementation of this technique, built on the theorem prover SPASS, [37].

In the rest of the paper, we first briefly review previous work in security protocol analysis, refutation of incorrect conjectures and proof by consistency (§2). This explains the motivation for our development of CORAL. The CORAL system is described in §3, and CORAL's protocol model in §4. We give a description of the Asokan–Ginzboorg protocol in §5. In §6, we explain how we modelled the Asokan–Ginzboorg protocol for a group of unbounded size. Then we show how we used CORAL to discover three attacks on the protocol in §7. At the end of §7, we propose a new improved version of the protocol. In §8, we compare our results using CORAL to other research on protocol analysis, paying particular attention to work on group protocol analysis. We suggest further work in §9, and summarise and conclude in §10.

2   Background

Security protocols were first proposed by Needham and Schroeder, [25]. The authors predicted that their protocols may be 'prone to extremely subtle errors that are unlikely to be detected in normal operation'. They turned out to be correct. Several attacks, i.e. sequences of messages leading to a breach in security, were found in subsequent years. Since then more protocols have been proposed, many of which also turned out to be flawed. In the last five years or so, the interest in the problem from formal methods researchers has greatly increased. The problem of deciding whether a protocol is secure or not is in general undecidable, [14], due to the unbounded number of agents and parallel runs of the protocol that must be considered, and the unbounded number of terms an intruder can generate. However, good results in terms of new attacks and security guarantees have been achieved by a variety of methods, e.g. [19,20,27, 11]. Techniques based on model checking, term rewriting, theorem proving and logic programming each have their advantages and their advocates. For example, model checking approaches can find flaws very quickly, but can only be applied to finite (and typically very small) instances of the protocol, [19]. This means that if no attack is found, there may still be an attack upon a larger instance. Other methods can find guarantees of security quickly, but provide no help in finding attacks on flawed protocols, [11], or require the user to find and prove lemmas in order to reduce the problem to a tractable finite search space, [20]. Recently, dedicated tools for protocol analysis such as Athena have been built, combining techniques from model checking and theorem proving with a special purpose calculus and representation, [31]. Though user interaction is sometimes required to ensure termination, in general Athena's results are impressive. In terms of analysing the standard corpus of two and three party protocols given in [10], the field can now be said to be saturated. Most research attention has now turned to trying to widen the scope of the techniques, e.g. to more precise models of encryption, [23], 'second-level' protocols, [7], and group protocols, [21].

One method for protocol analysis that has proved very flexible is Paulson's inductive method, [27]. Protocols are formalised in typed higher-order logic as the set of all possible traces. Security properties can be proved by induction on traces, using the mechanized theorem prover Isabelle/HOL, [26]. The inductive method deals directly with the infinite state model, and assumes an arbitrary number of protocol participants, allowing properties of group protocols to be proved. However, proofs are tricky and require days or weeks of expert effort. Proof attempts may break down, and as Paulson notes, this can be hard to interpret. Perhaps further lemmas may need to be proved, or a generalisation made, or the conjecture may in fact be incorrect, indicating a flaw in the protocol. CORAL was designed to automate the task of refuting incorrect inductive conjectures in such a scenario. Additionally, if a refutation is found, CORAL provides a counterexample, giving the user the trace required to exploit the protocol flaw.

The refutation of incorrect inductive conjectures is a problem of general interest in the automated theorem proving community. Tools have been proposed by Protzen, [29], Reif, [30], and Ahrendt, [1]. Ahrendt's method works by constructing a set of clauses to send to a model generation prover, and is restricted to free datatypes. Protzen's technique progressively instantiates terms in the formula to be checked using the recursive definitions of the function symbols involved. Rief's method instantiates the formula with constructor terms, and uses simplifier rules in the theorem prover KIV to evaluate truth or falsehood. These techniques have found many small counterexamples, but are too naïve for a situation like protocol checking. They are designed for datatypes that can be easily enumerated, e.g. types in which any combination of constructors is a valid member of the type. In a protocol analysis scenario, this would tend to generate many non-valid traces, for example, traces in which an honest agent sends message 3 in a protocol without having received message 2. This could be accounted for in the specification by using a predicate to specify valid traces, but given the complexity of the protocol analysis problem, it would seem too inefficient to keep generating invalid traces only to later reject them. A method more suited to inductive datatypes is required.

Proof by consistency was originally conceived by Musser, [24], as a method for proving inductive theorems by using a modified Knuth-Bendix completion procedure. The idea is to show that a conjecture is a theorem by proving consistency with the axioms in the intended semantics. It was developed by Bachmair, [4], Ganzinger and Stuber, [17], and Bouhoula and Rusinowitch, [9], amongst others. Interest waned as it seemed too hard to scale the technique up to proving larger conjectures. However, later versions of the technique did have the property of being *refutation complete*, that is able to detect false conjectures in finite time. Comon and Nieuwenhuis, [12], have shown that the previous techniques for proof by consistency can be generalised to the production of a first-order axiomatisation $\mathcal{A}$ of the minimal Herbrand model such that $\mathcal{A} \cup E \cup C$ is consistent if and only if conjecture C is an inductive consequence of axioms E. With $\mathcal{A}$ satisfying the properties they define as a *Normal I-Axiomatisation*, inductive proofs can be reduced to first-order consistency problems and so can be solved by any saturation based theorem prover. This allows all the techniques developed for improving the performance of automatic first-order provers, such as reduction rules, redundancy detection, and efficient memory management techniques, to be used to aid the search for a proof or refutation. We describe the method in the next section.

3   The Coral System

CORAL is an implementation of the Comon-Nieuwenhuis method for proof by consistency, [12], in the theorem prover SPASS, [37]. There is only room for a summary of the technique and how it is implemented here. More details are available in [32].

The Comon–Nieuwenhuis method relies on a number of theoretical results, but informally, a proof attempt involves two parts. In the first part, we pursue a *fair induction derivation*. This is a restricted kind of saturation, [5], where we need only consider overlaps between axioms and conjectures, and produce inferences from an adapted superposition rule. In the second part, every clause in the induction derivation is checked for consistency against an *I-Axiomatisation*. This is typically a set of clauses sufficient for deciding inequality of ground terms. If all the consistency checks succeed, and the induction derivation procedure terminates, the theorem is proved. If any consistency check fails, then the conjecture is incorrect. Comon and Nieuwenhuis have shown refutation completeness for this system, i.e. any incorrect conjecture will be refuted in finite time, [12]. Since the induction derivation procedure may not terminate, we must carry out the consistency checks in parallel to retain refutation completeness. CORAL uses a parallel architecture to achieve this, using a socket interface to send clauses for I-Axiomatisation checking to a parallel prover. For problems specified by a *reductive definition*, [12, p. 19], which includes most natural specifications of inductive datatypes, this strategy offers marked performance improvements over a standard superposition strategy without losing refutation completeness. More details in [32].

In the case where a refutation is found, we are able to extract a counterexample by means of a well-known method first proposed by Green, [18]. When CORAL refutes a security conjecture of the form $\forall trace.P(trace)$, it has proved in its superposition calculus that $\exists trace.\neg P(trace)$. We track the instantiations made to the trace variable using an *answer literal*, following Green's method. Green has shown that this will always yield a correct constructive answer for these types of proofs. We show how new attacks are discovered as counterexamples in §7.

4 Coral's Protocol Model

The aim of CORAL's model was a first-order version of Paulson's inductive model for protocol analysis. Though Paulson's formalism is encoded in higher-order logic, no fundamentally higher-order concepts are used – in particular there is no unification of functional objects. Objects have types, and sets and lists are used. All this we model in first-order logic. Our formalism is typed, though it is also possible to relax the types and search for type attacks. Like Paulson's, our model allows an indeterminate and unbounded number of agents to participate, playing any role, and using an arbitrary number of fresh nonces and keys. Freshness is modelled by the *parts* operator: a nonce N is fresh with respect to a trace *trace* if $in(N, parts(trace)) = false$. This follows Paulson's model, [27, p. 12].

A protocol is modelled as the set of all possible traces, i.e. all possible sequences of messages sent by any number of honest users under the specification of the protocol and, additionally, faked messages sent by the intruder. A trace of messages is modelled as a list. A distinct feature of our formalism is that the entire state of the system is encoded in the trace. Other models with similar semantics often encode information about the knowledge of principals and the intruder in separate predicates. The latter approach has advantages in terms of the time required to find attacks on standard 2 or 3 party protocols, but our approach allows us to add unbounded numbers of messages to the trace under a single first-order rule, which is the key feature required to model the Asokan–Ginzboorg protocol without predetermining group size.

The intruder has the usual capabilities specified by the Dolev-Yao model, [13], i.e. the ability to intercept all messages in the trace, and to break down and reassemble the messages he has seen. He can only open encrypted packets if he has the correct key, and is assumed to be accepted as an honest player by the other agents. We specify intruder knowledge in terms of sets. Given a trace of messages exchanged, xt, we define $analz(xt)$ to be the least set including xt closed under projection and decryption by known keys. This is accomplished by using exactly the same rules as the Paulson model, [27, p. 12]. Then we can define the terms the intruder may build, given a trace xt, as being members of the set $synth(analz(xt))$, where $synth(x)$ is the least set containing x, including agent names and closed under pairing and encryption by known keys. The intruder may send anything in $synth(analz(xt))$ provided it matches the template of one of the messages in the protocol. This last feature is an optimisation to make searching for attacks more efficient, since the spy gains nothing from sending a message that no honest agent will respond to, [33]. We use this feature to define a domain specific redundancy rule for clauses: a clause is considered redundant if it specifies that the intruder must use a subterm which does not occur in any protocol message. By using the term indexing built into SPASS we can make this check extremely efficiently, resulting in a marked performance improvement. CORAL has a further redundancy rule which eliminates clauses with an unsatisfiable *parts* constraint, i.e. a *parts* literal where the variable that is supposed to represent a fresh nonce appears in the trace referred to by the *parts* operator. These clauses would of course be eliminated eventually by the axioms defining *parts*, but by eagerly pruning them, we save time.

For a specific protocol, we generally require one axiom per protocol message, each one having the interpretation, 'if xt is a trace containing message n addressed to agent xa, then xt may be extended by xa responding with message $n + 1$'. The tests that a suitable message n has been sent are known as *control conditions*. After the axioms specifying the protocol have been written down, a script automatically produces the clauses needed to model the intruder sending and receiving these protocol messages. These two sets of rules can be added to a standard set of axioms describing types, the member function, and the intruder to complete the protocol specification. It should be possibly to automatically generate the message axioms from a protocol specification language such as HLSPL, [6].

Using this model, CORAL has rediscovered several known attacks on two and three party security protocols including Needham-Schroeder, Neuman-Stubblebine and BAN Otway-Rees. This latter is significant since it requires an honest agent to generate two fresh nonces and to play the role of both the initiator and the responder, things which previous first-order models have not allowed, [36]. Run times on a Pentium IV Linux box vary from 17 seconds for Needham-Schroeder public key to 15 minutes for Otway-Rees. This is significantly slower than our competitors, e.g. [6,31]. However, there are two rather more positive aspects to CORAL's performance. The first is that CORAL searches for a counterexample in the general model of a protocol, not just in a model involving particular principals being involved in particular sessions as specified by the user, as for example in [6]. The second is that although CORAL's backwards search proceeds quite slowly, primarily because of time spent doing subsumption checking for each new clause, this checking together with the domain-specific redundancy rules described above means that many states are eliminated as being redundant with respect to the states we have already explored. This leads to quite an efficient search, for example for the Needham-Schroeder public key protocol, CORAL generates 1452 clauses, keeps 610 after redundancy checking, and discovers the attack at the 411th clause it considers. This gave us the confidence that CORAL would scale up to a group protocol whose model has a much larger branching rate, like the group protocol considered in this paper.

5   The Asokan–Ginzboorg Protocol

Asokan and Ginzboorg of the Nokia Research Centre have proposed an application level protocol for use with Bluetooth devices, [2]. The scenario under consideration is this: a group of people are in a meeting room and want to set up a secure session amongst their Bluetooth-enabled laptops. However, their computers have no shared prior knowledge and there is no trusted third party or public key infrastructure available. The protocol proceeds by assuming a short group password is chosen and displayed, e.g. on a whiteboard. The password is assumed to be susceptible to a *dictionary attack*, but the participants in the meeting then use the password to establish a secure secret key.

Asokan and Ginzboorg describe two protocols for establishing such a key in their paper, [2]. We have analysed the first of these. Completing analysis of the second protocol using CORAL would require some further work (see §9). Here is a description of the first Asokan–Ginzboorg protocol (which we will hereafter refer to as simply 'the Asokan–Ginzboorg protocol'). Let the group be of size n for some arbitrary $n \in \mathbb{N}, n \geq 2$. We write the members of the group as M_i, $1 \leq i \leq n$, with M_n acting as group leader.

1. $M_n \rightarrow \text{ALL} : M_n, \{\!| E |\!\}_P$
2. $M_i \rightarrow M_n \quad : M_i, \{\!| R_i, S_i |\!\}_E \qquad\qquad\qquad i = 1, \ldots, n-1$
3. $M_n \rightarrow M_i \quad : \{\!| \{S_j, j = 1, \ldots, n\} |\!\}_{R_i} \qquad i = 1, \ldots, n-1$
4. $M_i \rightarrow M_n \quad : M_i, \{\!| S_i, h(S_1, \ldots, S_n) |\!\}_K \quad$ some i, $K = f(S_1, \ldots, S_n)$

What is happening here is:

1. M_n broadcasts a message containing a fresh public key, E, encrypted under the password, P, which she has written on the whiteboard.
2. Every other participant M_i, for $i = 1, \ldots, n-1$, sends M_n a contribution to the final key, S_i, and a fresh symmetric key, R_i, encrypted under public key E.
3. Once M_n has a response from everyone in the room, she collects together the S_i in a package along with a contribution of her own (S_n) and sends out one message to each participant, containing this package $S_1, \ldots, S_n$ encrypted under the respective symmetric key R_i.
4. One participant M_i responds to M_n with the package he just received passed through a one way hash function $h()$ and encrypted under the new group key $K = f(S_1, \ldots, S_n)$, with f a commonly known function.

Asokan and Ginzboorg argue that it is sufficient for each group member to receive confirmation that one other member knows the key: everyone except M_n receives this confirmation in step 3. M_n gets confirmation from some member of the group in step 4. Once this final message is received, the protocol designers argue,

agents $M_1, \ldots, M_n$ must all have the new key $K = f(S_1, \ldots, S_n)$. The protocol
has three goals: the first is to ensure that a spy eavesdropping on Bluetooth
communications from outside the room cannot obtain the key. Secondly, it aims
to be secure against *disruption attacks* – attacks whereby a spy prevents the
honest agents from completing a successful run of the protocol – by an attacker
who can add fake messages, but not block or delay messages. Thirdly, it aims by
means of contributory key establishment to prevent a group of dishonest players
from restricting the key to a certain range. We used these goals as the basis of
our investigation described in §7.

6   Modelling the Asokan–Ginzboorg Protocol

In CORAL's model we reason about traces as variables, which may later be
instantiated to any number of messages. This allows us to model the protocol
generally with respect to the size of the group. The use of a general model means
that we do not have to guess how many players are needed to achieve an attack –
CORAL will search for attacks involving any number of participants. Of course,
CORAL is more likely to find attacks with small number of participants first,
since this will involve reasoning with smaller clauses.

Our methodology for modelling the Asokan–Ginzboorg protocol was the same
as for fixed 2 or 3 principal protocols, i.e. to produce one rule for each protocol message describing how a trace may be extended by that message, taking
into account the control conditions. So for message 2, our rule express the fact
that an agent can send a message 2 if he has seen a message 1 in the trace,
and will only use fresh numbers S_i and R_i in his message. However, for the
Asokan–Ginzboorg protocol, message 3 posed a problem. For a group of n participants, $n-1$ different message 3s will be sent out at once. Moreover, the group
leader for the run must check the control condition that she has received $n-1$
message 2s. This could not be modelled by a single first-order clause without
predetermining the size of the group. This problem was solved by adding two
more clauses to the model. We recursively define a new function which checks
a trace to see if all message 2s have been sent for a particular group leader
and number of participants. It returns a new trace containing the response prescribed by the protocol[1]. It works in all modes of instantiation, allowing us to
use it to construct protocol runs for various sized groups while the refutation
search is going on. This shows an advantage of a theorem-prover based formalism – we have access to full first-order inference for control conditions like
these in our protocol model. A full description of this function is available via
http://homepages.inf.ed.ac.uk/s9808756/asokan-ginzboorg-model/.

---

[1] Note that this was only required for modelling *honest* agents sending message 3s,
since they have to conform to the protocol. The intruder can send any combination
of message 3s, no matter what message 2s have appeared in the trace. He is only
constrained by what knowledge he can extract from previous messages in the trace,
by the same rules as for regular protocols.

7   Attacking the Asokan–Ginzboorg Protocol

We decided to test the protocol against two attackers: one inside the room, and one outside. The spy outside tries to affect disruption attacks, and the spy inside tries to gain control of communication in the room, e.g. by making all participants agree on different keys that only he knows. As we are in a wireless scenario, both spies capabilities differ from the normal Dolev-Yao model in that they cannot prevent particular messages from being received, they can only insert additional fake messages of their own.

Finding attacks in CORAL results from finding counterexamples to security properties. These are formulated in a similar way to Paulson's method. We must formulate the required properties in terms of the set of possible traces of messages. The following conjecture was used to check for disruption attacks:

%% some honest xi has sent message 4, so has key $f(Package)$:
$eqagent(XI,spy)=false \land$
$member(sent(XI,XK,pair(principal(XI),$
$\quad encr(pair(nonce(SI),h(Package)),f(Package))))),Trace) = true\land$

%% genuine messages 3 and 1 are in the trace to some agent XJ:
$member(sent(MN,XJ,encr(Package,nonce(RJ)))),Trace)=true \land$
$member(sent(MN,all,pair(principal(MN),encr(key(E),key(P))))) \land$
$\quad ,Trace)=true$

%% but XJ never sent a message 2 under public key E with nonces SJ
%% (which is in $Package$) and RJ (which the message 3 meant for
%% him was sent under). That means he doesn't have RJ, and so can't
%% get the key from his message 3.
$member(nonce(SJ),Package)=true \land$
$member(sent(XJ,MN,pair(principal(XJ),encr(pair(nonce(RJ),nonce(SJ)),$
$\quad key(E))))),Trace)=false$
$\rightarrow$

This conjecture expresses that a run has been finished (i.e. a message 4 has been sent) and that there is some agent who does not now have the key (i.e. he cannot read his message 3). Note that conjecture is negative, i.e. it says that for all possible traces, no trace can have the combination of genuine messages 4,3 and 1 without a corresponding message 2. Note also that the conjecture is completely general is terms of the size of the group. When first run with this conjecture, CORAL produces the following counterexample for a group of size 2:

1. $M_2 \rightarrow ALL : M_2, \{\!| E |\!\}_P$
1'. $spy_{M_1} \rightarrow ALL : M_1, \{\!| E |\!\}_P$
2'. $M_2 \rightarrow M_1 \ : M_2, \{\!| R_2, S_2 |\!\}_E$
2. $spy_{M_1} \rightarrow M_2 \ : M_1, \{\!| R_2, S_2 |\!\}_E$
3. $M_2 \rightarrow M_1 \ : \{\!| S_2', S_2 |\!\}_{R_2}$
3'. $spy_{M_1} \rightarrow M_2 \ : \{\!| S_2', S_2 |\!\}_{R_2}$
4.' $M_2 \rightarrow M_1 \ : M_2, \{\!| S_2, h(S_2', S_2) |\!\}_{f(S_2', S_2)}$

At the end of the run, M_2 now accepts the key $f(S_2', S_2)$ as a valid group key, but it contains numbers known only to M_2, and not to M_1. Encrypting the agent identifier in message 1 stops the spy from sending the fake message 1', preventing the attack. However, when we made this correction to the protocol, and ran CORAL again with the same conjecture, CORAL found the following counterexample for a group of size 3:

1. $M_1 \rightarrow \text{ALL} : \{\!| M_1, E |\!\}_P$
2. $M_2 \rightarrow M_1$ $: M_2, \{\!| R_2, S_2 |\!\}_E$
2. $spy_{M_3} \rightarrow M_1$ $: M_3, \{\!| R_2, S_2 |\!\}_E$
3. $M_1 \rightarrow M_2$ $: \{\!| S_2, S_2, S_1 |\!\}_{R_2}$
3. $M_1 \rightarrow M_3$ $: \{\!| S_2, S_2, S_1 |\!\}_{R_2}$
4. $M_2 \rightarrow M_1$ $: M_2, \{\!| S_2, h(S_2, S_2, S_1) |\!\}_{f(S_2, S_2, S_1)}$

This is another disruption attack, where the spy eavesdrops on the first message 2 sent, and then fakes a message 2 from another member of the group. This results in the protocol run ending with only two of the three person group sharing the key. This attack can also be prevented by a small change to the protocol, this time by encrypting the agent identifier in message 2 (see §7.1 below). CORAL took about 2.5 hours to find the first attack, and about 3 hours to find the second. With these two attacks prevented, CORAL finds no further disruption attacks after three days run-time.

When considering a scenario where one of the agents inside the room is a spy, we decided to consider what might be possible when all the players in the room think they have agreed on a key, but have in fact agreed on different keys. If the spy knows all these keys, he could filter all the information exchanged, perhaps making changes to documents, such that the other agents in the room are none the wiser. The only change to CORAL's model for this scenario is to allow the spy to read the password on the whiteboard. We then checked for non-matching key attacks by giving CORAL the following conjecture:

% we have distinct honest agents XI and XJ
eqagent(XI,spy)=false ∧
eqagent(XJ,spy)=false ∧
eqagent(XJ,XI)=false ∧

% they both sent message 2s in response to the same message 1
member(sent(XI,MN,pair(principal(XI),
encr(pair(nonce(RI), nonce(SI)), key(E)))), Trace) = true∧
member(sent(XJ,MN,pair(principal(XJ),
encr(pair(nonce(RJ), nonce(SJ)), key(E)))), Trace) = true∧

% and received message 3s under the correct keys, RI and RJ
member(sent(MN,XI,encr(Package1,nonce(RI))),Trace)=true∧
member(sent(MN,XJ,encr(Package2,nonce(RJ))),Trace)=true∧

% but the packages they received were different
eq(Package1,Package2)=false →

Note again that the conjecture is negative, i.e. it states that there is no trace for which this combination of conditions can hold. CORAL refuted this property in about 3 hours, producing the counterexample trace:

1. $spy \to$ ALL : $spy, \{\!| E |\!\}_P$
2. $M_1 \to spy$: $M_1, \{\!| R_1, S_1 |\!\}_E$
2. $M_2 \to spy$: $M_2, \{\!| R_2, S_2 |\!\}_E$
3. $spy \to M_1$: $\{\!| S_1, S_2, S_{spy} |\!\}_{R_1}$
3. $spy \to M_2$: $\{\!| S_1, S_2, S'_{spy} |\!\}_{R_2}$
4. $M_1 \to spy$: $M_1, \{\!| S_1, h(S_1, S_2, S_{spy}) |\!\}_{f(S_1, S_2, S_{spy})}$

This attack is just a standard protocol run for three participants, except that in the first message 3, the spy switches in a number of his own (S'_{spy} in the place of S_2). This means that M_1 accepts the key as $f(S_{spy}, S_1, S'_{spy})$, whereas M_2 accepts $f(S_{spy}, S_1, S_2)$, and both of these keys are known to the spy.

7.1   An Improved Version of the Protocol

As mentioned above, we can prevent the disruption attacks by encrypting the agent identifiers in messages 1 and 2. To prevent the attack by the spy inside the room, we can require message 4 to be broadcast to all participants, so that everyone can check they have agreed on the same key. Here is the revised Asokan–Ginzboorg protocol, with boxes highlighting the changes:

1. $M_n \to$ ALL   : $\{\!| \boxed{M_n}, E |\!\}_P$
2. $M_i \to M_n$: $\{\!| \boxed{M_i}, R_i, S_i |\!\}_E, i = 1, \ldots, n-1$
3. $M_n \to M_i$: $\{\!| \{Sj, j = 1, \ldots, n\} |\!\}_{R_i}, i = 1, \ldots, n-1$
4. $M_i \to \boxed{\text{ALL}}$: $M_i, \{\!| S_i, h(S_1, \ldots, S_n) |\!\}_K$, some i.

8   Related Work

A protocol for group key management has also been modelled by Taghdiri and Jackson, [34]. In this case the protocol under consideration was the Pull-Based Asynchronous Rekeying Framework, [35]. The Taghdiri-Jackson model is general in terms of the number of members of the group, but does not model a malicious intruder trying to break the protocol. Instead, they just investigate correctness properties of the protocol. One correctness property is found not to hold, revealing a flaw whereby a member of the group may accept a message from an ex-member of the group as being current. However, though potentially serious, this flaw is somewhat trivial to spot, and is evident from a description of the rather weak protocol. Taghdiri and Jackson propose a fix for the protocol. However, without the modelling of a wilfully malicious member, who can eavesdrop on protocol traffic and construct fake messages, they failed to produce a secure protocol. We have recently modelled the improved protocol in CORAL and discovered attacks which are just as serious as the two they discovered. Details will appear in a future publication.

The most successful attempt at group protocol analysis in terms of finding new attacks was [28], where the case study was the CLIQUES protocol suite, [3]. Pereira and Quisquater discovered a number of new attacks, using a pen-and-paper approach and borrowing some ideas from the strand space model, [15]. Their attacks were quite subtle, involving properties of the Diffie-Hellman exponentiation operation widely used in the CLIQUES suite. They also involved the spy doing some quite imaginative things, like joining the group, leaving, and then forcing the remaining members to accept a compromised key. Interestingly, some of their attacks required the group to be of size four or more, backing up the case for taking a general approach in terms of group size. Pereira and Quisquater showed the value of by-hand analysis taking algebraic properties of cryptographic functions into account, but only when undertaken by experts.

Meadows made an attempt to extend the NRL protocol analysis tool to make it suitable for analysing group protocols, [21]. Again the CLIQUES protocols were used as an example. However, the NRL tool was not able to rediscover the attacks Pereira and Quisquater had discovered, because of the intricate series of actions the spy has to perform to effect the attack. The NRL tool is tied to quite constrained notions of secrecy and authenticity, which may be where the problem lay. It would be interesting to see whether these kinds of attacks could be found automatically by CORAL. We hope that the very flexible inductive model used might mean that these attacks are within CORAL's scope. However, some work would be required to model the associative and commutative properties of the exponentiation operation used. Modelling these properties is a topic which has recently started to attract more research interest [23]

Tools like Athena, [31], and the On-The-Fly Model Checker (OFMC), [6], can discover attacks on standard 2 and 3 party protocols much faster than CORAL, thanks to their special purpose representations and inference algorithms. However, both would have trouble modelling the Asokan–Ginzboorg protocol and similar group protocols in a general way, without setting the group size. Athena uses an extension of the strand space model, [15]. One of Athena's requirements is that a so-called semi-bundle should be finite. However, this would be impossible for semi-bundles in an Asokan–Ginzboorg protocol strand space, since a strand may contain an unbounded number of message 3s. The OFMC is constrained to a one to one relationship between transitions and messages sent. This would seem to make it impossible to model an arbitrary number of message 3s being added to the trace without some adaptation of the tool.

9   Further Work

Our work with CORAL is ongoing. We do not intend to compete for speed of attack finding on a large corpus of standard protocols, since other approaches are better suited to this. Instead we intend to focusing on the kind of flexible or group protocols that are not easy to attack using a model checking approach. One goal is to develop a corpus of group protocols and attacks like the Clark-Jacob corpus for standard protocols, [10]. We also intend to write a converter to allow protocols to be given to CORAL in an existing protocol specification language like HLSPL, [6], or MuCAPSL, [22]. This would ease further comparison between CORAL and other tools. It would be very interesting to see if the Asokan–Ginzboorg protocol analysed here could be specified in such a language without choosing a group size, and if not, what features we should have to add to these languages.

As CORAL is built on SPASS, a theorem prover capable of equational reasoning, we should be able to reason about some simple algebraic properties of the cryptosystems underlying protocols, such as Diffie-Hellman type operations. This would allow us to analyse the second Asokan–Ginzboorg protocol, which is quite different to the first and seems not to be susceptible to the same attacks, and also the CLIQUES protocols mentioned above. The main task here would be to devise a way of modelling the low-level cryptographic operations in such a way that the essential properties are captured, but without going into too fine a degree of detail which would make automated analysis infeasible.

Apart from the small number of domain specific redundancy rules described in §4, CORAL is a general counterexample finding tool for inductive theories. We would like to explore other situations where such a tool may be of use, such as the detection of false lemmas and generalisations in an inductive theorem prover.

A longer term aim is to adapt CORAL to be able to analyse security APIs of cryptographic hardware modules, such as are used for electronic point-of-sale devices and automated teller machines. Serious flaws have been found in these in recent years, [8]. Some of these attacks involve reducing the complexity of guessing a secret value by brute force search using some information leaked by the API as the result of some unexpected sequence of commands. These kinds of attacks are beyond the scope of current protocol analysis tools, and would require measures of guess complexity to be considered along with the usual protocol traces. We could perhaps accomplish this in CORAL by attaching complexity literals to each clause in the same way that we currently attach answer literals.

10   Conclusions

We have presented CORAL, our system for refuting incorrect inductive conjectures, and shown three new attacks on the Asokan–Ginzboorg protocol that it found. An advantage of looking for attacks on group protocols with CORAL is that we can model the protocol generally and so look for attacks on any size of group. The distinct attacks discovered on groups of size two and three demonstrate the efficacy of this approach. A model checking type approach would have required us to fix the size of the group in advance. Additionally, refuting conjectures about traces in an inductive formalism allows us to look for unconventional attacks, e.g. where two members of the group share different keys with the intruder which they both believe to be the group key.

In future we intend to attack more group protocols, particularly wireless group protocols. We aim eventually to be able to find attacks on security APIs that require brute force guess complexity considerations to be taken into account.

References

1. W. Ahrendt. Deductive search for errors in free data type specifications using model generation. In A. Voronkov, editor, *18th Conference on Automated Deduction*, volume 2392 of *LNCS*, pages 211–225. Springer, 2002.
2. N. Asokan and P. Ginzboorg. Key-agreement in ad-hoc networks. *Computer Communications*, 23(17):1627–1637, 2000.
3. G. Ateniese, M. Steiner, and G. Tsudik. New multiparty authentication services and key agreement protocols. *IEEE Journal on Selected Areas in Communications*, 18(4):628–639, April 2000.
4. L. Bachmair. *Canonical Equational Proofs.* Birkhauser, 1991.
5. L. Bachmair and H. Ganzinger. Completion of First-order clauses with equality by strict superposition (extended abstract). In *Proceedings 2nd International CTRS Workshop*, pages 162–180, Montreal, Canada, 1990.
6. D. Basin, S. Mödersheim, and L. Viganò. An on-the-fly model-checker for security protocol analysis. In *Proceedings of the 2003 European Symposium on Research in Computer Security*, pages 253–270, 2003. Extended version available as Technical Report 404, ETH Zurich.
7. G. Bella. Verifying second-level security protocols. In D. Basin and B. Wolff, editors, *Theorem Proving in Higher Order Logics*, volume 2758 of *LNCS*, pages 352–366. Springer, 2003.
8. M. Bond and R. Anderson. API level attacks on embedded systems. *IEEE Computer Magazine*, pages 67–75, October 2001.
9. A. Bouhoula and M. Rusinowitch. Implicit induction in conditional theories. *Journal of Automated Reasoning*, 14(2):189–235, 1995.
10. J. Clark and J. Jacob. A survey of authentication protocol literature: Version 1.0. Available via http://www.cs.york.ac.uk/jac/papers/drareview.ps.gz, 1997.
11. E. Cohen. TAPS a first-order verifier for cryptographic protocols. In *Proceedings of the 13th IEEE Computer Security Foundations Workshop*, pages 144–158, Cambridge, England, July 2000.
12. H. Comon and R. Nieuwenhuis. Induction = I-Axiomatization + First-Order Consistency. *Information and Computation*, 159(1-2):151–186, May/June 2000.
13. D. Dolev and A. Yao. On the security of public key protocols. *IEEE Transactions in Information Theory*, 2(29):198–208, March 1983.
14. N. Durgin, P. Lincoln, J. Mitchell, and A. Scedrov. Undecidability of bounded security protocols. In N. Heintze and E. Clarke, editors, *Proceedings of the Workshop on Formal Methods and Security Protocols — FMSP, Trento, Italy*, July 1999. Electronic proceedings available at http://www.cs.bell-labs.com/who/nch/fmsp99/program.html.
15. F. Fábrega, J. Herzog, and Guttman J. Strand spaces: Proving security protocols correct. *Journal of Computer Security*, 7:191–230, 1999.
16. H. Ganzinger, editor. *Automated Deduction – CADE-16, 16th International Conference on Automated Deduction*, volume 1632 of *LNAI*, Trento, Italy, July 1999. Springer-Verlag.
17. H. Ganzinger and J. Stuber. *Informatik — Festschrift zum 60. Geburtstag von Günter Hotz*, chapter Inductive theorem proving by consistency for first-order clauses, pages 441–462. Teubner Verlag, 1992.
18. C. Green. Theorem proving by resolution as a basis for question-answering systems. In B. Meltzer and D. Michie, editors, *Machine Intelligence*, volume 4, pages 183–208. Edinburgh University Press, 1969.

19. G. Lowe. Breaking and fixing the Needham Schroeder public-key protocol using FDR. In *Proceedings of TACAS*, volume 1055, pages 147–166. Springer Verlag, 1996.
20. C. Meadows. The NRL protocol analyzer: An overview. *Journal of Logic Programming*, 26(2):113–131, 1996.
21. C. Meadows. Extending formal cryptographic protocol analysis techniques for group protocols and low-level cryptographic primitives. In P. Degano, editor, *Proceedings of the First Workshop on Issues in the Theory of Security*, pages 87–92, Geneva, Switzerland, July 2000.
22. J. Millen and G. Denker. MuCAPSL. In *DISCEX III, DARPA Information Survivability Conference and Exposition*, pages 238–249. IEEE Computer Society, 2003.
23. J. Millen and V. Shmatikov. Symbolic protocol analysis with products and Diffie-Hellman exponentiation. In *16th IEEE Computer Security Foundations Workshop*, pages 47–61, 2003.
24. D. Musser. On proving inductive properties of abstract data types. In *Proceedings 7th ACM Symp. on Principles of Programming Languages*, pages 154–162. ACM, 1980.
25. R. Needham and M. Schroeder. Using encryption for authentication in large networks of computers. *Communications of the ACM*, 21(12):993–999, December 1978.
26. L.C. Paulson. *Isabelle: A Generic Theorem Prover*, volume 828 of *LNCS*. Springer, 1994.
27. L.C. Paulson. The Inductive Approach to Verifying Cryptographic Protocols. *Journal of Computer Security*, 6:85–128, 1998.
28. O. Pereira and J.-J. Quisquater. Some attacks upon authenticated group key agreement protocols. *Journal of Computer Security*, 11(4):555–580, 2003. Special Issue: 14th Computer Security Foundations Workshop (CSFW14).
29. M. Protzen. Disproving conjectures. In D. Kapur, editor, *11th Conference on Automated Deduction*, pages 340–354, Saratoga Springs, NY, USA, June 1992. Published as Springer Lecture Notes in Artificial Intelligence, No 607.
30. W. Reif, G. Schellhorn, and A. Thums. Flaw detection in formal specifications. In *IJCAR'01*, pages 642–657, 2001.
31. D. Song, S. Berezin, and A. Perrig. Athena: A novel approach to efficient automatic security protocol analysis. *Journal of Computer Security*, 9(1/2):47–74, 2001.
32. G. Steel, A. Bundy, and E. Denney. Finding counterexamples to inductive conjectures and discovering security protocol attacks. In *Proceedings of the Foundations of Computer Security Workshop, FLoC'02*, 2002. Also available as Informatics Research Report 141 http://www.inf.ed.ac.uk/publications/report/0141.html.
33. P. Syverson, C. Meadows, and I. Cerversato. Dolev-Yao is no better than machiavelli. In P. Degano, editor, *Proceedings of the First Workshop on Issues in the Theory of Security*, pages 87–92, Geneva, Switzerland, 2000.
34. M. Tagdhiri and D. Jackson. A lightweight formal analysis of a multicast key management scheme. In *Proceedings of Formal Techniques of Networked and Distributed Systems - FORTE 2003*, LNCS, pages 240–256, Berlin, 2003. Springer.
35. S. Tanaka and F. Sato. A key distribution and rekeying framework with totally ordered multicast protocols. In *Proceedings of the 15th International Conference on Information Networking*, pages 831–838, 2001.
36. C. Weidenbach. Towards an automatic analysis of security protocols in first-order logic. In Ganzinger [16], pages 314–328.
37. C. Weidenbach et al. System description: SPASS version 1.0.0. In Ganzinger [16], pages 378–382.

Decision Procedures for Recursive Data Structures with Integer Constraints

Ting Zhang, Henny B. Sipma, and Zohar Manna[*]

Computer Science Department
Stanford University
Stanford, CA 94305-9045
{tingz,sipma,zm}@theory.stanford.edu

Abstract. This paper is concerned with the integration of recursive data structures with Presburger arithmetic. The integrated theory includes a length function on data structures, thus providing a tight coupling between the two theories, and hence the general Nelson-Oppen combination method for decision procedures is not applicable to this theory, even for the quantifier-free case. We present four decision procedures for the integrated theory depending on whether the language has infinitely many constants and whether the theory has quantifiers. Our decision procedures for quantifier-free theories are based on Oppen's algorithm for acyclic recursive data structures with infinite atom domain.

1 Introduction

Recursively defined data structures are essential constructs in programming languages. Intuitively, a data structure is recursively defined if it is partially composed of smaller or simpler instances of the same structure. Examples include lists, stacks, counters, trees, records and queues. To verify programs containing recursively defined data structures we must be able to reason about these data structures. Decision procedures for several data structures exist. However, in program verification decision procedures for a single theory are usually not applicable as programming languages often involve multiple data domains, resulting in verification conditions that span multiple theories. Common examples of such "mixed" constraints are combinations of data structures with integer constraints on the size of those structures.

In this paper we consider the integration of Presburger arithmetic with an important subclass of recursively defined data structures known as recursive data structures. This class of structures satisfies the following properties of term algebras: (i) the data domain is the set of data objects generated exclusively by applying constructors, and (ii) each data object is uniquely generated. Examples of such structures include lists, stacks, counters, trees and records; queues do not

---

[*] This research was supported in part by NSF grants CCR-01-21403, CCR-02-20134 and CCR-02-09237, by ARO grant DAAD19-01-1-0723, by ARPA/AF contracts F33615-00-C-1693 and F33615-99-C-3014, and by NAVY/ONR contract N00014-03-1-0939.

D. Basin and M. Rusinowitch (Eds.): IJCAR 2004, LNAI 3097, pp. 152–167, 2004.

belong to this class as they are not uniquely generated: they can grow at both ends.

Our language of the integrated theory has two sorts; the integer sort $\mathbb{Z}$ and the data (term) sort λ. Intuitively, the language is the set-theoretic union of the language of recursive data structures and the language of Presburger arithmetic plus the additional length function $|.| : \lambda \to \mathbb{Z}$. Formulae are formed from data literals and integer literals using logical connectives and quantifications. Data literals are exactly those literals in the theory of recursive data structures. Integer literals are those that can be built up from integer variables (including the length function applied to data terms), addition and the usual arithmetic functions and relations.

We present four decision procedures for different variants of the theory depending on whether the language has infinitely many atoms and whether the theory is quantifier-free. Our decision procedures for quantifier-free theories are based on Oppen's algorithm for acyclic recursive data structures with infinite atom domain [17]. When integer constraints in the input are absent, our decision procedures can be viewed as an extension of Oppen's original algorithm to cyclic structures and to the structures with finite atom domain.

Related Work. Our component theories are both decidable. Presburger arithmetic was first shown to be decidable in 1929 by the quantifier elimination method [6]. Efficient algorithms were later discovered by Cooper et al [4,18]. It is well-known that recursive data structures can be modeled as term algebras which were shown to be decidable by the quantifier elimination method [13,11, 8]. Decision procedures for the quantifier-free theory of recursive data structures were discovered by Nelson, Oppen et al [15,17,5]. In [17] Oppen gave a linear algorithm for acyclic structures and a quadratic algorithm was given in [15] for cyclic structures. If the values of the selector functions on atoms are specified, then the problem is NP-complete [17].

A general combination method for decision procedures for quantifier-free theories was developed by Nelson and Oppen in 1979 [14]. However, this method is not applicable to the combination of our component theories. The method requires that component theories be loosely coupled, that is, have disjoint signatures, and are stably infinite[1]. The method is not applicable to tightly coupled theories such as single-sorted theories with shared signatures or, as in our case, multisorted theories with functions mapping elements in one sort to another.

The integration of Presburger arithmetic with recursive data structures was discussed by Bjørner [1] and an incomplete procedure was implemented in STeP (Stanford Temporal Prover) [2]. Zarba constructed decision procedures for the combined theory of sets and integers [22] and multisets and integers [21] by extending the Nelson-Oppen combination method.

Paper Organization. Section 2 provides the preliminaries: it introduces the notation and terminology. Section 3 defines recursive data structures and ex-

---

[1] A theory is stably infinite if a quantifier-free formula in the theory is satisfiable if and only if it is satisfiable in an infinite model.

plains Oppen's algorithm, the basis for our decision algorithm for quantifier-free theories. Section 4 introduces the combined theory of recursive data structures and Presburger arithmetic and outlines our approach for constructing the decision procedures. Sections 5-8 describe the four classes of decision procedures for the different variants of the theory in detail. Section 9 discusses complexity issues and Section 10 concludes with some ideas for future work. Because of space limitations most proofs have been omitted. They are available for reference in the extended version of this paper at http://theory.stanford.edu/~tingz/papers/ijcar04_extended.pdf.

2   Preliminaries

We assume the first-order syntactic notions of variables, parameters and quantifiers, and semantic notions of structures, satisfiability and validity as in [6]. A signature Σ is a set of parameters (function symbols and predicate symbols) each of which is associated with an arity. The function symbols with arity 0 are also called constants. The set of Σ-terms $\mathcal{T}(\Sigma, \mathcal{X})$ is recursively defined by: (i) every constant $c \in \Sigma$ or variable $x \in \mathcal{X}$ is a term, and (ii) if $f \in \Sigma$ is an n-place function symbol and $t_1, \dots, t_n$ are terms, then $f(t_1, \dots, t_n)$ is a term.

An atomic formula (atom) is a formula of the form $P(t_1, \dots, t_n)$ where P is an n-place predicate symbol and $t_1, \dots, t_n$ are terms (equality is treated as a binary predicate symbol). A literal is an atomic formula or its negation. A ground formula is a formula with no variables. A variable occurs free in a formula if it is not in the scope of a quantifier. A sentence is a formula in which no variable occurs free. A formula without quantifiers is called quantifier-free. Every quantifier-free formula can be put into disjunctive normal form, that is, a disjunction of conjunctions of literals.

A Σ-structure (or Σ-interpretation) $\mathfrak{A}$ is a tuple $\langle A, I \rangle$ where A is a non-empty domain and I is a function that associates each n-place function symbol f (resp. predicate symbol P) with an n-place function $f^{\mathfrak{A}}$ (resp. relation $P^{\mathfrak{A}}$) on A. We usually denote $\mathfrak{A}$ by $\langle A; \Sigma \rangle$ which is called the signature of $\mathfrak{A}$. We use Gothic letters (like $\mathfrak{A}$) for structures and Roman letters (like A) for the underlying domain. A variable valuation (or variable assignment) ν (w.r.t. $\mathfrak{A}$) is a function that assigns each variable an element of A. The truth value of a formula is determined when an interpretation and a variable assignment is given.

A formula φ is satisfiable (or consistent) if it is true under some variable valuation; it is unsatisfiable (or inconsistent) otherwise. A formula φ is valid if it is true under every variable valuation. A formula φ is valid if and only if $\neg\varphi$ is unsatisfiable. We say that $\mathfrak{A}$ is a model of a set T of sentences if every sentence in T is true in $\mathfrak{A}$. A sentence φ is (logically) implied by T (or T-valid), written $T \models \varphi$, if φ is true in every model of T. Similarly we say that φ is T-satisfiable if φ is true in some model of T and it is T-unsatisfiable otherwise. The notions of $(T$-$)$validity and $(T$-$)$satisfiability naturally extend to a set of formulae. A theory T is a set of sentences that is closed under logical implication, that is, if $T \models \varphi$, then $\varphi \in T$. By a theory of structure $\mathfrak{A}$, written $\mathsf{Th}(\mathfrak{A})$, we shall mean the set of all valid sentences in $\mathfrak{A}$. We write $\mathsf{Th}^\forall(\mathfrak{A})$ for the quantifier-free fragment of $\mathsf{Th}(\mathfrak{A})$.

A **term algebra** (TA) of Σ with basis $\mathcal{X}$ is the structure $\mathfrak{A}$ whose domain is $\mathcal{T}(\Sigma, \mathcal{X})$ and for any n-place function symbol $f \in \Sigma$ and $t_1, \dots, t_n \in \mathcal{T}(\Sigma, \mathcal{X})$, $f^{\mathfrak{A}}(t_1, \dots, t_n) = f(t_1, \dots, t_n)$. We assume that Σ does not contain any predicate symbols except equality.

Presburger arithmetic (PA) is the first-order theory of addition in the arithmetic of integers. The corresponding structure is denoted by $\mathfrak{A}_{\mathbb{Z}} = \langle \mathbb{Z}; 0, +, < \rangle$.

In this paper all decision procedures for quantifier-free theories are refutation-based; to determine the validity of a formula φ it suffices to determine the unsatisfiability of $\neg\varphi$, which further reduces to determining the unsatisfiability of each disjunct in the disjunctive normal form of $\neg\varphi$. Henceforth, in discussions related to quantifier-free theories, an input formula always refers to a conjunction of literals.

3   Recursive Data Structures and Oppen's Algorithm

We present a general language of recursive data structures. For simplicity, we do not distinguish syntactic terms in the language from semantic terms in the corresponding interpretation. The meaning should be clear from the context.

Definition 1. *A recursive data structure* $\mathfrak{A}_\lambda : \langle \lambda; \mathcal{A}, \mathcal{C}, \mathcal{S}, \mathcal{T} \rangle$ *consists of*

1. λ: *The* **data domain**, *which consists of all terms built up from constants by applying constructors. Elements in λ are called λ-terms (or data terms).*
2. $\mathcal{A}$: *A set of* **atoms** *(constants): a, b, c, ...*
3. $\mathcal{C}$: *A finite set of* **constructors**: $\alpha, \beta, \gamma, \dots$ *The arity of α is denoted by $\mathsf{ar}(\alpha)$. We say that an object is α-typed (or an α-term) if its outmost constructor is α.*
4. $\mathcal{S}$: *A finite set of* **selectors**. *For each constructor α with arity $k > 0$, there are k selectors $\mathsf{s}_1^\alpha, \dots, \mathsf{s}_k^\alpha$ in $\mathcal{S}$. We call s_i^α $(1 \le i \le k)$ the ith α-selector. For a term x, $\mathsf{s}_i^\alpha(x)$ returns the ith component of x if x is an α-term and x itself otherwise.*
5. $\mathcal{T}$: *A finite set of* **testers**. *For each constructor α there is a corresponding tester Is_α. For a term x, $\mathsf{Is}_\alpha(x)$ is true if and only if x is an α-term. In addition there is a special tester Is_A such that $\mathsf{Is}_A(x)$ is true if and only if x is an atom. Note that there is no need for individual atom testers as $x = a$ serves as $\mathsf{Is}_a(x)$.*

The theory of recursive data structures is essentially the theory of term algebras (with the empty variable basis) which is axiomatizable as follows.

Proposition 1 (Axiomatization of Recursive Data Structures [8]). *Let $\bar{z}_\alpha$ abbreviate $z_1, \dots, z_{\mathsf{ar}(\alpha)}$. The following formula schemes, in which variables are implicitly universally quantified over λ, axiomatize $\mathsf{Th}(\mathfrak{A}_\lambda)$.*

A. $t(x) \ne x$, *if t is built solely by constructors and t properly contains x.*

B. $a \neq b$, $a \neq \alpha(x_1 \dots, x_{\text{ar}(\alpha)})$, and $\alpha(x_1 \dots, x_{\text{ar}(\alpha)}) \neq \beta(y_1, \dots, y_{\text{ar}(\beta)})$, if a
   and b are distinct atoms and if α and β are distinct constructors.

C. $\alpha(x_1, \dots, x_{\text{ar}(\alpha)}) = \alpha(y_1, \dots, y_{\text{ar}(\alpha)}) \rightarrow \bigwedge_{1 \leq i \leq \text{ar}(\alpha)} x_i = y_i$.

D. $\text{Is}_\alpha(x) \leftrightarrow \exists \, \bar{z}_\alpha \alpha(\bar{z}_\alpha) = x$, $\text{Is}_A(x) \leftrightarrow \bigwedge_{\alpha \in C} \neg \text{Is}_\alpha(x)$.

E. $\text{s}_i^\alpha(x) = y \leftrightarrow \exists \bar{z}_\alpha \big(\alpha(\bar{z}_\alpha) = x \wedge y = z_i \big) \big) \vee \big(\forall \bar{z}_\alpha (\alpha(\bar{z}_\alpha) \neq x) \wedge x = y \big)$.

In general, selectors and testers can be defined by constructors and vice versa. One direction has been shown by (D) and (E), which are pure definitional axioms.

Example 1. Consider the LISP list structure $\mathfrak{A}_{\text{List}} = \langle \text{List}; \text{cons}, \text{car}, \text{cdr} \rangle$ where List denotes the domain, cons is the 2-place constructor (pairing function) and car and cdr are the corresponding left and right selectors (projectors) respectively. Let $\{\text{car}, \text{cdr}\}^+$ denote any nonempty sequence of car and cdr. The axiom schemas in Proposition 1 reduce to the following.

$(i) \, \text{Is}_A(x) \leftrightarrow \neg \text{Is}_{\text{cons}}(x)$, $(ii) \, \text{car}(\text{cons}(x, y)) = x$, $(iii) \, \text{cdr}(\text{cons}(x, y)) = y$,
$(iv) \, \text{Is}_A(x) \leftrightarrow \{\text{car}, \text{cdr}\}^+(x) = x$, $(v) \, \text{Is}_{\text{cons}}(x) \leftrightarrow \text{cons}(\text{car}(x), \text{cdr}(x)) = x$.

Decision Procedures for Recursive Data Structures

In [17] Oppen presented a decision procedure for acyclic data structures $\mathfrak{A}_\lambda$. The basic idea of the decision procedure is to generate all equalities between terms implied by asserted equalities in the formula and check for inconsistencies with the asserted disequalities in the formula.

The decision procedure relies on the fact that $\text{Th}(\mathfrak{A}_\lambda)$ is convex.

Definition 2 (Convexity). *A theory is **convex** if whenever a conjunction of literals implies a disjunction of atoms, it also implies one of the disjuncts.*

Let Φ be a conjunction of literals and Ψ a disjunction of equalities. The convexity of $\text{Th}(\mathfrak{A}_\lambda)$ can be rephrased as follows: if none of the disjuncts in Ψ is implied by Φ, then $\neg \Psi$ is Φ-satisfiable. Hence Φ is satisfiable if and only if for any terms s and t, whenever $s \neq t \in \Phi$, $\Phi \not\models s = t$. The idea of Oppen's algorithm is to discover all logically implied equalities (between terms in Φ) using the DAG representation and the bidirectional closure algorithm, which we introduce below.

Definition 3 (DAG Representation). *A term t can be represented by a tree T_t such that (i) if t is a constant or variable, then T_t is a leaf vertex labeled by t, and (ii) if t is in the form $\alpha(t_1, \dots, t_k)$, then T_t is the tree having the root labeled by t and having $T_{t_1}, \dots, T_{t_k}$ as its subtrees. A **directed acyclic graph (DAG)** G_t of t is obtained from T_t by "factoring out" the common subtrees (subterms).*

For a vertex u, let $\pi(u)$ denote the label, $\delta(u)$ the outgoing degree and $u[i]$ $(1 \leq i \leq \delta(u))$ the ith successor of u. The DAG of a formula is the DAG representing all terms in the formula. For example, Figure 1 shows the DAG for $\text{cons}(y, z) = \text{cons}(\text{cdr}(x), z) \wedge \text{cons}(\text{car}(x), y) \neq x$ under the assumption that x is not an atom.

Definition 4 (Bidirectional Closure). *Let R be a binary relation on a DAG and let u, v be any two vertices such that $\delta(u) = \delta(v)$. We say that R' is the* unification closure *of R (denoted by $R{\downarrow}$) if R' is the smallest equivalence relation extending R such that $\pi(u) = \pi(v)$ implies $(u[i], v[i]) \in R'$ for $1 \leq i \leq \delta(u)$. We say that R' is the* congruence closure *of R (denoted by $R{\uparrow}$) if R' is the smallest equivalence relation extending R such that $(u[i], v[i]) \in R'$ ($1 \leq i \leq \delta(u)$) implies $\pi(u) = \pi(v)$. If R' is both unification and congruence closed (w.r.t. R), we call it the* bidirectional closure, *denoted by $R{\updownarrow}$.*

Let R be the set of all pairs asserted equal in Φ. It has been shown that $R{\updownarrow}$ represents all equalities logically implied by Φ [17]. Therefore Φ is unsatisfiable if and only if there exists t and s such that $t \neq s \in \Phi$ and $(t, s) \in R{\updownarrow}$.

Algorithm 1 (Oppen's Decision Procedure for Acyclic $\mathfrak{A}_\lambda$ [17]).
Input: $\Phi : q_1 = r_1 \wedge \ldots \wedge q_k = r_k \wedge s_1 \neq t_1 \wedge \ldots \wedge s_l \neq t_l$

1. *Construct the DAG G of Φ.*
2. *Compute the bidirectional closure $R{\updownarrow}$ of $R = \{(q_i, r_i) \mid 1 \leq i \leq k\}$.*
3. *Return FAIL if $\exists i(s_i, t_i) \in R{\updownarrow}$; return SUCCESS otherwise.*

In our setting $\mathfrak{A}_\lambda$ is cyclic and values of α-selectors on non α-terms are specified, e.g., $s_i^\alpha(x) = x$ if x is not an α-term. It was shown that for such structures the decision problem is NP-complete [15]. The complication is that it is not known a priori whether $s(x)$ is a proper subterm of x and hence it is not possible to use the DAG representation directly. A solution to this problem is to guess the type information of terms occurring immediately inside selector functions before applying Algorithm 1.

Definition 5 (Type Completion). *Φ' is a* type completion *of Φ if Φ' is obtained from Φ by adding tester predicates such that for any term $s(t)$ either $\mathsf{Is}_\alpha(t)$ (for some constructor α) or $\mathsf{Is}_A(t)$ is present in Φ'.*

Example 2. A possible type completion for $y = \mathsf{car}(\mathsf{cdr}(x))$ is $y = \mathsf{car}(\mathsf{cdr}(x)) \wedge \mathsf{Is}_{\mathsf{cons}}(x) \wedge \mathsf{Is}_A(\mathsf{cdr}(x))$.

A type completion Φ' is *compatible* with Φ if the satisfiability of Φ implies that Φ' is satisfiable and if any solution of Φ' is a solution of Φ. Obviously Φ is satisfiable if and only if it has a satisfiable compatible completion. This leads to the following nondeterministic algorithm that relies on the successful guess of a satisfiable compatible completion if such completion exists.

Algorithm 2 (The Decision Procedure for Cyclic $\mathfrak{A}_\lambda$). *Input: Φ.*

1. *Guess a type completion Φ' of Φ and simplify selector terms accordingly.*
2. *Call Algorithm 1 on Φ'.*

Example 3. Figure 1 shows the DAG representation of

$$\mathsf{cons}(y, z) = \mathsf{cons}(\mathsf{cdr}(x), z) \wedge \mathsf{cons}(\mathsf{car}(x), y) \neq x \tag{1}$$

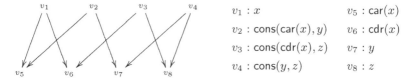

$$v_1 : x \qquad\qquad v_5 : \mathsf{car}(x)$$
$$v_2 : \mathsf{cons}(\mathsf{car}(x), y) \qquad v_6 : \mathsf{cdr}(x)$$
$$v_3 : \mathsf{cons}(\mathsf{cdr}(x), z) \qquad v_7 : y$$
$$v_4 : \mathsf{cons}(y, z) \qquad\qquad v_8 : z$$

Fig. 1. The DAG of (1) under the assumption $\mathsf{Is_{cons}}(x)$.

under the guess $\mathsf{Is_{cons}}(x)$. Initially $R = \{(v_3, v_4)\}$ as v_3 and v_4 are asserted equal in (1). (For simplicity reflexive pairs are not listed.) By the unification algorithm (v_6, v_7) are merged, which gives $R{\downarrow} = \{(v_3, v_4), (v_6, v_7)\}$. Then by the congruence algorithm (v_1, v_2) are merged, resulting in $R{\Updownarrow} = \{(v_1, v_2), (v_3, v_4), (v_6, v_7)\}$. Obviously this branch fails as $v_1 \neq v_2$ is asserted by (1). The remaining branch (with presence of $\mathsf{Is}_A(x)$) simplifies to $\mathsf{Is}_A(x) \wedge x = y$ which is clearly satisfiable, and therefore so is (1).

Note that the correctness of both Algorithm 1 and 2 relies on the (implicit) assumption that the atom domain is infinite, since otherwise the theory is not convex. As a counter example, for the structure $\mathfrak{A}_\lambda$ with only two atoms a and b, we have $\mathsf{Is}_A(x) \models x = a \vee x = b$, but neither $\mathsf{Is}_A(x) \models x = a$ nor $\mathsf{Is}_A(x) \models x = b$. We shall see (in Section 6) that our algorithm extends Oppen's original algorithm to structures with finite atom domain.

4   The Approach for the Integrated Theory

In this section we describe the different variants of the integrated theory and outline our approach for constructing the decision procedures for each of these variants. The details of each of these four decision procedures are presented in the next four sections.

Definition 6. *The structure of the integrated theory is* $\mathfrak{B}_\lambda = (\mathfrak{A}_\lambda; \mathfrak{A}_{\mathbb{Z}}; |.| : \lambda \to \mathbb{Z})$ *where* $\mathfrak{A}_\lambda$ *is a recursive data structure,* $\mathfrak{A}_{\mathbb{Z}}$ *is Presburger arithmetic, and* $|.|$ *denotes the length function defined recursively by: (i) for any atom* a, $|a| = 1$, *and (ii) for a term* $\alpha(t_1, \ldots, t_k)$, $|\alpha(t_1, \ldots, t_k)| = \sum_{i=1}^{k} |t_i|$.

Notice that we have chosen a length function that does not count the intermediate vertices. However, the algorithms can easily be modified for an alternative length function. We view terms of the form $|t|$ (where t is a non-ground λ-term) as (generalized) integer variables. If ν is an assignment of data variables, let $|\nu|$ denote the corresponding assignment of generalized integer variables. From now on, we use $\Phi_{\mathbb{Z}}$ (resp. Φ_λ) to denote a conjunction of integer literals (resp. a conjunction of data literals). Whenever it is clear from context, we write $\mathfrak{B}$ for $\mathfrak{B}_\lambda$. By $\mathfrak{B}^\omega$, $\mathfrak{B}^{<\omega}$ and $\mathfrak{B}^{=k}$ $(k > 0)$ we denote the structures with infinitely many atoms, with finitely many atoms and with exactly k atoms, respectively.

As was mentioned before, the general purpose combination method in [14] is not directly applicable due to the presence of the length function. The following example illustrates the problem.

Example 4. Consider $\mathfrak{B}^\omega_{\mathsf{List}}$. The constraints $\Phi_\lambda : x = \mathsf{cons}(\mathsf{car}(y), y)$ and $\Phi_\mathbb{Z} :$ $|x| < 2|\mathsf{car}(x)|$ are clearly satisfiable, respectively, in $\mathfrak{A}_{\mathsf{List}}$ and $\mathfrak{A}_\mathbb{Z}$. However, since Φ_λ implies that $\mathsf{car}(x) = \mathsf{car}(y)$, x contains two copies of $\mathsf{car}(y)$ and so its length should be at least two times the length of $\mathsf{car}(x)$. Therefore, $\Phi_\mathbb{Z} \wedge \Phi_\lambda$ is unsatisfiable.

A simple but crucial observation is that constraints of data structures impose "hidden" constraints on the lengths of those structures. Here we distinguish two types of integer constraints; "external integer constraints" that occur explicitly in input formulae, and "internal integer constraints" that are "induced" by satisfying assignments for the pure data structure constraints. In the following we will show that if we can express sound and complete length constraints (in the sense defined below) in Presburger arithmetic, we can derive decision procedures by utilizing the decision procedures for Presburger arithmetic and for recursive data structures.

Definition 7 (Induced Length Constraint). *A* length constraint *Φ_Δ with respect to Φ_λ is a Presburger formula in which only generalized integer variables occur free. Φ_Δ is* sound *if for any satisfying assignment ν_λ of Φ_λ, $|\nu_\lambda|$ is a satisfying assignment for Φ_Δ. Φ_Δ is* complete *if, whenever Φ_λ is satisfiable, for any satisfying assignment ν_Δ of Φ_Δ there exists a satisfying assignment ν_λ of Φ_λ such that $|\nu_\lambda| = \nu_\Delta$. We also say that Φ_Δ is* realizable *w.r.t. Φ_λ provided that Φ_λ is satisfiable. We say that Φ_Δ is* induced *by Φ_λ if Φ_Δ is both sound and complete.*

Example 5. Consider the formula $\Phi : \mathsf{cons}(x, y) = z$. The length constraint $|x| < |z| \wedge |y| < |z|$ is sound but it is not complete for Φ, as the integer assignment $\nu_\Delta : \{|x| = 3, |y| = 3, |z| = 4\}$ can not be realized. On the other hand, $|x| + |y| = |z| \wedge |x| > 5 \wedge |y| > 0$ is complete for Φ, but it is not sound because it does not satisfy the data assignment $\nu_\lambda : \{x = a, y = a, z = \mathsf{cons}(a, a)\}$. Finally, $|x| + |y| = |z| \wedge |x| > 0 \wedge |y| > 0$ is both sound and complete, and hence is the induced length constraint of Φ.

Main Theorem. *Let Φ be in the form $\Phi_\mathbb{Z} \wedge \Phi_\lambda$. Let Φ_Δ be the induced length constraint with respect to Φ_λ. Then Φ is satisfiable in $\mathfrak{B}$ if and only if $\Phi_\Delta \wedge \Phi_\mathbb{Z}$ is satisfiable in $\mathfrak{A}_\mathbb{Z}$ and Φ_λ is satisfiable in $\mathfrak{A}_\lambda$.*

Proof. ("$\Rightarrow$") Suppose that Φ is satisfiable and let ν be a satisfying assignment. ν divides into two disjoint parts: ν_λ for data variables and $\nu_\mathbb{Z}$ for integer variables. Φ_λ is obviously satisfiable under ν_λ. Let $\nu_\Delta = |\nu_\lambda|$. By soundness of Φ_Δ we know that ν_Δ satisfies Φ_Δ. Hence $\Phi_\Delta \wedge \Phi_\mathbb{Z}$ is satisfiable under the joint assignment of ν_Δ and $\nu_\mathbb{Z}$.
("$\Leftarrow$") Suppose that both $\Phi_\Delta \wedge \Phi_\mathbb{Z}$ and Φ_λ are satisfiable. Let ν be a satisfying assignment for $\Phi_\Delta \wedge \Phi_\mathbb{Z}$. ν divides into two disjoint parts: $\nu_\mathbb{Z}$ for pure integer

variables and ν_Δ for the generalized integer variables. By the completeness of Φ_Δ, ν_Δ can be realized by a satisfying assignment ν_λ such that $\nu_\Delta = |\nu_\lambda|$. Hence the joint assignment $\nu_\lambda \cup \nu_\mathbb{Z} \cup \nu_\Delta$ will satisfy Φ.   □

By the main theorem the decision problem for quantifier-free theories reduces to computing the induced length constraints in Presburger arithmetic. In the next two sections we show that this is possible for the two quantifier-free variants of our theory.

5   The Decision Procedure for $\mathsf{Th}^\forall(\mathfrak{B}^\omega)$

The first and easiest of the four variants considered is the quantifier-free combination with an infinite atom domain. In the structure $\mathfrak{B}^\omega$ the induced length constraints of a formula can be derived directly from the DAG for the formula. Before we present the algorithm we define the following predicates on terms in the DAG:

$$
\begin{aligned}
\mathsf{Tree}(t) \quad &: \exists x_1, \dots, x_n \geq 0 \ \left(\ |t| = \left(\textstyle\sum_{i=1}^{n}(d_i - 1)x_i \right) + 1 \right) \\
\mathsf{Node}^\alpha(t, \bar{t}_\alpha) &: |t| = \textstyle\sum_{i=1}^{\delta(\alpha)} |t_i| \\
\mathsf{Tree}^\alpha(t) \quad &: \exists \bar{t}_\alpha \left(\mathsf{Node}^\alpha(t, \bar{t}_\alpha) \wedge \bigwedge_{i=1}^{\delta(\alpha)} \mathsf{Tree}(t_i) \right)
\end{aligned}
$$

where $\bar{t}_\alpha$ stands for $t_1, \dots, t_{\delta(\alpha)}$ and and $d_1, \dots, d_n$ are the distinct arities of the constructors. The predicate $\mathsf{Tree}(t)$ is true iff $|t|$ is the length of a well-formed tree, since whenever a leaf expands one level with outgoing degree d, the length of the tree increases by $d - 1$. The second predicate forces the length of an α-typed node with known children to be the sum of the lengths of its children. The last predicate states the length constraint for an α-typed leaf. With these predicates the construction of the induced length constraint is given by the following algorithm.

Algorithm 3 (Construction of Φ_Δ in $\mathfrak{B}^\omega$). *Let Φ_λ be a (type-complete) data constraint, G_λ the DAG of Φ_λ and $R\mathbb{\Updownarrow}$ the bidirectional closure obtained by Algorithm 1. Initially set $\Phi_\Delta = \emptyset$. For each term t add the following to Φ_Δ.*

- $|t| = 1$, *if t is an atom;*
- $|t| = |s|$, *if $(t, s) \in R\mathbb{\Updownarrow}$.*
- $\mathsf{Tree}(t)$ *if t is an untyped leaf vertex.*
- $\mathsf{Node}^\alpha(t, \bar{t}_\alpha)$ *if t is an α-typed vertex with children $\bar{t}_\alpha$.*
- $\mathsf{Tree}^\alpha(t)$ *if t is an α-typed leaf vertex.*

Proposition 2. *Φ_Δ obtained by Algorithm 3 is expressible in a quantifier-free Presburger formula linear in the size of Φ.*

Theorem 1. *Φ_Δ obtained by Algorithm 3 is the induced length constraint of Φ_λ.*

Algorithm 4 (Decision Procedure for $\mathsf{Th}^\forall(\mathfrak{B}^\omega)$). *Input: $\Phi_\lambda \wedge \Phi_\mathbb{Z}$.*

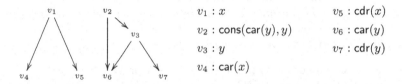

$v_1 : x$

$v_2 : \mathsf{cons}(\mathsf{car}(y), y)$

$v_3 : y$

$v_4 : \mathsf{car}(x)$

$v_5 : \mathsf{cdr}(x)$

$v_6 : \mathsf{car}(y)$

$v_7 : \mathsf{cdr}(y)$

Fig. 2. The DAG of (2) under the assumption $\mathsf{Is_{cons}}(y)$

1. *Guess a type completion Φ'_λ of Φ_λ.*
2. *Call Algorithm 1 on Φ'_λ.*
   - *Return FAIL if Φ'_λ is unsatisfiable; continue otherwise.*
3. *Construct Φ_Δ from G'_λ using Algorithm 3.*
   - *Return SUCCESS if Φ_Δ is satisfiable and $\Phi_{\mathbb{Z}}$ is satisfiable.*
   - *Return FAIL otherwise.*

The correctness of the algorithm follows from Main Theorem and Theorem 1.

Example 6. Figure 2 shows the DAG of

$$x = \mathsf{cons}(\mathsf{car}(y), y) \wedge |\mathsf{cons}(\mathsf{car}(y), y)| < 2|\mathsf{car}(x)|, \tag{2}$$

assuming that y is not an atom. The computed $R\mathord{\Uparrow}$ is $\{(v_1, v_2), (v_3, v_5), (v_4, v_6)\}$. By Algorithm 3 Φ_Δ includes the conjunction $|\mathsf{cons}(\mathsf{car}(y), y)| = |\mathsf{car}(y)| + |y| \wedge |\mathsf{car}(x)| = |\mathsf{car}(y)| \wedge |\mathsf{cdr}(x)| = |y|$ which implies $|\mathsf{cons}(\mathsf{car}(y), y)| \geq 2|\mathsf{car}(x)|$, contradicting $|\mathsf{cons}(\mathsf{car}(y), y)| < 2|\mathsf{car}(x)|$. If y is an atom, v_3, v_6, v_7 are merged. But still we have $|\mathsf{cons}(\mathsf{car}(y), y)| \geq 2|\mathsf{car}(x)|$. Therefore (2) is unsatisfiable.

6   The Decision Procedures for $\mathsf{Th}^{\forall}(\mathfrak{B}^{=k})$ and $\mathsf{Th}^{\forall}(\mathfrak{B}^{<\omega})$

Algorithm 3 cannot be used to construct the induced length constraints in structures with a finite number of atoms, $\mathfrak{B}^{=k}$, as illustrated by the following example.

Example 7. Consider $\mathfrak{B}^{=1}_{\mathsf{List}}$ (with atom a). The constraint

$$|x| = 3 \wedge \mathsf{Is}_A(y) \wedge x \neq \mathsf{cons}(\mathsf{cons}(y, y), y) \wedge x \neq \mathsf{cons}(y, \mathsf{cons}(y, y)) \tag{3}$$

is unsatisfiable while Φ_Δ obtained by Algorithm 3 is

$$|y| = 1 \wedge |\mathsf{cons}(y, y)| = 2 \wedge |\mathsf{cons}(\mathsf{cons}(y, y), y)| = 3 \wedge |\mathsf{cons}(y, \mathsf{cons}(y, y))| = 3$$

which is obviously satisfiable together with $|x| = 3$.

The reason is that if the atom domain is finite there are only finitely many terms of length n for any $n > 0$. If a term t is forced to be distinct from all of them, then t cannot have length n. Therefore Φ_Δ needs to include constraints that count the number of distinct terms of a certain length.

Definition 8 (Counting Constraint). *A* counting constraint *is a predicate* $\mathsf{CNT}^{\alpha}_{k,n}(x)$ *that is* true *if and only if there are at least* $n+1$ *different α-terms of length x in the language with exactly $k > 0$ distinct atoms.* $\mathsf{CNT}_{k,n}(x)$ *is similarly defined with α-terms replaced by λ-terms.*

The following two monotonicity properties are easily proven: for any $l \geq k > 0$ and $m \geq n > 0$, (i) $\mathsf{CNT}^{\alpha}_{k,n}(x) \rightarrow \mathsf{CNT}^{\alpha}_{l,n}(x)$ and (ii) $\mathsf{CNT}^{\alpha}_{k,m}(x) \rightarrow \mathsf{CNT}^{\alpha}_{k,n}(x)$.

Example 8. For $\mathfrak{B}^{=1}_{\mathsf{List}}$, $\mathsf{CNT}^{\mathsf{cons}}_{1,n}(x)$ is $x \geq m$ where m is the least number such that the m-th Catalan number $C_m = \frac{1}{m}\binom{2m-2}{m-1}$ is greater than n. This is not surprising as C_m gives the number of binary trees with m leaf vertices.

In general we have the following result.

Proposition 3. $\mathsf{CNT}^{\alpha}_{k,n}(x)$ *is expressible by a quantifier-free Presburger formula that can be computed in time $O(n)$.*

In order to construct counting constraints, we need equality information between terms.

Definition 9 (Equality Completion). *An* equality completion Φ'_{λ} *of Φ_{λ} is formula obtained from Φ_{λ} such that for any two terms u and v in Φ_{λ} either $u = v$ or $u \neq v$, and either $|u| = |v|$ or $|u| \neq |v|$ are in Φ'_{λ}.*

As before we present a nondeterministic algorithm; Φ is satisfiable if and only if at least one of the compatible (type and equality) completions of Φ is. By $\mathsf{NEQ}_n(x_0, x_1, \dots, x_n)$ we shall mean that $x_0, \dots, x_n$ have the same length but are pairwise distinct.

Algorithm 5 (Construction of Φ_{Δ} in $\mathfrak{B}^{=k}$).
Input: Φ_{λ} (type and equality complete), G_{λ} and $R\updownarrow$.

1. *Call Algorithm 3 to obtain Φ_{Δ}.*
2. *Add $\mathsf{CNT}^{\alpha}_{k,n}(|t|)$ to Φ_{Δ} for each t occurring in $\mathsf{NEQ}_n(t, t_1, \dots, t_n)$.*

Proposition 4. Φ_{Δ} *obtained by Algorithm 5 is expressible in a quantifier-free Presburger formula and the size of such a formula is linear in the size of Φ.*

Theorem 2. Φ_{Δ} *obtained by Algorithm 5 is the induced length constraint of Φ_{λ}.*

Algorithm 6 (Decision Procedure for $\mathsf{Th}^{\forall}(\mathfrak{B}^{=k})$). *Input : $\Phi_{\lambda} \wedge \Phi_{\mathbb{Z}}$.*

1. *Guess a type and equality completion Φ'_{λ} of Φ_{λ}.*
2. *Call Algorithm 1 on Φ'_{λ}.*
   - *Return* FAIL *if Φ'_{λ} is unsatisfiable; continue otherwise.*
3. *Construct Φ_{Δ} from G'_{λ} using Algorithm 5.*
   - *Return* SUCCESS *if Φ_{Δ} is satisfiable and $\Phi_{\mathbb{Z}}$ is satisfiable.*
   - *Return* FAIL *otherwise.*

The correctness of Algorithm 6 follows from the Main Theorem and Theorem 2. Notice that, when $\Phi_{\mathbb{Z}}$ is empty, the algorithm can be viewed as an extension of Oppen's original algorithm for structures with finite atom domain.

Example 9. Let us return to Example 7. Constraint (3) has exactly one compatible completion, namely $\mathsf{NEQ}_3(x, \mathsf{cons}(\mathsf{cons}(y, y), y), \mathsf{cons}(y, \mathsf{cons}(y, y)))$. This results in the counting constraint $|x| \geq 4$, contradicting $|x| = 3$.

Algorithm 4 is also a decision procedure for $\mathsf{Th}^\forall(\mathfrak{B}^{<\omega})$ according to the following theorem.

Theorem 3. $\mathsf{Th}^\forall(\mathfrak{B}^{<\omega}) = \mathsf{Th}^\forall(\mathfrak{B}^\omega)$ *in the languages with no constants.*

7   The Decision Procedure for $\mathsf{Th}(\mathfrak{B}^\omega)$

In this section and the next one we show the decidability of $\mathsf{Th}(\mathfrak{B}^\omega)$, $\mathsf{Th}(\mathfrak{B}^{=k})$ and $\mathsf{Th}(\mathfrak{B}^{<\omega})$ by extending the quantifier elimination procedure for the theory of term algebras [13,11,8] to our combined theory. A structure is said to "admit quantifier elimination" if any formula can be equivalently (and effectively) transformed into a quantifier-free formula. We demonstrate a procedure by which a sentence of $\mathfrak{B}$ is reduced to a ground quantifier-free formula (i.e., with no variable occurrence) whose validity can be easily checked. Our elimination procedures induce decision procedures for quantifier-free theories as satisfiability of a quantifier-free formula is the same as validity of its existential closure. However, unfortunately, the complexity lower bound of the theory of term algebras is non-elementary [3,7].

It is well-known that eliminating arbitrary quantifiers reduces to eliminating existential quantifiers from formulae in the form $\exists x(A_1(x) \wedge \ldots \wedge A_n(x))$, where $A_i(x)$ $(1 \leq i \leq n)$ are literals [8]. In the rest of the paper, we assume that all transformations are done on formulae of this form. We may also assume that $A_i's$ are not of the form $x = t$ as $\exists x(x = t \wedge \varphi(x, \bar{y}))$ simplifies to $\varphi(t, \bar{y})$, if x does not occur in t, to $\exists x \varphi(x, \bar{y})$ if $t \equiv x$, and to false by Axiom (A) if t properly contains x. As before, we present nondeterministic algorithms, but now a formula is valid if and only if it is true in every existential branch.

For our two-sorted language, we show how to eliminate quantifiers on integer variables as well as quantifiers on data variables. It is easy to see the soundness of transformations in the elimination procedures presented in this section and the next one. We leave the termination proofs to the extended version of this paper.

Eliminate Quantifiers on Integer Variables

We may assume formulae with quantifiers on integer variables are in the form

$$\exists x : \mathbb{Z} \ (\Phi_{\mathbb{Z}}(x, \bar{y}, \bar{z}) \wedge \Phi_\lambda(\bar{z})), \tag{4}$$

where $\bar{y}$ (resp. $\bar{z}$) denotes a sequence of integer variables (resp. data variables). Since $\Phi_\lambda(\bar{z})$ does not contain x, we can move it out of the scope of $\exists x$, and obtain

$$\exists x : \mathbb{Z} \ (\Phi_{\mathbb{Z}}(x, \bar{y}, \bar{z})) \wedge \Phi_\lambda(\bar{z}). \tag{5}$$

Notice that in $\Phi_{\mathbb{Z}}(x, \bar{y}, \bar{z})$, $\bar{z}$ only occurs inside generalized integer variables of the form $|t|$. Therefore $\exists x : \mathbb{Z}(\Phi_{\mathbb{Z}}(x, \bar{y}, \bar{z}))$ is essentially a Presburger formula and we can proceed to remove the quantifier using Cooper's method [4].

Eliminate Quantifiers on Data Variables

We may assume formulae with quantifiers on data variables are in the form

$$\exists x : \lambda \; (\Phi_{\mathbb{Z}}(x, \bar{y}, \bar{z}) \wedge \Phi_{\lambda}(x, \bar{z})), \qquad (6)$$

where $\bar{y}$ (resp. $\bar{z}$) denotes a sequence of integer variables (resp. data variables). As before in $\Phi_{\mathbb{Z}}(\bar{y}, x, \bar{z})$, $x, \bar{z}$ only occur inside integer terms. We may assume that (6) does not contain constructors (see Section 3).

First we make sure that x does not appear properly inside any terms. Suppose otherwise that $s(x)$ occurs for some selector s. As in [8], by guessing that x is α-typed, (6) becomes

$$\exists x, x_1, \dots, x_{\mathsf{ar}(\alpha)} : \lambda \; \left[\mathsf{Is}_{\alpha}(x) \wedge \bigwedge_{1 \le i \le \mathsf{ar}(\alpha)} s_i^{\alpha}(x) = x_i \wedge \Phi_{\lambda}(x, \bar{z}) \wedge \Phi_{\mathbb{Z}}(x, \bar{y}, \bar{z}) \right]. \quad (7)$$

We simplify (7) as follows: replace $x \ne t$ by $\bigvee_{1 \le i \le \mathsf{ar}(\alpha)} s_i^{\alpha}(t) \ne x_i \vee \neg \mathsf{Is}_{\alpha}(t)$ (which will cause disjunctive splittings), replace $s_i^{\alpha}(x)$ by x_i, replace $s_j^{\beta}(x)$ (for $\alpha \not\equiv \beta$) by x and replace $|x|$ by $\Sigma_{i=1}^{\mathsf{ar}(\alpha)} |x_i|$. After the simplification no x occurs in Φ_{λ} and $\Phi_{\mathbb{Z}}$, so we can remove $\bigwedge_{1 \le i \le \mathsf{ar}(\alpha)} s_i^{\alpha}(x) = x_i$, $\mathsf{Is}_{\alpha}(x)$ and the quantifier $\exists x$. Repeat the process if all of $x_1 \dots x_{\mathsf{ar}(\alpha)}$ occur inside selector functions.

Although the transformation from (6) to (7) introduces new quantified variables, those new variables appear in smaller terms (compared with x). Therefore the process will eventually terminate[2], producing formulae of the form

$$\exists x : \lambda \; (\bigwedge_{i < n} x \ne t_i \wedge \Phi_{\lambda}(\bar{z}) \wedge \Phi_{\mathbb{Z}}(x, \bar{y}, \bar{z})), \qquad (8)$$

where x does not appear in $\Phi_{\lambda}(\bar{z})$ and t_i ($i < n$). (8) says that there exists an x which is not equal to any terms of $t_0, \dots t_n$ and whose length is constrained by $\Phi_{\mathbb{Z}}(x, \bar{y}, \bar{z})$. But in $\mathfrak{B}^{\omega}$ for any $n > 0$ there are infinitely many terms of length n. It follows that $\bigwedge_{i < n} x \ne t_i$ can be ignored and thus (8) is equivalent to

$$\exists n : \mathbb{Z} \; \left[\mathsf{Tree}(x) \big[|x|/n \big] \wedge \Phi_{\lambda}(\bar{z}) \wedge \Phi_{\mathbb{Z}}(x, \bar{y}, \bar{z}) \big[|x|/n \big] \right], \qquad (9)$$

where $\varphi[|x|/n]$ stands for the formula obtained from φ by substituting all occurrences of $|x|$ for n. Now (9) can be handled by the elimination procedure for integer quantifiers. Note that if $\mathsf{Is}_{\alpha}(x)$ is in (8) we use $\mathsf{Tree}^{\alpha}(x)$ instead of $\mathsf{Tree}(x)$ in (9). Also, if we guessed that x is an atom, we would directly arrive at (8) by substituting x for $s(x)$, and then to (9) with n instantiated to 1.

8   The Decision Procedures for $\mathsf{Th}(\mathfrak{B}^{=k})$ and $\mathsf{Th}(\mathfrak{B}^{<\omega})$

In $\mathfrak{B}^{=k}$ the reduction from (8) to (9) is not sound for the same reason as in Section 6. However, the technique (of adding counting constraints) is still applicable here. We first introduce some new notations.

---

[2] To guarantee termination, we require the next round variable selection be restrained to $\{x_1 \dots x_{\mathsf{ar}(\alpha)}\}$. In other words, the variable selection is done in depth-first manner.

Definition 10 (Partitioning Formula). *Let x be a data variable and S be the set of data terms $\{t_1,\dots,t_n\}$. Let P be a partition of S and $Q \subseteq P$. By a partitioning formula, written* $\mathsf{PART}[x,S,P,Q]$, *we shall mean*

$$\mathsf{NE}_\lambda[x,S] \wedge \underbrace{\mathsf{NE}'_\lambda[S,P] \wedge \mathsf{EQ}_\lambda[S,P] \wedge \mathsf{NE}_\mathbb{Z}[x,S,Q] \wedge \mathsf{EQ}_\mathbb{Z}[x,S,Q]}_{\mathsf{PART}'[x,S,P,Q]}, \qquad (10)$$

where $\mathsf{NE}_\lambda[x,S] \equiv \bigwedge_{t\in S} x \neq t,$ $\mathsf{NE}_\mathbb{Z}[x,S,Q] \equiv \bigwedge_{t\in S_i \notin Q} |x| \neq |t|,$

   $\mathsf{NE}'_\lambda[S,P] \equiv \bigwedge_{t\in S_i \in P, t'\in S_j \in P, S_i \neq S_j} t \neq t',$ $\mathsf{EQ}_\mathbb{Z}[x,S,Q] \equiv \bigwedge_{t\in S_i \in Q} |x| = |t|,$

   $\mathsf{EQ}_\lambda[S,P] \equiv \bigwedge_{t,t'\in S_i \in P} t = t'.$

In fact $\mathsf{PART}[x,S,P,Q]$ is an equality completion of terms $\{x\} \cup S$.

Example 10. Let $S = \{t_1,t_2,t_3\}$, $P = \{\{t_1,t_2\},\{t_3\}\}$ and $Q = \{\{t_1,t_2\}\}$. Then
   $\mathsf{NE}_\lambda[x,S] : x \neq t_1 \wedge x \neq t_2 \wedge x \neq t_3$ $\mathsf{NE}_\mathbb{Z}[x,S,Q] : |x| \neq |t_3|$
   $\mathsf{NE}'_\lambda[S,P] : t_1 \neq t_3 \wedge t_2 \neq t_3$ $\mathsf{EQ}_\mathbb{Z}[x,S,Q] : |x| = |t_1| \wedge |x| = |t_2|$
   $\mathsf{EQ}_\lambda[S,P] : t_1 = t_2$

The elimination of quantifiers on integer variables is the same as before. We show how to eliminate quantifiers on data variables as follows. Starting from (8) we guess a partition P of $S = \{t_1,\dots,t_n\}$ and a set $Q \subseteq P$, resulting in

$$\exists x : \lambda \left[\mathsf{NE}_\lambda[x,S] \wedge \mathsf{PART}'[x,S,P,Q] \wedge \Phi_\lambda(\bar{z}) \wedge \Phi_\mathbb{Z}(x,\bar{y},\bar{z}) \right], \qquad (11)$$

which says that there exists an x such that (i) x is not equal to any terms in S, and (ii) S is partitioned into $|P|$ equivalence classes among which there are $|Q|$ classes whose members have the same length as x. But this is exactly what the counting constraint $\mathsf{CNT}_{k,|Q|}(|x|)$ states, and so we can transform (11) to

$$\exists n : \mathbb{Z} \left[\mathsf{Tree}(x)\left[|x|/n\right] \wedge \mathsf{CNT}_{k,|Q|}(n) \wedge \right.$$
$$\left. \mathsf{PART}'[x,S,P,Q]\left[|x|/n\right] \wedge \Phi_\lambda(\bar{z}) \wedge \Phi_\mathbb{Z}(|x|/n,\bar{y},\bar{z}) \right]. \qquad (12)$$

Again, we are left with the known task of eliminating integer quantifiers. As before, in case of presence of $\mathsf{Is}_\alpha(x)$, we use $\mathsf{Tree}^\alpha(x)$ and $\mathsf{CNT}^\alpha_{k,|Q|}(n)$ instead of $\mathsf{Tree}(x)$ and $\mathsf{CNT}_{k,|Q|}(n)$, respectively, in (12). If we had guessed that x is an atom, we would have (12) with n instantiated to 1.

Example 11. Consider, in $\mathfrak{B}^{=1}_{\mathsf{List}}$ (with atom a), $\exists x : \lambda(x \neq \mathsf{cons}(a,a) \wedge |x| = 2)$. The only compatible completion is $\exists x : \lambda(\mathsf{Is}_{\mathsf{cons}}(x) \wedge |x| = |\mathsf{cons}(a,a)| \wedge x \neq \mathsf{cons}(a,a) \wedge |x| = 2)$. This produces the counting constraint $\mathsf{CNT}^{\mathsf{cons}}_{1,1}(|x|)$ which is $|x| > 2$, contradicting $|x| = 2$.

We can also derive a decision procedure for $\mathsf{Th}(\mathfrak{B}^{<\omega})$ using the above elimination procedure.

Theorem 4. $\mathsf{Th}(\mathfrak{B}^{<\omega})$ *(in the languages with no constants) is decidable.*

9   Complexity

In this section we briefly discuss complexity of the quantifier theories. Let n be the input size of Φ. First it is not hard to see that both $\mathsf{Th}^{\forall}(\mathfrak{B}^{\omega})$ and $\mathsf{Th}^{\forall}(\mathfrak{B}^{=k})$ are NP-hard as they are super theories of $\mathsf{Th}^{\forall}(\mathfrak{A}_{\lambda})$ and $\mathsf{Th}^{\forall}(\mathfrak{A}_{\mathbb{Z}})$, either of which is NP-complete. Second, Algorithm 3 computes Φ_{Δ} in $O(n)$ (see Proposition 2) and so does Algorithm 5 (see Proposition 4). Third, the size of any type and equality completion of Φ is bounded by $O(n^2)$ as there are at most n^2 pairs of terms. By the nondeterministic nature of our algorithms, we see that each branch of computation (in Algorithm 4 and Algorithm 6 respectively) is in P. Therefore both $\mathsf{Th}^{\forall}(\mathfrak{B}^{\omega})$ (or $\mathsf{Th}^{\forall}(\mathfrak{B}^{<\omega})$) and $\mathsf{Th}^{\forall}(\mathfrak{B}^{=k})$ are NP-complete.

10   Conclusion

We presented four classes of decision procedures for recursive data structures integrated with Presburger arithmetic. Our technique is based on the extraction of sound and complete integer constraints from data constraints. We believe that this technique may apply to arithmetic integration of various other theories. We plan to extend our results in two directions. The first is to reason about the combination of recursive data structures with integers in richer languages such as the theory of recursive data structures with subterm relation $\preceq$ [20]. The second is to relax the restriction of unique construction of data objects to enable handling of structures in which a data object can be constructed in more than one way such as in the theory of queues [1,19] and word concatenation [12].

Recently it came to our attention that the combination of Presburger arithmetic and term algebras has been used in [9,10] to show that the quantifier-free theory of term algebras with Knuth-Bendix order is NP-complete. For quantifier-free theories in finite signatures, the decidability result is more general than ours, as the language also includes the (Knuth-Bendix) ordering predicate. But on the other hand, our quantifier elimination method presented in this paper can be readily extended to show the decidability of the first-order theory of Knuth-Bendix order [23]. We will compare the work with ours in the extended version of this paper.

References

1. Nikolaj S. Bjørner. *Integrating Decision Procedures for Temporal Verification*. PhD thesis, Computer Science Department, Stanford University, November 1998.
2. Nikolaj S. Bjørner, Anca Browne, Michael Colón, Bernd Finkbeiner, Zohar Manna, Henny B. Sipma, and Tomás E. Uribe. Verifying temporal properties of reactive systems: A STeP tutorial. *Formal Methods in System Design*, 16(3):227–270, June 2000.
3. K. J. Compton and C. W. Henson. A uniform method for proving lower bounds on the computational complexity of logical theories. *Annals of Pure and Applied Logic*, 48:1–79, 1990.

4. D. C. Cooper. Theorem proving in arithmetic without multiplication. In *Machine Intelligence*, volume 7, pages 91–99. American Elsevier, 1972.
5. J. Downey, R. Sethi, and R. E. Tarjan. Variations of the common subexpression problem. *Journal of ACM*, 27:758–771, 1980.
6. H. B. Enderton. *A Mathematical Introduction to Logic*. Academic Press, 2nd edition, 2001.
7. J. Ferrante and C. W. Rackoff. *The Computational Complexity of Logical Theories*. Springer, 1979.
8. Wilfrid Hodges. *Model Theory*. Cambridge University Press, Cambridge, UK, 1993.
9. Konstantin Korovin and Andrei Voronkov. A decision procedure for the existential theory of term algebras with the knuth-bendix ordering. In *Proceedings of 15th IEEE Symposium on Logic in Computer Science (LICS)*, pages 291 – 302, 2000.
10. Konstantin Korovin and Andrei Voronkov. Knuth-Bendix constraint solving is NP-complete. In *Proceedings of 28th International Colloquium on Automata, Languages and Programming (ICALP)*, volume 2076 of *Lecture Notes in Computer Science*, pages 979–992. Springer, 2001.
11. M. J. Maher. Complete axiomatizations of the algebras of finite, rational and infinite tree. In *Proceedings of the 3rd IEEE Symposium on Logic in Computer Science (LICS)*, pages 348–357. IEEE Press, 1988.
12. G. S. Makanin. The problem of solvability of equations in a free semigroup. *Math. Sbornik*, 103:147–236, 1977. English translation in Math. USSR Sb. 32, 129-198.
13. A. I. Mal'cev. Axiomatizable classes of locally free algebras of various types. In *The Metamathematics of Algebraic Systems, Collected Papers*, chapter 23, pages 262–281. North-Holland, 1971.
14. Greg Nelson and Derek C. Oppen. Simplification by cooperating decision procedures. *ACM Transactions on Programming Languages and Systems (TOPLAS)*, 1(2):245–257, October 1979.
15. Greg Nelson and Derek C. Oppen. Fast decision procedures based on congruence closure. *Journal of ACM*, 27(2):356–364, April 1980.
16. Derek C. Oppen. Elementary bounds for presburger arithmetic. In *Proceedings of 5th Annual ACM Symposium on Theory of Computing (STOC)*, pages 34–37. ACM Press, 1973.
17. Derek C. Oppen. Reasoning about recursively defined data structures. *Journal of ACM*, 27(3), July 1980.
18. C. R. Reddy and D. W. Loveland. Presburger arithmetic with bounded quantifier alternation. In *Proceedings of the 10th Annual Symposium on Theory of Computing (STOC)*, pages 320–325. ACM Press, 1978.
19. Tatiana Rybina and Andrei Voronkov. A decision procedure for term algebras with queues. *ACM Transactions on Computational Logic*, 2(2):155–181, 2001.
20. K. N. Venkataraman. Decidability of the purely existential fragment of the theory of term algebras. *Journal of ACM*, 34,2:492–510, 1987.
21. Calogero G. Zarba. Combining multisets with integers. In Andrei Voronkov, editor, *Proceedings of the 18th International Conference on Automated Deduction*, volume 2392 of *Lecture Notes in Artificial Intelligence*, pages 363–376. Springer, 2002.
22. Calogero G. Zarba. Combining sets with integers. In Alessandro Armando, editor, *Proceedings of the 4th International Workshop on Frontiers of Combining Systems (FroCoS)*, volume 2309 of *Lecture Notes in Artificial Intelligence*, pages 103–116. Springer, 2002.
23. Ting Zhang, Henny B. Sipma, and Zohar Manna. The decidability of the first-order theory of term algebras with Knuth-Bendix order. Submitted to CP'04. Extended version available at theory.stanford.edu/~tingz/papers/cp04_extended.pdf.

Modular Proof Systems for Partial Functions with Weak Equality

Harald Ganzinger, Viorica Sofronie-Stokkermans, and Uwe Waldmann

Max-Planck-Institut für Informatik, Stuhlsatzenhausweg 85,
66123 Saarbrücken, Germany, {hg,sofronie,uwe}@mpi-sb.mpg.de

Abstract. The paper presents a modular superposition calculus for the combination of first-order theories involving both total and partial functions. Modularity means that inferences are pure, only involving clauses over the alphabet of either one, but not both, of the theories. The calculus is shown to be complete provided that functions that are not in the intersection of the component signatures are declared as partial. This result also means that if the unsatisfiability of a goal modulo the combined theory does not depend on the totality of the functions in the extensions, the inconsistency will be effectively found. Moreover, we consider a constraint superposition calculus for the case of hierarchical theories and show that it has a related modularity property. Finally we identify cases where the partial models can always be made total so that modular superposition is also complete with respect to the standard (total function) semantics of the theories.

1 Introduction

This paper aims at providing new modularity results for refutational theorem proving in first-order logic with equality. In Nelson/Oppen-style combinations of two first-order theories $\mathcal{T}_1$ and $\mathcal{T}_2$ over signatures Σ_1 and Σ_2, inferences are *pure* in that all premises of an inference are clauses over only one of the signatures Σ_i where i depends on the inference. Therefore, no mixed formulas are ever generated when refuting goals represented by sets of pure formulas. What needs to be passed between the two theory modules are only universal formulas[1] over the intersection $\Sigma_1 \cap \Sigma_2$ of the two signatures. For stably infinite theories where, in addition, $\Sigma_1 \cap \Sigma_2$ consists of constants only, pure inference systems exist. This is one of the main consequences of Nelson's and Oppen's results [12] (also see, e.g., Tinelli and Harandi [14] for additional clarification). Ghilardi [8] has recently extended these completeness results for modular inference systems to a more general case of "compatibility" between the component theories $\mathcal{T}_i$. Future work might aim at liberating these compatibility requirements even further.

In this paper we take a different point of departure. We will consider arbitrary theory modules and investigate what one loses in terms of completeness when

---

[1] For Nelson/Oppen-style combination of theories, one even restricts the information exchange between theories to ground clauses over the intersection signature.

D. Basin and M. Rusinowitch (Eds.): IJCAR 2004, LNAI 3097, pp. 168–182, 2004.

superposition inferences are restricted to be pure. Superposition is refutationally complete for equational first-order logic, and by choosing term orderings appropriately (terms over $\Sigma_1 \cap \Sigma_2$ should be minimal in the term ordering), many, but not all, cases of impure inferences can be avoided. Impure inferences arise when one of the extensions $\Sigma_1 \setminus \Sigma_2$ or $\Sigma_2 \setminus \Sigma_1$ has additional non-constant function symbols. It is known that in such cases interpolants of implications of the form $\phi_1 \supset \phi_2$, with ϕ_i a Σ_i-formula, normally contain existential quantification. That means, that refutationally complete clausal theorem provers where existential quantifiers are skolemized, need to pass clauses from $\mathcal{T}_1$ to $\mathcal{T}_2$ [from $\mathcal{T}_2$ to $\mathcal{T}_1$] containing function symbols not in Σ_2 [Σ_1]. In other words, inference systems are either incomplete or impure necessarily. One of the main results of the paper is that if the extensions only introduce additional relations and partial functions,[2] a particular calculus of superposition for partial functions to be developed in this paper becomes a complete and modular proof system where inferences are pure. This result can be applied to problems where partial functions arise naturally. Alternatively we may think of this result as indicating what we lose if superposition is restricted to pure inferences. If a proof cannot be found in the pure system, a partial algebra model exists for the goal to be refuted. Conversely, if the inconsistency of a goal does not depend on the totality of the functions in the extensions, we will be able to find the inconsistency with the modular partial superposition calculus. There are interesting cases of problem classes where partial models can always be totalized and where the modular system is therefore in fact complete (cf. Section 5).

In Section 2 we will describe the logic of partial functions we are working with. The logic is that of weak equality in the sense of Evans [6]. This logic allows one to specify undefinedness, but not definedness, of a function. (However we may specify a kind of relative definedness as explained below.) Then, in Section 3, we state and prove sound and complete a superposition calculus for clauses over signatures where functions can be declared as either total or partial. The calculus might be of independent interest for problem domains where partial functions arise in a natural manner. (That aspect, however, will not be explored any further in this paper as we are mainly interested in modularity.) We show that the calculus only admits pure inferences in cases of theory combinations where all functions that are not in the intersection of the signatures are declared as partial. In Section 4 we consider a variant of the calculus, called constraint superposition, suitable for hierarchical extensions $\mathcal{T}_1$ of a base theory $\mathcal{T}_0$. It differs from the previous calculus in that unification is replaced by generating equality constraints over the base theory. This system is modular in that no inferences involving base clauses (over Σ_0) need to be made. Rather, we may integrate any refutationally complete prover for $\mathcal{T}_0$ accepting the base clauses generated from non-base inferences and returning falsum whenever the accumulated set of base

---

[2] A non-equational literal $p(t_1, \ldots, t_n)$ or $\neg p(t_1, \ldots, t_n)$, where p is a relation symbol, can be encoded as an equational literal $f_p(t_1, \ldots, t_n) \approx true$ or $\neg f_p(t_1, \ldots, t_n) \approx true$, where f_p is a partial function and $true$ a total constant. Thus we will in the sequel not mention relations anymore.

clauses is inconsistent with $\mathcal{T}_0$. In Section 5 we consider both shallow and local extensions of base theories, showing that for those classes of extensions constraint superposition is complete also with respect to the total algebra semantics of theories and goals. Finally Section 6 discusses related work.

2   Partial Functions with Weak Equality

A *signature* Σ is the disjoint union of a set Σ_{T} of total function symbols and a set Σ_{P} of partial function symbols. Each function symbol comes with a fixed arity.[3] Terms are built over Σ and a set V of variables.

A Σ-*algebra* A consists of a non-empty set U_A, a total function $f_A : U_A^n \to U_A$ for every $f/n \in \Sigma_{\mathrm{T}}$ and a partial function $g_A : U_A^n \to U_A$ for every $g/n \in \Sigma_{\mathrm{P}}$.

An *assignment* β into A is a mapping from the set of variables V to U_A. Given an algebra A and an assignment β into A, the value $(A, \beta)(t)$ of a Σ-term t is either a member of the universe U_A or one of the two special values $\perp_{\mathrm{u}}$ ("undefined") or $\perp_{\mathrm{i}}$ ("irrelevant"). It is defined as follows:

$(A, \beta)(x) = \beta(x)$
    if x is a variable.

$(A, \beta)(f(t_1, \ldots, t_n)) = f_A(a_1, \ldots, a_n)$
    if $(A, \beta)(t_i) = a_i \in U_A$ for all $i \in \{1, \ldots, n\}$ and $f_A(a_1, \ldots, a_n)$ is defined.

$(A, \beta)(f(t_1, \ldots, t_n)) = \perp_{\mathrm{u}}$
    if $(A, \beta)(t_i) = a_i \in U_A$ for all $i \in \{1, \ldots, n\}$ and $f_A(a_1, \ldots, a_n)$ is undefined.

$(A, \beta)(f(t_1, \ldots, t_n)) = \perp_{\mathrm{i}}$
    if $(A, \beta)(t_i) \in \{\perp_{\mathrm{u}}, \perp_{\mathrm{i}}\}$ for some $i \in \{1, \ldots, n\}$.

In other words, a term is irrelevant if one of its proper subterms is undefined.

To evaluate the truth of an equation, we use a three-valued logic with the values 1 (true), $\frac{1}{2}$ (undefined), and 0 (false). The truth values 1 and $\frac{1}{2}$ are called positive.

Given an algebra A and an assignment β into A, the truth value of a formula F w.r.t. A and β is denoted by $(A, \beta)(F)$. If F is an equation $s \approx t$, then $(A, \beta)(F) = 1$ if $(A, \beta)(s) = (A, \beta)(t) \in U_A$; $(A, \beta)(F) = \frac{1}{2}$ if $(A, \beta)(s) = (A, \beta)(t) = \perp_{\mathrm{u}}$ or $(A, \beta)(s) = \perp_{\mathrm{i}}$ or $(A, \beta)(t) = \perp_{\mathrm{i}}$; and otherwise $(A, \beta)(F) = 0$.

For complex formulas, we have

$$(A, \beta)(F \wedge G) = \min\{(A, \beta)(F), (A, \beta)(G)\},$$
$$(A, \beta)(F \vee G) = \max\{(A, \beta)(F), (A, \beta)(G)\},$$
$$(A, \beta)(\neg F) = 1 - (A, \beta)(F),$$
$$(A, \beta)(\forall x.F) = \min\{(A, \beta[x \mapsto a])(F) \mid a \in U_A\},$$
$$(A, \beta)(\exists x.F) = \max\{(A, \beta[x \mapsto a])(F) \mid a \in U_A\}.$$

---

[3] For simplicity, we restrict to the one-sorted case and to equality as the only predicate symbol. The extensions to a many-sorted framework where every function symbol has a unique declaration $f : s_1 \ldots s_n \to s_0$ and to additional predicate symbols are obvious.

We use $s \not\approx t$ as a shorthand for $\neg\, s \approx t$; in inference rules, the symbol $\overset{.}{\approx}$ denotes either $\approx$ or $\not\approx$.

An algebra A is a *model* of a formula F, if $(A, \beta)(F) \geq \frac{1}{2}$ for every β, or in other words, if F is positive (i.e., true or undefined) w.r.t. A and β; it is a model of a set N of formulas, if it is a model of every formula in N. If A is a model of F, we say that F *holds* in A. A formula F *follows* from a set N of formulas (denoted by $N \models F$), if every model of N is a model of F. Note that an algebra A is a model of a ground equation $s \approx t$ if both s and t are defined and equal in A, or if both are undefined, or if at least one of them is irrelevant; A is a model of $s \not\approx t$ unless both s and t are defined and different in A. It is easy to check that every ground clause C holds in an algebra A as soon as one term occurring in C is irrelevant in A. Intuitively, the ground instances of a clause that contain irrelevant terms are those instances which we choose to ignore.

Example 1. Let $\Sigma_T = \{\mathsf{nil}/0, \mathsf{cons}/2\}$, $\Sigma_P = \{\mathsf{car}/1, \mathsf{cdr}/1\}$, and let A be the algebra of finite lists with the usual interpretation of these symbols.

Then A is a model of $\forall x.\mathsf{cons}(\mathsf{car}(x), \mathsf{cdr}(x)) \approx x$: Suppose that x is mapped to some $a \in U_A$. Then either both $\mathsf{car}_A(a)$ and $\mathsf{cdr}_A(a)$ are undefined, hence the value of $\mathsf{cons}(\mathsf{car}(x), \mathsf{cdr}(x))$ is irrelevant, and the equation has the truth value $\frac{1}{2}$. Or $\mathsf{car}_A(a)$ and $\mathsf{cdr}_A(a)$ are defined; in this case $\mathsf{cons}_A(\mathsf{car}_A(a), \mathsf{cdr}_A(a)) = a$, so the equation has the truth value 1. The truth value of the universally quantified formula is $\min\{\frac{1}{2}, 1\} = \frac{1}{2}$, therefore A is a model of the formula.

Since $\mathsf{car}_A(\mathsf{nil}_A)$ and $\mathsf{cdr}_A(\mathsf{nil}_A)$ are undefined, A is a model of both the formula $\mathsf{car}(\mathsf{nil}) \approx \mathsf{cdr}(\mathsf{nil})$ and its negation $\mathsf{car}(\mathsf{nil}) \not\approx \mathsf{cdr}(\mathsf{nil})$. It is not a model of $\mathsf{car}(\mathsf{nil}) \approx \mathsf{nil}$ (the left-hand side is undefined, the right-hand side is defined), it is, however, a model of $\mathsf{car}(\mathsf{car}(\mathsf{nil})) \approx \mathsf{nil}$ (the left-hand side is irrelevant).

Note that explicit [un-]definedness predicates are not present in this logic. To express that a term t is not defined, one can simply state that $t \not\approx t$. Expressing that t (not containing partial function symbols below the top) is defined is only possible, if Σ_T contains appropriate total function symbols or can be extended by new symbols. For example, for an algebra B to be a model of $\forall x, y.\mathsf{car}(\mathsf{cons}(x, y)) \approx x$, car_B has to be defined for every b in the codomain of cons_B. Equations of this form implicitly express definedness requirements for partial functions.

From now on, we will consider only the clausal fragment of this logic. As usual, all variables in a clause are implicitly universally quantified. The theorem proving calculus described below will check whether a set N of clauses is inconsistent, that is, whether $N \models \bot$, where $\bot$ is the empty clause. The entailment problem "does a clause F follow from N" can be reduced to this refutation problem, but the reduction is a bit more complicated than in usual two-valued logic. The following example demonstrates the principal ideas of the reduction:

Example 2. Suppose that $\Sigma_T \supseteq \{a/0, b/0\}$ and $\Sigma_P \supseteq \{f/1, g/1\}$. We want to check whether $N \models f(a) \approx g(b)$ for some set N of clauses. One might think that this is equivalent to $N \cup \{f(a) \not\approx g(b)\} \models \bot$, but this is not true: If $N = \{f(a) \approx g(b)\}$, then $N \models f(a) \approx g(b)$, but still the set $N \cup \{f(a) \not\approx g(b)\}$ has a model, namely one in which $f(a)$ and $g(b)$ are undefined. The statement $N \models f(a) \approx g(b)$ holds if in each model of N either $f(a)$ and $g(b)$ are defined and equal, or both are undefined. Conversely, it does not hold, if there is a model of N in which $f(a)$ is defined and $g(b)$ is undefined or defined and different from $f(a)$, or vice versa. To translate the entailment problem into a *set of* refutation problems, we need therefore a new total function symbol $c/0$: $N \models f(a) \approx g(b)$ holds if and only if both $N \cup \{f(a) \approx c,\, g(b) \not\approx c\} \models \bot$ and $N \cup \{f(a) \not\approx c,\, g(b) \approx c\} \models \bot$.

3   Superposition for Partial Functions

The inference rules of the traditional superposition calculus (Bachmair and Ganzinger [1]) must be modified in several ways in order to be sound for our logic of partial functions. For instance, a literal $s \not\approx s$ may hold in an algebra – namely if s is undefined or irrelevant – so the equality resolution rule may be applied only if s is guaranteed to be defined. Similarly, replacement of equals by equals may be unsound: Assume that g is a partial function, $f(g(a))$ is irrelevant in some algebra A, and b is defined, then $f(g(a)) \approx b$ and $f(g(a)) \not\approx b$ hold in A, but $b \not\approx b$ does not. Consequently, a term that is replaced using some inference rule may contain a partial function symbol at the top, but not below the top (so that it is either defined or undefined, but not irrelevant).

For the same reason, substitutions that introduce partial function symbols must be ruled out: We say that a substitution is *total*, if no variable is mapped to a term containing a partial function symbol. If Q is a term/formula/inference and σ is a total substitution, then $Q\sigma$ is a total instance of Q.

Inference System 3. Let us start the presentation of the inference rules of the *partial superposition calculus* with a few general conventions.

The inference system is parameterized by a reduction ordering $\succ$ on terms that is total on ground terms and that has the property that every ground term over Σ_T is smaller than every ground term containing a symbol from Σ_P (for instance, a lexicographic path ordering where all symbols from Σ_P have higher precedence than symbols from Σ_T).[4]

To a positive literal $s \approx t$, we assign the multiset $\{s, t\}$, to a negative literal $\neg\, s \approx t$ the multiset $\{s, s, t, t\}$. The literal ordering $\succ_L$ compares these multisets using the multiset extension of $\succ$. The clause ordering $\succ_C$ compares clauses by comparing their multisets of literals using the multiset extension of $\succ_L$.

A literal that is involved in an inference must be maximal in the respective clause (except for the literal $s_0 \approx s'_0$ in *merging paramodulation* and the literals $t_i \approx t'_i$ $(i > 1)$ in *partial top-superposition*). A positive literal that is involved in a *superposition*, *partial top-superposition*, or *merging paramodulation* inference must be strictly maximal in the respective clause (with the exceptions above). In inferences with two premises, the left premise is not greater than or equal to the right premise.

---

[4] Since we are interested in total ground instances only, this implies that a variable may be considered as smaller than every term containing a symbol from Σ_P.

Equality Resolution

$$\frac{C' \vee s \not\approx s'}{C'\sigma}$$

if s does not contain partial function symbols and σ is a total most general unifier of s and s'.

Superposition

$$\frac{D' \vee t \approx t' \qquad C' \vee s[u] \dot{\approx} s'}{(D' \vee C' \vee s[t'] \dot{\approx} s')\sigma}$$

if u is not a variable, t does not contain partial function symbols below the top, σ is a total most general unifier of t and u, $t\sigma \not\preceq t'\sigma$, $s\sigma \not\prec s'\sigma$, and, if $s \dot{\approx} s'$ occurs positively or s is a Σ_{T}-term, then $s\sigma \not\preceq s'\sigma$.

Partial Top-Superposition

$$\frac{D' \vee t_1 \approx t'_1 \vee \ldots \vee t_n \approx t'_n \qquad C' \vee s \approx s'}{(D' \vee C' \vee s' \approx t'_1 \vee \ldots \vee s' \approx t'_n)\sigma}$$

if $n \geq 2$, s contains a partial function symbol at the top and no partial function symbols below the top, each t'_i contains a partial function symbol, σ is a total most general unifier of s and all t_i, $t_i\sigma \not\preceq t'_i\sigma$, and $s\sigma \not\preceq s'\sigma$.[5]

Merging Paramodulation

$$\frac{D' \vee t \approx t' \qquad C' \vee s_0 \approx s'_0 \vee s \approx s'[u]}{(D' \vee C' \vee s_0 \approx s'_0 \vee s \approx s'[t'])\sigma}$$

if u is not a variable, t does not contain partial function symbols below the top, σ is a total most general simultaneous unifier of t and u and of s_0 and s, $t\sigma \not\preceq t'\sigma$, $s\sigma \not\preceq s'\sigma$, $s\sigma \not\preceq s'_0\sigma$, and $s'\sigma \not\preceq s'_0\sigma$,

Factoring

$$\frac{C' \vee s \approx s' \vee t \approx t'}{(C' \vee s \approx s')\sigma}$$

if σ is a total most general simultaneous unifier of s and t and of s' and t'.

Theorem 4. *The inference rules of the partial superposition calculus are sound, that is, whenever the premises of an inference hold in some algebra A, then the conclusion holds in A.*

Proof. Let us consider first the *equality resolution* rule. Suppose that A is a model of the clause $C = C' \vee s \not\approx s'$, where s is a Σ_{T}-term; let σ be a total unifier of s and s' and let β be an arbitrary assignment. Since σ is total, $x\sigma$ is a Σ_{T}-term and $(A,\beta)(x\sigma) \in U_A$ for every variable x. Define the assignment γ by $\gamma(x) =$

---

[5] *Partial top-superposition* corresponds to iterated *superposition* into the right premise, except that the intermediate conclusions may not be eliminated if they are redundant as defined below; in fact, it can be implemented that way.

$(A, \beta)(x\sigma)$. By assumption, $\frac{1}{2} \leq (A, \gamma)(C) = (A, \beta)(C\sigma) = (A, \beta)(C'\sigma \lor s\sigma \not\approx s'\sigma)$. Now note that $s\sigma = s'\sigma$ is a Σ_{T}-term, hence $(A, \beta)(s\sigma) = (A, \beta)(s'\sigma) \in U_A$ and therefore $(A, \beta)(s\sigma \not\approx s'\sigma) = 0$. Consequently, $(A, \beta)(C'\sigma) \geq \frac{1}{2}$. Since β could be chosen arbitrarily, A is a model of $C'\sigma$.

For the *superposition* rule assume that A is a model of the clauses $D = D' \lor t \approx t'$ and $C = C' \lor s[u] \mathrel{\dot{\approx}} s'$, where t does not contain Σ_{P}-symbols below the top. W.l.o.g., C and D have no common variables. Let σ be a total unifier of t and u and let β be an arbitrary assignment. Since σ is total, $x\sigma$ is a Σ_{T}-term and $(A, \beta)(x\sigma) \in U_A$ for every variable x. Define the assignment γ by $\gamma(x) = (A, \beta)(x\sigma)$. By assumption, $\frac{1}{2} \leq (A, \gamma)(C) = (A, \beta)(C\sigma) = (A, \beta)(C'\sigma \lor s\sigma[u\sigma] \mathrel{\dot{\approx}} s'\sigma)$ and $\frac{1}{2} \leq (A, \gamma)(D) = (A, \beta)(D\sigma) = (A, \beta)(D'\sigma \lor t\sigma \approx t'\sigma)$. If $(A, \beta)(C'\sigma) \geq \frac{1}{2}$ or $(A, \beta)(D'\sigma) \geq \frac{1}{2}$, it is obvious that the conclusion is positive w.r.t. A and β. Otherwise $(A, \beta)(s\sigma[u\sigma] \mathrel{\dot{\approx}} s'\sigma) \geq \frac{1}{2}$ and $(A, \beta)(t\sigma \approx t'\sigma) \geq \frac{1}{2}$. Since t does not contain Σ_{P}-symbols below the top, $(A, \beta)(t\sigma) \in U_A \cup \{\perp_{\mathrm{u}}\}$. This leaves two possible reasons why $(A, \beta)(t\sigma \approx t'\sigma)$ is positive: If $(A, \beta)(t\sigma) = (A, \beta)(t'\sigma) \in U_A \cup \{\perp_{\mathrm{u}}\}$, then clearly $(A, \beta)(s\sigma[t'\sigma] \mathrel{\dot{\approx}} s'\sigma) = (A, \beta)(s\sigma[t\sigma] \mathrel{\dot{\approx}} s'\sigma) = (A, \beta)(s\sigma[u\sigma] \mathrel{\dot{\approx}} s'\sigma) \geq \frac{1}{2}$. Otherwise $(A, \beta)(t'\sigma) = \perp_{\mathrm{i}}$, then $(A, \beta)(s\sigma[t'\sigma]) = \perp_{\mathrm{i}}$, hence $(A, \beta)(s\sigma[t'\sigma] \mathrel{\dot{\approx}} s'\sigma) = \frac{1}{2}$.

The soundness of the *partial top-superposition* and *merging paramodulation* rules is proved analogously.

Finally we consider the *factoring* rule. Let A be a model of the clause $C = C' \lor s \approx s' \lor t \approx t'$; let σ be a total simultaneous unifier of s and t and of s' and t', and let β be an arbitrary assignment. Define the assignment γ by $\gamma(x) = (A, \beta)(x\sigma)$. By assumption, $\frac{1}{2} \leq (A, \gamma)(C) = (A, \beta)(C\sigma) = (A, \beta)(C'\sigma \lor s\sigma \approx s'\sigma \lor t\sigma \approx t'\sigma)$. Clearly, $(A, \beta)(s\sigma \approx s'\sigma) = (A, \beta)(t\sigma \approx t'\sigma)$, hence $(A, \beta)(C'\sigma \lor s\sigma \approx s'\sigma) = (A, \beta)(C\sigma) \geq \frac{1}{2}$. Since β could be chosen arbitrarily, A is a model of the conclusion.

As usual, the inference rules of the partial superposition calculus are accompanied by a redundancy criterion. A ground clause C is called redundant w.r.t. a set N of ground clauses, if it follows from clauses in N that are smaller than C. A ground inference is called redundant w.r.t. a set N of ground clauses, if its conclusion follows from clauses in N that are smaller than the largest premise. For general clauses and inferences, redundancy is defined by lifting: A clause or inference is redundant w.r.t. a set N of clauses, if all its total ground instances are redundant w.r.t the set of total ground instances of clauses in N.

A set N of clauses is called saturated up to redundancy, if all inferences between clauses in N are redundant w.r.t. N. We will show that the partial superposition calculus is refutationally complete, that is, that a saturated set of clauses has a model if and only if it does not contain the empty clause. The "only if" part of this proposition is of course trivial. For the "if" part, we have to construct a model of a saturated set N. This model is represented by a convergent term rewrite system or, equivalently, by an equational theory. The universe of the model consists of all ground normal forms of the rewrite system that are Σ_{T}-terms (or, equivalently, of the congruence classes of all ground Σ_{T}-terms). Given such a model, a ground term is defined, if its normal form is a Σ_{T}-term;

it is undefined, if all its immediate subterms have normal forms that are Σ_T-terms, but the term itself does not; it is irrelevant, if some of its subterms does not have a normal forms that is a Σ_T-term. The rewrite system is constructed from the set $\overline{N}$ of total ground instances of clauses in N. Starting with an empty interpretation all such instances are inspected in ascending order w.r.t. the clause ordering. If a reductive clause is false and irreducible in the interpretation constructed so far, its maximal equation is turned into a rewrite rule and added to the interpretation. If the original clause set is saturated and doesn't contain the empty clause, then the final interpretation is a model of all ground instances, and thus of the original clause set (Bachmair and Ganzinger [1]).

Let N be a set of clauses not containing $\bot$. Using induction on the clause ordering we define sets of rewrite rules E_C and R_C for all $C \in \overline{N}$ as follows:

Assume that E_D has already been defined for all $D \in \overline{N}$ with $D \prec_c C$. Then $R_C = \bigcup_{D \prec_c C} E_D$. The set E_C contains the rewrite rule $s \to s'$, if

    (a) $C = C' \vee s \approx s'$.
    (b) $s \approx s'$ is strictly maximal in C.
    (c) $s \succ s'$.
    (d) C is false in R_C.
    (e) C' is false in $R_C \cup \{s \to s'\}$.
    (f) s is irreducible w.r.t. R_C and contains no Σ_P-symbols below the top.
    (g) the R_C-normal form of s' contains no Σ_P-symbols.
    (h) no clause $D \in \overline{N}$ with $D \prec_c C$ is false in $R_C \cup \{s \to s'\}$.

In this case, C is called productive. Otherwise $E_C = \emptyset$. Finally, $R_\infty = \bigcup_{D \in \overline{N}} E_D$.

Lemma 5. *Let N be a set of clauses that is saturated up to redundancy and does not contain the empty clause. Then we have for every total ground instance $C\theta \in \overline{N}$:*

    (i) $E_{C\theta} = \emptyset$ if and only if $C\theta$ has positive truth value in $R_{C\theta}$.
    (ii) $C\theta$ has positive truth value in R_∞ and in R_D for every $D \succ_c C\theta$.

The proof can be found in the full version of the paper.

Theorem 6. *The partial superposition calculus is refutationally complete.*

Proof. We have to show that a saturated set N of clauses has a model if and only if does not contain the empty clause.

If N contains the empty clause, then obviously it does not have a model. Otherwise, the rewrite system R_∞ constructed above gives us a model A of N: The universe of A consists of all ground normal forms of R_∞ that are Σ_T-terms (or, equivalently, of the congruence classes of all ground Σ_T-terms). A function $f_A : U_A^n \to U_A$ maps the terms $t_1, \dots, t_n$ to the R_∞-normal form of $f(t_1, \dots, t_n)$ if this is a Σ_T-term, it is undefined otherwise.

There are alternative ways of dealing with partial functions in automated theorem proving, notably by encoding a partial function f/n as an $(n+1)$-ary relation r together with a clause $\neg r(x_1, \dots, x_n, y) \vee \neg r(x_1, \dots, x_n, y') \vee y \approx y'$.

One may ask whether partial superposition has any advantages over such an encoding. First, it is clear that the flattening of terms resulting from the relational encoding will generally make it more difficult to detect simplification opportunities. Second, the strengthened ordering restrictions of partial superposition reduce the number of possible inferences. The following trivial example illustrates this:

Example 7. Let $\Sigma_T = \{a/0,\ b/0,\ c/0\}$, let $\Sigma_P = \{f/1\}$, and suppose that N contains the clauses

$$f(a) \approx b$$
$$f(a) \approx c$$
$$b \not\approx c$$

where $a \succ b \succ c$. Partial superposition derives $b \approx c$ from the first two clauses, then $c \not\approx c$, and then the empty clause. This whole process is completely deterministic: no other inferences are possible. Besides, every superposition between a unit clause and a ground clause is necessarily a simplification of the second premise, so that the second premise $f(a) \approx b$ can be deleted from the set of clauses.

If we use relational encoding of partial functions, then N is turned into

$$r(a,b)$$
$$r(a,c)$$
$$\neg\, r(x,y) \ \vee\ \neg\, r(x,y') \ \vee\ y \approx y'$$
$$b \not\approx c$$

In contrast to partial superposition, where we had exactly one way to derive $b \approx c$, there are now two different hyperresolution inferences that produce this clause, plus two further hyperresolution inferences that produce the tautologies $b \approx b$ and $c \approx c$. Moreover, we need now one further computation step to see that $b \approx c$ and $r(a,c)$ make $r(a,b)$ redundant.

We now show that the partial superposition calculus is modular for combinations of theories where all total functions are in the intersection of their signatures. Assume that we have two signatures Σ_1 and Σ_2. Call an inference pure if its premises are either all clauses over Σ_1 or they are all clauses over Σ_2. Note that a pure inference of the partial superposition calculus, in particular, derives a pure Σ_1-clause or a pure Σ_2-clause.

Theorem 8. *Suppose that Σ_1 and Σ_2 are two signatures that share the set of total function symbols and have disjoint sets of partial function symbols. Let N be a set of clauses, such that every clause in N is either a pure Σ_1-clause or a pure Σ_2-clause. Then all inferences of the partial superposition calculus with premises in N are pure.*

Proof. For the inference rules with only one premise, the result is trivial, since the clauses in N are pure. For the binary inference rules observe that either the term t or t_1 in the first premise contains a partial symbol; then this symbol must also occur in the second premise so that both premises are pure clauses over the same Σ_i. Or t is a Σ_{T}-term, then by the properties of the ordering the first premise contains only total symbols, hence is both a Σ_1- and a Σ_2-clause. Again, the inference is pure.

4   Hierarchic Extensions

The inference system of the partial superposition calculus (and its completeness proof) can be turned – with slight modifications – into a calculus for hierarchic structures. We assume the following scenario: Let Σ_0 be a (total) signature, and let $\mathcal{T}_0$ be a first-order theory over Σ_0 such that some refutationally complete prover exists that is able to check the unsatisfiability of sets of Σ_0-clauses w.r.t. $\mathcal{T}_0$. Now we consider an extension $\Sigma_1 \supseteq \Sigma_0$ of the signature, such that all symbols in $\Sigma_1 \setminus \Sigma_0$ are partial, and a set N of Σ_1-clauses. The task is to check whether N is unsatisfiable relative to $\mathcal{T}_0$, that is, whether there is no model of N whose restriction to Σ_0 is a model of $\mathcal{T}_0$, using the prover for $\mathcal{T}_0$ as a black-box.

We will modify the rules of the partial superposition calculus as follows:

- Replace unification by lazy unification: Change the inference rules in such a way that total most general unifiers are not computed, but instead the unification constraint is turned into new antecedent literals.
- Do not perform inferences that do not involve partial function symbols. (This implies that the *equality resolution* rule is dropped completely.)
- Add a new inference rule that allows to derive a contradiction from any finite set of Σ_0-clauses that is inconsistent with $\mathcal{T}_0$.

Inference System 9. The *constraint superposition calculus* uses the inference rules defined below.

Term and literal orderings are defined as before. A literal that is involved in an inference must be maximal in the respective clause (except for the literal $s_0 \approx s_0'$ in *merging paramodulation* and the literals $t_i \approx t_i'$ ($i > 1$) in *partial top-superposition*). A positive literal that is involved in a *superposition*, *partial top-superposition*, or *merging paramodulation* inference must be strictly maximal in the respective clause (with the exceptions above). In inferences with two premises, the left premise is not greater than or equal to the right premise.

Superposition
$$\frac{D' \vee f(\vec{u}) \approx t' \qquad C' \vee s[f(\vec{v})] \mathrel{\dot{\approx}} s'}{D' \vee C' \vee \vec{u} \not\approx \vec{v} \vee s[t'] \mathrel{\dot{\approx}} s'}$$

if $f \in \Sigma_1 \setminus \Sigma_0$, $\vec{u}$ and $\vec{v}$ do not contain partial function symbols, $f(\vec{u}) \not\preceq t'$, and $s \not\prec s'$.

Partial Top-Superposition
$$\frac{D' \vee \bigvee_{1 \leq i \leq n} f(\vec{u}_i) \approx t_i' \qquad C' \vee f(\vec{v}) \approx s'}{D' \vee C' \vee \bigvee_{1 \leq i \leq n} \vec{u}_i \not\approx \vec{v} \vee \bigvee_{1 \leq i \leq n} s' \approx t_i'}$$

if $n \geq 2$, $f \in \Sigma_1 \setminus \Sigma_0$, $\vec{u}_i$ and $\vec{v}$ do not contain partial function symbols, each t'_i contains a partial function symbol, $f(\vec{u}_i) \not\preceq t'_i$, and $f(\vec{v}) \not\preceq s'$.

Merging Paramodulation
$$\frac{D' \vee f(\vec{u}) \approx t' \quad C' \vee g(\vec{q}) \approx s'_0 \vee g(\vec{r}) \approx s'[f(\vec{v})]}{D' \vee C' \vee \vec{q} \not\approx \vec{r} \vee \vec{u} \not\approx \vec{v} \vee g(\vec{q}) \approx s'_0 \vee g(\vec{r}) \approx s'[t']}$$

if $f, g \in \Sigma_1 \setminus \Sigma_0$, $\vec{u}$ and $\vec{v}$ do not contain partial function symbols, $f(\vec{u}) \not\preceq t'$, $g(\vec{r}) \not\preceq s'$, $g(\vec{r}) \not\preceq s'_0$, and $s' \not\preceq s'_0$,

Factoring
$$\frac{C' \vee s \approx s' \vee t \approx t'}{C' \vee \bigvee_{1 \leq i \leq n} x_i \not\approx t_i \vee s \approx s'}$$

if $\sigma = \{ x_i \mapsto t_i \mid 1 \leq i \leq n \}$ is a total most general simultaneous unifier of s and t and of s' and t'.

Constraint Refutation
$$\frac{M}{\bot}$$

if M is a finite set of Σ_0-clauses that is inconsistent with the base theory $\mathcal{T}_0$.

Theorem 10. *Let N be a saturated set of clauses. Then N has a model whose restriction to Σ_0 is a model of the base theory $\mathcal{T}_0$ if and only if N does not contain the empty clause.*

Proof. The proof proceeds in essentially the same way as for the partial superposition calculus. The main difference is that we do not start the model construction with the empty interpretation but with some convergent ground rewrite system that is contained in $\succ$ and represents an arbitrary term-generated $\mathcal{T}_0$-model of all Σ_0-clauses in N. Such a model does exist, otherwise, by compactness of $\mathcal{T}_0$, the *constraint refutation* rule would have been applied.

The constraint superposition calculus is related to a calculus presented by Bachmair, Ganzinger, and Waldmann [3], where a base theory is extended by *total* functions, but where sufficient completeness of the extension is necessary for the refutational completeness of the calculus.

5   Shallow and Local Extensions of a Base Theory

As shown in Section 4, constraint superposition is complete whenever all functions in the extension are declared as partial. In fact this result can be extended to the many-sorted case where only functions in the extension having a codomain of base sort have to be partial, whereas functions with a result of extension sort can be declared as either total or partial.

From our point of view, an important application of this result is to approximate refutational theorem proving in extensions of base theories for which refutationally complete black box theorem provers exist. If constraint superposition finds a contradiction for a set of clauses in the extended signature, the set is unsatisfiable in particular also with respect to total algebras. In that sense constraint superposition is a sound and modular approximation of refutational theorem proving for hierarchical first-order theories.

In this section we discuss cases for when this approximation is, in fact, complete. A particularly simple case is that of a shallow extension. Suppose $\mathcal{T}_0 \subseteq \mathcal{T}_1$ is a theory extension in which all functions in the extension $\Sigma_1 \setminus \Sigma_0$ having a codomain in the set S_0 of (base) sorts in Σ_0 are declared partial. A Σ_1-clause C is called *shallow* if partial function symbols occur in C only positively and only at the root of terms. The theory extension $\mathcal{T}_0 \subseteq \mathcal{T}_1$ is shallow if $\mathcal{T}_1 \setminus \mathcal{T}_0$ consists only of shallow clauses.

For an example, suppose we have the natural numbers (of sort nat) as base theory. Consider as an extension the two clauses

$$\mathsf{read}(\mathsf{write}(a, i, x), i) \approx x$$

$$x \approx y \lor \mathsf{read}(\mathsf{write}(a, i, x), j) \approx \mathsf{read}(a, j)$$

where array is a new sort, write : array $\times$ nat $\times$ nat $\to$ array is a total and read : array $\times$ nat $\to$ nat a partial function symbol, and a, i, j, x are variables of suitable sort. Under these assumptions the two clauses are shallow. This definition of read is tail-recursive, and tail-recursive definitions in general of a partial function will be shallow. Other kinds of recursive definitions will normally not be shallow, as exemplified by the case of append over nat-lists

$$\mathsf{append}(\mathsf{cons}(x, l), l') \approx \mathsf{cons}(x, \mathsf{append}(l, l'))$$

where append : list $\times$ list $\to$ list is partial and cons : nat $\times$ list $\to$ list is total.

Shallow extensions enjoy the property that any partial algebra model can be extended to total algebra model.

Theorem 11. *Suppose that $\mathcal{T}_0 \subseteq \mathcal{T}_1$ is a theory extension in which all functions in $\Sigma_1 \setminus \Sigma_0$ of a codomain in S_0 are declared partial. If all clauses in $\mathcal{T}_1 \setminus \mathcal{T}_0$ are shallow, then $\mathcal{T}_1$ has a partial model if, and only if, $\mathcal{T}_1$ has a total model.*

Proof. Suppose A is a partial Σ_1-algebra that is a model of $\mathcal{T}_1$. Pick, for each sort s, an element a_s from the carrier A_s associated with the sort s in A. Let B be the extension of A into a total algebra obtained by making f_B return a_s, wherever f_A is undefined in A, for every partial f of codomain s. It is easy to see that B is also a model of $\mathcal{T}_1$: Since all function symbols in Σ_0 are total, $B|_{\Sigma_0}$ coincides with $A|_{\Sigma_0}$ so that B is a model of $\mathcal{T}_0$. Suppose $\mathcal{T}_1 \setminus \mathcal{T}_0$ contains, say, an equation $f(\vec{s}) \approx g(\vec{t})$ with f partial and of sort s. Since the equation is shallow, neither $\vec{s}$ nor $\vec{t}$ contain any partial function symbol. Thus, for each assignment of the variables, the values $\vec{a}$ and $\vec{b}$ for $\vec{s}$ and $\vec{t}$, respectively, are defined. Therefore, in order for the equation to be satisfied in A, f_A is defined on $\vec{a}$ if, and only if, g_A is defined on $\vec{b}$. If $f_A(\vec{a})$ is defined, so is $g_A(\vec{b})$, and $f_B(\vec{a}) = f_A(\vec{a}) = g_A(\vec{b}) = g_B(\vec{b})$. If $f_A(\vec{a})$ is undefined so is $g_A(\vec{b})$, thus $f_B(\vec{a}) = a_s = g_B(\vec{b})$. For the case of general clauses also note that partial functions do not occur negatively in shallow clauses.

Note that any set of ground clauses can be turned into a set of shallow ground clauses by introducing new (total) constants for subterms that start with a partial function. That *flattening transformation* preserves [un-]satisfiability with respect to total algebra semantics. Therefore, a simple application of the preceding theorem is to the case where a base theory is extended by free function symbols and where we want to prove unsatisfiability of sets of ground clauses over this extension: flattening the clauses followed by applying constraint superposition is a sound and complete (with respect to total algebra semantics) and modular method for this problem. Constraint superposition can be used to prove that if the universal (clause) theory of the base theory $\mathcal{T}_0$ is decidable then the universal theory of any extension of $\mathcal{T}_0$ by free function symbols is also decidable. This provides a simple proof of a result established also in (Tinelli and Zarba [15]).

A more general, but related, case is that of local theories. Call a theory extension $\mathcal{T}_0 \subseteq \mathcal{T}_1$ *local* if, for every set G of ground Σ_1-clauses, whenever $\mathcal{T}_1 \cup G$ is unsatisfiable then also $\mathcal{T}_0 \cup \mathcal{T}_1[G] \cup G$ is unsatisfiable where $\mathcal{T}_1[G]$ is the set of those instances of clauses in $\mathcal{T}_1 \setminus \mathcal{T}_0$ in which each subterm starting with a partial function is a ground subterm appearing in G or in $\mathcal{T}_1 \setminus \mathcal{T}_0$. The set $\mathcal{T}_1[G]$ is finite and can be effectively computed from G. Flattening the terms in $\mathcal{T}_1[G] \cup G$ by introducing auxiliary total constants and then applying constraint superposition is, therefore, a sound and complete (with respect to total algebra semantics) and modular method for proving unsatisfiability of ground goals G with respect to $\mathcal{T}_1$.

For a local theory extension, constraint superposition is also complete (w.r.t. total semantics) when applied directly to $\mathcal{T}_1 \cup G_F$, with G_F the flattened form of G. Thus, we can avoid computing the ground instances $\mathcal{T}_1[G]$, which might be too expensive in many cases.

6   Related Work

In this section related work is summarized and compared with the results presented in the paper.

Validity of identities in partial algebras. There are many possibilities for defining validity of identities in partial algebras, from which we mention only a few (for further details we refer to [4]): "existential validity" $((A, \beta) \models t \overset{e}{=} t'$ if and only if $(A, \beta)(t)$ and $(A, \beta)(t')$ are both defined and equal); "strong validity" $((A, \beta) \models t \overset{s}{=} t'$ if and only if either both $(A, \beta)(t)$ and $(A, \beta)(t')$ are undefined or both are defined and equal); and validity in the sense of Evans [6,5,9], which is the form of validity we use in this paper. We have chosen Evans' definition, which we call weak equality, as it is too weak to allow one to define totality of a partial function (except by definitions involving previously defined total functions). On the other hand Evans validity is often related to properties of embeddability of partial algebras into total algebras [6,5,9]. This connection allows us to replace equational reasoning for total functions with reasoning about partial functions, or with relational reasoning. Evans validity was also used in (Ganzinger [7]) for establishing relationships between semantic and proof-theoretic approaches to polynomial time decidability for uniform word problems for quasi-varieties, in particular connections between embeddability and locality of equational theories. In the present paper we have extended Evans' validity of identities to a notion of validity for clauses, and have shown an embeddability result for shallow and for local extensions of base theories.

Resolution calculi for partial functions and partial congruences. An alternative way to dealing with undefinedness, which goes back to Kleene [11], is to use many-valued logic, with an additional truth value for "undefined". Kleene's logic has been used by various authors for giving logical systems for partial functions and for reasoning about partial functions in a many-valued framework. A resolution calculus for partial functions, where undefinedness is formalized using Kleene's strong three valued logic, was proposed by Kerber and Kohlhase in [10]. Refinements of resolutions such as paramodulation or superposition are not considered in [10].

Bachmair and Ganzinger [2] give a version of ordered chaining for partial equivalence and congruence axioms. This calculus is devised for *strong* or *existential validity*; consequently, reflexivity resolution is replaced with a rule which encodes partial reflexivity.

Modular theorem proving in combinations of theories. In Nelson/Oppen-style combinations of stably infinite theories $\mathcal{T}_1$ and $\mathcal{T}_2$ over signatures Σ_1 and Σ_2 which are disjoint or share only constants, inferences are always pure. Ghilardi [8] has recently extended the completeness results for modular inference systems for combinations of theories over non-disjoint signatures. Theorem 8, one of the main results of our paper, also provides a modular way of combining extensions $\mathcal{T}_1$ and $\mathcal{T}_2$ of an *arbitrary* base theory $\mathcal{T}_0$. The main difference between Ghilardi's approach and our work is that in (Ghilardi [8]) the component theories need to satisfy a rather strong compatibility condition with respect to the shared theory. On the other hand, our calculi are only complete with respect to the partial function semantics. We have shown, however, that for shallow or local extensions of base theories partial models can always be made total. Ghilardi's compatibility conditions [8] ensure, in addition, that the Craig interpolants consist of positive ground clauses whereas in the modular partial superposition calculus described in this paper clauses with variables need to be exchanged between the theory modules.

For Theorem 8 to be applicable, the theories $\mathcal{T}_1$ and $\mathcal{T}_2$ (regarded as theories with partial functions in Σ_1, Σ_2) have the same total function symbols. A similar situation was analyzed by Tinelli [13], who gives a method for cooperation of background reasoners for universal theories which have *the same function symbols*. However, we have shown that there are interesting problem classes where partial models can always be totalized, and therefore, in these cases the condition that the theories $\mathcal{T}_1$ and $\mathcal{T}_2$ have the same total function symbols can be relaxed. The superposition calculus for partial functions developed in this paper also allows to efficiently compute the (universal) Craig interpolant even in this more general case.

References

1. Leo Bachmair and Harald Ganzinger. Rewrite-based equational theorem proving with selection and simplification. *Journal of Logic and Computation*, 4(3):217–247, 1994.
2. Leo Bachmair and Harald Ganzinger. Ordered chaining calculi for first-order theories of transitive relations. *Journal of the ACM*, 45(6), 1998.
3. Leo Bachmair, Harald Ganzinger, and Uwe Waldmann. Refutational theorem proving for hierarchic first-order theories. *Applicable Algebra in Engineering, Communication and Computing*, 5(3/4):193–212, 1994.
4. Peter Burmeister. *A Model Theoretic Oriented Approach to Partial Algebras: Introduction to Theory and Application of Partial Algebras, Part I*, vol. 31 of *Mathematical Research*. Akademie-Verlag, Berlin, 1986.
5. Stanley Burris. Polynomial time uniform word problems. *Mathematical Logic Quarterly*, 41:173–182, 1995.
6. T. Evans. Embeddability and the word problem. *J. London Math. Soc.*, 28:76–80, 1953.
7. Harald Ganzinger. Relating semantic and proof-theoretic concepts for polynomial time decidability of uniform word problems. In *Sixteenth Annual IEEE Symposium on Logic in Computer Science*, 2001, pp. 81–92. IEEE Computer Society Press.
8. Silvio Ghilardi. Quantifier elimination and provers integration. *Electronic Notes in Theoretical Computer Science*, 86(1), 2003.
9. George Grätzer. *Universal algebra*. Springer Verlag, 2. edition, 1968.
10. Manfred Kerber and Michael Kohlhase. A mechanization of strong Kleene logic for partial functions. In Alan Bundy, ed., *Twelfth International Conference on Automated Deduction*, Nancy, France, 1994, LNAI 814, pp. 371–385. Springer-Verlag.
11. Stephen C. Kleene. *Introduction to Metamathematics*. D. Van Nostrand Company, Inc., Princeton, New Jersey, 1952.
12. Greg Nelson and Derek C. Oppen. Simplification by cooperating decision procedures. *ACM Transactions on Programming Languages and Systems*, 1(2):245–257, 1979.
13. Cesare Tinelli. Cooperation of background reasoners in theory reasoning by residue sharing. *Journal of Automated Reasoning*, 30(1):1–31, 2003.
14. Cesare Tinelli and Mehdi Harandi. A new correctness proof of the Nelson-Oppen combination procedure. In Franz Baader and Klaus U. Schulz, eds., *Frontiers of Combining Systems, First International Workshop*, Munich, Germany, 1996, Applied Logic Series, Vol. 3, pp. 103–119. Kluwer Academic Publishers.
15. Cesare Tinelli and Calogero Zarba. Combining non-stably infinite theories. *Electronic Notes in Theoretical Computer Science*, 86(1), 2003.

A New Combination Procedure for the Word Problem That Generalizes Fusion Decidability Results in Modal Logics

Franz Baader[1]*, Silvio Ghilardi[2], and Cesare Tinelli[3]

[1] Institut für Theoretische Informatik, TU Dresden
[2] Dipartimento di Scienze dell'Informazione, Università degli Studi di Milano
[3] Department of Computer Science, The University of Iowa

Abstract. Previous results for combining decision procedures for the word problem in the non-disjoint case do not apply to equational theories induced by modal logics—whose combination is not disjoint since they share the theory of Boolean algebras. Conversely, decidability results for the fusion of modal logics are strongly tailored towards the special theories at hand, and thus do not generalize to other equational theories. In this paper, we present a new approach for combining decision procedures for the word problem in the non-disjoint case that applies to equational theories induced by modal logics, but is not restricted to them. The known fusion decidability results for modal logics are instances of our approach. However, even for equational theories induced by modal logics our results are more general since they are not restricted to so-called normal modal logics.

1 Introduction

The combination of decision procedures for logical theories arises in many areas of logic in computer science, such as constraint solving, automated deduction, term rewriting, modal logics, and description logics. In general, one has two first-order theories T_1 and T_2 over the signatures Σ_1 and Σ_2, for which validity of a certain type of formulae (e.g., universal, existential positive, etc.) is decidable. The question is then whether one can combine the decision procedures for T_1 and T_2 into one for their union $T_1 \cup T_2$. The problem is usually much easier (though not at all trivial) if the theories do not share symbols, i.e., if $\Sigma_1 \cap \Sigma_2 = \emptyset$. For non-disjoint signatures, the combination of theories can easily lead to undecidability, and thus one must find appropriate restrictions on the theories to be combined.

In automated deduction, the Nelson-Oppen combination procedure [17,16] as well as the problem of combining decision procedures for the word problem [19,21,18,6] have drawn considerable attention. The Nelson-Oppen method combines decision procedures for the validity of quantifier-free formulae in so-called stably infinite theories. If we restrict the attention to equational theories,[1] then

---

* Partially supported by DFG under grant BA 1122/3-3.
[1] Equational theories are stably infinite if one adds the axiom $\exists x, y.\ x \not\approx y$ [6].

D. Basin and M. Rusinowitch (Eds.): IJCAR 2004, LNAI 3097, pp. 183–197, 2004.

it is easy to see that the validity of arbitrary quantifier-free formulae can be reduced to the validity of formulae of the form $s_1 \approx t_1 \wedge \ldots \wedge s_n \approx t_n \rightarrow s \approx t$ where $s_1, \ldots, t$ are terms. Thus, in this case the Nelson-Oppen method combines decision procedures for the *conditional word problem* (i.e., for validity of conditional equations of the above form). Though this may at first sight seem surprising, combining decision procedures for the *word problem* (i.e., for validity of equations $s \approx t$) is a harder task: the known combination algorithms for the word problem are more complicated than the Nelson-Oppen method, and the same applies to their proofs of correctness. The reason is that the algorithms for the component theories are then less powerful. For example, if one applies the Nelson-Oppen method to a word problem $s \approx t$, then it will generate as input for the component procedures conditional word problems, not word problems (see [6] for a more detailed discussion). Both the Nelson-Oppen method and the methods for combining decision procedures for the word problem have been generalized to the non-disjoint case [11,24,7,12]. The main restriction on the theories to be combined is that they share only so-called constructors.

In modal logics, one is interested in whether properties (like decidability, finite axiomatizability) of uni-modal logics transfer to multi-modal logics that are obtained as the fusion of uni-modal logics. For the decidability transfer, one usually considers two different decision problems, the *validity* problem (Is the formula φ a theorem of the logic?) and the *relativized validity* problem (Does the formula φ follow from the global assumption ψ?). There are strong combination results that show that in many cases decidability transfers from two modal logics to their fusion [15,23,25,4]. Again, transfer results for the harder decision problem, relativized validity,[2] are easier to show than for the simpler one, validity. In fact, for validity the results only apply to so-called *normal* modal logics,[3] whereas this restriction is not necessary for relativized validity.

There is a close connection between the (conditional) word problem and the (relativized) validity problem in modal logics. In fact, in so-called *classical* modal logics (which encompass most well-known modal logics), modal formulae can be viewed as terms, on which equivalence of formulae induces an equational theory. The fusion of modal logics then corresponds to the union of the corresponding equational theories, and the (relativized) validity problem to the (conditional) word problem. The union of the equational theories corresponding to two modal logics is over non-disjoint signatures since the Boolean operators are shared. Unfortunately, in this setting the Boolean operators are not shared constructors in the sense of [24,7] (see [12]), and thus the decidability transfer results for modal logics cannot be obtained as special cases of the results in [24,7,12].

Recently, a new generalization of the Nelson-Oppen combination method to non-disjoint theories was developed in [13,14]. The main restriction on the theories T_1 and T_2 to be combined is that they are *compatible* with their shared theory T_0, and that their shared theory is *locally finite* (i.e., its finitely generated models are finite). A theory T is compatible with a theory T_0 iff (i) $T_0 \subseteq T$; (ii)

---

[2] This is in fact a harder problem since in modal logics the deduction theorem typically does not hold.

[3] An exception is [4], where only the existence of "covering normal terms" is required.

T_0 has a model completion T_0^*; and (iii) every model of T embeds into a model of $T \cup T_0^*$. It is well-known that the theory BA of Boolean algebras is locally finite and that the equational theories induced by classical modal logics are compatible with BA (see [2] for details). Thus, the combination method in [14, 13] applies to (equational theories induced by) classical modal logics. However, since it generalizes the Nelson-Oppen method, it only yields transfer results for decidability of the conditional word problem (i.e., the relativized validity problem).

In the present paper, we address the harder problem of designing a combination method for the word problem in the non-disjoint case that has the known transfer results for decidability of validity in modal logics as instances.

In fact, we will see that our approach strictly generalizes these results since it does not require the modal logics to be normal. The question of whether such transfer results hold also for non-normal modal logics was a long-standing open problem in modal logics. In addition to the conditions imposed in [13,14], our method needs the shared theory T_0 to have *local solvers*. Roughly speaking, this is the case if in T_0 one can solve an arbitrary equation with respect to any of its variables (see Definition 3 for details).

In the next section, we introduce some basic notions for equational theories, and define the restrictions under which our combination approach applies. In Section 3, we describe the new combination procedure, and show that it is sound and complete. Section 4 shows that the restrictions imposed by our procedure are satisfied by all classical modal logics. In particular, we show there that the theory of Boolean algebras has local solvers. In this section, we also comment on the complexity of our combination procedure if applied to modal logics, and illustrate the working of the procedure on an example.

For space constraints we must forgo most of the proofs of the results presented here. The interested reader can find them in [2].

2   Preliminaries

In this paper we will use standard notions from equational logic, universal algebra and term rewriting (see, e.g., [5]). We consider only first-order theories (with equality $\approx$) over a functional signature. We use the letters Σ, Ω, possibly with subscripts, to denote signatures. Throughout the paper, we fix a countably-infinite set V of *variables* and a countably-infinite set C of *free constants*, both disjoint with any signature Σ and with each other. For any $X \subseteq V \cup C$, $T(\Sigma, X)$ denotes the set of Σ-*terms* over X, i.e., first-order terms with variables and free constants in X and function symbols in Σ.[4] First-order Σ-*formulae* are defined in the usual way, using equality as the only predicate symbol. A Σ-*sentence* is a Σ-formula without *free* variables, and a *ground* Σ-*formula* is a Σ-formula without variables. An equational theory E over Σ is a set of (implicitly universally quantified) Σ-identities of the form $s \approx t$, where $s, t \in T(\Sigma, V)$. As usual, first-order interpretations of Σ are called Σ-*algebras*. We denote algebras

---

[4] Note that Σ may also contain constants.

by calligraphic letters $(\mathcal{A}, \mathcal{B}, \ldots)$, and their carriers by the corresponding Roman letter $(A, B, \ldots)$. A Σ-algebra $\mathcal{A}$ is a *model* of a set T of Σ-sentences iff it satisfies every sentence in T. For a set Γ of sentences and a sentence φ, we write $\Gamma \models_E \varphi$ if every model of E that satisfies Γ also satisfies φ. When Γ is the empty set, we write just $\models_E \varphi$, as usual. We denote by $\approx_E$ the equational consequences of E, i.e., the relation $\approx_E = \{(s,t) \in T(\Sigma, V \cup C) \times T(\Sigma, V \cup C) \mid \models_E s \approx t\}$. The *word problem* for E is the problem of deciding the relation $\approx_E$.

If $\mathcal{A}$ is an Ω-algebra and $\Sigma \subseteq \Omega$, we denote by $\mathcal{A}^\Sigma$ the Σ-*reduct* of $\mathcal{A}$, i.e., the algebra obtained from $\mathcal{A}$ by ignoring the symbols in $\Omega \setminus \Sigma$. An *embedding* of a Σ-algebra $\mathcal{A}$ into a Σ-algebra $\mathcal{B}$ is an injective Σ-homomorphism from $\mathcal{A}$ to $\mathcal{B}$. If such an embedding exists then we say that $\mathcal{A}$ can be embedded into $\mathcal{B}$. If $\mathcal{A}$ is Σ-algebra and $\mathcal{B}$ is an Ω-algebra with $\Sigma \subseteq \Omega$, we say that $\mathcal{A}$ can be Σ-embedded into $\mathcal{B}$ if there is an embedding of $\mathcal{A}$ into $\mathcal{B}^\Sigma$. We call the corresponding embedding a Σ-embedding of $\mathcal{A}$ into $\mathcal{B}$. If this embedding is the inclusion function, then we say that $\mathcal{A}$ is a Σ-subalgebra of $\mathcal{B}$.

Given a signature Σ and a set X disjoint with $\Sigma \cup V$, we denote by $\Sigma(X)$ the signature obtained by adding the elements of X as constant symbols to Σ. When X is included in the carrier of a Σ-algebra $\mathcal{A}$, we can view $\mathcal{A}$ as a $\Sigma(X)$-algebra by interpreting each $x \in X$ by itself. The Σ-*diagram* $\Delta_X^\Sigma(\mathcal{A})$ of $\mathcal{A}$ (w.r.t. X) consists of all ground $\Sigma(X)$-literals that hold in $\mathcal{A}$. We write just $\Delta^\Sigma(\mathcal{A})$ when X coincides with the whole carrier of $\mathcal{A}$. By a result known as Robinson's Diagram Lemma [9] embeddings and diagrams are related as follows.

Lemma 1. *Let $\mathcal{A}$ be a Σ-algebra generated by a set X, and let $\mathcal{B}$ be an Ω-algebra for some $\Omega \supseteq \Sigma(X)$. Then $\mathcal{A}$ can be $\Sigma(X)$-embedded into $\mathcal{B}$ iff $\mathcal{B}$ is a model of $\Delta_X^\Sigma(\mathcal{A})$.*

A consequence of the lemma above, which we will use later, is that if two Σ-algebras $\mathcal{A}$, $\mathcal{B}$ are both generated by a set X and if one of them, say $\mathcal{B}$, satisfies the other's diagram w.r.t. X, then they are isomorphic.

Ground formulae are invariant under embeddings in the following sense.

Lemma 2. *Let $\mathcal{A}$ be a Σ-algebra that can be Σ-embedded into an algebra $\mathcal{B}$. For all ground $\Sigma(A)$-formulae φ, $\mathcal{A}$ satisfies φ iff $\mathcal{B}$ satisfies φ where $\mathcal{B}$ is extended to a $\Sigma(A)$-algebra by interpreting $a \in A$ by its image under the embedding.*

Given equational theories E_1, E_2 over their respective signatures Σ_1, Σ_2, we want to define conditions under which the decidability of the word problem for E_1 and E_2 implies the decidability of the word problem for their union.

First restriction: We will require that both E_1 and E_2 be *compatible* with a shared subtheory E_0 over the signature $\Sigma_0 := \Sigma_1 \cap \Sigma_2$. The definition of compatibility depends on the notion of a model completion. A first-order Σ-theory E^* is a *model completion* of an equational Σ-theory E iff it extends E and for every model $\mathcal{A}$ of E (i) $\mathcal{A}$ can be embedded into a model of E^*, and (ii) $E^* \cup \Delta^\Sigma(\mathcal{A})$ is a complete $\Sigma(A)$-theory, i.e., $E^* \cup \Delta^\Sigma(\mathcal{A})$ is satisfiable and for any $\Sigma(A)$-sentence φ, either φ or its negation follows from $E^* \cup \Delta^\Sigma(\mathcal{A})$.

Definition 1 (Compatibility). *Let E be an equational theory over the signature Σ, and let E_0 be an equational theory over a subsignature $\Sigma_0 \subseteq \Sigma$. We say that E is E_0-compatible iff (1) $\approx_{E_0} \subseteq \approx_E$; (2) E_0 has a model completion E_0^*; (3) every model of E embeds into a model of $E \cup E_0^*$.*

Examples of theories that satisfy this definition can be found in [2,13,14] and in Section 4. Here we just show two consequences that will be important when proving completeness of our combination procedure.

Lemma 3. *Assume that E_1 and E_2 are two equational theories over the respective signatures Σ_1 and Σ_2 that are both E_0-compatible for some equational theory E_0 with signature $\Sigma_0 = \Sigma_1 \cap \Sigma_2$. For $i = 0, 1, 2$, let $\mathcal{A}_i$ be a model of E_i such that $\mathcal{A}_0$ is a Σ_0-subalgebra of both $\mathcal{A}_1$ and $\mathcal{A}_2$. Then there are a model $\mathcal{A}$ of $E_1 \cup E_2$ and Σ_i-embeddings f_i of $\mathcal{A}_i$ into $\mathcal{A}$ whose restrictions to A_0 coincide.*

In the following, we call conjunctions of Σ-identities *e-formulae*. We will write $\varphi(\boldsymbol{x})$ to denote an e-formula φ all of whose variables are included in the tuple $\boldsymbol{x}$. If $\boldsymbol{x} = (x_1, \ldots, x_n)$ we will write $\varphi(\boldsymbol{a})$ to denote that $\boldsymbol{a}$ is a tuple of free constants of the form $(a_1, \ldots, a_n)$ and $\varphi(\boldsymbol{a})$ is the formula obtained from φ by replacing every occurrence of x_i by a_i for $i = 1, \ldots, n$.

Lemma 4. *Let E_1 be E_0-compatible where E_1 and E_0 are equational theories over the respective signatures Σ_1 and Σ_0 with $\Sigma_1 \supseteq \Sigma_0$. Let $\psi_1(\boldsymbol{x}, \boldsymbol{y})$ be an e-formula in the signature Σ_1 and $\psi_2(\boldsymbol{y}, \boldsymbol{z})$ an e-formula in the signature Σ_0 such that $\psi_1(\boldsymbol{a}_1, \boldsymbol{a}_0) \models_{E_1} \psi_2(\boldsymbol{a}_0, \boldsymbol{a}_2)$, where $\boldsymbol{a}_1$, $\boldsymbol{a}_0$ and $\boldsymbol{a}_2$ are pairwise disjoint tuples of fresh constants. Then, there is an e-formula $\psi_0(\boldsymbol{y})$ in the signature Σ_0, such that $\psi_1(\boldsymbol{a}_1, \boldsymbol{a}_0) \models_{E_1} \psi_0(\boldsymbol{a}_0)$ and $\psi_0(\boldsymbol{a}_0) \models_{E_0} \psi_2(\boldsymbol{a}_0, \boldsymbol{a}_2)$.*

Second restriction: We will require that all the finitely generated models of E_0 be finite. From a more syntactical point of view this means that if C_0 is a finite subset of C, then there are only finitely many E_0-equivalence classes of terms in $T(\Sigma_0, C_0)$. For our combination procedure to be effective, we must be able to compute representatives of these equivalence classes.

Definition 2. *An an equational theory E_0 over the signature Σ_0 is effectively locally finite iff for every (finite) tuple $\boldsymbol{c}$ of constants from C we can effectively compute a finite set of terms $R_{E_0}(\boldsymbol{c}) \subseteq T(\Sigma_0, \boldsymbol{c})$ such that*

1. *$s \not\approx_{E_0} t$ for all distinct $s, t \in R_{E_0}(\boldsymbol{c})$;*
2. *for all terms $s \in T(\Sigma_0, \boldsymbol{c})$, there is some $t \in R_{E_0}(\boldsymbol{c})$ such that $s \approx_{E_0} t$.*

Example 1. A well-known example of an effectively locally finite theory is the usual (equational) theory BA of Boolean algebras over the signature $\Sigma_{BA} := \{\cap, \cup, \overline{(\_)}, 1, 0\}$. In fact, if $\boldsymbol{c} = (c_1, \ldots, c_n)$, every ground Boolean term over the constants in $\boldsymbol{c}$ is equivalent in BA to a term in "conjunctive normal form," a meet of terms of the kind $d_1 \cup \cdots \cup d_n$, where each d_i is either c_i or $\overline{c}_i$. It is easy to see that the set $R_{BA}(\boldsymbol{c})$ of such normal forms is isomorphic to the powerset of the powerset of $\boldsymbol{c}$, which is effectively computable and has cardinality 2^{2^n}.

Third restriction: We will require that E_1 and E_2 be each a *conservative exten-sions* of E_0, i.e., for $i = 1, 2$ and for all $s, t \in T(\Sigma_0, V)$, $s \approx_{E_0} t$ iff $s \approx_{E_i} t$.

Fourth restriction: Finally, we will require the theory E_0 to have local solvers, in the sense that any finite set of equations can be *solved* with respect to any of its variables.

Definition 3 (Gaussian). *The equational theory E_0 is* Gaussian *iff for every e-formula $\varphi(\boldsymbol{x}, y)$ it is possible to compute an e-formula $C(\boldsymbol{x})$ and a term $s(\boldsymbol{x}, \boldsymbol{z})$ with fresh variables $\boldsymbol{z}$ such that*

$$\models_{E_0} \varphi(\boldsymbol{x}, y) \Leftrightarrow (C(\boldsymbol{x}) \wedge \exists \boldsymbol{z}.(y = s(\boldsymbol{x}, \boldsymbol{z}))) \tag{1}$$

We call the formula C the solvability condition *of φ w.r.t. y, and the term s a (local)* solver *of φ w.r.t. y in E_0.*

There is a close connection between the above definition and Gaussian elim-ination, which is explained in the following example.

Example 2. Let K be a fixed field (e.g., the field of rational or real numbers). We consider the theory of vector spaces over K whose signature consists of a symbol for addition, a symbol for additive inverse and, for every scalar $k \in K$, a unary function symbol $k \cdot (-)$. Axioms are the usual vector spaces axioms (namely, the Abelian group axioms plus the axioms for scalar multiplication). In this theory, terms are equivalent to linear homogeneous polynomials (with non-zero coefficients) over K. Every e-formula $\varphi(\boldsymbol{x}, y)$ can be transformed into a homogeneous system $t_1(\boldsymbol{x}, y) = 0 \wedge \cdots \wedge t_k(\boldsymbol{x}, y) = 0$ of linear equations with unknowns $\boldsymbol{x}, y$. If y does not occur in φ, then φ is its own solvability condition and any fresh variable z is a local solver of φ w.r.t. y.[5] If y occurs in φ, then (modulo easy algebraic transformations) we can assume that φ contains an equation of the form $y = t(\boldsymbol{x})$; this equation gives the local solver, which is $t(\boldsymbol{x})$ (the sequence of existential quantifiers $\exists \boldsymbol{z}$ in (1) is empty), whereas the solvability condition is the e-formula obtained from φ by eliminating y, i.e., replacing y by $t(\boldsymbol{x})$ everywhere in φ.

In Section 4 we will see that the theory of Boolean algebras introduced in Example 1 is not only Gaussian but also satisfies our other restrictions.

3   The Combination Procedure

In the following, we assume that E_1, E_2 are equational theories over the signa-tures Σ_1, Σ_2 with decidable word problems, and that there exists an equational theory E_0 over the signature $\Sigma_0 := \Sigma_1 \cap \Sigma_2$ such that

- E_0 is Gaussian and effectively locally finite;
- for $i = 1, 2$, E_i is E_0-compatible and a conservative extension of E_0.

---

[5]   Note that φ is trivially equivalent to $\varphi \wedge \exists z.(y = z)$.

Abstraction rewrite systems. Our combination procedure works on the following data structure (where C is again a set of free constants disjoint with Σ_1 and Σ_2).

Definition 4. *An* abstraction rewrite system (ARS) *is a finite ground rewrite system R that can be partitioned into $R = R_1 \cup R_2$ so that*

- *for $i = 1, 2$, the rules of R_i have the form $a \to t$ where $a \in C$, $t \in T(\Sigma_i, C)$, and every constant a occurs at most once as a left-hand side in R_i;*
- *$R = R_1 \cup R_2$ is terminating.*

The ARS R is an initial *ARS iff every constant a occurs at most once as a left-hand side in the whole R.*

In particular, for $i = 1, 2$, R_i is also terminating, and the restriction that every constant occurs at most once as a left-hand side in R_i implies that R_i is confluent. We denote the unique normal form of a term s w.r.t. R_i by $s\!\downarrow_{R_i}$.

Given a ground rewrite system R, an equational theory E, and an e-formula ψ, we write $R \models_E \psi$ to express that $\{l \approx r \mid l \to r \in R\} \models_E \psi$.

Lemma 5. *Let $R = R_1 \cup R_2$ be an ARS, and $s, t \in T(\Sigma_i, C)$ for some $i \in \{1, 2\}$. Then $R_i \models_{E_i} s \approx t$ iff $s\!\downarrow_{R_i} \approx_{E_i} t\!\downarrow_{R_i}$.*

If we want to decide the word problem in $E_1 \cup E_2$, it is sufficient to consider ground terms with free constants, i.e., terms $s, t \in T(\Sigma_1 \cup \Sigma_2, C)$. Given such terms s, t we can employ the usual abstraction procedures that replace subterms by new constants in C (see, e.g., [7]) to generate terms $u, v \in T(\Sigma_0, C)$ and an initial ARS $R = R_1 \cup R_2$ such that $s \approx_{E_1 \cup E_2} t$ iff $R \models_{E_1 \cup E_2} u \approx v$. Thus, to decide $\approx_{E_1 \cup E_2}$, it is sufficient to devise a procedure that can solve problems of the form "$R \models_{E_1 \cup E_2} u \approx v$?" where R is an initial ARS and $u, v \in T(\Sigma_0, C)$.

The combination procedure. The input of the procedure is an initial ARS $R = R_1 \cup R_2$ and two terms $u, v \in T(\Sigma_0, C)$. Let $>$ be a total ordering of the left-hand side (lhs) constants of R such that for all $a \to t \in R$, t contains only lhs constants smaller than a (this ordering exists since R is terminating). Given this ordering, we can assume that $R = \{a_i \to t_i \mid i = 1, \ldots, n\}$ for some $n \geq 0$ where $a_n > a_{n-1} > \cdots > a_1$.

Note that u, v and each t_i may also contain free constants from C that are not left-hand side constants. In the following, we use c to denote a tuple of all these constants. Furthermore, for $j = 1, 2$ and $i = 0, \ldots, n$, we denote by $R_j^{(i)}$ the restriction of R_j to the rules whose left-hand sides are smaller or equal to a_i—where, by convention, $R_j^{(0)}$ is the empty system.

The combination procedure is described in Figure 1. First, note that all of the steps of the procedure are effective. Step 1 of the for loop is trivially effective; Step 2 is effective because E_0 is effectively locally finite by assumption. Step 3 is effective because the test that $R_j^{(i)} \models_{E_j} t \approx t'$ can be reduced by Lemma 5 to testing that $t\!\downarrow_{R_j^{(i)}} \approx_{E_j} t'\!\downarrow_{R_j^{(i)}}$. The latter test is effective because,

Input: an initial ARS $R = R_1 \cup R_2 = \{a_i \to t_i \mid i = 1, \dots, n\}$ and
   terms $u, v \in T(\Sigma_0, C)$.

Let c collect the free constants in R, u, v that are not in $\{a_1, \dots, a_n\}$.

for $i = 1$ to n do

   1. Let j be such that $a_i \to t_i \in R_j$ and k such that $\{j, k\} = \{1, 2\}$.
   2. Let $T = R_{E_0}(a_1, \dots, a_i, c)$ (see Definition 2).
   3. For each pair of distinct terms $t, t' \in T$, test whether $R_j^{(i)} \models_{E_j} t \approx t'$.
   4. Let $\varphi(a_1, \dots, a_i, c)$ be the conjunction of those identities
      $t \approx t'$ for which the test succeeds.
   5. Let $s(a_1, \dots, a_{i-1}, c, d)$ be a local solver of φ w.r.t. a_i in E_0.
   6. Add to R_k the new rule $a_i \to s(a_1, \dots, a_{i-1}, c, d)$.

done

Output: "yes" if $R_1 \models_{E_1} u \approx v$, and "no" otherwise.

Fig. 1. The combination procedure.

(i) the word problem in E_j is decidable by assumption and (ii) $R_j^{(i)}$ is confluent
and terminating at each iteration of the loop. In Step 4 the formula φ can be
computed because T is finite and the local solver in Step 5 can be computed by
the algorithm provided by the definition of a Gaussian theory. Step 6 is trivial
and for the final test after the loop, the same observations as for Step 3 apply.

A few more remarks on the procedure are in order. In the fifth step of the
loop, d is a tuple of new constants introduced by the solver s. In the definition
of a local solver, we have used variables instead of constants, but this difference
will turn out to be irrelevant since free constants behave like variables. One may
wonder why the procedure ignores the solvability condition for the local solver.
The reason is that this condition follows from both R_1 and R_2, as will be shown
in the proof of completeness.

Adding the new rule to R_k in the sixth step of the loop does not destroy the
property of $R_1 \cup R_2$ being an ARS—although it will make it non-initial. In fact,
$s(a_1, \dots, a_{i-1}, c, d)$ contains only lhs constants smaller than a_i, and R_k before
did not contain a rule with lhs a_i because the input was an *initial* ARS.

The test after the loop is performed using R_1, E_1. The choice R_1 and E_1
versus R_2 and E_2 is arbitrary (see Lemma 6).

The correctness proof. Since the combinations procedure obviously terminates on any input, it is sufficient to show soundness and completeness. In the
proofs, we will use $R_{1,i}, R_{2,i}$ to denote the updated rewrite systems obtained after step i in the loop ($R_{1,0}$ and $R_{2,0}$ are the input systems R_1 and R_2). Soundness
is not hard to show (see [2]).

Proposition 1 (Soundness). *If the combination procedure answers "yes",
then* $R_1 \cup R_2 \models_{E_1 \cup E_2} u \approx v$.

The following lemma, which is used in the completeness proof, depends on our definition of a Gaussian theory (see [2] for details).

Lemma 6. *For every ground e-formula ψ in the signature $\Sigma_0 \cup \{a_1, .., a_n\} \cup c$, $R_{1,n} \models_{E_1} \psi$ iff $R_{2,n} \models_{E_2} \psi$.*

Proposition 2 (Completeness). *If $R_1 \cup R_2 \models_{E_1 \cup E_2} u \approx v$, then the combination procedure answers "yes".*

Proof. Since the procedure is terminating, it is enough to show that $R_{1,0} \cup R_{2,0} \not\models_{E_1 \cup E_2} u \approx v$ whenever the combination procedure answers "no". We do that by building a model of $R_{1,0} \cup R_{2,0} \cup E_1 \cup E_2$ that falsifies $u \approx v$. Let $\boldsymbol{a} := (a_1, \dots, a_n)$ and let $k \in \{1, 2\}$. Where $\boldsymbol{c}$ is defined as in Figure 1 and $\boldsymbol{d_k}$ is a tuple collecting all the new constants introduced in the rewrite system R_k during execution of the procedure (see Step 4 of the loop), let $\mathcal{A}_{k,0}$ be the initial model (see, e.g., [5] for a definition) of E_k over the signature $\Sigma_k \cup \boldsymbol{c} \cup \boldsymbol{d_k}$.

Observe that the final rewrite system $R_{k,n}$ contains (exactly) one rule of the form $a_i \rightarrow u_i$ for all $i = 1, \dots, n$. This is because either the rule $a_i \rightarrow t_i$ was already in $R_{k,0}$ to begin with (then $u_i = t_i$), or a rule of the form $a_i \rightarrow s_i$ for some solver s_i was added to $R_{k,i-1}$ at step i to produce $R_{k,i}$ (in which case $u_i = s_i$). Thus, we can use the rewrite rules of $R_{k,n}$ to define by induction on $i = 1, \dots, n$ an expansion $\mathcal{A}_{k,i}$ of $\mathcal{A}_{k,0}$ to the constants $a_1, \dots, a_i$. Specifically, $\mathcal{A}_{k,i}$ is defined as the expansion of $\mathcal{A}_{k,i-1}$ that interprets a_i as $u_i^{\mathcal{A}_{k,i-1}}$ where u_i is the term such that $a_i \rightarrow u_i \in R_{k,n}$. Note that $u_i^{\mathcal{A}_{k,i-1}}$ is well defined because u_i does not contain any of the constants $a_i, \dots, a_n$.

By induction on i it is easy to show (see [2]) for every ground e-formula $\varphi(a_1, \dots, a_i, \boldsymbol{c}, \boldsymbol{d_k})$ in the signature $\Sigma_k \cup \{a_1, \dots, a_i\} \cup \boldsymbol{c} \cup \boldsymbol{d_k}$, that

$$\mathcal{A}_{k,i} \text{ satisfies } \varphi(a_1, \dots, a_i, \boldsymbol{c}, \boldsymbol{d_k}) \text{ iff } R_{k,n}^{(i)} \models_{E_k} \varphi(a_1, \dots, a_i, \boldsymbol{c}, \boldsymbol{d_k}). \quad (2)$$

Let $\mathcal{A}_k = \mathcal{A}_{k,n}^{\Omega_k}$ where $\Omega_k = \Sigma_k \cup \boldsymbol{a} \cup \boldsymbol{c}$. As a special case of (2) above, we have that for every ground e-formula $\varphi(\boldsymbol{a}, \boldsymbol{c})$ in the signature $\Sigma_0 \cup \boldsymbol{a} \cup \boldsymbol{c}$,

$$\mathcal{A}_k \text{ satisfies } \varphi \text{ iff } R_{k,n} \models_{E_k} \varphi. \quad (3)$$

For $k = 1, 2$ let $\mathcal{B}_k$ be the subalgebra of $\mathcal{A}_k^{\Sigma_0}$ generated by (the interpretations in $\mathcal{A}_k$ of) the constants $\boldsymbol{a} \cup \boldsymbol{c}$. We claim that the algebras $\mathcal{B}_1$ and $\mathcal{B}_2$ satisfy each other's diagram. To see that, let ψ be a ground identity of signature $\Sigma_0 \cup \boldsymbol{a} \cup \boldsymbol{c}$. Then, $\psi \in \Delta_{\boldsymbol{a} \cup \boldsymbol{c}}^{\Sigma_0}(\mathcal{B}_k)$ iff $\mathcal{B}_k$ satisfies ψ (by definition of $\Delta_{\boldsymbol{a} \cup \boldsymbol{c}}^{\Sigma_0}(\mathcal{B}_k)$) iff $\mathcal{A}_k$ satisfies ψ (by construction of $\mathcal{B}_k$ and Lemma 2) iff $R_{k,n} \models_{E_k} \psi$ (by (3) above).

By Lemma 6, we can conclude that $\psi \in \Delta_{\boldsymbol{a} \cup \boldsymbol{c}}^{\Sigma_0}(\mathcal{B}_1)$ iff $\psi \in \Delta_{\boldsymbol{a} \cup \boldsymbol{c}}^{\Sigma_0}(\mathcal{B}_2)$. It follows from the observation after Lemma 1 that $\mathcal{B}_1$ and $\mathcal{B}_2$ are Σ_0-isomorphic, hence they can be identified with no loss of generality. Therefore, let $\mathcal{A}_0 = \mathcal{B}_1 = \mathcal{B}_2$ and observe that (i) $\mathcal{A}_k^{\Sigma_k}$ is a model of E_k by construction; (ii) $\mathcal{A}_0$ is a Σ_0-subalgebra of $\mathcal{A}_k^{\Sigma_k}$; and (iii) $\mathcal{A}_0$ is a model of E_0 because $\mathcal{A}_k^{\Sigma_0}$ is a model of E_0 and the set of models of an equational theory is closed under subalgebras.

By Lemma 3 it follows that there is a model $\mathcal{A}$ of $E_1 \cup E_2$ such that there are Σ_k-embeddings f_k of $\mathcal{A}_k^{\Sigma_k}$ into $\mathcal{A}$ $(i = 1, 2)$ satisfying $f_1(c^{\mathcal{A}_1}) = f_2(c^{\mathcal{A}_2})$ for all $c \in \boldsymbol{a} \cup \boldsymbol{c}$. Let then $\mathcal{A}'$ be the expansion of $\mathcal{A}$ to the signature $\Sigma_1 \cup \Sigma_2 \cup \boldsymbol{a} \cup \boldsymbol{c}$ such that $c^{\mathcal{A}'} = f_1(c^{\mathcal{A}_1})$ for every $c \in \boldsymbol{a} \cup \boldsymbol{c}$. It is not difficult to see that f_k is an Ω_k-embedding of $\mathcal{A}_k$ into $\mathcal{A}'$ for $k = 1, 2$. Observe that $\mathcal{A}'$, which is clearly a model of $E_1 \cup E_2$, is also a model of $R_{1,0} \cup R_{2,0}$. In fact, by construction of $R_{1,n}$ and $R_{2,n}$, for all $a \to t \in R_{1,0} \cup R_{2,0}$, there is a $k \in \{1, 2\}$ such that $a \to t \in R_{k,n}$. It follows immediately that $R_{k,n} \models_{E_k} a \approx t$, which implies by (3) above that $\mathcal{A}_k$ satisfies $a \approx t$. But then $\mathcal{A}'$ satisfies $a \approx t$ as well by Lemma 2.

In conclusion, we have that $\mathcal{A}'$ is a model of $R_{1,0} \cup R_{2,0} \cup E_1 \cup E_2$. Since the procedure returns "no" by assumption, it must be that $R_{1,n} \not\models_{E_1} u \approx v$. We then have that $\mathcal{A}_1$ falsifies $u \approx v$ by (3) above and $\mathcal{A}'$ falsifies $u \approx v$ by Lemma 2. □

From the total correctness of the combination procedure, we then obtain the following modular decidability result.

Theorem 1. *Let E_0, E_1, E_2 be three equational theories of respective signature $\Sigma_0, \Sigma_1, \Sigma_2$ such that*

- *$\Sigma_0 = \Sigma_1 \cap \Sigma_2$;*
- *E_0 is Gaussian and effectively locally finite;*
- *for $i = 1, 2$, E_i is E_0-compatible and a conservative extension of E_0.*

If the word problem in E_1 and in E_2 is decidable, then the word problem in $E_1 \cup E_2$ is also decidable.

4   Fusion Decidability in Modal Logics

First, we define the modal logics to which our combination procedure applies. Basically, these are modal logics that corresponds to equational extensions of the theory of Boolean algebras. A *modal signature* Σ_M is a set of operation symbols endowed with corresponding arities; from Σ_M propositional formulae are built up using countably many propositional variables, the operation symbols in Σ_M, the Boolean connectives, and the constant $\top$ for truth and $\bot$ for falsity. We use letters $x, x_1, \ldots, y, y_1, \ldots$ for propositional variables and letters $t, t_1, \ldots, u, u_1, \ldots$ as metavariables for propositional formulae. The following definition is adapted from [22].

Definition 5. *A classical modal logic L based on a modal signature Σ_M is a set of propositional formulae that (i) contains all classical tautologies; (ii) is closed under uniform substitution of propositional variables by propositional formulae; (iii) is closed under the modus ponens rule (from t and $t \Rightarrow u$ infer u); (iv) for each n-ary $o \in \Sigma_M$, is closed under the following replacement rule:*

$$\text{from } \; t_1 \Leftrightarrow u_1, \; \ldots, \; t_n \Leftrightarrow u_n \; \text{ infer } \; o(t_1, \ldots, t_n) \Leftrightarrow o(u_1, \ldots, u_n).$$

As classical modal logics (based on a given modal signature) are closed under intersections, it makes sense to speak of the least classical modal logic $[S]$ containing a certain set of propositional formulae S. If $L = [S]$, we say that S is a set of axiom schemata for L and write $S \vdash t$ for $t \in [S]$.

We say that a classical modal logic L is *decidable* iff L is a recursive set of propositional formulae; the *decision problem* for L is just the membership problem for L.

A classical modal logic L is said to be *normal* iff for every n-ary modal operator o in the signature of L and every argument position $i = 1, \ldots, n$, L contains the formulae $o(\boldsymbol{x}, \top, \boldsymbol{x}')$ and $o(\boldsymbol{x}, (y \Rightarrow z), \boldsymbol{x}') \Rightarrow (o(\boldsymbol{x}, y, \boldsymbol{x}') \Rightarrow o(\boldsymbol{x}, z, \boldsymbol{x}'))$. The least normal (classical modal, unary, unimodal) logic is the modal logic usually called **K** [8]. Most well-known modal logics considered in the literature (both normal and non-normal) fit Definition 5 (see [2] for some examples).

Let us call an equational theory *Boolean-based* if its signature includes the signature Σ_{BA} of Boolean algebras and its axioms include the Boolean algebras axioms BA (see Example 1). For notational convenience, we will assume that Σ_{BA} also contains the binary symbol $\supset$, defined by the axiom $x \supset y \approx \overline{x} \cup y$.

Given a classical modal logic L we can associate with it a Boolean-based equational theory E_L. Conversely, given a Boolean-based equational theory E we can associate with it a classical modal logic L_E. In fact, given a classical modal logic L with modal signature Σ_M, we define E_L as the theory having as signature $\Sigma_M \cup \Sigma_{BA}$ and as a set of axioms the set $BA \cup \{t_{BA} \approx 1 \mid t \in L\}$ where t_{BA} is obtained from t by replacing t's logical connectives $(\neg, \wedge, \vee, \Rightarrow)$ by the corresponding Boolean algebra operators $(\overline{(\_)}, \cap, \cup, \supset)$, and the logical constants $\top$ and $\bot$ by 1 and 0, respectively. Vice versa, given a Boolean-based equational theory E over the signature Σ, we define L_E as the classical modal logic over the modal signature $\Sigma \setminus \Sigma_{BA}$ axiomatized by the formulae $\{t_L \mid \models_E t \approx 1\}$ where t_L is obtained from t by the inverse of the replacement process above.

Classical modal logics (in our sense) and Boolean-based equational theories are equivalent formalisms, as is well-known from algebraic logic [20]. In particular, for our purposes, the following standard proposition is crucial, as it reduces the decision problem for a classical modal logic L to the word problem in E_L.

Proposition 3. *For every classical modal logic L and for every propositional formula t, we have that $t \in L$ iff $\models_{E_L} t_{BA} \approx 1$.*

Given two classical modal logics L_1, L_2 over two *disjoint* modal signatures Σ_M^1, Σ_M^2, the *fusion* of L_1 and L_2 is the classical modal logic $L_1 \oplus L_2$ over the signature $\Sigma_M^1 \cup \Sigma_M^2$ defined as $[L_1 \cup L_2]$. As $E_{L_1 \oplus L_2}$ is easily seen to be deductively equivalent to the theory $E_{L_1} \cup E_{L_2}$ (i.e., $\approx_{E_{L_1 \oplus L_2}} = \approx_{E_{L_1} \cup E_{L_2}}$), it is clear that the decision problem $L_1 \cup L_2 \vdash t$ reduces to the word problem $E_{L_1} \cup E_{L_2} \models t_{BA} \approx 1$. Our goal in the remainder of this section is to show that, thanks to the combination result in Theorem 1, this combined word problem for $E_{L_1} \cup E_{L_2}$ reduces to the single word problems for E_{L_1} and E_{L_2}, and thus to the decision problems for L_1 and L_2.

Note that, although the modal signatures Σ_M^1 and Σ_M^2 are disjoint, the signatures of E_{L_1} and E_{L_2} are no longer disjoint, because they share the Boolean operators. To show that our combination theorem applies to E_{L_1} and E_{L_2}, we thus must establish that the common subtheory BA of Boolean algebras matches the requirements for our combination procedure. To this end, we will restrict ourselves to component modal logics L_1 and L_2 that are *consistent*, that is, do not include $\perp$, (or, equivalently, do not contain all modal formulae over their signature). This restriction is without loss of generality because when either L_1 or L_2 are inconsistent $L_1 \oplus L_2$ is inconsistent as well, which means that its decision problem is trivial.

We have already shown in Section 2 that BA satisfies one of our requirements, namely effective local finiteness. As for the others, for every consistent classical modal logic L, the theory E_L is guaranteed to be a conservative extension of BA. The main reason is that there are no non-trivial equational extensions of the theory of Boolean algebras. In fact, as soon as one extends BA with an axiom $s \approx t$ for any s and t such that $s \not\approx_{BA} t$, the equation $0 \approx 1$ becomes valid.[6] By Proposition 3, this entails that if an equational theory E_L induced by a classical modal logic L is not a conservative extension of BA then $L \vdash \perp$. Hence L cannot be consistent.

Thus, it remains to be shown that BA is Gaussian and that E_L is BA-compatible for every consistent classical modal logic L. For space constraints we cannot do this here, but we refer the interested reader to [2] for complete proofs. Here, we just point out how the local solver looks like for BA. For each e-formula of the form $u(\boldsymbol{x}, y) \approx 1$ (and fresh variable z), the term

$$s(\boldsymbol{x}, z) := (u(\boldsymbol{x}, 1) \supset u(\boldsymbol{x}, z)) \supset (z \cap (u(\boldsymbol{x}, 0) \supset u(\boldsymbol{x}, z))) \tag{4}$$

is a local solver for $u(\boldsymbol{x}, y) \approx 1$ in BA w.r.t. y. Note that $s(\boldsymbol{x}, z)$ can be computed in linear time from $u(\boldsymbol{x}, y)$ and that the restriction to formulae of the form $u(\boldsymbol{x}, y) \approx 1$ can be made with no loss of generality because every Boolean e-formula can be (effectively) converted in linear time into a BA-equivalent e-formula of that form.

Combining Theorem 1 with the results above on the theories BA and E_L, we get the following general modular decidability result.

Theorem 2. *If L_1, L_2 are decidable classical modal logics, so is $L_1 \oplus L_2$.*

In [2] we also show that the complexity upper-bounds for the combined decision procedures obtained by applying our combination procedure to classical modal logics are not worse than the ones given in [4] for the case of normal modal logics. If the decision procedures for L_1 and for L_2 are in PSPACE, we get an EXPSPACE combined decision procedure for $L_1 \oplus L_2$. If instead the procedures are in EXPTIME, we get a 2EXPTIME combined decision procedure.

We close this section by giving an examples of our combination procedure at work.

Example 3. Consider the classical modal logic **KT** with modal signature $\{\Box\}$ and obtained by adding to **K** the axiom schema $\Box x \Rightarrow x$. Now let $\mathbf{KT}_1$ and

---

[6]   This is can be shown by a proper instantiation of the variables of $s \approx t$ by 0 and 1, followed by simple Boolean simplifications.

$\mathbf{KT}_2$ be two signature disjoint renamings of $\mathbf{KT}$ in which $\Box_1$ and $\Box_2$, respectively, replace $\Box$, and consider the fusion logic $\mathbf{KT}_1 \oplus \mathbf{KT}_2$. We can use our combination procedure to show that $\mathbf{KT}_1 \oplus \mathbf{KT}_2 \vdash \Box_2 x \Rightarrow \Diamond_1 x$ (where as usual $\Diamond_1 x$ abbreviates $\neg \Box_1 \neg x$). For $i = 1, 2$, let E_i be the equational theory corresponding to $\mathbf{KT}_i$. It is enough to show that

$$\models_{E_1 \cup E_2} (\Box_2(x) \supset \Diamond_1(x)) \approx 1$$

where now $\Diamond_1 x$ abbreviates $\overline{\Box_1(\overline{x})}$. After the abstraction process, we get the two rewrite systems $R_1 = \{a_1 \to \Diamond_1(c)\}$ and $R_2 = \{a_2 \to \Box_2(c)\}$ and the goal equation $(a_2 \supset a_1) \approx 1$ where a_1, a_2 and c are fresh constants.

As explained in [2], for the test in Step 3 of the procedure's loop we need to consider only identities of the form $t \approx 1$ where t is a term-clause over the set of constants under consideration.[7] During the first execution of the procedure's loop the constants in question are a_1 and c, therefore there are only four identities to consider: $\overline{a}_1 \cup \overline{c} \approx 1$, $\overline{a}_1 \cup c \approx 1$, $a_1 \cup \overline{c} \approx 1$, and $a_1 \cup c \approx 1$. The only identity for which the test is positive is $a_1 \cup \overline{c}$. In fact, $a_1 \cup \overline{c}$ rewrites to $\Diamond_1(c) \cup \overline{c}$, which is equivalent to $c \supset \Diamond_1(c)$. This is basically the contrapositive of (the translation of) the axiom schema $\Box_1(c) \supset c$.[8]

Using the formula (4) seen earlier, we can produce a solver for that identity, which reduces to $c \cup d_1$ after some simplifications, where d_1 is a fresh free constant. Hence, the following rewrite rule is added to R_2 in Step 6 of the loop: $a_1 \to c \cup d_1$.

Continuing the execution of the loop with the second—and final—iteration, we get the following. Among the eight term-clauses involving a_1, a_2, c, the test in Step 3 is positive for four of them. The conjunction of such term-clauses gives a Boolean e-formula that is equivalent to $(a_2 \supset c) \cap (c \supset a_1) \approx 1$. This e-formula, once solved with respect to a_2, gives (after simplifications) the rewrite rule $a_2 \to d_2 \cap ((c \supset a_1) \supset (d_2 \supset c))$, which is added to R_1 before quitting the loop. Using this R_1, the final test of the procedure $(R_1 \models_{E_1} a_2 \supset a_1 \approx 1)$ succeeds because the modal formula $d_2 \wedge ((c \Rightarrow \Diamond_1 c) \Rightarrow (d_2 \Rightarrow c)) \Rightarrow \Diamond_1 c$ is a theorem of $\mathbf{KT}_1$.

5   Conclusion

In this paper, we have described a new approach for combining decision procedures for the word problem in equational theories over *non-disjoint* signatures. Unlike the previous combination methods for the word problem [7,12] in the non-disjoint case, this approach has the known decidability transfer results for *validity* in the fusion of modal logics [15,25] as consequences. Our combination result is however more general than these transfer results since it applies also

---

[7] For a given set of constants $c_1, \dots, c_m$, a term-clause is a term of the form $b_1 \cup \cdots \cup b_m$ where each b_j is either c_j or $\overline{c}_j$.

[8] Another approach for checking this, and also that the tests for the other term-clauses are negative, is to translate the rewritten term-clauses into the corresponding modal formulae, and then check whether their complement is unsatisfiable in all Kripke structures with a reflexive accessibility relation (see [10], Fig. 5.1).

to *non-normal* modal logics—thus answering in the affirmative a long-standing open question in modal logics—and to equational theories not induced by modal logics (see, e.g., Example 2). Nevertheless, for the modal logic application, the complexity upper-bounds obtained through our approach are the same as for the more restricted approaches [25,4].

Our results are not consequences of combination results for the conditional word problem (the relativized validity problem) recently obtained by generalizing the Nelson-Oppen combination method [13,14]. In fact, there are modal logics (obtained by translating certain description logics into modal logic notation) for which the validity problem is decidable, but the relativized validity problem is not. This is, e.g, the case for description logics with feature agreements [1] or with concrete domains [3].

Our new combination approach is orthogonal to the previous combination approaches for the word problem in equational theories over non-disjoint signatures [7,12]. On the one hand, the previous results do not apply to theories induced by modal logics [12]. On the other hand, there are equational theories that (i) satisfy the restrictions imposed by the previous approaches, and (ii) are not locally finite [7], and thus do not satisfy our restrictions. Both the approach described in this paper and those in [7,12] have the combination results for the case of disjoint signatures as a consequence. For the previous approaches, this was already pointed out in [7,12]. For our approach, this is not totally obvious since some minor technical problems have to be overcome (see [2] for details).

References

1. F. Baader, H.-J. Bürckert, B. Nebel, W. Nutt, and G. Smolka. On the expressivity of feature logics with negation, functional uncertainty, and sort equations. *J. of Logic, Language and Information*, 2:1–18, 1993.
2. F. Baader, S. Ghilardi, and C. Tinelli. A new combination procedure for the word problem that generalizes fusion decidability results in modal logics. Technical Report 03-03, Department of Computer Science, The University of Iowa, December 2003.
3. F. Baader and P. Hanschke. Extensions of concept languages for a mechanical engineering application. In *Proc. of the 16th German Workshop on Artificial Intelligence (GWAI'92)*, volume 671 of *Lecture Notes in Computer Science*, pages 132–143, Bonn (Germany), 1992. Springer-Verlag.
4. F. Baader, C. Lutz, H. Sturm, and F. Wolter. Fusions of description logics and abstract description systems. *Journal of Artificial Intelligence Research*, 16:1–58, 2002.
5. F. Baader and T. Nipkow. *Term Rewriting and All That*. Cambridge University Press, United Kingdom, 1998.
6. F. Baader and C. Tinelli. A new approach for combining decision procedures for the word problem, and its connection to the Nelson-Oppen combination method. In W. McCune, editor, *Proceedings of the 14th International Conference on Automated Deduction (Townsville, Australia)*, volume 1249 of *Lecture Notes in Artificial Intelligence*, pages 19–33. Springer-Verlag, 1997.
7. F. Baader and C. Tinelli. Deciding the word problem in the union of equational theories. *Information and Computation*, 178(2):346–390, December 2002.

8. A. Chagrov and M. Zakharyaschev. *Modal Logic*, volume 35 of *Oxford Logic Guides*. Clarendon Press, Oxford, 1997.

9. C.-C. Chang and H. J. Keisler. *Model Theory*. North-Holland, Amsterdam-London, IIIrd edition, 1990.

10. B. F. Chellas. *Modal Logic, an Introduction*. Cambridge University Press, Cambridge, 1980.

11. E. Domenjoud, F. Klay, and C. Ringeissen. Combination techniques for non-disjoint equational theories. In A. Bundy, editor, *Proceedings of the 12th International Conference on Automated Deduction, Nancy (France)*, volume 814 of *Lecture Notes in Artificial Intelligence*, pages 267–281. Springer-Verlag, 1994.

12. Camillo Fiorentini and S. Ghilardi. Combining word problems through rewriting in categories with products. *Theoretical Computer Science*, 294:103–149, 2003.

13. S. Ghilardi. Model Theoretic Methods in Combined Constraint Satisfiability. Journal of Automated Reasoning, 2004. To appear.

14. S. Ghilardi and L. Santocanale. Algebraic and model theoretic techniques for fusion decidability in modal logics. In Moshe Vardi and Andrei Voronkov, editors, *Proceedings of the 10th International Conference on Logic for Programming, Artificial Intelligence, and Reasoning (LPAR 2003)*, volume 2850 of *Lecture Notes in Computer Science*, pages 152–166. Springer-Verlag, 2003.

15. M. Kracht and F. Wolter. Properties of independently axiomatizable bimodal logics. *The Journal of Symbolic Logic*, 56(4):1469–1485, December 1991.

16. G. Nelson. Combining satisfiability procedures by equality-sharing. In W. W. Bledsoe and D. W. Loveland, editors, *Automated Theorem Proving: After 25 Years*, volume 29 of *Contemporary Mathematics*, pages 201–211. American Mathematical Society, Providence, RI, 1984.

17. G. Nelson and D. C. Oppen. Simplification by cooperating decision procedures. *ACM Trans. on Programming Languages and Systems*, 1(2):245–257, October 1979.

18. T. Nipkow. Combining matching algorithms: The regular case. *Journal of Symbolic Computation*, 12:633–653, 1991.

19. D. Pigozzi. The join of equational theories. *Colloquium Mathematicum*, 30(1):15–25, 1974.

20. H. Rasiowa. *An Algebraic Approach to Non-Classical Logics*, volume 78 of *Studies in Logic and the Foundations of Mathematics*. North Holland, Amsterdam, 1974.

21. M. Schmidt-Schauß. Unification in a combination of arbitrary disjoint equational theories. *Journal of Symbolic Computation*, 8(1–2):51–100, July/August 1989. Special issue on unification. Part II.

22. K. Segerberg. *An Essay in Classical Modal Logic*, volume 13 of *Filosofiska Studier*. Uppsala Universitet, 1971.

23. E. Spaan. *Complexity of Modal Logics*. PhD thesis, Department of Mathematics and Computer Science, University of Amsterdam, The Netherlands, 1993.

24. C. Tinelli and C. Ringeissen. Unions of non-disjoint theories and combinations of satisfiability procedures. *Theoretical Computer Science*, 290(1):291–353, 2003.

25. F. Wolter. Fusions of modal logics revisited. In M. Kracht, M. de Rijke, H. Wansing, and M. Zakharyaschev, editors, *Advances in Modal Logic*. CSLI, 1998.

Using Automated Theorem Provers to Certify Auto-generated Aerospace Software

Ewen Denney[†], Bernd Fischer[‡], and Johann Schumann[‡]

[†]QSS / [‡]RIACS, NASA Ames Research Center,
{edenney,fisch,schumann}@email.arc.nasa.gov

Abstract. We describe a system for the automated certification of safety properties of NASA software. The system uses Hoare-style program verification technology to generate proof obligations which are then processed by an automated
first-order theorem prover (ATP). For full automation, however, the obligations
must be aggressively preprocessed and simplified. We discuss the unique requirements this application places on the ATPs and demonstrate how the individual
simplification stages, which are implemented by rewriting, influence the ability of
the ATPs to solve the proof tasks. Our results are based on 13 certification experiments that lead to more than 25,000 proof tasks which have each been attempted
by Vampire, Spass, and e-setheo.

1 Introduction

Software certification aims to show that the software in question satisfies a certain level
of quality, safety, or security. Its result is a *certificate*, i.e., independently checkable
evidence of the properties claimed. Certification approaches vary widely, ranging from
code reviews to full formal verification, but the highest degree of confidence is achieved
with approaches that are based on formal methods and use logic and theorem proving
to construct the certificates.

We have developed a certification approach which uses Hoare-style techniques to
demonstrate the safety of aerospace software which has been automatically generated
from high-level specifications. Our core idea is to extend the code generator so that it
simultaneously generates code *and* detailed annotations, e.g., loop invariants, that enable
a safety proof. A verification condition generator (VCG) processes the annotated code
and produces a set of *safety obligations*, which are provable if and only if the code
is safe. An automated theorem prover (ATP) then discharges these obligations and the
proofs, which can be verified by an independent proof checker, serve as certificates. This
approach largely decouples code generation and certification and is thus more scalable
than, e.g., verifying the generator or generating code and safety proofs in parallel.

In this paper, we describe and evaluate the application of ATPs to discharge the
emerging safety obligations. This is a crucial aspect of our approach since its practicability hinges on a very high degree of automation. Our first hypothesis is that the current
generation of high-performance ATPs is—in principle—already powerful enough for
practical applications. However, this is still a very demanding area because the number

D. Basin and M. Rusinowitch (Eds.): IJCAR 2004, LNAI 3097, pp. 198–212, 2004.
© Springer-Verlag Berlin Heidelberg 2004

of obligations is potentially very large and program verification is generally a hard problem domain for ATPs. Our second hypothesis is thus that the application still needs to carefully preprocess the proof tasks to make them more tractable for ATPs.

In our case, there are several factors which make a successful ATP application possible. First, we certify separate aspects of safety and not full functional correctness. This separation of concerns allows us to show non-trivial properties like matrix symmetry but results in more tractable obligations. Second, the extensions of the code generator are specific to the safety properties to be certified and to the algorithms used in the generated programs. This allows us to fine-tune the annotations which, in turn, also results in more tractable obligations. Third, we aggressively simplify the obligations before they are handed over to the prover, taking advantage of domain-specific knowledge.

We have tested our two hypotheses by running five high-performance provers on seven different versions of the safety obligations resulting from certifying five different safety policies for four different programs—in total more than 25,000 obligations per prover. In Section 2 we give an overview of the system architecture, describing the safety policies as well as the generation and preprocessing of the proof tasks. In Section 3, we outline the experimental set-up used to evaluate the theorem provers over a range of different preprocessing levels. The detailed results are given in Section 4; they confirm our hypotheses: the provers are generally able to certify all test programs for all polices but only after substantial preprocessing of the obligations. Finally, Section 5 draws some conclusions.

Conceptually, this paper continues the work described in [25,26] but the actual implementation of the certification system has been completely revised and substantially extended. We have expanded the range of both algorithms and safety properties which can be certified; in particular, our approach is now fully integrated with the AUTOFILTER system [27] as well as with the AUTOBAYES system [9] and the certification process is now completely automated. We have also implemented a new generic VCG which can be customized for a given safety policy and which directly processes the internal code representation instead of Modula-2 as in the previous version. All these improvements and extensions to the underlying logical framework result in a substantially larger experimental basis than reported before.

Related Work. KIV [17,18] is an interactive verification environment which can use ATPs but heavily relies on term rewriting and user guidance. Sunrise [11] is a fully automatic system but uses custom-designed tactics in HOL to discharge the obligations. Houdini [7] is a similar system. Here the generated proof obligations are discharged by ESC/Java but again, this relies on a significant amount of user interaction.

2   System Architecture

The certification tool is built as an extension to the AUTOBAYES and AUTOFILTER program synthesis systems. AUTOBAYES works in the statistical data analysis domain and generates parameter learning programs while AUTOFILTER generates state estimation code based on variants of the Kalman filter algorithm. The synthesis systems take as input a high-level problem specification (cf. Section 3.1 for informal examples). The code that implements the specification is then generated by a schema-based process.

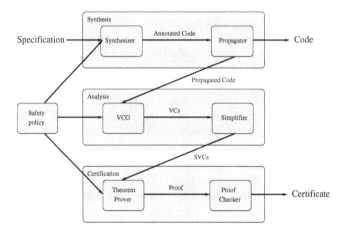

Fig. 1. Certification system architecture

Schemas are generic algorithms which are instantiated in a problem-specific way after their applicability conditions are proven to hold for the given problem specification. Both systems first generate C++-style intermediate code which is then compiled down into any of the different supported languages and runtime environments. Figure 1 gives an overview of the overall system architecture.

2.1   Safety Properties and Safety Policies

The certification tool automatically certifies that a program satisfies a given *safety property*, i.e., an operational characterization that the program "does not go wrong". It uses a corresponding *safety policy*, i.e., a set of Hoare-style proof rules and auxiliary definitions which are specifically designed to show that programs satisfy the safety property of interest. The distinction between safety properties and policies is explored in [2].

We further distinguish between *language-specific* and *domain-specific* properties and policies. Language-specific properties can be expressed in the constructs of the underlying programming language itself (e.g., array accesses), and are sensible for any given program written in the language. Domain-specific properties typically relate to high-level concepts outside the language (e.g., matrix multiplication), and must thus be expressed in terms of program fragments. Since these properties are specific to a particular application domain, the corresponding policies are not applicable to all programs.

We have defined five different safety properties and implemented the corresponding safety policies. Array-bounds safety (*array*) requires each access to an array element to be within the specified upper and lower bounds of the array. Variable initialization-before-use (*init*) asserts that each variable or individual array element has been assigned a defined value before it is used. Both are typical examples of language-specific properties. Matrix symmetry (*symm*) requires certain two-dimensional arrays to be symmetric. Sensor input usage (*in-use*) is a variation of the general *init*-property which guarantees that each sensor reading passed as an input to the Kalman filter algorithm is actually used during the computation of the output estimate. These two examples are specific to the Kalman

Table 1. Safety formulas for different policies

safety policy	safety condition	domain theory
array	$\forall a[i] \in c . a_{lo} \leq i \leq a_{hi}$	arithmetic
init	$\forall$ *read-var* $x \in c . init(x)$	propositional
in-use	$\forall$ *input-var* $x \in c . use(x)$	propositional
symm	$\forall$ *matrix-exp* $m \in c . \forall i, j . m[i, j] = m[j, i]$	matrices
norm	$\forall$ *vector* $v \in c . \Sigma_{i=1}^{\texttt{size}(v)} v[i] = 1$	arithmetic, summations

filter domain. The final example (*norm*) ensures that certain one-dimensional arrays represent normalized vectors, i.e., that their contents add up to one; it is specific to the data analysis domain.

The safety policies can be expressed in terms of two families of definitions. For each command the policy defines a safety condition and a substitution, which captures how the command changes the environmental information relevant to the safety policy. The rules of the safety policy can then be derived systematically from the standard Hoare rules of the underlying programming language [2].

From our perspective, the safety conditions are the most interesting aspect since they have the greatest bearing on the form of the proof obligations. Table 1 summarizes the different formulas and the domain theories needed to reason about them. Both variable initialization and usage as well as array bounds certification are logically simple and rely just on propositional and simple arithmetic reasoning, respectively, but can require a lot of information to be propagated throughout the program. The symmetry policy needs reasoning about matrix expressions expressed as a first-order quantification over all matrix entries. The vector norm policy is formalized in terms of the summation over entries in a one-dimensional array, and involves symbolic reasoning over finite sums.

2.2 Generating Proof Obligations

For certification purposes, the synthesis system *annotates* the code with mark-up information relevant to the selected safety policy. These annotations are part of the schema and thus instantiated in parallel with the code fragments. The annotations contain local information in the form of logical pre- and post-conditions and loop invariants, which is propagated throughout the code. The fully annotated code is then processed by the VCG, which applies the rules of the safety policy to the annotated code in order to generate the safety conditions. As usual, the VCG works backwards through the code. At each line, safety conditions are generated and the safety substitutions are applied. The VCG has been designed to be "correct-by-inspection", i.e., to be sufficiently simple so that it is straightforward to see that it correctly implements the rules of the logic. Hence, the VCG does not carry out any simplifications; in particular, it does not actually apply the substitutions (i.e., execute the specified replacements) but maintains explicit formal substitution terms. Consequently, the generated verification conditions (VCs) tend to be large and must be simplified separately; the more manageable simplified verification conditions (SVCs) which result are then processed by a first order theorem prover. The resulting proofs can be sent to a proof checker, e.g., Ivy [14]. However, since most ATPs

$$\ldots \forall x, y \cdot 0 \leq x \leq 5 \wedge 0 \leq y \leq 5 \Rightarrow sel(id\_init, x, y) = init \qquad \Big\} \text{ environmental in-}$$
$$\wedge \quad \forall x, y \cdot 0 \leq x \leq 5 \wedge 0 \leq y \leq 5 \Rightarrow sel(tmp1\_init, x, y) = init \qquad \Big\} \text{ formation}$$
$$\ldots \forall x, j \cdot 0 \leq x \leq i - 1 \wedge 0 \leq y \leq 5 \Rightarrow sel(tmp2\_init, x, y) = init$$
$$\wedge \quad \forall x, y \cdot 0 \leq y \leq 5 \wedge 0 \leq x \leq n\_p - 1 \Rightarrow$$
$$(x < i \Rightarrow sel(tmp2\_init, x, y) = init \wedge \qquad \qquad \Big\} \text{ invariants}$$
$$(y < j \wedge x = i \Rightarrow sel(tmp2\_init, x, y) = init)))$$
$$\ldots 0 \leq i \leq 5 \wedge 0 \leq j \leq 5 \qquad \qquad \qquad \qquad \quad \Big\} \text{ index bounds}$$
$$\Rightarrow \quad (sel(id\_init, i, j) = init \wedge sel(tmp1\_init, i, j) = init) \qquad \Big\} \text{ safety obligation}$$

Fig. 2. Structure of a safety obligation

do not produce explicit proofs—let alone in a standardized format—we will not focus on proof checking here but concentrate on the simplification and theorem proving steps.

The structure of a typical safety obligation (after substitution reduction and simplification) is given in Figure 2. It corresponds to the initialization safety of an assignment within a nested loop. Most of the hypotheses consist of annotations which have been propagated through the code and are irrelevant to the line at hand. The proof obligation also contains the local loop invariants together with bounds on for-loops. Finally, the conclusion is generated from the safety formula of the corresponding safety policy.

2.3 Processing Proof Obligations and Connecting the Prover

The simplified safety obligations are then exported as a number of individual proof obligations using TPTP first order logic syntax. A small script then adds the axioms of the domain theory, before the completed proof task is processed by the theorem prover. Parts of the domain theory are generated dynamically in order to facilitate reasoning with (small) integers. The domain theory is described in more detail in Section 3.3.

The connection to a theorem prover is straightforward. For provers that do not accept the TPTP syntax, the appropriate TPTP2X-converter was used before invoking the theorem prover. Run-time measurement and prover control (e.g., aborting provers) were performed with the same TPTP tools as in the CASC competition [22].

3 Experimental Setup

3.1 Program Corpus

As basis for the certification experiments we generated annotated programs from four different specifications which were written prior to and independently of the experiments. The size of the generated programs ranges from 431 to 1157 lines of commented C-code, including the annotations. Table 2 in Section 4 gives a more detailed breakdown. The first two examples are AUTOFILTER specifications. ds1 is taken from the attitude control system of NASA's Deep Space One mission [27]. iss specifies a component in a simulation environment for the Space Shuttle docking procedure at the International Space Station. In both cases, the generated code is based on Kalman filter algorithms, which make extensive use of matrix operations. The other two examples are AUTOBAYES specifications which are part of a more comprehensive analysis of planetary nebula images taken by the Hubble Space Telescope (see [5,8] for more details). segm describes an image segmentation problem for which an iterative (numerical) statistical clustering algorithm is synthesized. Finally, gauss fits an image against a two-dimensional Gaussian curve. This requires a multivariate optimization which is implemented by the Nelder-Mead simplex method. The code generated for these two examples has a substantially different structure from the state estimation examples. First, the numerical optimization code contains many deeply nested loops. Also, some of the loops are convergence loops which have no fixed upper bounds but are executed until a dynamically calculated error value gets small enough. In contrast, in the Kalman filter code, all loops are executed a fixed (i.e., known at synthesis time) number of times. Second, the numerical optimization code accesses all arrays element by element and contains no operations on entire matrices (e.g., matrix multiplication). The example specifications and all generated proof obligations can be found at http://ase.arc.nasa.gov/autobayes/ijcar.

3.2 Simplification

Proof task simplification is an important and integral part of our overall architecture. However, as observed before [10,6,20], simplifications—even on the purely propositional level—can have a significant impact on the performance of a theorem prover. In order to evaluate this impact, we used six different rewrite-based simplifiers to generate multiple versions of the safety obligations. We focus on rewrite-based simplifications rather than decision procedures because rewriting is easier to certify: each individual rewrite step $T \rightsquigarrow S$ can be traced and checked independently, e.g., by using an ATP to prove that $S \Rightarrow T$ holds.

Baseline. The baseline is given by the rewrite system $\mathcal{T}_\emptyset$ which eliminates the extralogical constructs (including explicit formal substitutions) which the VCG employs during the construction of the safety obligations. Our original intention was to axiomatize these constructs in first-order logic and then (ab-) use the provers for this elimination step, but that turned out to be infeasible. The main problem is that the combination with equality reasoning produces tremendous search spaces.

Propositional Structure. The first two proper simplification levels only work on the propositional structure of the obligations. $\mathcal{T}_{\forall,\Rightarrow}$ splits the few but large obligations generated by the VCG into a large number of smaller obligations. It consists of two rewrite rules $\forall x \cdot P \wedge Q \rightsquigarrow (\forall x \cdot P) \wedge (\forall x \cdot Q)$ and $P \Rightarrow (Q \wedge R) \rightsquigarrow (P \Rightarrow Q) \wedge (P \Rightarrow R)$ which distribute universal quantification and implication, respectively over conjunction. Each of the resulting conjuncts is then treated as an independent proof task. $\mathcal{T}_{\text{prop}}$ simplifies the propositional structure of the obligations more aggressively. It uses the rewrite rules

$$
\begin{array}{ll}
\neg\, true \rightsquigarrow false & \neg\, false \rightsquigarrow true \\
true \wedge P \rightsquigarrow P & false \wedge P \rightsquigarrow false \\
true \vee P \rightsquigarrow true & false \vee P \rightsquigarrow P \\
P \Rightarrow true \rightsquigarrow true & P \Rightarrow false \rightsquigarrow \neg P
\end{array}
$$

$$true \Rightarrow P \leadsto P \qquad\qquad false \Rightarrow P \leadsto true$$
$$P \Rightarrow P \leadsto true \qquad\qquad (P \wedge Q) \Rightarrow P \leadsto true$$
$$P \Rightarrow (Q \Rightarrow R) \leadsto (P \wedge Q) \Rightarrow R \qquad \forall x \cdot true \leadsto true$$

in addition to the two rules in $\mathcal{T}_{\forall,\Rightarrow}$. The rules have been chosen so that they preserve the overall structure of the obligations as far as possible; in particular, conjunction and disjunction are not distributed over each other and implications are not eliminated. Their impact on the clausifier should thus be minimal.

Ground Arithmetic. This simplification level additionally handles common extensions of plain first-order logic, i.e., equality, orders, and arithmetic. The rewrite system $\mathcal{T}_{eval}$ contains rules for the reflexivity of equality and partial orders as well as the irreflexivity of strict orders, although the latter rules are not invoked on the example obligations. In addition, it normalizes orders into $\leq$ and $>$ using the (obvious) rules

$$x \geq y \leadsto y \leq x \qquad \neg x > y \leadsto x \leq y$$
$$x < y \leadsto y > x \qquad \neg x \leq y \leadsto x > y$$

The choice of the specific orders is arbitrary; choosing for example $<$ instead of $>$ makes no difference. However, a further normalization by elimination of either the partial or the strict order (e.g., using a rule $x \leq y \leadsto x < y \vee x = y$) leads to a substantial increase in the formula size and thus proves to be counter-productive.

$\mathcal{T}_{eval}$ also contains rules to evaluate ground integer operations (i.e., addition, subtraction, and multiplication), equalities, and partial and strict orders. Moreover, it converts addition and subtraction with one small integer argument (i.e., $n \leq 5$) into Pressburger notation, using rules of the form $n + 1 \leadsto succ(n)$ and $n - 1 \leadsto pred(n)$. For many safety policies (e.g., *init*), such terms are introduced by relativized bounded quantifiers (e.g., $\forall x \cdot 0 \leq x \leq n - 1 \Rightarrow P(x)$) and contain the only occurrences of arithmetic operators. A final group of rules handles the interaction between *succ* and *pred*, as well as with the orders.

$$succ(pred(x)) \leadsto x \qquad\qquad pred(succ(x)) \leadsto x$$
$$succ(x) \leq y \leadsto x < y \qquad\qquad succ(x) > y \leadsto x \geq y$$
$$x \leq pred(y) \leadsto x < y \qquad\qquad x > pred(y) \leadsto x \geq y$$

Language-Specific Simplification. The next level handles constructs which are specific to the program verification domain, in particular array-expressions and conditional expressions, encoding the necessary parts of the language semantics. The rewrite system $\mathcal{T}_{array}$ adds rewrite formulations of McCarthy's array axioms [13], i.e., $sel(upd(a, i, v), j) \leadsto i = j ? v : sel(a, j)$ for one-dimensional arrays and similar forms for higher-dimensional arrays. Some safety policies are formulated using arrays of a given dimensionality which are uniformly initialized with a specific value. These are represented by a *constarray*-term, for which similar rules are required, e.g., $sel(constarray(v, d), i) \leadsto v$.

Nested *sel/upd*-terms, which result from sequences of individual assignments to the same array, lead to nested conditionals which in turn lead to an exponential blow-up during the subsequent language normalization step. $\mathcal{T}_{array}$ thus also contains two rules $true ? x : y \leadsto x$ and $false ? x : y \leadsto y$ to evaluate conditionals.

In order to evaluate the effect of these domain-specific simplifications properly, we also experimented with a rewrite system $\mathcal{T}_{array^*}$, which applies the two *sel*-rules in isolation.

Policy-Specific Simplification. The most aggressive simplification level $\mathcal{T}_{policy}$ uses a number of rules which are fine-tuned to handle situations that frequently arise with specific safety policies. The *init*-policy requires a rule

$$\forall x \cdot 0 \le x \le n \Rightarrow (x \ne 0 \land \ldots \land x \ne n \Rightarrow P) \rightsquigarrow true$$

which is derived from the finite induction axiom to handle the result of simplifying nested *sel/upd*-terms. For *in-use*, we need a single rule $def = use \rightsquigarrow false$, which follows from the fact that the two tokens *def* and *use* used by the policy are distinct. For *symm*, we make use of a lemma about the symmetry of specific matrix expressions: $A + BCB^T$ is already symmetric if (but not only if) the two matrices A and C are symmetric, regardless of the symmetry of B. The rewrite rule

$$sel(A + BCB^T, i, j) = sel(A + BCB^T, j, i)$$
$$\rightsquigarrow sel(A, i, j) = sel(A, j, i) \land sel(C, i, j) = sel(C, j, i)$$

formulates this lemma in an element-wise fashion.

For the *norm*-policy, the rules become a lot more specialized and complicated. Two rules are added to handle the inductive nature of finite sums:

$$\sum_{i=0}^{pred(0)} x \rightsquigarrow 0$$
$$P \land x = \sum_{i=0}^{pred(n)} Q(i) \Rightarrow x + Q(n) = \sum_{i'=0}^{n} Q(i')$$
$$\rightsquigarrow P \land x = \sum_{i=0}^{pred(n)} Q(i) \Rightarrow \sum_{i=0}^{n} Q(i) = \sum_{i=0}^{n} Q(i)$$

The first rule directly implements the base case of the induction; the second rule, which implements the step case, is more complicated. It requires alpha-conversion for the summations as well as higher-order matching for the body expressions. However, both are under explicit control of this specific rewrite rule and not the general rewrite engine, and are implemented directly as Prolog-predicates. A similar rule is required in a very specific situation to substitute an equality into a summation:

$$P \land (\forall i \cdot 0 \le i \le n \Rightarrow x = sel(f, i)) \Rightarrow \sum_{i=0}^{n} sel(f, i) = 1$$
$$\rightsquigarrow P \land (\forall i \cdot 0 \le i \le n \Rightarrow x = sel(f, i)) \Rightarrow \sum_{i=0}^{n} x = 1$$

The above rules capture the central steps of some of the proofs for the *norm*-policy and mirror the fact that these are essentially higher-order inferences.

Another set of rewrite rules handles all occurrences of the random number generator by asserting that the number is within its given range, i.e., $l \le rand(l, u) \le u$.

Normalization. The final preprocessing step transforms the obligations into pure first-order logic. It eliminates conditional expressions which occur as top-level arguments of predicate symbols, using rules of the form $P ? T : F = R \rightsquigarrow (P \Rightarrow T = R) \land (\neg P \Rightarrow F = R)$ and similarly for partial and strict orders. A number of congruence rules move nested occurrences of conditional expressions into the required positions. Finite sums, which only occur in obligations for the *norm*-policy, are represented with a de Bruijn-style variable-free notation.

Control. The simplifications are performed by a small but reasonably efficient rewrite engine implemented in Prolog (cf. Table 2 for runtime information). This engine does not support full AC-rewriting but flattens and orders the arguments of AC-operators. The rewrite rules, which are implemented as Prolog-clauses, then do their own list matching but can take the list ordering into account. The rules within each system are applied exhaustively. However, the two most aggressive simplification levels $\mathcal{T}_{array}$ and $\mathcal{T}_{policy}$ are followed by a "clean-up" phase. This consists of the language normalization followed by the propositional simplifications $\mathcal{T}_{prop}$ and the finite induction rule. Similarly, $\mathcal{T}_{array*}$ is followed by the language normalization and then by $\mathcal{T}_{\forall,\Rightarrow}$ to split the obligations.

3.3 Domain Theory

Each safety obligation is supplied with a first-order domain theory. In our case, the domain theory consists of a fixed part which contains 44 axioms, and a set of axioms which is generated dynamically for each proof task. The static set of axioms defines the usual properties of equality and the order relations, as well as axioms for simple Pressburger arithmetic and for the domain-specific operators (e.g., *sel/upd* or *rand*). The dynamic axioms are added because most theorem provers cannot calculate with integers, and to avoid the generation of large terms of the form $succ(\ldots(succ(0)\ldots))$. For all integer literals n, m in the proof task, we generate the corresponding axioms of the form $m > n$. For small integers (i.e., $n \leq 5$), we also generate axioms for explicit successor-terms, i.e., $n = succ^n(0)$ and add a finite induction schema of the form $\forall x \: : \: 0 \leq x \leq n \Rightarrow (x = 0 \vee x = 1 \vee \ldots \vee x = n)$. In our application domain, these axioms are needed for some of the matrix operations; thus n can be limited to the statically known maximal size of the matrices.

3.4 Theorem Provers

For the experiments, we selected several high-performance theorem provers for untyped first-order formulas with equality. Most of the provers participated at the CASC-19 [21] proving competition in the FOL-category. We used two versions of e-setheo [15] which were both derived from the CASC-version. For e-setheo-csp03F, the clausification module has been changed and instead of the clausifier provided by the TPTP toolset [22], FLOTTER V2.1 [23,24] was used to convert the formulas into a set of clauses. e-setheo-new is a recent development version with several improvements over the original e-setheo-csp03 version. Both versions of Vampire [19] have been taken directly "out of the box"—they are the versions which were running during CASC-19. Spass 2.1 was obtained from the developer's website [23].

In the experiments, we used the default parameter settings and none of the special features of the provers. For each proof obligation, we limited the run-time to 60 seconds; the CPU-time actually used was measured with the TPTP-tools on a 2.4GHz dual processor standard PC with 4GB memory.

Table 2. Results of generating safety obligations

example	loc	policy	loa	$T_\emptyset$		$T_{\forall,\Rightarrow}$		T_{prop}		T_{eval}		T_{array}		T_{array*}		T_{policy}	
ds1	431	array	0	5.5	11	5.3	103	5.4	55	5.5	1	5.5	1	5.6	103	5.5	1
		init	87	9.5	21	14.1	339	11.3	150	11.0	142	10.5	74	20.1	543	11.4	74
		in-use	61	7.3	19	12.9	453	7.7	59	7.6	57	7.4	21	16.2	682	8.1	21
		symm	75	4.8	17	5.7	101	4.7	21	4.9	21	66.7	858	245.6	2969	70.8	865
iss	755	array	0	24.6	1	28.1	582	24.8	114	24.2	4	24.0	4	27.9	582	24.7	4
		init	88	39.5	2	65.9	957	42.3	202	41.8	194	39.2	71	82.6	1378	39.7	71
		in-use	60	33.4	2	68.1	672	36.7	120	35.7	117	32.6	28	79.1	2409	31.6	1
		symm	87	33.0	1	34.9	185	28.1	35	27.9	35	71.0	479	396.8	3434	66.2	480
segm	517	array	0	3.0	29	3.3	85	2.9	8	2.9	3	3.0	3	3.3	85	3.0	1
		init	171	6.5	56	12.1	464	7.8	172	7.7	130	7.6	121	12.8	470	7.6	121
		norm	195	3.8	54	5.0	155	3.8	41	3.6	30	3.8	32	5.2	157	3.6	14
gauss	1039	array	20	21.0	69	24.9	687	21.2	98	21.0	20	20.9	20	24.3	687	21.3	20
		init	118	49.8	85	65.5	1417	54.1	395	53.2	324	53.9	316	66.2	1434	54.3	316

4 Empirical Results

4.1 Generating and Simplifying Obligations

Table 2 summarizes the results of generating the different versions of the safety obligations. For each of the example specifications, it lists the size of the generated programs (without annotations), the applicable safety policies, the respective size of the generated annotations (before propagation), and then, for each simplifier, the elapsed time and the number of generated obligations.

The elapsed times include synthesis of the programs as well as generation, simplification, and file output of the safety obligations; synthesis alone accounts for approximately 90% of the times listed under the *array* safety policy. In general, the times for generating and simplifying the obligations are moderate compared to both generating the programs and discharging the obligations. All times are CPU-times and have been measured in seconds using the Unix `time`-command.

Almost all of the generated obligations are valid, i.e., the generated programs are safe. The only exception is the *in-use*-policy which produces one invalid obligation for each of the `ds1` and `iss` examples. This is a consequence of the respective specifications which do not use all elements of the initial state vectors. The invalidity is confined to a single conjunct in one of the original obligations, and since none of the rewrite systems contains a distributive law, the number of invalid obligations does not change with simplification.

The first four simplification levels show the expected results. The baseline $T_\emptyset$ yields relatively few but large obligations which are then split up by $T_{\forall,\Rightarrow}$ into a much larger (on average more than an order of magnitude) number of smaller obligations. The next two levels then eliminate a large fraction of the obligations. Here, the propositional simplifier T_{prop} alone already discharges between 50% and 90% of the obligations while the additional effect of evaluating ground arithmetic (T_{eval}) is much smaller and generally well below 25%. The only significant difference occurs for the *array*-policy where more than 80% (and in the case of `ds1` even all) of the remaining obligations are reduced to true. This is a consequence of the large number of obligations which have the form $\neg n \leq n \Rightarrow P$ for an integer constant n representing the (lower or upper) bound of an array. The effect of the domain-specific simplifications is at first glance less clear. Using the array-rules only, $\mathcal{T}_{\text{array}^*}$, generally leads to an increase over $\mathcal{T}_{\forall, \Rightarrow}$ in the number of obligations; this even surpasses an order of magnitude for the *symm*-policy. However, in combination with the other simplifications ($\mathcal{T}_{\text{array}}$), most of these obligations can be discharged again, and we generally end up with less obligations than before; again, the *symm*-policy is the only exception. The effect of the final policy-specific simplifications is, as should be expected, highly dependent on the policy. For *in-use* and *norm* a further reduction is achieved, while the rules for *init* and *symm* only reduce the size of the obligations.

4.2   Running the Theorem Provers

Table 3 summarizes the results obtained from running the theorem provers on all proof obligations (except for the invalid obligations from the *in-use*-policy), grouped by the different simplification levels. Each line in the table corresponds to the proof tasks originating from a specific safety policy (*array, init, in-use, symm,* and *norm*). Then, for each prover, the percentage of solved proof obligations and the total CPU-time are given.

For the fully simplified version ($\mathcal{T}_{\text{policy}}$), all provers are able to find proofs for all tasks originating from at least one safety policy; e-setheo-csp03F can even discharge *all* the emerging safety obligations This result is central for our application since it shows that current ATPs can in fact be applied to certify the safety of synthesized code, confirming our first hypothesis.

For the unsimplified safety obligations, however, the picture is quite different. Here, the provers can only solve a relatively small fraction of the tasks and leave an unacceptably large number of obligations to the user. The only exception is the *array*-policy, which produces by far the simplest safety obligations. This confirms our second hypothesis: aggressive preprocessing is absolutely necessary to yield reasonable results.

Let us now look more closely at the different simplification stages. Breaking the large original formulas into a large number of smaller but independent proof tasks ($\mathcal{T}_{\forall, \Rightarrow}$) boosts the relative performance considerably. However, due to the large absolute number of tasks, the absolute number of failed tasks also increases. With each additional simplification step, the percentage of solved proof obligations increases further. Interestingly, however, $\mathcal{T}_{\forall, \Rightarrow}$ and $\mathcal{T}_{\text{array}}$ seem to have the biggest impact on performance. The reason seems to be that equality reasoning on deeply nested terms and formula structures can then be avoided, albeit at the cost of the substantial increase in the number of proof tasks. The results with the simplification strategy $\mathcal{T}_{\text{array}^*}$, which only contains the language-specific rules, also illustrates this behavior. The *norm*-policy clearly produces the most difficult proof obligations, requiring essentially inductive and higher-order reasoning. Here, all simplification steps are required to make the obligations go through the first-order ATPs.

The results in Table 3 also indicate there is no single best theorem prover. Even variants of the "same" prover can differ widely in their results. For some proof obligations,

Table 3. Certification results and times

			e-setheo03F		e-setheo-new		SPASS		Vampire6.0		Vampire5.0	
simp.	policy	N	%	T_{proof}	%	T_{proof}	%	T_{proof}	%	T_{proof}	%	T_{proof}
$T_\emptyset$	array	110	96.4	192.4	94.5	284.9	96.4	73.4	95.5	178.1	95.5	102.1
	init	164	76.8	3000.8	13.1	1759.8	75.0	2898.3	8.5	9224.9	8.5	8251.0
	in-use	19	57.9	610.8	44.4	612.2	68.4	512.8	57.9	773.1	47.4	645.5
	symm	18	50.0	387.7	8.3	266.1	38.9	555.3	16.7	744.9	16.7	723.6
	norm	54	51.9	1282.4	51.9	1341.0	51.9	1224.2	50.0	1316.5	48.1	1327.1
$T_{\forall,\Rightarrow}$	array	1457	99.0	903.4	94.2	5925.0	99.8	217.0	99.9	240.5	99.8	152.4
	init	3177	88.4	3969.4	91.7	20784.8	97.4	8732.2	95.0	14482.2	93.5	14203.4
	in-use	1123	59.3	819.1	96.4	4100.3	99.1	1733.5	95.3	4183.7	94.3	4206.8
	symm	286	93.4	1785.9	90.6	2341.0	88.5	3638.7	90.2	3315.8	91.3	1789.2
	norm	155	85.8	1422.1	73.5	2552.5	84.5	1572.0	87.7	1359.9	87.1	1276.0
T_{prop}	array	275	99.3	278.2	76.4	4080.8	99.3	157.5	99.3	187.5	99.3	132.6
	init	919	94.7	4239.4	73.0	17472.2	92.8	5469.7	84.9	10598.0	83.2	10546.8
	in-use	177	86.4	1854.0	77.4	2768.2	94.9	1008.3	70.1	3806.2	65.0	3960.6
	symm	56	66.1	1476.2	51.8	1944.4	48.2	1911.3	58.9	1596.7	58.9	1424.8
	norm	41	46.3	1361.2	41.5	1484.6	41.5	1478.2	53.7	1286.7	51.2	1275.3
T_{eval}	array	28	100.0	16.2	100.0	19.7	100.0	10.4	100.0	12.7	100.0	1.7
	init	790	94.6	3944.2	94.1	8288.0	93.3	4380.1	82.5	10239.0	82.0	9040.2
	in-use	172	86.0	1852.2	83.1	2305.2	94.8	1023.1	69.8	3718.1	67.4	3561.1
	symm	56	66.1	1451.1	66.1	1500.4	51.8	1716.0	62.5	1455.5	58.9	1389.8
	norm	30	53.3	859.4	13.3	1575.8	50.0	940.5	66.7	736.7	53.3	858.0
T_{array}	array	28	100.0	15.4	100.0	19.8	100.0	10.4	100.0	12.7	100.0	1.7
	init	582	100.0	527.6	100.0	823.9	99.7	875.8	100.0	1401.3	99.0	785.1
	in-use	47	100.0	323.9	100.0	343.2	100.0	171.3	100.0	262.6	87.2	525.2
	symm	1337	100.0	1104.3	99.9	1629.3	99.4	746.4	99.1	963.9	99.0	922.7
	norm	32	59.4	678.4	18.8	1583.1	59.4	709.7	62.5	791.7	50.0	858.6
T_{array*}	array	1457	99.9	916.4	94.2	5918.0	99.9	210.8	99.9	240.6	99.9	153.1
	init	3825	99.7	3412.3	96.3	13536.1	99.5	4574.9	99.8	4952.1	98.4	6000.1
	in-use	3089	99.8	2438.4	99.4	5139.0	99.8	889.2	99.8	793.5	99.6	925.9
	symm	6403	99.9	5317.4	99.7	11787.7	99.7	3385.1	99.6	3277.3	99.6	1807.0
	norm	157	86.0	1306.8	72.6	2670.8	86.0	1351.3	86.6	1449.9	86.0	1276.2
T_{policy}	array	26	100.0	15.0	100.0	17.7	100.0	9.9	100.0	12.0	100.0	1.6
	init	582	100.0	529.2	100.0	827.9	99.5	875.2	100.0	1418.9	99.0	782.5
	in-use	20	100.0	281.7	100.0	329.7	100.0	170.7	100.0	262.6	70.0	524.8
	symm	1345	100.0	1104.6	99.9	1640.5	99.4	760.0	99.1	1048.8	99.0	926.9
	norm	14	100.0	9.0	57.1	375.8	100.0	26.2	100.0	108.0	71.4	241.8

the choice of the clausification module makes a big difference. The TPTP-converter implements a straightforward algorithm similar to the one described in [12]. Flotter has a highly elaborate conversion algorithm which performs many simplifications and avoids exponential increase in the number of generated clauses. This effect is most visible on the unsimplified obligations (e.g., $T_\emptyset$ under *init*), where Spass and e-setheo-csp03F—which both use the Flotter clausifier—perform substantially better than the other provers.

Since our proof tasks are generated directly by a real application and are not hand-picked for certain properties, many of them are (almost) trivial—even in the unsimplified case. Figure 3 shows the resources required for the proof tasks as a series of pie charts

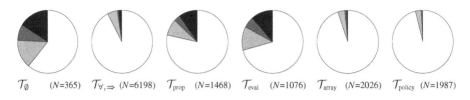

$\mathcal{T}_\emptyset$ (N=365) $\mathcal{T}_{\forall,\Rightarrow}$ (N=6198) $\mathcal{T}_{prop}$ (N=1468) $\mathcal{T}_{eval}$ (N=1076) $\mathcal{T}_{array}$ (N=2026) $\mathcal{T}_{policy}$ (N=1987)

Fig. 3. Distribution of easy ($T_{proof} < 1s$, white), medium ($T_{proof} < 10s$, light grey), difficult ($T_{proof} < 60s$, dark grey) proofs, and failing proof tasks (black) for the different simplification stages (prover: e-setheo-csp03F). N denotes the total number of proof tasks at each stage.

for the different simplification stages. All numbers are obtained with e-setheo-csp03F; the figures for the other provers look similar. Overall, the charts reflect the expected behavior: with additional preprocessing and simplification of the proof obligations, the number of easy proofs increases substantially and the number of failing proof tasks decreases sharply from approximately 16% to zero. The relative decrease of easy proofs from $\mathcal{T}_{\forall,\Rightarrow}$ to $\mathcal{T}_{prop}$ and $\mathcal{T}_{eval}$ is a consequence of the large number of easy proof tasks already discharged by the respective simplifications.

4.3   Difficult Proof Tasks

Since all proof tasks are generated in a uniform manner through the application of a safety policy by the VCG, it is obvious that many of the difficult proof tasks share some structural similarities. We have identified three classes of hard examples; these classes are directly addressed by the rewrite rules of the policy-specific simplifications.

Most safety obligations generated by the VCG are of the form $\mathcal{A} \Rightarrow \mathcal{B}_1 \wedge \ldots \wedge \mathcal{B}_n$ where the $\mathcal{B}_i$ are variable disjoint. These obligations can be split up into n smaller proof obligations of the form $\mathcal{A} \Rightarrow \mathcal{B}_i$ and most theorem provers can then handle these smaller independent obligations much more easily than the large original.

The second class contains formulas of the form $symm(r) \Rightarrow symm(diag\text{-}updates(r))$. Here, r is a matrix variable which is updated along its diagonal, and we need to show that r remains symmetric after the updates. For a 2x2 matrix and two updates (i.e., $r_{00} = x$ and $r_{11} = y$), we obtain the following simplified version of an actual proof task:

$$\forall i, j \cdot (0 \leq i, j \leq 1 \Rightarrow sel(r, i, j) = sel(r, j, i)) \Rightarrow$$
$$(\forall k, l \cdot (0 \leq k, l \leq 1 \Rightarrow$$
$$sel(upd(upd(r, 1, 1, y), 0, 0, x), k, l) = sel(upd(upd(r, 1, 1, y), 0, 0, x), l, k))).$$

This pushes the provers to their limits—e-setheo cannot prove this while Spass succeeds here but fails if the dimensions are increased to 3x3, or if three updates are made. In our examples, matrix dimensions up to 6x6 with 36 updates occur, yielding large proof obligations of this specific form which are not provable by current ATPs without further preprocessing.

Another class of trivial but hard examples, which frequently shows up in the *init*-policy, also results from the expansion of deeply nested *sel/upd*-terms. These problems have the form

$$\forall i, j \cdot 0 \leq i < n \wedge 0 \leq j \leq n \Rightarrow (i \neq 0 \wedge j \neq 0 \wedge \ldots i \neq n \wedge j \neq n \Rightarrow false)$$

and soon become intractable for the clausifier, even for small n ($n = 2$ or $n = 3$), although the proof would be easy after successful clausification.

5 Conclusions

We have described a system for the automated certification of safety properties of NASA state estimation and data analysis software. The system uses a generic VCG together with explicit safety policies to generate policy-specific safety obligations which are then automatically processed by a first-order ATP. We have evaluated several state-of-the-art ATPs on more than 25,000 obligations generated by our system. With "out-of-the-box" provers, only about two-thirds of the obligations could be proven. However, after aggressive simplification, most of the provers could solve all emerging obligations. In order to see the effects of simplification more clearly, we experimented with specific preprocessing stages.

It is well-known that, in contrast to traditional mathematics, software verification hinges on large numbers of mathematically shallow (in terms of the concepts involved) but structurally complex proof tasks, yet current provers are not well suited to this. Since the propositional structure of a formula is of great importance, we believe that clausification algorithms should integrate more simplification and split goal tasks into independent subtasks.

Certain application-specific constructs (e.g., *sel/upd*) can easily lead to proof tasks which cannot be handled by current ATPs. The reason is that simple manipulations on deep terms, when combined with equational reasoning, can result in a huge search space. Although specific parameter settings in a prover might overcome this problem, this would require a deep knowledge of the individual theorem provers. In our experiments, therefore, we did not use any specific features or parameter settings for the individual theorem provers.

With our approach to certification of auto-generated code, we are able to automatically produce safety certificates for code of considerable length and structural complexity. By combining rewriting with state-of-the-art automated theorem proving, we obtain a safety certification tool which compares favorably with tools based on static analysis (see [3] for a comparison). Our current efforts focus on extending the certification system in a number of areas. One aim is to develop a certificate management system, along the lines of the Programatica project [16]. We also plan to combine our work on certification with automated safety and design document generation [4] tools that we are developing. Finally, we continue to integrate additional safety properties.

References

[1] W. Bibel and P. H. Schmitt, (eds.). *Automated Deduction — A Basis for Applications*. Kluwer, 1998.

[2] E. Denney and B. Fischer. "Correctness of Source-Level Safety Policies". In *Proc. FM 2003: Formal Methods, LNCS 2805*, pp. 894–913. Springer, 2003.

[3] E. Denney, B. Fischer, and J. Schumann. "Adding Assurance to Automatically Generated Code". In *Proc. 8th IEEE Intl. Sympl. High Assurance System Engineering*, pp. 297–299. IEEE Comp. Soc. Press, 2004.

[4]   E. Denney and R. P. Venkatesan. "A generic software safety document generator". In *Proc. 10th AMAST*. To appear, 2004.

[5]   B. Fischer, A. Hajian, K. Knuth, and J. Schumann. Automatic Derivation of Statistical Data Analysis Algorithms: Planetary Nebulae and Beyond. In *Proc. 23rd MaxEnt*. To appear, 2004. http://ase.arc.nasa.gov/people/fischer/.

[6]   B. Fischer. *Deduction-Based Software Component Retrieval*. PhD thesis, U. Passau, Germany, 2001. http://elib.ub.uni-passau.de/opus/volltexte/2002/23/.

[7]   C. Flanagan and K. R. M. Leino. "Houdini, an Annotation Assistant for ESC/Java". In *Proc. FME 2001: Formal Methods for Increasing Software Productivity, LNCS 2021*, pp. 500–517. Springer, 2001.

[8]   B. Fischer and J. Schumann. "Applying AutoBayes to the Analysis of Planetary Nebulae Images". In *Proc. 18th ASE*, pp. 337–342. IEEE Comp. Soc. Press, 2003.

[9]   B. Fischer and J. Schumann. "AutoBayes: A System for Generating Data Analysis Programs from Statistical Models". *J. Functional Programming*, 13(3):483–508, 2003.

[10]   B. Fischer, J. Schumann, and G. Snelting. "Deduction-Based Software Component Retrieval". Volume II of Bibel and Schmitt [1], pp. 265–292. 1998.

[11]   P. Homeier and D. Martin. "Trustworthy Tools for Trustworthy Programs: A Verified Verification Condition Generator". In *Proc. TPHOLS 94*, pp. 269–284. Springer, 1994.

[12]   D. W. Loveland. *Automated Theorem Proving: A Logical Basis*. North–Holland, 1978.

[13]   J. McCarthy. "Towards a Mathematical Science of Computation". In *Proc. IFIP Congress 62*, pp. 21–28. North-Holland, 1962.

[14]   W. McCune and O. Shumsky. "System description: IVY". In *Proc. 17th CADE, LNAI 1831*, pp. 401–405. Springer, 2000.

[15]   M. Moser, O. Ibens, R. Letz, J. Steinbach, C. Goller, J. Schumann. and K. Mayr. "The Model Elimination Provers SETHEO and E-SETHEO". *J. Automated Reasoning*, 18:237–246, 1997.

[16]   The Programatica Team. "Programatica Tools for Certifiable, Auditable Development of High-assurance Systems in Haskell". In *Proc. High Confidence Software and Systems Conf.*, Baltimore, MD, April 2003.

[17]   W. Reif. "The KIV Approach to Software Verification". In *KORSO: Methods, Languages and Tools for the Construction of Correct Software, LNCS 1009*, pp. 339–370. Springer, 1995.

[18]   W. Reif, G. Schellhorn, K. Stenzel, and M. Balser. *Structured Specifications and Interactive Proofs with KIV*. Volume II of Bibel and Schmitt [1], pp. 13–40, 1998.

[19]   A. Riazanov and A. Voronkov. "The Design and Implementation of Vampire". *AI Communications*, 15(2–3):91–110, 2002.

[20]   J. Schumann. *Automated Theorem Proving in Software Engineering*. Springer, 2001.

[21]   G. Sutcliffe and C. Suttner. CASC Home Page. http://www.tptp.org/CASC.

[22]   G. Sutcliffe and C. Suttner. TPTP Home Page. http://www.tptp.org.

[23]   C. Weidenbach. SPASS Home Page. http://spass.mpi-sb.mpg.de.

[24]   C. Weidenbach, B. Gaede, and G. Rock. "Spass and Flotter version 0.42". In *Proc. 13th CADE, LNAI 1104*, pp. 141–145. Springer, 1996.

[25]   M. Whalen, J. Schumann, and B. Fischer. "AutoBayes/CC — Combining Program Synthesis with Automatic Code Certification (System Description)". In *Proc. 18th CADE, LNAI 2392*, pp. 290–294. Springer, 2002.

[26]   M. Whalen, J. Schumann, and B. Fischer. "Synthesizing Certified Code". In *Proc. FME 2002: Formal Methods—Getting IT Right, LNCS 2391*, pp. 431–450. Springer, 2002.

[27]   J. Whittle and J. Schumann. Automating the Implementation of Kalman Filter Algorithms, 2004. In review.

ARGO-LIB: A Generic Platform for Decision Procedures

Filip Marić and Predrag Janičić

Faculty of Mathematics, University of Belgrade
Studentski trg 16, 11 000 Belgrade, Serbia
{filip,janicic}@matf.bg.ac.yu

Abstract. ARGO-LIB is a C++ library that provides support for using decision procedures and for schemes for combining and augmenting decision procedures. This platform follows the SMT-LIB initiative which aims at establishing a library of benchmarks for satisfiability modulo theories. The platform can be easily integrated into other systems. It also enables comparison and unifying of different approaches, evaluation of new techniques and, hopefully, can help in advancing the field.

1 Introduction

The role of decision procedures is very important in theorem proving, model checking of real-time systems, symbolic simulation, etc. Decision procedures can reduce the search space of heuristic components of such systems and increase their abilities. Usually, decision procedures have to be combined, to communicate with heuristic components, or some additional hypotheses have to be invoked. There are several influential approaches for using decision procedures, some of them focusing on combining decision procedures [8,12,1] and some focusing on augmenting decision procedures [2,7]. We believe there is a need for an efficient generic platform for using decision procedures. Such a platform should provide support for a range of decision procedures (for a range of theories) and also for different techniques for using decision procedures. ARGO-LIB platform[1] is made motivated by the following requirements:

- it provides a flexible and efficient implementation of required primitives and also support for easily implementing higher-level procedures and different approaches for using decision procedures;
- it is implemented in the standard C++ programming language; the gains are in efficiency and in taking the full control over the proving process; the standard, plain C++ implementation is portable to different operating systems and there is no need for some specific operating system or some specific compiler; this also encourages and promotes sharing of code between different groups and enables converging implementation efforts in one direction;

---

[1] The web page for ARGO-LIB is at: www.matf.bg.ac.yu/~janicic/argo. The name ARGO comes from *Automated Reasoning GrOup* and from $A\rho\gamma\omega$, the name of the galley of argonauts (this galley was very light and very fast).

D. Basin and M. Rusinowitch (Eds.): IJCAR 2004, LNAI 3097, pp. 213–217, 2004.

- it is "light" (i.e., small in size and stand-alone, requires only any standard C++ compiler) and fast (capable of coping with complex real-world problems using the existing or new approaches);
- its architecture is flexible and modular; any user can easily extend it by new functionalities or reimplement some modules of the system;
- it can work stand-alone but can also be integrated into some other tool (e.g., a theorem prover, constraint solver, model checking system etc.);
- it enables comparison on an equal footing between schemes for using decision procedures (especially when the worst-case complexity can be misleading);
- it is publicly available and it has simple, documented interface to all functionalities of the platform;
- it supports a library of relevant conjectures and a support for benchmarking (here we have in mind motivations, ideas, and standards promoted by the SMT-LIB initiative [10]); this enables exchange of problems, results, ideas, and techniques between different groups.

Although some of the above requirements might seem conflicting, we do believe that we have built a system that meet them in a high degree. Indeed, instead of making a new programming language and/or a system that interprets proof methods and that takes care about "protecting" its user, we opt for C++ and give the user more freedom and responsibility. Namely, it is sensible to assume that the user interested in using efficient decision procedures is familiar with C++ and for him/her it is simpler to install and to use some C++ compiler then some specific theorem proving platform. Therefore, and also because of efficient code, we believe that C++ is one of the best (or the best) options for a generic platform for decision procedures which aims at *realistic* wider use (both in academia and in industry).

2 Background

Concerning the background logic (first order classical logic with equality), underlying theories, description of theories, and format, ARGO-LIB follows the SMT-LIB initiative [10]. The main goal of the SMT-LIB initiative, supported by a growing number of researchers world-wide is to produce a library of benchmarks for satisfiability modulo theories and all required standards and notational conventions. Such a library will facilitate the evaluation and the comparison of different approaches for using decision procedures and advance the state of the art in the field. The progress that has been made so far supports these expectations.

In terms of logical organisation and organisation of methods, ARGO-LIB follows the proof-planning paradigm [3] and the GS framework [6]. The GS framework for using decision procedures is built from a set of methods some of which are abstraction, entailment, congruence closure, and "lemma invoking". There are generic and theory-specific methods. Some methods use decision procedures as black boxes, while some also use functionalities like elimination of variables. The GS framework is flexible and general enough to cover a range of schemes for both combining and augmenting decision procedures, including Nelson and Oppen's [8], Shostak's [12], and the approach used in the TECTON system [7].

3   Architecture

Representation of expressions. ARGO-LIB contains a class hierarchy for representing first-order terms and formulae, with a wide set of low-level functions for manipulating them. Expressions are represented in a tree-like structure based on the Composite design pattern.

Unification and rewriting. ARGO-LIB provides support for first order unification. The unification algorithm is encapsulated in the class Unification, so it can be easily altered. A generic rewriting mechanism (including reduction ordering) for terms and formulae is also supported. A set of rewrite rules can be provided directly in the code or it can be read from an external file.

Theories. Each (decidable) theory inherits the base class Theory and is characterized by: *Signature, Decision procedure, Simplification procedure* (a function that performs theory-specific simplification of a given formula). At the moment, there is support for *PRA* (Presburger Rational Arithmetic), *FOLeq* (universally quantified fragment of equality), and *LISTS* (theory of lists).

Goals. A goal is an object of the class Goal and is represented by: *List of underlying theories, Conjecture* (a first order logic formula over the combination of given theories), *Status* of the conjecture (satisfiable, valid, unsatisfiable, invalid), *Extra signature* (the list of uninterpreted symbols). A number of iterator classes (for iterating over the subgoals) is also provided. For example, the class GoalTheoriesIterator is used to iterate through the subgoals of a goal whose conjecture is a conjunction of (abstracted) literals defined over several theories — a subgoal for each theory is created and its conjecture consists of all literals of the initial goal that are defined over that theory.

Methods. Each method inherits the abstract base class Method and transforms a given goal. A method has the following slots: *Name, Preconditions* (checks if the method is applicable to a given goal), *Effect* (a function that describes the effect of the method to a given goal), *Postconditions* (conditions that have to hold after the method application), *Parameters* (additional parameters, e.g., a variable to be eliminated). In ARGO-LIB, the methods don't have tactics attached and do not produce object-level proofs. A list of methods is constantly being expanded. Existing methods are divided in several groups: *general purpose methods* (e.g., Rewriting, Prenex, DNF, CNF, Skolemization, Negation, FOLSimplification), *theory specific methods* (e.g., NelsonOppenCCC, FourierMotzkinElimination, ListsSimplification), *combination of theories* (e.g., Abstraction, Partitioning, DeduceEqualityAndReplace, Simplification, UnsatModuloTheories), *combination schemes* (e.g., NelsonOppenCombinationScheme, TectonScheme).

Input format and parsing. The native input format for ARGO-LIB is the SMT-LIB format (for benchmarks). Description of theories and their signatures are also read from the files in the (slightly extended) SMT-LIB format.

Output generating. A number of classes (e.g., LaTeXOutput for LaTeX output, SMTOutput for SMT-LIB output, MathMLOutput for XML output) provide support for generating output in a readable and easy to understand format. The level and form of output information can be controlled.

4   Samples and Results

The example below shows the C++ code for Nelson-Oppen's scheme [8]. The procedure is entirely based on the use of the ARGO-LIB methods. Notice how the code faithfully reflects logical, high-level description of the procedure. Further below we show a part of one ARGO-LIB output. We have tested ARGO-LIB on a number of examples and the results are very good. For instance, CPU times are an order of magnitude less then for the PROLOG implementation reported in [6]. For instance, proving that $\forall x \forall y (x \le y \land y \le x + car(cons(0,x)) \land p(h(x) - h(y)) \Rightarrow p(0))$ is valid in combination of three theory takes 0.03s (on PC 700MHz).

```
void NelsonOppenCombinationScheme::Effect(Goal& g) {
    negation.ApplyTo(g);
    prenex.ApplyTo(g);
    skolemization.ApplyTo(g);
    dnf.ApplyTo(g);
    GoalDisjunctsIterator gdi(g);
    for (gdi.First(); !gdi.IsDone(); gdi.Next())
    {   Goal& current_disjunct=gdi.GetCurrent();
        abstraction_and_partitioning.ApplyTo(current_disjunct);
        while(1)
        {   simplify.ApplyTo(current_disjunct);
            unsat_modulo_theories.ApplyTo(current_disjunct);
            E_BOOL is_trivially_met=current_disjunct.IsTriviallyMet();
            if (is_trivially_met==TRUE)
            { current_disjunct.SetTrue(); break;  }
            else if (is_trivially_met==FALSE)
            { current_disjunct.SetFalse(); g.SetFalse(); return; }
            if (!deduce_equality_and_replace.ApplyTo(current_disjunct))
            { current_disjunct.SetFalse(); g.SetFalse(); return; }
        }
    }
    g.SetTrue();
}
```

1.	$(c_2 = h(c_3)) \land (c_1 + -2 * c_2 = 0) \land \neg(h(c_1 + -1 * c_2) = h(h(c_3)))$ is *unsatisfiable* in the theory $\langle$PRA, FOL=$\rangle$
	if and only if (by the method *AbstractionAndPartitioning*)
2.	$((c_1 + -2*c_2 = 0) \land (c_1 + -1*c_2 = c_4)) \land ((c_2 = h(c_3)) \land \neg(h(c_4) = h(h(c_3))))$ is *unsatisfiable* in the theory $\langle$PRA, FOL=$\rangle$
	if and only if (by the method *Deduce Equality and Replace* $(c_2=c_4)$)
3.	$((c_1 - 2 * c_4 = 0) \land (c_1 - c_4 = c_4)) \land ((c_4 = h(c_3)) \land \neg(h(c_4) = h(h(c_3))))$ is *unsatisfiable* in the theory $\langle$PRA, FOL=$\rangle$
	if and only if (by the method *Unsat Modulo Theory (FOL=)*)
4.	$\top$ The goal has been met

5 Related Work

The work presented in this paper is related to the long line of results and systems for using decision procedures [8,12,1,2,7]. Especially, the ARGO-LIB builds on the GS framework [6]. Also, our work is related to several recent systems — systems being developed with similar goals (light and efficient support for decision procedures), including *haRVey* [9], *CVC Lite* [4], *TSAT++* [14], *ICS* [5].

6 Conclusions and Future Work

In this paper we presented ARGO-LIB — a C++ library that provides support for using a range of decision procedures and mechanisms for their combination and augmentation. The library builds upon ideas from the GS framework and the SMT-LIB initiative. We hope that ARGO-LIB will promote exchange of ideas, techniques, benchmarks and code and so help advance in the field. For further work we are planning to extend ARGO-LIB by new decision procedures and also to further improve some low-level modules (e.g., unification). We are planning to test ARGO-LIB on wide sets of real-world problems and to compare it to the rival systems. We will also look for its applications in systems of industrial strength.

References

1. C. W. Barrett, D. L. Dill, and Aaron Stump. A Framework for Cooperating Decision Procedures. *CADE-17*, LNAI 1831. Springer, 2000.
2. R. S. Boyer and J S. Moore. Integrating Decision Procedures into Heuristic Theorem Provers: A Case Study of Linear Arithmetic. *Machine Intelligence 11*, 1988.
3. A. Bundy. The Use of Explicit Plans to Guide Inductive Proofs. *CADE-9*, Springer, 1988.
4. CVC Lite. on-line at: http://verify.stanford.edu/CVCL/.
5. ICS. on-line at: http://www.icansolve.com/.
6. P. Janičić and A. Bundy. A General Setting for the Flexible Combining and Augmenting Decision Procedures. *Journal of Automated Reasoning*, 28(3), 2002.
7. D. Kapur and M. Subramaniam. Using an induction prover for verifying arithmetic circuits. *Software Tools for Technology Transfer*, 3(1), 2000.
8. G. Nelson and D. C. Oppen. Simplification by cooperating decision procedures. *ACM Transactions on Programming Languages and Systems*, 1(2), 1979.
9. S. Ranise and D. Deharbe. Light-weight theorem proving for debugging and verifying units of code. *SEFM-03)*. IEEE Computer Society Press, 2003.
10. S. Ranise and C. Tinelli. The SMT-LIB Format: An Initial Proposal. 2003. on-line at: http://goedel.cs.uiowa.edu/smt-lib/.
11. H. Rueß and N. Shankar. Deconstructing Shostak. In *Proceedings of the Conference on Logic in Computer Science (LICS)*, 2001.
12. R. E. Shostak. Deciding combinations of theories. *Journal of the ACM*, 31(1), January 1984.
13. Aaron Stump, Arumugam Deivanayagam, Spencer Kathol, Dylan Lingelbach, and Daniel Schobel. Rogue deicision procedures. *Workshop PDPAR*. 2003,
14. TSAT++. on-line at: http://www.mrg.dist.unige.it/Tsat.

The ICS Decision Procedures
for Embedded Deduction*

Leonardo de Moura, Sam Owre, Harald Rueß, John Rushby, and
Natarajan Shankar

Computer Science Laboratory
SRI International, 333 Ravenswood Ave.
Menlo Park, CA 94025, USA
{demoura,owre,ruess,rushby,shankar}@csl.sri.com

Automated theorem proving lies at the heart of all tools for formal analysis of software and system descriptions. In formal verification systems such as PVS [10], the deductive capability is explicit and visible to the user, whereas in tools such as test case generators it is hidden and often ad-hoc. Many tools for formal analysis would benefit—both in performance and ease of construction—if they could draw on a powerful embedded service to perform common deductive tasks.

An embedded deductive service should be fully automatic, and this suggests that its focus should be restricted to those theories whose decision and satisfiability problems are decidable. However, there are some contexts that can tolerate incompleteness. For example, in extended static checking, the failure to prove a valid verification condition results only in a spurious warning message. In other contexts such as construction of abstractions, speed may be favored over completeness, so that undecidable theories (e.g., nonlinear integer arithmetic) and those whose decision problems are often considered infeasible in practice (e.g., real closed fields) should not be ruled out completely.

Most problems that arise in practice involve *combinations* of theories: the question whether $f(cons(4 \times car(x) - 2 \times f(cdr(x)), y)) = f(cons(6 \times cdr(x), y))$ follows from $2 \times car(x) - 3 \times cdr(x) = f(cdr(x))$, for example, requires simultaneously the theories of uninterpreted functions, linear arithmetic, and lists. The ground (i.e., quantifier-free) fragment of many combinations is decidable when the fully quantified combination is not, and practical experience indicates that automation of the ground case is adequate for most applications.

Practical experience also suggests several other desiderata for an effective deductive service. Some applications (e.g., construction of abstractions) invoke their deductive service a huge number of times in the course of a single calculation, so that performance of the service must be very good. Other applications such as proof search explore many variations on a formula (i.e., alternately asserting and denying various combinations of its premises), so the deductive service should not examine individual formulas in isolation, but should provide a rich application programming interface that supports incremental assertion, retraction, and querying of formulas. Other applications such as test case generation

* This work was supported by SRI International, by NSF grants CCR-ITR-0326540 and CCR-ITR-0325808, by NASA/Langley under contract NAS1-00079, and by NSA under contract MDA904-02-C-1196.

D. Basin and M. Rusinowitch (Eds.): IJCAR 2004, LNAI 3097, pp. 218–222, 2004.

generate propositionally complex formulas with thousands or millions of propositional connectives applied to terms over the decided theories, so that this type of proof search must be performed efficiently inside the deductive service.

We have developed a system called ICS (the name stands for *Integrated Canonizer/Solver*) that can be embedded in applications to provide deductive services satisfying the desiderata above. ICS includes functionality for

- deciding equality constraints in the combination of theories including arithmetic, lists, and other commonly used datatypes,
- for solving propositional combinations of constraints,
- for incrementally processing atomic formulas in an *online* manner, and for
- managing and manipulating a multitude of large assertional contexts in a functional way.

This makes ICS suitable for use in applications with highly dynamic environments such as proof search or symbolic simulation. With an interactive theorem prover such as PVS [10], ICS can be used as a backend verification engine that manages assertional contexts corresponding to open subgoals in a *multi-threaded* way, thereby supporting efficient context switching between open subgoals during proof search. ICS is also highly efficient and is able to deal with huge formulas generated by fully automated applications such as bounded model checking.

ICS is available free of charge for noncommercial use at

`ics.csl.sri.com`

ICS can be used as a standalone application that reads formulas interactively, and may also be included as a library in any application that requires embedded deduction. Binaries for Red Hat Linux, Mac OSX, and Cygwin are precompiled. This distribution also includes libraries for use with C, Ocaml, and Lisp. The source code of ICS is available under a license agreement.

1   Core ICS

The core algorithm of ICS is a corrected version of Shostak's combination procedure for equality and disequality with both uninterpreted and interpreted function symbols [11, 13]. The concepts of canonization and solving have been extended to include inequalities over arithmetic terms [12]. The theory supported by ICS includes rational and integer linear arithmetic (currently integer arithmetic is incomplete), tuples and projections from tuples, coproducts, Boolean constants, S-expressions, functional arrays, combinatory logic with case-splitting, bitvectors [8], and an (incomplete) extension to handle nonlinear multiplication.

Consider, for example, demonstrating the unsatisfiability of the conjunction of the literals `f(f(x)-f(y)) <> f(z)`, `y <= x`, `y >= x + z`, and `z > 0`. Using the ICS interactor, these literals are asserted from left to right.

```
ics> assert f(f(x)-f(y)) <> f(z).
:ok s1
```

This assertion causes ICS to add `f(f(x)-f(y)) <> f(z)` to the initially empty logical context, and the resulting context is named `s1`. These names can be used to arbitrarily jump between a multitude of contexts. The `show` command displays the current state of the decision procedure.

```
ics> show.
  d: {u!5 <> u!4}
  u: {u!1 = f(y), u!2 = f(x), u!4 = f(v!3), u!5 = f(z)}
  la: {v!3 = u!2 + -1 * u!1}
```

This context consists of variable disequalities in `d`, equalities over uninterpreted terms in `u`, and linear arithmetic facts in `la`. The equalities in `u` are flat equalities of the form `x = f(x1,...,xn)` with `f` an uninterpreted function symbol. Fresh variables such as `u!1` are introduced to flatten input terms. Equalities in `la` are all in solved form as explained in [13].

```
ics> assert y <= x; y >= x + z.
:ok s2
```

These inequalities are asserted using an online Simplex algorithm [12] which generates fresh slack variables such as `k!6` and `k!7` which are restricted to non-negative values.

```
ics> show.
  d: {u!5  <>  u!4}
  u: {u!1 = f(y), u!2 = f(x), u!4 = f(v!3), u!5 = f(z)}
  la: {y = x-k!6,  z = -k!7-k!6, v!3 = u!2-u!1}
```

Finally, the last assertion detects the inconsistency, and returns `:unsat`.

```
ics> assert z > 0.
:unsat {-y + x >=0,  z >0, -z + y - x >=0}
```

In such a case, ICS returns a *justification* in terms of a set of the asserted atoms that participate in demonstrating the inconsistency. This is not only useful for suggesting counterexamples, but is essential for efficient integration with a SAT solver (see below). Since there is a trade-off between the preciseness of justifications and the cost for computing them, the justification set provided by ICS might not be minimal.

The above sequence of commands can also be placed in a file `foo.ics` and ICS can be invoked in batch mode as a shell command Alternatively, using the `-server` command line flag, ICS interacts through the specified port instead of reading and writing on the standard input and output channels.

The integration of non-solvable theories such as functional arrays is obtained by an extension of the basic Shostak combination procedure. The basic idea behind this completion-like procedure is similar to the one described by Nelson [9].

```
ics> assert a[j:=x] = b[k:=y]; i <> j; i = k.
:ok s1
```

Assertion of the three literals above, for example, causes the ICS engine to explicitly generate new equalities based on forward chains. The resulting state `s1` is displayed using the `show` command.

```
ics> show.
... arr: {a!3=b[k:=y], y=a!3[k], a!3=a[j:=x], x=a!3[j], y=a[k]}
```

The representation of the array context is in terms of equalities with variables on the right-hand side and flat array terms on the left-hand side. Fresh variables such as a!3 are introduced to flatten terms. The equality a!3 = b[k := y], for example, causes the addition of the derived equality y = a!3[k]. With these completions, a canonizer can be defined similar to the case for solvable Shostak theories [13], and a[k] has canonical form y in the current context.

```
ics>  can a[k].
:term y
:justification {i = k,  b[k := y] = a[j := x], i <> j}
```

In addition, the ICS canonizer returns a subset of the input literals sufficient for proving validity of the equality a[k] = y. In general, a full case-split on array indices is required for completeness.

2 SAT-Based Constraint Satisfaction

ICS decides propositional satisfiability problems with literals drawn from the combination of theories of the core theory described above.

```
ics> sat [x > 2 | x < 4 | p] & 2 * x > 6.
:sat s2
:model [p |-> :true] [-6 + 2 * x >0]
```

This example shows satisfiability of a propositional formula with linear arithmetic constraints. In addition to the satisfying assignment to the propositional variable p, a set of assignments for the variable x is described by means of an arithmetic constraint.

The verification engine underlying the sat command combines the ICS ground decision procedures with a non-clausal propositional SAT solver using the paradigm of *lazy theorem proving* [6]. Let ϕ be the formula whose satisfiability is being checked. Let L be an injective map from fresh propositional variables to the atomic subformulas of ϕ such that $L^{-1}[\phi]$ is a propositional formula. We can use the SAT solver to check that $L^{-1}[\phi]$ is satisfiable, but the resulting truth assignment, say $l_1 \wedge \ldots \wedge l_n$, might be spurious, that is $L[l_1 \wedge \ldots \wedge l_n]$ might not be ground-satisfiable. If that is the case, we can repeat the search with the added clause $(\neg l_1 \vee \ldots \vee \neg l_n)$ and invoke the SAT solver on $(\neg l_1 \vee \ldots \vee \neg l_n) \wedge L^{-1}[\phi]$. This ensures that the next satisfying assignment returned is different from the previous assignment that was found to be ground-unsatisfiable.

The sat command implements several crucial optimizations. First, the SAT solver notifies the ground decision procedures of every variable assignment during its search, and the ground decision procedure might therefore trigger non-chronological backtracking and determine adequate backtracking points. Second, the justifications of inconsistencies provided by the ground decision procedure are used to further prune the search space as described in [6]. Note that for the combination to be effective, both the ground decision procedures and the SAT solver must support the incremental introduction of information.

3 Applications

One of our main applications of ICS is within SAL [4] where it is used for bounded model checking of infinite-state systems ($BMC(\infty)$) [6] and induction proofs [7]. Transition systems are encoded in the SAL language, and $BMC(\infty)$ problems are generated in terms of satisfiability problems of propositional constraint formulas. Currently, we support verification backends for UCLID [3], CVC [2], SVC [1], and ICS for discharging these satisfiability problems. In comparison with these other systems, ICS performs favorably on a wide range of benchmarks [5].

4 Outlook

We plan to enlarge the services provided by ICS so that even less deductive glue will be required in future. In particular, we intend to add quantifier elimination, rewriting, and forward chaining. Other planned enhancements include generation of concrete solutions to satisfiability problems, and generation of proof objects. We expect that the latter will also improve the interaction between core ICS and its SAT solver, and thereby further increase the performance of ICS.

References

1. C.W. Barrett, D. Dill, and J. Levitt. Validity checking for combinations of theories with equality. *LNCS*, 1166:187–201, 1996.
2. C.W. Barrett, D.L. Dill, and A. Stump. Checking satisfiability of first-order formulas by incremental translation to SAT. *LNCS*, 2404:236–249, 2002.
3. R.E. Bryant, S. K. Lahiri, and S. A. Seshia. Deciding CLU logic formulas via boolean and pseudo-boolean encodings. *LNCS*, 2003.
4. L. de Moura, S. Owre, H. Rueß, J. Rushby, N. Shankar, M. Sorea, and A. Tiwari. SAL 2. In R. Alur and D. Peled, editors, *Computer-Aided Verification, CAV'2004*, LNCS, Boston, MA, July 2004. Springer Verlag.
5. L. de Moura and H. Rueß. An experimental evaluation of ground decision procedures. In R. Alur and D. Peled, editors, *Computer-Aided Verification, CAV'2004*, LNCS, Boston, MA, July 2004. Springer Verlag.
6. L. de Moura, H. Rueß, and M. Sorea. Lazy theorem proving for bounded model checking over infinite domains. *LNCS*, 2392:438–455, 2002.
7. L. de Moura, H. Rueß, and M. Sorea. Bounded model checking and induction: From refutation to verification. *LNCS*, 2725:14–26, 2003.
8. O. Möller and H. Rueß. Solving bit-vector equations. *LNCS*, 1522:36–48, 1998.
9. G. Nelson. Techniques for program verification. Technical Report CSL-81-10, Xerox Palo Alto Research Center, Palo Alto, Ca., 1981.
10. S. Owre, J. Rushby, N. Shankar, and F. von Henke. Formal verification for fault-tolerant architectures: Prolegomena to the design of PVS. *IEEE Transactions on Software Engineering*, 21(2):107–125, February 1995.
11. H. Rueß and N. Shankar. Deconstructing Shostak. In *16th LICS*, pages 19–28. IEEE Computer Society, 2001.
12. H. Rueß and N. Shankar. Solving Linear Arithmetic Constraints. Technical Report SRI-CSL-04-01, CSL, SRI International, Menlo Park, CA, 94025, March 2004.
13. N. Shankar and H. Rueß. Combining Shostak theories. In S. Tison, editor, *RTA'02*, volume 2378 of *LNCS*, pages 1–18. Springer, 2002.

System Description: E 0.81

Stephan Schulz

RISC Linz, Johannes Kepler Universität Linz, Austria,
`schulz@informatik.tu-muenchen.de`

Abstract. E is an equational theorem prover for clausal logic with
equality. We describe the latest version, E 0.81 *Tumsong*, with special
emphasis on the important aspects that have changed compared to previously described versions.

1 Introduction

E is a high-performance theorem prover for full clausal logic with equality. It has
consistently been ranked among the top systems of the UEQ and MIX categories
of the yearly CADE ATP system competitions, and has found applications in
hard- and software verification as well as mathematics. It has been previously
described in [Sch01] and in some more detail in [Sch02].

The latest published release is E 0.81 *Tumsong*. It contains a number of
important improvements and additions on various levels. The system is available
under the GNU GPL at [Sch04c]. The source distribution is easy to install and
known to compile without any problems under most modern UNIX dialects,
including recent versions of GNU/Linux, Solaris, and MacOS-X.

2 Calculus

E uses a saturating refutation procedure, i.e. it attempts to show the unsatisfiability of a set of clauses by deriving the empty clause. It implements the superposition calculus **SP** [Sch02], a variant of the calculus described in [BG94], enhanced with a variety of explicit redundancy elimination techniques and pseudo-
splitting [RV01]. E is based on a purely equational view, with non-equational
atoms encoded as equalities with a reserved constant $\$true$. To maintain correct-
ness, first-order terms and atoms are represented by two disjoint sorts.

The major generating inference rule is *superposition* (restricted paramodulation), which typically is responsible for more than 99% of all generated clauses.
The most important simplification techniques are tautology elimination, rewriting, clause normalization, subsumption, and AC redundancy elimination, all of
which have been described in [Sch02]. We have recently added *equational definition unfolding* and *contextual literal cutting*.

An equational definition is a unit clause of the form $f(X_1, \ldots, X_n) = t$ (where
f does not occur in t and t contains no additional variables). Such a clause fully
defines f. We can apply this equation to replace f with its definition throughout

D. Basin and M. Rusinowitch (Eds.): IJCAR 2004, LNAI 3097, pp. 223–228, 2004.

the formula, and then delete it, completely getting rid of the function symbol f. Currently, E unfolds equational definitions in a preprocessing step. We found that this makes it much easier to find good term orderings, in particular for many unit equational problems.

Contextual literal cutting, also known as *subsumption resolution*, views a clause as a conditional fact, i.e. it considers one literal as the *active* literal, and the other literals as *conditions*. If the conditions are implied in the context of another clause, we can use the active literal to simplify that clause. Currently, we only use the active literal for cutting off an implied literal of opposite polarity:

$$\text{(CLC)} \quad \frac{\sigma(C) \vee \sigma(R) \vee \sigma(l) \qquad C \vee \bar{l}}{\sigma(C) \vee \sigma(R) \qquad\qquad C \vee \bar{l}} \qquad \text{where } \bar{l} \text{ is the negation of } l \text{ and } \sigma \text{ is a substitution}$$

Note that this is a simplification rule, i.e. the clauses in the precondition are *replaced* by the clauses in the conclusion.

3 Implementation

E is implemented in ANSI C, based on a layered set of libraries. There have been three significant recent changes to the core inference engine: A simplified proof procedure, a much improved rewriting engine, and the addition of a new indexing technique for forward and backward subsumption. It is perhaps interesting to note that while we fixed a couple of bugs in the last 3 years, we found only one problem with completeness of the prover (an over-eager subsumption function), and no problem with correctness. Also, stability of the prover has been consistently excellent in practice, with no reported problems.

3.1 Proof Procedure

E uses the DISCOUNT variant of the given clause algorithm. The proof state S is represented by two sets of clauses, $S = P \cup U$. Clauses in P are called *processed* or *active*. P is maximally inter reduced, and all generating inferences between clauses in P have been performed. Moreover, normally only clauses in P are used for simplification. U is the set of *unprocessed* clauses, which, in out case, are also *passive*, i.e. they are not used in any generating inferences and normally only as the simplified partner in simplifications.

The proof procedure selects a clause g from U for processing, simplifies it, uses it to back-simplify the clauses in P (simplified clauses in P get moved back into U), adds it to P, and performs all generating inferences between g and clauses from P. The newly generated clauses are simplified (primarily to enable a better heuristic evaluation) and added to U. This process is repeated until either the empty clause (and hence an explicit refutation) is found, or all clauses have been processed.

In the original version, E generalized a technique for unfailing completion to avoid the reprocessing of some simplified clauses in P. In particular, if the set of maximal terms in inference literals could not possibly be changed by a simplification, the affected clause was not moved into U. Instead, a separate

while $U \neq \{\}$	while $U \neq \{\}$
$\quad g = $ delete_best(U)	$\quad g = $ delete_best(U)
$\quad g = $ simplify(g,P)	$\quad g = $ simplify(g,P)
$\quad$ if $g == \square$	$\quad$ if $g == \square$
$\quad\quad$ SUCCESS, Proof found	$\quad\quad$ SUCCESS, Proof found
$\quad$ if g is not redundant w.r.t. P	$\quad$ if g is not redundant w.r.t. P
$\quad\quad P = P\backslash\{c \in P | c$ redundant w.r.t. $g\}$	$\quad\quad P = P\backslash\{c \in P | c$ redundant w.r.t. $g\}$
$\quad\quad T = \{c \in P | c$ simplifiable with $g\}$	$\quad\quad T = \{c \in P | c$ crit.-simplifiable with $g\}$
	$\quad\quad P = \{$simplify$(p, \{g\}) | p \in P\}$
$\quad\quad P = (P\backslash T) \cup \{g\}$	$\quad\quad P = (P\backslash T) \cup \{g\}$
$\quad\quad T = T \cup$ generate(g, P)	$\quad\quad T = T \cup$ generate(g, P)
$\quad\quad$ foreach $c \in T$	$\quad\quad$ foreach $c \in T$
$\quad\quad\quad c = $ cheap_simplify(c, P)	$\quad\quad\quad c = $ cheap_simplify(c, P)
$\quad\quad\quad$ if c is not trivial	$\quad\quad\quad$ if c is not trivial
$\quad\quad\quad\quad U = U \cup \{c\}$	$\quad\quad\quad\quad U = U \cup \{c\}$
SUCCESS, original U is satisfiable	SUCCESS, original U is satisfiable

Fig. 1. Simplified current (left) and original E proof procedure

interreduction of P ensured the invariant that P was maximally simplified. The most obvious example is the rewriting of the minimal term in an orientable unit clause (i.e. the right hand side of a rewrite rule).

In theory, this optimization avoids repetition of work. In practice, however, we found that the special case was very rarely triggered. Moreover, it complicates the code significantly. In addition to the detection and the special interreduction stage, clauses in P can change in the original version of the code. This makes *indexing* of clauses in P unnecessarily hard.

It was primarily for this reason that we changed to the simpler, new version. We could find significant changes in the search behavior for only very few examples, and no change in overall performance. Fig 1 shows a comparison of the two different algorithms.

3.2 Rewriting

One of the design goals of E was the exploration of *shared* rewriting, i.e. an implementation of rewriting where a single rewrite step would affect all occurrences of a given term. Up until E 0.63, this was implemented by *destructive* global rewriting on an aggressively shared *term bank*, a data structure for the administration of terms. If a term was rewritten, it was replaced in the term bank, and the change was propagated to all superterms and literals in which it occurred. This required each term to carry information about all references to it, incurring a significant memory penalty. Moreover, special care had to be taken in cases where rewriting would make previously different terms equal. Additional complications arose from the fact that terms in any clause could change due to rewriting of terms in unrelated clauses.

Performance was generally adequate. However, in [LS01] we found that it was not qualitatively different from a well-implemented conventional rewrite engine. The reason for this disappointment is the following: By far most terms occur in the set U of unprocessed clauses. However, nearly all rewriting takes place on terms from the temporary term set T containing the clauses newly generated during each traversal of the main loop. If we *destructively* replace terms in T, the information about the rewrite steps is lost for the future. The small benefit from rewriting multiple instances of terms in T barely compensates for the overhead.

As a consequence, we now moved to a *cached* rewriting model. Instead of destructively changing terms that are rewritten, we just add a link to the new normal form. This annotation persists until the term cell is no longer referenced by any clause and eventually collected by our garbage collector. Typically, rewrite links for frequently occurring terms will never expire, while links for rare terms will still persist for some time. In clause normalization, we just follow rewrite links until we find a term not yet known to be rewritable. Only in that case do we apply standard matching and, if possible, add a new rewrite link. Actual references in a clause are only changed explicitly, never as a side effect.

Our new interpretation of shared rewriting has resulted in much simpler, more reliable and maintainable code. It also has improved performance and decreased memory consumption significantly.

3.3   Subsumption

Subsumption is one of the most important redundancy elimination techniques for saturating theorem provers. In versions earlier than E 0.8, we used simple sequential search for nearly all subsumption attempts, in particular for all non-unit subsumption. Despite the fact that E's proof procedure actually de-emphasizes subsumption compared to Otter-loop provers, the total cost of subsumption frequently was much as 20% in terms of CPU time. In version 0.8 we integrated a newly-developed indexing technique for subsumption, *feature vector indexing* [Sch04a]. This uses the same small and simple index for both forward- and backward subsumption, typically reducing the time for subsumption to the limit of measurability. This became especially important since contextual literal cutting is implemented via subsumption, and can increase subsumption cost significantly. Without an adequate indexing technique, that cost can well be higher than the benefit.

3.4   Heuristic Control

Heuristic control has always been one of the strong points of E compared to other provers. Changes in recent versions include much better generation of term orderings (based primarily on the frequency of symbols in the input formula), a better support for set-of-support based strategies, new clause selection heuristics, better literal selection, and a much better automatic mode integrating most of the new features.

Table 1. Performance of recent versions of E

Version	Proofs	Models	Notes
E 0.63	2673	339	Essentially equivalent to E 0.62dev described in [Sch02]
E 0.7	2833	372	Cached rewriting, gc for terms, improved literal selection
E 0.71	2978	410	Frequency-based ordering generation
E 0.8	3017	410	Feature-vector indexing, preliminary contextual unit cutting
E 0.81	3051	413	Test results with unit cutting and set of support, watch list

Of particular interest for some users might be the integration of a *watch list* of clauses that can be used to either guide the proof search (watch list clauses can be preferred for processing) or just to get direct deductive proofs (which, despite the theoretical incompleteness, often works quite well in practice). Details about the new heuristic options are described in [Sch04b].

4 Performance

We have evaluated recent versions of E on the 5180 clause normal form problems of the TPTP problem library, version 2.5.1, following the guidelines for the use of that library. Test runs were performed with a time limit of 300 seconds and a memory limit of 192 MB on a cluster of SUN Ultra-60 workstations.

Table 1 shows the results and notes on the introduction of the most important new features. Note that each version also features new search heuristics at least partially adapted to or based on the other changes. The NEWS file in the E distribution lists changes more extensively.

5 Conclusion

We have described the progress of the theorem prover E during the last 3 years. This includes advances in the calculus and heuristic control, along with more significant changes to the inference engine. As a result, E is now more stable and powerful than ever, and we are confident that further improvements will follow in the future.

At the moment we are testing a parser for full first-order format and a clause normal form translator, work on new in- and output formats, and on tools for post-processing and lemma detection. We are also planning to use more indexing techniques, especially for paramodulation and backward rewriting, and to speed up ordering constraint checks.

References

[BG94]   L. Bachmair and H. Ganzinger. Rewrite-Based Equational Theorem Proving with Selection and Simplification. *Journal of Logic and Computation*, 3(4):217–247, 1994.

[LS01]   B. Löchner and S. Schulz. An Evaluation of Shared Rewriting. In H. de Nivelle and S. Schulz, editors, *Proc. 2nd IWIL*, MPI Preprint, 33–48, Saarbrücken, 2001. Max-Planck-Institut für Informatik.

[RV01]   A. Riazanov and A. Voronkov. Splitting without Backtracking. In B. Nebel, editor, *Proc. 17th IJCAI, Seattle*, 1:611–617. Morgan Kaufmann, 2001.

[Sch01]   S. Schulz. System Abstract: E 0.61. In R. Goré, A. Leitsch, and T. Nipkow, editors, *Proc. 1st IJCAR, Siena, LNAI 2083*, 370–375. Springer, 2001.

[Sch02]   S. Schulz. E – A Brainiac Theorem Prover. *Journal of AI Communications*, 15(2/3):111–126, 2002.

[Sch04a]   S. Schulz. Simple and Efficient Clause Subsumption with Feature Vector Indexing. 2004. (to be published).

[Sch04b]   S. Schulz. *The E Equational Theorem Prover – User Manual*. See [Sch04c].

[Sch04c]   S. Schulz. The E Web Site. http://www.eprover.org, 2004.

Second-Order Logic over Finite Structures – Report on a Research Programme*

Georg Gottlob

Institut für Informationssysteme, Technische Universität Wien
Favoritenstraße 9-11, A-1040 Wien, Austria
gottlob@dbai.tuwien.ac.at

Abstract. This paper reports about the results achieved so far in the context of a research programme at the cutting point of logic, formal language theory, and complexity theory. The aim of this research programme is to classify the complexity of evaluating formulas from different prefix classes of second-order logic over different types of finite structures, such as strings, graphs, or arbitrary structures. In particular, we report on classifications of second-order logic on strings and of existential second-order logic on graphs.

1 Introduction

Logicians and computer scientists have been studying for a long time the relationship between fragments of predicate logic and the solvability and complexity of decision problems that can be expressed within such fragments. Among the studied fragments, quantifier prefix classes play a predominant role. This can be explained by the syntactical simplicity of such prefix classes and by the fact that they form a natural hierarchy of increasingly complex fragments of logic that appears to be deeply related to core issues of decidability and complexity. In fact, one of the most fruitful research programs that kept logicians and computer scientists busy for decades was the exhaustive solution of Hilbert's classical Entscheidungsproblem (cf. [3]), i.e., of the problem of determining those prefix classes of first-order logic for which formula-satisfiability (resp. finite satisfiability of formulas) is decidable.

Quantifier prefixes emerged not only in the context of decidability theory (a common branch of recursion theory and theoretical computer science), but also in core areas of computer science such as formal language and automata theory, and later in complexity theory. In automata theory, Büchi [5,4] and Trakhtenbrot [32] independently proved that a language is regular iff it can be described by a sentence of monadic second-order logic, in particular, by a sentence of monadic existential second-order logic. In complexity theory, Fagin [11] showed that a problem on finite structures is in NP iff it can be described by a sentence of existential second-order logic (ESO). These fundamental results have engendered a large number of further investigations and results on characterizing language and complexity classes by fragments of logic (see, e.g. the monographs [27,24,7, 16]).

---

* Most of the material contained in this paper stems, modulo editorial adaptations, from the much longer papers [9,10,12].

D. Basin and M. Rusinowitch (Eds.): IJCAR 2004, LNAI 3097, pp. 229–243, 2004.

While the classical research programme of determining the prefix characterizations of decidable fragments of first-order logic was successfully completed around 1984 (cf. [3]), until recently little was known on analogous problems on finite structures, in particular, on the tractability/intractability frontier of the model checking problem for prefix classes of second-order logic (SO), and in particular, of existential second order logic (ESO) over finite structures. In the late nineties, a number of scientists, including Thomas Eiter, Yuri Gurevich, Phokion Kolaitis, Thomas Schwentick, and the author started to attack this new research programme in a systematic manner.

By *complexity of a prefix class* C we mean the complexity of the following model-checking problem: Given a fixed sentence Φ in C, decide for variable finite structures A whether A is a model of Φ, which we denote by $A \models \Phi$. Determining the complexity of all prefix classes is an ambitious research programme, in particular the analysis of various types of finite structures such as *strings*, i.e., finite word structures with successor, *trees*, *graphs*, or arbitrary finite relational structures (corresponding to relational datavases). Over strings and trees, one of the main goals of this classification is to determine the *regular* prefix classes, i.e., those whose formulas express regular languages only; note that by Büchi's theorem, regular fragments over strings are (semantically) included in monadic second-order logic.

In the context of this research programme, three systematic studies were carried out recently, that shed light on the prefix classes of the existential fragment ESO (also denoted by Σ_1^1) of second-order logic:

- In [9], the ESO prefix-classes over strings are exhaustively classified. In particular, the precise frontier between regular and nonregular classes is traced out, and it is shown that every class that expresses some nonregular language also expresses some NP-complete language. There is thus a huge complexity gap in ESO: some prefix classes can express only regular languages (which are well-known to have extremely low complexity), while all others are intractable. The results of [9] are briefly reviewed in Section 2.
- In [10] this line of research was continued by by systematically investigating the syntactically more complex prefix classes $\Sigma_k^1(\mathcal{Q})$ of second-order logic for each integer $k > 1$ and for each first-order quantifier prefix $\mathcal{Q}$. An exhaustive classification of the regular and nonregular prefix classes of this form was given, and complexity results for the corresponding model-checking problems were derived.
- In [12], the complexity of all ESO prefix-classes over graphs and arbitrary relational structures is analyzed, and the tractability/intractability frontier is completely delineated. Unsurprisingly, several classes that are regular over strings become NP-hard over graphs. Interestingly, the analysis shows that one of the NP-hard classes becomes polynomial for the restriction to undirected graphs without self-loops. A brief account of these results is given in Section 4.

1.1 Preliminaries and Classical Results

We consider second-order logic with equality (unless explicitly stated otherwise) and without function symbols of positive arity. Predicates are denoted by capitals and individual variables by lower case letters; a bold face version of a letter denotes a tuple of corresponding symbols.

A *prefix* is any string over the alphabet $\{\exists, \forall\}$, and a *prefix set* is any language $\mathcal{Q} \subseteq \{\exists, \forall\}^*$ of prefixes. A prefix set $\mathcal{Q}$ is *trivial*, if $\mathcal{Q} = \emptyset$ or $\mathcal{Q} = \{\lambda\}$, i.e., it consists of the empty prefix. In the rest of this paper, we focus on nontrivial prefix sets. We often view a prefix Q as the prefix class $\{Q\}$. A *generalized prefix* is any string over the extended prefix alphabet $\{\exists, \forall, \exists^*, \forall^*\}$. A prefix set $\mathcal{Q}$ is *standard*, if either $\mathcal{Q} = \{\exists, \forall\}^*$ or $\mathcal{Q}$ can be given by some generalized prefix.

For any prefix Q, the class $\Sigma_0^1(Q)$ is the set of all prenex first-order formulas (which may contain free variables and constants) with prefix Q, and for every $k \geq 0$, $\Sigma_{k+1}^1(Q)$ (resp., Π_{k+1}^1) is the set of all formulas $\exists \mathbf{R} \varPhi$ (resp., $\forall \mathbf{R} \varPhi$) where $\varPhi$ is from Π_k^1 (resp., Σ_k^1). For any prefix set $\mathcal{Q}$, the class $\Sigma_k^1(\mathcal{Q})$ is the union $\Sigma_k^1(\mathcal{Q}) = \bigcup_{Q \in \mathcal{Q}} \Sigma_k^1(Q)$. We write also ESO for Σ_1^1. For example, $\mathrm{ESO}(\exists^* \forall \exists^*)$ is the class of all formulas $\exists \mathbf{R} \exists \mathbf{y} \forall x \exists \mathbf{z} \varphi$, where φ is quantifier-free; this is the class of ESO-prefix formulas, whose first-order part is in the well-known Ackermann class with equality.

Let $A = \{a_1, \dots, a_m\}$ be a finite alphabet. A *string* over A is a finite first-order structure $W = \langle U, C_{a_1}^W, \dots, C_{a_m}^W, Succ^W, min^W, max^W \rangle$, for the vocabulary $\sigma_A = \{C_{a_1}, \dots, C_{a_m}, Succ, min, max\}$, where

- U is a nonempty finite initial segment $\{1, 2, \dots, n\}$ of the positive integers;
- each $C_{a_i}^W$ is a unary relation over U (i.e., a subset of U) for the unary predicate C_{a_i}, for $i = 1, \dots, m$, such that the $C_{a_i}^W$ are pairwise disjoint and $\bigcup_i C_{a_i}^W = U$.
- $Succ^W$ is the usual successor relation on U and min^W and max^W are the first and the last element in U, respectively.

Observe that this representation of a string is a *successor structure* as discussed e.g. in [8]. An alternative representation uses a standard linear order $<$ on U instead of the successor $Succ$. In full ESO or second-order logic, $<$ is tantamount to $Succ$ since either predicate can be defined in terms of the other.

The strings W for A correspond to the nonempty finite words over A in the obvious way; in abuse of notation, we often use W in place of the corresponding word from A^* and vice versa.

A *SO sentence* $\varPhi$ over the vocabulary σ_A is a second-order formula whose only free variables are the predicate variables of the signature σ_A, and in which no constant symbols except min and max occur. Such a sentence defines a language over A, denoted $\mathcal{L}(\varPhi)$, given by $\mathcal{L}(\varPhi) = \{W \in A^* \mid W \models \varPhi\}$. We say that a language $L \subseteq A^*$ is *expressed* by $\varPhi$, if $\mathcal{L}(\varPhi) = L \cap A^+$ (thus, for technical reasons, without loss of generality we disregard the empty string); L is *expressed by a set S of sentences*, if L is expressed by some $\varPhi \in S$. We say that S *captures* a class C of languages, if S expresses all and only the languages in C.

Example 1.1. Let us consider some languages over the alphabet $A = \{a, b\}$, and how they can be expressed using logical sentences.

- $L_1 = \{a, b\}^* b \{a, b\}^*$: This language is expressed by the simple sentence

$$\exists x . C_b(x).$$

- $L_2 = a^*b$:   This language is expressed by the sentence

$$C_b(max) \wedge \forall x \neq max.C_a(x).$$

- $L_3 = (ab)^*$:   Using the successor predicate, we can express this language by

$$C_a(min) \wedge C_b(max) \wedge \forall x, y.Succ(x, y) \to (C_a(x) \leftrightarrow \neg C_a(y)).$$

- $L_4 = \{w \in \{a, b\}^* \mid |w| = 2n, n \geq 1\}$: We express this language by the sentence

$$\exists E \, \forall x, y.\neg E(min) \wedge E(max) \wedge Succ(x, y) \to (E(x) \leftrightarrow \neg E(y)).$$

Note that this a monadic ESO sentence. It postulates the existence of a monadic predicate E, i.e., a "coloring" of the string such that neighbored positions have different color, and the first and last position are uncolored and colored, respectively.
- $L_5 = \{a^n b^n \mid n \geq 1\}$:   Expressing this language is more involved:

$$\exists R \, \forall x, x^+, y, y^-.R(min, max) \wedge [R(x, y) \to (C_a(x) \wedge C_b(y))] \wedge$$
$$[(Succ(x, x^+) \wedge Succ(y^-, y) \wedge R(x, y)) \to R(x^+, y^-)].$$

Observe that this sentence is not monadic. Informally, it postulates the existence of an arc from the first to the last position of the string W, which must be an a and a b, respectively, and recursively arcs from the i-th to the $(|W| - i + 1)$-th position.

Let A be a finite alphabet. A sentence Φ over σ_A is called *regular*, if $\mathcal{L}(\Phi)$ is a regular language. A set of sentences S (in particular, any ESO-prefix class) is *regular*, if for every finite alphabet A, all sentences $\Phi \in S$ over σ_A are regular.

Büchi [5] has shown the following fundamental theorem, which was independently found by Trakhtenbrot [32]. Denote by MSO the fragment of second-order logic in which all predicate variables have arity at most one,[1] and let REG denote the class of regular languages.

Proposition 1.1 (Büchi's Theorem). MSO *captures* REG.

That MSO can express all regular languages is easy to see, since it is straightforward to describe runs of a finite state automaton by an existential MSO sentence. In fact, this is easily possible in monadic ESO($\forall\exists$) as well as in monadic ESO($\forall\forall$). Thus, we have the following lower expressiveness bound on ESO-prefix classes over strings.

Proposition 1.2. *Let* $\mathcal{Q}$ *be any prefix set. If* $\mathcal{Q} \cap \{\exists, \forall\}^* \forall \{\exists, \forall\}^+ \neq \emptyset$*, then* ESO($\mathcal{Q}$) *expresses all languages in* REG.

On the other hand, with non-monadic predicates allowed ESO has much higher expressivity. In particular, by Fagin's result [11], we have the following.

Proposition 1.3 (Fagin's Theorem). ESO *captures* NP.

---

[1] Observe that we assume MSO allows one to use nullary predicate variables (i.e., propositional variables) along with unary predicate variables. Obviously, Büchi's Theorem survives.

This theorem can be sharpened to various fragments of ESO. In particular, by Leivant's results [19,8], in the presence of a successor and constants min and max, the fragment ESO($\forall^*$) captures NP; thus, ESO($\forall^*$) expresses all languages in NP.

Before proceeding with the characterization of the regular languages by nonmonadic fragments of ESO, we would like to note that many papers cover either extensions or restrictions of MSO or REG, and cite some relevant results.

Lynch [20], has studied the logic over strings obtained from existential MSO by adding addition. He proved that model checking for this logic lies in $\text{NTIME}(n)$, i.e., in nondeterministic linear time. Grandjean and Olive [14,22] obtained interesting results related to those of Lynch. They gave logical representations of the class NLIN, i.e., linear time on random access machines, in terms of second-order logic with unary functions instead of relations (in their setting, also the input string is represented by a function).

Lautemann, Schwentick and Thérien [18] proved that the class CFL of context-free languages is characterized by ESO formulas of the form $\exists B \varphi$ where φ is first-order, B is a binary predicate symbol, and the range of the second-order quantifier is restricted to the class of *matchings*, i.e., pairing relations without crossover. Note that this is not a purely prefix-syntactic characterization of CFL. From our results and the fact that some languages which are not context-free can be expressed in the minimal nonregular ESO-prefix classes, it follows that a syntactic characterization of CFL by means of ESO-prefix classes is impossible.

Several restricted versions of REG where studied and logically characterized by restricted versions of ESO. McNaughton and Papert [21] showed that first-order logic with a linear ordering precisely characterizes the *star-free* regular languages. This theorem was extended by Thomas [29] to ω-languages, i.e., languages with infinite words. Later several hierarchies of the star-free languages were studied and logically characterized (see, e.g. [29,23,24,25]). Straubing, Thérien and Thomas [28] showed that first-order logic with modular counting quantifiers characterize the regular languages whose syntactic monoids contain only solvable groups. These and many other related results can be found in the books and surveys [27,29,23,24,25].

2 Recent Results on ESO over Strings

Combining and extending the results of Büchi and Fagin, it is natural to ask: What about (nonmonadic) prefix classes ESO($\mathcal{Q}$) over finite strings? We know by Fagin's theorem that all these classes describe languages in NP. But there is a large spectrum of languages contained in NP ranging from regular languages (at the bottom) to NP-hard languages at the top. What can be said about the languages expressed by a given prefix class ESO($\mathcal{Q}$)? Can the expressive power of these fragments be characterized? In order to clarify these issues, the following particular problems where investigated in [9]:

• Which classes ESO($\mathcal{Q}$) express only regular languages? In other terms, for which fragments ESO($\mathcal{Q}$) is it true that for any sentence $\Phi \in$ ESO($\mathcal{Q}$) the set $Mod(\Phi) = \{W \in A^* \mid W \models \Phi\}$ of all finite strings (over a given finite alphabet A) satisfying Φ constitutes a regular language? By Büchi's Theorem, this question is identical to the following: Which prefix classes of ESO are (semantically) included in MSO?

Note that by Gurevich's classifiability theorem (cf. [3]) and by elementary closure properties of regular languages, it follows that there is *a finite number of maximal regular prefix classes* ESO($\mathcal{Q}$), and similarly, of minimal nonregular prefix classes; the latter are, moreover, standard prefix classes (cf. Section 1.1). It was the aim of [9] to determine the maximal regular prefix classes and the minimal nonregular prefix classes.

- What is the complexity of model checking (over strings) for the *nonregular* classes ESO($\mathcal{Q}$), i.e., deciding whether $W \models \Phi$ for a given W (where Φ is fixed)?

Model checking for regular classes ESO($\mathcal{Q}$) is easy: it is feasible by a finite state automaton. We also know (e.g. by Fagin's Theorem) that some classes ESO($\mathcal{Q}$) allow us to express NP-complete languages. It is therefore important to know (i) which classes ESO($\mathcal{Q}$) can express NP-complete languages, and (ii) whether there are prefix classes ESO($\mathcal{Q}$) of intermediate complexity between regular and NP-complete classes.

- Which classes ESO($\mathcal{Q}$) *capture* the class REG? By Büchi's Theorem, this question is equivalent to the question of which classes ESO($\mathcal{Q}$) have exactly the expressive power of MSO over strings.

- For which classes ESO($\mathcal{Q}$) is finite satisfiability decidable, i.e., given a formula $\Phi \in$ ESO($\mathcal{Q}$), decide whether Φ is true on some finite string ?

Reference [9] answers all the above questions exhaustively. Some of the results are rather unexpected. In particular, a surprising dichotomy theorem is proven, which sharply classifies all ESO($\mathcal{Q}$) classes as either regular or intractable. Among the main results of [9] are the following findings.

(1) The class ESO($\exists^*\forall\exists^*$) is regular. This theorem is the technically most involved result of [9]. Since this class is nonmonadic, it was not possible to exploit any of the ideas underlying Büchi's proof for proving it regular. The main difficulty consists in the fact that relations of higher arity may connect elements of a string that are very distant from one another; it was not *a priori* clear how a finite state automaton could guess such connections and check their global consistency. To solve this problem, new combinatorial methods (related to hypergraph transversals) were developed.

Interestingly, model checking for the fragment ESO($\exists^*\forall\exists^*$) is NP-complete over *graphs*. For example, the well-known set-splitting problem can be expressed in it. Thus the fact that our input structures are monadic *strings* is essential (just as for MSO).

(2) The class ESO($\exists^*\forall\forall$) is regular. The regularity proof for this fragment is easier but also required the development of new techniques (more of logical than of combinatorial nature). Note that model checking for this class, too, is NP-complete over graphs.

(3) Any class ESO($\mathcal{Q}$) not contained in the union of ESO($\exists^*\forall\exists^*$) and ESO($\exists^*\forall\forall$) is not regular.

Thus ESO($\exists^*\forall\exists^*$) and ESO($\exists^*\forall\forall$) are the *maximal regular standard prefix classes*. The unique maximal (general) regular ESO-prefix class is the union of these two classes, i.e, ESO($\exists^*\forall\exists^*$) $\cup$ ESO($\exists^*\forall\forall$) $=$ ESO($\exists^*\forall(\forall \cup \exists^*)$).

As shown in [9], it turns out that there are three minimal nonregular ESO-prefix classes, namely the standard prefix classes ESO($\forall\forall\forall$), ESO($\forall\forall\exists$), and ESO($\forall\exists\forall$). All these classes express nonregular languages by sentences whose list of second-order variables consists of a single binary predicate variable.

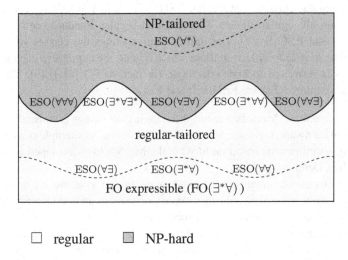

Fig. 1. Complete picture of the ESO-prefix classes on finite strings

Thus, 1.-3. give a *complete characterization* of the regular $ESO(\mathcal{Q})$ classes.

(4) The following dichotomy theorem is derived: Let $ESO(\mathcal{Q})$ be any prefix class. Then, either $ESO(\mathcal{Q})$ is regular, or $ESO(\mathcal{Q})$ expresses some NP-complete language. This means that model checking for $ESO(\mathcal{Q})$ is either possible by a deterministic finite automaton (and thus in constant space and linear time) or it is already NP-complete. Moreover, for all NP-complete classes $ESO(\mathcal{Q})$, NP-hardness holds already for sentences whose list of second-order variables consists of a single binary predicate variable. There are no fragments of intermediate difficulty between REG and NP.

(5) The above dichotomy theorem is paralleled by the solvability of the finite satisfiability problem for ESO (and thus FO) over strings. As shown in [9], over finite strings satisfiability of a given $ESO(\mathcal{Q})$ sentence is decidable iff $ESO(\mathcal{Q})$ is regular.

(6) In [9], a precise characterization is given of those prefix classes of ESO which are equivalent to MSO over strings, i.e. of those prefix fragments that *capture* the class REG of regular languages. This provides new logical characterizations of REG. Moreover, in [9] it is established that any regular ESO-prefix class is over strings either equivalent to full MSO, or is contained in first-order logic, in fact, in $FO(\exists^*\forall)$.

It is further shown that $ESO(\forall^*)$ is the unique minimal ESO prefix class which captures NP. The proof uses results in [19,8] and well-known hierarchy theorems.

The main results of [9] are summarized in Figure 1. In this figure, the ESO-prefix classes are divided into four regions. The upper two contain all classes that express nonregular languages, and thus also NP-complete languages. The uppermost region contains those classes which capture NP; these classes are called NP-*tailored*. The region next below, separated by a dashed line, contains those classes which can express some NP-hard languages, but not all languages in NP. Its bottom is constituted by the minimal nonregular classes, ESO($\forall\forall\forall$), ESO($\forall\exists\forall$), and ESO($\forall\forall\exists$). The lower two regions contain all regular classes. The maximal regular standard prefix classes are ESO($\exists^*\forall\exists^*$) and ESO($\exists^*\forall\forall$). The dashed line separates the classes which capture REG(called *regular-tailored*), from those which do not; the expressive capability of the latter classes is restricted to first-order logic (in fact, to FO($\exists^*\forall$)) [9]. The minimal classes which capture REG are ESO($\forall\exists$) and ESO($\forall\forall$).

Potential Applications. Monadic second-order logic over strings is currently used in the verification of hardware, software, and distributed systems. An example of a specific tool for checking specifications based on MSO is the MONA tool developed at the BRICS research lab in Denmark [1,15].

Observe that certain interesting desired properties of systems are most naturally formulated in *nonmonadic* second-order logic. Consider, as an unpretentious example[2], the following property of a ring P of processors of different types, where two types may either be compatible or incompatible with each other. We call P *tolerant*, if for each processor p in P there exist two other distinct processors $backup_1(p) \in P$ and $backup_2(p) \in P$, both compatible to p, such that the following conditions are satisfied:

1. for each $p \in P$ and for each $i \in \{1, 2\}$, $backup_i(p)$ is not a neighbor of p;
2. for each $i, j \in \{1, 2\}$, $backup_i(backup_j(p)) \notin \{p, backup_1(p), backup_2(p)\}$.

Intuitively, we may imagine that in case p breaks down, the workload of p can be reassigned to $backup_1(p)$ or to $backup_2(p)$. Condition 1 reflects the intuition that if some processor is damaged, there is some likelihood that also its neighbors are (e.g. in case of physical affection such as radiation), thus neighbors should not be used as backup processors. Condition 2 states that the backup processor assignment is antisymmetric and anti-triangular; this ensures, in particular, that the system remains functional, even if two processors of the same type are broken (further processors of incompatible type might be broken, provided that broken processors can be simply bypassed for communication).

Let T be a fixed set of processor types. We represent a ring of n processors numbered from 1 to n where processor i is adjacent to processor $i+1$ $(mod\ n)$ as a string of length n from T^* whose i-th position is τ if the type of the i-th processor is τ; logically, $C_\tau(i)$ is then true. The property of P being tolerant is expressed by the following second order sentence Φ:

$$\Phi : \exists R_1, R_2, \forall x \exists y_1, y_2.\ compat(x, y_1) \wedge compat(x, y_2) \wedge$$

$$R_1(x, y_1) \wedge R_2(x, y_2) \wedge$$
$$\bigwedge_{i=1,2} \bigwedge_{j=1,2} \left(\neg R_i(y_j, x) \wedge \neg R_1(y_j, y_i) \wedge \neg R_2(y_j, y_i) \right) \wedge$$
$$x \neq y_1 \wedge x \neq y_2 \wedge y_1 \neq y_2 \wedge$$
$$\neg Succ(x, y_1) \wedge \neg Succ(y_1, x) \wedge \neg Succ(x, y_2) \wedge \neg Succ(y_2, x) \wedge$$
$$\left((x = max) \rightarrow (y_1 \neq min \wedge y_2 \neq min) \right) \wedge$$

---

[2] Our goal here is merely to give the reader some intuition about a possible type of application.

$$\Big((x = min) \rightarrow (y_1 \neq max \wedge y_2 \neq max)\Big),$$

where $compat(x, y)$ is the abbreviation for the formal statement that processor x is compatible to processor y (which can be encoded as a simple boolean formula over C_τ atoms).

Φ is the natural second-order formulation of the tolerance property of a ring of processors. This formula is in the fragment $ESO(\exists^*\forall\exists^*)$; hence, by our results, we can immediately classify tolerance as a regular property, i.e., a property that can be checked by a finite automaton.

In a similar way, one can exhibit examples of $ESO(\exists^*\forall\forall)$ formulas that naturally express interesting properties whose regularity is not completely obvious *a priori*. We thus hope that our results may find applications in the field of computer aided verification.

3   Recent Results on SO over Strings

In [10], we further investigated the extension of the results in the previous section from ESO to full second-order logic over strings. Our focus of attention were the *SO* prefix classes which are regular vs. those which are not. In particular, we were interested to know how adding second-order variables affects regular fragments.

Generalizing Fagin's theorem, Stockmeyer [26] has shown that full SO captures the polynomial hierarchy (PH). Second-order variables turn out to be quite powerful. In fact, already two first-order variables, a single binary predicate variable, and further monadic predicate variables are sufficient to express languages that are complete for the levels of PH.

We were able to precisely characterize the regular and nonregular standard Σ^1_k prefix classes. The maximal standard Σ^1_k prefix classes which are regular and the minimal standard Σ^1_k prefix classes which are non-regular are summarized in Figure 2.

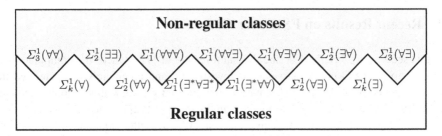

Fig. 2. Maximal regular and minimal non-regular SO prefix classes on strings.

Thus, no $\Sigma^1_k(Q)$ fragment, where $k \geq 3$ and Q contains at least two variables, is regular, and for $k = 2$, only two such fragments ($Q = \exists\forall$ or $Q = \forall\exists$) are regular. Note that Grädel and Rosen have shown [13] that $\Sigma^1_1(FO^2)$, i.e., existential second order logic with two first-order variables, is over strings regular. By our results, $\Sigma^1_k(\mathrm{FO}^2)$, for $k \geq 2$, is intractable.

Figure 3 shows inclusion relationship between the classes $\Sigma^1_2(Q)$ where Q contains 2 quantifiers. Similar relationships hold for $\Sigma^1_k(Q)$ classes. Furthermore, as shown in [10], we have that $\Sigma^1_2(\forall\exists) = \Sigma^1_1(\bigwedge\forall\exists)$ and $\Sigma^1_3(\exists\forall) = \Sigma^1_2(\bigvee\exists\forall)$, where $\Sigma^1_k(\bigvee Q)$ (resp., $\Sigma^1_k(\bigwedge Q)$) denote the class of Σ^1_k sentences where the first-order part is a finite disjunction (resp., conjunction) of prefix formulas with quantifier in Q.

$$\begin{array}{ccc} \Sigma^1_2(\exists\forall) & & \Sigma^1_2(\exists\exists) \\ \supseteq & & \subseteq \\ & \Sigma^1_2(\forall\forall) = \Sigma^1_2(\forall\exists) & \end{array}$$

Fig. 3. Semantic inclusion relations between $\Sigma^1_2(Q)$ classes over strings, $|Q| = 2$

As for the complexity of model checking, the results from [9] and above imply that deciding whether $W \models \Phi$ for a fixed formula Φ and a given string W is intractable for all prefix classes $\Sigma^1_k(Q)$ which are (syntactically) not included in the maximal regular prefix classes shown in Figure 2, with the exception $\Sigma^1_2(\exists\forall)$, for which the tractability is currently open.

In more detail, the complexity of SO over strings increases with the number of SO quantifier alternations. In particular, $\Sigma^1_2(\exists\exists)$ can express Σ^p_2-complete languages, where $\Sigma^p_2 = \mathrm{NP}^{\mathrm{NP}}$ is from the second level of the polynomial hierarchy. Other fragments of Σ^1_2 have lower complexity; for example, $\Sigma^1_2(\forall^*\exists)$ is merely NP-complete. By adding further second-order variables, we can encode evaluating Σ^p_k-complete QBFs into $\Sigma^1_k(Q)$, where $Q = \exists\exists$ if $k > 2$ is even and $Q = \forall\forall$ if $k > 1$ is odd.

4   Recent Results on ESO over Graphs

In this section, we briefly describe the main results of [12], where the computational complexity of ESO-prefix classes of is investigated and completely characterized in three contexts: over (1) directed graphs, (2) undirected graphs with self-loops, and (3) undirected graphs without self-loops. A main theorem of [12] is that a *dichotomy* holds in these contexts, that is to say, each prefix class of ESO either contains sentences that can express NP-complete problems or each of its sentences expresses a polynomial-time solvable problem. Although the boundary of the dichotomy coincides for 1. and 2. (which we refer ot as *general graphs* from now on), it changes if one moves to 3. The key difference is that a certain prefix class, based on the well-known *Ackermann class*, contains sentences that can express NP-complete problems over general graphs, but becomes tractable over undirected graphs without self-loops. Moreover, establishing the dichotomy in case 3. turned out to be technically challenging, and required the use of sophisticated machinery from graph theory and combinatorics, including results about graphs of bounded tree-width and Ramsey's theorem.

In [12], a special notation for ESO-prefix classes was used in order to describe the results with the tightest possible precision involving both the number of SO quantifiers and their arities.[3] Expressions in this notation are built according to the following rules:

- E (resp., E_i) denotes the existential quantification over a single predicate of arbitrary arity (arity $\leq i$).

- a (resp., e) denotes the universal (existential) quantification of a single first-order variable.

- If η is a quantification pattern, then η^* denotes all patterns obtained by repeating η zero or more times.

An expression $\mathcal{E}$ in the special notation consists of a string of ESO quantification patterns (E-patterns) followed by a string of first-order quantification patterns (a or e patterns); such an expression represents the class of all prenex ESO-formulas whose quantifier prefix corresponds to a (not-necessarily contiguous) substring of $\mathcal{E}$.

For example, $E_1^* eaa$ denotes the class of formulas $\exists P_1 \cdots \exists P_r \exists x \forall y \forall z \varphi$, where each P_i is monadic, x, y, and z are first-order variables, and φ is quantifier-free.

A prefix class C is NP-*hard* on a class $\mathcal{K}$ of relational structures, if some sentence in C expresses an NP-hard property on $\mathcal{K}$, and C is *polynomial-time* (PTIME) on $\mathcal{K}$, if for each sentence $\Phi \in C$, model checking is polynomial. Furthermore, C is called *first-order* (FO), if every $\Phi \in C$ is equivalent to a first-order formula.

The first result of [12] completely characterizes the computational complexity of ESO-prefix classes on general graphs. In fact, the same characterization holds on the collection of all finite structures over any relational vocabulary that contains a relation symbol of arity ≥ 2. This characterization is obtained by showing (assuming P $\neq$ NP) that there are four *minimal* NP-hard and three *maximal* PTIME prefix classes, and that these seven classes combine to give complete information about all other prefix classes. This means that every other prefix either contains one of the minimal NP-hard prefix classes as a substring (and, hence, is NP-hard) or is a substring of a maximal PTIME prefix class (and, hence, is in PTIME). Figure 4 depicts the characterization of the NP-hard and PTIME prefix classes of ESO on general graphs.

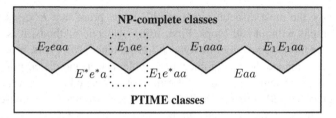

Fig. 4. ESO on arbitrary structures, directed graphs and undirected graphs with self-loops.

---

[3] For ESO over strings [9], the same level of precision was reached with simpler notation.

As seen in Figure 4, the four minimal NP-hard classes are E_2eaa, E_1ae, E_1aaa, and E_1E_1aa, while the three maximal PTIME classes are E^*e^*a, E_1e^*aa, and Eaa. The NP-hardness results are established by showing that each of the four minimal prefix classes contains ESO-sentences expressing NP-complete problems. For example, a SAT encoding on general graphs can be expressed by an E_1ae sentence. Note that the first-order prefix class ae played a key role in the study of the classical decision problem for fragments of first-order logic (see [3]). As regards the maximal PTIME classes, E^*e^*a is actually FO, while the model checking problem for fixed sentences in E_1e^*aa and Eaa is reducible to 2SAT and, thus, is in PTIME (in fact, in NL).

The second result of [12] completely characterizes the computational complexity of prefix classes of ESO on undirected graphs without self-loops. As mentioned earlier, it was shown that a dichotomy still holds, but its boundary changes. The key difference is that E^*ae turns out to be PTIME on undirected graphs without self-loops, while its subclass E_1ae is NP-hard on general graphs. It can be seen that interesting properties of graphs are expressible by E^*ae-sentences. Specifically, for each integer $m > 0$, there is a E^*ae-sentence expressing that a connected graph contains a cycle whose length is divisible by m. This was shown to be decidable in polynomial time by Thomassen [30]. E^*ae constitutes a maximal PTIME class, because all four extensions of E_1ae by any single first-order quantifier are NP-hard on undirected graphs without self-loops [12]. The other minimal NP-hard prefixes on general graphs remain NP-hard also on undirected graphs without self-loops. Consequently, over such graphs, there are seven minimal NP-hard and four maximal PTIME prefix classes that determine the computational complexity of all other ESO-prefix classes (see Figure 5).

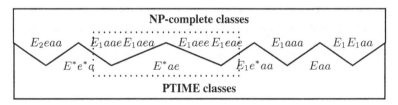

Fig. 5. ESO on undirected graphs without self-loops. The dotted boxes in Figures 4 and 5 indicate the difference between the two cases.

Technically, the most difficult result of [12] is the proof that E^*ae is PTIME on undirected graphs without self-loops. First, using syntactic methods, it is shown that each E^*ae-sentence is equivalent to some E_1^*ae-sentence. After this, it is shown that for each E_1^*ae-sentence the model-checking problem over undirected graphs without self-loops is is equivalent to a natural coloring problem called the *saturation problem*. This problem asks whether there is a particular mapping from a given undirected graph without self-loops to a fixed, directed *pattern graph P* which is extracted from the E_1^*ae-formula under consideration. Depending on the labelings of cycles in P, two cases of the saturation problem are distinguished, namely *pure pattern graphs* and *mixed pattern graphs*. For each case, a polynomial-time algorithm is designed. In simplified terms and focussed on the case of connected graphs, the one for pure pattern graphs has three main ingredients. First, adapting results by Thomassen [30] and using a new graph coloring method, it is shown that if a $E_1^* ae$-sentence Φ gives rise to a pure pattern graph, then a fixed integer k can be found such that every undirected graph without self-loops and having tree-width bigger than k satisfies Φ. Second, Courcelle's theorem [6] is used by which model-checking for MSO sentences is polynomial on graphs of bounded tree-width. Third, Bodlaender's result [2] is used that, for each fixed k, there is a polynomial-time algorithm to check if a given graph has tree-width at most k.

The polynomial-time algorithm for mixed pattern graphs has a similar architecture, but requires the development of substantial additional technical machinery, including a generalization of the concept of graphs of bounded tree-width. The results of [12] can be summarized in the following theorem.

Theorem 4.1. *Figures 4 and 5 provide a complete classification of the complexity of all ESO prefix classes on graphs.*

5 Open Problems

Let us conclude this paper by pointing out a few interesting (and in our opinion important) issues that should eventually be settled.

• While the work on word structures concentrated so far on strings with a successor relation $Succ$, one should also consider the case where in addition a predefined linear order $<$ is available on the word structures, and the case where the successor relation $Succ$ is replaced by such a linear order. While for full ESO or SO, $Succ$ and $<$ are freely interchangeable, because eiter predicate can be defined in terms of the other, this is not so for many of the limited ESO-prefix classes. Preliminary results suggest that most of the results in this paper carry over to the $<$ case.

• Delineate the tractability/intractability frontier for all SO prefix classes over graphs.

• Study SO prefix classes over trees and other interesting classes of structures (e.g. planar graphs).

• The scope of [9] and of Section 2 in this paper are *finite* strings. However, infinite strings or ω-words are another important area of research. In particular, Büchi has shown that an analogue of his theorem (Proposition 1.1) also holds for ω-words [4]. For an overview of this and many other important results on ω-words, we refer the reader to the excellent survey paper [31]. In this context, it would be interesting to see which of the results established so far survive for ω-words. For some results, such as the regularity of $\text{ESO}(\exists^*\forall\forall)$ this is obviously the case since no finiteness assumption on the input word structures was made in the proof. For determining the regularity or nonregularity of some other clases such as $\text{ESO}(\exists^*\forall\exists^*)$, further research is needed.

Acknowledgments. This work was supported by the Austrian Science Fund (FWF) Project Z29-N04.

References

1. D. Basin and N. Klarlund Hardware verification using monadic second-order logic. *Proc. 7th Intl. Conf. on Computer Aided Verification (CAV'95)*, LNCS 939, pp. 31–41, 1995.
2. H. L. Bodlaender. A linear-time algorithm for finding tree-decompositions of small treewidth. *SIAM Journal on Computing*, 25(6):1305–1317, 1996.
3. E. Börger, E. Grädel, and Y. Gurevich. *The Classical Decision Problem*. Springer, 1997.
4. J. R. Büchi. On a decision method in restriced second-order arithmetic. In E. Nagel et al., editor, *Proc. International Congress on Logic, Methodology and Philosophy of Science*, pp. 1–11, Stanford, CA, 1960. Stanford University Press.
5. J. R. Büchi. Weak second-order arithmetic and finite automata. *Zeitschrift für mathematische Logik und Grundlagen der Mathematik*, 6:66–92, 1960.
6. B. Courcelle. The monadic second-order logic of graphs I: recognizable sets of finite graphs. *Information and Computation*, 85:12–75, 1990.
7. H.-D. Ebbinghaus and J. Flum. *Finite Model Theory*. Springer, 1995.
8. T. Eiter, G. Gottlob, and Y. Gurevich. Normal forms for second-order logic over finite structures, and classification of NP optimization problems. *Annals of Pure and Applied Logic*, 78:111–125, 1996.
9. T. Eiter, G. Gottlob, and Y. Gurevich. Existential second-order logic over strings. *Journal of the ACM*, 47(1):77–131, 2000.
10. T. Eiter, G. Gottlob, and T. Schwentick. Second Order Logic over Strings: Regular and Non-Regular Fragments. *Developments in Language Theory, 5th Intl. Conference (DLT'01)*, Vienna, Austria, July 16-21, 2001, Revised Papers, Springer LNCS 2295, pp.37-56, 2002.
11. R. Fagin. Generalized first-order spectra and polynomial-time recognizable sets. In R. M. Karp, editor, *Complexity of Computation*, pp. 43–74. AMS, 1974.
12. G. Gottlob, P. Kolaitis, and T. Schwentick. Existential second-order logic over graphs: charting the tractability frontier. In *Journal of the ACM*, 51(2):312–362, 2004.
13. E. Grädel and E. Rosen. Two-variable descriptions of regularity. In *Proc. 14th Annual Symposium on Logic in Computer Science (LICS-99)* , pp. 14–23. IEEE CS Press, 1999.
14. E. Grandjean. Universal quantifiers and time complexity of random access machines. *Mathematical Systems Theory*, 13:171–187, 1985.
15. J.G. Henriksen, J. Jensen, M. Jørgensen, N. Klarlund, B. Paige, T. Rauhe, and A. Sandholm Mona: Monadic Second-order Logic in Practice. in Proceedings of *Tools and Algorithms for the Construction and Analysis of Systems, First Intl. Workshop, TACAS'95*, Springer LNCS 1019, pp. 89–110, 1996.
16. N. Immerman. *Descriptive Complexity*. Springer, 1997.
17. P. Kolaitis and C. Papadimitriou. Some computational aspects of circumscription. *Journal of the ACM*, 37(1):1–15, 1990.
18. C. Lautemann, T. Schwentick, and D. Thérien. Logics for context-free languages. In *Proc. 1994 Annual Conference of the EACSL*, pages 205–216, 1995.
19. D. Leivant. Descriptive characterizations of computational complexity. *Journal of Computer and System Sciences*, 39:51–83, 1989.
20. J. F. Lynch. The quantifier structure of sentences that characterize nondeterministic time complexity. *Computational Complexity*, 2:40–66, 1992.
21. R. McNaughton and S. Papert. *Counter-Free Automata*. MIT Press, Cambridge, Massachusetts, 1971.
22. F. Olive. A Conjunctive Logical Characterization of Nondeterministic Linear Time. In *Proc. Conference on Computer Science Logic (CSL '97)*, LNCS. Springer, 1998 (to appear).

23. J.-E. Pin. *Varieties of Formal Languages*. North Oxford, London and Plenum, New York, 1986.
24. J.-E. Pin. Logic On Words. *Bulletin of the EATCS*, 54:145–165, 1994.
25. J.-E. Pin. Semigroups and Automata on Words. *Annals of Mathematics and Artificial Intelligence*, 16:343–384, 1996.
26. L. J. Stockmeyer. The polynomial-time hierarchy. *Theoretical Comp. Sc.*, 3:1–22, 1977.
27. H. Straubing. *Finite Automata, Formal Logic, and Circuit Complexity*. Birkhäuser, 1994.
28. H. Straubing, D. Thérien, and W. Thomas. Regular Languages Defined with Generalized Quantifiers. *Information and Computation*, 118:289–301, 1995.
29. W. Thomas. Languages, Automata, and Logic. In G. Rozenberg and A. Salomaa, editors, *Handbook of Formal Language Theory*, volume III, pages 389–455. Springer, 1996.
30. C. Thomassen. On the presence of disjoint subgraphs of a specified type. *Journal of Graph Theory*, 12:1, 101-111, 1988.
31. W. Thomas. Automata on infinite objects. In J. van Leeuwen, editor, *Handbook of Theoretical Computer Science*, volume B, chapter 4. Elsevier Science Pub., 1990.
32. B. Trakhtenbrot. Finite automata and the logic of monadic predicates. *Dokl. Akad. Nauk SSSR*, 140:326–329, 1961.

Efficient Algorithms for Constraint Description Problems over Finite Totally Ordered Domains*

Extended Abstract

Ángel J. Gil[1], Miki Hermann[2], Gernot Salzer[3], and Bruno Zanuttini[4]

[1] Universitat Pompeu Fabra, Barcelona, Spain. angel.gil@upf.edu
[2] LIX (FRE 2653), École Polytechnique, France. hermann@lix.polytechnique.fr
[3] Technische Universität Wien, Austria. salzer@logic.at
[4] GREYC (UMR 6072), Université de Caen, France. zanutti@info.unicaen.fr

Abstract. Given a finite set of vectors over a finite totally ordered domain, we study the problem of computing a constraint in conjunctive normal form such that the set of solutions for the produced constraint is identical to the original set. We develop an efficient polynomial-time algorithm for the general case, followed by specific polynomial-time algorithms producing Horn, dual Horn, and bijunctive constraints for sets of vectors closed under the operations of conjunction, disjunction, and median, respectively. We also consider the affine constraints, analyzing them by means of computer algebra. Our results generalize the work of Dechter and Pearl on relational data, as well as the papers by Hébrard and Zanuttini. They also complete the results of Hähnle *et al.* on multivalued logics and Jeavons *et al.* on the algebraic approach to constraints. We view our work as a step toward a complete complexity classification of constraint satisfaction problems over finite domains.

1 Introduction and Summary of Results

Constraint satisfaction problems constitute nowadays a well-studied topic on the frontier of complexity, logic, combinatoric, and artificial intelligence. It is indeed well-known that this framework allows us to encode many natural problems or knowledge bases. In principle, an instance of a constraint satisfaction problem is a finite set of variable vectors associated with an allowed set of values. A *model* is an assignment of values to all variables that satisfy every constraint. When a constraint satisfaction problem encodes a decision problem, the models represent its *solutions*. When it encodes some knowledge, the models represent possible combinations that the variables can assume in the described universe.

The constraints are usually represented by means of a set of variable vectors associated with an allowed set of values. This representation is not always well-suited for our purposes, therefore other representations have been introduced. The essence of the most studied alternative is the notion of of a *relation*, making it easy to apply it within the database or knowledge base framework. An

---

* Dedicated to the memory of Peter Ružička (1947 – 2003).

D. Basin and M. Rusinowitch (Eds.): IJCAR 2004, LNAI 3097, pp. 244–258, 2004.

instance of a constraint satisfaction problem is then represented as a conjunction of relation applications. We study in this paper the constraint *description* problem, i.e., that of converting a constraint from a former set representation to the latter one by means of conjunction over relations. We consider this problem first in its general setting without any restrictions imposed on the initial set of vectors. We continue by imposing several closure properties on the initial set, like the closure under the minimum, maximum, median, and affine operations. We subsequently discover that these closure properties induce the description by Horn, dual Horn, bijunctive, and affine constraints, respectively.

The motivation to study constraint description problems is numerous. From the artificial intelligence point of view description problems formalize the notion of exact acquisition of knowledge from examples. This means that they formalize situations where a system is given access to a set of examples and it is asked to compute a constraint describing it exactly. Satisfiability poses a keystone problem in artificial intelligence, automated deduction, databases, and verification. It is well-known that the satisfiability problem for the general constraints is an NP-complete problem. Therefore it is important to look for restricted classes of constraints that admit polynomial algorithms deciding satisfiability. Horn, dual Horn, bijunctive, and affine constraints constitute exactly these tractable classes, as it was mentioned by Schaefer [17] for the case of Boolean constraints. Thus the description problem for these four classes can be seen as storing a specific knowledge into a knowledge base while we are required to respect its format. This problem is also known as *structure identification*, studied by Dechter with Pearl [7] and by Hébrard with Zanuttini [11,19], both for the Boolean case. Another motivation for studying description problems comes from combinatorics. Indeed, since finding a solution for an instance of a constraint satisfaction problem is difficult in general but tractable in the four aforementioned cases, it is important to be able to recognize constraints belonging to these tractable cases.

The study of Boolean constraint satisfaction problems, especially their complexity questions, was started by Schaefer in [17], although he did not yet consider constraints explicitly. During the last ten years, constraints gained considerable interest in theoretical computer science. An excellent complexity classification of existing Boolean constraint satisfaction problems can be found in the monograph [6]. Jeavons *et al.* [5,13,14] started to study constraint satisfaction problems from an algebraic viewpoint. Feder, Kolaitis, and Vardi [8,16] posed a general framework for the study of constraint satisfaction problems.

There has not been much progress done on constraint satisfaction problems on domains with larger cardinality. Hell and Nešetřil [12] studied constraint satisfaction problems by means of graph homomorphisms. Bulatov [4] made a significant breakthrough with a generalization of Schaefer's result to a three-element domain. On the other hand, Hähnle *et al.* [2,3,10] studied the complexity of satisfiability problems for many-valued logics that present yet another viewpoint of constraint satisfaction problems. We realized reading the previous articles on many-valued logics that in the presence of a total order the satisfiability problem for the Horn, dual Horn, bijunctive, and affine many-valued formulas of signed logic are decidable in polynomial time. We saw also that Jeavons and Cooper [15] studied some aspects of tractable constraints on finite ordered domains from an algebraic standpoint. This lead us to the idea to look more carefully on constraint description problems over finite totally ordered domains, developing a new formalism for constraints based on an already known concept of inequalities. The purpose of our paper is manifold. We want to generalize the work of Dechter and Pearl [7], based on the more efficient algorithms for Boolean description problems by Hébrard and Zanuttini [11,19]. We also want to complement the work of Hähnle *et al.* on many-valued logics. We also want to pave the way for a complete complexity classification of constraint satisfaction problems over finite totally ordered domains[1].

2 Preliminaries

Let D be a finite, totally ordered domain, say $D = \{0, \dots, n-1\}$, and let V be a set of variables. For $x \in V$ and $d \in D$, the inequalities $x \geq d$ and $x \leq d$ are called positive and negative *literal*, respectively. The set of *constraints* over D and V is defined as follows: the logical constants *false* and *true* are constraints; literals are constraints; if φ and ψ are constraints, then the expressions $(\varphi \wedge \psi)$ and $(\varphi \vee \psi)$ are constraints. We write $\varphi(x_1, \dots, x_\ell)$ to indicate that constraint φ contains exactly the variables $x_1, \dots, x_\ell$. For convenience, we use the following shorthand notation, as usual: $x > d$ means $x \geq d + 1$ for $d \in \{0, \dots, n-2\}$, and *false* otherwise; $x < d$ means $x \leq d - 1$ for $d \in \{1, \dots, n-1\}$, and *false* otherwise; $x = d$ means $x \geq d \wedge x \leq d$; $\neg false$ and $\neg true$ mean *true* and *false*, respectively; $\neg(x \geq d)$, $\neg(x \leq d)$, $\neg(x > d)$, and $\neg(x < d)$ mean $x < d$, $x > d$, $x \leq d$, and $x \geq d$, respectively; $\neg(x = d)$ and $x \neq d$ both mean $x < d \vee x > d$; $\neg(\varphi \wedge \psi)$ and $\neg(\varphi \vee \psi)$ mean $\neg\varphi \vee \neg\psi$ and $\neg\varphi \wedge \neg\psi$, respectively. Note that $x = d$ and $x \neq d$ asymptotically require the same space as their alternative notations, i.e., $O(\log n)$. Since d is bounded by n, its binary coding has length $O(\log n)$.

A *clause* is a disjunction of literals. It is a *Horn* clause if it contains at most one positive literal, *dual Horn* if it contains at most one negative literal, and *bijunctive* if it contains at most two literals. Following Schaefer [17], we extend the notion of constraints by introducing affine clauses. An *affine* clause is an equation $a_1 x_1 + \cdots + a_\ell x_\ell = b \pmod{n}$ where $x_1, \dots, x_\ell \in V$ and $a_1, \dots, a_\ell, b \in D$. A constraint is in *conjunctive normal form* (CNF) if it is a conjunction of clauses. It is a Horn, a dual Horn, a bijunctive, or an affine constraint if it is a conjunction of Horn, dual Horn, bijunctive, or affine clauses, respectively.

A *model* for a constraint $\varphi(x_1, \dots, x_\ell)$ is a mapping $m: \{x_1, \dots, x_\ell\} \to D$ assigning a domain element $m(x)$ to each variable x. The *satisfaction relation* $m \models \varphi$ is inductively defined as follows: $m \models true$ and $m \not\models false$; $m \models x \leq d$ if $m(x) \leq d$, and $m \models x \geq d$ if $m(x) \geq d$; $m \models \varphi \wedge \psi$ if $m \models \varphi$ and $m \models \psi$; $m \models \varphi \vee \psi$ if $m \models \varphi$ or $m \models \psi$. An affine clause is satisfied by a model m if $a_1 m(x_1) + \cdots + a_\ell m(x_\ell) = b \pmod{n}$. The set of all models satisfying φ is denoted by $\mathrm{Sol}(\varphi)$. If we arrange the variables in some arbitrary but fixed order, say as

---

[1] See http://www.lix.polytechnique.fr/~hermann/publications/cdesc04.ps.gz for the proofs, since many of them are omitted here due to lack of space.

a vector $x = (x_1, \ldots, x_\ell)$, then the models can be identified with the vectors in D^ℓ. The j-th component of a vector m, denoted by $m[j]$, gives the value of the j-th variable, i.e., $m(x_j) = m[j]$. The operations of conjunction, disjunction, addition, and median on vectors $m, m', m'' \in D^\ell$ are defined as follows:

$$m \wedge m' = (\min(m[1], m'[1]), \ldots, \min(m[\ell], m'[\ell]))$$
$$m \vee m' = (\max(m[1], m'[1]), \ldots, \max(m[\ell], m'[\ell]))$$
$$m + m' = (m[1] + m'[1] \ (\mathrm{mod} \ |D|), \ldots, m[\ell] + m'[\ell] \ (\mathrm{mod} \ |D|))$$
$$\mathrm{med}(m, m', m'') = (\mathrm{med}(m[1], m'[1], m''[1]), \ldots, \mathrm{med}(m[\ell], m'[\ell], m''[\ell])$$

The ternary *median* operator is defined as follows: for each choice of three values $a, b, c \in D$ such that $a \leq b \leq c$, we have $\mathrm{med}(a, b, c) = b$. Note that the median can also be defined by $\mathrm{med}(a, b, c) = \min(\max(a, b), \max(b, c), \max(c, a))$.

We say that a set of vectors is *Horn* if it is closed under conjunction, *dual Horn* if it is closed under disjunction, *bijunctive* if it is closed under median, and *affine* if it is a Cartesian product of affine spaces, i.e., of vector spaces translated by some vector.

3 Constraints in Conjunctive Normal Form

We investigate first the description problem for arbitrary sets of vectors.

Problem: DESCRIPTION
Input: A finite set of vectors $M \subseteq D^\ell$ over a finite totally ordered domain D.
Output: A constraint $\varphi(x_1, \ldots, x_\ell)$ over D in CNF such that $\mathrm{Sol}(\varphi) = M$.

The usual approach to this problem in the literature is to compute first the complement set $\bar{M} = D^\ell \setminus M$, followed by a construction of a clause $c(\bar{m})$ for each vector $\bar{m} \in \bar{M}$ missing from M such that $\bar{m}$ is the unique vector falsifying $c(\bar{m})$. The constraint φ is then the conjunction of the clauses $c(\bar{m})$ for all missing vectors $\bar{m} \in \bar{M}$. However, this algorithm is essentially exponential, since the complement set $\bar{M}$ can be exponentially bigger than the original set of vectors M. We present a new algorithm running in polynomial time and producing a CNF constraint of polynomial length with respect to $|M|$, ℓ, and $\log |D|$.

To construct the constraint φ we proceed in the following way. We arrange the set M as an ordered tree T_M, with branches corresponding to the vectors in M. In case M contains all possible vectors, i.e. $M = D^\ell$, T_M is a complete tree of branching factor $|D|$ and depth ℓ. Otherwise, some branches are missing, leading to gaps in the tree. We characterize these gaps by conjunctions of literals. Their disjunction yields a complete description of all vectors that are *missing* from M. Finally, by negation and de Morgan's law we obtain φ.

Let T_M be an ordered tree with edges labeled by domain elements such that each path from the root to a leaf corresponds to a vector in M. The tree T_M contains a path labeled $d_1. \cdots .d_i$ from the root to some node if there is a vector $m \in M$ such that $m[j] = d_j$ holds for every $j = 1, \ldots, i$. The level of a node is its distance to the root plus 1, i.e., the root is at level 1 and a node reachable via

$d_1. \cdots .d_i$ is at level $i+1$ (Fig. 1(a)). Note that all leaves are at level $\ell + 1$. If the edges between a node and its children are sorted in ascending order according to their labels, then traversing the leaves from left to right enumerates the vectors of M in lexicographic order, say $m_1, \ldots, m_{|M|}$. A vector m is lexicographically smaller than a vector m', if there is a level i such that $m[i] < m'[i]$ holds, and for all $j < i$ we have $m[j] = m'[j]$.

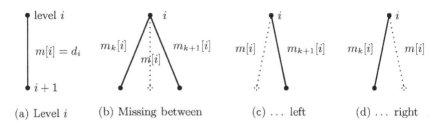

(a) Level i (b) Missing between   (c) ... left   (d) ... right

Fig. 1. Tree representation of vectors

Suppose that m_k and m_{k+1} are immediate neighbors in the lexicographic enumeration of M and let m be a vector lexicographically in between, thus missing from M. There are three possibilities for the path corresponding to m. It may either leave the tree at the fork *between* m_k and m_{k+1} (Fig. 1(b)), or at the fork to the *left* of m_{k+1} (Fig. 1(c)), or at the fork to the *right* of m_k (Fig. 1(d)). Therefore the missing vector m can be characterized by the constraints $\text{middle}(k,i)$, $\text{left}(k+1,i)$, and $\text{right}(k,i)$ defined as follows.

$$\text{middle}(k,i) = \bigwedge_{j<i}(x_j = m_k[j]) \quad \wedge \quad (x_i > m_k[i]) \wedge (x_i < m_{k+1}[i])$$

$$\text{left}(k+1,i) = \bigwedge_{j<i}(x_j = m_{k+1}[j]) \quad \wedge \quad (x_i < m_{k+1}[i])$$

$$\text{right}(k,i) = \bigwedge_{j<i}(x_j = m_k[j]) \quad \wedge \quad (x_i > m_k[i])$$

The situation depicted in Fig. 1 is a snapshot at level i of the tree T_M. Of course, if the situations in Fig. 1(c) or Fig. 1(d) occur at a level i, then subsequent forks of a missing vector to the left or to the right, respectively, occur also at each level $i' > i$.

To describe *all* vectors missing from M we form the disjunction of the above constraints for appropriate values of k and i. We need to determine the levels at which neighboring models fork by means of the following function.

$$\text{fork}(k) = \begin{cases} 0 & \text{for } k = 0 \\ \min\{i \mid m_k[i] \neq m_{k+1}[i]\} & \text{for } k = 1, \ldots, |M| - 1 \\ 0 & \text{for } k = |M| \end{cases}$$

The values fork(0) and fork($|M|$) correspond to imaginary models m_0 and $m_{|M|+1}$ forking at a level above the root. They allow to write the conditions below in a concise way at the left and right border of the tree. The three situations in Fig. 1 can now be specified by the following conditions.

$$
\begin{array}{llll}
i = \mathrm{fork}(k) & \wedge & m_k[i] + 1 < m_{k+1}[i] & \text{(edges missing in between)} \\
\mathrm{fork}(k) < i & \wedge & m_{k+1}[i] > 0 & (\ \ldots\ \text{to the left}) \\
\mathrm{fork}(k) < i & \wedge & m_k[i] < |D| - 1 & (\ \ldots\ \text{to the right})
\end{array}
$$

The second condition in each line ensures that there is at least one missing edge. It avoids that the constraints middle(k, i), left($k + 1, i$), and right(k, i) evaluate to false. The disjunction of clauses middle(k, i), left($k+1, i$), and right(k, i) that satisfy the first, second, and third condition, respectively, for all models and all levels represent a disjunctive constraint satisfied by the models *missing* from M. After applying negation and de Morgan's law, we arrive at the constraint

$$
\begin{aligned}
\varphi(M) \\
= \quad & \bigwedge \{\, \neg\,\mathrm{middle}(k, i) \mid 0 < k < |M|,\ i = \mathrm{fork}(k),\ m_k[i] + 1 < m_{k+1}[i]\,\} \\
\wedge \quad & \bigwedge \{\, \neg\,\mathrm{left}(k + 1, i) \mid 0 \le k < |M|,\ \mathrm{fork}(k) < i \le \ell,\ m_{k+1}[i] > 0\,\} \\
\wedge \quad & \bigwedge \{\, \neg\,\mathrm{right}(k, i) \quad \mid 0 < k \le |M|,\ \mathrm{fork}(k) < i \le \ell,\ m_k[i] < |D| - 1\,\}.
\end{aligned}
$$

The constraint φ is in conjunctive normal form and the condition $\mathrm{Sol}(\varphi) = M$ holds. Note that we use negation not as an operator on the syntax level but as a meta-notation expressing that the constraint following the negation sign has to be replaced by its dual. Note also that the conjunct left($k + 1, i$) is defined and used with the shifted parameter $k + 1$. This is necessary because the conjunct left characterizes a gap lexicographically before the vector m_{k+1}.

It follows directly from the construction that the constraint $\varphi(M)$ contains at most $3\,|M|\,\ell$ clauses. Each clause contains at most 2ℓ literals, namely at most one positive and one negative for each variable. Each literal has the length $O(\log|D|)$, since the domain elements are written in binary notation. Hence, the overall length of the constraint $\varphi(M)$ is $O(|M|\,\ell^2 \log|D|)$.

The vectors in M can be lexicographically sorted in time $O(|M|\,\ell \log|D|)$ using radix sort. The factor $\log|D|$ stems from the comparison of domain elements. The fork levels can also be computed in time $O(|M|\,\ell \log|D|)$, in parallel with sorting the set M. The constraint $\varphi(M)$ is produced by two loops, where the outer loop is going through each vector in M and the inner loop through the variables. The clauses $\neg\,\mathrm{middle}(k, i)$, $\neg\,\mathrm{left}(k + 1, i)$, and $\neg\,\mathrm{right}(k, i)$ are potentially written in each step inside the combined loops. This makes an algorithm with time complexity $O(|M|\,\ell^2 \log|D|)$.

Theorem 1. *For each set of vectors $M \subseteq D^\ell$ over a finite ordered domain D there exists a constraint φ in conjunctive normal form such that $M = \mathrm{Sol}(\varphi)$. It contains at most $3\,|M|\,\ell$ clauses and its length is $O(|M|\,\ell^2 \log|D|)$. The algorithm constructing φ runs in time $O(|M|\,\ell^2 \log|D|)$ and space $O(|M|\,\ell \log|D|)$.*

4   Horn Constraints

Horn clauses and formulas constitute a frequently studied subclass. We investigate in this section the description problem for a generalization of Horn formulas to ordered finite domains, namely for sets of vectors closed under conjunction.

Problem: DESCRIPTION[HORN]
Input: A finite set of vectors $M \subseteq D^\ell$, closed under conjunction, over a finite totally ordered domain D.
Output: A Horn constraint φ over D such that $\text{Sol}(\varphi) = M$.

The general construction in Section 3 does not guarantee that the final constraint is Horn whenever the set M is closed under conjunction. Therefore we must shorten the clauses of the constraint φ, produced in Section 3, to obtain only Horn clauses. For this, we will modify a construction proposed by Jeavons and Cooper in [15]. Their method is exponential, since it proposes to construct a Horn clause for each vector in the complement set $D^\ell \setminus M$. Contrary to the method of Jeavons and Cooper, our new proposed method is polynomial with respect to $|M|$, ℓ and $\log |D|$.

Let $\varphi(M)$ be a constraint produced by the method of Section 3 and let c be a clause from $\varphi(M)$. We denote by c^- the disjunction of the negative literals in c. The vectors in M satisfying a negative literal in c satisfy also the restricted clause c^-. Hence we have only to care about the vectors that satisfy a positive literal but no negative literals in c, described by the set

$$M_c \quad = \quad \{m \in M \mid m \not\models c^-\} \ .$$

If M_c is empty, we can replace the clause c by $h(c) = c^-$ in the constraint $\varphi(M)$ without changing the set of models $\text{Sol}(\varphi)$. Otherwise, note that M_c is closed under conjunction, since M is already closed under this operation. Indeed, if the vectors m and m' falsify every negative literal $x \leq d$ of c^- then the conjunction $m \wedge m'$ falsifies the same negative literals. Hence M_c contains a unique minimal model $m_* = \bigwedge M_c$. Every positive literal in c satisfied by m_* is also satisfied by all the vectors in M_c. Let l be a positive literal from c and satisfied by m_*. There exists at least one such literal since otherwise m_* would satisfy neither c^- nor any positive literal in c, hence it would not be in M_c. Then c can be replaced with the Horn clause $h(c) = l \vee c^-$, without changing the set of models $\text{Sol}(\varphi)$. We obtain a Horn constraint $h(M)$ for a Horn set M by replacing every non-Horn clause c in $\varphi(M)$ by its Horn restriction $h(c)$.

The length of $h(M)$ is basically the same as that of $\varphi(M)$. The number of clauses is the same and the length of clauses is $O(\ell \log |D|)$ in both cases. There are at most 2ℓ literals in each clause of $\varphi(M)$ (one positive and one negative literal per variable) versus $\ell + 1$ literals in each clause of $h(M)$ (one negative literal per variable plus a single positive literal).

The construction of each Horn clause $h(c)$ requires time $O(|M| \ell \log |D|)$. For every vector $m \in M$ we have to evaluate at most ℓ negative literals in c to find out whether m belongs to M_c. The evaluation of a literal takes time

$O(\log |D|)$. Hence the computation of the set M_c takes time $O(|M|\ell \log |D|)$. To obtain $m_* = \bigwedge M_c$, we have to compute $|M_c| - 1$ conjunctions between vectors of length ℓ, each of the ℓ conjunctions taking time $O(\log |D|)$. Therefore m_* can also be computed in time $O(|M|\ell \log |D|)$. Since there are at most $3|M|\ell$ clauses in $\varphi(M)$, the transformation of $\varphi(M)$ into $h(M)$ can be done in time $O(|M|^2 \ell^2 \log |D|)$. Hence, the whole algorithm producing the Horn constraint $h(M)$ from the set of vectors M runs in time $O(|M|^2 \ell^2 \log |D|)$.

Theorem 2. *For each set of vectors $M \subseteq D^\ell$ over a finite totally ordered domain D that is closed under conjunction, there exists a Horn constraint φ such that $M = \mathrm{Sol}(\varphi)$. The constraint φ contains at most $3|M|\ell$ clauses and its length is $O(|M|\ell^2 \log |D|)$. The algorithm constructing the constraint φ runs in time $O(|M|^2 \ell^2 \log |D|)$ and space $O(|M|\ell \log |D|)$.*

Note that the result of Theorem 2 is not optimal. We can indeed derive a better algorithm with time complexity $O(|M|\ell(|M| + \ell) \log |D|)$ from the one for the Boolean case presented in [11]. For that purpose, notice that the conjuncts $\mathrm{middle}(k,i)$ and $\mathrm{left}(k+1,i)$ behave like the first case studied in the aforementioned paper, whereas the conjunct $\mathrm{right}(k,i)$ can be treated by means of the other cases, taking advantage of the following sets defined as generalizations of the corresponding ones from [11]:

$$I(k,i) = \{m \in M \mid m[i] > m_k[i] \text{ and } \forall j < i,\ m[j] \geq m_k[j]\},$$
$$\mathrm{sim}(k,i) = \max_{1 \leq j \leq k} \{j \mid \exists m \in I(k,i) \text{ such that } \forall l,\ l < j \text{ implies } m[l] \leq m_k[l]\}.$$

Using Theorem 2, we are able to prove a generalization of a well-known characterization of Horn sets. A related characterization in a different setting can be found in [9].

Proposition 3. *A set of vectors M over a finite totally ordered domain is closed under conjunction if and only if there exists a Horn constraint φ satisfying the identity $\mathrm{Sol}(\varphi) = M$.*

If we interchange conjunctions with disjunctions of models, as well as positive and negative literals throughout Section 4, we obtain identical results for dual Horn constraints.

Theorem 4. *A set of vectors $M \subseteq D^\ell$ over a finite ordered domain D is closed under disjunction if and only if there exists a dual Horn constraint φ satisfying the identity $M = \mathrm{Sol}(\varphi)$. Given M closed under disjunction, the dual Horn constraint φ contains at most $3|M|\ell$ clauses and its length is $O(|M|\ell^2 \log |D|)$. It can be constructed in time $O(|M|\ell(|M|+\ell) \log |D|)$ and space $O(|M|\ell \log |D|)$.*

5 Bijunctive Constraints

Bijunctive clauses and formulas present another frequently studied subclass of propositional formulas. We investigate in this section the description problem for a generalization of bijunctive formulas to ordered finite domains, namely for sets of vectors closed under the median operation.

Problem: DESCRIPTION[BIJUNCTIVE]
Input: A finite set of vectors $M \subseteq D^\ell$, closed under median, over a finite totally ordered domain D.
Output: A bijunctive constraint φ over D such that $\text{Sol}(\varphi) = M$.

The general construction in Section 3 does not guarantee that the final constraint is bijunctive whenever the set M is closed under median. Therefore we add a post-processing step that transforms the constraint into a bijunctive one. Let $\varphi(M)$ be the constraint produced by the method of Section 3 and let c be a clause from $\varphi(M)$. We construct a bijunctive restriction $b(\varphi)$ by removing appropriate literals from φ such that no more than two literals remain in each clause. Since φ is a conjunctive normal form, any model of $b(\varphi)$ is still a model of φ. The converse does not hold in general. However, if $\text{Sol}(\varphi)$ is closed under median, the method presented below preserves the models, i.e., every model of φ remains a model of $b(\varphi)$. In the proof we need a simple lemma.

Lemma 5. *The model* $\text{med}(m_1, m_2, m_3)$ *satisfies a literal l if and only if at least two of the models m_1, m_2, and m_3 satisfy l.*

We say that a literal l is *essential* for a clause c if there is a model $m \in M$ that satisfies l, but no other literal in c; we also say that m is a *justification* for l. Obviously, we may remove non-essential literals from c without losing models. It remains to show that no clause from φ contains more than two essential literals.

To derive a contradiction, suppose that c is a clause from φ containing at least three essential literals, say l_1, l_2, and l_3. Let m_1, m_2, and m_3 be their justifications, i.e., for each i we have $m_i \models l_i$ and m_i does not satisfy any other literal in c. According to Lemma 5, in this case the model $\text{med}(m_1, m_2, m_3)$ satisfies no literal at all. Hence $\text{med}(m_1, m_2, m_3)$ satisfies neither c nor φ, which contradicts the assumption that $\text{Sol}(\varphi)$ is closed under median.

The preceding discussion suggests applying the following algorithm to every clause c of φ. For every literal l in $c = c' \vee l$, check whether the remaining clause c' is still satisfied by all models in M. If yes, the literal is not essential and can be removed. Otherwise it is one of the two literals in the final bijunctive clause $b(c)$.

Proposition 6. *Given a finite set of vectors $M \subseteq D^\ell$ over a finite totally ordered domain D, closed under median, and a constraint φ in conjunctive normal form such that $M = \text{Sol}(\varphi)$, an equivalent bijunctive constraint $b(M)$ with the same number of clauses can be computed in time $O(d\,|M|\,\ell \log |D|)$ and space $O(|M|\,\ell \log |D|)$, where d is the number of clauses in φ.*

Theorem 7. *For each set of vectors $M \subseteq D^\ell$ over a finite ordered domain D that is closed under median, there exists a bijunctive constraint φ such that $M = \text{Sol}(\varphi)$. Its length is $O(|M|\,\ell(\log \ell + \log |D|))$ and it contains at most $3\,|M|\,\ell$ clauses. The algorithm constructing φ runs in time $O(|M|^2\,\ell^2 \log |D|)$ and space $O(|M|\,\ell \log |D|)$.*

Proposition 8. *A set of vectors M over a finite totally ordered domain is closed under median if and only if there exists a bijunctive constraint φ satisfying the identity $M = \mathrm{Sol}(\varphi)$.*

We wish to point out that reducing the length of clauses also allows us to compute a *Horn* constraint whenever the set M is closed under conjunction. In this case a *minimal* clause in a constraint φ with $M = \mathrm{Sol}(\varphi)$ is always Horn.

Contrary to the Horn case, the simplest known algorithm for the bijunctive description problem does not seem to lift well from the Boolean to the finite domain. Dechter and Pearl [7] showed that in the Boolean case this problem can be solved in time $O(|M|\,\ell^2)$, which is better than our result even when ignoring the unavoidable factor $\log|D|$. Their algorithm generates first all the $O(\ell^2)$ bijunctive clauses built from the variables of the formula, followed by an elimination of those falsified by a vector from M, where the bijunctive formula is the conjunction of the retained clauses. However, there are $O(\ell^2\,|D|^2)$ bijunctive clauses for a finite domain D yielding an algorithm with time complexity $O(|M|\,\ell^2\,|D|^2)$, which is exponential in the size $O(\log|D|)$ of the domain elements.

Another idea, not applicable efficiently in the finite domain case, is that of projecting M onto each pair of variables, then computing a bijunctive constraint for each projection. This requires time $O(|M|\,\ell^2)$ in the Boolean case, since we only need to compute a CNF for each projection. A CNF for a projection is always bijunctive, thus only the general, efficient algorithm of Theorem 1 has to be used. However, in the finite domain case, computing a constraint with the algorithm of Theorem 1 does not necessarily yield a bijunctive one. Each clause can contain up to four literals, a positive and a negative one for each variable. Thus we need to use an algorithm for computing a *bijunctive* CNF, like that of Theorem 7, yielding an overall time complexity of $O(|M|^2\,\ell^2\,\log|D|)$.

6   Affine Constraints

Recall that a set of vectors over D is *affine* if it is a Cartesian product of affine spaces (one for each prime factor of n), i.e., of vector spaces translated by some vector. For sake of completeness this section summarizes some results from the theory of finite fields relevant for the analysis of affine constraints. An interested reader can find more information e.g. in the monograph [18].

The main question in the case of affine constraints and affine sets is whether the cardinality n of the domain is prime or not. According to this, we distinguish three cases: (1) n is prime, which will be studied in full detail; (2) n is a product of prime factors, which will reduced by means of the Chinese Remainder Theorem to the previous case; and (3) n is a prime power, which will be reduced by means of Hensel lifting to the first case.

6.1 n Is Prime

If n is prime then the working domain D is the finite field $\mathbb{Z}_n$, sometimes denoted also $\mathbb{F}_n$ or $\mathbb{Z}/n\mathbb{Z}$. Recall first some necessary results from linear algebra. Let

$\varphi(x)$ be an affine constraint equivalent to the affine system $Ax = b$ over the finite field $\mathbb{Z}_n$, where A is a $p \times q$ matrix over $\mathbb{Z}_n$. If the system $Ax = b$ is consistent, i.e., it does not imply an equation $a_1 = a_2$ for two different values $a_1 \neq a_2$ from $\mathbb{Z}_n$, and of full row rank, then it has n^{q-p} solutions. The solutions of the system $Ax = b$ form an affine space, i.e., a vector space translated by a vector. The set of solutions M of the system $Ax = b$ can be written as a direct sum of the solutions M^* of the homogeneous system $Ax = 0$ and a particular solution m of the system $Ax = b$. The set of solutions M^* form a vector space and the particular solution m is the translating vector.

This section investigates the description problem for sets of vectors representing an affine space.

Problem: DESCRIPTION[AFFINE], n PRIME
Input: A finite set of vectors $M \subseteq D^\ell$, representing an affine space over a finite totally ordered domain D.
Output: An affine constraint φ over D such that $\mathrm{Sol}(\varphi) = M$.

The first point to check in an affine description problem is whether the cardinality of a set of vectors M over $\mathbb{Z}_n$ is a power of n, otherwise M cannot be an affine space. Then we choose an arbitrary vector $m^* \in M$ and form the set $M^* = \{m - m^* \mid m \in M\}$. It is clear that M is an affine space if and only if M^* is a vector space. If the cardinality of M^* is equal to n^q for some q, then there must exist a homogeneous linear system $(I\ B)(u\ v) = 0$ over $\mathbb{Z}_n$, where I is an $(\ell - q) \times (\ell - q)$ identity matrix and B is a $(\ell - q) \times q$ matrix of full row rank over $\mathbb{Z}_n$. We will concentrate on the construction of the matrix B.

The i-th row of the system $(I\ B)(u\ v) = 0$ is $u_i + b_i^1 v_1 + \cdots + b_i^q v_q = 0$, or $u_i + b_i v = 0$, where b_i is the i-th row of the matrix B. Let m be a vector from the set M^*. We substitute $m[i]$ for u_i and $m[\ell - q + j]$ for v_j for each $j = 1, \dots, q$. This implies the equation $m[i] + b_i^1 m[\ell - q + 1] + \cdots + b_i^q m[\ell] = 0$, or $(e_i\ b_i)m = 0$, for each $m \in M^*$, where e_i is the corresponding unit vector. This means that we construct the homogeneous system $S_i \colon M^*(e_i\ b_i) = 0$, where M^* is the matrix whose rows are the vectors of the set M^*. The system S_i is an inhomogeneous system over $\mathbb{Z}_n$ with q variables b_i, since the dot product of the unit vector e_i with each vector $m \in M$ produces the constant $m[i]$. If M^* is a vector space, then the system S_i has exactly one solution, which constitutes the i-th row of the matrix B. This is because the vector space M^* of cardinality n^q has the dimension q, i.e., each basis of M^* contains exactly q linearly independent vectors. If S_i has no solution, then M^* is not a vector space. The system S_i cannot have more than one solution, which follows from the cardinality of the set M^* and the construction of the system S_i.

Once all the systems S_i have been solved, we have determined the coefficients of the matrix B in the homogeneous system $(I\ B)(u\ v) = 0$. To determine the inhomogeneous system $(I\ B)(u\ v) = b$ that describes the original set of vectors M, we substitute the vector m^* for the variables $(u\ v)$ and derive the values of the vector b. This implies the following result.

Theorem 9. *If $M \subseteq \mathbb{Z}_n^\ell$ is a set of vectors representing an affine space then there exists an affine system $Ax = b$ with $\ell - \log|M| / \log|D|$ rows over $\mathbb{Z}_n$,* *such that* $\text{Sol}(Ax = b) = M$, *that can be computed in time* $O(|M|^{mx} (\ell \log |D| + \log |M|) \log |D|)$ *and space* $O((|M| + \ell)\ell \log |D| - \ell \log |M|)$, *where* mx *is the exponent in the asymptotic complexity for matrix multiplication.*

Following the preceding discussion, there exists an easy way to determine whether a set of vectors M over $\mathbb{Z}_n$ is an affine space. Since we work in a ring, we can also use the subtraction operation over $\mathbb{Z}_n$, making the characterization more compact.

Proposition 10. *A set of vectors* $M \subseteq \mathbb{Z}_n^\ell$ *is an affine space if and only if it is closed under the affine operation* $\text{aff}(x, y, z) = x - y + z \pmod{n}$, *i.e., for any choice of three not necessarily distinct vectors* $m, m', m'' \in M$ *the vector* $m - m' + m''$ *also belongs to* M.

6.2 n Is Not Prime

If n is not prime, then it can be written as a product of relatively prime factors $n = n_1 \cdots n_q$. We assume that this factorization is *a priori* known, since it does not make sense to factorize n every time we need to solve an affine problem over $\mathbb{Z}_n$. The affine description problem is then formulated as follows.

Problem: DESCRIPTION[AFFINE], n COMPOSED
Input: A finite set of vectors $M \subseteq D^\ell$, representing a Cartesian product of affine spaces over a finite totally ordered domain D.
Output: An affine constraint φ over D such that $\text{Sol}(\varphi) = M$.

A lot of results can be reused from the previous case, when n is prime, but we need to use the Chinese Remainder Theorem to solve the affine systems.

Theorem 11 (Chinese Remainder Theorem). *Let* $n_1, \ldots, n_q$ *be pairwise relatively prime and* $n = n_1 \cdots n_q$. *Consider the mapping* $f: a \leftrightarrow (a_1, \ldots, a_q)$, *where* $a \in \mathbb{Z}_n$, $a_i \in \mathbb{Z}_{n_i}$, *and* $a_i = a \mod n_i$, *for each* $i \in \{1, \ldots, q\}$. *Then the mapping* f *is a bijection between* $\mathbb{Z}_n$ *and the Cartesian product* $\mathbb{Z}_{n_1} \times \cdots \times \mathbb{Z}_{n_q}$.

The Chinese Remainder Theorem says that operations performed on the elements of $\mathbb{Z}_n$ can be equivalently performed on the corresponding q-tuples by performing the operations independently in each coordinate position in the appropriate system. Instead of working with an affine system of equations $Ax = b$ over $\mathbb{Z}_n$, we work with q affine systems $Ax = b$ over $\mathbb{Z}_{n_k}$. The factorization $n = n_1 \cdots n_q$ implies that $q \leq \log n$, since the inequality $n_k \geq 2$ holds for each k. This means that we have only a logarithmic number of subsystems to consider. Since the length of n is $\log n$, this means that we have only a polynomial number (with respect to the length of input) of systems $Ax = b$ over $\mathbb{Z}_{n_i}$ to consider.

The constraint description algorithm uses the Chinese Remainder Theorem in the other direction. In practice, we compute first the coefficients of the matrix B_k, as in Section 6.1, modulo n_k for each $k = 1, \ldots, q$. By means of the mapping f in the Chinese Remainder Theorem we determine from $B_1, \ldots, B_q$ the matrix B of the homogeneous system $(I \; B)(u \; v) = \mathbf{0}$. The vector b of the final inhomogeneous system $(I\ B)(u\ v) = b$ is determined as in Section 6.1. An application of the Chinese Remainder Theorem requires $O(\text{im}(\log |D|) \log \log |D|)$ operations [18, Theorem 10.25], where $\text{im}(r)$ denotes the time for multiplication of two integers of length r.

Theorem 12. *If $M \subseteq \mathbb{Z}_n^\ell$ is a set of vectors representing a Cartesian product of affine spaces then there exists a system $Ax = b$ over $\mathbb{Z}_n$, where $\text{Sol}(Ax = b) = M$.*

A factor n_i in the previous factorization $n = n_1 \cdots n_q$ need not be prime, it can also be a prime power. Therefore we need to consider a third case.

6.3 n Is a Prime Power

If $n = p^q$ for some prime p and an integer exponent $q > 1$, then we use Hensel lifting to solve our problem. Since the Hensel lifting is by far beyond the scope of this paper, we only state the main result without any particular presentation of the lifting method. An interesting reader is strongly encouraged to consult the part on Hensel lifting in the monograph [18]. We assume that the power p^q is known to us in binary notation, therefore its length is $O(q \log p)$. We compute first the system $(I\ B)(u\ v) = b \pmod{p}$, as in Section 6.1, followed by a Hensel lifting to the system $A'x = b' \pmod{p^{2^{\lceil \log_2 k \rceil}}}$. Each coefficient $B(i,j)$ and b_i of the matrix B and the vector b, respectively, is lifted separately, similarly to the application of the Chinese Remainder Theorem in Section 6.2. Finally, we cut the result down to modulo $m = p^q$. Usually, Hensel lifting is presented for polynomials over a ring R. To adapt our approach to the usual presentation, we use the same trick as that applied for cyclic codes: a number $a = a_d a_{d-1} \cdots a_1 a_0$ is interpreted as a polynomial $a_d X^d + a_{d-1} X^{d-1} + \cdots + a_1 X + a_0$, where X is a formal variable. The description problem for affine constraints is formulated as follows.

Problem: DESCRIPTION[AFFINE], n PRIME POWER
Input: A finite set of vectors $M \subseteq D^\ell$, representing a power of an affine space over a finite totally ordered domain D.
Output: An affine constraint φ over D such that $\text{Sol}(\varphi) = M$.

Each Hensel step requires $O(\text{im}(\log |D|) \text{im}(\log p))$ operations, where $\text{im}(r)$ is the time for multiplying two integers of length r. There are $O(\log q)$ iterations of the Hensel step. Hence, there exists a polynomial-time algorithm to compute the system $A'x = b' \pmod{p^q}$ from the system $(I\ B)(u\ v) = b \pmod{p}$.

Theorem 13. *If $M \subseteq \mathbb{Z}_n^\ell$ is a set of vectors representing a power of an affine space then there exists an affine system $Ax = b$ over $\mathbb{Z}_n$, where $\text{Sol}(Ax = b) = M$.*

7   Changing the Literals

If we change the underlying notion of literals, using $x = d$ and $x \neq d$ as basic building blocks, the situation changes drastically. Former positive literal $x \geq d$ becomes a shorthand for the disjunction $(x = d) \vee (x = d + 1) \vee \cdots \vee (x =$

$n-1$), whereas the former negative literal $x \leq d$ now represents the disjunction $(x = 0) \vee (x = 1) \vee \cdots \vee (x = d)$. Even if we compress literals containing the same variable into a bit vector, the new representation still needs n bits, i.e., its size is $O(n)$. Compared to the former literals of size $O(\log n)$, this amounts to an exponential blow-up. As an immediate consequence the algorithms given in the preceding sections become exponential, since we have to replace literals like $x_i < m_k[i]$, $x_i > m_k[i]$, and $x_i < m_{k+1}[i]$ by disjunctions of equalities.

The satisfiability problem for constraints in CNF over finite totally ordered domains with basic operators $\leq$ and $\geq$ is defined similarly to Boolean satisfiability. The complexity of these problems was studied for fixed domain cardinalities, from the standpoint of many-valued logics, by Hähnle et $al.$ [3,10]. The NP-completeness proof for Boolean satisfiability generalizes uniformly to finite ordered domains. Hähnle et $al.$ [3,10] proved that the satisfiability problems restricted to Horn, dual Horn, and bijunctive constraints, are decidable in polynomial time for a fixed domain cardinality. The tractability of the affine restriction is a consequence of Section 6.

The satisfiability of constraints in conjunctive normal form is also affected when switching to $=$ and $\neq$ as basic operators. While the satisfiability problem for general constraints remains NP-complete, the restrictions to Horn, dual Horn, and bijunctive constraints change from polynomially solvable to NP-complete for $|D| \geq 3$. This can be shown by encoding for example the graph problem of k-COLORING [1,5]. When we use the Horn and bijunctive clause $(u \neq d \vee v \neq d)$, we can express by $C(u,v) = (u \neq 0 \vee v \neq 0) \wedge \cdots \wedge (u \neq k-1 \vee v \neq k-1)$ that the adjacent vertices of the edge (u,v) are "colored" by different "colors". On the other hand, Beckert et $al.$ [2] proved that bijunctive constraints restricted to positive literals can be solved in linear time.

8 Concluding Remarks

The studied constraint description problems constitute a generalization of the Boolean structure identification problems, studied by Dechter and Pearl [7], with more efficient algorithms as a byproduct. Our paper presents a complement to the work of Hähnle et $al.$ [10] on the complexity of the satisfiability problems in many-valued logics. It also completes the study of tractable constraints [5,14,15] by Jeavons and his group.

We have constructed efficient polynomial-time algorithms for constraint description problems over a finite totally ordered domain, where the produced constraint is in conjunctive normal form. If the original set of vectors is closed under the operation of conjunction, disjunction, or median, we have presented specific algorithms that produce a Horn, a dual Horn, or a bijunctive constraint, respectively. In all three cases, the constraint contains at most $3\,|M|\,\ell$ clauses. It is interesting to note that the produced algorithms are compatible, with respect to asymptotic complexity, with known algorithms for the Boolean case presented in [11,19]. This means that the restriction of the new algorithms presented in our paper to domains D with cardinality $|D| = 2$ produces the aforementioned algorithms for the Boolean case. However, the presented algorithms are not just straightforward extensions of the previous ones for the Boolean case, but they required the development of new methods.

It would be interesting to know if more efficient algorithms exist or whether our algorithms are asymptotically optimal. Certainly a more involved lower bound analysis is necessary to answer this open question. A possible extension of our work would be a generalization of our algorithms to partially ordered domains and to domains with a different structure, like lattices.

References

1. C. Ansótegui and F. Manyà. New logical and complexity results for signed-SAT. In *Proc. 33rd ISMVL 2003, Tokyo (Japan)*, pages 181–187. 2003.
2. B. Beckert, R. Hähnle, and F. Manyà. The 2-SAT problem of regular signed CNF formulas. In *Proc. 30th ISMVL, Portland (OR, USA)*, pages 331–336. 2000.
3. R. Béjar, R. Hähnle, and F. Manyà. A modular reduction of regular logic to classical logic. In *Proc. 31st ISMVL, Warsaw (Poland)*, pages 221–226. 2001.
4. A. A. Bulatov. A dichotomy theorem for constraints on a three-element set. In *Proc. 43rd FOCS, Vancouver (BC, Canada)*, pages 649–658, 2002.
5. M. C. Cooper, D. A. Cohen, and P. Jeavons. Characterising tractable constraints. *Artificial Intelligence*, 65(2):347–361, 1994.
6. N. Creignou, S. Khanna, and M. Sudan. *Complexity Classifications of Boolean Constraint Satisfaction Problems*. SIAM Monographs on Discrete Mathematics and Applications. SIAM, Philadelphia (PA), 2001.
7. R. Dechter and J. Pearl. Structure identification in relational data. *Artificial Intelligence*, 58(1-3):237–270, 1992.
8. T. Feder and M. Y. Vardi. The computational structure of monotone monadic SNP and constraint satisfaction: a study through Datalog and group theory. *SIAM Journal on Computing*, 28(1):57–104, 1998.
9. R. Hähnle. Exploiting data dependencies in many-valued logics. *Journal of Applied Non-Classical Logics*, 6(1), 1996.
10. R. Hähnle. Complexity of many-valued logics. In *Proc. 31st ISMVL, Warsaw (Poland)*, pages 137–148. 2001.
11. J.-J. Hébrard and B. Zanuttini. An efficient algorithm for Horn description. *Information Processing Letters*, 88(4):177–182, 2003.
12. P. Hell and J. Nešetřil. On the complexity of H-coloring. *Journal of Combinatorial Theory, Series B*, 48:92–110, 1990.
13. P. Jeavons. On the algebraic structure of combinatorial problems. *Theoretical Computer Science*, 200(1-2):185–204, 1998.
14. P. Jeavons, D. Cohen, and M. Gyssens. Closure properties of constraints. *Journal of the Association for Computing Machinery*, 44(4):527–548, 1997.
15. P. Jeavons and M. C. Cooper. Tractable constraints on ordered domains. *Artificial Intelligence*, 79(2):327–339, 1995.
16. P. G. Kolaitis and M. Y. Vardi. Conjunctive-query containment and constraint satisfaction. *Journal of Computer and System Science*, 61(2):302–332, 2000.
17. T. J. Schaefer. The complexity of satisfiability problems. In *Proc. 10th STOC, San Diego (CA, USA)*, pages 216–226, 1978.
18. J. von zur Gathen and J. Gerhard. *Modern Computer Algebra*. Cambridge University Press, 1999.
19. B. Zanuttini and J.-J. Hébrard. A unified framework for structure identification. *Information Processing Letters*, 81(6):335–339, 2002.

PDL with Negation of Atomic Programs

Carsten Lutz[1] and Dirk Walther[2]

[1] Inst. for Theoretical Computer Science
TU Dresden, Germany
lutz@tcs.inf.tu-dresden.de
[2] Dept. of Computer Science
University of Liverpool, UK
dwalther@csc.liv.ac.uk

Abstract. Propositional dynamic logic (PDL) is one of the most successful variants of modal logic. To make it even more useful for applications, many extensions of PDL have been considered in the literature. A very natural and useful such extension is with negation of programs. Unfortunately, as long-known, reasoning with the resulting logic is undecidable. In this paper, we consider the extension of PDL with negation of atomic programs, only. We argue that this logic is still useful, e.g. in the context of description logics, and prove that satisfiability is decidable and EXPTIME-complete using an approach based on Büchi tree automata.

1 Introduction

Propositional dynamic logic (PDL) is a variant of propositional modal logic that has been developed in the late seventies as a tool for reasoning about programs [1, 2,3,4,5]. Since then, PDL was used rather successfully in a large number of application areas such as reasoning about knowledge [6], reasoning about actions [7, 8], description logics [9], and others. Starting almost with its invention around 1979 [3], many extensions of PDL have been proposed with the goal to enhance the expressive power and make PDL even more applicable; see e.g. [10,4,5]. Some of these extensions are tailored toward specific application areas, such as the *halt* predicate that allows to state termination in the context of reasoning about programs [11]. The majority of proposed extensions, however, is of a general nature and has been employed in many different application areas—for instance, the extension of PDL with the widely applied converse operator [12].

Among the general purpose extensions of PDL, two of the most obvious ones are the addition of program intersection "$\cap$" and of program negation "$\neg$" [13, 4,5]. Since PDL already provides for program union "$\cup$", the latter is more general than the former: $\alpha \cap \beta$ can simply be expressed as $\neg(\neg\alpha \cup \neg\beta)$. The main obstacle for using these two extensions in practical applications is that they are problematic w.r.t. their computational properties: first, adding intersection destroys many of the nice model-theoretic properties of PDL. The only known algorithm for reasoning in the resulting logic PDL$^\cap$ is the quite intricate one given in [13]. Up to now, it is unknown whether the provided 2-EXPTIME upper bound is tight—in contrast to EXPTIME-complete reasoning in PDL. Second,

D. Basin and M. Rusinowitch (Eds.): IJCAR 2004, LNAI 3097, pp. 259–273, 2004.
© Springer-Verlag Berlin Heidelberg 2004

the situation with PDL extended with negation ($\text{PDL}^{\neg}$) is even worse: it was observed quite early in 1984 that reasoning in $\text{PDL}^{\neg}$ is undecidable [4].

This undecidability was often regretted [4,10,14], in particular since reasoning in $\text{PDL}^{\neg}$ would be quite interesting for a number of application areas. To illustrate the usefulness of this logic, let us give three examples of its expressive power: first, it was already noted that negation can be employed to express intersection. Intersection, in turn, is very useful for reasoning about programs since it allows to capture the parallel execution of programs. Second, program negation allows to express the universal modality $\Box_U \varphi$ by writing $[a]\varphi \wedge [\neg a]\varphi$, with a an arbitrary atomic program. The universal modality is a very useful extension of modal logics that comes handy in many applications; see e.g. [15]. Third, program negation can be used to express the window operator $\boxminus_a$ [16, 17,18], whose semantics is as follows: $\boxminus_a \varphi$ holds at a world w iff φ holding at a world w' implies that w' is a-accessible from w. In $\text{PDL}^{\neg}$, we can thus just write $[\neg a]\neg\varphi$ instead of $\boxminus_a \varphi$. The window operator can be viewed as expressing sufficiency in contrast to the standard box operator of modal logic, which expresses necessity. Moreover, the window operator has important applications, e.g. in description logics [19].

Due to the usefulness of program negation, it is natural to attempt the identification of fragments of $\text{PDL}^{\neg}$ that still capture some of the desirable properties of program negation, but are well-behaved in a computational sense. One candidate for such a fragment is $\text{PDL}^{\cap}$. As has already been noted, this fragment is indeed decidable, but has a quite intricate model theory. The purpose of this paper is to explore another interesting option: $\text{PDL}^{(\neg)}$, the fragment of $\text{PDL}^{\neg}$ that allows the application of program negation to *atomic* programs, only. Indeed, we show that reasoning in $\text{PDL}^{(\neg)}$ is decidable, and ExpTime-complete—thus not harder than reasoning in PDL itself. Moreover, $\text{PDL}^{(\neg)}$ has a simpler model theory than $\text{PDL}^{\cap}$: we are able to use a decision procedure that is an extension of the standard automata-based decision procedure for PDL [20], and of the standard automata-based decision procedure for Boolean modal logic [21]. Finally, we claim that $\text{PDL}^{(\neg)}$ is still useful for applications: while intersection cannot be expressed any more, the universal modality and the window operator are still available.

To give some more concrete examples of the practicability of $\text{PDL}^{(\neg)}$, let us take a description logic perspective. Description logics are a family of logics that originated in artificial intelligence as a tool for the representation of conceptual knowledge [22]. It is well-known that many description logics (DLs) are notational variants of modal logics [23,9]. In particular, the description logic $\mathcal{ALC}_{\text{reg}}$, which extends the basic DL $\mathcal{ALC}$ with regular expressions on roles, corresponds to PDL [9]. More precisely, DL concepts can be understood as PDL formulas, and DL roles as PDL programs. Thus, the extension $\mathcal{ALC}_{\text{reg}}^{(\neg)}$ of $\mathcal{ALC}_{\text{reg}}$ with negation of atomic (!) roles is a notational variant of $\text{PDL}^{(\neg)}$. We give two examples of knowledge representation with $\mathcal{ALC}_{\text{reg}}^{(\neg)}$. These examples, which use DL syntax rather than PDL syntax, illustrate that the combination of regular expressions on roles and of atomic negation of roles is a very useful one.

1. Some private universities prefer to admit students whose ancestors donated money to the university. Using $\mathcal{ALC}_{reg}$, the class of all applicants having a donating ancestor can be described with the concept $\exists$parent$^+$.Donator. To describe the set of preferred students, we can now combine this concept with the window operator: the $\mathcal{ALC}_{reg}^{(\neg)}$-concept

$$\text{UniversityX} \rightarrow \forall\text{prefer.Applicant} \sqcap \forall\neg\text{prefer.}\neg(\exists\text{parent}^+.\text{Donator})$$

states that, in the case of University X, only people who actually applied are preferred, and all applicants with donating ancestors are preferred.

2. Suppose that we want to use $\mathcal{ALC}_{reg}^{(\neg)}$ to talk about trust and mistrust among negotiating parties. Also assume that we have a very strong notion of trust, namely that it is transitive: if I trust x, and x trusts y, then I trust y as well. An analogous assumption for mistrust should clearly not be made. Then, we can model mistrust by using an atomic role mistrust, and trust by using $(\neg\text{mistrust})^*$ and say, e.g., that I trust some politicians and never mistrust a family member :

$$\exists(\neg\text{mistrust})^*.\text{Politician} \sqcap \forall\text{mistrust.}\neg\text{Familymember.}$$

Note that reversing the roles of trust and mistrust does not work: first, to achieve transitivity of trust, we'd have to introduce an atomic direct-trust relation. And second, we could then only speak about the negation of direct-trust, but not about the negation of direct-trust*, which corresponds to mistrust.

2   PDL with Negation

In this section, we introduce propositional dynamic logic (PDL) with negation of programs. We start with defining full PDL$^\neg$, i.e. PDL extended with negation of (possibly complex) programs. Then, the logics PDL and PDL$^{(\neg)}$, are defined as fragments of PDL$^\neg$.

Definition 1 (PDL$^\neg$ Syntax). *Let Φ_0 and Π_0 be countably infinite and disjoint sets of propositional letters and atomic programs, respectively. Then the set $\Pi^\neg$ of PDL$^\neg$-programs and the set $\Phi^\neg$ of PDL$^\neg$-formulas are defined by simultaneous induction, i.e., they are the smallest sets such that:*

- *$\Phi_0 \subseteq \Phi^\neg$;*
- *$\Pi_0 \subseteq \Pi^\neg$;*
- *if $\varphi, \psi \in \Phi^\neg$, then $\{\neg\varphi, \varphi \wedge \psi, \varphi \vee \psi\} \subseteq \Phi^\neg$;*
- *if $\pi_1, \pi_2 \in \Pi^\neg$, then $\{\neg\pi_1, \pi_1 \cup \pi_2, \pi_1; \pi_2, \pi_1^*\} \subseteq \Pi^\neg$;*
- *if $\pi \in \Pi^\neg$, and $\varphi \in \Phi^\neg$, then $\{\langle\pi\rangle\varphi, [\pi]\varphi\} \subseteq \Phi^\neg$;*
- *if $\varphi \in \Phi^\neg$, then $\varphi? \in \Pi^\neg$*

We use $\top$ as abbreviation for an arbitrary propositional tautology, and $\bot$ as abbreviation for $\neg\top$. Moreover, for $\pi, \pi' \in \Pi^\neg$ we use $\pi \cap \pi'$ as abbreviation for $\neg(\neg\pi \cup \neg\pi')$.

A formula $\varphi \in \Phi^\neg$ is called a PDL$^{(\neg)}$-formula (PDL-formula) if, in φ, negation occurs only in front of atomic programs and formulas (only in front of formulas).

Throughout this paper, the operator $\langle \pi \rangle$ is called the diamond operator, $[\pi]$ is called the box operator, and programs of the form $\psi?$ are called *tests*. Let us note how formulas of PDL$^\neg$ can be converted into concepts of the description logic $\mathcal{ALC}^{(\neg)}_{\text{reg}}$ mentioned in the introduction: simply replace $\wedge$, $\vee$, $\langle \pi \rangle \psi$, and $[\pi]\psi$ with $\sqcap$, $\sqcup$, $\exists \pi.\psi$, and $\forall \pi.\psi$, respectively.

Definition 2 (PDL$^\neg$ Semantics). *Let* $\mathcal{M} = (W, \mathcal{R}, V)$ *be a* Kripke structure *where* W *is the* set of worlds, $\mathcal{R}$ *is a family of* accessibility relations *for atomic programs* $\{R_\pi \subseteq W^2 \mid \pi \in \Pi_0\}$, *and* $V : \Phi_0 \to 2^W$ *is a* valuation function. *In the following, we define accessibility relations for compound programs and the satisfaction relation* $\models$ *by simultaneous induction, where* $\cdot^*$ *denotes the reflexive-transitive closure:*

$$
\begin{aligned}
R_{\varphi?} &:= \{(u, u) \in W^2 \mid \mathcal{M}, u \models \varphi\} \\
R_{\neg\pi} &:= W^2 \backslash R_\pi \\
R_{\pi_1 \cup \pi_2} &:= R_{\pi_1} \cup R_{\pi_2} \\
R_{\pi_1 ; \pi_2} &:= R_{\pi_1} \circ R_{\pi_2} \\
R_{\pi^*} &:= (R_\pi)^* \\
\mathcal{M}, u \models p \quad &\textit{iff} \quad u \in V(p) \text{ for any } p \in \Phi \\
\mathcal{M}, u \models \neg\varphi \quad &\textit{iff} \quad \mathcal{M}, u \not\models \varphi \\
\mathcal{M}, u \models \varphi_1 \vee \varphi_2 \quad &\textit{iff} \quad \mathcal{M}, u \models \varphi_1 \text{ or } \mathcal{M}, u \models \varphi_2 \\
\mathcal{M}, u \models \varphi_1 \wedge \varphi_2 \quad &\textit{iff} \quad \mathcal{M}, u \models \varphi_1 \text{ and } \mathcal{M}, u \models \varphi_2 \\
\mathcal{M}, u \models \langle\pi\rangle\varphi \quad &\textit{iff} \quad \text{there is a } v \in W \text{ with } (u, v) \in R_\pi \text{ and } \mathcal{M}, v \models \varphi \\
\mathcal{M}, u \models [\pi]\varphi \quad &\textit{iff} \quad \text{for all } v \in W, (u, v) \in R_\pi \text{ implies } \mathcal{M}, v \models \varphi
\end{aligned}
$$

If $\mathcal{M}, u \models \varphi$ *for some formula* $\varphi \in \Phi^\neg$ *and world* $u \in W$, *then* φ *is* true at u in $\mathcal{M}$, *and* $\mathcal{M}$ *is called* model *of* φ. *A formula is* satisfiable *if it has a model.*

It is well-known that satisfiability of PDL$^\neg$-formulas is undecidable [4]. Since this can be established in a very simple way, we give a proof for illustrative purposes.

The proof is by reduction of the undecidable word-problem for finitely presented semi-groups [24]: given a set of word identities $\{u_1 = v_1, \ldots, u_k = v_k\}$, the task is to decide whether they imply another word identity $u = v$. To reduce this problem to PDL$^\neg$-satisfiability, we need to introduce the universal modality $\Box_U \varphi$, which has the following semantics:

$$
\mathcal{M}, u \models \Box_U \varphi \quad \textit{iff} \quad \mathcal{M}, v \models \varphi \text{ for all } v \in W.
$$

Clearly, in PDL$^\neg$ we can replace $\Box_U \varphi$ with the equivalent $[a]\varphi \wedge [\neg a]\varphi$, where $a \in \Pi_0$ is an arbitrary atomic program. Using the universal modality, the reduction is now easy: we assume that, for every generator of the semi-group, there is an atomic program of the same name, and then note that $\{u_1 = v_1, \ldots, u_k = v_k\}$ implies $u = v$ if and only if the following formula is unsatisfiable:

$$
\Big(\langle u \cap \neg v \rangle \top \vee \langle \neg u \cap v \rangle \top \Big) \wedge \Box_U \Big(\bigwedge_{i=1..k} [u_i \cap \neg v_i]\bot \wedge [v_i \cap \neg u_i]\bot \Big).
$$

Here, we assume that the symbols of the words u_i and v_i (and of u and v) are separated by program composition ";".

Since PDL$^\neg$ is a very useful logic for a large number of purposes, this unde-cidability result is rather disappointing. As has been argued in the introduction, it is thus a natural idea to search for decidable fragments of PDL$^\neg$ that still extend PDL in a useful way. In the remainder of this paper, we will prove that PDL$^{(\neg)}$ is such a fragment. Note that, in PDL$^{(\neg)}$, we can still define the uni-versal modality as described above. Also note that we can use negated atomic programs nested inside other program operators.

3   An Automata-Based Variant of PDL$^{(\neg)}$

Similar to some related results in [20], our decidability proof is based on Büchi-automata on infinite trees. It has turned out that, for such proofs, it is rather convenient to use variants of PDL in which complex programs are described by means of automata on finite words, rather than by regular expressions. Therefore, in this section we define a corresponding variant APDL$^{(\neg)}$ of PDL$^{(\neg)}$.

Definition 3 (Finite automata). *A (nondeterministic) finite automaton (NFA) $\mathcal{A}$ is a quintuple $(Q, \Sigma, q_0, \Delta, F)$ where*

- Q *is a finite set of* states,
- Σ *is a finite* alphabet,
- q_0 *is an* initial state,
- $\Delta : Q \times \Sigma \to 2^Q$ *is a (partial)* transition function, *and*
- $F \subseteq Q$ *is the set of* accepting states.

The function Δ can be inductively extended to a function from $Q \times \Sigma^*$ to 2^Q in a natural way:

- $\Delta(q, \varepsilon) := \{q\}$, *where ε is the* empty word;
- $\Delta(q, wa) := \{q'' \in Q \mid q'' \in \Delta(q', a) \text{ for some } q' \in \Delta(q, w)\}$.

A sequence $p_0, \ldots, p_n \in Q$, $n \geq 0$, is a run of $\mathcal{A}$ on the word $a_1 \cdots a_n \in \Sigma^*$ if $p_0 = q_0$, $p_i \in \Delta(p_{i-1}, a_i)$ for $0 < i \leq n$, and $p_n \in F$. A word $w \in \Sigma^*$ is accepted by $\mathcal{A}$ if there exists a run of $\mathcal{A}$ on w. The language accepted by $\mathcal{A}$ is the set $\mathcal{L}(\mathcal{A}) := \{w \in \Sigma^* \mid w \text{ is accepted by } \mathcal{A}\}$.

To obtain APDL$^{(\neg)}$ from PDL$^{(\neg)}$, we replace complex programs (i.e. regular expressions) inside boxes and diamonds with automata. For the sake of exactness, we give the complete definition.

Definition 4 (APDL$^{(\neg)}$ Syntax). *The set $\Pi_0^{(\neg)}$ of program literals is de-fined as $\{a, \neg a \mid a \in \Pi_0\}$. The sets $A\Pi^{(\neg)}$ of program automata and $A\Phi^{(\neg)}$ of APDL$^{(\neg)}$-formulas are defined by simultaneous induction, i.e., $A\Pi^{(\neg)}$ and $A\Phi^{(\neg)}$ are the smallest sets such that:*

- $\Phi_0 \subseteq A\Phi^{(\neg)}$;
- *if $\varphi, \psi \in A\Phi^{(\neg)}$, then $\{\neg\varphi, \varphi \vee \psi, \varphi \wedge \psi\} \subseteq A\Phi^{(\neg)}$;*
- *if $\alpha \in A\Pi^{(\neg)}$ and $\varphi \in A\Phi^{(\neg)}$, then $\{\langle\alpha\rangle\varphi, [\alpha]\varphi\} \subseteq A\Phi^{(\neg)}$;*

– if α is a finite automaton with alphabet $\Sigma \subseteq \Pi_0^{(\neg)} \cup \{\psi? \mid \psi \in A\Phi^{(\neg)}\}$, then $\alpha \in A\Pi^{(\neg)}$

Note that the alphabet of program automata is composed of atomic programs, of negated atomic programs, and of tests.

Definition 5 (APDL$^{(\neg)}$ Semantics). *Let $\mathcal{M} = (W, \mathcal{R}, V)$ be a Kripke structure as in Definition 2. We inductively define a relation R mapping each program literal, each test, and each program automaton to a binary relation over W. This is done simultaneously with the definition of the satisfaction relation $\models$:*

$$
\begin{aligned}
R(a) &:= R_a \text{ for each } a \in \Pi_0 \\
R(\neg a) &:= W^2 \setminus R_a \text{ for each } a \in \Pi_0 \\
R(\psi?) &:= \{(u,u) \in W^2 \mid \mathcal{M}, u \models \psi\} \\
R(\alpha) &:= \{(u,v) \in W^2 \mid \text{there is a word } w = w_1 \cdots w_m \in \mathcal{L}(\alpha), \\
&\qquad m \geq 0, \text{ and worlds } u_0, \ldots, u_m \in W \text{ such that} \\
&\qquad u = u_0 R(w_1) u_1 R(w_2) \cdots u_{m-1} R(w_m) u_m = v\}
\end{aligned}
$$

$$
\begin{aligned}
\mathcal{M}, u &\models p & \text{iff} \quad & u \in V(p) \text{ for any } p \in \Phi, \\
\mathcal{M}, u &\models \neg\varphi & \text{iff} \quad & \mathcal{M}, u \not\models \varphi, \\
\mathcal{M}, u &\models \varphi_1 \vee \varphi_2 & \text{iff} \quad & \mathcal{M}, u \models \varphi_1 \text{ or } \mathcal{M}, u \models \varphi_2, \\
\mathcal{M}, u &\models \varphi_1 \wedge \varphi_2 & \text{iff} \quad & \mathcal{M}, u \models \varphi_1 \text{ and } \mathcal{M}, u \models \varphi_2, \\
\mathcal{M}, u &\models \langle\alpha\rangle\varphi & \text{iff} \quad & \text{there is a } u' \in W \text{ with } (u, u') \in R(\alpha) \text{ and } \mathcal{M}, u' \models \varphi, \\
\mathcal{M}, u &\models [\alpha]\varphi & \text{iff} \quad & \text{for all } u' \in W, (u, u') \in R(\alpha) \text{ implies } \mathcal{M}, u' \models \varphi.
\end{aligned}
$$

Since every language defined by a regular expression can also be accepted by a finite automaton and vice versa [25], it is straightforward to verify that PDL$^{(\neg)}$ and APDL$^{(\neg)}$ have the same expressive power. Moreover, upper complexity bounds carry over from APDL$^{(\neg)}$ to PDL$^{(\neg)}$ since conversion of regular expressions to finite automata can be done with at most a polynomial blow-up in size (the converse does not hold true).

It is interesting to note that, in many automata-based decision procedures for variants of PDL, a *deterministic* version of APDL is used, i.e. a variant of APDL in which there may be at most one successor for each world and each atomic program [20]. In a second step, satisfiability in the non-deterministic APDL-variant is then reduced to satisfiability in the deterministic one. We cannot take this approach here since we cannot w.l.o.g. assume that both atomic programs *and their negations* are deterministic. Indeed, this would correspond to limiting the size of Kripke structures to only two worlds.

4   Hintikka-Trees

This section provides a core step toward using Büchi-tree automata for deciding the satisfiability of APDL$^{(\neg)}$-formulas. The intuition behind this approach is as follows: to decide the satisfiability of an APDL$^{(\neg)}$-formula φ, we translate it into a Büchi-tree automaton $\mathcal{B}_\varphi$ such that the trees accepted by the automaton correspond in some way to models of the formula φ. To decide satisfiability of

φ, it then remains to perform a simple emptiness-test on the automaton $\mathcal{B}_\varphi$: the accepted language will be non-empty if and only if φ has a model.

In the case of $APDL^{(\neg)}$, one obstacle to this approach is that $APDL^{(\neg)}$ does not enjoy the *tree model property (TMP)*, i.e., there are $APDL^{(\neg)}$-formulas that are satisfiable only in non-tree models. For example, for each $n \in \mathbb{N}$ the following $PDL^{(\neg)}$-formula enforces a cycle of length n:

$$\psi_1^n \wedge \langle a \rangle (\psi_2^n \wedge \langle a \rangle (\cdots (\psi_n^n \wedge [\neg a]\neg\psi_1^n) \cdots)),$$

where, for $1 \le i \le n$, $\psi_i^n = p_1 \wedge \cdots \wedge \neg p_i \wedge \cdots \wedge p_n$ with propositional variables $p_1, \ldots, p_n$. Note that the formula inside the diamond simulates the window operator and in this way closes the cycle. Thus, we have to invest some work to obtain tree-shaped representations of (possibly non-tree) models that can then be accepted by Büchi-automata.

As a preliminary, we assume that all $APDL^{(\neg)}$-formulas are in *negation normal form (NNF)*, i.e. that negation occurs only in front of propositional letters. This assumption can be made w.l.o.g. since each formula can be converted into an equivalent one in NNF by exhaustively eliminating double negation, applying DeMorgan's rules, and exploiting the duality between diamonds and boxes. For the sake of brevity, we introduce the following notational conventions:

- for each $APDL^{(\neg)}$-formula φ, $\dot{\neg}\varphi$ denotes the NNF of $\neg\varphi$;
- for each program literal π, $\overline{\pi}$ denotes $\neg\pi$ if π is an atomic program, and a if $\pi = \neg a$ for some atomic program a;
- for each program automaton α, we use Q_α, Σ_α, q_α, Δ_α, and F_α to denote the components of $\alpha = (Q, \Sigma, q_0, \Delta, F)$;
- for each program automaton α and state $q \in Q_\alpha$, we use α_q to denote the automaton $(Q_\alpha, \Sigma_\alpha, q, \Delta_\alpha, F_\alpha)$, i.e. the automaton obtained from α by using q as the new initial state.

Before we can develop the tree-shaped abstraction of models, we need to fix a *closure*, i.e. a set of formulas $\mathsf{cl}(\varphi)$ relevant for deciding the satisfiability of an input formula φ. This is done analogous to [3,20]. In the following, when we talk of a subformula ψ of a formula φ, we mean that ψ can be obtained from φ by decomposing only formula operators, but not program operators. For example, a is a subformula of $\langle b? \rangle a$, while b is not.

Definition 6 (Closure). *Let φ be a $APDL^{(\neg)}$-formula. The set $\mathsf{cl}(\varphi)$ is the smallest set which is closed under the following conditions:*

    (C1) $\varphi \in \mathsf{cl}(\varphi)$

    (C2) if ψ is a subformula of $\psi' \in \mathsf{cl}(\varphi)$, then $\psi \in \mathsf{cl}(\varphi)$

    (C3) if $\psi \in \mathsf{cl}(\varphi)$, then $\dot{\neg}\psi \in \mathsf{cl}(\varphi)$

    (C4) if $\langle \alpha \rangle \psi \in \mathsf{cl}(\varphi)$, then $\psi' \in \mathsf{cl}(\varphi)$ for all $\psi'? \in \Sigma_\alpha$

    (C5) if $\langle \alpha \rangle \psi \in \mathsf{cl}(\varphi)$, then $\langle \alpha_q \rangle \psi \in \mathsf{cl}(\varphi)$ for all $q \in Q_\alpha$

    (C6) if $[\alpha]\psi \in \mathsf{cl}(\varphi)$, then $\psi' \in \mathsf{cl}(\varphi)$ for all $\psi'? \in \Sigma_\alpha$

    (C7) if $[\alpha]\psi \in \mathsf{cl}(\varphi)$, then $[\alpha_q]\psi \in \mathsf{cl}(\varphi)$ for all $q \in Q_\alpha$

It is standard to verify that the cardinality of $\mathsf{cl}(\varphi)$ is polynomial in the length of φ, see e.g. [5]. We generally assume the diamond formulas (i.e. formulas of the form $\langle \alpha \rangle \psi$) in $\mathsf{cl}(\varphi)$ to be linearly ordered and use ϵ_i to denote the i-th diamond formula in $\mathsf{cl}(\varphi)$, with ϵ_1 being the first one. Note that a changed initial state of an automaton results in a different diamond formula.

To define *Hintikka-trees*, the tree-shaped abstraction of models underlying our decision procedure, we proceed in three steps. First, we introduce *Hintikka-sets* that will be used as (parts of) node labels. Intuitively, each node in the tree describes a world of the corresponding model, and its label contains the formulas from the closure of the input formula φ that are true in this world. Second, we introduce a *matching relation* that describes the possible "neighborhoods" that we may find in Hintikka-trees, where a neighborhood consists of a labeled node and its labeled successors. And third, we use these ingredients to define Hintikka-trees.

Definition 7 (Hintikka-set). *Let $\psi \in \Phi^{(\neg)}$ be an $APDL^{(\neg)}$-formula, and $\alpha \in A\Pi^{(\neg)}$ a program automaton. The set $\Psi \subseteq \mathsf{cl}(\varphi)$ is a Hintikka-set for φ if*

> *(H1) if $\psi_1 \wedge \psi_2 \in \Psi$, then $\psi_1 \in \Psi$ and $\psi_2 \in \Psi$*
>
> *(H2) if $\psi_1 \vee \psi_2 \in \Psi$, then $\psi_1 \in \Psi$ or $\psi_2 \in \Psi$*
>
> *(H3) $\psi \in \Psi$ iff $\dot{\neg}\psi \notin \Psi$*
>
> *(H4) if $[\alpha]\psi \in \Psi$ and $q_\alpha \in F_\alpha$, then $\psi \in \Psi$*
>
> *(H5) if $[\alpha]\psi \in \Psi$ then, for any state $q \in Q_\alpha$ and test $\theta? \in \Sigma_\alpha$,*
>
> $$q \in \Delta_\alpha(q_\alpha, \theta?) \text{ implies that } \dot{\neg}\theta \in \Psi \text{ or } [\alpha_q]\psi \in \Psi$$

The set of all Hintikka-sets for φ is designated by $\mathcal{H}_\varphi$.

The conditions (H1) to (H3) are standard, with one exception: (H3) is stronger than usual since it enforces maximality of Hintikka-sets by stating that, for each formula $\psi \in \mathsf{cl}(\varphi)$, either ψ or $\dot{\neg}\psi$ must be in the Hintikka-set. This will be used later on to deal with negated programs. The last two conditions (H4) and (H5) deal with the "local" impact of box formulas.

Next, we define the matching relation. The purpose of this relation can be understood as follows: in the Hintikka-tree, each node has exactly one successor for every diamond formula in $\mathsf{cl}(\varphi)$. The matching relation helps to ensure that all diamond formulas in a node's label can be satisfied "via" the corresponding successor in the Hintikka-tree, and that none of the box formulas is violated via any successors. We talk of "via" here since going to an immediate successor corresponds to travelling along a *single* program literal. Since programs in $APDL^{(\neg)}$ are automata that may only accept words of length greater one, in general we cannot satisfy diamonds by going only to the immediate successor, but rather we must perform a sequence of such moves.

Before we define the matching relation formally, let us fix the structure of node labels of Hintikka-trees. For reasons that will be discussed below, node labels not only contain a Hintikka-set, but also two additional components. More precisely, if φ is an APDL$^{(\neg)}$-formula and cl(φ) contains k diamond formulas, then we use

- $\Pi_\varphi^{(\neg)}$ to denote the set of all program literals occurring in φ; and
- Λ_φ to abbreviate $\mathcal{H}_\varphi \times (\Pi_\varphi^{(\neg)} \cup \{\bot\}) \times \{0, \ldots, k\}$, i.e. the set of triples containing a Hintikka-set for φ, a program literal of φ or $\bot$, and a number at most k.

The elements of Λ_φ will be used as node labels in Hintikka-trees. Intuitively, the first component lists the formulas that are true at a node, the second component fixes the program literal with which the node can be reached from its predecessor (or $\bot$ if this information is not important), and the third component will help to ensure that diamond formulas are eventually satisfied when moving through the tree. For a triple $\lambda \in \Lambda_\varphi$, we refer to the first, second and third triple component with λ^1, λ^2, and λ^3, respectively. For the following definition, recall that we use ϵ_i to denote the i-th diamond in cl(φ).

Definition 8 (Matching). *Let φ be a formula and k the number of diamond formulas in* cl(φ). *A $k + 1$-tuple of Λ_φ-triples $(\lambda, \lambda_1, \ldots, \lambda_k)$ is* matching *if, for $1 \leq i \leq k$ and all automata $\alpha \in A\Pi^{(\neg)}$, the following holds:*

(M1) if $\epsilon_i = \langle \alpha \rangle \psi \in \lambda^1$, then there is a word $w = \psi_1? \cdots \psi_n? \in \Sigma_\alpha^*$, $n \geq 0$,

and a state $q_1 \in Q_\alpha$ such that $\{\psi_1, \ldots, \psi_n\} \subseteq \lambda^1$, $q_1 \in \Delta_\alpha(q_\alpha, w)$,

and one of the following holds:

(a) q_1 is a final state, $\psi \in \lambda^1$, $\lambda_i^2 = \bot$, and $\lambda_i^3 = 0$

(b) there is a program literal $\pi \in \Sigma_\alpha$ and a state $q_2 \in Q_\alpha$ such that

$$q_2 \in \Delta_\alpha(q_1, \pi), \ \epsilon_j = \langle \alpha_{q_2} \rangle \psi \in \lambda_i^1, \ \lambda_i^2 = \pi, \ and \ \lambda_i^3 = j.$$

(M2) if $[\alpha]\psi \in \lambda^1$, $q \in Q_\alpha$, and $\pi \in \Sigma_\alpha$ a program literal such that

$$q \in \Delta_\alpha(q_\alpha, \pi), \ then \ \pi = \lambda_i^2 \ implies \ [\alpha_q]\psi \in \lambda_i^1.$$

As already noted, the purpose of the matching relation is to describe the possible neighborhoods in Hintikka-trees. To this end, think of λ as the label of a node, and of $\lambda_1, \ldots, \lambda_k$ as the labels of its successors. The purpose of Conditions (M1) and (M2) is to ensure that diamonds are satisfied and that boxes are not violated, respectively. Let us consider only (M1). If a diamond $\epsilon_i = \langle \alpha \rangle \psi$ is in the first component of λ, it can either be satisfied in the node labeled with λ itself (Condition (a)) or we can "delay" its satisfaction to the i-th successor node that is reserved specifically for this purpose (Condition (b)). In Case (a), it is not important over which program literal we can reach the i-th successor, and thus the second component of λ_i can be set to $\bot$. In the second case, we must choose a suitable program literal π and a suitable state q of α, make sure that the i-th successor is reachable over π via its second λ_i-component, and guarantee that the first component of λ_i contains the diamond under consideration with the automata α "advanced" to initial state q.

The remaining building block for ensuring that diamonds are satisfied is to enforce that the satisfaction of diamonds is not delayed forever. This is one of the two core parts of the definition of Hintikka-trees, the other being the proper treatment of negation. Before we can discuss the prevention of infinitely delayed diamonds in some more detail, we have to introduce some basic notions.

Let M be a set and $k \in \mathbb{N}$. An *(infinite) k-ary M-tree* T is a mapping $T : [k]^* \to M$, where $[k]$ is used (now and in the following) as an abbreviation for the set $\{1, \ldots, k\}$. Intuitively, the node αi is the i-th child of α. We use ε to denote the empty word (corresponding to the root of the tree). An infinite *path* in a k-ary M-tree is an infinite word γ over the alphabet $[k]$. We use $\gamma[n]$, $n \geq 0$, to denote the prefix of γ up to the n-th element of the sequence (with $\gamma[0]$ yielding the empty sequence).

Now back to the prevention of infinitely delayed diamonds. Given a formula φ with k diamond formulas in $\mathsf{cl}(\varphi)$, a Hintikka-tree will be defined as a k-ary Λ_φ-tree in which every neighborhood is matching and some additional conditions are satisfied. To detect infinite delays of diamonds in such trees, it does *not* suffice to simply look for infinite sequences of nodes that all contain the same diamond: firstly, diamonds are evolving while being "pushed" through the tree since their initial state might be changed. Secondly, such a sequence does not necessarily correspond to an infinite delay of diamond satisfaction: it could as well be the case that the diamond is satisfied an infinite number of times, but always immediately "regenerated" by some other formula. Also note that we cannot use the standard technique from [20] since it only works for deterministic variants of PDL.

Precisely for this purpose, the easy detection of infinitely delayed diamonds, we have introduced the third component of node labels in Hintikka trees: if a diamond was pushed to the current node x from its predecessor, then by (M1) the third component of x's label contains the number of the pushed diamond. Moreover, if the pushed diamond is not satisfied in x, we again use the third component of x: it contains the number of the successor of x to which the diamond's satisfaction is (further) delayed. If no diamond was pushed to x, its third component is simply zero. Thus, the following definition captures our intuitive notion of infinitely delayed diamonds.

Definition 9 (Diamond Starvation). *Let φ be an $APDL^{(\neg)}$-formula with k diamond formulas in $\mathsf{cl}(\varphi)$, T a k-ary Λ_φ-tree, $x \in [k]^*$ a node in T, and $\epsilon_i = \langle \alpha \rangle \psi \in T(x)^1$. Then the diamond formula $\langle \alpha \rangle \psi$ is called* starving *in x if there exists a path $\gamma = \gamma_1 \gamma_2 \cdots \in [k]^\omega$ such that*

1. $\gamma_1 = i$,
2. $T(x\gamma[n])^3 = \gamma_{n+1}$ *for $n \geq 1$.*

We have now gathered all ingredients to define Hintikka-trees formally.

Definition 10 (Hintikka-tree). *Let φ be an $APDL^{(\neg)}$-formula with k diamond formulas in $\mathsf{cl}(\varphi)$. A k-ary Λ_φ-tree T is a* Hintikka-tree *for φ if T satisfies,* *for all nodes* $x, y \in [k]^*$, *the following conditions:*

(T1) $\varphi \in T(\varepsilon)^1$

(T2) *the* $k + 1$-*tuple* $(T(x), T(x1), \dots, T(xk))$ *is matching*

(T3) *no diamond formula from* $\mathsf{cl}(\varphi)$ *is starving in* x

(T4) *if* $[\alpha]\psi, [\beta]\theta \in T(x)^1$, $\pi \in \Pi_0^{(\neg)}$, $q'_\alpha \in Q_\alpha$, *and* $q'_\beta \in Q_\beta$ *such that*
$q'_\alpha \in \Delta_\alpha(q_\alpha, \pi)$ *and* $q'_\beta \in \Delta_\beta(q_\beta, \overline{\pi})$, *then*
$[\alpha_{q'_\alpha}]\psi \notin T(y)^1$ *implies* $[\beta_{q'_\beta}]\theta \in T(y)^1$.

Conditions (T1) to (T3) are easily understood. The purpose of Condition (T4) is to deal with negated programs. In particular, for each atomic program a we have to ensure that any pair of nodes x, y of a Hintikka-tree T can be related by one of a and $\neg a$ without violating any boxes. This is done by (T4) together with (H3)—indeed, this is the reason for formulating (H3) stronger than usual. Intuitively, the treatment of negation can be understood as follows: suppose that $[\alpha]\psi \in T(x)^1$, let $q \in \Delta_\alpha(q_\alpha, a)$ for some atomic program a, and let y be a node. By (H3), we have either $[\alpha_q]\psi \in T(y)^1$ or $\dot{\neg}[\alpha_q]\psi \in T(y)^1$. In the first case, x and y can be related by a. In the second case, (T4) ensures that they can be related by $\neg a$. This technique is inspired by [21], but generalized to program automata.

The following proposition shows that Hintikka-trees are indeed proper abstractions of models. A proof can be found in [26].

Proposition 1. *An* $APDL^{(\neg)}$-*formula* φ *is satisfiable iff it has a Hintikka-tree.*

5   Büchi Automata for Hintikka-Trees

In this section, we show that it is possible to construct, for every $APDL^{(\neg)}$-formula φ, a Büchi tree automaton $\mathcal{B}_\varphi$ that accepts exactly the Hintikka-trees for φ. By Proposition 1, since the size of $\mathcal{B}_\varphi$ is at most exponential in the length of φ, and since the emptiness of Büchi-tree automata can be verified in quadratic time [20], this yields an EXPTIME decision procedure for the satisfiability of $APDL^{(\neg)}$-formulas. We start with introducing Büchi tree automata.

Definition 11 (Büchi Tree Automaton). *A* Büchi tree automaton $\mathcal{B}$ *for* k-*ary* M-*trees is a quintuple* (Q, M, I, Δ, F), *where*

- Q *is a finite set of* states,
- M *is a finite* alphabet,
- $I \subseteq Q$ *is the set of* initial states,
- $\Delta \subseteq Q \times M \times Q^k$ *is the* transition relation, *and*
- $F \subseteq Q$ *is the set of* accepting states.

Let M *be a set of labels, and* T *a* k-*ary* M-*tree. Then, a* run *of* $\mathcal{B}$ *on* T *is a* k-*ary* Q-*tree* r *such that*

1. $r(\varepsilon) \in I$, and
2. $(r(x), T(x), r(x1), \ldots, r(xk)) \in \Delta$ for all nodes $x \in [k]^*$.

Let $\gamma \in [k]^\omega$ be a path. The set $\inf_r(\gamma)$ contains the states in Q that occur infinitely often in run r along path γ. A run r of $\mathcal{B}$ on T is accepting if, for each path $\gamma \in [k]^\omega$, we have $\inf_r(\gamma) \cap F \neq \emptyset$. The language accepted by $\mathcal{B}$ is the set $\mathcal{L}(\mathcal{B}) = \{T \mid \text{there is an accepting run of } \mathcal{B} \text{ on } T\}$.

Given a Büchi automaton $\mathcal{B}$, the problem whether its language is empty, i.e., whether it holds that $\mathcal{L}(\mathcal{B}) = \emptyset$, is called the *emptiness problem*. This problem is solvable in time quadratic in the size of the automaton [20].

We now give the translation of APDL$^{(\neg)}$-formulas φ into Büchi-automata $\mathcal{B}_\varphi$. To simplify the notation, we write $\mathcal{P}_\square(\varphi)$ to denote the set of sets $\{\{[\alpha]\psi, [\beta]\theta\} \mid [\alpha]\psi, [\beta]\theta \in \mathrm{cl}(\varphi)\}$. We first introduce our automata formally and then explain the intuition.

Definition 12. *Let φ be an APDL$^{(\neg)}$-formula with $\mathrm{cl}(\varphi)$ containing k diamond formulas. The Büchi tree automaton $\mathcal{B}_\varphi = (Q, \Lambda_\varphi, I, \Delta, F)$ on k-ary Λ_φ-trees is defined as follows:*

- Q *contains those triples* $((\Psi, \pi, \ell), P, d) \in \Lambda_\varphi \times 2^{\mathcal{P}_\square(\varphi)} \times \{\emptyset, \uparrow\}$ *that satisfy the following conditions:*
  (1) if $\{[\alpha]\psi, [\beta]\theta\} \subseteq \Psi$, *then* $\{[\alpha]\psi, [\beta]\theta\} \in P$
  (2) if $\{[\alpha]\psi, [\beta]\theta\} \in P$, $\pi \in \Pi^{(\neg)}$, $q'_\alpha \in \Delta_\alpha(q_\alpha, \pi)$, $q'_\beta \in \Delta_\beta(q_\beta, \overline{\pi})$, *and* $[\alpha_{q'_\alpha}]\psi \notin \Psi$, *then* $[\beta_{q'_\beta}]\theta \in \Psi$

- $I := \{((\Psi, \pi, \ell), P, d) \in Q \mid \varphi \in \Psi, \text{ and } d = \emptyset\}$.

- $((\lambda_0, P_0, d_0), (\Psi, \pi, \ell), (\lambda_1, P_1, d_1), \ldots, (\lambda_k, P_k, d_k)) \in \Delta$ *if and only if, for each* $i \in [k]$, *the following holds:*
  1. $\lambda_0 = (\Psi, \pi, \ell)$,
  2. $P_0 = P_i$,
  3. *the tuple* $(\lambda_0, \ldots, \lambda_k)$ *is matching,*
  4. $d_i = \begin{cases} \uparrow & \text{if } d_0 = \emptyset, \lambda_i^3 \neq 0 \text{ and } \epsilon_i \in \Psi \\ \uparrow & \text{if } d_0 = \uparrow, \lambda_0^3 = i, \text{ and } \lambda_i^3 \neq 0 \\ \emptyset & \text{otherwise.} \end{cases}$

- *The set F of accepting states is* $F := \{(\lambda, P, d) \in Q \mid d = \emptyset\}$.

While it is not hard to see how the set of initial states enforces (T1) of Hintikka-trees and how the transition relation enforces (T2), Conditions (T3) and (T4) are more challenging. In the following, we discuss them in detail.

Condition (T3) is enforced with the help of the third component of states, which may take the values "$\emptyset$" and "$\uparrow$". Intuitively, the fourth point in the definition of Δ ensures that, whenever the satisfaction of a diamond is delayed in a node x and r is a run, then r assigns states with third component $\uparrow$ to all nodes on the path that "tracks" the diamond delay. Note that, for this purpose, the definition of Δ refers to the third component of Λ_φ-tuples, which is "controlled" by (M1) in the appropriate way. All nodes that do not appear on delayed diamond paths are labeled with $\oslash$. Then, the set of accepting states ensures that there is no path that, from some point on, is constantly labeled with ↑. Thus, we enforce that no diamonds are delayed infinitely in trees accepted by our automata, i.e. no starvation occurs.

There is one special case that should be mentioned. Assume that a node x contains a diamond $\epsilon_i = \langle\alpha\rangle\psi$ that is not satisfied "within this node" (Case (a) of (M1) does not apply). Then there is a potential starvation path for ϵ_i that starts at x and goes through the node xi: (M1) "advances" the automaton α to α_q, and ensures that $\epsilon_j = \langle\alpha_q\rangle\psi \in T(xi)^1$ and that $T(xi)^3 = j$. Now suppose that $T(xi)^1$ contains another diamond $\epsilon_k = \langle\beta\rangle\theta$ with $\epsilon_j \neq \epsilon_k$. If ϵ_k is not satisfied within xi, there is a potential starvation path for ϵ_k starting at xi and going through xik. Since the starvation path for ϵ_i and the starvation path for ϵ_k are for different diamonds, we must be careful to separate them—failure in doing this would result in some starvation-free Hintikka-trees to be rejected. Thus, the definition of Δ ensures that runs label xik with $\oslash$, and the constant ↑-labeling of the starvation path for ϵ_k is delayed by one node: it starts only at the *successor* of xik on the starvation path for ϵ_k.

Now for Condition (T4). In contrast to Conditions (T1) and (T2), this condition has a global flavor in the sense that it does not only concern a node and its successors. Thus, we need to employ a special technique to enforce that (T4) is satisfied: we use the second component of states as a "bookkeeping component" that allows to propagate global information. More precisely, Point (1) of the definition of Q and Point (1) of the definition of Δ ensure that, whenever two boxes appear in a Hintikka-set labeling a node x in a Hintikka-tree T, then this joint occurrence is recorded in the second component of the state that any run assigns to x. Via the definition of the transition relation (second point), we further ensure that all states appearing in a run share the same second component. Thus, we may use Point (2) of the definition of Q and Point (1) of the definition of Δ to ensure that any node y satisfies the property stated by Condition (T4).

The following proposition shows that the Büchi tree automaton $\mathcal{B}_\varphi$ indeed accepts precisely the Hintikka-trees for APDL$^{(\neg)}$-formula φ. A proof can be found in [26].

Proposition 2. *Let φ be an APDL$^{(\neg)}$-formula and T a k-ary Λ_φ-tree. Then T is a Hintikka-tree for φ iff $T \in \mathcal{L}(\mathcal{B}_\varphi)$.*

Putting together Propositions 1 and 2, it is now easy to establish decidability and ExpTime-complexity of APDL$^{(\neg)}$ and thus also of PDL$^{(\neg)}$.

Theorem 1. *Satisfiability of PDL$^{(\neg)}$-formulas is ExpTime-complete.*

Proof. From Propositions 1 and 2, it follows that an APDL$^{(\neg)}$-formula φ is satisfiable if and only if $\mathcal{L}(\mathcal{B}_\varphi) \neq \emptyset$. The emptiness problem for Büchi automata is decidable in time quadratic in the size of the automaton [20]. To show that APDL$^{(\neg)}$-formula satisfiability is in ExpTime, it thus remains to show that the size of $\mathcal{B}_\varphi = (Q, \Lambda_\varphi, I, \Delta, F)$ is at most exponential in φ.

Let n be the length of φ. Since the cardinality of $\mathsf{cl}(\varphi)$ is polynomial in n, the cardinality of $\mathcal{H}_\varphi$ (the set of Hintikka-sets for φ) is at most exponential in n. Thus, it is readily checked that the same holds for Λ_Φ and Q. The exponential upper bound on the cardinalities of I and F is trivial. It remains to determine the size of Δ: since the size of Q is exponential in n and the out-degree of trees accepted by automata is polynomial in n, we obtain an exponential bound.

Thus, $\mathrm{APDL}^{(\neg)}$-formula satisfiability and hence also $\mathrm{PDL}^{(\neg)}$-formula satisfiability are in EXPTIME. For the lower bound, it suffices to recall that PDL-formula satisfiability is already EXPTIME-hard [3]. □

6 Conclusion

This paper introduces the propositional dynamic logic $\mathrm{PDL}^{(\neg)}$, which extends standard PDL with negation of atomic programs. We were able to show that this logic extends PDL in an interesting and useful way, yet retaining its appealing computational properties. There are some natural directions for future work. For instance, it should be simple to further extend $\mathrm{PDL}^{(\neg)}$ with the converse operator without destroying the EXPTIME upper bound. It would be more interesting, however, to investigate the interplay between (full) negation and PDL's program operators in some more detail. For example, to the best our our knowledge it is unknown whether the fragment of $\mathrm{PDL}^\neg$ that has only the program operators "¬" and ";" is decidable.

References

1. Pratt, V.: Considerations on floyd-hoare logic. In: FOCS: IEEE Symposium on Foundations of Computer Science (FOCS). (1976)
2. Fischer, M.J., Ladner, R.E.: Propositional modal logic of programs. In: Conference record of the ninth annual ACM Symposium on Theory of Computing, ACM Press (1977) 286–294
3. Fischer, M.J., Ladner, R.E.: Propositional dynamic logic of regular programs. Journal of Computer and System Sciences **18** (1979) 194–211
4. Harel, D.: Dynamic logic. In Gabbay, D.M., Guenthner, F., eds.: Handbook of Philosophical Logic, Volume II. D. Reidel Publishers (1984) 496–604
5. Harel, D., Kozen, D., Tiuryn, J.: Dynamic Logic. MIT Press (2000)
6. Fagin, R., Halpern, J.Y., Moses, Y., Vardi, M.Y.: Reasoning About Knowledge. MIT Press (1995)
7. De Giacomo, G., Lenzerini, M.: PDL-based framework for reasoning about actions. In: Proceedings of the 4th Congress of the Italian Association for Artificial Intelligence (AI*IA'95). Volume 992., Springer (1995) 103–114
8. Prendinger, H., Schurz, G.: Reasoning about action and change: A dynamic logic approach. Journal of Logic, Language, and Information **5** (1996) 209–245
9. Giacomo, G.D., Lenzerini, M.: Boosting the correspondence between description logics and propositional dynamic logics. In: Proceedings of the Twelfth National Conference on Artificial Intelligence (AAAI'94). Volume 1, AAAI Press (1994) 205–212

10. Passy, S., Tinchev, T.: An essay in combinatory dynamic logic. Information and Computation **93** (1991)
11. Harel, D., Pratt, V.: Nondeterminism in logics of programs. In: Proceedings of the Fifth Symposium on Principles of Programming Languages, ACM (1978) 203–213
12. Vardi, M.Y.: The taming of converse: Reasoning about two-way computations. In Parikh, R., ed.: Proceedings of the Conference on Logic of Programs. Volume 193 of LNCS., Springer (1985) 413–424
13. Danecki, S.: Nondeterministic propositional dynamic logic with intersection is decidable. In Skowron, A., ed.: Proceedings of the Fifth Symposium on Computation Theory. Volume 208 of LNCS., Springer (1984) 34–53
14. Broersen, J.: Relativized action complement for dynamic logics. In Philippe Balbiani, Nobu-Yuki Suzuki, F.W., Zakharyaschev, M., eds.: Advances in Modal Logics Volume 4, King's College Publications (2003) 51–69
15. Goranko, V., Passy, S.: Using the universal modality: Gains and questions. Journal of Logic and Computation **2** (1992) 5–30
16. Humberstone, I.L.: Inaccessible worlds. Notre Dame Journal of Formal Logic **24** (1983) 346–352
17. Gargov, G., Passy, S., Tinchev, T.: Modal environment for Boolean speculations. In Skordev, D., ed.: Mathematical Logic and Applications, New York, USA, Plenum Press (1987) 253–263
18. Goranko, V.: Modal definability in enriched languages. Notre Dame Journal of Formal Logic **31** (1990) 81–105
19. Lutz, C., Sattler, U.: Mary likes all cats. In Baader, F., Sattler, U., eds.: Proceedings of the 2000 International Workshop in Description Logics (DL2000). Number 33 in CEUR-WS (http://ceur-ws.org/) (2000) 213–226
20. Vardi, M.Y., Wolper, P.: Automata-theoretic techniques for modal logic of programs. Journal of Computer and System Sciences **32** (1986) 183–221
21. Lutz, C., Sattler, U.: The complexity of reasoning with boolean modal logics. In Wolter, F., Wansing, H., de Rijke, M., Zakharyaschev, M., eds.: Advances in Modal Logics Volume 3, CSLI Publications, Stanford, CA, USA (2001)
22. Baader, F., McGuiness, D.L., Nardi, D., Patel-Schneider, P.: The Description Logic Handbook: Theory, implementation and applications. Cambridge University Press (2003)
23. Schild, K.D.: A correspondence theory for terminological logics: Preliminary report. In Mylopoulos, J., Reiter, R., eds.: Proceedings of the Twelfth International Joint Conference on Artificial Intelligence (IJCAI-91), Morgan Kaufmann (1991) 466–471
24. Matijasevich, Y.: Simple examples of undecidable associative calculi. Soviet mathematics (Doklady) (1967) 555–557
25. Kleene, S.: Representation of events in nerve nets and finite automata. In C.E.Shannon, J.McCarthy, eds.: Automata Studies. Princeton University Press (1956) 3–41
26. Lutz, C., Walther, D.: PDL with negation of atomic programs. LTCS-Report 03-04, Technical University Dresden (2003) Available from http://lat.inf.tu-dresden.de/research/reports.html.

Counter-Model Search in Gödel-Dummett Logics

Dominique Larchey-Wendling

LORIA – CNRS
Campus Scientifique, BP 239
Vandœuvre-lès-Nancy, France

Abstract. We present a new method for deciding Gödel-Dummett logic
LC. We first characterize the validity of irreducible sequents of LC by the
existence of r-cycles in bi-colored graphs and we propose a linear algorithm to detect r-cycles and build counter-models. Then we characterize
the validity of formulae by the existence of r-cycles in boolean constrained
graphs. We also give a parallel method to detect r-cycles under boolean
constraints. Similar results are given for the finitary versions LC_n.

1 Introduction

Gödel-Dummett logic LC and its finitary versions $(\mathsf{LC}_n)_{n>0}$ are the intermediate
logics (between classical and intuitionistic logics) characterized by linear Kripke
models. LC was introduced by Gödel in [10] and later axiomatized by Dummett in [6]. It is now one of the most studied intermediate logics and has been
recognized recently as one of the fundamental *t-norm based fuzzy logics* [11].
Proof-search in LC has benefited from the development of proof-search in intuitionistic logic IL with two important seeds: the *contraction-free calculus* of
Dyckhoff [7,8,1] and the *hyper-sequent* calculus of Avron [2,13]. Two of the
most recent contributions propose a similar approach based on a set of *local*
and *strongly invertible* proof rules (for either sequent [12] or hyper-sequent [2]
calculus,) and a semantic criterion to decide *irreducible (hyper)-sequents* and
eventually build a counter-model.

We are interested in studying combination of proof-search and counter-model
construction to provide decision procedures for intermediate logics. We have already proposed such a combination for LC [12], but here we investigate deeper
counter-model search to obtain a new system with the following fundamental
property: all the irreducible sequents arising from a proof-search can be represented in a shared semantic structure. The semantic criterion that decides
irreducible sequents can be computed in parallel on this shared structure. Instead of simply combining proof-search and counter-model construction, we are
now at the frontier of these two techniques and so provide a calculus of *parallel
counter-model search.*

In section 3, we recall the indexation technique of [12] and introduce a new
proof-system with the property that irreducible sequents are now composed only
of atomic implications. In section 4 we associate a *bi-colored graph* to each irreducible sequent. We show that irreducible sequents validity in LC can be decided

D. Basin and M. Rusinowitch (Eds.): IJCAR 2004, LNAI 3097, pp. 274–288, 2004.
© Springer-Verlag Berlin Heidelberg 2004

from the existence of *r-cycles* in the graph. We propose a linear time algorithm to detect r-cycles and eventually build counter-models. We prove similar results for LC_n. In section 5, we show that proof-search can be viewed as a *non-deterministic choice* of arrows in a bi-colored graph corresponding to left or right premises of proof-rules. We postpone non-deterministic choices by introducing *boolean selectors* and represent proof-search by a *conditional bi-colored graph*. We prove that validity is characterized by the existence of r-cycles in every instance of this graph and discuss r-cycle detection combined with boolean constraint solving. Moreover, we characterize the smallest n for which a given formula is invalid in LC_n. In section 6, we detail some implementation techniques for a constrained r-cycle detection algorithm based on *matrices of binary decision diagrams* and briefly present our *parallel counter-model search* system both for LC and for its finitary versions LC_n. Due to space considerations, the proofs of section 3 are not included is this document.[1]

2 Introduction to Gödel-Dummett Logics LC_n

In this section, we present the algebraic semantics of the family of propositional Gödel-Dummett logics LC_n. The value n belongs to the set $\overline{\mathbb{N}}^* = \{1, 2, \dots\} \cup \{\infty\}$ of strictly positive natural numbers with its natural order $\leqslant$, augmented with a greatest element ∞. In the case $n = \infty$, the logic LC_∞ is also denoted by LC: this is the usual Gödel-Dummett logic. The set of propositional *formulae*, denoted Form is defined inductively, starting from a set of propositional *variables* denoted by Var with an additional bottom constant $\perp$ denoting *absurdity* and using the connectives $\wedge$, $\vee$ and $\supset$. IL will denote the set of formulae that are provable in any intuitionistic propositional calculus (see [7]) and CL will denote the classically valid formulae. As usual an *intermediate propositional logic* [1] is a set of formulae $\mathcal{L}$ satisfying $\mathsf{IL} \subseteq \mathcal{L} \subseteq \mathsf{CL}$ and closed under the rule of modus ponens and under arbitrary substitution. For any $n \in \overline{\mathbb{N}}^*$, the Gödel-Dummett logic LC_n is an intermediate logic. On the semantic side, it is characterized by the linear Kripke models of size n (see [6].) The following strictly increasing sequence holds: $\mathsf{IL} \subset \mathsf{LC} = \mathsf{LC}_\infty \subset \cdots \subset \mathsf{LC}_n \subset \cdots \subset \mathsf{LC}_1 = \mathsf{CL}$ In the particular case of LC, the logic has a simple Hilbert axiomatic system: $(X \supset Y) \vee (Y \supset X)$ added to the axioms of IL.

In this paper, we will use the algebraic semantics characterization of LC_n [2] rather than Kripke semantics. Let us fix a particular $n \in \overline{\mathbb{N}}^*$. The algebraic model is the set $\overline{[0, n)} = [0, \dots, n[\cup\{\infty\}$ composed of $n + 1$ elements. An interpretation of propositional variables $[\![\cdot]\!] : \mathsf{Var} \to \overline{[0, n)}$ is inductively extended to formulae: $\perp$ interpreted by 0, the conjunction $\wedge$ is interpreted by the *minimum* function denoted $\wedge$, the disjunction $\vee$ by the *maximum* function $\vee$ and the implication $\supset$ by the operator $\twoheadrightarrow$ defined by $a \twoheadrightarrow b = $ if $a \leqslant b$ then ∞ else b. A formula D is *valid* for the interpretation $[\![\cdot]\!]$ if the equality $[\![D]\!] = \infty$ holds. This interpretation is complete for LC. A *counter-model* of a formula D is an interpretation $[\![\cdot]\!]$ such

---

[1] See http://www.loria.fr/~larchey/LC for an extended version including those proofs and a prototype implementation of the counter-model search engine in Ocaml.

that $\llbracket D \rrbracket < \infty$. A *sequent* is a pair $\Gamma \vdash \Delta$ where Γ and Δ are multisets of formulae. Γ, Δ denotes the sum of the two multisets. Given a sequent $\Gamma \vdash \Delta$ and an interpretation $\llbracket \cdot \rrbracket$ of variables, we interpret $\Gamma \equiv A_1, \dots, A_n$ by $\llbracket \Gamma \rrbracket = \llbracket A_1 \rrbracket \wedge \cdots \wedge \llbracket A_n \rrbracket$ and $\Delta \equiv B_1, \dots, B_p$ by $\llbracket \Delta \rrbracket = \llbracket B_1 \rrbracket \vee \cdots \vee \llbracket B_p \rrbracket$. This sequent is *valid* with respect to the interpretation $\llbracket \cdot \rrbracket$ if $\llbracket \Gamma \rrbracket \leqslant \llbracket \Delta \rrbracket$ holds. On the other hand, a *counter-model* to this sequent is an interpretation $\llbracket \cdot \rrbracket$ such that $\llbracket \Delta \rrbracket < \llbracket \Gamma \rrbracket$, i.e. for any pair (i, j), the inequality $\llbracket B_j \rrbracket < \llbracket A_i \rrbracket$ holds.

3   Proof-Search and Sequent Calculus for LC_n

In this section we present a refinement of the decision procedure described in [12]. As before, we have the three following steps: first a linear reduction of a formula D into a *flat sequent* $\delta^-(D) \vdash \diamond \supset \mathcal{X}_D$, then a sequent based proof-search that reduces this flat sequent into a set of *implicational sequents* $\Gamma_i \vdash \Delta_b$ and finally a semantic algorithm that builds counter-models of $\Gamma_i \vdash \Delta_b$.

In [12], the proof-search process produced what we called pseudo-atomic sequents, which are sequents composed only of formulae of the form X and $X \supset Y$ where X and Y are propositional variables. In our new system, we further constraint our proof-search procedure to only produce atomic implications $X \supset Y$. We do not integrate the $\bot$ constant but it can be treated as in [12]. So atomic formulae are variables. We introduce a new special variable $\diamond \notin \mathsf{Var}$. From now, (propositional) variables are elements of $\mathsf{Var}_\diamond = \mathsf{Var} \cup \{\diamond\}$ but we require that $\diamond$ does not occur in the formulae we are trying to decide, so it has a special meaning during proof-search: a *b-context* denoted Δ_b is a non-empty multiset of implications such that if $A \supset B \in \Delta_b$ then $\diamond \supset B \in \Delta_b$.

We refine our indexation technique and the notion of flat sequent. A *flat formula* is of one of the following forms: $X \supset Y$ or $Z \supset (X \otimes Y)$ or $(X \otimes Y) \supset Z$ where X, Y and Z are propositional variables and $\otimes \in \{\wedge, \vee, \supset\}$. $\Gamma \vdash \Delta_b$ is a *flat sequent* if all formulae of Γ are flat and Δ_b is a b-context of atomic implications. Let us fix a particular formula D (not containing $\diamond$.) We introduce a new variable $\mathcal{X}_A$ (not occurring in D and different from $\diamond$) for every occurrence A of subformula of D. So the variables $\mathcal{X}_A$ represent the nodes of the decomposition tree of D. Then, we define the multisets $\delta^+(K)$ and $\delta^-(K)$ by induction on the *occurence* of the subformula K (the comma represents the sum of multisets.) These two multisets are composed of only flat formulae with variables of the form $\mathcal{X}_A$ or V (where V is a variable of D):

$$\delta^+(V) = \mathcal{X}_V \supset V \text{ when } V \text{ is a variable}$$
$$\delta^+(A \otimes B) = \delta^+(A), \delta^+(B), \mathcal{X}_{A \otimes B} \supset (\mathcal{X}_A \otimes \mathcal{X}_B) \text{ when } \otimes \in \{\wedge, \vee\}$$
$$\delta^+(A \supset B) = \delta^-(A), \delta^+(B), \mathcal{X}_{A \supset B} \supset (\mathcal{X}_A \supset \mathcal{X}_B)$$

$$\delta^-(V) = V \supset \mathcal{X}_V \text{ when } V \text{ is a variable}$$
$$\delta^-(A \otimes B) = \delta^-(A), \delta^-(B), (\mathcal{X}_A \otimes \mathcal{X}_B) \supset \mathcal{X}_{A \otimes B} \text{ when } \otimes \in \{\wedge, \vee\}$$
$$\delta^-(A \supset B) = \delta^+(A), \delta^-(B), (\mathcal{X}_A \supset \mathcal{X}_B) \supset \mathcal{X}_{A \supset B}$$

$$\frac{\Gamma, A \supset C \vdash \Delta \qquad \Gamma, B \supset C \vdash \Delta}{\Gamma, (A \wedge B) \supset C \vdash \Delta} \ [\supset_2] \qquad \frac{\Gamma, A \supset B, A \supset C \vdash \Delta}{\Gamma, A \supset (B \wedge C) \vdash \Delta} \ [\supset_2']$$

$$\frac{\Gamma, A \supset C, B \supset C \vdash \Delta}{\Gamma, (A \vee B) \supset C \vdash \Delta} \ [\supset_3] \qquad \frac{\Gamma, A \supset B \vdash \Delta \qquad \Gamma, A \supset C \vdash \Delta}{\Gamma, A \supset (B \vee C) \vdash \Delta} \ [\supset_3']$$

$$\frac{\Gamma, A \supset C \vdash \Delta \qquad \Gamma, B \supset C \vdash \Delta}{\Gamma, A \supset (B \supset C) \vdash \Delta} \ [\supset_4']$$

Fig. 1. Proof rules for implications in LC_n.

We just recall the proposition 1 already proved in [12]. We also introduce the proposition 2 which is the semantic counterpart of the substitution property of intermediate logics.

Proposition 1. *Both* $\delta^+(D), \mathcal{X}_D \vdash D$ *and* $\delta^-(D), D \vdash \mathcal{X}_D$ *are valid in* LC_n.

Proposition 2. *If* $[\![\cdot]\!]$ *is such that* $[\![\mathcal{X}_K]\!] = [\![K]\!]$ *holds for any occurrence* K *of subformula of* D, *then equation* $\lfloor \delta^-(D) \rfloor = \lfloor \delta^+(D) \rfloor = \infty$ *holds.*

Both proofs are left to the reader (trivial induction on K, see [12]). Then we are able to propose our linear reduction of a formula into a flat sequent. We recall that $\Diamond$ is just another variable but is required not to occur elsewhere in the sequent, and so does not occur in D or as one of the $\mathcal{X}_K$ introduced during the computation of $\delta^-(D)$.

Theorem 1. *In* LC_n, D *is valid if and only if* $\delta^-(D) \vdash \Diamond \supset \mathcal{X}_D$ *is valid.*

In figure 1, we recall proof rules for LC_n [12]. All of them are sound and strongly invertible. We replace the "old" rule $[\supset_4]$ (not included in the figure) with a new version:

$$\frac{\Gamma, B \supset C \vdash A \supset B, \Diamond \supset B, \Delta_b \qquad \Gamma, \Diamond \supset C \vdash \Delta_b}{\Gamma, (A \supset B) \supset C \vdash \Delta_b} \ [\supset_4] \quad \Delta_b \text{ is a b-context}$$

Theorem 2. *In* LC_n, *rule* $[\supset_4]$ *is sound and strongly invertible.*

The aim for such a new rule is that applied backward, all proof rules only introduce implications. Applying rules $[\supset_2], [\supset_3], [\supset_4]$ and $[\supset_2'], [\supset_3'], [\supset_4']$ bottom-up, it is easy to see that the proof-search process applied to the flat sequent $\delta^-(D) \vdash \Diamond \supset \mathcal{X}_D$ terminates and produces a set of *irreducible sequents,* meaning that no more rules can be applied to them. We characterize these sequents as *implicational,* i.e. composed only of implications between variables like in $X_1 \supset Y_1, \ldots, X_k \supset Y_k \vdash A_1 \supset B_1, \ldots, A_l \supset B_l$. Indeed, as there is one rule for each case of $(X \otimes Y) \supset Z$ or $Z \supset (X \otimes Y)$ and as the corresponding rule transforms the compound formula into one, two or three atomic implications in each premise, these irreducible sequents are implicational sequents. Moreover, if $\Gamma \vdash \Delta$ is any of those irreducible sequents then Δ is also a b-context. Indeed, applied backward, all proof-rules preserve flat sequents and the starting sequent is itself flat. We point out that there is no more a distinction between $\Diamond$ and other variables in the notion of implicational sequent: the notion of b-context is only used during proof-search. Section 5.1 discusses the proof-search process in further details.

4   Counter-Models of Implicational Sequents

In this section, we present a new and visual criterion to decide implicational sequents. In particular, the sequents that arise as irreducible during the previously described proof-search process are implicational.

4.1   Bi-colored Graph of an Implicational Sequent

Let $\mathcal{S} = X_1 \supset Y_1, \ldots, X_k \supset Y_k \vdash A_1 \supset B_1, \ldots, A_l \supset B_l$ be an implicational sequent. We build a *bi-colored graph* $\mathcal{G}_\mathcal{S}$ which has nodes in the set $\{X_i\} \cup \{Y_i\} \cup \{A_i\} \cup \{B_i\}$ and has two kinds of arrows, green (denoted by $\rightarrow$) and red (denoted by $\Rightarrow$.) The set of arrows is $\{X_1 \rightarrow Y_1, \ldots, X_k \rightarrow Y_k\} \cup \{B_1 \Rightarrow A_1, \ldots, B_l \Rightarrow A_l\}$. The red arrow $B_i \Rightarrow A_i$ is in the direction *opposite* to that in the implication $A_i \supset B_i$.

We will often use the symbols $\rightarrow$ and $\Rightarrow$ to denote the corresponding incidence relation in the graph. So for example, $\rightarrow\Rightarrow$ denotes the composition of the two relations and $X \rightarrow\Rightarrow Y$ means there exists a chain $X \rightarrow Z \Rightarrow Y$ in $\mathcal{G}_\mathcal{S}$. Also $\rightarrow^\star$ is the reflexive and transitive closure of $\rightarrow$, i.e. the accessibility for the $\rightarrow$ relation, And $\rightarrow + \Rightarrow$ is the union of relations.

4.2   Heights in Bi-colored Graphs

We define a notion of bi-height in bi-colored graphs. The idea is very simple. A green arrow $\rightarrow$ weighs 0 and a red arrow $\Rightarrow$ weighs at least 1.

Definition 1 (Bi-height in a bi-colored graph). *Let $\mathcal{G}$ be a bi-colored graph. A* bi-height *is a function $h : \mathcal{G} \rightarrow \mathbb{N}$ such that for any $x, y \in \mathcal{G}$, if $x \rightarrow y \in \mathcal{G}$ then $h(x) \leqslant h(y)$ and if $x \Rightarrow y \in \mathcal{G}$ then $h(x) < h(y)$.*

As we will see, bi-height can be used to compute counter-models. We characterize graphs that admit bi-heights by the notion of r-cycle. This notion is similar to the notion of G-cycle in [3] but here we give a simple and efficient algorithm to find cycles and compute counter-models.

Definition 2 (R-cycle). *A r-cycle is a chain of the form $x(\rightarrow + \Rightarrow)^\star \Rightarrow x$.*

It is clear that if a graph has a r-cycle, then there is no bi-height. We give a linear[2] algorithm to compute a bi-height when there is no r-cycle.

Theorem 3. *Let $\mathcal{G}$ be a bi-colored graph. It is possible to decide if $\mathcal{G}$ has r-cycles in linear time and if not, to compute a bi-height h for $\mathcal{G}$ in linear time.*

---

[2] Linearity is measured w.r.t. the number of vertexes and arrows in the graph.

Proof. Even if it has no r-cycle, $\mathcal{G}$ may still contain green ($\rightarrow$) cycles. To remove all cycles, we introduce the contracted graph $\mathcal{G}'$ of $\mathcal{G}$: let $\mathcal{C}$ be the set of strongly connected components for the "green" sub-graph of $\mathcal{G}$ (i.e. $\mathcal{G}_\rightarrow$), $\mathcal{C} = \{[x] \mid x \in \mathcal{G}\}$ and $[x]$ is the strongly connected component of x. $\mathcal{G}'$ has $\mathcal{C}$ as set of nodes, and the set of arrows is described by:

$$[x] \rightarrow [y] \text{ iff } [x] \neq [y] \text{ and } \exists x', y' \text{ s.t. } [x] = [x'], [y] = [y'] \text{ and } x' \rightarrow y' \in \mathcal{G}$$
$$[x] \Rightarrow [y] \text{ iff } \exists x', y' \text{ s.t. } [x] = [x'], [y] = [y'] \text{ and } x' \Rightarrow y' \in \mathcal{G}$$

$\mathcal{G}'$ is computed in linear time by standard depth first search algorithms. $\mathcal{G}'$ has no green ($\rightarrow$) cycle (because they collapse into a strongly connected component) and so $\mathcal{G}'$ has a cycle (with either $\rightarrow$ or $\Rightarrow$ arrows) if and only if $\mathcal{G}$ has a r-cycle. Finding a cycle in $\mathcal{G}'$ takes linear time in the size of $\mathcal{G}'$ thus of $\mathcal{G}$.

Now suppose that $\mathcal{G}'$ has no cycle (i.e. no r-cycle in $\mathcal{G}$). The relation $(\rightarrow + \Rightarrow)^*$ is a finite partial order and we can define $h' : \mathcal{G}' \rightarrow \mathbb{N}$ inductively by:

$$h'([y]) = \max \left\{ \begin{array}{l} h'([x]) \text{ for } [x] \rightarrow [y] \in \mathcal{G}' \\ h'([x]) + 1 \text{ for } [x] \Rightarrow [y] \in \mathcal{G}' \end{array} \right\}$$

We compute the whole function h' in linear time by sorting the nodes of $\mathcal{G}'$ along $(\rightarrow + \Rightarrow)^*$, again by depth first search. We define $h(x) = h'([x])$ and prove that h is a bi-height in $\mathcal{G}$. If $x \rightarrow y \in \mathcal{G}$: first case $[x] = [y]$ and then $h(x) = h(y)$, second case $[x] \neq [y]$ and then $[x] \rightarrow [y] \in \mathcal{G}'$ thus $h'([x]) \leqslant h'([y])$, so $h(x) \leqslant h(y)$. If $x \Rightarrow y \in \mathcal{G}$ then $[x] \Rightarrow [y] \in \mathcal{G}'$ and $h'([x]) + 1 \leqslant h'([y])$ so $h(x) < h(y)$.

Theorem 4. *If $\mathcal{G}$ has no chain of type $(\rightarrow^* \Rightarrow)^n$ then the height h of theorem 3 satisfies $\forall x \in \mathcal{G}$, $h(x) < n$.*

Proof. In $\mathcal{G}'$, if $h'([y]) = n$ then there exists a chain $[x](\Rightarrow \rightarrow^*)^n[y]$ in $\mathcal{G}'$. This result is straightforwardly proved by induction on n. Then suppose that there exists y such that $h(y) = m \geqslant n$. We obtain a chain of type $(\Rightarrow \rightarrow^*)^m$ in $\mathcal{G}'$. Expanding the "green" strongly connected components of $\mathcal{G}$, we obtain a chain of type $(\rightarrow^*(\Rightarrow \rightarrow^*)\rightarrow^*)^m$ in $\mathcal{G}$. It contains a subchain of type $(\rightarrow^* \Rightarrow)^n$.

Theorem 5. *Let $\mathcal{G}$ be a bi-colored graph with no r-cycle. Then, for n greater than the number of nodes of $\mathcal{G}$, the graph $\mathcal{G}$ has no chain of the form $(\rightarrow^* \Rightarrow)^n$.*

Proof. Let s be the number of nodes of $\mathcal{G}$. Let $n \geqslant s$. Suppose that $\mathcal{G}$ has a chain of the form $x_0 \rightarrow^* \Rightarrow x_1 \rightarrow^* \Rightarrow \cdots \rightarrow^* \Rightarrow x_n$. If all the x_i are different then the set $\{x_0, x_1, \ldots, x_n\}$ contains $n + 1 > s$ nodes of $\mathcal{G}$. This is not possible. So let $i < j$ be such that $x_i = x_j$. The chain $x_i \rightarrow^* \Rightarrow \cdots \rightarrow^* \Rightarrow x_j$ is a r-cycle.

4.3 Counter-Models Versus Chains in Bi-colored Graphs

Lemma 1. *Let $\mathcal{S}$ be an _implicational sequent_ and $\mathcal{G}_\mathcal{S}$ its associated bi-colored graph. Let $[\![\cdot]\!] : \mathrm{Var}_\diamond \rightarrow \overline{[0, n)}$ be a counter-model of $\mathcal{S}$ in LC_n and $X_1 \rightarrow \cdots \rightarrow X_k \Rightarrow Y$ a chain in $\mathcal{G}_\mathcal{S}$. Then $[\![X_1]\!] \leqslant \cdots \leqslant [\![X_k]\!] < [\![Y]\!]$ holds.*

Proof. Let $\mathcal{S} \equiv \Gamma \vdash \Delta$. As $[\![\cdot]\!]$ is a counter-model, the relation $\lceil\Delta\rceil < \lfloor\Gamma\rfloor$ holds. As $X_k \Rightarrow Y \in \mathcal{G}_\mathcal{S}$, the formula $Y \supset X_k$ is an element of Δ. So we deduce $[\![Y \supset X_k]\!] \leqslant \lceil\Delta\rceil < \infty$. Thus we obtain $[\![Y]\!] > [\![X_k]\!]$ and $[\![X_k]\!] = [\![Y \supset X_k]\!] \leqslant \lceil\Delta\rceil$. Also $X_{k-1} \to X_k \in \mathcal{G}_\mathcal{S}$, so $X_{k-1} \supset X_k$ belongs to Γ. Thus $[\![X_k]\!] \leqslant \lceil\Delta\rceil < \lfloor\Gamma\rfloor \leqslant [\![X_{k-1} \supset X_k]\!]$ holds. So it is necessary that $[\![X_{k-1}]\!] \leqslant [\![X_k]\!]$ (because otherwise, $[\![X_{k-1} \supset X_k]\!] = [\![X_k]\!]$ holds) and we deduce $[\![X_{k-1}]\!] \leqslant [\![X_k]\!] \leqslant \lceil\Delta\rceil$. By descending induction on i for $k-1, \ldots, 2$, we prove $[\![X_{i-1}]\!] \leqslant [\![X_i]\!] \leqslant \lceil\Delta\rceil$

Theorem 6. *For $n < \infty$, the implicational sequent $\mathcal{S}$ has a counter-model in* LC_n *if and only if its associated graph $\mathcal{G}_\mathcal{S}$ contains no chain of type* $(\to^\star\Rightarrow)^{n+1}$.

Proof. Let $\mathcal{S} \equiv \Gamma \vdash \Delta$. First we prove the if part. We suppose that $\mathcal{G}_\mathcal{S}$ contains no chain of the form $(\to^\star\Rightarrow)^{n+1}$. Then by theorem 4, there exists a bi-height function $h : \mathcal{G}_\mathcal{S} \to [0, n]$. We define the semantic function $[\![\cdot]\!] : \mathrm{Var}_\diamond \to \overline{[0, n)}$ by $[\![X]\!] = h(X)$ if $h(X) < n$ and $[\![X]\!] = \infty$ if $h(X) = n$ if X occurs in $\mathcal{S}$ (i.e. is a node of $\mathcal{G}_\mathcal{S}$,) and $[\![X]\!] = \infty$ (or any other value) when X does not occur in $\mathcal{S}$. Now, let us prove that $[\![\cdot]\!]$ is a counter-model of $\mathcal{S}$. Indeed, if $X \supset Y \in \Gamma$ then $X \to Y \in \mathcal{G}_\mathcal{S}$ and then $h(X) \leqslant h(Y)$. It follows that $[\![X]\!] \leqslant [\![Y]\!]$ and so $[\![X \supset Y]\!] = \infty$. We have $\lfloor\Gamma\rfloor = \infty$. If $X \supset Y \in \Delta$ then $Y \Rightarrow X \in \mathcal{G}_\mathcal{S}$. Thus $h(Y) < h(X)$ and $[\![X \supset Y]\!] = [\![Y]\!] = h(Y) < h(X) \leqslant n$. $[\![X \supset Y]\!] \leqslant n - 1$ holds and so $\lceil\Delta\rceil \leqslant n - 1$ holds. Finally, $\lceil\Delta\rceil \leqslant n - 1 < \infty = \lfloor\Gamma\rfloor$ so $[\![\cdot]\!]$ is a counter-model of $\mathcal{S}$.

Now we prove the only if part of the theorem. Let $[\![\cdot]\!] : \mathrm{Var}_\diamond \to \overline{[0, n)}$ be a counter-model of $\mathcal{S}$. Suppose there is a chain of the form $(\to^\star\Rightarrow)^{n+1}$ in $\mathcal{G}_\mathcal{S}$: $X_0 \to^\star\Rightarrow X_1 \to^\star\Rightarrow X_2 \to^\star\Rightarrow \cdots \to^\star\Rightarrow X_n \to^\star\Rightarrow X_{n+1}$. So for any i, there is a chain $X_i \to^\star\Rightarrow X_{i+1}$ and by lemma 1, we obtain $[\![X_i]\!] < [\![X_{i+1}]\!]$. Then, $[\![X_0]\!] < [\![X_1]\!] < \cdots < [\![X_{n+1}]\!]$ is a strictly increasing sequence of $n+2$ elements in $\overline{[0, n)}$. As this set has $n+1$ elements, we get a contradiction.

Theorem 7. *An implicational sequent $\mathcal{S}$ has a counter-model in* LC *if and only if its associated graph $\mathcal{G}_\mathcal{S}$ has no r-cycle.*

Proof. For the if part, if $\mathcal{G}_\mathcal{S}$ has no r-cycle, by theorem 3, there is a height $h : \mathcal{G}_\mathcal{S} \to \mathbb{N}$. We define $[\![X]\!] \in \overline{\mathbb{N}}$ by $[\![X]\!] = h(X)$ and obtain a counter-model of $\mathcal{S}$ in LC. For the only if part, the existence of a chain $X \to^\star\Rightarrow \to^\star\Rightarrow \cdots \to^\star\Rightarrow X$ would lead to $[\![X]\!] < [\![X]\!]$ by the same argument as before.

4.4   Algebraic Criteria and the Limit Counter-Model

We have seen that the existence of counter-models of implicational sequents is equivalent to the lack of r-cycles in the associated bi-colored graphs. Now we present an algebraic formula that expresses the existence of r-cycles. Let $\mathcal{G}$ be a bi-colored graph of k nodes with its incidence relations $\to$ and $\Rightarrow$. The relation $\to$ (or $\Rightarrow$) can be viewed as an *incidence $k \times k$-matrix* whose rows and columns are indexed by the nodes of $\mathcal{G}$. The cells of these matrices take their value in the boolean algebra $\{0, 1\}$. So there is a 1 at cell (x, y) in the matrix of $\to$ if and only if $x \to y \in \mathcal{G}$. We define $+$ as the disjunction (or logical "or") and $\cdot$ as the conjunction (or logical "and") in the boolean algebra $\{0, 1\}$. These operations extend naturally to sum and multiplication of square boolean matrices.

Now if we identify the relations $\to$ and $\Rightarrow$ with their respective matrices, the composed relation $\to\Rightarrow$ has a corresponding matrix $\to \cdot \Rightarrow$ and the union of relations $\to$ and $\Rightarrow$ has a corresponding matrix $\to + \Rightarrow$. The relation $\to^{\star}$ corresponds to a matrix $\sum_{i \geqslant 0} \to^i$. So $((\to + \Rightarrow)^{\star} \Rightarrow)_{x,x} = 1$ means that there exists a chain of the form $x(\to + \Rightarrow)^{\star} \Rightarrow x$ in the graph $\mathcal{G}$. Let $\mathrm{tr}(\cdot)$ denote the trace of matrices defined by $\mathrm{tr}(M) = \sum_x M_{x,x}$.

Proposition 3. $\mathcal{G}$ has a r-cycle if and only if $\mathrm{tr}\big((\to + \Rightarrow)^{\star} \Rightarrow\big) = 1$.

In section 6, we will explain how to compute this trace efficiently. We conclude this section by a criterion to determine the minimal n for which a given sequent $\mathcal{S}$ has a counter-model in LC_n. Let $\sum M$ denote the sum of all the elements of the matrix M defined by $\sum M = \sum_{x,y} M_{x,y}$.

Proposition 4. Let $\mathcal{S}$ be an implicational sequent not valid in LC. There exists a minimal n such that $\mathcal{S}$ has a counter-model in LC_n and it is the first n s.t. $\sum(\to^{\star}\Rightarrow)^{n+1} = 0$.

Proof. By theorem 6, $\mathcal{S}$ has a counter-model in LC_n, iff $\mathcal{G}_{\mathcal{S}}$ has no chain of the form $(\to^{\star}\Rightarrow)^{n+1}$. Having no chain of the form $(\to^{\star}\Rightarrow)^{n+1}$ means the matrix of this relation is the zero matrix, i.e. $\sum(\to^{\star}\Rightarrow)^{n+1} = 0$. As $\mathcal{S}$ is not valid in LC, by theorem 7, its bi-colored graph $\mathcal{G}_{\mathcal{S}}$ has no r-cycle. Then, there is an n s.t. $\mathcal{G}_{\mathcal{S}}$ has no chain of the form $(\to^{\star}\Rightarrow)^n$ by theorem 5. So there exists a minimal one. $\square$

5   Parallel Counter-Model Search

Combining the results of the preceding sections provides an algorithm to decide LC_n by indexation followed by proof-search to obtain a set of irreducible sequents (which are implicational sequents in our setting) and then counter-model construction for these irreducible sequents. This combination of proof-search and counter-model construction can now be viewed in a common graph-theoretic setting that will lead us to efficient parallel counter-model construction. First we present a graph-theoretic approach for proof-search. Then we combine proof-search and counter-model construction into a r-cycle search problem on a *conditional bi-colored graph*, i.e. a bi-colored graph where arrows might be indexed with *boolean selectors*. In section 6, we present a global technique which efficiently solves the conditional r-cycle search problem.

5.1   Proof-Search as Bi-colored Graph Construction

Let us fix a particular formula D. Using the results of section 3, D is indexed into an equivalent flat sequent $\delta^-(D) \vdash \Diamond \supset \mathcal{X}_D^-$.[3] Then, let us study the formulae occurring in $\delta^-(D)$. From the definition of δ (see section 3), there is exactly one formula in $\delta^-(D)$ for each occurrence of a subformula of D:

---

[3] The polarity of occurrences of subformulae is suffixed by their indexes like in $\mathcal{X}_V^+$.

Table 1. Proof rules and bi-colored graph construction

$\dfrac{\Gamma, \mathcal{X}_A^- \supset \mathcal{X}_{A \wedge B}^- \vdash \Delta_b \qquad \Gamma, \mathcal{X}_B^- \supset \mathcal{X}_{A \wedge B}^- \vdash \Delta_b}{\Gamma, (\mathcal{X}_A^- \wedge \mathcal{X}_B^-) \supset \mathcal{X}_{A \wedge B}^- \vdash \Delta_b}$ $[\supset_2]$	$\begin{array}{c} \wedge^- \\ \nearrow \\ A^- \qquad B^- \end{array}$	$\begin{array}{c} \wedge^- \\ \nwarrow \\ A^- \qquad B^- \end{array}$
$\dfrac{\Gamma, \mathcal{X}_{A \wedge B}^+ \supset \mathcal{X}_A^+, \mathcal{X}_{A \wedge B}^+ \supset \mathcal{X}_B^+ \vdash \Delta_b}{\Gamma, \mathcal{X}_{A \wedge B}^+ \supset (\mathcal{X}_A^+ \wedge \mathcal{X}_B^+) \vdash \Delta_b}$ $[\supset_2']$	$\begin{array}{c} \wedge^+ \\ \swarrow \quad \searrow \\ A^+ \qquad B^+ \end{array}$	
$\dfrac{\Gamma, \mathcal{X}_A^- \supset \mathcal{X}_{A \vee B}^-, \mathcal{X}_B^- \supset \mathcal{X}_{A \vee B}^- \vdash \Delta_b}{\Gamma, (\mathcal{X}_A^- \vee \mathcal{X}_B^-) \supset \mathcal{X}_{A \vee B}^- \vdash \Delta_b}$ $[\supset_3]$	$\begin{array}{c} \vee^- \\ \nearrow \quad \nwarrow \\ A^- \qquad B^- \end{array}$	
$\dfrac{\Gamma, \mathcal{X}_{A \vee B}^+ \supset \mathcal{X}_A^+ \vdash \Delta_b \qquad \Gamma, \mathcal{X}_{A \vee B}^+ \supset \mathcal{X}_B^+ \vdash \Delta_b}{\Gamma, \mathcal{X}_{A \vee B}^+ \supset (\mathcal{X}_A^+ \vee \mathcal{X}_B^+) \vdash \Delta_b}$ $[\supset_3']$	$\begin{array}{c} \vee^+ \\ \swarrow \\ A^+ \qquad B^+ \end{array}$	$\begin{array}{c} \vee^+ \\ \searrow \\ A^+ \qquad B^+ \end{array}$
$\dfrac{\begin{array}{c} \Gamma, \mathcal{X}_B^- \supset \mathcal{X}_{A \supset B}^- \vdash \mathcal{X}_A^+ \supset \mathcal{X}_B^-, \Diamond \supset \mathcal{X}_B^-, \Delta_b \\ \vdots \\ \Gamma, \Diamond \supset \mathcal{X}_{A \supset B}^- \vdash \Delta_b \end{array}}{\Gamma, (\mathcal{X}_A^+ \supset \mathcal{X}_B^-) \supset \mathcal{X}_{A \supset B}^- \vdash \Delta_b}$ $[\supset_4]$	$\begin{array}{c} \supset^- \qquad \Diamond \\ \nwarrow \quad \nearrow\!\!\!\!\nearrow \\ A^+ \Longleftarrow B^- \end{array}$	$\begin{array}{c} \supset^- \longleftarrow \Diamond \\ \\ A^+ \qquad B^- \end{array}$
$\dfrac{\Gamma, \mathcal{X}_{A \supset B}^+ \supset \mathcal{X}_B^+ \vdash \Delta_b \qquad \Gamma, \mathcal{X}_A^- \supset \mathcal{X}_B^+ \vdash \Delta_b}{\Gamma, \mathcal{X}_{A \supset B}^+ \supset (\mathcal{X}_A^- \supset \mathcal{X}_B^+) \vdash \Delta_b}$ $[\supset_4']$	$\begin{array}{c} \supset^+ \\ \searrow \\ A^- \qquad B^+ \end{array}$	$\begin{array}{c} \supset^+ \\ \\ A^- \longrightarrow B^+ \end{array}$

- if V is a variable occurring positively (resp. negatively), the formula in $\delta^-(D)$ is $\mathcal{X}_V^+ \supset V$ (resp. $V \supset \mathcal{X}_V^-$) which is already an atomic implication which will not be decomposed further during proof search;

- if $A \supset B$ occurs negatively, $(\mathcal{X}_A^+ \supset \mathcal{X}_B^-) \supset \mathcal{X}_{A \supset B}^-$ will appear in $\delta^-(D)$ and could be decomposed once using rule $[\supset_4]$.

and so on... There is exactly one logical rule for each case: $[\supset_2]$ for $\wedge^-$, $[\supset_2']$ for $\wedge^+$, $[\supset_3]$ for $\vee^-$, $[\supset_3']$ for $\vee^+$, $[\supset_4]$ for $\supset^-$, and $[\supset_4']$ for $\supset^+$. The left column of table 1 presents all these cases.

What are the atomic implications occurring in the sequents generated by a full proof-search from starting $\delta^-(D) \vdash \Diamond \supset \mathcal{X}_D^-$? Some appear already at the beginning like the formulae $\mathcal{X}_V^+ \supset V$ or $V \supset \mathcal{X}_V^-$ for V occurrence of a variable of D and are not changed by proof rules. Also $\Diamond \supset \mathcal{X}_D^-$ appears at the beginning and is not changed by proof rules. So in the corresponding bi-colored graph, there are arrows $\mathcal{X}_V^+ \to V$, $V \to \mathcal{X}_V^-$ and $\mathcal{X}_D^- \Rightarrow \Diamond$. The situation is summarized at the right hand side.

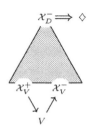

The other atomic implications are generated by the decompositions of formulae $\mathcal{X}_{A \otimes B} \supset (\mathcal{X}_A \otimes \mathcal{X}_B)$ and $(\mathcal{X}_A \otimes \mathcal{X}_B) \supset \mathcal{X}_{A \otimes B}$ occurring during backward proof rules application. As an example, we consider the case of a negative occurrence of $A \supset B$. Then $(\mathcal{X}_A^+ \supset \mathcal{X}_B^-) \supset \mathcal{X}_{A \supset B}^-$ occurs in $\delta^-(D)$, the corresponding rule is

$$\frac{\ldots, \mathcal{X}_B^- \supset \mathcal{X}_{A \supset B}^- \vdash \mathcal{X}_A^+ \supset \mathcal{X}_B^-, \Diamond \supset \mathcal{X}_B^-, \ldots \qquad \ldots, \Diamond \supset \mathcal{X}_{A \supset B}^- \vdash \ldots}{\ldots, (\mathcal{X}_A^+ \supset \mathcal{X}_B^-) \supset \mathcal{X}_{A \supset B}^- \vdash \ldots} \; [\supset_4]$$

Let us consider a completed proof-search branch ending with an implicational sequent. Its associated graph either contains $\mathcal{X}_B^- \to \mathcal{X}_{A \supset B}^-$, $\mathcal{X}_B^- \Rightarrow \mathcal{X}_A^+$ and $\mathcal{X}_B^- \Rightarrow \Diamond$ if we choose left premise of rule $[\supset_4]$ or $\Diamond \to \mathcal{X}_{A \supset B}^-$ if we choose right premise. This could be summarized by the two arrow introduction rules displayed on the rhs.

The complete set of arrow introduction rules is given in the right column of table 1. With this set of rules, each internal occurrence of a subformula introduces arrows in the bi-colored graph, depending on the choice of the left premise or right premise for the cases $\wedge^-$, $\vee^+$, $\supset^-$, and $\supset^+$. Since rules $\wedge^+$ and $\vee^-$ only have one premise, there is no choice in these cases and proof-search does not branch. So the end-sequent of a completed proof-search branch and its corresponding bi-colored graph is characterized by a choice of left or right premise for internal nodes of the shape $\wedge^-$, $\vee^+$, $\supset^-$, and $\supset^+$. The reader is reminded that the proof rules implied in proof-search are all strongly invertible and so the order of application of those rules does not influence the validity of the generated sequents: all the rules are permutable. Let us illustrate these ideas on an example.

5.2 An Example: The Peirce's Formula

$D = ((A \supset B) \supset A) \supset A$ is the Peirce's formula. We index it:

$$((A_5^+ \supset_3^- B_6^-) \supset_1^+ A_4^+) \supset_0^- A_2^-$$

The decomposition tree of this formula has three internal nodes $\supset_0^-$, $\supset_1^+$ and $\supset_3^-$. The leaves of the decomposition tree introduce the arrows $A_5^+ \to A$, $A_4^+ \to A$, $A \to A_2^-$ and $B \to B_6^-$ in the bi-colored graph. And of course, there is also the arrow $\supset_0^- \Rightarrow \Diamond$ corresponding to $\ldots \vdash \Diamond \supset \mathcal{X}_D^-$.

Then we choose one proof-search branch: left premise for $\supset_0^-$ and $\supset_3^-$ and right premise for $\supset_1^+$. We obtain the bi-colored graph presented on the

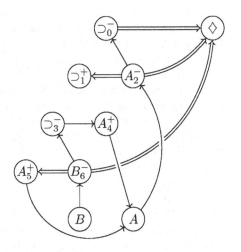

rhs. We explain the remaining arrows: the left premise for $\supset_0^-$ (rule $[\supset_4]$) introduces $A_2^- \to \supset_0^-$, $A_2^- \Rightarrow \supset_1^+$ and $A_2^- \Rightarrow \Diamond$; the right premise for $\supset_1^+$ (rule $[\supset_4']$) introduces $\supset_3^- \to A_4^+$; the left premise for $\supset_3^-$ (rule $[\supset_4]$) introduces $B_6^- \to \supset_3^-$, $B_6^- \Rightarrow A_5^+$ and $B_6^- \Rightarrow \Diamond$. This graph is the bi-colored graph associated with the implicational end-sequent of the proof-search branch characterized by the choice: left for $\supset_0^-$ and $\supset_3^-$ and right for $\supset_1^+$.

To decide if this sequent is valid in LC, we look for r-cycles. For this, we redesign the graph so that $\rightarrow$ arrows are horizontal or go up and $\Rightarrow$ go up strictly. It appears that this graph has a bi-height h defined by $h(x) = 0$ for $x \in \{B, 6, 3, 4\}$, $h(x) = 1$ for $x \in \{5, A, 2, 0\}$ and $h(x) = 2$ for $x \in \{\Diamond, 1\}$. Then $\llbracket \cdot \rrbracket$ defined by $\llbracket A \rrbracket = 1$ and $\llbracket B \rrbracket = 0$ is a counter-model of the Peirce's formula since then $\llbracket ((A \supset B) \supset A) \supset A \rrbracket = ((1 \rightarrow 0) \rightarrow 1) \rightarrow 1 = (0 \rightarrow 1) \rightarrow 1 = \infty \rightarrow 1 = 1 < \infty$. $\llbracket \cdot \rrbracket$ is a counter-model of Peirce's formula not only in LC but also in LC_2 (just under $\mathsf{CL} = \mathsf{LC}_1$.)

5.3   Postponing Proof-Search Branch Selection

Our system has a very important property that the others lack: proof-search can be seen as the incremental construction of a semantic graph. The nodes of this graph do not depend on the proof-search branch; only the choices of arrows depend on the branch chosen. In section 5.2, we have chosen a "good" proof-search branch from which we can extract a counter-model. There are many other branches that lead to other bi-colored graphs which might also lack r-cycles, leading to other counter-models. Clearly, a proof-search branch lacking r-cycles corresponds to a counter-model. Is it possible to find all the counter-models and thus, all such branches?

An idea for that is to postpone the choice of premises characterizing proof-search branches, detect r-cycles and then select the branches for which no r-cycle exists.

Instead of using either left or right premise, let us use both. Of course this would not be sound unless we keep track of the fact that these two choices cannot coexist in the same branch. This is done by introducing boolean selectors (x and its negation $\overline{x}$) and indexing arrows with those selectors. For example, we obtain the following transformation of rules for a positive occurrence of disjunction. Let us reconsider the case of the Peirce's formula. We apply all the possible rules, postponing all the choices between left or right premises and introducing a new boolean selector for each potential choice. All three rules have two premises and we introduce the selectors

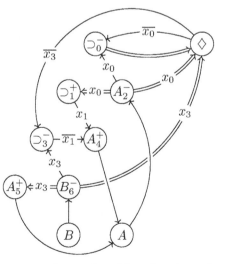

x_0 for $\supset_0^-$, x_1 for $\supset_1^+$ and x_3 for $\supset_3^-$. The corresponding indexed bi-colored graph is represented at the right. The reader might notice that the graph of section 5.2 is just an instance of this graph with $x_0 = 1$, $x_1 = 0$ and $x_3 = 1$.

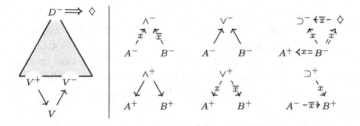

Fig. 2. Counter-model search system for LC_n

We want to discover all the possible valuations of selectors such that the corresponding instance graph has no r-cycle. For example, $0 \Rightarrow \Diamond \rightarrow 0$ is a r-cycle. To "break" this cycle, the constraint $\overline{x_0} = 0$ has to be satisfied, i.e. x_0 has to be satisfied. Let us do this for all the basic r-cycles, i.e. the r-cycles that do not repeat nodes. It is sufficient to "break" basic r-cycles for all the r-cycles to be removed because any r-cycle contains at least one basic r-cycle.

Let us look for basic r-cycles: no r-cycle passes through B so we can remove it from the graph. Then no r-cycle passes through B_6^-. Then no r-cycle passes through A_5^+. Then, we

$0 \Rightarrow \Diamond \rightarrow 0$	x_0
$0 \Rightarrow \Diamond \rightarrow 3 \rightarrow 4 \rightarrow A \rightarrow 2 \rightarrow 0$	$x_3 + x_1 + \overline{x_0}$
$2 \Rightarrow \Diamond \rightarrow 3 \rightarrow 4 \rightarrow A \rightarrow 2$	$\overline{x_0} + x_3 + x_1$
$2 \Rightarrow 1 \rightarrow 4 \rightarrow A \rightarrow 2$	$\overline{x_0} + \overline{x_1}$

obtain four basic r-cycles and their associated constraints expressing the condition at which the corresponding r-cycle is broken: x_0, $x_3 + x_1 + \overline{x_0}$, $\overline{x_0} + x_3 + x_1$ and $\overline{x_0} + \overline{x_1}$. Then we solve these constraints altogether, i.e. we search a valuation satisfying the conjunction of all these constraints. There is only one solution which is $x_0 = 1$, $x_1 = 0$ and $x_3 = 1$, corresponding to the branch characterized by left for $\supset_0^-$ and $\supset_3^-$ and right for $\supset_1^+$. This is the proof-search branch we have chosen in section 5.2. With any other valuation, one of the four basic r-cycles would exist, and so, it is the only one for which the instanced bi-colored graph has no r-cycle.

5.4   A Parallel Counter-Model Search System for LC_n

We introduce our final counter-model search system which is a combination of the graphical system of table 1 and the idea of postponing choice of left or right premise with the help of boolean selectors. The word parallel means that all the search branches are explored simultaneously with the help of selectors.

The system is presented in figure 2. The principle is to build a conditional bi-colored graph (where arrows might be indexed by boolean selectors,) notion which formalizes the concept we sketched in section 5.3. We fix a formula D. We start from the nodes of the decomposition tree of D composed of internal nodes ($\mathcal{X}_{A \otimes B}^+$ or $\mathcal{X}_{A \otimes B}^-$) and leaves ($\mathcal{X}_V^+$ or $\mathcal{X}_V^-$). We add one new node for each variable V occurring in D.[4] We also add one node for $\Diamond$. Then, we add the arrow

---

[4] For a variable V occurring in D, there are some nodes $\mathcal{X}_V^+$ (resp. $\mathcal{X}_V^-$) where V is a positive (resp. negative) occurrence in D, and one node for the variable V.

$D^- \Rightarrow \Diamond$ (D^- is the root of the decomposition tree) and arrows $V^+ \to V$ (resp. $V \to V^-$) for each positive (resp. negative) occurrence of the variable V. This is summarized by the left side of figure 2. All these arrows are not annotated: they are present in every proof-search branch, i.e. for any valuation on selectors.

Then for each internal node, we add arrows according to the schemes of the right part of figure 2. There is one scheme for each case in $\{\wedge, \vee, \supset\} \times \{+, -\}$. For the schemes $\wedge^-$, $\vee^+$, $\supset^+$ and $\supset^-$, we introduce conditional arrows, i.e arrows indexed with a boolean selector of the form x or $\bar{x}$. This selector has to be new: there is exactly one selector for each instance of a scheme. Each scheme introduces two or four arrows so the construction of the conditional bi-colored graph is a process which take linear time w.r.t. the size of D.

Theorem 8. *Let D be a formula and $\mathcal{G}$ be the corresponding conditional bi-colored graph obtained by the previously described process. Then D has a counter-model in* LC *if and only if there exists a valuation v on the selectors such that the instance graph $\mathcal{G}_v$ has no r-cycle.*

Proof. Because all the proof rules are strongly invertible, $[\![\cdot]\!]$ is a counter-model of D if and only if it is a counter-model of at least one of the irreducible (and implicational) end-sequent of a completed proof-search branch. This end-sequent is characterized by a choice of left or right premises and it corresponds to a choice of a valuation on selectors.

By theorem 7, the end-sequent of a proof-search branch has a counter-model if and only if its associated bi-colored graph has no r-cycle. But this graph is exactly the instance $\mathcal{G}_v$ where v is the valuation on selectors corresponding to the proof-search branch.

Theorem 9. *D has a counter-model in* LC_n *iff there exists a valuation v on the selectors such that the instance graph $\mathcal{G}_v$ has no chain of the form $(\to^\star \Rightarrow)^{n+1}$.*

Proof. For LC_n with $n \neq \infty$, we just have to apply theorem 6 instead of theorem 7. The proof rules are common for all the family LC_n (including $n = \infty$.)

Computing the conditional bi-colored graph from a given formula takes a linear time but finding all the r-cycles in a graph and solving the boolean constraints system have both exponential complexity. Indeed, there exists formulae for which the conditional bi-colored graph has exponentially many cycles. So on the complexity side, finding cycles and then building a constraint system is not a good idea. An approach to solve this problem would be to try to mix r-cycle detection and boolean constraint solving to obtain a kind of conditional r-cycle detection but this is not an easy task. After reflection, we can observe that we do not need the list of all the r-cycles to decide a formula: we only need to compute and solve the boolean condition characterizing the existence of r-cycles.

6   Practical r-Cycle Detection in Conditional Graphs

In section 5.4, we have introduced the notion of conditional bi-colored graph which is a graph in which arrows might be indexed with boolean constraints. To implement conditional graphs, we have chosen to represent the relations $\rightarrow$ and $\Rightarrow$ as generalized incidence matrices. In an incidence matrix, there is a 1 in a cell if the corresponding arrow exists in the graph. Generalized incidence matrix cells might contain not only 0 or 1 (for unconditional arrows) but also arbitrary boolean expressions built upon atomic boolean selectors. These expressions are considered up to boolean equivalence.

Definition 3 (Conditional matrix). *A conditional matrix on set $\mathcal{S}$ of size k is a $k \times k$-array with values in the free boolean algebra over the set of selectors.*

So a conditional bi-colored graph is viewed as a pair $(\rightarrow, \Rightarrow)$ of conditional matrices. The algebraic operations we have defined like $\rightarrow + \Rightarrow$, $\rightarrow \cdot \Rightarrow$, $\rightarrow^\star$, $\mathrm{tr}(\cdot)$ and $\sum(\cdot)$ extend naturally to conditional matrices because they rely only on the boolean operators $\cdot$ and $+$. If v is a valuation of boolean variables in $\{0,1\}$, we call an *instance graph* and denote by $\mathcal{G}_v$ the bi-colored graph obtained by instantiating boolean expressions in the cells of the matrices $\rightarrow$ and $\Rightarrow$ with the valuation v. Instantiation commutes with algebraic operations on matrices because it commutes with the boolean operators $\cdot$ and $+$. For example $[\rightarrow + \Rightarrow]_v = \rightarrow_v + \Rightarrow_v$.

Theorem 10. *Let $\mathcal{G} = (\rightarrow, \Rightarrow)$ be a conditional bi-colored graph. There exists a r-cycle in every instance $\mathcal{G}_v$ of $\mathcal{G}$ if and only if $\mathrm{tr}\big((\rightarrow + \Rightarrow)^\star \Rightarrow\big) = 1$ holds.*

The proof is straightforward: $\big[\mathrm{tr}\big((\rightarrow + \Rightarrow)^\star \Rightarrow\big)\big]_v = \mathrm{tr}\big((\rightarrow_v + \Rightarrow_v)^\star \Rightarrow_v\big)$ and then theorem 8. The reader is reminded that the equation $\mathrm{tr}\big((\rightarrow + \Rightarrow)^\star \Rightarrow\big) = 1$ is an equivalence of boolean expressions. The result for the finitary version LC_n also holds. To compute the smallest n such that a formula is not valid, we compute sequentially until $\sum(\rightarrow^\star \Rightarrow)^{n+1} < 1$ and then, find a valuation v such that the instance is refuted: $\sum(\rightarrow_v^\star \Rightarrow_v)^{n+1} = 0$.

We have implemented a prototype of counter-model search engine for LC based on these principles in Objective Caml. We give a brief description of the practical choices we have made. A $k \times k$ conditional matrix is represented by a *sparse matrix* of integers. These integers represent nodes of a *shared RO-BDD* [5] so they uniquely encode boolean expressions up to equivalence and algebraic operations may be performed efficiently. How can we compute traces efficiently? Let α be a $k \times k$ conditional matrix. Let I be the identity $k \times k$ matrix ($I_{x,x} = 1$ and $I_{x,y} = 0$ otherwise.) Then one could check that $\alpha^\star = (I + \alpha)^k$. It is even sufficient to compute the sequence $I, I + \alpha, (I + \alpha)^2, \ldots$ until it stabilizes, which happens in at most k steps. So let $A = \rightarrow + \Rightarrow$ and $B = \Rightarrow$. We want to compute $\mathrm{tr}(A^\star B)$. We compute the trace column by column. Let B_i be the i-th column of B and T be the column with 1 on each cell. Then we compute the sequence $t_0, \ldots, t_k$: $t_0 = 0$ and $t_i = \big[A^\star(t_{i-1}T + B_i)\big]_i$ for $i = 1, \ldots, k$ by induction and we obtain $t_k = \mathrm{tr}(A^\star B)$. The sequence $t_0, \ldots, t_k$ is increasing and the computation can be stopped as soon as $t_i = 1$ in which case, the formula is provable. Previously generated constraints are used to accelerate the computation by adding t_{i-1} to each cell of B_i (in $t_{i-1}T + B_i$).

7   Conclusion

Compared to the work initiated by Avron [2,3] that focuses mainly on proof-search and even logic programming [13], we develop the idea of mixing proof-search and counter-model construction. Then, we propose a new system for counter-model search in LC and LC_n, mainly based on the notion of r-cycles in conditional graphs, and thus an efficient algorithm to decide these logics and provide counter-models. In further work, we will deeper investigate the relationships between the notion of r-cycle and the G-cycles of [3] and analyze if our conditional graphs also fit in the hyper-sequent setting. We will also investigate the relationships between our parallel counter-model search and other approaches based for example on parallel dialogue games [4,9].

References

[1] Alessendro Avellone, Mauro Ferrari, and Pierangelo Miglioli. Duplication-Free Tableau Calculi and Related Cut-Free Sequent Calculi for the Interpolable Propositional Intermediate Logics. *Logic Journal of the IGPL*, 7(4):447–480, 1999.

[2] Arnon Avron. A Tableau System for Gödel-Dummett Logic Based on a Hypersequent Calculus. In *TABLEAUX 2000*, volume 1847 of *LNAI*, pages 98–111, 2000.

[3] Arnon Avron and Beata Konikowska. Decomposition Proof Systems for Gödel-Dummett Logics. *Studia Logica*, 69(2):197–219, 2001.

[4] Matthias Baaz and Christian Fermüller. Analytic Calculi for Projective Logics. In *TABLEAUX'99*, volume 1617 of *LNCS*, pages 36–50, 1999.

[5] Randal E. Bryant. Graph-based algorithms for Boolean function manipulation. *IEEE Transactions on Computers*, C-35(8):677–691, 1986.

[6] Michael Dummett. A Propositional Calculus with a Denumerable matrix. *Journal of Symbolic Logic*, 24:96–107, 1959.

[7] Roy Dyckhoff. Contraction-free Sequent Calculi for Intuitionistic Logic. *Journal of Symbolic Logic*, 57(3):795–807, 1992.

[8] Roy Dyckhoff. A Deterministic Terminating Sequent Calculus for Gödel-Dummett logic. *Logical Journal of the IGPL*, 7:319–326, 1999.

[9] Christian Fermüller. Parallel Dialogue Games and Hypersequents for Intermediate Logics. In *TABLEAUX 2003*, volume 2796 of *LNAI*, pages 48–64, 2003.

[10] Kurt Gödel. Zum intuitionistischen Aussagenkalkül. In *Anzeiger Akademie des Wissenschaften Wien*, volume 69, pages 65–66. 1932.

[11] Petr Hajek. *Metamathematics of Fuzzy Logic*. Kluwer Academic Publishers, 1998.

[12] Dominique Larchey-Wendling. Combining Proof-Search and Counter-Model Construction for Deciding Gödel-Dummett Logic. In *CADE-18*, volume 2392 of *LNAI*, pages 94–110, 2002.

[13] George Metcalfe, Nicolas Olivetti, and Dov Gabbay. Goal-Directed Calculi for Gödel-Dummett Logics. In *CSL*, volume 2803 of *LNCS*, pages 413–426, 2003.

Generalised Handling of Variables in Disconnection Tableaux

Reinhold Letz and Gernot Stenz

Institut für Informatik
Technische Universität München
D-85748 Garching, Germany
{letz,stenzg}@in.tum.de

Abstract. Recent years have seen a renewed interest in instantiation based theorem proving for first-order logic. The disconnection calculus is a successful approach of this kind, it integrates clause linking into a tableau guided proof procedure. In this paper we consider extensions of both the linking concept and the treatment of variables which are based on generalised notions of unifiability. These extensions result in significantly enhanced search space pruning and may permit the reduction of the length of refutations below Herbrand complexity which normally is a lower bound to the proof lengths of instantiation based methods. We present concise proofs of soundness and completeness of the new methods and show how these concepts relate to other instantiation based approaches.

1 Introduction

Instantiation based approaches to automated deduction have found a lot of attention recently [8,17,7,6,1,2]. One of the most successful such frameworks is the disconnection tableau calculus [4,11,15,16,20]. In contrast to most instantiation based methods, this framework does not interleave an instance generation subroutine with a separate propositional decision procedure. Instead, the detection of (partial) unsatisfiability is directly integrated into the instantiation procedure by using a tableau as a data structure for guiding the search. One of the main advantages of the disconnection tableau calculus is that it permits branch saturation and hence model generation for certain formula classes, which is in contrast to the family of so-called free-variable tableaux like, e.g., model elimination. On the other hand, the framework fully integrates unification in a controlled manner without the need for any form of blind instantiation like in Smullyan's γ-rule (which is explicitly or implicitly needed, for example, in all forms of hyper tableau calculi). The high potential of the disconnection tableau approach was convincingly demonstrated in the CADE-19 system competition where our disconnection tableau prover DCTP [19] came out best in the competition category devoted to the Bernays-Schönfinkel class.

It is well-known, however, that plain tableau procedures have weaknesses concerning the lengths of minimal refutations, which is because of their cutfreeness. Consequently, controlled forms of cut have been integrated into tableau

D. Basin and M. Rusinowitch (Eds.): IJCAR 2004, LNAI 3097, pp. 289–306, 2004.

methods like *folding up, folding down, factorisation* or *complement splitting* [14]. But those methods typically are most effective on the propositional level.

Another reason for the fact that tableau proofs tend to be relatively long is purely first-order in nature. Instantiation based proof methods, to which the plain disconnection calculus belongs, are based directly on Herbrand's theorem by showing the unsatisfiability of a first-order formula through reduction to ground instances. As a consequence, those methods have the natural limitation that the minimal size of a refutation is bounded from below by the *Herbrand complexity* of the formula. The use of so-called *local variables* presented in this paper represents an attempt to shorten tableau proofs in such a manner that Herbrand complexity is no more a lower bound. Additionally, this concept can be used to restrict the number of necessary clause instances in such a manner that certain forms of subsumption deletion become viable which normally are not compatible with instantiation based approaches.

In [16], a first attempt was pursued to extend the exploitation of local variables from unit clauses to clauses of arbitrary lengths. The completeness of the method, however, turned out to be a delicate problem and had to be left open in [16]. For example, constructive methods for proving completeness by means of proof transformations turned out to fail. In the meantime, as an additional complication, we detected that the new inference rule is no more *unitary* and hence may introduce an additional form of indeterminism. With the new concept of ∀-linking developed in this paper, this problem simply vanishes.

As a further main contribution of this paper, we address the so-called regularity condition, which excludes the duplication of literals on a tableau branch. With the method of local variables we can achieve a significant sharpening of regularity, thus bringing it closer to subsumption. This results in a further strong reduction of the search space.

The investigation of these new concepts showed that the fundamental means of instance generation used by instantiation based methods can be significantly improved. This is achieved by generalising standard unification in manners suited to the notions of complementarity employed in these approaches. These generalised methods are also applicable to other instantiation based calculi.

The paper is organised as follows. The next section briefly defines the basic notation and describes the disconnection calculus. Then, in Section 3 we introduce the new unification and clause linking concepts. In Section 4 we illustrate a fundamental weakness of standard disconnection tableaux and show how this weakness can be remedied by our integration of local variables. In Section 5, we prove the completeness of the new method. Section 6 discusses the regularity concept and its modification and we show that completeness can be preserved. Section 7 compares our work to related instantiation based approaches. We conclude with an assessment of this work and mention future perspectives in Section 8.

2   The Disconnection Tableau Calculus

Since its first presentation in [4], properties and extensions of the disconnection tableau calculus were discussed in [11,15,16]. Essentially, the method can be viewed as an integration of the *clause linking* method [8] into a tableau control structure. Given a set S of input clauses, clause linking iteratively produces instances of the clauses in the input set. This is done by applying substitutions σ to the clauses which result from *connections* or *links* in S, i.e., pairs of literals l, k such that $l\sigma$ and $\neg k\sigma$ are contradictory. In clause linking the set S is augmented with those instances and periodically tested for propositional unsatisfiability with the proviso that all variables are treated as one and the same constant.

In the disconnection tableau framework, the generation of new clause instances and the testing for propositional unsatisfiability (modulo the substitution of all variables) is performed in a uniform procedure, which avoids among other things the interfacing problems inherent to linking methods.

2.1   Terminological Prerequisites

Before we come to the definition of the method, we clarify our usage of the needed terminology. As usual, a *literal* is an atomic formula or a negated atomic formula, and a *clause* is a finite set of literals, sometimes displayed as a disjunction of literals. We use the terms *substitution*, *unifier* and *most general unifier (mgu)*, as well as *variant* and *(variable) renaming of a clause* in their standard meanings. Given a set of clauses S, its *Herbrand universe* $\mathcal{H}_U(S)$ and its *Herbrand base* $\mathcal{H}_B(S)$ are the set of all ground terms resp. the set of all ground atoms over the signature of S (if necessary augmented with a constant). The *Herbrand set S^** of S is the set of all ground instances of clauses in S with terms from $\mathcal{H}_U(S)$. A *total interpretation* $\mathcal{I}$ for a set of clauses S is a set of ground literals such that for every atom $A \in \mathcal{H}_B(S)$ exactly one of A or $\neg A$ is in $\mathcal{I}$. An *interpretation* $\mathcal{I}$ for S is any subset of a total interpretation for S. An interpretation $\mathcal{I}$ is termed a *model* for S if every clause in the Herbrand set S^* of S contains a literal which is in $\mathcal{I}$.[1]

Now, we come to notions more specific to the disconnection tableau method.

Definition 1 (Literal occurence, link, linking instance). *A literal occurence is any pair $\langle c, l \rangle$ where c is a clause and l is a literal in c. Given two literal occurences $\langle c, l \rangle$ and $\langle c', \neg l' \rangle$ with c and c' variable-disjoint, if there is a most general unifier σ of l and l', then the set $\ell = \{\langle c, l \rangle, \langle c', \neg l' \rangle\}$ is called a* connection *or* link *(between the clauses c and c'). The clauses $c\sigma$ and $c'\sigma$ are* linking instances *of c resp. of c' w.r.t. ℓ.*

Definition 2 (Path). *A path through a clause set S is any total mapping P : $S \rightarrow \bigcup S$ with $P(c) \in c$, i.e., a set containing exactly one literal occurence $\langle c, l \rangle$*

---

[1] So we have a more general view of interpretations and models, which normally would be rather called *partial* interpretations and models.

for every $c \in S$.[2] *With* the (set) of clauses *resp.* literals on *or in* P *we mean the domain resp. the range of the mapping* P. *A path* P *is* complementary *if it contains two literal occurences of the form* $\langle c, l \rangle$ *and* $\langle c', \neg l \rangle$, *otherwise* P *is called* open.

According to these definitions, we have the following.

Proposition 1. *For any set of clauses* S, *if* P *is an open path through the Herbrand set* S^* *of* S, *then the range of* P, *i.e., the set of its literals, is a model for* S.

Definition 3 (Tableau). *A* tableau *is a (possibly infinite) downward tree with all tree nodes except the root labeled with literal occurences. Given a clause set* S, *a* tableau for S *is a tableau in which, for every tableau node* N, *the literal occurences at its immediate successor nodes* $N_1, \ldots, N_m$ *are* $\langle c, l_1 \rangle, \ldots, \langle c, l_m \rangle$, *resp., and* $c = l_1 \vee \cdots \vee l_m$ *is an instance of a clause in* S. *Furthermore, a* branch *of a tableau* T *is any maximal sequence* $B = N_1, N_2, N_3, \ldots$ *of nodes in* T *such that* N_1 *is an immediate successor of the root node and any* N_{i+1} *is an immediate successor of* N_i, *i.e., every tableau branch is a path through the clauses on the branch.*

2.2 Details of the Disconnection Tableau Approach

A specific feature of the disconnection tableau calculus is that the construction of a tableau for a clause set S is started w.r.t. a path through S, which we call the *initial path*. An initial path is needed in order to be able to start the tableau construction. The initial path can be chosen arbitrarily, but remains fixed throughout the entire tableau construction. The disconnection tableau calculus consists of a single complex inference rule, the so-called *linking rule*.

Definition 4 (Linking rule). *Given an initial path[3]* P *and a tableau branch* B *with two literal occurences* $\langle c, l \rangle$ *and* $\langle c', \neg l' \rangle$ *in* $P \cup B$, *such that* $\ell = \{\langle c, l \rangle, \langle c', \neg l' \rangle\}$ *is a connection with most general unifier* σ, *then*

1. *expand the branch* B *with a variable-disjoint renaming[4] of a linking instance w.r.t.* ℓ *of one of the two clauses, say, with* $c\sigma$,
2. *below the node labelled with* $l\sigma$, *expand the branch with a variable-disjoint renaming of a linking instance w.r.t.* ℓ *of* $c'\sigma$.

In other words, we perform a clause linking step and attach the coupled renamed linking instances below the leaf node N of the current tableau branch B. Afterwards, the respective connection need not be used any more below N on any extension of B, thus "disconnecting" the connected literals. This last feature explains the term *disconnection tableau calculus* for the proof method.

---

[2] I.e., we treat an n-ary mapping as an $n+1$-ary relation which is unique on the last argument position, as usual.

[3] An initial path need not be open, since closures do not apply to the initial path.

[4] The renaming achieves that all clauses in a tableau are variable-disjoint.

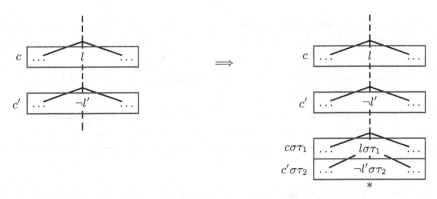

Fig. 1. Illustration of a linking step (with renamings τ_1, τ_2).

The standard form of tableau closure is not sufficient as a branch closure condition for the disconnection tableau calculus, but the same notion as employed in the clause linking methods [8] will do.

Definition 5 ($\forall$-variant, $\forall$–closure). *We call two literals (or clauses) l and l' $\forall$-variants, denoted by $l =_\forall l'$, iff $l\sigma = l'\sigma$, where σ is a substitution mapping all variables to one and the same variable. A tableau branch B is $\forall$-closed if there are two literals l and $\neg l'$ on B such that $l =_\forall l'$. A tableau is $\forall$-closed if all its branches are $\forall$-closed.*

Applied to the tableau in Figure 1, this means that after the linking step at least the middle branch is $\forall$-closed, as indicated by an asterisk below the leaf.

The *disconnection tableau calculus* then simply consists of the linking rule plus a rule for the selection of an initial path applicable merely once at the beginning of the proof construction. The calculus is sound and complete for any initial path selection, i.e., a clause set S is unsatisfiable if and only if, for any initial path P, there exists a finite disconnection tableau T for S and P such that all branches of T are $\forall$-closed. Completeness is preserved if we require *variant-freeness* along a branch, i.e., two different nodes on a branch must not be labeled with two literal occurences whose clauses are variants of each other. Note, however, that this condition does not extend to the initial path of a tableau. Variant-freeness implies that a link can be used only once on each branch, resulting in a decision procedure for certain formula classes, most notably the Bernays-Schönfinkel class (which cannot be decided by resolution). As shown in [11], we may also extract a model from a finitely failed open branch.

3   A Generalisation of the Linking Rule

An analysis of the clause linking rules used in the literature [8,17,6,7,11] shows that none of the paradigms used exploits the $\forall$-closure concept to its full power. We can achieve this by introducing the following generalisation of a unifier.

Definition 6 (∀-unifier). *Given two literals l and l', a substitution σ is called a ∀-unifier of l and l' if $l\sigma =_\forall l'\sigma$. σ is called a most general ∀-unifier of l and l' if it is more general than every ∀-unifier of l and l'.*

The question how such most general ∀-unifiers can be computed is answered by the following proposition.

Proposition 2. *Let σ be any most general unifier of two literals l and l'. Let σ' be the substitution resulting from σ by iteratively renaming each occurence of a variable in the range of σ to a new variable. Then σ' is a most general ∀-unifier of l and l'.*

Definition 7 (∀-linking instance, -linking rule). *A clause c_0 is a ∀-linking instance of a clause c w.r.t. a link $\ell = \{\langle c, l\rangle, \langle c', l'\rangle\}$ if $c_0 = c\sigma$ where σ is a most general ∀-unifier of l and l'. The ∀-linking rule generalises the linking rule in such a manner that most general ∀-unifiers are used instead of most general unifiers.*

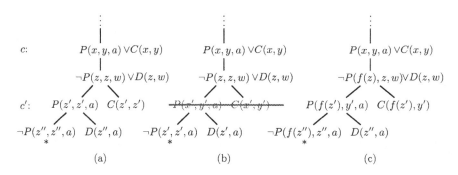

Fig. 2. Linking instance generation for standard linking (a), ∀-linking with subsequent variant deletion (b) and with ∀-linking generating a more general linking instance than standard linking (c).

Example 1. Figure 2 illustrates the differences between standard linking and ∀-linking. Whereas in partial tableau (a) the standard linking rule enforces a proper specialisation of clause $c = P(x, y, a) \vee C(x, y)$ that also affects the remainder of the linking instance, the use of ∀-linking in partial tableau (b) maintains the full generality of the linking instance c' of c which, being a variant of the original clause, can be deleted. The partial tableau (c) demonstrates a case where ∀-linking still produces two proper linking instances. However, the upper linking instance is strictly more general than a linking instance generated by the standard linking rule.

As the example shows, with this new inference rule either more general clause instances or even less clause instances are produced, yet the closure condition is the same as for the standard linking rule.

Example 2. Consider the partial tableau in Figure 3 (a). The negative unit clauses each have $n - 1$ variables with one variable occuring in two adjoining positions. Then, $n - 1$ different clause instances can be generated, none of which are variants of each other. Using $\forall$-linking on the other hand, in Figure 3 (b), all possible $\forall$-linking instances are variants of each other, therefore only one instance will be placed on the tableau.

This generalisation of linking instances is also applicable to other contemporary frameworks based on instance generation [8,17,7,6] and may result in an improved search behaviour.

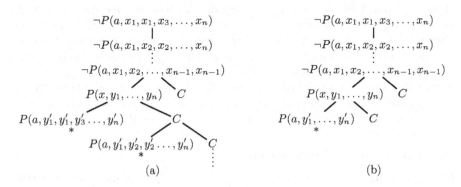

Fig. 3. Reduction of linking instances by $\forall$-linking (b) compared to standard linking (a).

In order for this new linking concept to even further reduce the number of clauses to be generated during the proof search, a new and also generalised method of variant deletion can be applied.

Definition 8 ($\forall$-variant freeness). *A tableau branch B is $\forall$-variant free if for no literal occurence $L = \langle c, l \rangle$ on B there is an $L' = \langle c', l' \rangle$ above L on B such that $c =_\forall c'$ and c is an instance of c'.*

4   Local Variables

4.1   Instantiation Based Methods and Herbrand Complexity

By their very nature, instantiation based methods demonstrate the unsatisfiability of a first-order formula through a reduction to the ground case. Therefore, the minimal size of a refutation in such a method cannot be less than the so-called *Herbrand complexity* of the formula. Recall that the Herbrand complexity of a set S of clauses is the minimal complexity of an unsatisfiable set containing only ground instances of clauses in S. It is well-known that Herbrand complexity is not a lower bound for resolution. More precisely, resolution may have exponentially shorter refutations (see, e.g., [12] p. 157 for an example). The essential reason for this fact is that resolution permits the generation of arbitrarily many renamed copies of resolvents. [5] Renaming of resolvents amounts to having a form of a *quantifier introduction rule* when viewed from the perspective of the sequent system; standard tableaux and standard instantiation based methods, in contrast, have *quantifier elimination* rules only.

A trivial case where this difference already shows up is when one considers unit hyper resolution, which is one of the most successful approaches for Horn clause sets. It inherently relies on the fact that resolvents can be used in more than one non-ground instance, which is not compatible with a plain instantiation based approach. So, in order to have a generally successful method, the paradigm of instantiation basedness has to be relaxed. This has been recognised since the very first implementations of our disconnection prover DCTP, which treats unit clauses in a special manner. For example, unit clauses are exempted from the instantiation mechanism of the framework and unit clauses can be freely used for subsumption in tableaux. It is important to emphasize that this special handling of unit clauses is one of the main reasons for the success of the system DCTP. Here, we consider the problems occuring when extending such a special handling from literals in unit clauses to certain literals in clauses of arbitrary lengths.

4.2   Universal and Local Variables

In a general free-variable tableau a branch formula F containing free variables has often to be used more than once on the branch, but with different instances for some of the variables, say, $u_1, \ldots, u_n$. This may lead to multiple occurences of similar subproofs. Such a duplication can be avoided if one can show that the respective variables are *universal* wrt. the formula F, i.e., when $\forall u_1 \cdots u_n F$ holds on the branch. A general description of this property is given, e.g., in [3]. Since proving this property is undecidable in general, efficient sufficient conditions for universality have been developed, e.g., the concept of locality in [9]. We will use a similar approach.

Definition 9 (Local and shared variables and literals). *A variable u is called* local *in a literal l of a clause $l \vee c$ if u does occur in l but not in c. A variable u is* local *in a clause c if u is local in a literal of c. All other variables in a clause are called* shared. *A literal l is called* local *if it contains no shared variables. l is called* mixed *if it contains both local and shared variables.*

Obviously, since all clauses in a disconnection tableau are variable-disjoint, any local variable in a literal on a tableau branch is universal wrt. the literal. In order to illustrate a fundamental weakness of standard disconnection tableaux and the potential of the use of local variables, we discuss a very simple example. [6]

---

[5] As a side-effect, the number of *inference steps* of a resolution derivation may not be a representative complexity measure for the derivation (see [13] or [12] pp. 160ff.).

[6] In practice, such clauses will normally not occur in the respective input set, but, as our experiments show, clauses with local variables may be dynamically generated during the proof process as a result of simplification and factoring steps [20].

Example 3. Let S be a set consisting of the four clauses $\neg R$, $Q(x) \vee P(x,u)$, $\neg Q(c) \vee R$, and $\neg P(x,a) \vee \neg P(x,b)$ where the u denote local variables, the x denote shared variables and a and b denote constants.

A minimal closed disconnection tableau for this set is displayed in Figure 4(a).

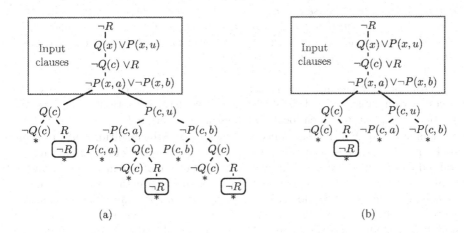

(a)   (b)

Fig. 4. Closed tableau for the clause set from Example 3.

The redundancy in the proof is obvious. Without making use of the local variable u in the second clause, the subgoal R needs to be solved repeatedly, which can result in a much larger proof if the refutation of R is not a simple unit clause but a complex sub-proof.

Different methods have been developed to avoid this redundancy (see [11] for a short review). Our approach is a very direct one. We extend the closure rule in such a manner that a literal containing local variables can be used for different branch closures. This feature may have a significant proof shortening effect. Note that it also destroys the *ground instance property* of tableaux, which guarantees that any closed free-variable tableau remains closed when all variables are instantiated with ground terms. As a result, Herbrand complexity need no more be a lower bound to the complexity of such tableau refutations.

Furthermore, the presence of local variables can also be used for search pruning by reducing the number of necessary clause instances, as follows. We simply forbid that local literals be instantiated when the linking instances are generated.

We describe first the generalisation of the closure rule. In order to simplify the presentation, we partition the set of variables of our first-order language into two subsets, the set of *local* variables and the set of *shared* variables.

Definition 10 (L-variant, l-closure). *We call two literals (or clauses) k and k' l-variants, denoted by $k =_l k'$, if $k\sigma =_{\forall} k'\sigma$ where σ is a substitution on local variables only. A tableau branch B is l-closed if there are two literals k and $\neg k'$ on B such that $k =_l k'$. A tableau T is l-closed if all its branches are l-closed.*

Definition 11 (L-unifier). *Given two literals k and k', a substitution σ is called an l-unifier of k and k', if $k\sigma =_l k'\sigma$; σ is called a* most general l-unifier *of k and k' if it is more general than every l-unifier of k and k'.*

Definition 12 (L-linking instance, l-linking rule). *A clause c' is an l-linking instance of a clause c w.r.t. a link ℓ if $c' = c\sigma$ where σ is the restriction of a most general l-unifier for the literals occuring in ℓ to shared variables. The l-linking rule generalises the $\forall$-linking rule by using most general l-unifiers and l-linking instances instead of the $\forall$-concepts.*

With the new closure condition and the new linking rule, we achieve the disconnection tableau proof displayed in Figure 4(b). Note that no clauses are put on the tableau in which local variables are instantiated.

It should be noted that the concept of locality and the resulting optimisations are limited to variables occurring on a *single branch* and therefore cannot be generalised to variables occurring in more than one literal of a clause. Furthermore, as illustrated by the above example, the search space reduction obtained cannot be achieved through techniques like *explicit clause splitting*, which replaces a clause $l \vee c$ by two new clauses $s(\vec{x}) \vee l$ and $\neg s(\vec{x}) \vee c$ connected by the new split literal $s(\vec{x})$, where $\vec{x}$ denotes the list of shared variables of l. Applied to $Q(x) \vee P(x, u)$ clause splitting would produce the clauses $S(x) \vee P(x, u)$ and $\neg S(x) \vee Q(x)$ and hence not eliminate the local variable u.

Using standard linking instead of $\forall$-linking as a basis for l-linking creates a particular problem, the non-uniqueness of most general unifiers. Then, the result of a unification between literals containing both local and shared variables can be different depending on the way the variables of the literals are substituted, as shown in the following example:

Example 4. Consider the pair of clauses $P(u, x, f(u)) \vee R(x)$ and $\neg P(v(f(v), y) \vee S(y)$. Using local variables, the link between $P(u, x, f(u))$ and $\neg P(v(f(v), y)$ can be unified in two ways, resulting in either $P(u, f(v), f(v))$ and $\neg P(v, f(v), f(v))$ or $P(u, f(u), f(u))$ and $\neg P(v, f(v), f(u))$. In either way one of the linking instances no longer has a local variable.

As a matter of fact, this indeterminism might turn out to be critical for the practicality of the method in automated deduction. The problem, however, completely vanishes when the concept of $\forall$-linking is used instead of regular linking. The resulting unique $\forall$-linking instances for the above example then are $P(v', f(v''), f(v')) \vee R(f(v''))$ and $\neg P(v, f(v), f(u')) \vee S(f(u'))$.

Concerning the *soundness* of the new method, obviously only the l-closure rule need be considered, since the l-linking rule produces clause instances only. The soundness of l-closure can be recognised by a straightforward proof transformation similar to the one used in [16].

Proposition 3 (Soundness of l-closure). *If there is an l-closed disconnection tableau T for a set of clauses S, then S is unsatisfiable.*

Proof. We show the soundness by transforming T into a tableau which can be closed according to general tableau rules [5]. First, we substitute all shared variables in T by the same arbitrary constant, which does not affect the l-closedness. Then, we replace every literal k containing the local variables $u_1, \ldots, u_n$ with the finite conjunction $\mathcal{C}_k$ of all instances $k\theta_i$ of k in which k is used in l-closure steps; this is sound, because every pair of a variable assignment and an interpretation which satisfies $\forall u_1 \cdots u_n k$ also satisfies $\mathcal{C}_k$. Now, we can simulate any l-closure step using k and leaf literals k_i by an application of the standard α-rule in tableaux by selecting the respective literals $k\theta_i$ from the conjunction $\mathcal{C}_k$. Then the respective branch is complementary in the standard sense. qed.

5 Completeness of Disconnection Tableaux with Local Variables

In the rest of the paper, we consider as disconnection tableau calculus the method described in Section 2 using the l-linking and the l-closure rules instead of the standard rules. We also assume that the $\forall$-variant-freeness condition is respected in the tableau construction.

Showing *completeness* of the new method is much more difficult than showing soundness. The feature critical for completeness is the l-linking rule, for the following reason. The disconnection approach is an instantiation oriented approach, which works by generating instances of input clauses. With the l-linking rule the generation of clause instances is weakened, since certain clauses are not put on the tableau in their full specialisation. The question is whether the l-closure rule is powerful enough to compensate for this and to preserve completeness. Note that our first attempt to show completeness by means of a proof transformation failed even for a less restricted version of the calculus not incorporating $\forall$-unification. The problem here is that it is not possible to cast a tableau with standard linking into a tableau with l-linking without avoiding a complete restructuring of the tableau, possibly including an increase in proof length.

The central concept we have developed in [11] is that of an *instance preserving enumeration*. This notion bridges the gap between tableau branches and Herbrand paths in a natural manner and thus facilitates a concise completeness argument, but the original concept must be slightly adapted to encompass $\forall$-linking.

Definition 13 (Instance preserving enumeration (ipe)). *Let P be a path and S the set of clauses in the elements of P. An* instance preserving enumeration (ipe) *of P is any sequence $E = L_1, L_2, L_3, \ldots$ in which exactly the elements of P occur and in such an order that, for any $L_i = \langle c_i, l_i \rangle$ and $L_j = \langle c_j, l_j \rangle$, when $c_i \theta$ is a proper instance of $c_j \theta$ where θ is a substitution mapping all variables to one and the same variable, then $i > j$. Let, furthermore, S^* be the Herbrand set of S. Now, for any instance preserving enumeration E of P, there is a path P^* through S^* satisfying the following property:*

- a literal occurence $L = \langle c, l \rangle$ is in P^* if and only if there is a literal occurence $L_k = \langle c_k, l_k \rangle$ in E and a substitution σ with $\langle c_k \sigma, l_k \sigma \rangle = \langle c, l \rangle$, and there is no $L_j = \langle c_j, l_j \rangle$ with $j > k$ in E such that c is an instance of c_j.

We term L_k the minimal matcher in E of the clause c resp. literal occurence $\langle c, l \rangle$. Such a path P^* is called an Herbrand path of E and of P.[7]

We will prove that the existence of a tableau branch which cannot be l-closed entails the existence of a model. In contrast to tableau calculi like connection tableaux [10] or Fitting's free-variable tableaux [5], the disconnection tableau calculus is *non-destructive*, i.e., any inference step performs just an expansion of the previous tableau. This is exploited as follows.

Definition 14 (Disconnection tableau sequence). *Given a set of clauses S and an initial path P through S, a* disconnection tableau sequence *for S and P is any (possibly infinite) sequence $\mathcal{T} = T_0, T_1, T_2, \dots$ satisfying the following properties:*

- T_0 *is the trivial tableau consisting just of the root node,*
- *any T_{i+1} in $\mathcal{T}$ can be obtained from T_i by application of an l-linking step.*

Any tableau in $\mathcal{T}$ is called a disconnection tableau *for S and P.*

Because of the non-destructiveness of the disconnection approach, any tableau in such a sequence $\mathcal{T}$ contains all tableaux with smaller indices in the sequence as initial segments. Therefore we can form the tree union $\bigcup \mathcal{T}$ of all tableaux in $\mathcal{T}$. In the case of a finite sequence, $\bigcup \mathcal{T}$ is just the last element in the sequence. In general, $\bigcup \mathcal{T}$ may be an infinite tableau. We call $\bigcup \mathcal{T}$ the *limit tableau* of the sequence $\mathcal{T}$.

Definition 15 (L-saturation). *A branch B in a (possibly infinite) tableau is called* l-saturated *if B is not l-closed and, for any link $\ell = \{\langle c_1, l_1 \rangle, \langle c_2, l_2 \rangle\}$ on B, the branch B contains a literal occurence $\langle c_3, l_3 \rangle$ such that there is an l-linking instance c_4 of c_1 or c_2 w.r.t. ℓ which is no $\forall$-variant of c_1 or c_2, resp., $c_4 =_\forall c_3$ and c_4 is an instance of c_3. The limit tableau of a disconnection tableau sequence is called* l-saturated *if either all its branches are l-closed or if it has an l-saturated branch.*

We have to consider a particularity of the calculus concerning the rôle of the initial path. Since the initial path is just used to get the tableau construction process going, the initial path is not subject to the variant-freeness condition. So, possibly we may have two literal occurences $\langle c, l \rangle$ and $\langle c', l' \rangle$ with c and c' being variants, one on the initial branch and the other on the tableau branch. But l and l' may not be variants, i.e., the clauses c and c' are passed at different positions, thus leading to an ambiguity concerning the corresponding Herbrand path. This is resolved by giving preference to the tableau branch over the initial path and by deleting the respective literal occurences from the initial path.

---

[7] Note that we could associate a *unique* Herbrand path P^* with an ipe E by using a total order R on the ground atoms of S^* and taking as $P^*(c)$, from all literals in c which are instances of the literal l_k in the minimal matcher, just the minimal one in R.

Definition 16 (Adjusted branch). *Let B be a branch in a disconnection tableau with initial path P. P^{-B} is obtained from P by deleting all literal occurrences $\langle c, l \rangle$ from P with $\langle c', l' \rangle$ in B, $c' =_\forall c$ and c' being more general than c. We call $B^+ = B \cup P^{-B}$ the* adjusted branch *of P and B.*

Proposition 4 (Model characterisation). *Let B be a branch of the limit tableau of a disconnection tableau sequence for a clause set S and an initial path P. If the adjusted branch B^+ is l-saturated, then the set of literals I of any Herbrand path P^* of B^+ is an Herbrand model for S.*

Proof. First, P^* must be a Herbrand path through the entire Herbrand set S^* of S, since all clauses in S are in the domain of B^+. Then, the main part of the proof is to demonstrate that P^* is an open Herbrand path through S^*. Assume, indirectly, that P^* be not. This means that P^* contains two literal occurences $\langle c, l \rangle$ and $\langle c', \neg l \rangle$. Since P^* is an Herbrand path of an instance preserving enumeration E of B^+, B^+ must contain two literal occurences which are more general than $\langle c, l \rangle$ and $\langle c', \neg l \rangle$, respectively. We select their minimal matchers $L_1 = \langle c_1, l_1 \rangle$ and $L_2 = \langle c_2, \neg l_2 \rangle$ in E. Obviously, l_1 and l_2 must be unifiable. Let σ be a most general l-unifier of l_1 and l_2. By the definition of l-saturation, a literal occurence $L_3 = \langle c_3, l_3 \rangle$ must be on the branch B where c_3 is a more general $\forall$-variant of an l-linking instance but no $\forall$-variant of c_1 or c_2 w.r.t. the link $\{L_1, L_2\}$. Then, by construction, c or c' must be an instance of c_3, and hence L_3 a matcher of c or c'. Additionally, $c_3 \theta$ is a proper instance of $c_1 \theta$ or $c_2 \theta$ where θ is a substitution identifying all variables. Therefore, according to the definition of an instance preserving enumeration (Definition 13), the index of L_3 in E must be greater than the index of one of L_1 or L_2. So one of L_1 and L_2 cannot be a minimal matcher, which contradicts our assumption. Consequently, P^* cannot have a link and hence must be an open Herbrand path through S^*. Finally, by Proposition 1, the set of literals I of P^* is an Herbrand model for S.   qed.

With the model characterisation at hand, the completeness of the disconnection tableau calculus can be recognised with a standard technique which is traditional for tableaux. The key approach is to use a *systematic* or *fair* inference strategy f which guarantees that, for any clause set S, the limit tableau for S under strategy f is saturated [18]. There are simple and effective fair inference strategies for the disconnection tableau calculus [11]. A sufficient condition is not to delay links on a branch arbitrarily long.

Proposition 5 (Completeness). *If S is an unsatisfiable clause set and f is a fair inference strategy, then the limit tableau T for S under strategy f is a finite l-closed disconnection tableau for S.*

Proof. First, we show that every branch of T is l-closed. Assume a branch be not l-closed. Now, because of the fairness of the strategy f, T would be l-saturated and contain an l-saturated branch. Then, by Proposition 4, S would be satisfiable, contradicting our assumption. The l-closedness entails the finiteness of every branch. Since T is finitely branching, by König's Lemma, it must be finite.
        qed.

6   Completeness with l-Regularity

The condition of *regularity*, which avoids that a formula appears more than once on a tableau branch, is one of the most effective search pruning methods for tableau calculi. In the disconnection framework, however, it applies only to ground literals, since all clauses on a tableau branch are variable-disjoint. The idea of sharpening regularity is basically to perform forward subsumption with l-substitutions.

Definition 17 (L-regularity). *A tableau branch B is called* l-regular *if, for any literal occurence $\langle c, k \rangle$ on B, there is no literal occurence $\langle c', k' \rangle$ below $\langle c, k \rangle$ on the branch such that c' contains an instance of k which is an l-variant of k.*

Note that since all clauses on a branch are variable-disjoint, the subsuming literal k must be a local literal, thus not containing any shared variables. In fact, we will apply this concept in an even stronger form. Whenever we perform a linking step with link ℓ on a branch B producing two clause instances c_1 and c_2, we may avoid putting *any* of the clauses on the branch if there is a literal occurence $\langle c, k \rangle$ above on B not in the link ℓ such that c_1 or c_2 contains an l-instance of k. This is a form of *coupled elimination* [4], since such an l-regularity violation renders the entire inference redundant. Consequently, we use the following generalised notion of the saturation of a branch.

Definition 18 (Weak l-saturation). *A branch B in a (possibly infinite) tableau is called* weakly l-saturated *if B is not l-closed and, for any link $\ell = \{\langle c_1, l_1 \rangle, \langle c_2, l_2 \rangle\}$ on B, the branch contains a literal occurence $\langle c_3, l_3 \rangle$ not in ℓ such that*

(1) either there is an l-linking instance c_4 of c_1 or c_2 w.r.t. ℓ which is no $\forall$-variant of c_1 or c_2, resp., $c_4 =_\forall c_3$ and c_4 is an instance of c_3
(2) or l_3 is a local literal and subsumes an l-linking-instance c_4 of ℓ, i.e., there is a substitution τ such that $l_3\tau$ is contained in c_4.

The limit tableau of a disconnection tableau sequence is called weakly l-saturated *if either all its branches are l-closed or if it has a weakly l-saturated branch.*

The idea of the model characterisation argument for the l-regular case is the following. We try to pursue the same main line as in Section 5, but we assume that all literal occurences $\langle c, l \rangle$ in a path where c is subsumed by a local literal l' are ignored on the branch. In order to guarantee that every clause in the Herbrand set S^* is passed after this deletion, we define that such a deleted clause c is passed at a literal occurence $\langle c, k \rangle$ where $k\tau = l'$ for some l-substitution τ. Since there may be more than one such subsuming literal, the determination of a unique Herbrand path is achieved by imposing an order on the local literals on a branch.

Definition 19 (Local literal (enumeration)). *Given an instance preserving enumeration E, i.e., a sequence of literal occurences, a local literal enumeration (lle) of E is any sequence $\mathcal{L} = l_1, l_2, l_3, \ldots$ in which exactly the local literals in* *the members of E occur and in such an order that, for any l_i and l_j, when l_i is a proper instance of l_j, then $i > j$. Let, furthermore, S^* be the Herbrand set of the clauses occuring in the elements of an ipe E. An Herbrand path P^* of E under a local literal enumeration $\mathcal{L}$ of E has to satisfy the following conditions. For any clause $c \in S^*$, if $P^*(c) = l$, then:*[8]

- *if c contains a literal which is an instance of a literal in the lle $\mathcal{L}$, then l must be an instance of such a literal l_i in $\mathcal{L}$ with smallest index i,*
- *otherwise, l must be an instance of the literal in the minimal matcher (Definition 13) of the clause c in E.*

Proposition 6 (Model characterisation with l-regularity). *Given a branch B of the limit tableau of a sequence of l-regular disconnection tableaux for a clause set S and an initial path P. If the adjusted branch B^+ (Definition 16) is weakly l-saturated and P^* is the Herbrand path of an ipe E of B^+ under an lle $\mathcal{L}$ of E, then the set of literals in the elements of P^* is an Herbrand model for S.*

Proof. Again, we demonstrate that P^* is an open Herbrand path through S^*. Assume, indirectly, that P^* be not, i.e., P^* contains two literal occurences $\langle c, l \rangle$ and $\langle c', \neg l \rangle$. We may distinguish three cases

Case 1. Both l and $\neg l$ are instances of literals occuring in the local literal enumeration $\mathcal{L}$. But then B^+ would be l-closed, contradicting our assumption that B is weakly l-saturated.

Case 2. None of l and $\neg l$ is an instance of a literal in the lle $\mathcal{L}$. This case is already captured in the proof of model characterisation (Proposition 4).

Case 3. Exactly one of l or $\neg l$ is an instance of a literal in $\mathcal{L}$, w.l.o.g. l. Let l_1 be the first literal in $\mathcal{L}$ which is more general than l. According to the l-regularity assumption, there must be exactly one literal occurence $\langle c_1, l_1 \rangle$ on B^+ with literal l_1. Let further $\langle c_2, \neg l_2 \rangle$ be the minimal matcher of c' on B^+. Now, since B^+ is weakly l-saturated, there must be a literal occurence $\langle c_3, l_3 \rangle$ not in $\ell = \{\langle c_1, l_1 \rangle, \langle c_2, \neg l_2 \rangle\}$ on B^+ which satisfies one of the conditions in Definition 18:

(1) Either there is an l-linking instance c_4 of c_2 w.r.t. ℓ which is no $\forall$-variant of c_2, $c_4 =_\forall c_3$ and c_4 is an instance of c_3. The fact that c' must be an instance of c_3 contradicts the assumption of $\langle c_2, \neg l_2 \rangle$ being the minimal matcher of c'.

(2) Or l_3 is a local literal and subsumes an l-linking instance c_4 of the link ℓ, we select the first suitable l_3 in $\mathcal{L}$.

(a) Either $c_4 = c_1\sigma$: now l_1 is a local literal, i.e., contains no shared variable, hence the only l-linking instance of c_1 w.r.t. ℓ is c_1, so $c_4 = c_1$ and l_3 subsumes c_1. By Definition 19 the path P^* must pass c at a literal subsumed by l_3. Therefore l_1 is subsumed by l_3. By l-regularity, l_3 is no variant of l_1, and

---

[8] For convenience we assume that if more than one literal in a ground clause is an instance of a literal in L, then the minimal literal according to some total order is used.

since l_1 is before l_3 in the lle $\mathcal{L}$, l_3 must be an instance of l_1, which is a contradiction.

(b) Or, $c_4 = c_2\sigma$: then, by assumption $\neg l$ and hence $\neg l_2\sigma$ are not subsumed by l_3. So, l_3 must subsume another literal in $c_2\sigma$. According to Definition 19, the clause c' must not be passed at literal $\neg l$, which again contradicts our assumption.

Therefore, P^* must be an open Herbrand path through S^*. Finally, by Proposition 1, the set of literals I in P^* is an Herbrand model for S.   qed.

7   Related Work

In the last years, a number of other instantiation based proof methods have been studied. The contemporary approaches can be classified into two main categories. On the one hand there are methods based on clause linking which require a separate procedure for detecting unsatisfiablity on the propositional or higher level. On the other hand, there are methods which integrate the detection of unsatisfiability directly into the instantiation procedure.

Representatives of the first category can be found in [8,17,7,6]. Lee and Plaisted [8] extend the clause linking method to hyperlinking which requires that a simultaneous unifier exists which unifies each literal in a clause with another literal in the formula. Hyperlinking is not compatible with the disconnection method because of the path restriction [20]. More recently, this hyperlinking concept has been enhanced with forms of semantic guidance in [17]. The approaches presented by Hooker et al. [7] focus on the selection of links according to term depths, which is also reflected in their completeness proofs. Furthermore, they try to optimise the interleaving with the propositional decision procedure in order to reduce the duplication of work. Ganzinger and Korovin [6] integrate literal selection functions and extended concepts of redundancy w.r.t. instances as well as inferences. Also, they describe generalisations of the methods in [7] in order to achieve more flexibility. Finally, they try to tailor their instance generation process towards the use of non-propositional decision procedures. All the methods mentioned here might profit from the search pruning achievable through the new ∀-linking rule developed in this paper. Interestingly, the methods developed here for the use of local variables, by their very deviation from the plain instantiation based paradigm, seem difficult to integrate into instantiation based methods with separate decision procedures. In order to fully exploit local variables, one would have to extend the used propositional decision procedures with a special non-ground handling of local variables, but this would significantly reduce the performance of the decision procedures.

The second category of instantiation based calculi, to which our approach belongs, has been the subject of [1,2]. Both the FDPLL [1] and the Model Evolution [2] frameworks are based on forms of semantic trees which have a fixed binary branching rate. The newer Model Evolution approach directly integrates the unit propagation mechanism lifted from the propositional to a higher level. Also, a limited form of local variables is used which is currently generalised to encompass literals containing both local and shared variables. There, shared variables correspond to so-called *parameters* while only local variables are actually denoted as *variables*. Interestingly, Model Evolution requires a form of runtime Skolemisation which dynamically increases the signature of the problem by new constants. It has been argued that a semantic tree approach can be superior to a tableau based approach, because the number of instances of formulae necessary in the proof search might be significantly smaller. However, when one works with clauses of fixed length, this difference can be polynomial only.[9]

Unfortunately, none of the described other approaches has an implementation which is generally competitive in performance with state-of-the-art theorem provers.

8 Conclusion and Outlook

In this paper, we have presented generalisations of the inference and deletion rules in the disconnection tableau calculus. These generalisations may significantly reduce both the lengths of minimal proofs and the sizes of the search spaces. Furthermore, the methods can also be applied to other instance based approaches. We have given concise completeness proofs of the new calculi and developed new meta-theoretic notions which can be useful for future enhancements. Note also that, in our framework, the occurence of mixed literals poses no problems at all. We have also compared our method with the most important other instantiation based methods and explained the main differences.

Experiments with a prototypical implementation show a significant gain in performance for problems containing local variables, e.g. *Andrew's Challenge* [16].

The framework of disconnection tableaux is by far not exploited to its full potential. We consider the following issues to be of high importance for the future. First, the new treatment of variables should be integrated with the equality handling described in [15]. Then, the regularity concept, which is incompatible with equality handling (as described in [20]) should be relaxed to a compatible version. Next, efficient implementations of the new concepts should be developed; note that both the ∀- and the l-linking rule require significant changes in the design of the prover. Finally, unit propagation should be integrated in a more direct manner similar to the approach in [2].

References

1. Peter Baumgartner. FDPLL – A First-Order Davis-Putnam-Logeman-Loveland Procedure. In David McAllester, editor, *Proceedings of the 17th International Conference on Automated Deduction (CADE-17), Pittsburgh, USA*, volume 1831 of *Lecture Notes in Artificial Intelligence*, pages 200–219. Springer, 2000.

---

[9] Obviously, it is possible to translate every set of clauses into a set containing clauses of length at most three.

2. Peter Baumgartner and Cesare Tinelli. The Model Evolution Calculus. In Franz Baader, editor, *Automated Deduction – CADE-19*, volume 2741 of *Lecture Notes in Artificial Intelligence*. Springer, 2003.
3. B. Beckert and R. Hähnle. Analytic tableaux. In *Automated Deduction — A Basis for Applications*, volume I: Foundations, pages 11–41. Kluwer, Dordrecht, 1998.
4. J.-P. Billon. The disconnection method: a confluent integration of unification in the analytic framework. In *Proceedings, 5th TABLEAUX*, volume 1071 of *LNAI*, pages 110–126, Berlin, 1996. Springer.
5. Melvin C. Fitting. *First-Order Logic and Automated Theorem Proving*. Springer, second revised edition, 1996.
6. Harald Ganzinger and Konstantin Korovin. New directions in instantiation-based theorem proving. In *Proceedings of the eightteenth Annual IEEE Syposium on Logic in Computer Science (LICS-03)9*, pages 55–64, Los Alamitos, CA, June 22–25 2003. IEEE Computer Society.
7. J. Hooker, G. Rago, V. Chandru, and A. Shrivastava. Partial instantiation methods for inference in first-order logic. *Journal of Automated Reasoning*, 28(5):371–396, 2002.
8. S.-J. Lee and D. Plaisted. Eliminating duplication with the hyper-linking strategy. *Journal of Automated Reasoning*, pages 25–42, 1992.
9. R. Letz. Clausal tableaux. In Wolfgang Bibel and Peter H. Schmitt, editors, *Automated Deduction — A Basis for Applications*, volume I: Foundations, pages 43–72. Kluwer, Dordrecht, 1998.
10. R. Letz, K. Mayr, and C. Goller. Controlled integration of the cut rule into connection tableau calculi. *Journal of Automated Reasoning*, 13(3):297–337, December 1994.
11. R. Letz and G. Stenz. Proof and Model Generation with Disconnection Tableaux. In Andrei Voronkov, editor, *Proceedings, 8th LPAR, Havanna, Cuba*, pages 142–156. Springer, Berlin, December 2001.
12. Reinhold Letz. *First-order calculi and proof procedures for automated deduction*. PhD thesis, TH Darmstadt, June 1993.
13. Reinhold Letz. On the polynomial transparency of resolution. In Ruzena Bajcsy, editor, *Proceedings of the 13th International Joint Conference on Artificial Intelligence (IJCAI), Chambery, France*, pages 123–129. Morgan Kaufmann, 1993.
14. Reinhold Letz, Klaus Mayr, and C. Goller. Controlled integration of the cut rule into connection tableau calculi. *Journal of Automated Reasoning*, 13(3):297–338, December 1994.
15. Reinhold Letz and Gernot Stenz. Integration of Equality Reasoning into the Disconnection Calculus. In *Proceedings, TABLEAUX-2002, Copenhagen, Denmark*, volume 2381 of *LNAI*, pages 176–190. Springer, Berlin, 2002.
16. Reinhold Letz and Gernot Stenz. Universal Variables in Disconnection Tableaux. In *Proceedings, TABLEAUX-2003, Rome, Italy*, volume 2796 of *LNAI*, pages 117–133. Springer, Berlin, 2003.
17. David A. Plaisted and Yunshan Zhu. Ordered semantic hyper linking. *Journal of Automated Reasoning*, 25(3):167–217, 2000.
18. Raymond Smullyan. *First-Order Logic*. Springer, 1968.
19. G. Stenz. DCTP 1.2 – System Abstract. In *Proceedings, TABLEAUX-2002, Copenhagen, Denmark*, volume 2381 of *LNAI*, pages 335–340. Springer, Berlin, 2002.
20. G. Stenz. *The Disconnection Calculus*. Logos Verlag, Berlin, 2002. Dissertation, Fakultät für Informatik, Technische Universität München.

Chain Resolution for the Semantic Web

Tanel Tammet

Tallinn Technical University
tammet@staff.ttu.ee

Abstract. We investigate the applicability of classical resolution-based
theorem proving methods for the Semantic Web. We consider several
well-known search strategies, propose a general schema for applying resolution provers and propose a new search strategy "chain resolution"
tailored for large ontologies. Chain resolution is an extension of the standard resolution algorithm. The main idea of the extension is to treat
binary clauses of the general form $A(x) \vee B(x)$ with a special chain resolution mechanism, which is different from standard resolution used otherwise. Chain resolution significantly reduces the size of the search space
for problems containing a large number of simple implications, typically
arising from taxonomies. Finally we present a compilation-based schema
for practical application of resolution-based methods as inference engines
for Semantic Web queries.

1 Introduction

Building efficient inference engines for large practical examples is one of the
crucial tasks for the Semantic Web project as a whole (see an early paper
[SemanticWeb01] and the project page [SemanticWebURL]). Since the current
mainstream languages RDF and OWL are based on classical first order logic (or
subsets of first order logic) it is particularly important to investigate specialised
methods for the kinds of first order logic (FOL) inferencing tasks likely to arise
in practical examples.

It is common knowledge that although RDF is a very simple subset of FOL,
and OWL Lite is also severely restricted, already OWL Full is undecidable, hence
in a certain sense equivalent to full first order logic (see [RDF], [OWL]). It is possible that as the field progresses, several new languages will appear. In particular,
unless the use of the common equality predicate is tightly restricted, equality
will lead to undecidable classes. Inevitably some parts of the knowledge on the
web will be presented in decidable and some parts in undecidable fragments of
FOL.

The nature of undecidability makes it impossible to obtain an "optimal" or
even "close to optimal" algorithm. Inevitably a good inference system will need
to employ a variety of specialised algorithms and strategies, depending on the
nature of the inference task at hand.

The majority of research in building inference engines for the Semantic Web
is carried out in the context of description logics, which have also heavily influenced the development of OWL (see [Handbook03]). The inference systems

D. Basin and M. Rusinowitch (Eds.): IJCAR 2004, LNAI 3097, pp. 307–320, 2004.
© Springer-Verlag Berlin Heidelberg 2004

for description logics are very well optimised for the specific decidable classes of FOL.

However, in the light of a likelihood of encountering undecidable classes of formulas among the information on the Web, we argue that it is important to also investigate methods suited for undecidable classes. Even in case of decidable classes the high complexity of the decision problem diminishes the likelihood that one or two "optimal" algorithms could be found.

Automated theorem proving in full FOL is a classical area of computer science. There is a significant amount of theory and a number of evolving, powerful provers (see [Casc]: an annual prover competition, [TPTP]: the largest currently existing collection of examples).

However, the majority of existing theory and provers have been targeting mathematically-oriented types of problems, which are very different from the kinds of problems we expect to encounter in the industrial use of Semantic Web systems. The prover examples from mathematics (see [TPTP] for the largest currently existing collection) are relatively small, yet encode highly complex structures and employ equality extensively. On the other hand, typical examples from the Semantic Web are ontology-based, with very large taxonomies and even larger databases of relatively simple facts.

Similarly to the description logic community we argue that the standard methods of classical theorem proving in FOL are not well suited for Semantic Web tasks. However, the "standard methods" like resolution are not fixed, specific algorithms, but rather frameworks of algorithms which can be heavily modified, specialised with search strategies and extended by additional methods.

The particular importance of the classical resolution-based theorem proving methods stems from the research and experience collected for full, undecidable FOL.

We suggest that it will be useful to combine the knowledge and experience from the field of description logics with the field of Datalog and the field of resolution theorem proving.

The current paper is targeted at presenting relatively simple, yet important extensions to the standard resolution method, with the goal to make resolution-based methods a practically useful addition to the current alternatives.

Most of the current high-level theorem provers for full FOL are based on the resolution method. Although the basic resolution method as introduced by Robinson is not particularly efficient, it has proved to be a flexible framework for introducing efficient, specialised strategies. For example, relatively simple clause ordering strategies produce efficient decision algorithms for most of the well-known, classical decidable classes of FOL, see [Fermüller93], [Handbook01]. It is easy to devise strategies for top-down and bottom-up search directions, additional mechanisms for answering database-style queries etc.

However, there are several application areas (for example, propositional logic and arithmetical theories) for which the basic strategies of resolution have not produced search algorithms with efficiency comparable to specialised algorithms significantly different from resolution.

In practical applications it is fairly common to arrive to a hard situation where a proof task contains both an "ordinary" FOL part well suited for resolution and a large special theory not well suited for resolution. One approach in such situations is to extend the resolution-based methods with a specialised "black box" system. Unfortunately, for complex "black boxes" it is often either very hard or impossible to prove completeness of the combined system.

We propose a simple "black box" system called "chain resolution" extending standard resolution in the context of answering queries in case large ontologies are present in the theory, as is common in the applications of semantic web and terminological reasoning.

We show that chain resolution is sound and complete. More importantly, it can be combined with several crucial strategies of resolution: set of support and ordering strategies.

An important feature of chain resolution is that it is easy to implement and easy to incorporate into existing theorem provers. The presence of a chain resolution component should not seriously degrade the performance of a theorem prover for tasks where chain resolution is useless, while significantly improving performance for tasks where it can be employed.

In the last sections of the paper we present a general scheme for rule database compilation, employing both chain resolution and ordering strategies.

The presented scheme along with the chain resolution algorithm is currently being implemented as a layer on top of a high-perfomance resolution prover Gandalf, see [Tammet97], [Tammet98], [Tammet02].

2   Informal Introduction to Chain Resolution

We assume familiarity with standard concepts of resolution, see [Handbook01]. We will use the following notions in our presentation:

Definition 1. *A* signed predicate symbol *is either a predicate symbol or a predicate symbol prefixed by a negation. A* literal *is a signed atom.*

Definition 2. *A* chain clause *is a clause of the form*

$$A(x_1, \ldots, x_n) \vee B(x_1, \ldots, x_n)$$

where $x_1, \ldots, x_n$ are different variables, A and B are signed predicate symbols.

Definition 3.
A clause C' is a propositional variation *of a clause C iff C' is derivable by a sequence of binary resolution steps from C and a set of chain clauses.*

Definition 4.
A search state *of a resolution prover is a set of clauses kept by the prover at some point during proof search.*

Definition 5.
A passive list of a resolution prover is a subset of a search state containing exactly these clauses which have not been resolved upon yet.

Throughout the paper we will only consider chain clauses of a simpler form $A(x) \vee B(x)$ where A and B are signed unary predicates. All the presented rules, algorithms and proofs obviously hold also for the general chain clauses as defined above.

We will bring a small toy example of a taxonomy containing only chain clauses: "a mammal is an animal", "a person is a mammal", "a person thinks", "a man is a person", "a woman is a person". The following clause set encodes this knowledge as implications in FOL: { $\neg mammal(x) \vee animal(x)$, $\neg person(x) \vee mammal(x)$, $\neg person(x) \vee thinks(x)$, $\neg man(x) \vee person(x)$, $\neg woman(x) \vee person(x)$ }. Chain clauses can also capture negated implications like "a man is not a woman". Observe that the chain resolution method is applicable for general, unrestricted FOL, not just the case where we only have chain clauses present.

We expect that ontologies typically contain a significant number of chain clauses, plus a number of clauses which are not chain clauses. In the extreme case where there are no chain clauses present (nor derivable from) in the initial clause set, chain resolution will be simply reduced to the ordinary resolution method.

We note that the concept of the chain clause can in specific circumstances be extended to cases where the literals $A(x_1, \ldots, x_n) \vee B(x_1, \ldots, x_n)$ contain constant arguments in addition to the distinct variables.

In order for this to be possible it must be shown that the constants occur in positions where there are either no variable occurrences in the corresponding literals in non-chain clauses or such clauses can be finitely saturated to the point that the offending variable occurrences disappear.

In such cases we can essentially treat a predicate symbol together with some constant occurrences as a new, specialised version of the predicate symbol.

We will not consider this extension in the current paper.

The basic idea of chain resolution is simple. Instead of keeping chain clauses in the search state of the prover and handling them similarly to all the other clauses, the information content of the chain clauses is kept in a special data structure which we will call *the chain box* in the paper.

Chain clauses are never kept or used similarly to regular clauses. At each stage of the proof search each signed unary predicate symbol P is associated with a set S of signed unary predicate symbols $P'_1, P'_2, \ldots, P'_m$ so that the set S contains all these and exactly these signed unary predicate symbols P'_i such that an implication $\neg P(x) \vee P'_i(x)$ is derivable using only the chain clauses in the initial clause set plus all chain clauses derived so far during proof search.

The "black" chain box of chain resolution encodes implications of the form $A(x) \Rightarrow B(x)$ where A and B may be positive or negative signed predicates. The chain box is used during ordinary resolution, factorisation and subsumption checks to perform these operations modulo chain clauses as encoded in the box. For example, literals $A(t)$ and $B(t')$ are resolvable upon in case $A(t)$ and $\neg A(t')$ are unifiable and $\neg B$ is a member of the set S associated with the signed predicate symbol $\neg A$ inside the chain box. Due to the way the chain box is constructed, no search is necessary during the lookup of whether $\neg B$ is contained in a corresponding chain: this operation can be performed in constant, logarithmic or linear time, depending on the way the chain box is implemented.

A suggested way to implement the chain box is using a bit matrix of size $4 * n^2$ where n is a number of unary predicate symbols in the problem at hand. Even for a large n, say, 10000, the bit matrix will use up only a small part of a memory of an ordinary computer.

The main effect of chain resolution comes from the possibility to keep only one propositional variation of each clause in the search state. In order to achieve this the subsumption algorithm of the prover has to be modified to subsume modulo information in the chain box, analogously to the way the resolution is modified.

Since the main modification of an "ordinary" resolution prover for implementing chain resolution consists of modifying the equality check of predicate symbols during unification and matching operations of literals, it would be easy to modify most existing provers to take advantage of chain resolution.

3   Motivation for Chain Resolution

During experimental resolution-based proof searches for query answering in large ontologies it has become clear to us that:

- New chain clauses are produced during proof search.
- Some chain clauses are typically present in the proof, making a proof larger and hence harder to find.
- Chain clauses produce a large number of propositional variations of non-chain clauses, making the search space larger.

We will elaborate upon the last, critical item. Suppose we have a clause C of the form $A_1(t_1) \vee \ldots \vee A_n(t_n) \vee \Gamma$ in the search state. Suppose search state also contains m_i chain clauses of the form $\neg A_i(x) \vee A'_1(x), \ldots \neg A_i(x) \vee A'_{m_i}(x)$ for each i such that $1 \leq i \leq n$. Then the resolution method will derive $m_1 * m_2 * \ldots * m_n$ new clauses (propositional variations) from C using chain clauses alone.

Although the propositional variations of a clause are easy to derive, the fact that they significantly increase the search space will slow down the proof search. Hence we decided to develop a special "black box" implementation module we present in this paper. Chain resolution module allows the prover to keep only one propositional variation of a clause in the search state and allows the prover to never use chain clauses in ordinary resolution steps.

An important aspect of the semantic web context is that a crucial part of the prospective formalisations of knowledge consist of formalisations of ontologies. Taxonomies, ie hierarchies of concepts, typically form a large part of an ontology. Most prospective applications of the semantic web envision formalisations where the ontologies are not simple taxonomies, but rather more general, ordered acyclic graphs.

A full ontology may also contain knowledge which is not representable by simple implications. Prospective applications and typical queries formalise knowledge which is also not representable by chain clauses. However, ontologies and simple implications are currently seen as a large, if not the largest part of prospective semantic web applications.

Chain resolution, although a very simple mechanism, makes efficient query answering possible also for the resolution-type methods.

4   Details of the Chain Resolution

We will present the chain resolution for a "general resolution algorithm", without bringing out specific details of different resolution algorithms, which are inessential for chain resolution.

4.1   Moving Chain Clauses to the Chain Box

Remember that the chain box contains a set of signed unary predicate symbols, each associated with a set of signed unary predicate symbols. Hence, the chain box is a set of pairs.

Definition 6.
A signed predicate symbol P is called a key *of a chain box in case P has a set of signed predicate symbols associated with it in the chain box. The associated set is called the* chain *of a key P, also denoted as $chain(P)$. The pair of a key P and the chain set of P is called the* row *of a chain box.*

The chain resolution mechanism consists of two separate parts invoked during the proof search process:

– Moving chain clauses to the chain box. We will later show that moving chain clauses to the chain box may also produce new unit clauses which have to be added to the search state.
– Using the chain box in ordinary resolution, factorisation and subsumption operations.

The general algorithm of moving clauses to the chain box during proof search:

– Chain box initialisation. For each unit predicate symbol P in the initial clause two *chain box rows* are formed: one with they key P, one with the key $\neg P$. For each chain box key the initial chain set of the key will contain the key itself as a single element of the set.
– Before the proof search starts, all the chain clauses present in the initial clause set are moved to the chain box, one by one. Each chain clause moved to the chain box is removed from the initial clause set. In case a move to the chain box produces a set of new unit clauses S, these clauses are added to the initial clause set.

– During the proof search process each time the proof search produces a new chain clause, this clause is moved to the chain box and removed from the search state. All the unit clauses produced by the moving operation are added to the passive list part of the search state, analogously to clauses derived by ordinary resolution steps.

Observe that the chains in the chain box correspond to propositional implications from keys seen as propositional atoms to corresponding chain elements, also seen as propositional atoms. After moving a chain to the chain box we want all the implicit implications derivable from the chain box to be explicitly present as chains. A chain clause $A(x) \lor B(x)$ is seen both as a propositional implication $\neg A \Rightarrow B$ and an equivalent implication $\neg B \Rightarrow A$.

The algorithm we present for adding a chain clause to the chain box is one of the several possible algorithms for adding chain clauses. The main element of the algorithm adding a chain clause to the chain box is the following recursive procedure $addchain(k, p)$ which adds a signed predicate p to the chain of the key k:

– do $addchainaux(k, p)$
– do $addchainaux(\neg p, \neg k)$

where $addchainaux(k, p)$ is defined as

– If $p \notin chain(k)$ then do:
  • $chain(k) := chain(k) \cup \{p\}$
  • for all $l \in chain(p)$ do $addchain(k, l)$
  • for each key $r \in chainbox$ where $k \in chain(r)$ do:
    * for all $m \in chain(k)$ do $addchain(r, m)$

Regardless of the concrete details of adding a predicate to a chain, the algorithm must propagate all the new implicit implications (for example, a case where $\neg A$ is a member of some chain and the chain of B contains more than one element) to explicit implications by adding elements of chains to corresponding chains, until no more changes can be made.

The full algorithm for moving one chain clause $A(x) \lor B(x)$ to the chain box:

– Do $addchain(\neg A, B)$ (observe that this also does $addchain(\neg B, A)$).
– Construct a set S of unit clauses derivable from the chain box. A unit clause $p(x)$ is derivable from the chain box in case a chain c with a key $\neg p$ contains both r and $\neg r$ for some predicate symbol r.
– Add elements of S to the passive list in the search state of the prover.

A concrete implementation of chain resolution should optimize the creation of a set of S of unit clauses in the previous algorithm so that the same unit clauses are not repeatedly produced during several chain clause additions to the chain box.

Observe that chain box may become contradictory. As shown later, for the completeness properties of chain resolution it is not necessary to check whether the chain box is contradictory or not: derivation of unit clauses suffices for completeness.

4.2   Chain Resolution, Factorisation, and Subsumption

In this subsection we describe the modifications to the standard resolution, factorisation and subsumption operations. The easiest way to modify these operations is to modify literal unification and literal subsumption algorithms. Hence we will not bring formal rules for chain resolution, factorisation and subsumption. Instead we assume familiarity with standard rules and suggest each implementer to modify the concrete forms of these rules he/she is using.

In chain resolution the presented chain operations replace standard resolution, factorisation and subsumption operations.

The *chain resolution algorithm* for resolution allows to resolve upon two literals $A(t_1, \ldots, t_n)$ and $B(t'_1, \ldots, t'_n)$ iff $A(t_1, \ldots, t_n)$ and $A(t'_1, \ldots, t'_n)$ are unifiable using standard unification and either $B = \neg A$ or $B \in chain(\neg A)$. Observe that if $B \in chain(\neg A)$ then by the construction of the chain box also $A \in chain(\neg B)$. Chain resolution does not introduce any changes to the substitution computed by a successful unification operation.

The *chain factorisation* algorithm requires a more substantial modification of the standard factorisation algorithm. When factorising two literals $A(t_1, \ldots, t_n)$ and $B(t'_1, \ldots, t'_n)$ in a clause the predicate symbol in the resulting factorised literal depends on the implications encoded in the chain box. First, chain factorisation can be successful only if $A(t_1, \ldots, t_n)$ and $A(t'_1, \ldots, t'_n)$ are unifiable with standard unification. The resulting literal of the factorisation before substitution is applied is obtained in the following way:

- if $A = B$ then the resulting literal is $A(t_1, \ldots, t_n)$.
- if $A \in chain(B)$ then the resulting literal is $A(t_1, \ldots, t_n)$.
- if $B \in chain(A)$ then the resulting literal is $B(t_1, \ldots, t_n)$.

If both $A \in chain(B)$ and $B \in chain(A)$ hold, then the choice of the resulting literal between $A(t_1, \ldots, t_n)$ and $B(t_1, \ldots, t_n)$ does not matter and it is also not necessary to derive two different resulting literals. Chain factorisation does not introduce any changes to the substitution computed by a successful unification operation.

By the *chain subsumption* algorithm of literals a literal $A(t_1, \ldots, t_n)$ subsumes a literal $B(t'_1, \ldots, t'_n)$ iff $A(t_1, \ldots, t_n)$ subsumes $B(t'_1, \ldots, t'_n)$ using standard subsumption and either $B = A$ or $B \in chain(A)$. Chain subsumption does not introduce any changes to the substitution computed by a successful subsumption operation. Neither does it introduce any changes to the clause subsumption algorithm, except modifying an underlying literal subsumption algorithm.

Observe that while chain resolution and chain factorisation may give advantages for proof search efficiency, the main efficiency-boosting effect for chain resolution is obtained by replacing a standard subsumption algorithm with chain subsumption.

5   Correctness and Completeness of Chain Resolution

Both the correctness and completeness proofs are relatively simple. Since the paper is application-oriented, we will not explicate the obvious details of the proofs.

Lemma 1. *Chain resolution is correct: if chain resolution derives a contradiction from a clause set S, then S is not satisfiable.*

Proof. The proof uses correctness of standard resolution as its base. We show that any clause derivable from S using chain resolution and factorisation rules is derivable using standard resolution and factorisation rules. Proof is constructed by the induction on the number of chain resolution and factorisation rules in a derivation D of a contradiction from S using chain resolution.

The applications of the resolution and factorisation rules of chain resolution in D can be replaced by derivations using corresponding standard rules of resolution along with the chain clauses encoded in the chain box. All the chain clauses encoded in the chain box are either present in S or derivable from S using standard resolution.

Lemma 2. *Chain resolution is complete: if a clause set S is not satisfiable, then chain resolution will derive a contradiction from S.*

Proof. We will first show that chain resolution without a chain subsumption rule is complete. Proof is by induction on the derivation D of contradiction using standard resolution rules. We transform D to a derivation D' of contradiction using chain resolution rules.

- Indeed, every resolution step between a chain clause C in D and a non-chain clause N in D can only result with a modification N' of N where one of the signed predicate symbols in N is replaced. Every further use of N' in D for factorisation or resolution with another non-chain clause can be replaced by either a corresponding chain factorisation or chain resolution step, since C would be in the chain box at the moment of the rule application. Further uses of N' in D for resolution with other chain clauses are treated in the similar manner.
- A resolution step between two chain clauses in D results either with a unit clause or a chain clause. A unit clause U in D obtained from two chain clauses would be present in the search state of D, by the result of moving a chain clause to a chain box. A chain clause C' in D obtained from two chain clauses would be encoded in the chain box inside two chains, by the result of moving a chain clause to a chain box.
- Factorisation steps as well as resolution steps between non-chain clauses in D are unmodified.

Second, we will show that chain subsumption does not destroy completeness either. The proof is based on the completeness of ordinary subsumption. Indeed, if a clause C chain-subsumes a clause C' using a chain rule R encoded in the chain box, then any use of C' in D can be replaced by a use of either C or C along with R with the help of the chain factorisation or chain resolution rule. Let $C = (A(t) \vee \Gamma)$, $C' = (B(t') \vee \Gamma')$, $R = (\neg A(x) \vee B(x))$. Then any application of a standard rule to the literal $B(t')$ in C' can be replaced by an application of the chain rule to the literal $A(t)$ in C along with a chain clause R.

6   Chain Resolution in Combination with Other Strategies

In the following we will show completeness of chain resolution in combination with the following important strategies:

- Set of support strategy. We show that a naive combination of chain strategy with the set of a support strategy is incomplete, and present an alternative weak combination, which preserves completeness.
- A family of strategies for ordering literals in a clause, based on ordering the term structures in the literals.

6.1   Set of Support Strategy

The set of support (sos), see [Handbook01] is a classical, completeness-preserving resolution strategy, employed for focusing search to the query. As such it is very useful in cases where the clause set consists of a large set of facts and rules, known to be consistent, and a relatively small query. A typical semantic web query, similarly to typical datalog and prolog queries, is likely to have such a structure. The sos strategy can be seen as one of the ways to make a resolution search behave similarly to a tableaux search.

The idea of the sos strategy is following. An initial clause set S is split into two disjoint parts:

- R: a set of clauses, assumed to be consistent. Typically R consists of a set of known facts and rules.
- Q: another set of clauses, typically obtained from the query.

Newly derived clauses are added to the initially empty set Q'. *Sos condition*: one of the two clauses C_1 and C_2 in any resolution rule application is required to be either present in Q or Q'. In other words, derivations from R alone are prohibited.

Sos strategy is not compatible with most of the other resolution strategies.

By the *naive combination* of sos strategy with chain resolution we mean the following strategy:

- Resolution is restricted by the sos condition.
- Chain clauses are moved to chain box both from R, Q and Q'.
- It is always allowed to use a chain clause encoded in the chain box, regardless of whether the chain clause originates from R, Q or Q'.

It is easy to see that the naive combination is incomplete. Consider sets $R=\{\neg A(c), \neg B(c)\}$ and $Q=\{A(x) \vee B(x)\}$. The union of R and Q is obviously inconsistent. The sos strategy will derive a contradiction from these sets. However, if we move the chain clause $\{A(x) \vee B(x)\}$ to the chain box before we start any resolution steps, the set Q will be empty and due to the sos condition we cannot derive any clauses at all.

We will define an alternative *weak combination* of sos strategy and chain resolution as follows:

- Resolution is restricted by the sos condition.
- Chain clauses are moved to the chain box from R, but not from Q or Q'.
- It is always allowed to use a chain clause encoded in the chain box.
- A clause in R is not allowed to chain subsume a clause in Q or Q'.
- A newly derived clause C may be chain subsumed by existing clauses in Q or Q'.

Lemma 3. *Chain resolution in weak combination with the sos strategy preserves completeness: if an empty clause can be derived from the clause sets R and Q using sos strategy, then it can be also derived using sos strategy weakly combined with the chain strategy.*

Proof. Observe that the previously presented proofs of the completeness of chain resolution hold, unmodified, in the situation where resolution is weakly combined with the sos strategy. Indeed, the sos condition is preserved during the transformation of an original derivation D to the derivation D'.

The completeness of the weak combination of sos and chain strategies is crucial for the compilation framework presented later in the paper.

6.2 Ordering Strategies

By the *ordering strategies* we mean resolution strategies which:

- Define an ordering $\prec_o$ on literals in a clause.
- Ordering condition: prohibit resolution with a literal B in a clause C iff C contains a literal A such that $B \prec_o A$.

Ordering strategies form a powerful, well-studied family of resolution strategies. In particular, several completeness-preserving ordering strategies of resolution have been shown to always terminate on clause sets from several known decidable subclasses of first order logic, thus giving a resolution-based decision algorithm for these classes. Ordering strategies have been used to find new decidable classes. Some ordering strategies have been shown to decide different classes of description logics, including extensions of well-known description logics, see [HustadtSchmidt00], [Handbook01], [Fermüller93].

In this section we will consider a subfamily of ordering strategies, which we will call *term-based orderings*. An ordering $\prec_t$ is term-based iff it is nonsensitive to variable and predicate names and polarities, ie iff the following conditions hold:

- If $A \prec_t B$ then $A' \prec_t B'$ where A' and B' are obtained from A and B, respectively, by renaming some variables simultaneously in A and B.
- If $A \prec_t B$ then $A' \prec_t B'$ where A' and B' are obtained from A and B, respectively, by replacing predicate names (and/or) polarity in A and B.

Term-based orderings generate term-based ordering strategies for resolution.

By a combination of a term-based ordering strategy with chain resolution we mean a following strategy:

- Both ordinary and chain resolution are restricted by the ordering condition.
- Both the initial and generated chain clauses are moved to the chain box.
- Chain subsumption is allowed.

We can prove the completeness lemma similar to the lemmas proved previously.

Lemma 4. *Chain resolution combined with any completeness-preserving term-based ordering strategy is completeness-preserving: if a clause set S is inconsistent, an empty clause can be derived using the combination of the term-based ordering strategy and the chain strategy.*

Proof. Observe that the previously presented proofs of the completeness of chain resolution hold, unmodified, in the situation where resolution is weakly combined with the sos strategy. Since chain resolution rule applications only replace the predicates present in clauses, and the ordering is assumed to be insensitive to the predicates and their polarities, the ordering condition is preserved during the transformation of an original derivation D to the derivation D'.

In order to use the combined chain and ordering strategies as decision procedures we would have to show that the known termination properties of ordering strategies are preserved.

In this paper we will not tackle the termination properties of the chain strategy combination with ordering strategies. However, intuition indicates that the termination properties are likely preserved for most, if not all important term-based ordering strategies.

7   A Resolution Framework for the Semantic Web Inference Algorithms

We will now present a general compilation-based scheme for the practical usage of resolution-based inference engines for solving queries arising in the semantic web context. We are currently working on an actual implementation of the presented scheme on top of the resolution prover Gandalf. The scheme we present is obviously only one of the ways to use the resolution-based inference engines in this context. Many open issues already known and continuously arising will have to be solved for the eventual industrial use of such systems.

Let us consider a likely situation where an inference engine will be used in our context. We will have a very large database of known facts and rules. A

large part of the rules form a taxonomy, or rather, several taxonomies. A large
part, if not all (depending on the task domain) of the other rules fall into a
decidable fragment. However, some rules may be present (for example, rules
involving equality, in case the use of equality is not tightly restricted) which do
not fall into known decidable classes.

The system will regularly ask queries against such a database. The database
will change over time and will be different for different task domains. However,
it is likely that the number of queries asked over time is much higher than the
number of times the rule database is changed during the same period.

Hence it will normally pay out to compile the database before queries are
asked against it.

Hence, for the resolution-based inference systems we suggest the following
scheme:

- *Compilation phase.*
  - *Analysis.* Analyse the rule base R to determine whether R or a large
    subset R_d of R falls into a known decidable class. Determine a suitable
    ordering-based strategy os for R or R_d.
  - *Terminating strategy run.* Run resolution with the strategy os combined
    with the chain strategy on R or R_d, for N seconds, where N is a pre-
    determined limit. This will build both a chain box C_b and a set of new
    clauses R_1. In case inconsistency was found, stop, otherwise continue.
  - *First filtering.* If R was not in a decidable class, then the previous step
    probably did not terminate. Even if R was decidable, N may have been
    too small for the previous step to terminate in time. Hence, in case the
    strategy did not terminate, the set of new clauses R_1 is likely to be very
    large. It is sensible to employ heuristics to keep only some part R_f of
    the derived clauses R_1 for the future.
  - *Chain box building.* Regardless of whether the terminating strategy run
    terminated in time or not, it will be advantageous to increase the size
    of a chain box by performing another run of resolution on $R \cup R_f$ plus
    C_b, this time using a different strategy, optimised for fast derivation of
    chain clauses. After a predetermined time of M seconds the run will be
    stopped (or it may terminate by itself earlier).
  - *Final filtering and storage.* The set of clauses R_2 obtained after the pre-
    vious run will be heuristically filtered, again storing a subset R_{f2} of R_2.
    Finally, both the chain box constructed so far as well as the new clauses
    R_{f2} along with the (possibly) simplified initial clause set R will be stored
    as a *compilation* of the initial clause set R.
- *Query phase.* The main motivation for the compilation phase is a likely
  situation where one compilation can be used for a number of different queries
  and the recompilations are not too frequent.

  In the query phase the clauses obtained from the query will be added to the
  compilation and several resolution runs will be conducted with different
  strategies. The obvious initial choice in case of a large inital R is the set
  of support resolution weakly combined with the chain strategy. For the set

of support strategy the R part of clauses will be formed of the compilation along with the chain box built. The query clauses will form the Q part of clauses for the strategy.

Observe that the compilation probably contains a large chain box, thus the "weakness" of the combination does not prohibit efficient chain box usage.

8   Summary

We have presented a practical extension called chain resolution to the standard resolution method, essentially combining resolution with a simple and efficient propositional algorithm. The extension is shown to be both correct and complete. Chain resolution is compatible with several well-known resolution strategies and is useful for solving queries in case large ontologies are present in the theory. It enables resolution provers to work efficiently even in case a problem at hand contains a large ontology, thus combining some ideas from terminological reasoning with the classical theorem provers for FOL. We have also presented a general scheme for practical compilation of large rule databases for resolution-based provers.

References

[Casc] http://www.cs.miami.edu/~tptp/CASC/

[Fermüller93] Fermüller, C., Leitsch, A., Tammet, T., Zamov, N. Resolution methods for decision problems. Lecture Notes in Artificial Intelligence 679, Springer Verlag, 1993.

[Handbook01] Robinson, A.J., Voronkov, A., editors. Handbook of Automated Reasoning. MIT Press, 2001.

[Handbook03] Baader, F., Calvanese, V., McGuinness, V., Nardi, D., Patel-Schneider, P., editors. The Description Logic Handbook. Cambridge University Press, 2003.

[HustadtSchmidt00] Hustadt, U., Schmidt, R. A. Issues of decidability for description logics in the framework of resolution. In G. Salzer and R. Caferra, editors, Automated Deduction in Classical and Non-Classical Logics, pp. 192-206. LNAI 1761, Springer.

[OWL] OWL Web Ontology Language. Semantics and Abstract Syntax. W3C Working Draft 31 March 2003. W3C. Patel-Schneider, P.F, Hayes, P., Horrocks, I., eds. http://www.w3.org/TR/owl-semantics/

[RDF] RDF Semantics. W3C Working Draft 23 January 2003. W3C. Hayes, P. ed. http://www.w3.org/TR/rdf-mt/

[SemanticWeb01] Berners-Lee, T., Hendler, J., Lassila, O. The Semantic Web. Scientific American, May 2001.

[SemanticWebURL] Semantic web. http://www.w3.org/2001/sw/. W3C.

[TPTP] http://www.cs.miami.edu/~ tptp/

[Tammet97] Tammet, T. Gandalf. Journal of Automated Reasoning vol 18 No 2 (1997) 199–204.

[Tammet98] Tammet, T. Towards Efficient Subsumption. In CADE-15, pages 427-441, Lecture Notes in Computer Science vol. 1421, Springer Verlag, 1998.

[Tammet02] Tammet, T. Gandalf. http://www.ttu.ee/it/gandalf/, Tallinn Technical University.

SONIC — Non-standard Inferences Go OILED

Anni-Yasmin Turhan and Christian Kissig*

TU Dresden, Germany
lastname@tcs.inf.tu-dresden.de

Abstract. SONIC[1] is the first prototype implementation of non-standard inferences for Description Logics usable via a graphical user interface. The contribution of our implementation is twofold: it extends an earlier implementation of the least common subsumer and of the approximation inference to number restrictions, and it offers these reasoning services via an extension of the graphical ontology editor OILED [3].

1 Introduction and Motivation

Description Logics (DLs) are a family of formalisms used to represent terminological knowledge of a given application domain in a structured and well-defined way. The basic notions of DLs are *concept descriptions* and *roles*, representing unary predicates and binary relations, respectively. The inference problems for DLs can be divided into so-called standard and non-standard ones. Well known standard inference problems are satisfiability and subsumption of concept descriptions. For a great range of DLs, sound and complete decision procedures for these problems could be devised and some of them are put into practice in state of the art DL systems as FACT [11] and RACER [9].

Prominent non-standard inferences are the least common subsumer (lcs), and approximation. Non-standard inferences resulted from the experience with real-world DL ontologies, where standard inference algorithms sometimes did not suffice for building and maintaining purposes. For example, the problem of how to structure the application domain by means of concept definitions may not be clear at the beginning of the modeling task. This kind of difficulties can be alleviated by non-standard inferences [1,7].

Given two concept descriptions A and B in a description logic $\mathcal{L}$, the *lcs* of A and B is defined as the least (w.r.t. subsumption) concept description in $\mathcal{L}$ subsuming A and B. The idea behind the lcs inference is to extract the commonalities of the input concepts. It has been argued in [1,7] that the lcs facilitates a "bottom-up"-approach to the modeling task: a domain expert can select a number of intuitively related concept descriptions already existing in an ontology and use the lcs operation to automatically construct a new concept description representing the closest generalization of them.

---

* This work has been supported by the Deutsche Forschungsgemeinschaft, DFG Project BA 1122/4-3.
[1] SONIC stands for "Simple OILED Non-standard Inference Component".

D. Basin and M. Rusinowitch (Eds.): IJCAR 2004, LNAI 3097, pp. 321–325, 2004.

Approximation was first mentioned as a new inference problem in [1]. The *approximation* of a concept description C from a DL $\mathcal{L}_1$ is defined as the least concept description (w.r.t. subsumption) in a DL $\mathcal{L}_2$ that subsumes C. The idea underlying approximation is to translate a concept description from one DL into a typically less expressive DL. Approximation can be used to make non-standard inferences accessible to more expressive DLs so that at least an approximate solution can be computed. In case the DL $\mathcal{L}$ provides disjunction, the lcs of C_1 and C_2 is just the disjunction $(C_1 \sqcup C_2)$. Thus, a user inspecting this concept does not learn anything about the commonalities of C_1 and C_2. Using approximation, however, one can make the commonalities explicit to some extent by first approximating C_1 and C_2 in a sublanguage of $\mathcal{L}$ which does not provide disjunction, and then compute the lcs of the approximations in $\mathcal{L}$. Another application of approximation lies in user-friendly DL systems, such as OILED [3], that offer a simplified frame-based view on ontologies defined in an expressive background DL. Here approximation can be used to compute simple frame-based representations of otherwise very complicated concept descriptions.

OILED is a widely accepted ontology editor and it can be linked to both state of the art DL systems, RACER [9] and FACT [11]. Hence this editor is a good starting point to provide users from practical applications with non-standard inference reasoning services. The system SONIC is the first system that provides some of these reasoning services via a graphical user interface. SONIC can be downloaded from `http://lat.inf.tu-dresden.de/systems/sonic.html`.

2   The SONIC Implementation

Let us briefly recall the DLs covered by SONIC. The DL $\mathcal{ALE}$ offers the top- and bottom-concept $(\top, \bot)$, conjunction $(C \sqcap D)$, existential $(\exists r.C)$, value restrictions $(\forall r.C)$, and primitive negation $(\neg C)$. The DL $\mathcal{ALC}$ extends $\mathcal{ALE}$ by disjunction $(C \sqcup D)$ and full negation. Extending each of these DLs by number restrictions $((\leq n\ r), (\geq n\ r))$ one obtains $\mathcal{ALEN}$ and $\mathcal{ALCN}$, respectively. For the definition of the syntax and semantics of these DLs, refer to [5,6]. A *TBox* is a finite set of concept definitions of the form $A \doteq C$, where A is a concept name and C is a concept description. Concept names occurring on the left-hand side of a definition are called *defined concepts*. All other concept names are called *primitive concepts*. SONIC can only process TBoxes that are acyclic and do not contain multiple definitions.

2.1   Implementing the Inferences

SONIC implements the lcs for $\mathcal{ALEN}$-concept descriptions and the approximation of $\mathcal{ALCN}$- by $\mathcal{ALEN}$-concept descriptions in Lisp. The algorithm for computing the lcs in $\mathcal{ALEN}$ was devised and proven correct in [12]. This algorithm consists of three main steps: first recursively replace defined concepts by their definitions from the TBox, then normalize the descriptions to make implicit information explicit, and finally make a recursive structural comparison of each role-level of the descriptions. In $\mathcal{ALEN}$ the last two steps are much more involved than in $\mathcal{ALE}$

since the number restrictions for a role, more precisely the at-most restrictions, necessitates merging of role-successors. The lcs algorithm for $\mathcal{ALEN}$ takes double exponential time in the worst case. Nevertheless, the lcs for $\mathcal{ALEN}$ realized in SONIC is a plain implementation of this algorithm. Surprisingly, a first evaluation shows that for concepts of an application ontology with only integers from 0 to 7 used in number restrictions the run-times remained under a second (on a Pentium IV System, 2 GHz). The implementation of the lcs for $\mathcal{ALE}$ as described in [2] uses unfolding only on demand—a technique known as lazy unfolding. Due to this technique shorter and thus more easily comprehensible concept descriptions can be obtained more quickly, see [2]. To implement lazy unfolding also for $\mathcal{ALEN}$ is yet future work.

The algorithm for computing the $\mathcal{ALCN}$ to $\mathcal{ALEN}$ approximation was devised and proven correct in [5]. The idea underlying it is similar to the lcs algorithm in $\mathcal{ALEN}$. For approximation the normalization process additionally has to "push" the disjunctions outward on each role-level before the commonalities of the disjuncts are computed by applying the lcs on each role-level. The $\mathcal{ALCN}$ to $\mathcal{ALEN}$ approximation was implemented in Lisp using our $\mathcal{ALEN}$ lcs implementation. An implementation of the $\mathcal{ALC}$ to $\mathcal{ALE}$ approximation is described in [6]. It was the basis for the implementation presented here. The worst case complexity of approximation in both pairs of DLs is double exponential time, nevertheless this is not a tight bound. A first evaluation of approximating randomly generated concept descriptions show that, unfortunately, both implementations run out of memory already for concepts that contain several disjunctions with about 6 disjuncts. The implementation of the algorithms for both inferences are done in a straightforward way without code optimizations or sophisticated data structures. This facilitated testing and debugging of SONIC.

Let us illustrate the procedure of lcs and approximation by an example. Consider a $\mathcal{ALCN}$-TBox with role r, primitive concepts A, B, and concept definitions: $C_1 \doteq \exists r.A \sqcap \forall r.B \sqcap (\geq 3\, r)$, $C_2 \doteq \forall r.(A \sqcap B) \sqcap (\geq 2\, r)$, $C \doteq C_1 \sqcup C_2$ and $D \doteq \exists r.(A \sqcap B) \sqcap \exists r.(\neg A \sqcap B)$. If we want to find the commonalities between C and D, we first compute the $\mathcal{ALEN}$-approximation of C and then the $\mathcal{ALEN}$-lcs of D and $approx(C)$. We compute $approx(C)$ by first unfolding C and then extracting the commonalities of C_1 and C_2. Both have a value restriction and the lcs of these restrictions is $\forall r.B$. Both C_i induce the number restriction $(\geq 2\, r)$, since $(\geq 2\, r)$ subsumes $(\geq 3\, r)$. C_1 has a value and an existential restriction inducing the existential restriction $\exists r.(A \sqcap B)$, whereas in C_2 the number restriction requires at least two distinct r-successors which in addition to the value restriction also induces the restriction $\exists r.(A \sqcap B)$. Thus we obtain $approx(C) = \exists r.(A \sqcap B) \sqcap \forall r.B \sqcap (\geq 2\, r)$. In the concept definition of D the occurrence of A and $\neg A$ induce that at least two r-successors exist. Thus the commonalities of C and D are $lcs(approx(C), D) = \exists r.(A \sqcap B) \sqcap (\geq 2\, r)$.

2.2   Linking the Components

In order to provide the lcs and approximation to OILED users, SONIC does not only have to connect to the editor OILED, but also to a DL system since both, lcs and approximation, use subsumption tests during their computation. A connection from SONIC to the editor OILED, is realized by a plug-in. Like OILED itself, this plug-in is implemented in Java. SONIC's

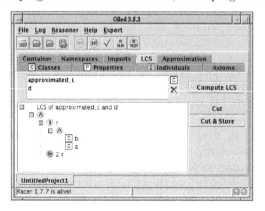

Fig. 1. Interface of SONIC

plug-in is implemented for OILED Version 3.5.3 and realizes mainly the graphical user interface of SONIC—its lcs tab is shown in Figure 1. SONIC's Java plug-in connects via the JLinker interface by Franz Inc. to the Lisp implementation to pass concepts between the components.

To classify an ontology from within OILED, the user can either connect OILED to the reasoner FACT (via CORBA) or to any DL reasoner supporting the DIG ("Description Logic Implementation Group") protocol. The DIG protocol is an XML-based standard for DL systems with a tell/ask syntax, see [4]. DL developers of most systems have committed to implement it in their system making it a promising standard for future DL related software.

SONIC must have access to the same instance of the reasoner that OILED is connected to in order to have access to the information from the ontology, more precisely, to make use of stored concept definitions and of cached subsumption relations obtained during classification by the DL reasoner. Obtaining the concept definitions from OILED directly, would result in storing the ontology in all of the three components and, moreover, the results for lcs and approximation might be incorrect, if OILED and the DL reasoner do not have consistent data.

Since SONIC needs to retrieve the concept definition of a defined concept in order to perform unfolding—a functionality that RACER provides—we decided to use RACER in our implementation. SONIC connects to RACER Version 1.7.7 via the TCP socket interface described in [10]. Note, that in this setting the RACER system need not run locally, but may even be accessed via the web by OILED and SONIC.

2.3   SONIC at Work

Starting the OILED editor with SONIC, the lcs and approximation inferences are available on extra tabs—as shown in Figure 1. Once the OILED user has defined some concepts in the OILED ontology, has connected to the DIG reasoner RACER and classified the ontology, she can use, for example, the lcs reasoning service to add a new super-concept of a number of concepts to the ontology. On the lcs tab she can select some concept names from all concept names in ontology. When the lcs button is clicked, the selected names are transmitted to SONIC's Lisp component and the lcs is computed based on the current concept definitions stored in RACER. The obtained lcs concept description is send to the plug-in and displayed on the lcs tab in OilEd. Since the returned concept descriptions can become very large, Sonic displays them in a tree representation, where uninteresting subconcepts can be folded away by the user and inspected later. In Figure 1 we see how the concept description obtained from the example in Section 2.1 is displayed in Sonic. Based on this representation Sonic also provides limited editing functionality. The OilEd user can cut subdescriptions from the displayed lcs concept description or cut and store (a part of) it under a new concept name in the ontology.

3 Outlook

Developing Sonic is ongoing work. Our next step is to optimize the current implementation of reasoning services and to implement minimal rewriting to obtain more concise result concept descriptions. Future versions of Sonic will comprise the already completed implementations of the difference operator [6] and of matching for $\mathcal{ALE}$ [8].

We would like to thank Ralf Möller and Sean Bechhofer for their help on how to implement Sonic's linking to Racer and to OilEd.

References

1. F. Baader, R. Küsters, and R. Molitor. Computing least common subsumers in description logics with existential restrictions. In, *Proceedings of IJCAI-99*, Stockholm, Sweden. Morgan Kaufmann, 1999.
2. F. Baader and A.-Y. Turhan. On the problem of computing small representations of least common subsumers. In *Proceedings of KI'02*, LNAI. Springer–Verlag, 2002.
3. S. Bechhofer, I. Horrocks, C. Goble, and R. Stevens. OilEd: a Reason-able Ontology Editor for the Semantic Web. In *Proceedings of KI'01*, *LNAI*, Springer-Verlag, 2001.
4. S. Bechhofer, R. Möller, and P. Crowther. The DIG description logic interface. In *Proceedings of DL 2003*, Rome, Italy, CEUR-WS, 2003.
5. S. Brandt, R. Küsters, and A.-Y. Turhan. Approximating $\mathcal{ALCN}$-concept descriptions. In *Proceedings of DL 2002*, nr. 53 in CEUR-WS. RWTH Aachen, 2002.
6. S. Brandt, R. Küsters, and A.-Y. Turhan. Approximation and difference in description logics. In *Proceedings of KR-02*, Morgan Kaufmann, 2002.
7. S. Brandt and A.-Y. Turhan. Using non-standard inferences in description logics – what does it buy me? In *Proc. of KIDLWS'01*, CEUR-WS. RWTH Aachen, 2001.
8. S. Brandt. Implementing matching in $\mathcal{ALE}$—first results. In *Proceedings of DL2003*, Rome, Italy, CEUR-WS, 2003.
9. V. Haarslev and R. Möller. RACER system description. In *Proceedings of the Int. Joint Conference on Automated Reasoning IJCAR'01*, LNAI. Springer Verlag, 2001.
10. V. Haarslev and R. Möller. *RACER User's Guide and Manual, Version 1.7.7*, Sept, 2003. available from: http://www.sts.tu-harburg.de/~r.f.moeller/racer/racer-manual-1-7-7.pdf.
11. I. Horrocks. Using an expressive description logic: FaCT or fiction? In *Proceedings of KR-98*, Trento, Italy, 1998.
12. R. Küsters and R. Molitor. Computing Least Common Subsumers in $\mathcal{ALEN}$. In *Proceedings of IJCAI-01*, Morgan Kaufman, 2001.

TeMP: A Temporal Monodic Prover[*]

Ullrich Hustadt[1], Boris Konev[1][**], Alexandre Riazanov[2], and Andrei Voronkov[2]

[1] Department of Computer Science, University of Liverpool, UK
{U.Hustadt, B.Konev}@csc.liv.ac.uk
[2] Department of Computer Science, University of Manchester, UK
{riazanov, voronkov}@cs.man.ac.uk

1 Introduction

First-Order Temporal Logic, FOTL, is an extension of classical first-order logic by temporal operators for a discrete linear model of time (isomorphic to $\mathbb{N}$, that is, the most commonly used model of time). Formulae of this logic are interpreted over structures that associate with each element n of $\mathbb{N}$, representing a moment in time, a first-order structure (D_n, I_n) with its own non-empty domain D_n. In this paper we make the *expanding domain assumption*, that is, $D_n \subseteq D_m$ if $n < m$. The set of valid formulae of this logic is not recursively enumerable. However, the set of valid *monodic* formulae is known to be finitely axiomatisable [13].

A formula ϕ in a FOTL language *without equality and function symbols* (constants are allowed) is called *monodic* if any subformula of ϕ of the form $\bigcirc\psi$, $\Box\psi$, $\Diamond\psi$, $\psi_1 \cup \psi_2$ or $\psi_1 \mathsf{W} \psi_2$ contains at most one free variable. For example, the formulae $\forall x \Box \exists y P(x,y)$ and $\forall x \Box P(x,c)$ are monodic, while $\forall x,y(P(x,y) \Rightarrow \Box P(x,y))$ is not monodic. The monodic fragment has a wide range of novel applications, for example in spatio-temporal logics [5] and temporal description logics [1].

In this paper we describe **TeMP**, the first automatic theorem prover for the monodic fragment of FOTL. The prover implements *fine-grained temporal resolution*, described in the following section, while Section 3 provides an overview of our implementation. Finally, Section 4 describes some preliminary experiments with **TeMP**.

2 Monodic Fine-Grained Temporal Resolution

Our temporal prover is based on *fine-grained temporal resolution* [9] which we briefly describe in this section. Every monodic temporal formula can be transformed in a satisfiability equivalence preserving way into a clausal form. The calculus operates on four kinds of temporal clauses, called *initial*, *universal*, *step*, and *eventuality* clauses. Essentially, initial clauses hold only in the initial moment in time, all other kinds of clauses hold in every moment in time. Initial and universal clauses are ordinary first-order clauses, containing no temporal operators. *Step* clauses in the clausal form of monodic temporal formulae are of the form $p \Rightarrow \bigcirc q$, where p and q are propositions, or of the form $P(x) \Rightarrow \bigcirc Q(x)$, where P and Q are unary predicate symbols and x a variable. During

---

[*] Work supported by EPSRC grant GR/L87491.
[**] On leave from Steklov Institute of Mathematics at St.Petersburg

D. Basin and M. Rusinowitch (Eds.): IJCAR 2004, LNAI 3097, pp. 326–330, 2004.
© Springer-Verlag Berlin Heidelberg 2004

a derivation more general *step* clauses can be derived, which are of the form $C \Rightarrow \bigcirc D$, where C is a *conjunction* of propositions, atoms of the form $P(x)$ and ground formulae of the form $P(c)$, where P is a unary predicate symbol and c is a constant such that c occurs in the input formula, D is a *disjunction* of arbitrary literals, such that C and D have at most one free variable in common. The *eventuality* clauses are of the form $\Diamond L(x)$, where $L(x)$ is a literal having at most one free variable.

Monodic fine-grained temporal resolution consists of the *eventuality resolution rule*:

$$\frac{\forall x(\mathcal{A}_1(x) \Rightarrow \bigcirc(\mathcal{B}_1(x))) \quad \ldots \quad \forall x(\mathcal{A}_n(x) \Rightarrow \bigcirc(\mathcal{B}_n(x))) \quad \Diamond L(x)}{\forall x \bigwedge_{i=1}^{n} \neg \mathcal{A}_i(x)} \; (\Diamond_{res}^{\mathcal{U}}),$$

where $\forall x(\mathcal{A}_i(x) \Rightarrow \bigcirc \mathcal{B}_i(x))$ are complex combinations of step clauses, called *full merged step clauses* [9], such that for all $i \in \{1, \ldots, n\}$, the *loop* side conditions $\forall x(\mathcal{U} \wedge \mathcal{B}_i(x) \Rightarrow \neg L(x))$ and $\forall x(\mathcal{U} \wedge \mathcal{B}_i(x) \Rightarrow \bigvee_{j=1}^{n}(\mathcal{A}_j(x)))$, with $\mathcal{U}$ being the current set of all universal clauses, are both valid; and the following five rules of *fine-grained step resolution*:

1. *First-order resolution between two universal clauses and factoring on a universal clause.* The result is a universal clause.
2. *First-order resolution between an initial and a universal clause, between two initial clauses, and factoring on an initial clause.* The result is again an initial clause.
3. *Fine-grained (restricted) step resolution.*

$$\frac{C_1 \Rightarrow \bigcirc(D_1 \vee L) \quad C_2 \Rightarrow \bigcirc(D_2 \vee \neg M)}{(C_1 \wedge C_2)\sigma \Rightarrow \bigcirc(D_1 \vee D_2)\sigma} \qquad \frac{C_1 \Rightarrow \bigcirc(D_1 \vee L) \quad D_2 \vee \neg M}{C_1\sigma \Rightarrow \bigcirc(D_1 \vee D_2)\sigma}$$

4. *(Step) factoring.*

$$\frac{C_1 \Rightarrow \bigcirc(D_1 \vee L \vee M)}{C_1\sigma \Rightarrow \bigcirc(D_1 \vee L)\sigma} \qquad \frac{(C_1 \wedge L \wedge M) \Rightarrow \bigcirc D_1}{(C_1 \wedge L)\sigma \Rightarrow \bigcirc D_1\sigma}$$

5. *Clause conversion.*
   A step clause of the form $C \Rightarrow \bigcirc\mathbf{false}$ is rewritten into the *universal clause* $\neg C$.

In rules 1 to 5, we assume that different premises and conclusions of the deduction rules have no variables in common; variables may be renamed if necessary. In rules 3 and 4, σ is a most general unifier of the literals L and M such that σ does not map variables from C_1 or C_2 into a constant or a functional term.

The input formula is unsatisfiable over expanding domains if and only if fine-grained temporal resolution derives the empty clause (see [9], Theorem 8).

3   Implementation

The deduction rules of fine-grained step resolution are close enough to classical first-order resolution to allow us to use first-order resolution provers to provide an implementation of our calculus.

Let **S** be a temporal problem in clausal form. For every k-ary predicate, P, occurring in **S**, we introduce a new $(k+1)$-ary predicate $\widetilde{P}$. We will also use the constant 0 (representing the initial moment in time), and unary function symbols s (representing the successor function on time) and h, which we assume not to occur in **S**. Let ϕ be a first-order formula in the vocabulary of **S**. We denote by $[\phi]^T$ the result of replacing all occurrences of predicates in ϕ by their "tilded" counterparts with T as the first argument (e.g. $P(x,y)$ is replaced with $\widetilde{P}(T,x,y)$). The term T will either be the constant 0 or the variable t (intuitively, t represents a moment in time). The variable t is assumed to be universally quantified.

Now, in order to realise fine-grained step resolution by means of classical first-order resolution, we define a set of first-order clauses **FO(S)** as follows.

- For every initial clause C from **S**, the clause $[C]^0$ is in **FO(S)**.
- For every universal clause D from **S**, the clause $[D]^t$ is in **FO(S)**.
- For every step clause $p \Rightarrow \bigcirc q$ from **S**, the clause $\neg\widetilde{p}(t) \vee \widetilde{q}(s(t))$ is in **FO(S)**, and for every step clause $P(x) \Rightarrow \bigcirc Q(x)$, the clause $\neg P(t,x) \vee Q(s(t),h(x))$ is in **FO(S)**.

The key insight is that fine-grained step resolution on **S**, including (implicitly) the clause conversion rule, can be realised using classical ordered first-order resolution with selection (see, e.g. [2]) on **FO(S)**. For rules 1 and 2, this is obvious. For step resolution and (step) factoring, we observe that if a clause contains a *next-state* literal, i.e. a literal whose first argument starts with the function symbol s, a factoring or resolution inference can only be performed on such a literal. This requirement can be enforced by an appropriate literal selection strategy. Note that standard redundancy deletion mechanisms, such as subsumption and tautology deletion, are also compatible with fine-grained step resolution (for details see [9]). As for the eventuality resolution rule, note that finding full merged clauses which satisfy the side conditions of the eventuality resolution rule is a non-trivial problem [9]. We find such merged clauses by means of a search algorithm presented in [9] which is again based on step resolution. Hence, the performance of the step resolution inference engine is critical for the overall performance of our system.

In our implementation, we extended the propositional temporal prover, **TRP**++ [6], to deal with monodic formulae. The main procedure of our implementation of this calculus consists of a loop where at each iteration (i) the set of temporal clauses is saturated under application of the step resolution rules, and (ii) then for every eventuality clause in the clause set, an attempt is made to find a set of premises for an application of the eventuality resolution rule. If we find such a set, the set of clauses representing the conclusion of the application is added to the current set of clauses. The main loop terminates if the empty clause is derived, indicating that the initial set of clauses is unsatisfiable, or if no new clauses have been derived during the last iteration of the main loop, which in the absence of the empty clause indicates that the initial set of clauses is satisfiable.

The task of saturating clause sets with classical resolution simulating step resolution is delegated to the **Vampire** kernel [11], which is linked to the whole system as a C++ library. Minor adjustments have been made in the functionality of **Vampire** to accommodate step resolution: a special mode for literal selection has been introduced such that in a clause containing a next-state literal only next-state literals can be selected. At the moment, the result of a previous saturation step, augmented with the result of an eventuality resolution application, is resubmitted to the **Vampire** kernel, although no inferences are performed between the clauses from the already saturated part. This is only a temporary solution, and in the future **Vampire** will support incremental input in order to reduce communication overhead.

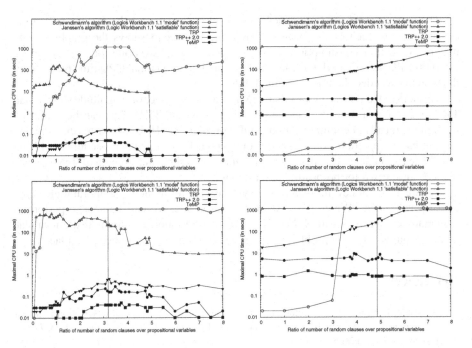

Fig. 1. Performance of the systems on $\mathcal{C}_{ran}^1$ (left) and $\mathcal{C}_{ran}^2$ (right)

4   Performance Evaluation

It is difficult to evaluate the performance of **TeMP** for two reasons. First, there are no established monodic benchmark sets. Second, there is no other monodic temporal prover to compare with.

However, we find it worthwhile to compare the performance of **TeMP** to that of **TRP++ 2.0** [6] on *propositional* temporal logic (PLTL) formulae. Both provers perform essentially the same inference steps on such formulae, since the rules of fine-grained temporal resolution presented in Section 2 coincide with those of propositional temporal resolution [3] on propositional temporal logic formulae.

Besides **TeMP** and **TRP++ 2.0** we have also included two tableau-based procedures for PLTL implemented in the Logics Workbench 1.1, described respectively in [8] and [12], and **TRP**, a prototype implementation of temporal resolution in SICStus Prolog by the first author.

We have compared the systems on two classes of semi-randomly generated PLTL-formulae, called $\mathcal{C}_{ran}^1$ and $\mathcal{C}_{ran}^2$, introduced in [7], for parameters $n = 12$, $k = 3$, and $p = 0.5$. The tests were performed on a PC with a 1.3GHz AMD Athlon processor, 512MB main memory, and 1GB virtual memory running Red Hat Linux 7.1. For each individual satisfiability test of a formula a time-limit of 1000 CPU seconds was used.

The left- and right-hand sides of Figure 1 depict the behaviour of the systems on $\mathcal{C}_{ran}^1$ and on $\mathcal{C}_{ran}^2$, respectively. All the graphs contain a vertical line. On the left of the line most formulae are satisfiable, on the right most formulae are unsatisfiable.

The upper part of the figure shows the resulting graphs for the median CPU time consumption of each of the systems, while the lower part shows the graphs for the maximal CPU time consumption. In all performance graphs, a point for a system above the 1000 CPU second mark indicates that the median or maximal CPU time required by the system exceeded the imposed time limit.

We can see from these graphs that **TeMP** is about an order of magnitude slower than **TRP++** 2.0, but still faster than the prototypical system **TRP**. This can be explained by high overheads in communications with **Vampire**. **TeMP** is also faster than the two tableau-based procedures on $\mathcal{C}_{ran}^1$ and is only outperformed by the procedure of [12] on satisfiable formulae in $\mathcal{C}_{ran}^2$. In our opinion, the results show the strength of **TeMP**, since it is not specialised for propositional reasoning.

We are aware of new *tableau-based* systems [10] for monodic temporal logic being under development. When these systems are available, we will be able to perform a systematic comparison with **TeMP**. We also intend to look at more realistic formulae coming from verification problems [4], instead of randomly generated formulae.

References

1. A. Artale, E. Franconi, F. Wolter, and M. Zakharyaschev. A temporal description logic for reasoning over conceptual schemas and queries. In *Proc. JELIA'02*, volume 2424 of *LNCS*, pages 98–110. Springer, 2002.
2. L. Bachmair and H. Ganzinger. Resolution theorem proving. In A. Robinson and A. Voronkov, editors, *Handbook of Automated Reasoning*, pages 19–99. Elsevier, 2001.
3. M. Fisher, C. Dixon, and M. Peim. Clausal temporal resolution. *ACM Transactions on Computational Logic*, 2(1):12–56, 2001.
4. M. Fisher and A. Lisitsa. Temporal verification of monodic abstract state machines. Technical Report ULCS-03-011, Department of Computer Science, University of Liverpool, 2003.
5. D. Gabelaia, R. Kontchakov, A. Kurucz, F. Wolter, and M. Zakharyaschev. On the computational complexity of spatio-temporal logics. In *Proc. FLAIRS 2003*, pages 460–464. AAAI Press, 2003.
6. U. Hustadt and B. Konev. TRP++ 2.0: A temporal resolution prover. In *Proc. CADE-19*, volume 2741 of *LNAI*, pages 274–278. Springer, 2003.
7. U. Hustadt and R. A. Schmidt. Scientific benchmarking with temporal logic decision procedures. In *Proc. KR2002*, pages 533–544. Morgan Kaufmann, 2002.
8. G. Janssen. *Logics for Digital Circuit Verification: Theory, Algorithms, and Applications*. PhD thesis, Eindhoven University of Technology, The Netherlands, 1999.
9. B. Konev, A. Degtyarev, C. Dixon, M. Fisher, and U. Hustadt. Mechanising first-order temporal resolution. Technical Report ULCS-03-023, University of Liverpool, Department of Computer Science, 2003. http://www.csc.liv.ac.uk/research/.
10. R. Kontchakov, C. Lutz, F. Wolter, and M. Zakharyaschev. Temporalising tableaux. *Studia Logica*, 76(1):91–134, 2004.
11. A. Riazanov and A. Voronkov. The design and implementation of Vampire. *AI Communications*, 15(2-3):91–110, 2002.
12. S. Schwendimann. *Aspects of Computational Logic*. PhD thesis, Universität Bern, Switzerland, 1998.
13. F. Wolter and M. Zakharyaschev. Axiomatizing the monodic fragment of first-order temporal logic. *Annals of Pure and Applied logic*, 118:133–145, 2002.

Dr.Doodle: A Diagrammatic Theorem Prover

Daniel Winterstein, Alan Bundy, and Corin Gurr

Edinburgh University

Abstract. This paper presents the Dr.Doodle system, an interactive theorem prover that uses diagrammatic representations. The assumption underlying this project is that, for some domains (principally geometry), diagrammatic reasoning is easier to understand than conventional algebraic approaches – at least for a significant number of people. The Dr.Doodle system was developed for the domain of metric-space analysis (a geometric domain, but traditionally taught using a dry algebraic formalism). Pilot experiments were conducted to evaluate its potential as the basis of an educational tool, with encouraging results.

1 Introduction

Diagrams are commonly used in virtually all areas of representation and reasoning. In particular – although current theorem-provers make very little use of diagrams – they are invaluable in mathematics texts. They are used in a variety of ways, including to give examples showing why a theorem is true, to give counter-examples, to explain the structure of a proof, and to prove a theorem outright. Insight is often more clearly perceived in these diagrammatic proofs than in the corresponding algebraic proofs. We have developed a system for producing such proofs in the domain of *metric-space analysis*, based upon a new diagrammatic logic.

As well as contributing to the development of more accessible theorem provers, this work also opens the exciting possibility of developing computer-based diagrams in new directions. If we consider the very real differences between text and hypertext, we see that diagrammatic reasoning on computers need not be just a straight conversion of diagrammatic reasoning on paper. Our work has led to the development of animated diagrams with a formal semantics as a meaningful representation for quantifiers (see [8]). The rigour required for doing mathematics forces a thorough investigation of the mechanics of such reasoning.

1.1 Our Domain: Metric-Space Analysis

Euclidean plane geometry has always been taught using diagrammatic reasoning. Traditionally though, only algebraic proofs are allowed in the slippery realms of more abstract geometries. We have investigated using diagrams in such a domain, that of *metric-space analysis*. This is a hard domain, and even great mathematicians such as Cauchy have made mistakes in this subject [5]. Students typically find it daunting, and we conjecture that the dry algebraic formalism

D. Basin and M. Rusinowitch (Eds.): IJCAR 2004, LNAI 3097, pp. 331–335, 2004.

used in the domain is partially responsible for these difficulties. Currently the system only covers a fraction of the domain, but this was sufficient to run some tutorials on the concept of *open sets*. This allowed us to experimentally compare the use of diagrams with an equivalent algebraic approach.

2   The Dr.Doodle System

The Dr.Doodle system is an interactive theorem prover for non-inductive reasoning, with diagrammatic representations for metric-space and real-line analysis concepts (involving objects such as functions, sets and lengths, and properties such as open, closed and continuous). The user selects which rule to apply at each step. In places this involves drawing, and the interaction sometimes resembles using a graphics program.

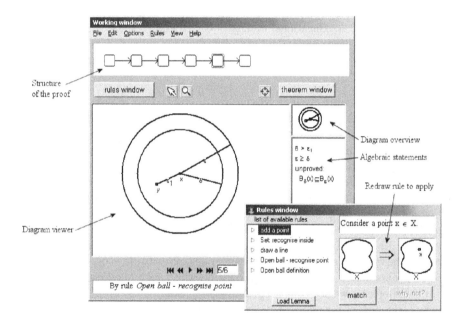

Fig. 1. Screenshot of Dr.Doodle.

2.1   Dr.Doodle Diagrams

The diagrams in our system are so-called *heterogenous representations*, combining graphical and textual elements. They consist of example objects with conjunctions of relation statements. Where possible, the example objects should constitute a model for the statements; if no model can be found (e.g. because the statements are inconsistent) then a *near* model is used. Disjunctions are represented using multiple diagrams (an intuitive representation, but with scaling issues). Statements can be represented in three different ways: *implicitly* (where the relation is true for the objects drawn, e.g. $a \in B$ for $a = \frac{1}{2}$, $B = [0,1]$), *graphically* (using conventions such as 'a dotted border indicates an open set') or *algebraically*. These methods are used as appropriate (e.g. relations such as $A \subset B$ are usually implicit, but can also be stated algebraically if necessary).

The system produces diagrams from an internal proof state as follows:

1. The example objects are drawn.
2. The drawing is tested to see which relations are implicitly represented (note that these tests take into account drawing resolution).
3. Often, a diagram will represent relations that are not specified in the internal proof state. Sometimes this is desirable. For example, if we represent $A \subset B$ by drawing A inside B, then a diagram for $A \subset B$, $B \subset C$ will inevitably also represent $A \subset C$ – which is, of course, true. Such 'free rides' are one advantage of using diagrams (c.f. [7]).
   Any extra relations found in the previous step are tested to see if they follow from the specified relations as a result of free rides. Unwanted extra relations are then explicitly removed by the system.
4. Relations which are not implicitly represented by drawing the objects (e.g. $y = f(x)$) are explicitly added to the diagram.

This process involves important design choices at several stages:

– How to choose and draw the example objects.
– Which relations can be represented implicitly.
– How to represent the explicit relations (e.g. arrows for function relations, but predicates for properties such as 'surjective').

There is a trade off here between simplicity, flexibility, and intuitive appeal, and our choices are tailored to the domain considered.

2.2   Dynamic Diagram Logic

We can only give a brief overview here of the logic implemented in Dr.Doodle. For more information, please refer to [8]. Often diagrammatic reasoning is presented as a question of interpreting static diagrams. Here we consider *dynamic* diagrammatic reasoning, where the process of drawing is important, as opposed to just the finished diagram.

The inference rules are specified using *redraw rules*, which are a visual adaptation of rewrite rules. Redraw rules are defined by an example diagram transformation; figure 2 shows an example. When a rule specifies creating new objects, these are typically drawn-in by the user (automated drawing, whilst interesting, requires both model-generation and aesthetic judgement to select 'good' examples). Theorems are stated in the same manner. A proof consists of a demonstration that the theorem antecedent can always be redrawn to give the consequent diagram using an accepted set of rules.This paradigm - whereby inference rules are mainly presented as drawing acts - restricts users to constructing forward reasoning proofs; future work includes handling other proof structures.

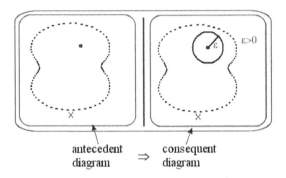

Fig. 2. A redraw rule for "X an open set, $x \in X \Rightarrow \exists \epsilon > 0$ s.t. $\{x' : |x' - x| < \epsilon\} \subset X$"

An Equivalent Algebraic Logic. Dr.Doodle also features an algebraic reasoning mode, where redraw rules are converted into natural-deduction rewrite rules. Although the two types of proof appear very different, the reasoning used is equivalent, and the Dr.Doodle system can convert a diagrammatic proof into an algebraic one (though not vice versa at present).

3 Experimental Evaluation

Pilot experiments were conducted to test the use of Dr.Doodle in a practical setting (teaching mathematics undergraduates). These experiments compared performance using Dr.Doodle in algebraic and diagrammatic modes (hence evaluating the potential of the diagrammatic logic – since this is the innovative aspect of the system – independently of the interface), and are described in [9]. The results show statistically significant improvements – both in exercise scores and 'efficiency' – when using diagrammatic reasoning. Informal feedback gave comments such as: "The pictures were useful for helping understand what was going on. Better than written explanations a lot of the time."

These positive results are not surprising. As the domain is a geometric one, we would expect visual representations to be useful. We conclude that diagrammatic reasoning is a useful tool in this field. However further experiments are desirable, especially as these experiments did not look at the interesting questions of *how* and *why* diagrams are useful here (and hence how general these findings are).

4 Related Work

Diagrammatic reasoning is a relatively unexplored research area, and there are few comparable systems. There are several powerful geometric theorem provers (e.g. Cinderella [4]), but these use proof techniques that are very hard for humans to follow. The work by Howse *et al* on spider diagrams is more relevant. These diagrams are used to represent set-theoretic statements, and can also support reasoning [2].

Barwise, Etchemendy *et al*'s HyperProof arguably sets the standard for educational applications of theorem provers [1]. It is a much more developed system than Dr.Doodle, but handles a very different domain. Aimed at philosophy students learning logic, it uses the blocksworld domain to give concrete visual meaning to predicate logic statements. Diagrammatic inferences in HyperProof involve reading information from the diagram, or testing propositions against the diagram. Psychometric studies by Stenning *et al* show that this is beneficial for some students [6]. Moreover, which students will benefit may be predictable using simple aptitude tests for spatial reasoning.

The most closely related system is Jamnik's Diamond, which uses geometric reasoning about area to prove natural number arithmetic theorems [3]. The user supplies proofs for example cases of a theorem, from which a general proof is extracted and checked. This project was initially conceived as extending Jamnik's work to a continuous domain whilst exploring practical applications. However the differences between countable domain and continuous domain reasoning have led to a very different system.

5   Conclusion

We have described the Dr.Doodle system and given an overview of the logic it uses based on diagrammatic representations. The aim of such work is to produce theorem provers whose proofs can be readily understood, based on the idea that diagrammatic representations are easier and more intuitive for some domains. Our pilot experiments support this idea. We now intend to develop Dr.Doodle further to produce a tutorial system for mathematical analysis.

This work is described in more detail in the first author's forthcoming Ph.D. thesis. A demonstration version of the Dr.Doodle system is available from the first author.

References

1. J.Barwise & J.Etchemendy "Heterogeneous Logic" in *Diagrammatic Reasoning: Cognitive and Computational Perspectives*, AAAI Press/MIT Press, 1995.
2. J.Flower & G.Stapleton "Automated Theorem Proving with Spider Diagrams" in *Computing: the Australasian Theory Symposium (CATS)*, 2004.
3. M.Jamnik "Mathematical Reasoning with Diagrams"CSLI Press, 2001.
4. U.Kortenkamp & J.Richter-Gebert "The Next Version of Cinderella" in *First International Congress Of Mathematical Software*, World Scientific, 2002.
5. E.Maxwell "Fallacies in Mathematics" Cambridge University Press, 1959.
6. J.Oberlander, R.Cox & K.Stenning "Proof styles in multimodal reasoning" in *Logic, Language and Computation*, CSLI Press, 1996.
7. A.Shimojima "Operational Constraints in Diagrammatic Reasoning" in *Logical Reasoning with Diagrams*, OUP, 1996.
8. D.Winterstein, A.Bundy, C.Gurr & M.Jamnik "Using Animation in Diagrammatic Theorem Proving" in *Diagrams 2002*, Springer-Verlag, 2002.
9. D.Winterstein, A.Bundy, C.Gurr & M.Jamnik "An Experimental Comparison of Diagrammatic and Algebraic Logics" in *Diagrams 2004*, Springer-Verlag, 2004.

Solving Constraints by Elimination Methods

Volker Weispfenning

University of Passau, D-94030 Passau, Germany,
weispfen@uni-passau.de,
http://www.fmi.uni-passau.de/algebra/staff/weispfen.php3

Abstract. We give an overview of some current variable elimination techniques in constraint solving and their applications. We focus on numerical constraints, with variables ranging over the reals, the integers, or the p-adics, but include also constraints of a more combinatorial nature, with Boolean variables of variables in free term structures.

1   Basics of Variable Elimination

Constraints in a general sense are systems of conditions on objects in a more or less specified domain. Major algorithmic goals of their study are the following
1. Decide the satisfiability of constraints.
2. Determine sample solutions.
3. Determine the structure of the solution set.
4. Find suitable equivalent transformations and normal forms for constraints.

Here we restrict our attention to constraints formalized by formulas of first-order logic over some specific first-order structure. In most cases these formulas will be quantifier-free, i.e. propositional combinations of atomic formulas, but this is not essential.

The essence of elimination methods is to eliminate the variables in constraints successively one-by-one. Thus each variable elimination deals only with a univariate situation which is much easier than the original multivariate constraint. The price to pay for this enormous simplification is that the remaining variables have to be treated as *parameters*, i.e. as fixed but unknown objects in the given domain. Conditions for the solvability of such a univariate parametric constraint in the given structure have be expressed *uniformly* as quantifier-free formulas about the parameters. If this succeeds, these elimination steps for a single variable can be repeated successively for other variables that have served previously as parameters. This iteration is performed until no parameters are left over and a decision about solvability of the original constraint is obtained. This procedure is called *quantifier elimination* for the given structure. *Extended quantifier elimination* specifies in addition in each elimination step finitely many candidate expressions for solutions wrt. the current variable depending uniformly on the parameters. By successive back substitution of these parametric expressions one obtains then sample solutions of the original constraint. For an overview over quantifier elimination in algebraic theories consult [28]; for a corresponding overview over general solution methods for numerical constraints see [2].

D. Basin and M. Rusinowitch (Eds.): IJCAR 2004, LNAI 3097, pp. 336–341, 2004.

An additional benefit of quantifier elimination methods is the possibility to cope also with *unspecified predicates* in constraints, provided their range can be coded by first-order parameters: Consider e.g. polynomial real order constraints. Then one may include several unspecified predicates ranging e.g. over open intervals by coding each of these predicates by their paramteric endpoints.

A simple and classical illustration of extended quantifier elimination in fields is *Gauss elimination* for conjunctions of linear equations. Another well-known illustration is *Fourier-Motzkin elimination* for conjunctions of linear inequalities in an ordered field [15,18]. These examples differ, however, vastly in efficiency: While Gauss elimination (with reduction of fractions) is quite efficient, Fourier-Motzkin elimination is quite inefficient, both in theory and in practice.

So in order to render elimination methods a useful tool in practical applications of constraint solving, we have to watch out for the theoretical and practical complexity of the methods.

2 Elimination by Modified Substitution of Parametric Test Points

As a paradigm for an a flexible and efficient elimination tool we consider in more detail *variable elimination by modified substitution of parametric test points*. This method also provides sample solutions as a byproduct. In this approach the variable to be eliminated is disjunctively replaced by finitely many expressions in the parameters that serve as test points for the satisfiability of the constraint wrt. to this variable. These expressions need not be terms of the given language. The corresponding 'substitutions' are done in such a way that the resulting expression is again a quantifier-free formula in the parameters. This fact then guarantees that the method can be iterated for other variables.

We illustrate this method for

1. linear real constraints [29,16,34]
2. quadratic real constraints [33]
3. linear integer constraints [30,32]
4. mixed real-integer linear constraints [36]
5. linear p-adic constraints [21,12]

In all these cases the method is close to optimal in theoretical worst-case complexity, and of great practical relevance. Implementations are realized in the REDLOG-package of the widely used computer algebra system REDUCE [10]. Using these implementations we will illustrate the method and its applications by examples from automatic theorem proving, geometry, optimization, scheduling, motion planning, solving congruence systems [13,14,22,34,5,6,17,40,39].

Besides the areas mentioned above the method has also proved to be useful in purely combinatorial settings. As examples mention boolean constraint solving [19] and extended quantifier elimination in free term algebras for arbitrary finite first-order languages. In the latter case the decision problem for constraints is extremely complex; nervertheless the method is successful and yields a finer complexity analysis[23].

This elimination method is, however, in no way the only approach to quantifier elimination. The most important alternative and supplement for real constraints is the quantifier elimination by cylindrical algebraic decomposition [4], [3], [9].

3   Solving Transcendental Constraints

Many interesting constraint solving problems e.g. in the reals involve besides polynomials also transcendental functions such as the exponential function and/or trigonometric functions. Problems of this type arise naturally e.g. in connection with solutions of differential equations. For general constraints involving such functions elimination methods fail in principle [26].

Nevertheless there are interesting subclasses of transcendental constraints for which elimination methods work. They involve typically only a single univariate transcendental function in combination with multivariate polynomials. In this situation all variables except the critical one occurring in the transcendental function can be eliminated essentially as in Tarski algebra. This works due to the fact that the transcendental expressions can simply be treated as parameters during the elimination of all uncritical variables. The resulting univariate polynomial-transcendental problems can then be treated by geometric arguments using the decisive fact that all the transcendental functions admitted satisfy some kind of Lindemann theorem in transcendental number theory.

Instances are the linear-transcendental problems treated in [1,37,38]. Here linear multivariate polynomials are additively combined with a single univariate transcendental function such as $\exp, \sin, \cos, \arctan$. The case of order constraints involving polynomials in $\exp(y)$ for a single variable y with coefficients that are arbitrary non-linear multivariate polynomials was studied in [27] by very sophisticated methods. Recently elementary, implementable algorithms for the solution of this type of constraint have been found.

4   REDLOG

Most of the algorithms described above are implemented in the REDLOG package of the computer algebra system REDUCE and are publically available as part of REDUCE 3.7. Some implementations are under way and will be available with the next release of REDUCE. REDLOG is well-documented and professionally maintained [10]; for more information and a collection of benchmark examples consult `http://www.fmi.uni-passau.de/ redlog/`. REDLOG stands for REDUCE logic system. It provides an extension of the computer algebra system REDUCE to a computer logic system implementing symbolic algorithms for first-order formulas wrt. temporarily fixed first-order languages and theories, refered in REDLOG as 'contexts'. REDLOG originated from implementations of quantifier elimination procedures. Successfully applying such methods to both academic and real-world problems, the authors have developed over more than a decade a large set of formula-manipulating tools for input, normal form computation and simplification in different contexts, many of which are meanwhile interesting in their own right [11].

At present the following contexts are available in REDLOG:

1. ACFSF for algebraically closed fields, in particular the complex field, with quantifier elimination based on comprehensive Gröbner bases [31].
2. OFSF for real closed ordered fields, in particular the field of real numbers. Three methods of real quantifier elimination are implemented. First, the virtual substitution method. Second, partial cylindrical algebraic decomposition. Here, next to a classic version, a generic quantifier elimination variant and a strategy for finding an efficient projection order is implemented [20, 9]. Third, Hermitian quantifer elimination [35,8,7].
3. DVFSF for discretely valued fields, in particular p-adic fields, with a linear quantifier elimination and algorithms for univariate polynomial constraint solving [21,12,24,25].
4. PASF for Presburger Arithmetic, i.e., the linear theory of integers with congruences modulo m for fixed natural numbers m. The syntax is extended to allow for large ANDs and ORs in form of bounded quantifiers. Simplification techniques for the extended syntax and context-specific simplification are implemented and used by a dedicated quantifier elimination method [30, 32].
5. IBALP for Initial Boolean Algebras. This is a framework to integrate propositional logic with first-order logic in such a way that it may be construed either as first-order logic over the two-element Boolean algebra or as propositional logic including Boolean quantification. Quantifier elimination and simplification is implemented; as a byproduct it can find uniform solutions for parametric QSAT problems [19].

References

1. Hirokazu Anai and Volker Weispfenning. Deciding linear-trigonometric problems. In C. Traverso, editor, *ISSAC'2000*, pages 14–22. ACM-Press, 2000.
2. A. Bockmayr and W. Weispfenning. *Solving Numerical Constraints*, volume I, chapter 12, pages 751–842. Elsevier Science and MIT Press, 2001.
3. Georg E. Collins. Quantifier elimination for real closed fields by cylindrical algebraic decomposition. In B.F. Caviness and J.R. Johnson, editors, *Quantifier Elimination and Cylindrical Algebraic Decomposition*, Texts and Monographs in Symbolic Computation, pages 85–121. Springer, Wien, New York, 1998.
4. George E. Collins. Quantifier elimination by cylindrical algebraic decomposition - twenty years of progress. In B.F. Caviness and J.R. Johnson, editors, *Quantifier Elimination and Cylindrical Algebraic Decomposition*, Texts and Monographs in Symbolic Computation, pages 8–23. Springer, Wien, New York, 1998.
5. Andreas Dolzmann. Solving geometric problems with real quantifier elimination. In Xiao-Shan Gao, Dongming Wang, and Lu Yang, editors, *Automated Deduction in Geometry*, volume 1669 of *Lecture Notes in Artificial Intelligence (Subseries of LNCS)*, pages 14–29. Springer-Verlag, Berlin Heidelberg, 1999.

6. Andreas Dolzmann. *Algorithmic Strategies for Applicable Real Quantifier Elimination*. Doctoral dissertation, Department of Mathematics and Computer Science. University of Passau, Germany, D-94030 Passau, Germany, March 2000.
7. Andreas Dolzmann and Lorenz A. Gilch. Generic hermitian quantifier elimination. Technical report, University of Passau, 2004. To appear.
8. Andreas Dolzmann and Lorenz A. Gilch. Hermitian quantifier elimination. Technical report, University of Passau, 2004. To appear.
9. Andreas Dolzmann, Andreas Seidl, and Thomas Sturm. Efficient projection orders for cad. In Jaime Gutierrez, editor, *Proceedings of the 2004 International Symposium on Symbolic and Algebraic Computation (ISSAC 2004)*, Santander, Spain, July 2004. ACM.
10. Andreas Dolzmann and Thomas Sturm. Redlog: Computer algebra meets computer logic. *ACM SIGSAM Bulletin*, 31(2):2–9, June 1997.
11. Andreas Dolzmann and Thomas Sturm. Simplification of quantifier-free formulae over ordered fields. *Journal of Symbolic Computation*, 24(2):209–231, August 1997.
12. Andreas Dolzmann and Thomas Sturm. P-adic constraint solving. In Sam Dooley, editor, *Proceedings of the 1999 International Symposium on Symbolic and Algebraic Computation (ISSAC 99), Vancouver, BC*, pages 151–158. ACM Press, New York, NY, July 1999.
13. Andreas Dolzmann, Thomas Sturm, and Volker Weispfenning. A new approach for automatic theorem proving in real geometry. *Journal of Automated Reasoning*, 21(3):357–380, 1998.
14. Andreas Dolzmann, Thomas Sturm, and Volker Weispfenning. Real quantifier elimination in practice. In B. H. Matzat, G.-M. Greuel, and G. Hiss, editors, *Algorithmic Algebra and Number Theory*, pages 221–247. Springer, Berlin, 1998.
15. J.B.J. Fourier. Solution d'une question particulière du calcul des inègalités. *Nouveau Bulletin des Sciences par la Socieétée Philomathique de Paris*, pages 99–100, 1826.
16. Rüdiger Loos and Volker Weispfenning. Applying linear quantifier elimination. *The Computer Journal*, 36(5):450–462, 1993. Special issue on computational quantifier elimination.
17. Isolde Mazzucco. Symopt: Symbolic parametric mathematical programming. In V.G. Ganzha and E.W. Mayr, editors, *Computer Algebra in Scientific Computing 2001*, pages 417–429, Konstanz, Germany, Sept 2001.
18. Theodore S. Motzkin. *Beiträge zur Theorie der linearen Ungleichungen*. Doctoral dissertation, Universität Zürich, 1936.
19. Andreas Seidl and Thomas Sturm. Boolean quantification in a first-order context. In V. G. Ganzha, E. W. Mayr, and E. V. Vorozhtsov, editors, *Computer Algebra in Scientific Computing (CASC 2003)*, pages 329–345, Passau, Germany, September 2003.
20. Andreas Seidl and Thomas Sturm. A generic projection operator for partial cylindrical algebraic decomposition. Technical Report MIP-0301, FMI, Universität Passau, D-94030 Passau, Germany, January 2003.
21. Thomas Sturm. Linear problems in valued fields. Technical Report MIP-9715, FMI, Universität Passau, D-94030 Passau, Germany, November 1998.
22. Thomas Sturm and Volker Weispfenning. Rounding and blending of solids by a real elimination method. In Achim Sydow, editor, *Proceedings of the 15th IMACS World Congress on Scientific Computation, Modelling, and Applied Mathematics (IMACS 97)*, volume 2, pages 727–732, Berlin, August 1997. IMACS, Wissenschaft & Technik Verlag.

23. Thomas Sturm and Volker Weispfenning. Quantifier elimination in term algebras. In *Computer Algebra in Scientific Computation - CASC 2002*, pages 285–30. TU München, 2002.

24. Thomas Sturm and Volker Weispfenning. P-adic root isolation. *RACSAM, Rev. R. Acad. Cien. Serie A. Mat.*, 2004. to appear.

25. Thomas Sturm and Volker Weispfenning. Solving univariate p-adic constraints. Technical Report MIP-040?, FMI, Universität Passau, D-94030 Passau, Germany, Apr 2004.

26. Lou van den Dries. Remarks on tarski's problem concerning $(R, +, \cdot, exp)$. In G. Lolli, G. Longo, and A. Marcja, editors, *Logic Colloqium '82*, volume 112 of *Studies in Logic and the Foundations of Mathematics*, pages 97–121, Florence, August 1984.

27. N.N. Vorobjov. The complexity of deciding consistency of systems of polynomials in exponent inequalities. *J. Symbolic Computation*, 13(2):139–173, 1992.

28. Volker Weispfenning. Aspects of quantifier elimination in algebra. In *Universal Algebra and its links with logic...*, volume 25, pages 85–105, Berlin, 1984. Heldermann Verlag.

29. Volker Weispfenning. The complexity of linear problems in fields. *Journal of Symbolic Computation*, 5(1–2):3–27, February–April 1988.

30. Volker Weispfenning. The complexity of almost linear Diophantine problems. *Journal of Symbolic Computation*, 10(5):395–403, November 1990.

31. Volker Weispfenning. Comprehensive Gröbner bases. *Journal of Symbolic Computation*, 14:1–29, July 1992.

32. Volker Weispfenning. Complexity and uniformity of elimination in Presburger arithmetic. In Wolfgang W. Küchlin, editor, *International Symposium on Symbolic and Algebraic Computation, Maui, Hawaii*, pages 48–53, New York, July 1997. ACM Press.

33. Volker Weispfenning. Quantifier elimination for real algebra—the quadratic case and beyond. *Applicable Algebra in Engineering Communication and Computing*, 8(2):85–101, February 1997.

34. Volker Weispfenning. Simulation and optimization by quantifier elimination. *Journal of Symbolic Computation*, 24(2):189–208, August 1997. Special issue on applications of quantifier elimination.

35. Volker Weispfenning. A new approach to quantifier elimination for real algebra. In B.F. Caviness and J.R. Johnson, editors, *Quantifier Elimination and Cylindrical Algebraic Decomposition*, Texts and Monographs in Symbolic Computation, pages 376–392. Springer, Wien, New York, 1998.

36. Volker Weispfenning. Mixed real-integer linear quantifier elimination. In S. Dooley, editor, *ISSAC'99*, pages 129–136. ACM-Press, 1999.

37. Volker Weispfenning. Deciding linear-transcendental problems. In V.G. Ganzha, E.W. Mayr, and E.V. Vorozhtshov, editors, *Computer Algebra in Scientific Computation - CASC 2000*, pages 423–438. Springer, 2000.

38. Volker Weispfenning. Solving linear-transcendental problems. In *EACA '2000*, pages 81–87. Universitat Polytechnica de Catalunya, 2000.

39. Volker Weispfenning. Semilinear motion planning among moving objects in REDLOG. In V.G. Ganzha, E.W. Mayr, and E.V. Vorozhtsov, editors, *CASC 2001*, pages 541–553. Springer Verlag, 2001.

40. Volker Weispfenning. Semilinear motion planning in REDLOG. *AAECC*, 12(6):455–475, December 2001.

Analyzing Selected Quantified Integer Programs

K. Subramani*

LCSEE,
West Virginia University,
Morgantown, WV
ksmani@csee.wvu.edu

Abstract. In this paper, we introduce a problem called Quantified Integer Programming, which generalizes the Quantified Satisfiability problem (QSAT). In a Quantified Integer Program (QIP), the program variables can assume arbitrary integral values, as opposed to the boolean values that are assumed by the variables of an instance of QSAT. QIPs naturally represent 2-person integer matrix games. The Quantified Integer Programming problem is PSPACE-hard in general, since the QSAT problem is PSPACE-complete. We focus on analyzing various special cases of the general problem, with a view to discovering subclasses that are tractable. Subclasses of the general QIP problem are obtained by restricting either the constraint matrix or the quantifier specification. We show that if the constraint matrix is totally unimodular, the problem of deciding a QIP can be solved in polynomial time.

1 Introduction

Quantified decision problems are useful in modeling situations, wherein a policy (action) can depend upon the effect of imposed stimuli. A typical such situation is a 2– person game. Consider a board game comprised of an initial configuration and two players A and B, each having a finite set of moves. We focus on the following decision problem: *Given the initial configuration, does Player A have a first move (policy), such that for all possible first moves of Player B (imposed stimulus), Player A has a second move, such that for all possible second moves of Player B,... , Player A eventually wins?*. If this question can be answered affirmatively, then Player A has a winning strategy; otherwise, Player B has one. The board configuration can be represented as a boolean expression or a constraint matrix; the expressiveness of the board configuration typically determines the complexity of the decision problem. The use of quantified boolean expressions to capture problem domains has been widespread within the AI and Planning communities [CGS98]. Our work in this paper is concerned with a problem closely related to Quantified Satisfiability called Quantified Integer Programming. Quantified Integer Programming generalizes the QSAT problem, in that program variables can assume arbitrary integral values. It follows that

---

* This research has been supported in part by a research grant from the Department of Computer Science, Carnegie Mellon University.

D. Basin and M. Rusinowitch (Eds.): IJCAR 2004, LNAI 3097, pp. 342–356, 2004.
© Springer-Verlag Berlin Heidelberg 2004

the Quantified Integer Programming problem is at least as hard as QSAT. QIPs naturally represent 2-person, integer matrix games. An interesting line of research is to focus on restrictions of the general problem, in order to discover useful subclasses that are tractable. In this paper, we study a number of special subclasses of QIPS that are obtained by restricting either the constraint matrix or the quantifier specification.

The rest of this paper is organized as follows: Section §2 describes the QIP problem and some of the special subclasses that we shall be studying in this paper. In Section §3, we motivate the study of our special cases; a detailed description of related work will be available in the journal version of this paper. We commence our analysis in Section §4, by deriving properties of arbitrary QIPs; these properties are then used in the succeeding sections. Section §5 demonstrates the existence of a polynomial time procedure for deciding totally unimodular QIPs. Although we define Planar QIPs and Box QIPs, their analyses has been relegated to the journal version of this paper. We conclude in Section §6 by summarizing our results and outlining problems for future research.

2   Statement of Problem(s)

Definition 1. *Let $\{x_1, x_2, \ldots, x_n\}$ be a set of n boolean variables. A literal is either the variable x_i or its complement $\bar{x}_i$. A disjunction of literals is called a clause, represented by C_i. A boolean expression of the form:*

$$Q_1 x_1 \in \{\textbf{true}, \textbf{false}\} \; Q_2 x_2 \in \{\textbf{true}, \textbf{false}\}$$
$$\ldots Q_n x_n \in \{\textbf{true}, \textbf{false}\} \quad C$$

where each Q_i is either $\exists$ or $\forall$ and $C = C_1 \wedge C_2 \ldots \wedge C_m$, is called a Quantified CNF formula (QCNF) and the problem of deciding whether it is true is called the Quantified Satisfiability problem (QSAT).

QSAT has been shown to be `PSPACE-complete`, even when there are at most 3 literals per clause (Q3SAT) [Pap94]. [Sch78] argued the existence of polynomial time algorithms for Q2SAT, although no constructive procedure was given. [APT79] provided the first polynomial time algorithm for the Q2SAT problem. In [Gav93], it was shown that the Q2SAT problem is in the parallel complexity class NC_2. Experimental work on algorithms for QSAT problems has been the thrust of [CGS98].

Definition 2. *Let $x_1, x_2, \ldots x_n$ be a set of n variables with integral ranges. A mathematical program of the form*

$$Q_1 \, x_1 \in \{a^1 - b^1\} \; Q_2 \, x_2 \in \{a^2 - b^2\} \ldots$$
$$Q_n \, x_n \in \{a^n - b^n\} \quad \mathbf{A} \cdot \boldsymbol{x} \leq \boldsymbol{b} \tag{1}$$

where each Q_i is either $\exists$ or $\forall$ is called a Quantified Integer Program (QIP).

The matrix $\mathbf{A}$ is called the constraint matrix of the QIP.

The PSPACE-hardness of QIPs follows immediately from the PSPACE-completeness of QSAT; in fact the reduction from QSAT to QIP is identical to that from SAT to 0/1 Integer Programming.

Without loss of generality, we assume that the quantifiers are strictly alternating and that $Q_1 = \exists$ (by using dummy variables, if necessary); further we denote the existentially quantified variables using $x_i \in \{a^i - b^i\}$, a^i, b^i integral, $i = 1, 2, \ldots, n$ and the universally quantified variables using $y_i \in \{c^i - d^i\}, c^i$, d^i integral, $i = 1, 2, \ldots, n$. Thus we can write an arbitrary QIP as :

$$
\begin{aligned}
\mathbf{QIPG} : \exists x_1 &\in \{a^1 - b^1\} \; \forall y_1 \in \{c^1 - d^1\} \\
\exists x_2 &\in \{a^2 - b^2\} \; \forall y_2 \in \{c^2 - d^2\} \\
\ldots \exists x_n &\in \{a^n - b^n\} \; \forall y_n \in \{c^n - d^n\} \\
&\mathbf{A} \cdot [\boldsymbol{x} \; \boldsymbol{y}]^{\mathbf{T}} \leq \boldsymbol{b}
\end{aligned}
\tag{2}
$$

for suitably chosen $\boldsymbol{x}, \boldsymbol{y}, \mathbf{A}, \boldsymbol{b}, n$.

The specification $\exists x_1 \in \{a^1 - b^1\} \; \forall y_1 \in \{c^1 - d^1\} \exists x_2 \in \{a^2 - b^2\} \; \forall y_2 \in \{c^2 - d^2\} \ldots \exists x_n \in \{a^n - b^n\} \; \forall y_n \in \{c^n - d^n\}$ is called the quantifier specification or quantifier string of the QIP and is denoted by $\mathbf{Q}(\mathbf{x}, \mathbf{y})$. We note that $\mathbf{Q}(\mathbf{x}, \mathbf{y})$ imposes a linear ordering on the program variables of the QIP. System (2) is referred to as a *QIP in general form or a general QIP*, on account of the unbounded alternation in the quantifier specification $\mathbf{Q}(\mathbf{x}, \mathbf{y})$.

The following assertions hold for the rest of this paper:

1. $\mathbf{A}$ and $\boldsymbol{b}$ are integral.
2. The intervals of the variables are defined by integers, i.e., a^i, b^i, c^i, d^i are integers, for all i.
3. We say that $x_i \in \{a^i - b^i\}, a^i \leq b^i$ to mean that the valid values for x_i are the integers in the set $\{a^i, a^i + 1, \ldots, b^i\}$.
4. The range constraints on the existentially quantified variables can be made part of the constraint system $\mathbf{A} \cdot [\boldsymbol{x} \; \boldsymbol{y}]^{\mathbf{T}} \leq \boldsymbol{b}$.

We now describe the various restrictions to System (2) that we shall analyze in this paper.

Definition 3. *A matrix $\mathbf{A}$ is said to be totally unimodular (TUM), if every square sub-matrix of $\mathbf{A}$ has determinant 0, 1 or −1.*

Remark 1. Definition (3) forces every entry in $\mathbf{A}$ to belong to $\{0, 1, -1\}$.

TUM matrices arise in Network Flow problems [AMO93], scheduling problems [Pin95] and a whole host of situations in which only strict difference constraints are permitted between program variables [Sch87].

Definition 4. *A totally unimodular QIP (TQIP) is a QIP in which the constraint matrix $(\mathbf{A})$ is totally unimodular.*

The first problem that we consider is:

P$_1$: *Is there a polynomial time procedure to decide an arbitrary TQIP as described in Definition (4)?*

Definition 5. *A Planar QIP (PQIP) is a QIP, in which all constraints exist between at most 2 variables, i.e., every constraint is a half-plane in the $x_1 - y_1$ plane.*

Note that all constraints are between the same 2 variables (x_1 and y_1). Planar QIPs are also known as 2−dimensional QIPs and should not be confused with Quantified Integer Programming with at most 2 non-zero variables per constraint, i.e., QIP(2). In the former, the dimension of the constraint matrix is 2, whereas in the latter the dimension of the constraint matrix is $2 \cdot n$. The second problem that we consider is:

P$_2$: *Is there a polynomial time procedure to decide an arbitrary PQIP as described in Definition (5)?*

Definition 6. *A Box QIP (BQIP) is a QIP, in which every quantifier is universal, i.e., a QIP of the form:*

$$\forall y_1 \in \{c^1 - d^1\}\forall y_2 \in \{c^2 - d^2\} \ldots \forall y_n \in \{c^n - d^n\}$$
$$\mathbf{A} \cdot \boldsymbol{y} \leq \boldsymbol{b} \qquad (3)$$

The third problem that we consider is:

P$_3$: *Is there a polynomial time procedure to decide an arbitrary BQIP as described in Definition (6)?*

We reiterate that only problem **P$_1$** is analyzed in this paper; problems **P$_2$** and **P$_3$** will be discussed in an extended version of this paper.

2.1   Model Verification

In this section, we formally specify what it means for a vector $\boldsymbol{x}$ to be a solution or a "model" of a QIP, such as **QIPG**. The specification involves the notion of a 2-person integer matrix game.

Let **X** denote the existential player and **Y** denote the universal player. The game consists of n rounds; in round i, **X** guesses a value for x_i, which may depend upon $\{y_1, y_2, \ldots, y_{i-1}\}$, while **Y** guesses a value for y_i which may depend upon $\{x_1, x_2, \ldots, x_i\}$. At the end of n rounds, we construct the vectors $\boldsymbol{x} = [x_1, x_2, \ldots, x_n]^T$ and $\boldsymbol{y} = [y_1, y_2, \ldots, y_n]^T$. If $\mathbf{A} \cdot [\boldsymbol{x} \ \boldsymbol{y}]^T \leq \boldsymbol{b}$, the game is said to be a win for **X**; otherwise it is said to be a win for **Y**. The guesses made by both players are non-deterministic, in that if **X** can win the game, his guesses will lead to a win; likewise if **Y** can thwart **X**, then the guesses made by **Y**, will lead to **X** losing. If **X** wins the game, **QIPG** is said to have a model or be *true*.

Observe that a solution to a QIP, such as **QIPG**, is in general, a strategy for the existential player **X** and not a numeric vector. Thus a solution vector will have the form $\boldsymbol{x_s} = [x_1, x_2, \ldots, x_n]^T = [c_0, f_1(y_1), f_2(y_1, y_2), \ldots, f_{n-1}(y_1, y_2, \ldots, y_{n-1})]^T$, where the f_i are the Skolem functions capturing the dependence of x_i on $y_1, y_2, \ldots, y_{i-1}$ and c_0 is a constant in $\{a^1 - b^1\}$. Likewise, the universal player $\mathbf{Y}$ also makes its moves, according to some strategy. Given a strategy $\boldsymbol{x_s}$ for $\mathbf{X}$ and a strategy $\boldsymbol{y_s}$ for $\mathbf{Y}$, we say that $\boldsymbol{x_s}$ is winning against $\boldsymbol{y_s}$, if the consequence of $\mathbf{X}$ playing according to $\boldsymbol{x_s}$ and $\mathbf{Y}$ playing according to $\boldsymbol{y_s}$ is that $\mathbf{X}$ wins the game. A strategy $\boldsymbol{x_s}$ for $\mathbf{X}$ is said to be *always-winning* or a model for System (2) (**QIPG**), if the consequence of $\mathbf{X}$ playing according to $\boldsymbol{x_s}$ is that $\mathbf{X}$ wins, regardless of the strategy employed by $\mathbf{Y}$. Thus the Quantified Integer Programming problem can be described as the problem of checking whether the existential player in a 2-person integer matrix game has an always-winning strategy. For the rest of the paper, we will use the phrase "a winning strategy for the existential player" to mean an always-winning strategy for the existential player, when there is no confusion.

3   Motivation

One of the principal areas in which QIPs find application in, is the modeling of uncertainty [Sub02b]. In most application models, there is the inherent assumption of constancy in data, which is neither realistic nor accurate. For instance, in scheduling problems [Pin95], it is standard to assume that the execution time of a job is fixed and known in advance. While this simplifying assumption leads to elegant models, the fact remains that in real-time systems, such an assumption would lead to catastrophic consequences [SSRB98]. Of late, there has been some interest in problems such as parametric flow [McC00] and Selective Assembly [IMM98], in which the capacities are assumed to be variable. Such flow problems are easily and naturally expressed as TQIPs.

Whereas Integer Programming is `NP-complete` [GJ79], there are some interesting restrictions to the constraint matrix (other than total unimodularity) that have polynomial time decision procedures. Foremost among them is Planar IP or Integer Programming in the plane. Planar IPs have been used to model and solve knapsack problems that arise in certain applications [Kan80,HW76]. Problem $\mathbf{P_2}$ generalizes the planar IP problem to the quantified case; once again, the applications are motivated by the uncertainty in knapsack parameters.

Box QIPs are useful models to check validity of Clausal Systems and Integer Programs, since the only way in which a BQIP is false, is if there is a witness attesting to this fact. One of the more important application areas of BQIPs is the field of Constraint Databases [Rev98a,Rev98b]. Queries of the form: *Enumerate all people between ages* 20 *and* 30 *who earn between* 20K *and* 30k *annually*, are naturally expressible as Box QIPs over the appropriate domain. Problem $\mathbf{P_3}$ generalizes the work in [Rev98c], in that the solutions that we seek are quantified lattice points and not rationals.

Additional applications of QIPs can be found in the areas of logical inference [CH99], Computer Vision [vH] and Compiler Construction [Pug92b,Pug92a].

4 Properties of General QIPs

In this section, we derive properties that are true of any general QIP. These properties when combined with constraint specific properties aid us in the design of polynomial time algorithms for specialized classes of QIPs.

Definition 7. *A mathematical program of the form*

$$\exists x_1 \in [a^1, b^1] \; \forall y_1 \in [c^1, d^1] \; \exists x_2 \in [a^2, b^2]$$
$$\forall y_2 \in [c^2, d^2] \dots \exists x_n \in [a^n, b^n] \; \forall y_n \in [c^n, d^n]$$
$$\mathbf{A} \cdot [\boldsymbol{x} \; \boldsymbol{y}]^{\mathrm{T}} \leq \boldsymbol{b} \qquad (4)$$

is called a Quantified Linear Program (QLP).

In System (4), the range of the each program variable is a continuous real interval and not a discrete interval as is the case with QIPs. However, the entities defining the ranges, viz., $\{a^i, b^i, c^i, d^i\}$, $i = 1, 2, \dots, n$ are integral.

Definition 8. *A TQLP is a QLP in which the constraint matrix is totally unimodular.*

The complexity of deciding QLPs is not known [Joh], although the class of TQLPs can be decided in polynomial time (See [Sub03]).

Definition 9. *A Quantified Integer Program in which some variables have discrete ranges, while the rest have continuous ranges is called a Mixed QIP (MQIP).*

Theorem (1) argues the equivalence of certain MQIPs and QLPs. The consequences of this equivalence, when the constraint matrix is totally unimodular, are pointed out in Corollary (2).

Theorem 1.

$$\mathbf{L} : \exists x_1 \in [a^1, b^1] \; \forall y_1 \in \{c^1 - d^1\}$$
$$\exists x_2 \in [a^2, b^2] \; \forall y_2 \in \{c^2 - d^2\}$$
$$\dots \exists x_n \in [a^n, b^n] \; \forall y_n \in \{c^n - d^n\}$$
$$\mathbf{A} \cdot [\boldsymbol{x} \; \boldsymbol{y}]^{\mathrm{T}} \leq \boldsymbol{b}$$
$$\Leftrightarrow$$
$$\mathbf{R} : \exists x_1 \in [a^1, b^1] \; \forall y_1 \in [c^1, d^1]$$
$$\exists x_2 \in [a^2, b^2] \; \forall y_2 \in [c^2, d^2]$$
$$\dots \exists x_n \in [a^n, b^n] \; \forall y_n \in [c^n, d^n]$$
$$\mathbf{A} \cdot [\boldsymbol{x} \; \boldsymbol{y}]^{\mathrm{T}} \leq \boldsymbol{b} \qquad (5)$$

In other words, the existential player of the game **L** *has a winning strategy, if and only if the existential player of the game* **R** *has a winning strategy.*

Proof: Let $\mathbf{X_L}$ and $\mathbf{Y_L}$ denote the existential and universal players of the game $\mathbf{L}$ respectively. Likewise, let $\mathbf{X_R}$ and $\mathbf{Y_R}$ denote the existential and universal players of the game $\mathbf{R}$.

Observe that $\mathbf{R}$ is a QLP, while $\mathbf{L}$ is an MQIP.

$\mathbf{R} \Rightarrow \mathbf{L}$ is straightforward. Suppose that $\mathbf{X_R}$ has a strategy that is winning when the universal player $\mathbf{Y_R}$ can choose his i^{th} move from the continuous interval $[c^i, d^i]$; then clearly the strategy will also be winning, when the universal player has to choose his i^{th} move from the discrete interval $\{c^i - d^i\}$. Thus $\mathbf{X_L}$ can adopt the same strategy as $\mathbf{X_R}$ and win against any strategy employed by $\mathbf{Y_L}$.

We now focus on proving $\mathbf{L} \Rightarrow \mathbf{R}$. Our proof uses induction on the length of the quantifier string and therefore on the dimension of $\mathbf{A}$. Note that as described in System (2), the quantifier string is always of even length, with the existentially quantified variables and the universally quantified variables strictly alternating. Further, the first variable is always existentially quantified and the last variable is always universally quantified.

In the base case of the induction, the length of the quantifier string is 2; accordingly, we have to show that:

$$\mathbf{L} : \exists x_1 \in [a^1, b^1] \; \forall y_1 \in \{c^1 - d^1\} \; \mathbf{A} \cdot [x_1 \; y_1]^T \le \mathbf{b}$$
$$\Rightarrow \mathbf{R} : \exists x_1 \in [a^1, b^1] \; \forall y_1 \in [c^1, d^1] \; \mathbf{A} \cdot [x_1 \; y_1]^T \le \mathbf{b}$$

Let us say that $\mathbf{L}$ is true and let $x_1 = c_0$ be a solution, where $c_0 \in [a^1, b^1]$. We can write the constraint system $\mathbf{A} \cdot [x_1 \; y_1]^T \le \mathbf{b}$ as: $x_1 \cdot \mathbf{g_1} + y_1 \cdot \mathbf{h_1} \le \mathbf{b}$. Note that c_0 is a fixed constant, independent of y_1 and holds for all integral values of y_1 in $\{c^1 - d^1\}$. Accordingly, we have:

$$c_0 \cdot \mathbf{g_1} \le \mathbf{b} - c^1 \cdot \mathbf{h_1}$$
$$c_0 \cdot \mathbf{g_1} \le \mathbf{b} - d^1 \cdot \mathbf{h_1} \tag{6}$$

Now consider the (real) parametric point $y_1 = \lambda \cdot c^1 + (1 - \lambda) \cdot d^1$, $0 \le \lambda \le 1$. Observe that

$$\mathbf{b} - (\lambda \cdot c^1 + (1 - \lambda) \cdot d^1) \cdot \mathbf{h_1}$$
$$= \lambda \cdot \mathbf{b} + (1 - \lambda) \cdot \mathbf{b} - \lambda \cdot c^1 \cdot \mathbf{h_1}$$
$$-(1 - \lambda) \cdot d^1 \cdot \mathbf{h_1}$$
$$= \lambda \cdot (\mathbf{b} - c^1 \cdot \mathbf{h_1}) + (1 - \lambda) \cdot (\mathbf{b} - d^1 \cdot \mathbf{h_1})$$
$$\ge \lambda \cdot (c_0 \cdot \mathbf{g_1}) + (1 - \lambda) \cdot (c_0 \cdot \mathbf{g_1})$$
$$= c_0 \cdot \mathbf{g_1}$$

In other words, $x_1 = c_0$ holds for all values of y_1 in the continuous range $[c^1, d^1]$, thereby proving the base case.

Assume that Theorem (1) always holds, when the quantifier string has length $2 \cdot m$; we need to show that it holds when the quantifier string has length $2 \cdot m + 2$, i.e., we need to show that:

$$\mathbf{L} : \exists x_1 \in [a^1, b^1] \; \forall y_1 \in \{c^1 - d^1\}$$

$$\exists x_2 \in [a^2, b^2] \ \forall y_2 \in \{c^2 - d^2\}$$
$$\ldots \exists x_{m+1} \in [a^{m+1}, b^{m+1}]$$
$$\forall y_{m+1} \in \{c^{m+1} - d^{m+1}\}$$
$$\mathbf{A} \cdot [\boldsymbol{x} \ \boldsymbol{y}]^{\mathbf{T}} \leq \boldsymbol{b}$$
$$\Rightarrow$$
$$\mathbf{R} : \exists x_1 \in [a^1, b^1] \ \forall y_1 \in [c^1, d^1]$$
$$\exists x_2 \in [a^2, b^2] \ \forall y_2 \in [c^2, d^2]$$
$$\ldots \exists x_{m+1} \in [a^{m+1}, b^{m+1}]$$
$$\forall y_{m+1} \in [c^{m+1}, d^{m+1}]$$
$$\mathbf{A} \cdot [\boldsymbol{x} \ \boldsymbol{y}]^{\mathbf{T}} \leq \boldsymbol{b} \tag{7}$$

Let $\boldsymbol{x_s} = [x_1, x_2, \ldots, x_{m+1}]^T$ be a model for $\mathbf{L}$, in the manner described in Section §2.1. We consider 2 distinct games $\mathbf{L_1}$ and $\mathbf{L_2}$ to decide $\mathbf{L}$; one in which $\mathbf{Y_L}$ is forced to choose c^{m+1} for y_{m+1} and another in which $\mathbf{Y_L}$ is forced to pick $y_{m+1} = d^{m+1}$.

Consider the complete set of moves made in $(m+1)$ rounds by $\mathbf{X_L}$ and $\mathbf{Y_L}$ to decide $\mathbf{L}$ in both games; let $\boldsymbol{x_L}$ denote the numeric vector guessed by $\mathbf{X_L}$, while $\boldsymbol{y_{L1}}$ denotes the vector guessed by $\mathbf{Y_L}$ for game $\mathbf{L_1}$ and $\boldsymbol{y_{L2}}$ denotes the vector guessed by $\mathbf{Y_L}$ for game $\mathbf{L_2}$. Note that the moves made by $\mathbf{X_L}$ cannot depend on y_{m+1} and hence the vector guessed by $\mathbf{X_L}$ is the same for both games; further $\boldsymbol{y_{L1}}$ and $\boldsymbol{y_{L2}}$ differ only in their $(m+1)^{th}$ component. We denote the m-vector (numeric) corresponding to the first m components of the 2 vectors $\boldsymbol{y_{L1}}$ and $\boldsymbol{y_{L2}}$ by $\boldsymbol{y_1}$. We rewrite the constraint system $\mathbf{A} \cdot [\boldsymbol{x} \ \boldsymbol{y}]^{\mathbf{T}} \leq \boldsymbol{b}$ as $\mathbf{G} \cdot \boldsymbol{x} + \mathbf{H'} \cdot \boldsymbol{y'} + y_{m+1} \cdot \boldsymbol{h_{m+1}} \leq \boldsymbol{b}$, where $\boldsymbol{x} = [x_1, x_2, \ldots, x_{m+1}]^T$ and $\boldsymbol{y'} = [y_1, y_2, \ldots, y_m]^T$.

Since $\boldsymbol{x_s}$ is a model for $\mathbf{L}$, we must have

$$\mathbf{G} \cdot \boldsymbol{x_L} + \mathbf{H'} \cdot \boldsymbol{y_1} + c^{m+1} \cdot \boldsymbol{h_{m+1}} \leq \boldsymbol{b} \tag{8}$$

and

$$\mathbf{G} \cdot \boldsymbol{x_L} + \mathbf{H'} \cdot \boldsymbol{y_1} + d^{m+1} \cdot \boldsymbol{h_{m+1}} \leq \boldsymbol{b} \tag{9}$$

Now consider the (real) parametric point

$$y_{m+1} = \lambda \cdot c^{m+1} + (1 - \lambda) \cdot d^{m+1}.$$

Observe that:

$$\mathbf{G} \cdot \boldsymbol{x_L} + \mathbf{H'} \cdot \boldsymbol{y_1} + y_{m+1} \cdot \boldsymbol{h_{m+1}}$$
$$= \mathbf{G} \cdot \boldsymbol{x_L} + \mathbf{H'} \cdot \boldsymbol{y_1} + (\lambda \cdot c^{m+1} + (1 - \lambda) \cdot d^{m+1}) \cdot \boldsymbol{h_{m+1}}$$
$$= \lambda \cdot (\mathbf{G} \cdot \boldsymbol{x_L} + \mathbf{H'} \cdot \boldsymbol{y_1} + c^{m+1} \cdot \boldsymbol{h_{m+1}})$$
$$+ (1 - \lambda) \cdot (\mathbf{G} \cdot \boldsymbol{x_L} + \mathbf{H'} \cdot \boldsymbol{y_1} + d^{m+1} \cdot \boldsymbol{h_{m+1}})$$
$$\leq \lambda \cdot \boldsymbol{b} + (1 - \lambda) \cdot \boldsymbol{b}$$
$$= \boldsymbol{b}$$

In other words, x_s serves as a winning strategy for $\mathbf{X_L}$ for all values of y_{m+1} in the continuous range $[c^{m+1}, d^{m+1}]$. Accordingly, we are required to show that

$$\mathbf{L} : \exists x_1 \in [a^1, b^1] \; \forall y_1 \in \{c^1 - d^1\}$$
$$\exists x_2 \in [a^2, b^2] \; \forall y_2 \in \{c^2 - d^2\}$$
$$\ldots \exists x_{m+1} \in [a^{m+1}, b^{m+1}]$$
$$\forall y_{m+1} \in [c^{m+1}, d^{m+1}] \; \mathbf{A} \cdot [\boldsymbol{x} \; \boldsymbol{y}]^{\mathbf{T}} \le \boldsymbol{b}$$
$$\Rightarrow$$
$$\mathbf{R} : \exists x_1 \in [a^1, b^1] \; \forall y_1 \in [c^1, d^1]$$
$$\exists x_2 \in [a^2, b^2] \; \forall y_2 \in [c^2, d^2]$$
$$\ldots \exists x_{m+1} \in [a^{m+1}, b^{m+1}]$$
$$\forall y_{m+1} \in [c^{m+1}, d^{m+1}] \; \mathbf{A} \cdot [\boldsymbol{x} \; \boldsymbol{y}]^{\mathbf{T}} \le \boldsymbol{b} \tag{10}$$

Observe that both y_{m+1} and x_{m+1} are continuous in $\mathbf{L}$ and $\mathbf{R}$ and hence can be eliminated using the quantifier elimination techniques used to eliminate the variables of a Quantified Linear Program [Sub03]; y_{m+1} is eliminated using variable substitution, while x_{m+1} is eliminated using the Fourier-Motzkin elimination technique. Accordingly, the constraint system $\mathbf{A} \cdot [\boldsymbol{x} \; \boldsymbol{y}]^{\mathbf{T}} \le \boldsymbol{b}$ is transformed into the system $\mathbf{A_1} \cdot [\boldsymbol{x_1} \; \boldsymbol{y_1}]^{\mathbf{T}} \le \boldsymbol{b_1}$, where $\boldsymbol{x_1} = [x_1, x_2, \ldots, x_m]^T$ and $\boldsymbol{y_1} = [y_1, y_2, \ldots, y_m]^T$. Since the quantifier string is now of length $2 \cdot m$, we can use the inductive hypothesis to conclude that $\mathbf{L} \Rightarrow \mathbf{R}$.

It follows that Theorem (1) is proven. □

Corollary 1. *If all the existentially quantified variables of a QIP have continuous ranges, then the discrete ranges of the universally quantified variables can be relaxed into continuous ranges.*

Theorem 2. *Let*

$$\mathbf{L} : \exists x_1 \in \{a^1 - b^1\} \; \forall y_1 \in \{c^1 - d^1\}$$
$$\exists x_2 \in \{a^2 - b^2\} \; \forall y_2 \in \{c^2 - d^2\}$$
$$\ldots \exists x_n \in \{a^n - b^n\} \; \forall y_n \in \{c^n - d^n\}$$
$$\mathbf{A} \cdot [\boldsymbol{x} \; \boldsymbol{y}]^{\mathbf{T}} \le \boldsymbol{b}$$

and

$$\mathbf{R} : \exists x_1 \in \{a^1 - b^1\} \; \forall y_1 \in \{c^1 - d^1\}$$
$$\exists x_2 \in \{a^2 - b^2\} \; \forall y_2 \in \{c^2 - d^2\}$$
$$\ldots \exists x_n \in \{a^n - b^n\} \; \forall y_n \in [c^n, d^n]$$
$$\mathbf{A} \cdot [\boldsymbol{x} \; \boldsymbol{y}]^{\mathbf{T}} \le \boldsymbol{b}$$

Then $\mathbf{L} \Leftrightarrow \mathbf{R}$.

Proof: As in Theorem (1), let $\mathbf{X_L}$ and $\mathbf{Y_L}$ denote the existential and universal players of the game $\mathbf{L}$ respectively. Likewise, let $\mathbf{X_R}$ and $\mathbf{Y_R}$ denote the existential and universal players of the game $\mathbf{R}$.

Once again note that $\mathbf{R} \Rightarrow \mathbf{L}$ is straightforward, since the strategy that enabled $\mathbf{X_R}$ to win when $\mathbf{Y_R}$ is allowed to choose his n^{th} move from the continuous interval $[c^n, d^n]$ will also be winning when $\mathbf{Y_R}$ is allowed to choose his n^{th} move from the restricted (discrete) interval $\{c^n - d^n\}$. In other words, a strategy that is winning for $\mathbf{X_R}$ is also winning for $\mathbf{X_L}$.

We focus on proving $\mathbf{L} \Rightarrow \mathbf{R}$.

Let $\boldsymbol{x_s} = [x_1, x_2, \ldots, x_n]^T$ be a model for $\mathbf{L}$, in the manner described in Section §2.1. We consider 2 distinct games $\mathbf{L_1}$ and $\mathbf{L_2}$ to decide $\mathbf{L}$; one in which $\mathbf{Y_L}$ is forced to choose c^n for y_n and another in which $\mathbf{Y_L}$ is forced to pick $y_n = d^n$.

Consider the complete set of moves made in n rounds by $\mathbf{X_L}$ and $\mathbf{Y_L}$ to decide $\mathbf{L}$ in both games; let $\boldsymbol{x_L}$ denote the numeric vector guessed by $\mathbf{X_L}$, while $\boldsymbol{y_{L1}}$ denotes the vector guessed by $\mathbf{Y_L}$ for game $\mathbf{L_1}$ and $\boldsymbol{y_{L2}}$ denotes the vector guessed by $\mathbf{Y_L}$ for game $\mathbf{L_2}$. Note that the moves made by $\mathbf{X_L}$ cannot depend on y_n and hence the vector guessed by $\mathbf{X_L}$ is the same for both games; further $\boldsymbol{y_{L1}}$ and $\boldsymbol{y_{L2}}$ differ only in their n^{th} component. We denote the $(n-1)$-vector (numeric) corresponding to the first $(n-1)$ components of the 2 vectors $\boldsymbol{y_{L1}}$ and $\boldsymbol{y_{L2}}$ by $\boldsymbol{y_1}$. We rewrite the constraint system $\mathbf{A} \cdot [\boldsymbol{x}\ \boldsymbol{y}]^T \leq \boldsymbol{b}$ as $\mathbf{G} \cdot \boldsymbol{x} + \mathbf{H}' \cdot \boldsymbol{y}' + y_n \cdot \boldsymbol{h_{m+1}} \leq \boldsymbol{b}$, where $\boldsymbol{x} = [x_1, x_2, \ldots, x_n]^T$ and $\boldsymbol{y}' = [y_1, y_2, \ldots, y_{n-1}]^T$.

Since $\boldsymbol{x_s}$ is a model for $\mathbf{L}$, we must have

$$\mathbf{G} \cdot \boldsymbol{x_L} + \mathbf{H}' \cdot \boldsymbol{y_1} + c^n \cdot \boldsymbol{h_n} \leq \boldsymbol{b} \tag{11}$$

and

$$\mathbf{G} \cdot \boldsymbol{x_L} + \mathbf{H}' \cdot \boldsymbol{y_1} + d^n \cdot \boldsymbol{h_n} \leq \boldsymbol{b} \tag{12}$$

Now consider the (real) parametric point

$$y_n = \lambda \cdot c^n + (1 - \lambda) \cdot d^n.$$

Observe that:

$$\mathbf{G} \cdot \boldsymbol{x_L} + \mathbf{H}' \cdot \boldsymbol{y_1} + y_n \cdot \boldsymbol{h_n}$$
$$= \mathbf{G} \cdot \boldsymbol{x_L} + \mathbf{H}' \cdot \boldsymbol{y_1} + (\lambda \cdot c^n + (1 - \lambda) \cdot d^n) \cdot \boldsymbol{h_n}$$
$$= \lambda \cdot (\mathbf{G} \cdot \boldsymbol{x_L} + \mathbf{H}' \cdot \boldsymbol{y_1} + c^n \cdot \boldsymbol{h_n})$$
$$+ (1 - \lambda) \cdot (\mathbf{G} \cdot \boldsymbol{x_L} + \mathbf{H}' \cdot \boldsymbol{y_1} + d^n \cdot \boldsymbol{h_n})$$
$$\leq \lambda \cdot \boldsymbol{b} + (1 - \lambda) \cdot \boldsymbol{b}$$
$$= \boldsymbol{b}$$

In other words, $\boldsymbol{x_s}$ serves as a winning strategy for $\mathbf{X_L}$ for all values of y_n in the continuous range $[c^n, d^n]$, i.e., the strategy that is winning for $\mathbf{X_L}$ is also winning for $\mathbf{X_R}$, thereby proving that $\mathbf{L} \Rightarrow \mathbf{R}$.

$\square$

5   TUM QIPs

In this section, we handle the first restriction to QIPs, viz., total unimodularity of the constraint matrix. Our techniques are based on showing that TQIPs can be relaxed to polynomial time solvable TQLPs, while preserving the solution space. Relaxing a QIP to a QLP involves replacing the discrete intervals $\{a^i - b^i\}$ with continuous intervals $[a'^i, b'^i]$, for suitably chosen a'^i, b'^i, in case of both the universally and existentially quantified variables of the QIP. In the rest of this section, we argue that such a solution-preserving relaxation exists, when the constraint matrix $\mathbf{A}$ of the QIP is TUM.

Theorem 3. *Let*

$$\mathbf{R} : \exists x_1 \in [a^1, b^1] \; \forall y_1 \in [c^1, d^1]$$
$$\exists x_2 \in [a^2, b^2] \; \forall y_2 \in [c^2, d^2]$$
$$\ldots \exists x_n \in [a^n, b^n] \; \forall y_n \in [c^n, d^n]$$
$$\mathbf{A} \cdot [\boldsymbol{x} \; \boldsymbol{y}]^{\mathbf{T}} \leq \boldsymbol{b} \tag{13}$$

have a model, where $\mathbf{A}$ is totally unimodular. Then x_1 can always be chosen integral, by the existential player (say $\mathbf{X_R}$), without affecting the outcome of the game.

Proof: (Recall that $a_i, b_i, c_i, d_i, \; i = 1, 2, \ldots n$ are integral and $\boldsymbol{b}$ is integral.) Observe that $\mathbf{R}$ is a QLP, hence we can use the algorithm developed in [Sub03] to decide it. The algorithm therein, eliminates a universally quantified variable as follows: The vector $\boldsymbol{b}$ is replaced with a new integral vector, say $\boldsymbol{b}'$, obtained by subtracting an appropriate integral vector from $\boldsymbol{b}$. The only change to the $\mathbf{A}$ matrix is that a column is eliminated and hence it stays totally unimodular. Existentially quantified variables are eliminated using Fourier-Motzkin elimination, which is a variation of pivoting and hence their elimination also preserves total unimodularity (See [NW99]). Since $\mathbf{R}$ has a model, the algorithm in [Sub03] determines a range for x_1 of the form $a \leq x_1 \leq b$. Since $\mathbf{A}$ is totally unimodular and stays so under the elimination operations, and $\boldsymbol{b}$ is integral and stays so under the elimination operations, there is at least one integer in this range. □

Corollary 2. *Let*

$$\mathbf{R} : \exists x_1 \in [a^1, b^1] \; \forall y_1 \in \{c^1 - d^1\}$$
$$\exists x_2 \in [a^2, b^2] \; \forall y_2 \in \{c^2 - d^2\}$$
$$\ldots \exists x_n \in [a^n, b^n] \; \forall y_n \in \{c^n - d^n\}$$
$$\mathbf{A} \cdot [\boldsymbol{x} \; \boldsymbol{y}]^{\mathbf{T}} \leq \boldsymbol{b} \tag{14}$$

be true, where $\mathbf{A}$ is TUM. Then x_1 can always be chosen integral.

Proof: Follows from Theorem (1) and Theorem (3). □

Theorem 4.

$$\mathbf{L} : \exists x_1 \in \{a^1 - b^1\} \ \forall y_1 \in \{c^1 - d^1\}$$
$$\exists x_2 \in \{a^2 - b^2\} \ \forall y_2 \in \{c^2 - d^2\}$$
$$\dots \exists x_n \in \{a^n - b^n\} \ \forall y_n \in \{c^n - d^n\}$$
$$\mathbf{A} \cdot [\boldsymbol{x} \ \boldsymbol{y}]^{\mathbf{T}} \leq \boldsymbol{b}$$

$$\Leftrightarrow$$

$$\mathbf{R} : \exists x_1 \in [a^1, b^1] \ \forall y_1 \in \{c^1 - d^1\}$$
$$\exists x_2 \in [a^2, b^2] \ \forall y_2 \in \{c^2 - d^2\}$$
$$\dots \exists x_n \in [a^n, b^n] \ \forall y_n \in \{c^n - d^n\}$$
$$\mathbf{A} \cdot [\boldsymbol{x} \ \boldsymbol{y}]^{\mathbf{T}} \leq \boldsymbol{b} \tag{15}$$

where $\mathbf{A}$ *is TUM.*

Proof: Let $\mathbf{X_L}$ and $\mathbf{Y_L}$ denote the existential and universal players of the game $\mathbf{L}$ respectively. Likewise, let $\mathbf{X_R}$ and $\mathbf{Y_R}$ denote the existential and universal players of the game $\mathbf{R}$.

$\mathbf{L} \Rightarrow \mathbf{R}$ is straightforward. If there exists a winning strategy for $\mathbf{X_L}$ against $\mathbf{Y_L}$, when $\mathbf{X_L}$ is forced to choose his i^{th} move from the discrete interval $\{a^i - b^i\}$, then the same strategy will also be winning for $\mathbf{X_L}$, when he is allowed to choose his i^{th} move from the continuous interval $[a^i - b^i]$. Thus $\mathbf{X_R}$ can adopt the same strategy against $\mathbf{Y_R}$ and win.

We focus on proving $\mathbf{R} \Rightarrow \mathbf{L}$. Let $\mathbf{X_R}$ have a winning strategy against against $\mathbf{Y_R}$ in the game $\mathbf{R}$.

From the hypothesis, we know that there exists a rational $x_1 \in [a_1, b_1]$, for all integral values of $y_1 \in \{c_1 - d_1\}$, there exists a rational $x_2 \in [a_2, b_2]$ (which could depend on y_1), for all integral values of $y_2 \in \{c_2 - d_2\}$, ..., there exists a rational $x_n \in [a_n, b_n]$ (which could depend upon $y_1, y_2, \dots, y_{n-1}$), for all integral values of $y_n \in \{c_n - d_n\}$ such that $\mathbf{A} \cdot [\boldsymbol{x} \ \boldsymbol{y}]^{\mathbf{T}} \leq \boldsymbol{b}$. From Corollary (2), we know that x_1 can always be guessed integral (say p_1), since $\mathbf{A}$ is TUM. Since y_1 must be guessed integral (say q_1), at the end of round 1, we have guessed 2 integers and the constraint system $\mathbf{A} \cdot [\boldsymbol{x} \ \boldsymbol{y}]^{\mathbf{T}} \leq \boldsymbol{b}$ is transformed into $\mathbf{A}' \cdot [\boldsymbol{x}' \ \boldsymbol{y}']^{\mathbf{T}} \leq \boldsymbol{b}'$, where $\mathbf{A}'$ is obtained by deleting the first column $(\boldsymbol{a_1})$ and the $(n+1)^{th}$ column $(\boldsymbol{a_{n+1}})$ of $\mathbf{A}$, $\boldsymbol{x}' = [x_2, x_3, \dots, x_n]^T$, $\boldsymbol{y}' = [y_2, y_3, \dots, y_n]^T$ and $\boldsymbol{b}' = \boldsymbol{b} - \boldsymbol{p_1} - \boldsymbol{q_1}$, where $\boldsymbol{p_1}$ is the m−vector $(p_1 \cdot \boldsymbol{a_1})$ and $\boldsymbol{q_1}$ is the m−vector $(q_1 \cdot \boldsymbol{a_{n+1}})$. Note that once again $\mathbf{A}'$ is TUM and $\boldsymbol{b}'$ is integral. So x_2 can be guessed integral and y_2 must be integral. Thus, the game can be played with the $\mathbf{X}$ constantly guessing integral values and $\mathbf{Y}$ being forced to make integral moves. From the hypothesis $\mathbf{R}$ is true; it follows that $\mathbf{L}$ is true. $\square$

Theorem 5. *TQIPs can be decided in polynomial time.*

Proof: Use Theorem (4) to relax the ranges of the existentially quantified variables and Theorem (1) to relax the ranges of the universally quantified variables to get a TQLP; then use the algorithm in [Sub03] to decide the TQLP in polynomial time. $\square$

We have thus shown that when the constraint matrix representing a system of linear inequalities is totally unimodular, a Quantified Integer Program can be relaxed to a Quantified Linear Program, in exactly the same way that a traditional Integer Program can be relaxed to a traditional Linear Program.

6 Conclusion

In this paper, we studied a number of interesting classes of the Quantified Integer Programming problem. Our principal contribution has been to demonstrate that QIPs can be relaxed to QLPs, when the constraint matrix is totally unimodular. As discussed in Section §3, Quantified Integer Programming is a versatile tool that is used to model a number of real-world problems. The restrictions to QIPs discussed in this paper seem to capture fairly large subclasses of problems that occur in practical situations. Some of the important open problems are:

1. Are there other restrictions to the quantifier specification or the constraint matrix that can be solved in polynomial time? For instance, in [Sub02a], we showed that the Q2SAT problem could be modeled as a polynomial-time solvable QIP; our result there is more general than the result in [APT79] and uses only polyhedral elimination techniques.
2. What is the simplest restriction to the constraint matrix for which the language of QIPs becomes hard, assuming an arbitrary quantifier string (i.e., unbounded alternation)? An extension to the planar case studied in this paper, is QIP(2), which is defined as the class of QIPs in which every constraint has at most 2 existentially quantified variables. IP(2), i.e., Integer Programming with at most 2 variables per constraint is known to be NP-complete [HN94,Lag85]; we believe that QIP(2) is PSPACE-complete.
3. A problem that we are currently working on regards characterizing the structure of *Parametric Lattices*, i.e., lattices whose lattice points are integer functions and not integers. Understanding the structure of these lattices will be crucial towards designing algorithms for general cases of QIPs.

References

[AMO93] R. K. Ahuja, T. L. Magnanti, and J. B. Orlin. *Network Flows: Theory, Algorithms and Applications*. Prentice-Hall, 1993.

[APT79] Bengt Aspvall, Michael F. Plass, and Robert Tarjan. A linear-time algorithm for testing the truth of certain quantified boolean formulas. *Information Processing Letters*, 8(3):121–123, 1979.

[CGS98] M. Cadoli, A. Giovanardi, and M. Schaerf. An algorithm to evaluate quantified boolean formulae. In *AAAI-98*, July 1998.

[CH99] V. Chandru and J. N. Hooker. *Optimization Methods for Logical Inference*. Series in Discrete Mathematics and Optimization. John Wiley & Sons Inc., New York, 1999.

[Gav93] F. Gavril. An efficiently solvable graph partition, problem to which many problems are reducible. *Information Processing Letters*, 45(6):285–290, 1993.

[GJ79]   M. R. Garey and D. S. Johnson. *Computers and Intractability: A Guide to the Theory of NP-Completeness*. W. H. Freeman Company, San Francisco, 1979.

[HN94]   Dorit S. Hochbaum and Joseph (Seffi) Naor. Simple and fast algorithms for linear and integer programs with two variables per inequality. *SIAM Journal on Computing*, 23(6):1179–1192, December 1994.

[HW76]   D.S. Hirschberg and C.K. Wong. A polynomial algorithm for the knapsack problem in 2 variables. *JACM*, 23(1):147–154, 1976.

[IMM98]   S. Iwata, T. Matsui, and S.T. McCormick. A fast bipartite network flow algorithm for selective assembly. *OR Letters*, pages 137–143, 1998.

[Joh]   D.S. Johnson. Personal Communication.

[Kan80]   R. Kannan. A polynomial algorithm for the two-variable integer programming problem. *JACM*, 27(1):118–122, 1980.

[Lag85]   J. C. Lagarias. The computational complexity of simultaneous Diophantine approximation problems. *SIAM Journal on Computing*, 14(1):196–209, 1985.

[McC00]   S.T. McCormick. Fast algorithms for parametric scheduling come from extensions to parametric maximum flow. *Operations Research*, 47:744–756, 2000.

[NW99]   G. L. Nemhauser and L. A. Wolsey. *Integer and Combinatorial Optimization*. John Wiley & Sons, New York, 1999.

[Pap94]   Christos H. Papadimitriou. *Computational Complexity*. Addison-Wesley, New York, 1994.

[Pin95]   M. Pinedo. *Scheduling: theory, algorithms, and systems*. Prentice-Hall, Englewood Cliffs, 1995.

[Pug92a]   W. Pugh. The definition of dependence distance. Technical Report CS-TR-2292, Dept. of Computer Science, Univ. of Maryland, College Park, November 1992.

[Pug92b]   W. Pugh. The omega test: A fast and practical integer programming algorithm for dependence analysis. *Comm. of the ACM*, 35(8):102–114, August 1992.

[Rev98a]   P. Revesz. Safe query languages for constraint databases. *ACM Transactions on Database Systems*, 23(1):58–99, 1998.

[Rev98b]   Peter Revesz. The evaluation and the computational complexity of datalog queries of boolean constraints. *International Journal of Algebra and Computation*, 8(5):553–574, 1998.

[Rev98c]   Peter Revesz. Safe datalog queries with linear constraints. In *Proceedings of the Fourth International Conference on the Principles and Practice of Constraint Programming*, pages 355–369, Pisa, Italy, October 1998. Springer-Verlag. LNCS 1520.

[Sch78]   T.J. Schaefer. The complexity of satisfiability problems. In Alfred Aho, editor, *Proceedings of the 10th Annual ACM Symposium on Theory of Computing*, pages 216–226, New York City, NY, 1978. ACM Press.

[Sch87]   Alexander Schrijver. *Theory of Linear and Integer Programming*. John Wiley and Sons, New York, 1987.

[SSRB98]   John A. Stankovic, Marco Spuri, Krithi Ramamritham, and Giorgio C. Buttazzo, editors. *Deadline Scheduling for Real-Time Systems*. Kluwer Academic Publishers, 1998.

[Sub02a]   K. Subramani. On identifying simple and quantified lattice points in the 2sat polytope. In et. al. Jacques Calmet, editor, *Proceedings of the 5th International Conference on Artificial Intelligence and Symbolic Computation (AISC)*, volume 2385 of *Lecture Notes in Artificial Intelligence*, pages 217–230. Springer-Verlag, July 2002.

[Sub02b]   K. Subramani. A specification framework for real-time scheduling. In W.I. Grosky and F. Plasil, editors, *Proceedings of the 29th Annual Conference on Current Trends in Theory and Practice of Informatics (SOFSEM)*, volume 2540 of *Lecture Notes in Computer Science*, pages 195–207. Springer-Verlag, November 2002.

[Sub03]   K. Subramani. An analysis of quantified linear programs. In C.S. Calude, et. al., editor, *Proceedings of the 4th International Conference on Discrete Mathematics and Theoretical Computer Science (DMTCS)*, volume 2731 of *Lecture Notes in Computer Science*, pages 265–277. Springer-Verlag, July 2003.

[vH]   Pascal van Hentenryck. Personal Communication.

Formalizing O Notation in Isabelle/HOL

Jeremy Avigad and Kevin Donnelly

Carnegie Mellon University

Abstract. We describe a formalization of asymptotic O notation using the Isabelle/HOL proof assistant.

1 Introduction

Asymptotic notions are used to characterize the approximate long-term behavior of functions in a number of branches of mathematics and computer science, including analysis, combinatorics, and computational complexity. Our goal here is to describe an implementation of one important asymptotic notion — "big O notation" — using the Isabelle/HOL proof assistant.

Developing a library to support such reasoning poses a number of interesting challenges. First of all, ordinary mathematical practice involving O notation relies on a number of *conventions*, some determinate and some ambiguous, so deciding on an appropriate formal representation requires some thought. Second, we will see that a natural way of handling the notation is inherently *higher-order*; thus the implementation is a way of putting the higher-order features of a theorem prover to the test. Finally, O notation is quite *general*, since many of the definitions and basic properties make sense for the analysis of any domain of functions $A \Rightarrow B$ where B has the structure of an ordered ring (or even, more generally, a ring with a valuation); in practice, A is often either the set of natural numbers or an interval of real numbers, and B may be $\mathbb{N}, \mathbb{Q}, \mathbb{R}$, or $\mathbb{C}$.

On the positive side, uses of O notation can have a very computational flavor, and making the right simplification rules available to an automated reasoner can yield effective results. Given the range of applications, then, this particular case study is a good test of the ability of today's proof assistants to support an important type of mathematical analysis.

Section 2 provides a quick refresher course in O notation. Section 3 then reviews some of the specific features of Isabelle that are required for, and well-suited to, the task of formalization. In Sections 4 and 5 we describe our implementation. In Section 6 we describe our initial application, that is, deriving certain identities used in analytic number theory. Future work is described in Section 7. Piotr Rudnicki has drawn our attention to a similar formalization of asymptotic notions in Mizar [3,4], which we also discuss in Section 7.

We are grateful to Clemens Ballarin for help with some of the type declarations in Section 4.

D. Basin and M. Rusinowitch (Eds.): IJCAR 2004, LNAI 3097, pp. 357–371, 2004.
© Springer-Verlag Berlin Heidelberg 2004

2   *O* Notation

If f and g are functions, the notation $f(x) = O(g(x))$ is used to express the fact that f's rate of growth is no bigger than that of g, in the following sense:

Definition 1. $f(x) = O(g(x))$ *means* $\exists c \, \forall x \, (|f(x)| \leq c \cdot |g(x)|)$.

Here it is assumed that f and g are functions with the same domain and codomain. The definition further assumes that notions of multiplication and absolute value are defined on the codomain; but otherwise the definition is very general.

Henceforth we will assume that the codomain is an ordered ring, giving us addition and subtraction as well. (We will also assume that rings are nondegenerate, i.e. satisfy $0 \neq 1$.) Absolute value is then defined by

$$|x| = \begin{cases} x & \text{if } 0 \leq x \\ -x & \text{otherwise} \end{cases}$$

This covers $\mathbb{Z}$, $\mathbb{Q}$, and $\mathbb{R}$. Here are some examples:

- As functions from the positive natural numbers, $\mathbb{N}^+$, to $\mathbb{Z}$, $4x^2 + 3x + 12 = O(x^2)$.
- Similarly, $4x^2 + 3x + 12 = 4x^2 + O(x)$.
- If $f(n) = \sum_{k=0}^{n} \binom{3n}{k}$ is viewed as a function from $\mathbb{N}^+$ to $\mathbb{Q}$, we have

$$f(n) = \binom{3n}{n} \left(2 - \frac{4}{n} + O(\frac{1}{n^2}) \right).$$

This last example is taken from Graham et al. [1, Chapter 9], which is an excellent reference for asymptotic notions in general.

Some observations are in order. First, the notation in the examples belie the fact that O notation is about functions; thus, the more accurate rendering of the first example is

$$\lambda x. \, 4x^2 + 3x + 12 = O(\lambda x. \, x^2).$$

The corresponding shorthand is used all the time, even though it is ambiguous; for the example, the expression

$$ax^2 + bx + c = O(x^2)$$

is not generally true if one interprets the terms as functions of a, b, or c, instead of x.

The next thing to notice is that we have already stretched the notation well beyond Definition 1. For example, we read the second example above as the assertion that for some function f such that $f = O(x)$, $4x^2 + 3x + 12 = 4x^2 + f$.

The last thing to note is that in these expressions, "=" does not behave like equality. For example, although the assertion

$$x^2 + O(x) = O(x^3)$$

is correct, the symmetric assertion

$$O(x^3) = x^2 + O(x)$$

is not.

The approach described by Graham et al. [1] is to interpret $O(f)$ as denoting the *set* of all functions that have that rate of growth; that is,

$$O(f) = \{g \mid \exists c\, \forall x\ (|g(x)| \le c \cdot |f(x)|)\}.$$

Then one reads equality in the expression $f = O(g)$ as the *element-of* relation, $f \in O(g)$. One can read the sum $g + O(h)$ as

$$g + O(h) = \{g\} + O(h) = \{g + k \mid k \in O(h)\},$$

and then interpret equality in the expression

$$g + O(h) = O(l)$$

as the subset relation, $g + O(h) \subseteq O(l)$. This is the reading we have followed.

The O operator can be extended to sets as well, by defining

$$O(S) = \bigcup_{f \in S} O(f) = \{g \mid \exists f \in S\ (f = O(g))\}.$$

Thus, if $f = g + O(h)$ and $h = k + O(l)$, we can make sense of the expression

$$f = g + O(k + O(l)).$$

There are various extensions to O notation, as we have described it. First of all, one often restricts the domain under consideration to some subset S of the domain of the relevant functions. For example, if f and g are functions with domain $\mathbb{R}$, we might say "$f = O(g)$ on the interval $(0, 1)$," thereby restricting the quantifier in the definition of $O(g)$. Second, one is often only concerned with the long-term behavior of functions, in which case one wants to ignore the behavior on any initial segment of the domain. The relation $f = O(g)$ is therefore often used instead to express the fact that f's rate of growth is *eventually* dominated by that of g, i.e. that Definition 1 holds provided the universal quantifier is restricted to sufficiently large inputs. As an example, notice that the assertion $x + 1 \in O(x^2)$ is false on our strict reading if x is assumed to range over the natural numbers, since no constant satisfies the definition for $x = 0$. The statement *is* true, however, on the interval $[1, \infty)$, as well as eventually true.

There is an elegant way of modeling these variants in our framework. If A is any set of functions from σ to τ and S is any set of elements of type σ, define "A on S" to be the set

$$\{f \mid \exists g \in A\ \forall x \in S\ (f(x) = g(x))\}$$

of all functions that agree with some function in A on elements of S. The second variant assumes that there is an ordering on the domain, but when this is the case, we can define "A eventually" to be the set

$$\{f \mid \exists k\, \exists g \in A\ \forall x \ge k\ (f(x) = g(x))\}$$

of all functions that eventually agree with some function in A. With these definitions, "$f = O(g)$ on S" and "$f = O(g)$ eventually" have the desired meanings.

Finally, as noted in the introduction, O notation makes sense for any ring with a suitable *valuation*, such as the modulus function $|\cdot| : \mathbb{C} \Rightarrow \mathbb{R}$ on the complex numbers. Such a variation would require only minor changes to the definitions and lemmas we give below; conversely, if we define $O'(f) = O(\lambda x.|f(x)|)$, properties of the more general version can be derived from properties of the more restricted one. Below, we will only treat the most basic version of O notation, described above.

3   Isabelle Preliminaries

Isabelle [12] is a generic proof assistant developed under the direction of Larry Paulson at Cambridge University and Tobias Nipkow at TU Munich. The HOL instantiation [6] provides a formal framework based on Church's simple type theory. In addition to basic types like *nat* and *bool* and type constructors for product types, function types, set types, and list types, one can turn any set of objects into a new type using a **typedef** declaration. Isabelle also supports polymorphic types and overloading. Thus, if $'a$ is a variable ranging over types, then in the term $(x::'a) + y$ it is inferred that y has type $'a$, and that $'a$ ranges over types for which an operation $+$ has been defined.

Moreover, in Isabelle/HOL types are grouped together into *sorts*. These sorts are ordered, and types of a sort inherit properties of any of the sort's ancestors. Thus the sorts, also known as *type classes*, form a hierarchy, with the predefined *logic* class, containing all types, at the top. This, in particular, supports subtype polymorphism; for example, the term

$$(\lambda x \; y. \; x \leq y)::('a::order\,) \Rightarrow 'a \Rightarrow bool$$

represents an object of type $'a \Rightarrow 'a \Rightarrow bool$, where $'a$ ranges over types of the sort *order*. Here the notation $t::(T::S)$ means that the term t has type T, which is a member of sort S.

Isabelle also allows one to impose axiomatic requirements on a sort; that is, an *axiomatic type class* is simply a class of types that satisfy certain axioms. Axiomatic type classes make it possible to introduce terms, concepts, and associated theorems at a useful level of generality. For example, the classes

axclass *plus* $<$ *term*
axclass *zero* $<$ *term*
axclass *one* $<$ *term*

are used to define the classes of types, after which polymorphic constants

$+::('a::plus) \Rightarrow 'a \Rightarrow 'a$
$0::('a::zero)$
$1::('a::one)$

can be declared to exist simultaneously for all types $'a$ in the respective classes. The following declares the class *plus-ac0* to be a class of types for which 0 and + are defined and satisfy appropriate axioms:

> **axclass** *plus-ac0 < plus, zero*
> *commute: $x + y = y + x$*
> *assoc: $(x + y) + z = x + (y + z)$*
> *zero: $0 + x = x$*

One can then derive theorems that hold for any type in this class, such as

> **theorem** *right-zero: $x + 0 = x$::($'a$::plus-ac0)*

One can also define operations that make sense for any member of a class. For example, the Isabelle HOL library defines a general summation operator

> *setsum::$('a \Rightarrow 'b) \Rightarrow (('a\ set) \Rightarrow ('b::plus\text{-}ac0))$*

Assuming $'b$ is any type in the class *plus-ac0*, A is any set of objects of type $'a$, and f is any function from $'a$ to $'b$, *setsum A f* denotes the sum $\sum_{x \in A} f(x)$. (This operator is defined to return 0 if A is infinite.) The assertion

> **instance** *nat::plus-ac0*

declares the particular type *nat* to be an instance of *plus-ac0*, and requires us to prove that elements of *nat* satisfy the defining axioms. Once this is done, however, we can use operators like *setsum* and the general results proved for *plus-ac0* freely for *nat*.

More details on Isabelle's mechanisms for handling axiomatic type classes and overloading can be found in [8,9].

4   The Formalization

Our formalization makes use of Isabelle 2003's HOL-Complex library. This includes a theory of the real numbers and the rudiments of real analysis (including nonstandard analysis), developed by Jacques Fleuriot (with contributions by Larry Paulson). It also includes an axiomatic development of rings by Markus Wenzel and Gertrud Bauer, which was important to our formalization.

Aiming for generality, we designed our library to deal with functions from any set into a nondegenerate ordered ring. O sets are defined as described by Definition 1 in Section 2:

$$O(g) = \{h \mid \exists c\ \forall x\ (|h(x)| \leq c \cdot |g(x)|)\}$$

When f and g are elements of a function type such that addition is defined on the codomain, we would like define $f + g$ to be their pointwise sum. Similarly, if A and B are sets of elements of a type for which addition is defined, we would like to define the sum $A + B$ to be the set of sums of their elements. The first step is to declare the appropriate function and set types to be appropriate for such overloading:

instance *set* :: (*plus*)*plus*
instance *fun* :: (*type,plus*)*plus*

We may then define the corresponding operations:

defs (overloaded)
  func-plus: $f + g == (\lambda x.\ f\ x + g\ x)$
  set-plus: $A + B == \{c.\ \exists a \in A.\ \exists b \in B.\ c = a + b\}$

We can define multiplication and subtraction similarly, assuming these operations are supported by the relevant types. We can also lift a zero element:

instance *fun* :: (*type,zero*)*zero*
instance *set* :: (*zero*)*zero*
defs (overloaded)
  func-zero: $0::(('a::type) \Rightarrow ('b::zero)) == \lambda x.\ 0$
  set-zero: $0::('a::zero)set == \{0\}$

In other words, the 0 function is the constant function which returns 0, and the 0 set is the singleton containing 0. Similarly, one can define appropriate notions of 1.

Now, asssuming the types in question are elements of the sort *plus-ac0*, we can show that the liftings to the function and set types are again elements of *plus-ac0*.

instance *fun* :: (*type,plus-ac0*)*plus-ac0*
instance *set* :: (*plus-ac0*)*plus-ac0*

This declaration requires justification: we have to show that the resulting addition operation is associative and commutative, and that the zero element is an additive identity. Similarly, if the relevant starting type is a ring, then the resulting function type is a ring:

instance *fun* :: (*type,ring*)*ring*

Similarly, if the underlying type is a ring, multiplication of sets is commutative and associative, distributes over addition, and has an identity, 1.

We can now define the operation O, which maps a suitable function f to the set of functions, $O(f)$:

constdefs
  bigo :: $('a \Rightarrow 'b::oring\text{-}nd) \Rightarrow ('a \Rightarrow 'b)\ set$ (($10'(\text{-}')$))
  $O(f::('a \Rightarrow 'b::oring\text{-}nd)) == \{h.\ \exists\,c.\ \forall x.\ abs\ (h\ x) \leq c * abs\ (f\ x)\}$
  bigoset :: $('a \Rightarrow 'b::oring\text{-}nd)\ set \Rightarrow ('a \Rightarrow 'b)\ set$ (($10'(\text{-}')$))
  $O(S::('a \Rightarrow 'b::oring\text{-}nd)\ set) == \{h.\ \exists f \in S.\ \exists\,c.\ \forall x.$
    $abs(h\ x) \leq (c * abs(f\ x))\}$

Recall that when it comes to O notation we also want to be able to deal with terms of the form $f + O(g)$, where f and g are functions. Thus we define an operation "$+o$" that takes, as argument, an element of a type and a set of elements of the same type:

constdefs
> *elt-set-plus* :: $'a::plus \Rightarrow 'a\ set \Rightarrow 'a\ set$ (**infixl** $+o$ 70)
> $a +o\ B == \{c.\ \exists b{\in}B.\ c = a + b\}$

The operation $*o$ is defined similarly. The following declaration indicates that these operations should be displayed as "+" and "*" in all forms of output from the theorem prover:

syntax (output)
> *elt-set-plus* :: $'a \Rightarrow 'a\ set \Rightarrow 'a\ set$ (**infix** + 70)
> *elt-set-times* :: $'a \Rightarrow 'a\ set \Rightarrow 'a\ set$ (**infix** * 80)

Remember that according to the conventions described in Section 2, we would like to interpret $x = O(f)$ as $x \in O(f)$, and, for example, $f + O(g) = O(h)$ as $f + O(g) \subseteq O(h)$. Thus we declare symbols "=o" and "=s" to denote these two "equalities":

syntax
> *elt-set-eq* :: $'a \Rightarrow 'a\ set \Rightarrow bool$ (**infix** $=o$ 50)
> *set-set-eq* :: $'a\ set \Rightarrow 'a\ set \Rightarrow bool$ (**infix** $=s$ 50)
translations
> $x =o\ A => x \in A$
> $A =s\ B => A \subseteq B$

Because they are translated immediately, the symbols $=o$ and $=s$ are displayed as $\in$ and $\subseteq$, respectively, in the prover's output. In a slightly underhanded way, however, we can arrange to have $=o$, $=s$, $+o$, and $*o$ appear as $=$, $=$, $+$, and $*$, respectively, in proof scripts: we introduce new symbols to denote the former, and then configure our editor, document generator, etc. to display these symbols as the latter.

The advantage to having these translations in place is that our formalizations come closer to textbook expressions. The following table gives some examples of ordinary uses of O notation, paired with our Isabelle representations:

$f = g + O(h)$	$f = g + O(h)$
$x^2 + 3x + 1 = x^2 + O(x)$	$(\lambda x.\ x\char`\^2 + 3 * x + 1) = (\lambda x.\ x\char`\^2) + O(\lambda x.\ x)$
$x^2 + O(x) = O(x^2)$	$(\lambda x.\ x\char`\^2) + O(\lambda x.\ x) = O(\lambda x.\ x\char`\^2)$

The equality symbol should be used to denote $\in$ and $\subseteq$, however, only when the context makes this usage clear. For example, it is ambiguous as to whether the expression $O(f) = O(g)$ refers to set equality or the subset relation, that is, whether the equality symbol is the ordinary one or a translation of the equality symbol we have defined for O notation. For that reason, we will use $\in$ and $\subseteq$ when describing properties of the notation in Section 5. However, we resort to the common textbook uses of equality when we describe some particular identities in Section 6.

5   Basic Properties

In this section we indicate some of the theorems in the library that we have developed to support O notation. As will become clear in Section 6, a substantial part of reasoning with the notation involves reasoning about relationships between sets with respect to addition of sets, and addition of elements and sets. Here are some of the basic properties:

set-plus-intro	$[\![a \in C, b \in D]\!] \Rightarrow a + b \in C + D$
set-plus-intro2	$b \in C \Rightarrow a + b \in a + C$
set-plus-rearrange	$(a + C) + (b + D) = (a + b) + (C + D)$
set-plus-rearrange2	$a + (b + C) = (a + b) + C$
set-plus-rearrange3	$(a + C) + D = a + (C + D)$
set-plus-rearrange4	$C + (a + D) = a + (C + D)$
set-zero-plus	$0 + C = C$

Here a and b range of elements of a type of sort *plus-ac0*, C and D range over arbitrary sets of elements of such a type, and in an expression like $a + C$, the symbol $+$ really denotes the $+o$ operator. The bracket notation in *set-plus-intro* means that the conclusion $a + b \in C + D$ follows from the two hypotheses $a \in C$ and $b \in D$. If we use the four rearrangments above to simplify a term built up from the three types of addition, the result is a term consisting of a sum of elements on the left "added" to a sum of sets on the right. This makes it easy verify identities that enable one to rearrange terms in a calculation.

Since reasoning about expressions involving O notation essentially boils down to reasoning about inclusions between the associated sets, the following monotonicity properties are central:

set-plus-mono	$C \subseteq D \Rightarrow a + C \subseteq a + D$
set-plus-mono2	$[\![C \subseteq D, E \subseteq F]\!] \Rightarrow C + E \subseteq D + F$
set-plus-mono3	$a \in C \Rightarrow a + D \subseteq C + D$
set-plus-mono4	$a \in C \Rightarrow a + D \subseteq D + C$

These are declared to the automated reasoner. We will see in Section 6 that, in conjunction with the properties below, this provides useful support for asymptotic calculations.

Analogous lists of properties holds for set multiplication, under the assumption that the multiplication on the underlying type is commutative and associative, with an identity. Assuming the underlying type is a ring, the distributivity of multiplication over addition lifts in various ways:

set-times-plus-distrib	$a * (b + C) = a * b + a * C$
set-times-plus-distrib2	$a * (C + D) = a * C + a * D$
set-times-plus-distrib3	$(a + C) * D \subseteq a * D + C * D$

The following theorem relates ordinary subtraction to element-set addition:

set-minus-plus	$(a - b \in C) = (a \in b + C)$

Note that equality here denotes propositional equivalence.

Turning now to O notation proper, the following properties follow more or less from the basic set-theoretic properties of the definitions:

bigo-elt-subset	$f \in O(g) \Rightarrow O(f) \subseteq O(g)$
bigoset-elt-subset	$f \in O(A) \Rightarrow O(f) \subseteq O(A)$
bigoset-mono	$A \subseteq B \Rightarrow O(A) \subseteq O(B)$
bigo-refl	$f \in O(f)$
bigoset-refl	$A \subseteq O(A)$
bigo-bigo-eq	$O(O(f)) = O(f)$

Here, variables like f and g range over functions whose codomain is a nondegenerate ordered ring, and A and B range over sets of such functions.

In the definition of $O(f)$, we can assume without loss of generality that c is strictly positive. (Here we assume $0 \neq 1$.) It is easier not to have to show this when demonstrating that the definitions are met; on the other hand, the next three theorems allow us to use this fact where convenient.

bigo-pos-const	$(\exists c \, \forall x \, (|h(x)| \leq c * |f(x)|)) =$
	$\quad (\exists c \, (0 < c \wedge \forall x \, (|h(x)| \leq c * |f(x)|)))$
bigo-alt-def	$O(f) = \{h \mid \exists c \, (0 < c \wedge \forall x \, (|h(x)| \leq c * |g(x)|)\}$
bigoset-alt-def	$O(A) = \{h \mid \exists f \in A \, \exists c \, (0 < c \wedge \forall x$
	$\quad (|h(x)| \leq c * |g(x)|))\}$

The following is an alternative characterization of the O operator applied to sets.

bigoset-alt-def2	$O(A) = \{g \mid \exists f \in A \, (h \in O(f))\}$

The following properties are useful for calculations. Expressed at this level of generality, their proofs can only rely on properties, like the triangle inequality, that are true in every nondegenerate ordered ring.

bigo-plus-idemp	$O(f) + O(f) = O(f)$
bigo-plus-subset	$O(f + g) \subseteq O(f) + O(g)$
bigo-plus-subset2	$O(f + A) \subseteq O(f) + O(A)$
bigo-plus-subset3	$O(A + B) \subseteq O(A) + O(B)$
bigo-plus-subset4	$[|\forall x \, (0 \leq f(x)), \forall x \, (0 \leq g(x))|] \Rightarrow$
	$\quad O(f + g) = O(f) + O(g)$
bigo-plus-absorb	$f \in O(g) \Rightarrow f + O(g) = O(g)$
bigo-plus-absorb2	$[|f \in O(g), A \subseteq O(g)|] \Rightarrow f + A \subseteq O(g)$

To see that the subset relation cannot be replaced by equality in *bigo-plus-subset*, consider what happens when $g = -f$. For most calculations, the subset relation is sufficient; but when the relevant functions are positive, *bigo-plus-subset4* can simplify matters.

The following group of theorems is also useful for calculations.

bigo-mult	$O(f) * O(g) \subseteq O(f * g)$
bigo-mult2	$f * O(g) \subseteq O(f * g)$
bigo-minus	$f \in O(g) \Rightarrow -f \in O(g)$
bigo-minus2	$f \in g + O(h) \Rightarrow -f \in -g + O(h)$
bigo-minus3	$O(-f) = O(f)$
bigo-add-commute	$(f \in g + O(h)) = (g \in f + O(h))$

Showing that a particular function is in a particular O set is often just a matter of unwinding definitions. The following theorems offer shortcuts:

bigo-bounded	$[|\forall x\ (0 \leq f(x)), \forall x\ (f(x) \leq g(x))|] \Rightarrow f \in O(g)$
bigo-bounded2	$[|\forall x\ (g(x) \leq f(x)), \forall x\ (f(x) \leq g(x) + h(x))|] \Rightarrow$ $f \in g + O(h)$

The next two theorems only hold for ordered rings with the additional property that for every nonzero c there is a d such that $cd \geq 1$. They are therefore proved under this hypothesis, which holds for any field, as well as for any archimedian ring.

bigo-const	$c \neq 0 \Rightarrow O(\lambda x.c) = O(1)$
bigo-const-mult	$c \neq 0 \Rightarrow O((\lambda x.c) * f) = O(f)$
bigo-const-mult2	$c \neq 0 \Rightarrow (\lambda x.c) * O(f) = O(f)$

In all three cases, the $\subseteq$ direction holds more generally for ordered rings, without the requirement that $c \neq 0$.

Additional properties of O notation can be shown to hold in more specialized situations. For example, the HOL-Complex library includes a function *sumr m n f*, intended to denote $\sum_{m \leq i < n} f(i)$, where m and n are elements of $\mathbb{N}$ and f is a function from $\mathbb{N}$ to $\mathbb{R}$. Using the more perspicuous notation for sums, we have the following theorems:

bigo-sumr-pos	$[|\forall x\ (0 \leq h(x)), f \in O(h)|] \Rightarrow$ $\lambda x.\ \sum_{i<x} f(i) \in O(\lambda x.\ \sum_{i<x} h(i))$
bigo-sumr-pos2	$[|\forall x\ (0 \leq h(x)), f \in g + O(h)|] \Rightarrow$ $\lambda x.\ \sum_{i<x} f(i) \in \lambda x.\ \sum_{i<x} g(i) +$ $O(\lambda x.\ \sum_{i<x} h(i))$

6   Application: Arithmetic Functions

In this section, we will describe an initial application of our library. All of the
following facts are used in analytic number theory:

$$\ln(1 + 1/(n + 1)) = 1/(n + 1) + O(1/(n + 1)^2) \tag{1}$$

$$\sum_{i<n+1} \frac{1}{i+1} = \ln(n + 1) + O(1) \tag{2}$$

$$\sum_{i<n+1} \ln(i + 1) = (n + 1)\ln(n + 1) - n + O(\ln(n + 1)) \tag{3}$$

$$\sum_{i<n} \frac{\ln(i + 1)}{i + 1} = \frac{1}{2}\ln^2(n + 1) + O(1) \tag{4}$$

$$\ln^2(n + 2) - \ln^2(n + 1) = 2\frac{\ln(n + 1)}{n + 1} + O\left(\frac{\ln(n + 1) + 1}{(n + 1)^2}\right) \tag{5}$$

$$\sum_{i<n+1} \ln^2(i + 1) = \frac{(n + 1)\ln^2(n + 1) - 2(n + 1)\ln(n + 1)+}{2n + O(\ln^2(n + 1))} \tag{6}$$

$$\sum_{i<n+1} \ln^2(\frac{n + 1}{i + 1}) = 2n + O(\ln^2(n + 1)) \tag{7}$$

In these identities, the relevant terms are to be viewed as functions $f(n)$ from the
natural numbers $\mathbb{N}$ to the real numbers $\mathbb{R}$, and all the sums range over natural
numbers. Keep in mind that here "equality" really means "element of"!
    A more natural formulation of the first identity, for example, would be

$$\ln(1 + 1/n) = 1/n + O(1/n^2) \quad \text{for } n \geq 1,$$

and a more natural formulation of the second identity would be

$$\sum_{i=1}^{n} 1/i = \ln(n) + O(1) \quad \text{for } n \geq 1.$$

We have found it convenient to replace n by $n + 1$ instead of dealing with the
side condition or using the type of positive natural numbers instead.
    All of the identities above have been formalized in Isabelle. Our formaliza-
tions of the first three look as follows:

```
(λn::nat. ln (1 + 1 / (real n + 1))) = (λn. 1 / (real n + 1)) +
   O(λn. 1 / ((real n) + 1)^2)

(λn. sumr 0 (n+1) (λi. 1/(real i + 1))) = (λn. ln(real n + 1)) + O(λn. 1)

(λn. sumr 0 (n+1) (λi. ln(real i + 1))) =
   ((λn. (real n + 1) * ln(real n + 1)) - (λn. real n)) +
   O(λn. ln (real n + 1))
```

These are reasonable approximations to the usual mathematical notation, but for readability, we will use the latter below.

In an ordinary mathematical development, the first identity has to be proved, somehow, from the definition of ln. (It can be seen immediately, for example, from the Taylor series expansion of $\ln(1+x)$ at $x = 0$.) The others are typically derived from this using direct calculation, as well as, sometimes, basic properties of integration; see, for example, [5,7]. For example, the second identity reflects the fact that $\ln x$ is equal to $\int_1^x (1/y)dy$ and the third reflects the identity $\int \ln x = x \ln x + C$, which may be obtained using integration by parts.

Since Isabelle's current theory of integration was not always robust enough to handle the standard arguments, we had to replace these arguments by more direct proofs. This was an illuminating exercising in "unwinding" the methods of calculus. Thus, (5) above was useful in our low-tech proofs of the (4) and (6). Even *with* calculus, however, a good deal of ordinary asymptotic calculations are involved. To illustrate the use of our library, we will show how the theorems described in Section 5 are used to obtain (6) from (2), (3), (4), and (5).

To prove (6), first note that it suffices to establish the slightly weaker identity with $O(\ln^2(n+1)+1)$ in place of $O(\ln^2(n+1))$. The stronger version can then be obtained by unwinding definitions, using the fact that the terms on each side of the equation are equal when $n = 0$, and $\ln^2(n+1) \geq \ln^2 2$ when $n > 0$.

First of all, using the technique of partial summation [5,7], we have

$$\sum_{i<n+1} \ln^2(i+1) = (n+1)\ln^2(n+1) - \sum_{i<n}(i+1)(\ln^2(i+2) - \ln^2(i+1)). \quad (8)$$

This identity can be verified directly using induction on n. To esimate the second term on the right side, we start by multipling (5) through by $n+1$, and using *set-times-plus-distrib* and *bigo-mult2* to obtain

$$(n+1)(\ln^2(n+2) - \ln^2(n+1)) = 2\ln(n+1) + O\left(\frac{\ln(n+1)+1}{n+1}\right).$$

Using *bigo-sumr-pos2*, we obtain

$$\sum_{i<n}(i+1)(\ln^2(i+2) - \ln^2(i+1)) = 2\sum_{i<n}\ln(i+1) + O\left(\sum_{i<n}\frac{\ln(i+1)+1}{i+1}\right)$$

$$= 2\sum_{i<n}\ln(i+1) + O\left(\sum_{i<n}\frac{\ln(i+1)}{i+1}\right) + O\left(\sum_{i<n}\frac{1}{i+1}\right) \quad (9)$$

Here, the second "equality" is really set inclusion, and is obtained using properties of sums, *bigo-plus-subset*, and *set-plus-mono*.

Now, let us estimate each of the three terms on the right-hand side, keeping in mind that we only care about equality up to $O(\ln^2(n+1)+1)$. Considering the first term, we have

$$\sum_{i<n} \ln(i+1) = \sum_{i<n+1} \ln(i+1) - \ln(n+1)$$
$$= -\ln(n+1) + ((n+1)\ln(n+1) - n + O(\ln(n+1)))$$
$$= ((n+1)\ln(n+1) - n) + (-\ln(n+1) + O(\ln(n+1)))$$
$$= (n+1)\ln(n+1) - n + O(\ln^2(n+1) + 1).$$

The second equality is obtained by applying *set-plus-intro2* to identity (3). In the third equality, we use *set-plus-rearrange2* and ordinary properties of real addition to group the terms appropriately. The last equality uses the fact that $\ln(n+1)$ is in $O(\ln^2(n+1)+1)$, and so, by *bigo-minus-eq*, $-\ln(n+1)$ is in $O(\ln^2(n+1)+1)$ as well; using *bigo-elt-subset* and *bigo-plus-absorb*, the second parenthetical expression is a subset of $O(\ln^2(n+1)+1)$, and the equality follows from *set-plus-mono*. Calculations like these work well with Isabelle/Isar's support for calculational reasoning [10], since intermediate identities can often be verified automatically. For example, the second-to-last equality above is verified by simplifying terms; the last equality is also confirmed by the automated reasoner, given only the fact that $\ln(n+1)$ is in $O(\ln^2(n+1)+1)$. Mutiplying through by 2, and using *set-times-plus-distrib* and *bigo-const-mult2*, we obtain

$$2\sum_{i<n} \ln(i+1) = (2(n+1)\ln(n+1) - 2n) + O(\ln^2(n+1) + 1).$$

Turning to the second term, we have

$$O\left(\sum_{i<n} \frac{\ln(i+1)}{i+1}\right) = O\big(\ln^2(n+1)/2 + O(1)\big)$$
$$= O(\ln^2(n+1)/2) + O(1)$$
$$= O(\ln^2(n+1) + 1) + O(\ln^2(n+1) + 1)$$
$$= O(\ln^2(n+1) + 1).$$

The first equality is obtained by applying *bigo-elt-subset* to identity (4). The second equality uses *bigo-plus-subset3* and *bigo-bigo-eq*. Finally, we use *bigo-elt-subset*, monotonicity properties of set addition, and *bigo-plus-idemp*, as before, to simplify the resulting O set. (Alternatively, one can use (4) to show $\sum_{i<n} \ln(i+1)/(i+1) = O(\ln^2(n+1)+1)$, and then use *bigo-elt-subset*.)

Turning to the third term, we have

$$O\left(\sum_{i<n} \frac{1}{i+1}\right) = O\left(\sum_{i<n+1} \frac{1}{i+1} + (-1/(n+1))\right)$$
$$= O\big(\ln(n+1) + O(1)\big) + O(1/(n+1))$$
$$= O(\ln(n+1)) + O(1) + O(1)$$
$$= O(\ln^2(n+1) + 1)$$

using reasoning similar to that above.

Returning to equation (9), grouping terms appropriately, and using mononocity of addition, we obtain

$$\sum_{i<n}(i+1)(\ln^2(i+2)-\ln^2(i+1)) = 2(n+1)\ln(n+1) - 2n + O(\ln^2(n+1)+1).$$

Using *bigo-minus2* to negate both sides, substituting the result into (8), and applying *set-plus-intro2*, we have

$$\sum_{i<n+1}\ln^2(i+1) = (n+1)\ln^2(n+1) - 2(n+1)\ln(n+1) + 2n + O(\ln^2(n+1)+1),$$

as required.

7   Future Work

The library described here is being used to obtain a fully formalized proof of the prime number theorem, which states that the the the number of primes $\pi(x)$ less than or equal to x is asymptotic to $x/\ln x$. As the development of this proof proceeds, we intend to improve our library and the interactions with Isabelle's automated methods, as well as the interactions with Isar's calculational reasoning. This will ensure that calculations using O notation can be carried out smoothly and naturally. Independently, we still need to formalize the variations on O notation described in Section 2, as well as other asymptotic notions.

A similar treatment of asymptotic notions has been given in Mizar [3,4] under the "eventually" reading of O notation given in Section 2. This implementation includes Θ and Ω notation as well. But the treatment is less general than the one described here, in that the notions apply only to eventually nonnegative sequences of real numbers.

It seems appropriate to mention here a perennial problem with simple type theory. Our treatment of O notation applies to any type of functions, that is, any collection of functions whose domain and codomain are also types. Since Isabelle's **typedef** command allows us to turn the real interval $(0,1)$, say, into a type, we can use O notation directly for functions defined on this interval. *But simple type theory does not allow one to define types that depend on a parameter*, so, for example, one cannot prove theorems involving O notation for functions defined more generally on a real interval (x,y), where x and y are variables. To do so, one has three options:

1. Work around the problem, for example, using the "on S" variant of O notation described in Section 2.
2. Replace simple type theory with a formalism that allows dependent types. For example, Coq [11] is based on the Coquand-Huet *calculus of constructions*.
3. Use locales instead of types to fix the subset and parameters that are of interest. (See [2].)

Each option has its drawbacks. The first can be notationally cumbersome; the second involves a more elaborate type theory, with more complex type-theoretic behavior; the third involves reducing one's reliance on the underlying type theory, but giving up the associated notational and computational advantages. We do not yet have a sense what approach, in the long term, is best suited to developing complex mathematical theories.

References

1. Ronald L. Graham, Donald E. Knuth, and Oren Patashnik. *Concrete mathematics: a foundation for computer science*. Addison-Wesley Publishing Company, Reading, MA, second edition, 1994.
2. F. Kammüller, M. Wenzel, and L. C. Paulson. Locales – a sectioning concept for Isabelle. In Y. Bertot, G. Dowek, A. Hirschowitz, C. Paulin, and L. Théry, editors, *Proceeding of Theorem Proving in Higher Order Logics, 12th International Conference, TPHOLs'99, Nice, France, September 14 - 17*, volume 1690 of *LNCS*, 1999.
3. Richard Krueger, Piotr Rudnicki, and Paul Shelley. Asymptotic notation. Part I: theory. Journal of Formalized Mathematics, Volume 11, 1999.
   http://mizar.org/JFM/Vol11/asympt_0.html.
4. Richard Krueger, Piotr Rudnicki, and Paul Shelley. Asymptotic notation. Part II: examples and problems. Journal of Formalized Mathematics, Volume 11, 1999.
   http://mizar.org/JFM/Vol11/asympt_1.html.
5. Melvyn B. Nathanson. *Elementary Methods in Number Theory*. Springer, New York, 2000.
6. Tobias Nipkow, Lawrence C. Paulson, and Markus Wenzel. *Isabelle/HOL. A proof assistant for higher-order logic*, volume 2283 of *Lecture Notes in Computer Science*. Springer Verlag, Berlin, 2002.
7. Harold N. Shapiro. *Introduction to the theory of numbers*. Pure and Applied Mathematics. John Wiley & Sons Inc., New York, 1983. A Wiley-Interscience Publication.
8. Markus Wenzel. Using axiomatic type classes in Isabelle, 1995.
   http://www.cl.cam.ac.uk/Research/HVG/Isabelle/docs.html.
9. Type classes and overloading in higher-order logic. In E. Gunther and A. Felty, editors, *Proceedings of the 10th international conference on thoerem provings in higher-order logic (TPHOLs '97)*, pages 307-322, Murray Hill, New Jersey, 1997.
10. Markus Wenzel and Gertrud Bauer. Calculational reasoning revisited (an Isabelle/Isar experience). In R. J. Boulton and P. B. Jackson, editors, *Theorem Proving in Higher Order Logics: 14th International Conference, TPHOLs 2001, Edinburgh, Scotland, UK, September 3-6, 2001, Proceedings*, volume 2152 of *Lecture Notes in Computer Science*, pages 75-90, 2001.
11. The Coq proof assistant. Developed by the LogiCal project.
    http://pauillac.inria.fr/coq/coq-eng.html.
12. The Isabelle theorem proving environment. Developed by Larry Paulson at Cambridge University and Tobias Nipkow at TU Munich.
    http://www.cl.cam.ac.uk/Research/HVG/Isabelle/index.html.

Experiments on Supporting Interactive Proof Using Resolution

Jia Meng and Lawrence C. Paulson

Computer Laboratory, University of Cambridge
15 JJ Thomson Avenue, Cambridge CB3 0FD (UK)
{jm318,lp15}@cam.ac.uk

Abstract. Interactive theorem provers can model complex systems, but require much effort to prove theorems. Resolution theorem provers are automatic and powerful, but they are designed to be used for very different applications. This paper reports a series of experiments designed to determine whether resolution can support interactive proof as it is currently done. In particular, we present a sound and practical encoding in first-order logic of Isabelle's type classes.

1 Introduction

Interactive proof tools such as HOL [4], Isabelle [8] and PVS [10] have been highly successful. They have been used for verifying hardware, software, protocols, and so forth. Unfortunately, interactive proof requires much effort from a skilled user. Many other tools are completely automatic, but they cannot be used to verify large systems. Can we use automatic tools to improve the automation of interactive provers?

In this paper, we report a year's experiments aimed at assessing whether resolution theorem provers can assist interactive provers. For such an integration to be effective, we must bridge the many differences between a typical interactive theorem prover and a resolution theorem prover:

- higher-order logic versus first-order logic
- polymorphically typed (or other complicated type system) versus untyped
- natural deduction or sequent formalism versus clause form

Particularly difficult is the problem of coping with large numbers of previously proved lemmas. In interactive proof, the user typically proceeds by proving hundreds or thousands of lemmas that support later parts of the verification. Ideally, the user should not have to select the relevant lemmas manually, but too many irrelevant facts may overwhelm the automatic prover.

A number of other researchers have attempted to combine interactive and automatic theorem provers. The HOL system has for many years included a model elimination theorem prover, which recently Hurd has attempted to improve upon [6]. The Coq system has also been interfaced with an automatic first-order prover [2]. These approaches expect the user to pick relevant lemmas

D. Basin and M. Rusinowitch (Eds.): IJCAR 2004, LNAI 3097, pp. 372–384, 2004.

manually. Another attempt, using the KIV system [1], includes an automatic mechanism for discarding irrelevant lemmas. This approach has attractions, and the overall performance might be improved by using a more powerful automatic prover.

Closest to our conception is the Ωmega system [18]. One key idea we share is that assistants should run as background processes. The interactive user should not have to notice that a certain tool may prove a certain subgoal: it should be attempted automatically. However, there are important differences. Ωmega is designed to support working mathematicians, and it has been combined with a large number of other reasoning tools. Our aim is to support formal verification, and we are trying to achieve the best possible integration with one or two other reasoning tools. Creative mathematics and verification are different applications: the mathematician's main concern is to arrive at the right definitions, while the verifier's main concern is to cope with fixed but enormous definitions.

Our work is by no means complete. Our experiments are designed to identify the main obstacles to an effective integration between an interactive and automatic prover. We have taken typical problems that can be solved in Isabelle, either using Isabelle's automatic tools or by short sequences of proof commands. We have converted these problems into clause form and supplied them to resolution provers (Vampire or SPASS). Crucially, we have given the resolution prover a large set of axioms, corresponding to previous lemmas that would by default be available to Isabelle's own automatic tools. One of our findings is that even the best resolution provers sometimes founder when given large sets of irrelevant axioms. We have been able to add several hard problems to the TPTP library [19].[1] We hope that the engineers of resolution provers will make progress on the problem of relevance, and we have already had excellent cooperation from the Vampire team. We have also developed a way of modelling Isabelle's type class system in first-order logic.

Paper outline. We present background information on Isabelle and Vampire (§2). We briefly describe earlier experiments involving an untyped formalism, namely set theory (§3). We then describe new experiments involving a polymorphically typed formalism, namely higher-order logic (§4). We finally offer some conclusions (§5).

2   Background: Isabelle and Vampire

Isabelle [8] is an interactive theorem prover. Unusually, Isabelle is generic: it supports a multiplicity of logics. The most important of these is higher-order logic, which is also the basis of the HOL system and PVS. Isabelle also supports Zermelo-Fraenkel set theory [14], which is an untyped formalism based upon first-order logic.

Isabelle provides substantial automation. Its reasoning tactics include the following:

---

[1] SET787-1.p, SET787-2.p, COL088-1.p to COL0100-2.p

- *simp* is the simplifier, which performs conditional rewriting augmented by other code, including a decision procedure for linear arithmetic.
- *blast* is a sort of generic tableaux theorem prover. It performs forward and backwards chaining using any lemmas supplied by the user [13].
- *auto* is a naive combination of the previous two tactics. It interleaves rewriting and chaining. However, this treatment of equality is primitive compared with that provided by a good resolution prover.

Many other Isabelle tactics are variants of those described above. Although these tactics are powerful, the user has to choose which one is appropriate and invoke it manually. A resolution prover can perform all the types of reasoning done by these tactics. Can a resolution prover, running in the background, replace all the calls to *simp*, *blast* and *auto*?

Isabelle's tactics have one major advantage: they let the user declare lemmas for them to use, and they easily cope with hundreds of such lemmas. After proving an equality, the user can declare it as a *simplification rule*. If a lemma looks suitable for forwards chaining, the user can declare it as such, and similarly for backwards chaining. The accumulation of many such declarations greatly improves the automation available to the user.

Certain other lemmas are not declared permanently, but supplied when needed for particular proofs. For example, distributive laws like $x*(y+z) = x*y + x*z$ should not usually be declared as simplification rules because they can cause an exponential blow up. Such lemmas can be named explicitly in an invocation of *simp*. Similarly, a transitivity law like $x < y \implies y < z \implies x < z$ should not usually be declared as suitable for backward chaining because it produces an explosion of subgoals. It can be mentioned in a call to *blast*.

If a resolution prover is to replace the role played by *simp*, *blast* and *auto*, then it must be able to do something appropriate with such lemma declarations. Equalities can be recognized by their syntactic form. The Vampire developers have kindly extended Vampire: version 6.03 allows annotations on literals to specify that a clause should be used for forward or backward chaining. This extension improves Vampire's performance on our examples. The main problem, which appears to affect most resolution provers, is that the sheer number of default lemmas makes the search space explode.

It is not enough to prove the theorems: we must also convince Isabelle that they have been proved. Isabelle, like HOL, minimises the risk of soundness bugs by allowing only a small kernel of the system to assert theorems. The translations between Isabelle and the resolution provers could easily contain errors, even if the provers themselves are sound. Therefore, we would like possible to translate the resolution proof back into a native one, as Hurd has already done between Gandalf and HOL [5]. This could be done by implementing Isabelle versions of the rules used in automatic proofs, such as resolution and paramodulation. Isabelle could then re-play the proof found by the automatic tool. The translated proof could also be stored, allowing future executions of the proof script to run without the automatic tool. Another use of the resolution proof is to identify the relevant Isabelle lemmas; that information might be valuable to users.

Vampire [16] and SPASS [20] are the provers we have used for our experiments. These are leading resolution provers that have done well in recent CADE ATP System Competitions. Our objective to support integration with any resolution prover that outputs explicit proofs. We convert Isabelle's formalisms into untyped first-order logic rather than expecting the resolution prover to support them.

3   Formalising Untyped Isabelle/ZF in FOL

As Meng has reported elsewhere [7], our first experiments concerned translating Isabelle/ZF formulas into first-order logic (FOL) in order to examine whether the use of resolution was practial. These experiments consisted of taking existing Isabelle proof steps performed by `simp`, `blast`, `auto`, etc., trying to reproduce them using Vampire. All simplification rules and backward/forward chaining rules of the current Isabelle context were translated to clause form. The goals were negated, converted to clauses and finally sent to Vampire.

3.1   Translation Issues

We translate those Isabelle/ZF formulas that are already in FOL form into conjunctive normal form (CNF) directly. However, there are some ZF formulas that lie outside first-order logic. We need to reformulate them into FOL form before CNF transformation. Some of these issues are particular to Isabelle alone. For example, set intersection satisfies the equality

$$(c \in A \cap B) = (c \in A \wedge c \in B)$$

An equation between Boolean terms is obviously not first-order. Moreover, Isabelle represents the left-to-right implication in a peculiar fashion related to its encoding of the sequent calculus [11]. Our translation has to recognize this encoding and translate it to the corresponding implication, which in this case is

$$\forall c\, A\, B\, [c \in A \cap B \to (c \in A \wedge c \in B)]$$

We also need to remove ZF terms, such as $\bigcup_{x \in A} B(x)$, since they are not present in first-order logic. Let $\phi(Z)$ be a formula containing a free occurrence of some term Z, and let $\phi(v)$ be the result of replacing Z by the variable v. Then $\phi(Z)$ is equivalent (in ZF) to

$$\exists v\, [\phi(v) \wedge \forall u\, [u \in v \leftrightarrow u \in Z]]$$

This transformation allows any occurrence of the term Z to be forced into a context of the form $u \in Z$. In this case, Z is $\bigcup_{x \in A} B(x)$, and translations can then replace $u \in \bigcup_{x \in A} B(x)$ by $\exists x\, [x \in A \wedge u \in B(x)]$.

Some other translation issues also arose during our experiments. The subset relation $A \subseteq B$ is equivalent to $\forall x\, (x \in A \to x \in B)$: this reduces the subset relation to the membership relation. Experiments [7] showed that Vampire can find a proof much more quickly if the subset relation is replaced by its equivalent membership relation. This is probably because during most of the complex proofs, subset relations have to be reduced to equivalent membership relations anyway.

3.2 Experimental Results

An aim of these experiments was to find out whether the Isabelle/ZF-Vampire integration can prove goals that were proved by Isabelle's built-in tools such as `simp`, `blast` and `auto`. (Obviously, our ultimate goal is to supersede these tools, not merely to equal them.) We used three theory files:

- `equalities.thy`: proofs of many simple set equalities
- `Comb.thy`: a development of combinatory logic similar to the Isabelle/HOL version described by Paulson [11]
- `PropLog.thy`: a development of propositional logic

Thirty-seven lemmas (63 separate goals) were taken from Isabelle/ZF's theory files. The resolution prover could prove 52 goals out of 63 in the presence of a large set of axioms: 129 to 160 axiom clauses.

We also tried to examine if this integration can prove goals that cannot be proved by Isabelle's built-in tools and hence reduce user interaction. These goals were originally proved by a series of proof steps. Fifteen lemmas from `Comb.thy` and `PropLog.thy` were examined. Vampire proved ten lemmas.

Meng [7] gives a more detailed presentation of the experimental results.

3.3 Performance on Large Axiom Sets

An issue that arose from our ZF experiments is that many problems could only be proved for a minimal set of axioms and not with the full set of default axioms. Recall that one of our objectives is to preserve Isabelle's policy of not usually requiring the user to identify which previous lemmas should be used.

We took fifteen problems that seemed difficult in the presence of the full axiom set. We offered them to Geoff Sutcliffe for inclusion in the TPTP Library [19]. He kindly ran experiments using three provers (E, SPASS and Vampire) together with a tool he was developing for the very purpose of eliminating redundant axioms. Gernot Stenz ran the same problems on E-SETHEO, because that system is not available for downloading. Finally, Meng herself attempted the problems using both Vampire and SPASS. Thus, we made $15 \times 6 = 90$ trials altogether. These trials were not uniform, as they involved different hardware and different resource limits, but they are still illustrative of our difficulty.

Of the fifteen problems, only five could be proved. Two of them were proved in under two seconds by E-SETHEO. Sutcliffe proved the same two using SPASS, as well as a third problem; these took 43, 75 and 154 seconds, respectively. Meng was able to prove two other problems using a new version of Vampire (6.03, which supports literal annotations; version 5.6 could not prove them). Thus, only seven of the ninety proof attempts succeeded.

Unfortunately, the hardest problems arose from proofs using the technique of *rule inversion*, which is important for reasoning about operational semantics. Rule inversion is a form of case analysis that involves identifying which of the many rules of an operational semantics definition may have caused a given event. Isabelle's `blast` method handles such proofs easily, but converting the case analysis rule to clause form yields an explosion: 135 clauses in one simple case. We have been able to reduce this number by various means, but the proofs remain difficult.

4   Formalising Typed Isabelle/HOL in FOL

Our previous experiments were based on Isabelle/ZF in order to avoid the complications of types. However, few Isabelle users use ZF. If our integration is to be useful, it must support higher-order logic, which in turn requires a sound modelling of the intricacies of Isabelle's type system.

4.1   Types and Sorts in Isabelle/HOL

Isabelle/HOL [8] supports *axiomatic type classes* [21]. A *type class* is a set of types for which certain operations are defined. An *axiomatic* type class has a set of axioms that must be satisfied by its instances: types belonging to that class. If a type τ belongs to a class C then it is written as $\tau :: C$.

A type class C can be a subclass of another type class D, if all axioms of D can be proved in C. If a type τ is an instance of C then it is an instance of D as well. Furthermore, a type class may have more than one direct superclass. If C is a subclass of both D_1 and D_2 then C is subset of intersection of D_1 and D_2. The intersection of type classes is called a *sort*. For example, Isabelle/HOL defines the type class of linear orders (`linorder`) to be a subclass of the type class of partial orders (`order`).

```
axclass linorder < order
   linorder_linear: "x≤ y ∨ y≤ x"
```

Now, to assert that type `real` is an instance of class `linorder`, we must show that the corresponding instance of the axiom `linorder_linear` holds for that type.

```
instance real :: linorder
   proof ... qed
```

Axiomatic type classes allows meaningful overloading of both operators and theorems about those operators. We can prove theorems for class `linorder`, and they will hold for all types in that class, including types defined in future Isabelle sessions. We can prove a theorem such as $(-a) \times (-b) = a \times b$ in type class `ring` and declare it as a simplification rule, where it will affect numeric types such as `int`, `rat`, `real` and `complex`. Type checking remains decidable, for it is the user's responsibility to notice that a type belongs to a certain class and to declare it as such, providing the necessary proofs. Of course, full type checking must be performed, including checking of sorts, since a theorem about linear orderings cannot be assumed to hold for arbitrary orderings.

FOL automatic provers usually do not support types. However, when a HOL formula is translated to FOL clauses, this type and class information should be kept. Not only is the type information essential for soundness, but it can also reduce the search space significantly.

We can formalise axiomatic type classes in terms of FOL predicates, with types as FOL terms. Each type class corresponds to a unary predicate. If a type τ is an instance of a class C, then $C(\tau)$ will be true. Therefore, the subclass relation can be expressed by predicate implications. Taking the class `linorder` as an example, we have the formula

$$\forall \tau \, [linorder(\tau) \rightarrow order(\tau)]$$

Sort information for multiple inheritance can be handled similarly. τ :: $C_1, \ldots, C_n$ can be expressed as $C_1(\tau) \wedge \ldots \wedge C_n(\tau)$.

A further complication of type classes concerns type constructors such as `list`. A list can be linearly ordered (by the usual lexicographic ordering) if we have a linear ordering of the list element types. This statement can be formalised in Isabelle as

```
instance list :: (linorder) linorder
  proof ... qed
```

We represent this in FOL by $\forall \tau \, [linorder(\tau) \rightarrow linorder(list(\tau))]$.

4.2   Polymorphic Operators Other than Equality

Many predicates and functions in Isabelle/HOL are polymorphic. In our type class example, the relation $\leq$ is polymorphic. Its type is $\alpha \rightarrow \alpha \rightarrow bool$, where α is a type variable of class `ord`. The effect is to allow $\leq$ to be applied only if its arguments belong to type class `ord`. This polymophic type can be specialised when $\leq$ is applied to different arguments. When applied to sets, its type will be $\alpha \, set \rightarrow \alpha \, set \rightarrow bool$, whereas for natural numbers its type will be $nat \rightarrow nat \rightarrow bool$.

To formalise this type information of operators, for each operator (except constants and equality), we include its type as an additional argument. Constants do not need to carry type information: their types should be inferred automatically.

For example, axiom `linorder_linear` will be translated to

$$\forall \tau \, x \, y \, [linorder(\tau) \rightarrow (le(F(\tau, F(\tau, bool)), x, y) \vee le(F(\tau, F(\tau, bool)), y, x))]$$

where le is the predicate $\leq$ in prefix form and F is the function type constructor.

4.3 Types for Equalities

In Isabelle/HOL, equality is polymophic. When we prove the equality $A = B$, we may have to use different inference rules depending on the type of A and B. However, Vampire's built-in equality `equal` does not allow us to include the type of A and B as an extra argument.

There could be several ways to solve the problem. We could define a new equality predicate taking three arguments instead of two. However, automatic provers usually treat equality specially, giving much better performance than could be achieved with any user defined equality literal. Experiments showed that Vampire needed excessive time to find proofs involving this new equality predicate.

Therefore, we decided to use the built-in equality predicate while embedding the type information in its arguments. Instead of writing the formula $equal(A, B)$, we can include the type as

$$equal(typeinfo(A, \tau), typeinfo(B, \tau))$$

where τ is the type of A and B. The value of τ will determine which inference rule should be used to prove an equality. Equalities in previously proved lemmas and conjectures will be translated into $equal$ in this format with all type information included. In addition, the axiom

$$equal(typeinfo(A, \tau), typeinfo(B, \tau)) \rightarrow equal(A, B)$$

is sometimes required for paramodulation. Its effect is to strip the types away. In experiments, this type embedding gave a reasonable performance.

Equalities between boolean-valued terms are simply viewed as two implications in the transformation to clause form.

4.4 Vampire Settings

In Isabelle, rules usually have information indicating whether they should be used for forward chaining (elimination rules) or backward chaining (introduction rules). We would like this information to be preserved after rules are translated into clauses. We attempted to accomplish this by assigning weights and precedences to functions and predicates to indicate which literals should be eliminated sooner; this information gives an ordering on literals, which Vampire computes using the Knuth-Bendix Ordering (KBO) [15]. However, since the resulting KBO is a partial ordering on terms with variables, it does not match our requirements exactly.

The Vampire developers gave us a new version of Vampire (v6.03), with syntax to specify which literal in a clause should be selected for resolution. This syntax is an extension of TPTP syntax. Any positive literal that should be resolved first will be tagged with +++, and similarly a negative literal should be tagged with ---. Many lemmas could only be proved with this facility, which supports Isabelle's notions of forward and backward chaining.

Furthermore, Vampire provides many settings that can be specified by users to fine tune its performance according to different types of problems. We carried out many experiments in order to find the best combination of these settings. The experimental results [7] showed that six combinations of settings worked well under various circumstances. They were written to six setting files so that we can run six processes in parallel if necessary. We intend to experiment with different strategies for making the best use of the available processors. For example, if we have only one processor, then we may adopt a time-sharing mechanism, assigning suitable priorities and time limits to processes.

One particular issue is Set-of-Support (SOS). This strategy, which according to Chang and Lee [3] dates from 1965, forbids resolution steps that take all their clauses from a part of the clause set known to be satisfiable. This strategy seems ideal for problems such as ours that have a large axiom set, since it prevents deductions purely involving the axioms. However, it is incomplete in the presence of modern ordering heuristics and therefore is normally switched off. We have found that some lemmas can be proved only if SOS is on.

4.5 Experimental Results

We used the same general approach as we did in our earlier experiments on untyped ZF formulas. Isabelle/HOL lemmas were chosen, each of them usually presenting more than one goal to Vampire. The combination of the six setting files was used. The time limit for each proof attempt was 60sec. We used *formula renaming* [9] before the CNF transformation in order to minimize the number of clauses.

For typed Isabelle/HOL formulas, the inclusion of type information also helps to cut down the search space significantly. For instance if we want to prove subset relation between two sets: $X \leq Y$, its typed formula will be

$$le(F(set(\tau), F(set(\tau), bool)), X, Y)$$

Clearly inference rules such as

$$le(F(nat, F(nat, bool)), A, B)$$

will be ignored as the types of le will not match. Some experiments have been carried out to demonstrate the benefit of using such type information.

The primary aim of these experiments was to investigate whether this encoding of type information is practical for resolution proofs. This set of experiments took 56 lemmas (108 goals) from Isabelle/HOL theory files and tried to reproduce the proofs. We used the following theories:

- Multiset.thy: a development of multisets
- Comb.thy: combinatory logic formalized in higher-order logic
- List_Prefix.thy: a prefixing relation on lists
- Message.thy: a theory of messages for security protocol verification [12]

Table 1. Number of Goals Proved for Typed Lemmas

Theory	Number of Lemmas	Number of Goals	Number of Goals Proved
Multiset	3	3	3
Comb	18	29	24
List_Prefix	7	8	8
Message	28	68	62

Around 70 to 130 axiom clauses were used. Ninety-seven goals were proved using this typed formalism, as shown in Table 1.

Eleven lemmas from `Message.thy` cannot be proved by Isabelle's classical reasoners directly. Either they consist of more than one proof command or they explicitly indicate how rules should be used. Vampire proved seven of these lemmas and three more once some irrelevant axioms were removed. Only one lemma could not be proved at all, and we came close: only one of its seven subgoals could not be proved. Although this is a small sample, it suggests that Vampire indeed surpasses Isabelle's built-in tools in many situations.

Further experiments on those failed proof attempts were carried out. Among the six failed proof attempts on goals from `Message.thy`, three were made provable by removing some irrelevant axiom clauses.

Moreover, during the experiments with Isabelle/HOL's `Comb.thy` we also tried to translate HOL conjectures into FOL clauses without the inclusion of type information in order to compare the performance between the typed proofs and untyped proofs. These untyped experiments took the same set of goals (29 goals) from `Comb.thy`. Among these 29 goals, there were three Isabelle goals where Vampire found proofs more quickly when given untyped input and four goals where Vampire proved faster when given typed inputs. In particular, there was a case where Vampire proved a lot faster when given typed input (0.4 sec compared with 36 sec for untyped input). However, for those goals where untyped input required less time to be proved, the difference in time taken for typed and untyped input was not very significant. When typed input gave better performance, it should be explained by the restriction of search space. For the cases where untyped input outperformed typed input, it could be caused by the large literals in typed input due to the inclusion of types. Large literals may slow down proof search to some extent.

We also performed more specific experiments in order to examine whether the use of type information on overloaded operators can indeed cut down the search space. These experiments involved proving lemmas about subset relations taken from Isabelle/HOL's theory file `Set.thy`. In addition to the relevant axioms about subset properties, many irrelevant axioms about natural numbers were also included in the axiom set as they share the overloaded operator $\leq$. We first ran the experiments by not including the axioms of natural numbers and then ran the experiments again while adding those natural number axioms. Vampire spent the same amount of time in proofs regardless whether the natural number axioms were added or not. Clearly this demonstrates the benefits of including type information of overloaded operators, since without these information, Vampire may pick up irrelevant natural number axioms in the proof search and hence would slow down the proof procedure.

Furthermore, several experiments on the formalisation of polymorphic equalities have been carried out. Among the goals from `Message.thy`, eight goals required the use of equality clauses (on sets), in either lemmas or conjectures. Five of them were proved by switching `SOS` off and were not proved otherwise. One of them was proved by turning `SOS` off and removing some irrelevant axioms. The other two could only be proved if the equality literals were replaced by two directional subset relations.

There were also some goals from `Multiset.thy` and `Message.thy` that were proved by applying results from several axiomatic type classes. This finding suggests that our formalisation of types and sorts is practical, while preventing the application of lemmas when the type in question does not belong to the necessary type class.

We have noticed that more proofs are found for Isabelle/HOL's lemmas than for Isabelle/ZF's lemmas. We believe that type information deserves the credit for this improvement: it reduces the search space.

5 Conclusion and Future Work

This paper has described the translation between Isabelle/HOL and first-order logic, with particular emphasis on the treatment of types. It has also reviewed some issues arising from our previous work involving Isabelle/ZF (an untyped formalism) and FOL.

The ZF transformation does not require an encoding of types: everything is represented by sets. Although it is based on FOL, there are many terms that are outside the scope of FOL, such as λ-terms, which need to be translated into FOL formulas.

Typed HOL formulas involve type and sort information, which must be preserved when translated into FOL. This paper has outlined the encoding of types and sorts in FOL. The use of such information should help automatic provers eliminate irrelevant axioms from consideration during a proof search. Moreover, it is essential in order to ensure soundness of proofs.

Experimental results for both ZF and HOL demonstrated the potential benefit of our integration. In particular, the results showed the encoding of HOL types and sorts was useful and practical. More experiments on compacting the encoding could improve Vampire's proof search speed as smaller literals should require less time for proofs.

There are some general issues that apply to both ZF and HOL. The treatment of equality is an example. Lemmas involving equality (typed or untyped) seem harder to prove and usually require us to turn off `SOS`. We tried to replace set equality by two subset relations before translating them into clauses, replacing $A = B$ by the conjunction of $A \subseteq B$ and $B \subseteq A$. We found that Vampire proved theorems much faster as a result. However, this approach is inflexible: proving a pair of set inclusions is but one way of proving an equation. Therefore it would be harder to decide how we should replace equalities. Further investigation on this should be useful.

Our experimental results also gave us a hint on the proof strategy. Vampire's default settings are usually good, but sometimes proofs can only be found using other settings. We can either run several processes in parallel or run with the default settings first.

One aim of this integration is to relieve users of the task of identifying which of their previously proved lemmas are relevant. Our results show that existing proof procedures still do not cope well with large numbers of irrelevant axioms. More research on this issue is essential.

Soundness of the translation between different formalisms (for example HOL and FOL) will be ensured by various means. We intend to perform most of the translation inside Isabelle's logic rather than using raw code. We are modelling the types and sorts of HOL in full detail. We intend to execute the proofs found by the resolution prover within Isabelle.

Acknowledgements. We are grateful to the Vampire team (Alexandre Riazanov and Andrei Voronkov) for their co-operation and to Gernot Stenz and Geoff Sutcliffe for running some of our problems on their reasoning systems. Research was funded by the EPSRC grant GR/S57198/01 *Automation for Interactive Proof.*

References

1. Woldgang Ahrendt, Bernhard Beckert, Reiner Hähnle, Wolfram Menzel, Wolfgang Reif, Gerhard Schellhorn, and Peter H. Schmitt. Integrating automated and interactive theorem proving. In Wolfgang Bibel and Peter H. Schmitt, editors, *Automated Deduction— A Basis for Applications*, volume II. Systems and Implementation Techniques, pages 97–116. Kluwer Academic Publishers, 1998.
2. Marc Bezem, Dimitri Hendriks, and Hans de Nivelle. Automatic proof construction in type theory using resolution. *Journal of Automated Reasoning*, 29(3-4):253–275, 2002.
3. C.-L. Chang and R. C.-T. Lee. *Symbolic Logic and Mechanical Theorem Proving*. Academic Press, 1973.
4. M. J. C. Gordon and T. F. Melham. *Introduction to HOL: A Theorem Proving Environment for Higher Order Logic*. Cambridge University Press, 1993.
5. Joe Hurd. Integrating Gandalf and HOL. In Yves Bertot, Gilles Dowek, André Hirschowitz, Christine Paulin, and Laurent Théry, editors, *Theorem Proving in Higher Order Logics: TPHOLs '99*, LNCS 1690, pages 311–321. Springer, 1999.
6. Joe Hurd. An LCF-style interface between HOL and first-order logic. In Andrei Voronkov, editor, *Automated Deduction — CADE-18 International Conference*, LNAI 2392, pages 134–138. Springer, 2002.
7. Jia Meng. Integration of interactive and automatic provers. In Manuel Carro and Jesus Correas, editors, *Second CologNet Workshop on Implementation Technology for Computational Logic Systems*, 2003. On the Internet at http://www.cl.cam.ac.uk/users/jm318/papers/integration.pdf.

8. Tobias Nipkow, Lawrence C. Paulson, and Markus Wenzel. *Isabelle/HOL: A Proof Assistant for Higher-Order Logic*. Springer, 2002. LNCS Tutorial 2283.

9. Andreas Nonnengart and Christoph Weidenbach. Computing small clause normal forms. In Robinson and Voronkov [17], chapter 6, pages 335–367.

10. S. Owre, S. Rajan, J.M. Rushby, N. Shankar, and M.K. Srivas. PVS: Combining specification, proof checking, and model checking. In Rajeev Alur and Thomas A. Henzinger, editors, *Computer Aided Verification: 8th International Conference, CAV '96*, LNCS 1102, pages 411–414. Springer, 1996.

11. Lawrence C. Paulson. Generic automatic proof tools. In Robert Veroff, editor, *Automated Reasoning and its Applications: Essays in Honor of Larry Wos*, chapter 3. MIT Press, 1997.

12. Lawrence C. Paulson. The inductive approach to verifying cryptographic protocols. *Journal of Computer Security*, 6:85–128, 1998.

13. Lawrence C. Paulson. A generic tableau prover and its integration with Isabelle. *Journal of Universal Computer Science*, 5(3):73–87, 1999.

14. Lawrence C. Paulson. Isabelle's isabelle's logics: FOL and ZF. Technical report, Computer Laboratory, University of Cambridge, 2003. On the Internet at http://isabelle.in.tum.de/dist/Isabelle2003/doc/logics-ZF.pdf.

15. A. Riazanov and A. Voronkov. Efficient checking of term ordering constraints. Preprint CSPP-21, Department of Computer Science, University of Manchester, February 2003.

16. Alexander Riazanov and Andrei Voronkov. Vampire 1.1 (system description). In Rajeev Goré, Alexander Leitsch, and Tobias Nipkow, editors, *Automated Reasoning — First International Joint Conference, IJCAR 2001*, LNAI 2083, pages 376–380. Springer, 2001.

17. Alan Robinson and Andrei Voronkov, editors. *Handbook of Automated Reasoning*. Elsevier Science, 2001.

18. Jörg Siekmann, Christoph Benzmüller, Armin Fiedler, Andreas Meier, Immanuel Normann, and Martin Pollet. Proof development with ωmega: The irrationality of $\sqrt{2}$. In Fairouz Kamareddine, editor, *Thirty Five Years of Automating Mathematics*, pages 271–314. Kluwer Academic Publishers, 2003.

19. Geoff Sutcliffe and Christian Suttner. The TPTP problem library: CNF Release v1.2.1. *Journal of Automated Reasoning*, 21(2):177–203, October 1998.

20. Christoph Weidenbach. Combining superposition, sorts and splitting. In Robinson and Voronkov [17], chapter 27, pages 1965–2013.

21. Markus Wenzel. Type classes and overloading in higher-order logic. In Elsa L. Gunter and Amy Felty, editors, *Theorem Proving in Higher Order Logics: TPHOLs '97*, LNCS 1275, pages 307–322. Springer, 1997.

A Machine-Checked Formalization of the Generic Model and the Random Oracle Model

Gilles Barthe[1], Jan Cederquist[2]*, and Sabrina Tarento[1]**

[1] INRIA Sophia-Antipolis, France
{Gilles.Barthe,Sabrina.Tarento}@sophia.inria.fr
[2] CWI, Amsterdam, The Netherlands
Jan.Cederquist@cwi.nl

Abstract. Most approaches to the formal analyses of cryptographic protocols make the perfect cryptography assumption, i.e. the hypothese that there is no way to obtain knowledge about the plaintext pertaining to a ciphertext without knowing the key. Ideally, one would prefer to rely on a weaker hypothesis on the computational cost of gaining information about the plaintext pertaining to a ciphertext without knowing the key. Such a view is permitted by the Generic Model and the Random Oracle Model which provide non-standard computational models in which one may reason about the computational cost of breaking a cryptographic scheme. Using the proof assistant Coq, we provide a machine-checked account of the Generic Model and the Random Oracle Model.

1 Introduction

Background. Cryptographic protocols are widely used in distributed systems as a means to authenticate data and principals and more generally as a fundamental mechanism to guarantee such security goals as the confidentiality and the integrity of sensitive data. Yet their design is particularly error prone [2], and serious flaws have been uncovered, even in relatively simple and carefully designed protocols such as the Needham-Schroeder protocol [20]. In light of the difficulty of achieving correct designs, numerous frameworks have been used for modeling and analyzing cryptographic protocols formally, just to name a few: belief logics [10], type systems [1,17], model checkers [20], proof assistants [8,26], and frameworks that integrate several approaches [21]. Many of these frameworks have been used to good effect for discovering subtle flaws or for validating complex cryptographic protocols such as IKE [22] or SET [6]; due to space constraints, we refer to the recent article by Meadows [23] for a more detailed account of the history and applications of formal methods to cryptographic protocol analysis.

* Supported by an ERCIM Fellowship. Most of this work was performed at INRIA Sophia-Antipolis.
** Part of this work was supported by the INRIA Color project "Acces: Algorithmes Cryptographiques Certifiés".

D. Basin and M. Rusinowitch (Eds.): IJCAR 2004, LNAI 3097, pp. 385–399, 2004.
© Springer-Verlag Berlin Heidelberg 2004

Although these approaches differ in their underlying formalisms, they share the perfect cryptography assumption, i.e. the hypothesis that there is no way to obtain knowledge about the plaintext pertaining to a ciphertext without knowing the key; see [12,13] for some mild extensions of these models. Ideally, one would prefer to rely on a weaker hypothesis about the probability and computational cost of gaining information about the plaintext pertaining to a ciphertext without knowing the key. Such a view is closer to the prevailing view in cryptographic research, and there have been a number of recent works that advocate bridging the gap between the formal and computational views [3,23]. Yet in contrast to the formal view of cryptography, there is little work on the machine-checked formalization of the computational view of cryptography.

Generic Model and Random Oracle Model. The Generic Model, or GM for short, and the Random Oracle Model, or ROM for short, provide non-standard computational models in which one may reason about the probability and computational cost of breaking a cryptographic scheme.

ROM was introduced by Bellare and Rogaway [7], but its idea originates from earlier work by Fiat and Shamir [16]. GM was introduced by Shoup [29] and Nechaev [24], and further studied e.g. by Jakobsson and Schnorr [27,28], who also considered the combination of GM and ROM. The framework of GM and ROM is expressive enough to capture a variety of problems, such as the complexity of the discrete logarithm or the decisional Diffie-Hellman problem. More generally, GM and ROM provide an overall guarantee that a cryptographic scheme is not flawed [27,28,31], and have been used for establishing the security of many cryptographic schemes, including interactive schemes for which traditional security proofs do not operate [27,28].[1]

The basic idea behind GM and ROM is to constrain the behavior of an attacker so that he can only launch attacks that do not exploit any specific weakness from the group or the hash function underlying the cryptographic scheme under scrutiny. More precisely, GM and ROM define a class of generic algorithms that operate on an ideal group G of prime order q, and using an ideal hash function H that behaves as an oracle providing for each query a random answer. Then, one considers the probability of an attacker, taken among generic algorithms, to gain information about some secret, e.g. a cyphertext or a key. As we assume an ideal group, the attacker cannot obtain knowledge from a specific representation of group elements, and information about the secret may only be gained through queries to the oracle, or through testing group equalities. The main results of GM and ROM are upper bounds to the probability for an attacker of finding the secret: these upper bounds, which depend on the order q of the group, and on the number t of computations performed by the attacker, show that the probability is negligible for a sufficiently large q and a reasonable t.

---

[1] On the other hand, there is an ongoing debate about the exact significance of results that build upon GM and ROM; part of the debate arises from signatures scheme which can be proven secure under ROM, but which cannot implemented in a secure way, see e.g. [11] and e.g. [15] for a similar result about GM. Yet there is a consensus about the interest of these models.

This paper reports on preliminary work to provide, within the proof assistant Coq [14], a machine-checked formalization of GM and ROM and of their applications. In particular, we use our model of GM to give a machine-checked proof of the intractability of the discrete logarithm [29] (we do not prove anything about interactive algorithms here).

The formalization of GM and ROM within a proof assistant poses a number of interesting challenges. In particular, these models appeal to involved mathematical concepts from probabilities, such as uniform probability distribution, or statistical independence which are notoriously hard to formalize in a proof assistant, and to advanced results on multivariate polynomials, such as Schwartz Lemma, see Lemma 1, which provides an upper bound to the probability for a vector x over a finite field of order q to be the root of a multivariate polynomial of degree d. Furthermore, proofs about GM and ROM are carried out on examples, rather than in the general case. In order to increase the applicability of our formalization, we thus had to generalize the results of [27,28,29]. In particular, our main results, i.e. Propositions 1 and 2, provide for an arbitrary non-interactive generic algorithm an upper bound to the probability of breaking the cryptographic scheme, whereas previous works only provide such results for specific algorithms, such as the Discrete Logarithm [29].

Finally, proofs about cryptographic schemes are carried out in a very informal style, which makes their formalization intricate. In particular, proofs about GM and ROM often ignore events that occur with a negligible probability, i.e. events whose probability tends to 0 when the size of the group tends to ∞. When formalizing such proofs, we thus had to make statements and proofs about GM and ROM more accurate; an alternative would have been to formalize the notion of negligible probability using limits, but proofs would have become more complex.

Contents of the paper. The remainder of the paper is organised as follows. Section 2 provides a brief account of the Coq proof assistant, and presents our formalization of probabilities and polynomials, which are required to prove our main results. Section 3 describes our formal model of GM, whereas Section 4 provides a brief account of our formal model of ROM. We conclude in Section 5.

Formal proofs. For the sake of readability, we do not use Coq notation and adopt instead an informal style to present our main definitions and results. Formal developments can be retrieved from

$$\texttt{http://www-sop.inria.fr/everest/soft/Acces}$$

2   Preliminaries in Coq

This section provides a brief overview of the proof assistant Coq, and discusses some of issues with the formalization of algebra. Further, it describes our formalization of probabilities and of multivariate polynomials.

2.1   Coq

Coq [14] is a general purpose proof assistant based the Calculus of Inductive Constructions, which extends the Calculus of Constructions with a hierarchy of universes and mechanisms for (co)inductive definitions.

Coq integrates a very rich specification language. For example, complex algebraic structures can be defined and manipulated as first-class objects; in our formalization we rely on the formalization of basic algebraic structures by L. Pottier, available in the Coq contributions. Likewise, complex data structures are definable as inductive types; for the purpose of this paper however, we make a limited use of inductive definitions and mostly use first-order parameterized datatypes such as the type $list_X$ of X-lists.

In order to reason about specifications, Coq also integrates (through to the Curry-Howard isomorphism) a higher-order predicate logic. One particularity of the underlying logic is to be intuitionistic, hence types need not have a decidable equality, and predicates need not be decidable. For the sake of readability, we gloss over this issue in our presentation, but our formalization addresses decidability by making appropriate assumptions in definitions and results.

Further, logical statements can be used in specifications, e.g. in order to form the "subset" of prime numbers as the type of pairs $\langle n, \phi \rangle$ where n is a natural number and ϕ is a proof that n is prime. There are, however, some limitations to the interaction between specifications and propositions. In particular, dependent type theories such as the Calculus of Inductive Constructions lack intensional constructs that allow the formation of subsets or quotients. In order to circumvent this problem, formalizations rely on *setoids* [5], that is mathematical structures packaging a carrier, the "set"; its equality, the "book equality"; and a proof component ensuring that the book equality is well-behaved. For the sake of readability, we avoid in as much as possible mentioning setoids in our presentation, although they are pervasive in our formalizations.

Finally, let us introduce some Coq notation: **Prop** denotes the universe of propositions, and **Type** denotes the universe of types.

2.2   The Field $\mathbb{Z}_q$

Integers modulo q are modeled as a setoid: given $q \in \mathbb{N}$, we formalize $\mathbb{Z}_q$ as a setoid with underlying type $\mathbb{Z}$ (defined in the Coq libraries), and with underlying equivalence relation $\equiv q$ where $\equiv$ is defined as

$$\lambda q \in \mathbb{N}. \ \lambda a, b \in \mathbb{Z}. \ \exists k \in \mathbb{Z}. \ a - b = k \times q$$

We have shown that $\mathbb{Z}_q$ is a commutative field for q prime. All ring operations are defined in the obvious way. Interestingly, the multiplicative inverse $\cdot^{-1} : \mathbb{Z}_q \to \mathbb{Z}_q$ is defined as $\lambda x. (\text{mod } x \ q)^{q-2}$ where $\text{mod } x \ q$ is the remainder of the Euclidean division of x by q, and we have used Fermat's little theorem, available from the Coq contributions, to prove that $\forall a \in \mathbb{Z}_q. \ a \not\equiv 0 \ [q] \to a^{-1} * a \equiv a * a^{-1} \equiv 1 \ [q]$.

2.3 Probabilities

As there is no appropriate library for probabilities in the reference libraries and contributions in Coq, we have developed a collection of basic definitions and results for probabilities over finite sets. Due to lack of space, we only provide the definition of probabilities and conditional probabilities, and the statement of one illustrative result.

Before delving into details, let us point out that there are several possible approaches for defining probabilities over finite setoids. One possibility is to assume that the setoid is finite, i.e. isomorphic to some initial segment of $\mathbb{N}$, for a suitable notion of isomorphism of setoids. We have found slightly more convenient to define probabilities w.r.t. an arbitrary type V and a finite subset E of V, given as a (non-repeating) V-list.

Given a fixed type V and a fixed enumeration $E : list_V$, we define an event to be a predicate over V, i.e. $Event : Type := V \to \mathbf{Prop}$. Then, we define the probability of an event A being true as the ratio between the number of elements in E for which A is true and the total number of elements in E, i.e.

$$Pr_E[A] = \frac{\mathsf{length}\ (\mathsf{filter}\ E\ A)}{\mathsf{length}\ E}$$

where length and filter are the usual functions on lists, i.e. length l computes the length of the list l, and filter l P removes from the list l all its elements that do not satisfy the predicate P.

Then, one can check that Pr_E satisfies the properties of a probability measure, e.g.:

- for every event A, $0 \le Pr_E[A] \le 1$;
- if *True* is the trivial proposition, which always holds, then $Pr_E[\lambda a.\,True] = 1$;
- for any sequence A_i of disjoint events $Pr_E[\bigcup_{1 \le i \le n} A_i] = \sum_{1 \le i \le n} Pr_E[A_i]$, where $\bigcup_{1 \le i \le n} A_i = \lambda a.\ A_1(a) \vee \cdots \vee A_n(a)$.

Conditional probabilities are defined in the usual way, i.e.

$$Pr_E[A|B] = \frac{Pr_E[\lambda a.(Aa) \wedge (Ba)]}{Pr_E[B]}$$

Then, one can check that Pr_E satisfies properties such as

$$Pr_E[A] = Pr_E[A|B]\ Pr[B] + Pr_E[A|\neg B]\ (1 - Pr[B])$$

In the sequel, E will often be omitted to adopt the notation $Pr[A]$.

Discussion. It would be useful for a number of purposes, including the formalization of GM and ROM, to develop a comprehensive library about probabilities. In particular, we believe that a formal treatment of negligible events is required, if we want to continue with machine-checked accounts of computational cryptography. To this end, it will also be necessary to develop a theory of limits. In this respect, it will be useful to consider existing works on machine-checked probabilities in HOL [19] or Mizar [25].

2.4   Polynomials

We have formalized a collection of basic definitions and results for multivariate polynomials. Due to lack of space, we only provide the definitions of polynomials and focus on Schwartz Lemma and its corollaries, which are crucial to prove Propositions 1 and 2. As in [4,18], we use a list-based representation of polynomials; other possibilities for defining polynomials are discussed at the end of the paragraph.

Definition. Given a fixed ring R with underlying carrier R_{carr}, and a fixed set X of indeterminates, a monomial is a function that associates an exponent to each variable, so the type Mon_X of monomials is defined as $X \to \mathbb{N}$ (note that X is a type, not a setoid). Moreover, a polynomial is modeled as a list of coefficient-monomial pairs, so the type $Pol_{R,X}$ of polynomials is defined as $list_{(R_{carr} \times Mon_X)}$. The setoid structure of $Pol_{R,X}$ is defined using the book equality over R. Using the underlying operations of R, one can easily endow $Pol_{R,X}$ with a ring structure; in particular, addition of two polynomials is defined as list concatenation, negation of a polynomial is defined pointwise, and multiplication of polynomials is defined from multiplication of monomials using the map function.

The definition above allows an easy definition of the degree of a polynomial, which is needed in the proof of Schwartz Lemma below. To compute the degree of a polynomial, we must first assume that X is a finite type $\mathbb{X}_k$, defined inductively in the usual way (and with inhabitants $var_1, \dots, var_k$), and define the degree of a monomial as the sum of the degrees of the variables, i.e. $deg_{mon} m = \sum_{1 \leq i \leq k} (m\, var_i)$ where the sum operation can be defined using the elimination principle for finite sets. Then, the degree of a polynomial is defined as the maximum degree of its monomials, i.e.

$$deg_{pol} p = \mathsf{max}\ (\mathsf{map}\ (\lambda \langle a, m \rangle.\ deg_{mon}\ m)\ p)$$

where max is the function that computes the maximum element of a $\mathbb{Z}_q$-list.

Interpretation. The possibility of evaluating polynomials from $Pol_{R,X}$ into R is crucial to our approach to modeling GM and ROM. In this paragraph, we briefly describe how to evaluate polynomials.

Assume that X is a finite type $\mathbb{X}_k$, and let R be a ring with underlying carrier R_{carr}. Then we can define an evaluation function $Eval$ that, given an interpretation function $f : X \to R_{carr}$ and a polynomial $p : Pol_{R,X}$, returns an element of R_{carr} that corresponds to the value of p under the interpretation f. The evaluation function $Eval$ is defined by structural recursion over lists, i.e. $Eval\ f\ (\mathsf{nil}) = 0_R$, where 0_R is the neutral element w.r.t. addition, and

$$Eval\ f\ (\mathsf{cons}\ \langle a, m \rangle\ l) = (a \times_R (Eval_{Mon}\ f\ m)) +_R (Eval\ f\ l)$$

where $\times_R$ and $+_R$ respectively denote the ring multiplication and addition, and $Eval_{Mon} : (X \to R_{carr}) \to Mon_X \to R_{carr}$ is an auxiliary function that computes the interpretation of monomials, i.e.

$$Eval_{Mon} \ f \ (x_1^{l_1} \ldots x_k^{l_k}) = (f \ x_1)^{l_1} \times_R \ldots \times_R (f \ x_k)^{l_k}$$

Schwartz Lemma. In order to prove Schwartz Lemma, we assume the fundamental theorem of algebra for integers.

Theorem 1. *Let p be a polynomial in $\mathbb{Z}_q$, in one variable, of degree n, not identical to 0. Then there are at most n roots of $p(x) = 0$.*

We have not proved this result yet. However, we do not expect any difficulty in performing the proof, which proof proceeds by induction over the degree, using the division algorithm in the inductive step[2].

Lemma 1 (Schwartz Lemma). *Let $p(x_1, \ldots, x_k)$ be a polynomial in k variables, not identical to 0, with degree at most d, and the values chosen uniformly and independently in $[0, q-1]$. Then $Pr[p(x_1, \ldots, x_k) = 0] \leq d/q$.*

Proof (sketch). By induction on k. For $k = 0$ we just note that p is constant distinct from 0, thus the probability is 0. For the induction step, we rewrite $p(x_1, \ldots, x_{k+1})$ according to the powers of x_{k+1}

$$p(x_1, \ldots, x_{k+1}) = p'(x_1, \ldots, x_{k+1}) + p_l(x_1, \ldots, x_k)x_{k+1}^l$$

where $l \leq d$ is the degree of x_{k+1} in p, and p' is a polynomial in which the degree of x_{k+1} is less than l and p_l is a polynomial in $x_1, \ldots, x_k$ not identical to 0.

In our formalization, we consider as type of events the type $(\mathbb{X}_k \to \mathbb{Z}) \to \textbf{Prop}$, and as enumeration list the non-repeating list of all functions $g : \mathbb{X}_k \to \mathbb{Z}$ s.t. $0 \leq g \ x \leq q - 1$ for all $x : \mathbb{X}_k$. Note that functions are treated up to extensional equality, i.e. for every function $g : \mathbb{X}_k \to \mathbb{Z}$ s.t. $0 \leq g \ x \leq q - 1$ for all $x : \mathbb{X}_k$, there exists $g' : \mathbb{X}_k \to \mathbb{Z}$ in V s.t. $g \ x = g' \ x$ for all $x : \mathbb{X}_k$ (which does not imply that $g = g'$ in Coq). Using this approach, then the interpretation of variables in X becomes random and uniformly distributed. This observation is crucial for the validity of our models.

We now state some corollaries of Schwartz Lemma that are used in Section 3. The first corollary follows rather directly from Schwartz Lemma, while the second corollary is proved by induction.

Lemma 2. *Let $p_1, \ldots, p_n$ be a sequence of polynomials, not identical to 0, with degree at most d, and the values are chosen uniformly and independently in $[0, q-1]$. Then $Pr[p_n = 0 | \forall j < n.p_j \neq 0] \leq \frac{d}{q-nd}$.*

Lemma 3. *Let $p_1, \ldots, p_n$ be a sequence of polynomials, not identical to 0, with degree at most d, values of the variables chosen uniformly and independently in $[0, q-1]$ and $nd < q$. Then $Pr[p_1 = 0 \vee \cdots \vee p_n = 0] \leq \frac{nd}{q-nd}$.*

---

[2] In fact, the result has been proved already by Geuvers et al. in their formalization of FTA and is available in the Coq contributions. However, it may not be straightforward to use their result in our work; indeed, the setting is slightly different, e.g. the underlying notion of set is provided by constructive setoids that feature an apartness relation, and algebraic structures are not formalized in exactly the same way.

Discussion. There are many possible definitions of polynomials. To cite a few that have been used in Coq developments, Geuvers *et al.*'s formalization of FTA use the Hörner representation for polynomials in one variable, whereas Théry's formalization of Buchberger's algorithm uses an order on the monomials in order to avoid repeated entries of monomials, and Pottier's formalization of algebraic structures uses an inductive definition of polynomial expressions. In a very recent (unpublished) work, Grégoire and Mahboubi have explored alternative representations of multivariate polynomials that allow for efficient reflexive tactics.

However, there lacks a standard and comprehensive library about polynomials. To provide a solid basis for further work about GM and ROM, it seems relevant to develop such a library, possibly taking advantage of the isomorphism between the different representations of polynomials.

3   Non-interactive Generic Algorithms

The framework of non-interactive generic algorithms is useful for establishing the security of the discrete logarithm problem, as well as several other related problems, most notably the Diffie-Hellman problem. This section presents our formalization of non-interactive generic algorithms.

3.1   Informal Account

Let G be a cyclic group of prime order q with generator g. A generic algorithm $\mathcal{A}$ over G is given by:

- its input $l_1, \ldots, l_{t'} \in \mathbb{Z}_q$, which depends upon a set of secrets, typically secret keys, say $s_1, \ldots, s_k \in \mathbb{Z}_q$. In the sequel, we define the group input $f_1, \ldots, f_{t'} \in G$ of the algorithm by $f_k = g^{l_k}$;
- a run, i.e. a sequence of t steps. A step can either be an input step, or a multivariate exponentiation (mex) step. An input step reads some input from the group input; for simplicity, we assume that all inputs are read exactly once at the beginning, for $1 \leq i \leq t'$, the algorithm at step i reads f_i from the group input. For $t' < i \leq t$, we assume that the algorithm at step i takes a mex step, i.e. selects arbitrarily $a_1^i, \ldots, a_{t'}^i \in \mathbb{Z}_q$ and computes $f_i = \prod_{1 \leq j \leq t'} f_j^{a_j^i}$.

The output of the generic algorithm $\mathcal{A}$ is the list $f_1, \ldots, f_t$. Further, we define *collisions* to be equalities $f_j = f_{j'}$ with $1 \leq j < j' \leq t$, and say that a collision $f_j = f_{j'}$ is *trivial* if it holds with probability 1, i.e. if it holds for all choices of secret data. In the sequel, we write $\mathcal{CO}(\mathcal{A})$ if the algorithm $\mathcal{A}$ finds non-trivial collisions.

The generic model considers an attacker as a generic algorithm $\mathcal{A}$, which tries to find information about secrets through testing equalities between outputs, i.e. through non-trivial collisions (trivial collisions do not reveal any information about secrets), and expresses a random guess for secrets if it fails to find them by the first method. Hence, the probability of $\mathcal{A}$ finding the secret s_j is deduced from the probability $ProbColl(\mathcal{A})$ that the algorithm discovers a non-trivial collision, i.e. that $\mathcal{CO}(\mathcal{A})$ holds.

In order to give an upper bound for $ProbColl(\mathcal{A})$, the Generic Model relies on Schwartz Lemma. To this end, the Generic Model assumes that $\mathcal{A}$ is a generic algorithm whose group inputs f_j are of the form $g^{m_j(s_1,\dots,s_k)}$ where $m_j(x_1,\dots,x_k)$ is a multivariate monomial over the set $X = \{x_1,\dots,x_k\}$ of secret parameters, and $s_1,\dots,s_k$ are the actual secrets.

Example 1 (Discrete logarithm). The algorithm is given as input the group generator $g \in G$ and the public key $h = g^r \in G$, and outputs a guess y for $\log_g h = r$. Observe that any non-trivial collision reveals the value of r: indeed, every f_i will be of the form $g^{a_i}(g^r)^{a_i'} = g^{(a_i + ra_i')}$. Hence for any collision $f_i = f_j$, we have $g^{(a_i + ra_i')} = g^{(a_j + ra_j')}$, and so $r(a_i' - a_j') \equiv a_j - a_i \ [q]$. If the collision is non-trivial, then $a_i' - a_j' \neq 0$ and we can deduce the value of r.

In this example, there is a single secret r and the formal inputs are the monomials $1 = r^0$ and r.

Example 2 (Decisional Diffie-Hellman Problem). The algorithm is given as input the group generator $g \in G$, the group elements g^x and g^y, and the group elements g^{xy} and g^z in random order, where x, y, z are random in $\mathbb{Z}_q$, and outputs a guess for g^{xy} (or equivalently, for the order of g^{xy} and g^z).

In this example, there are three secrets x and y and z, and the formal inputs are the monomials 1 and x and y and xy and z.

3.2 Formal Account

The main difficulty in formalizing generic algorithms is to capture formally the idea of a secret. We choose to model secrets by introducing a type Sec of formal secret parameters and an interpretation function $\sigma : Sec \to \mathbb{Z}_q$ that maps formal secret inputs to actual secrets.

Further, we assume given a non-repeating list of monomials $input : list_{mon_{Sec}}$ of length t', and let $m_1, \dots, m_{t'}$ be the elements of $input$. These monomials constitute the formal inputs of the algorithm; the actual inputs can be defined as $\mathsf{map}\ (Eval_{mon}\ \sigma)\ input : list_{\mathbb{Z}_q}$.

Finally, the type of generic algorithms is defined as the record type

$$GA = \{run : list_{list_{\mathbb{Z}_q}};\ ok : \dots\}$$

where run is the list of exponents selected by the algorithm at each step (the exponents are themselves gathered in a list), and ok is a predicate that guarantees some suitable properties on run, in particular that:

- all elements of run also have length t';
- for $1 \leq j \leq t'$, the j-th element of run is the list whose j-th element is 1, and whose remaining elements are 0;
- run is a non-repeating list, so as to avoid trivial collisions, see below.

The output of a generic algorithm is obtained by computing from the exponents $a_1^i, \ldots, a_{t'}^i$ the polynomial $p_i = \sum_{1 \leq j \leq t'} a_j^i\, m_j$, then evaluating each polynomial p_i with σ, finally obtaining in each case an element f_i of $\mathbb{Z}_q$ (as compared to the informal account, we find it more convenient to outputs as elements of $\mathbb{Z}_q$, which is legitimate since $\mathbb{Z}_q$ and G are isomorphic).

Formally, the output of the generic algorithm is modeled as

$$output : list_{\mathbb{Z}_q} := \mathsf{map}\ (\mathsf{eval\_pol}\ \sigma)\ (\mathsf{map}\ (\lambda l.\ \mathsf{zip}\ l\ input)\ run)$$

where zip is the obvious function of type $\forall A, B : \textbf{Type}.\ list_A \to list_B \to list_{A \times B}$.

Then, $\mathcal{CO}(\mathcal{A})$ is defined as doubles $output$, where doubles is the boolean-valued function that checks whether there are repeated entries in a list. Note that collisions occur iff there exist pairwise distinct i and i' s.t.

$$\mathsf{eval\_pol}\ \sigma\ p_i =_{\mathbb{Z}_q} \mathsf{eval\_pol}\ \sigma\ p_{i'}$$

Furthermore, trivial collisions are defined to satisfy the stronger property

$$\forall I : Sec \to \mathbb{Z}_q.\ \mathsf{eval\_pol}\ I\ p_i =_{\mathbb{Z}_q} \mathsf{eval\_pol}\ I\ p_{i'}$$

Such collisions never occur in our setting, since we assume that $input$ and run are non-repeating lists, and hence that $p_i - p_{i'}$ cannot be identical to 0.

Finally, a necessary condition $SecFound$ for an algorithm $\mathcal{A}$ to find a secret is that, either there was a collision or, there were no collisions but the algorithm happens to guess the correct value. Formally, given an algorithm $\mathcal{A}$, a secret $x : Sec$, and $g : \mathbb{Z}_q$ expressing the guess of the algorithm in case no collision is found, we define the relation $SecFound_{\mathcal{A}}(g, x)$ by the clauses:

$$\frac{Coll(\mathcal{A})}{SecFound_{\mathcal{A}}(g,x)} \qquad\qquad \frac{\neg Coll(\mathcal{A}) \quad \sigma x =_{\mathbb{Z}_q} g}{SecFound_{\mathcal{A}}(g,x)}$$

3.3   Properties of Generic Algorithms

In this section, we let $\mathcal{A}$ be a generic algorithm. Further, we let d be the maximal degree of the monomials m_j for $1 \leq j \leq t'$, let t be the number of steps $\mathcal{A}$ performs.

Proposition 1. $ProbColl(\mathcal{A}) \leq \dfrac{\binom{t}{2}d}{q - \binom{t}{2}d}$

Proof. All outputs are of the form $f_i = \sum_{1 \leq j \leq t'} a_j^i\, m_j(s_1, \ldots, s_k)$, where $p_i = \sum_{1 \leq j \leq t'} a_j^i\, m_j(x_1, \ldots, x_k)$ is a polynomial of degree d. Hence there exists a collision $f_i = f_{i'}$ iff $(s_1, \ldots, s_k)$ is a root of $p_i - p_{i'}$. There are $\binom{t}{2}$ equalities of the form $f_i = f_{i'}$ to test, hence $\binom{t}{2}$ polynomials of the form $p_i - p_{i'}$, each of which is not identical to 0 (as there are non-trivial collisions), and has degree $\leq d$. So we can apply Lemma 3 to deduce the expected result.

In the sequel, we let $ProbSecFound_{\mathcal{A}}(x) = Pr[\lambda g.\ SecFound(g, x)]$.

Proposition 2. *Let* $\mu = \frac{\binom{t}{2}d}{q-\binom{t}{2}d}$. *For any secret* $x : Sec$,

$$ProbSecFound_{\mathcal{A}}(x) \leq \mu + \frac{1-\mu}{q}$$

Proof. Immediate from the definition of $SecFound(g, x)$.

We can now instantiate the proposition to specific cryptographic schemes.

Example 3 (Discrete Logarithm, continued). Here $d = 1$ so the probability of finding the secret is $\mu + \frac{1-\mu}{q}$, where $\mu = \frac{\binom{t}{2}}{q-\binom{t}{2}}$.

Note that Proposition 2 only holds for a secret $x : Sec$ ranging uniformly over $\mathbb{Z}_q$. For some problems however, such as the Decisional Diffie-Hellman problem below, x ranges uniformly over a subset of $\mathbb{Z}_q$. In this case, the probability of finding a secret is $\mu + \frac{1-\mu}{q'}$, where $\mu = \frac{\binom{t}{2}d}{q-\binom{t}{2}d}$ and q' is the cardinal of the set of possible values for $x : Sec$.

Example 4 (Decisional Diffie-Hellman problem, continued). Here $q' = 2$ and $d = 2$ so the probability of finding the secret is $\frac{1+\mu}{2}$, where $\mu = \frac{2\binom{t}{2}}{q-2\binom{t}{2}}$.

We conclude this section with some brief remarks:

– our estimates are higher than the ones given for example in [27,28], since we must take negligible events into account;
– in our presentation, we have followed [27,28] in that attackers that do not find non-trivial collisions provide a random guess without taking advantage of the computations that they have performed. However, in our formalisation we also consider more powerful attackers that express a random guess over the set of values that have not yet been considered by the run, as every step that does not induce a collision reduces the set of possible values for the secret. In any case, the probability of such attackers is only negligibly higher than our probabilities.
– for the sake of readability, formal inputs are taken to be monomials. However, in our formalisation we also consider more general forms of formal inputs, namely polynomials. The main difficulty introduced by this more general form of formal inputs is the possibility for trivial collisions to occur, and hence one must add a new conjunct in the definition of *ok* that ensures the absence of such collisions.

4   Formalization of Interactive Algorithms

In this section, we present the main steps towards a formalization of interactive generic algorithms. However, we have not proven any result about interactive generic algorithms at this stage.

4.1 Informal Account

An interactive generic algorithm can read an input, or take a mex-step, or perform an interaction. We consider two common forms of interactions in cryptographic algorithms: queries to hash functions and decryptors. These forms of interaction are used in particular in the signed ElGamal encryption protocol. Note that interactions provide the algorithm with values, and that, in this setting, mex-steps select perform computations of the form $f_i = \prod_{1 \leq j \leq t'} f_j^{a_j^i}$, where for $1 \leq j \leq t'$, f_j is an input of the algorithm, and where $a_1^i, \ldots, a_{t'}^i$ are arbitrary but may depend on values that the algorithm received through interactions. More generally, values obtained from oracle interactions can be used in mex-steps as well as in future interactions.

Example 5. Let $x \in \mathbb{Z}_q$ and $h = g^x$ be the private and public keys for encryption, $m \in G$ the message to be encrypted. For encryption, pick random $r \in \mathbb{Z}_q$, (g^r, mh^r) is the ElGamal ciphertext. To add Schnorr signatures, pick random $s \in \mathbb{Z}_q$, compute $c = H(g^s, g^r, mh^r)$ and $z = s + cr$, then (g^r, mh^r, c, z) is the signed ciphertext. A decryptor Dec takes a claimed ciphertext $(\bar{h}, \bar{f}, c, z)$ and computes

$$F = (if\ H(g^z \bar{h}^{-c}, \bar{h}, \bar{f}) = c\ then\ \bar{h}^x\ else\ ?)$$

where ? is a random value, and then returns $\frac{\bar{f}}{F}$ which is the original message, if $(\bar{h}, \bar{f}, c, z)$ is a valid ciphertext.

As in the non-interactive model, an attacker is a generic algorithm that seeks to gain knowledge about secrets through testing equalities between the group elements it outputs, possibly through interactions. However, the attacker has now access to oracles for computing hash values and for decryption. Note that each operation performed by the attacker, i.e. reading an input, performing an interaction, or taking a mex-step, counts as a step in the run. However, as in the non-interactive case, testing equality is free.

Let us conclude this section by pointing that the fundamental assumption of ROM is that the hash function $H : G \rightarrow G \rightarrow G \rightarrow \mathbb{Z}_q$ is chosen at random with uniform probability distribution over all functions of that type. This assumption must of course be reflected in our formalization.

4.2 Formalization

The main difficulty in formalizing interactive algorithms is of course to capture the idea of hash function. In our formalization, we introduce a type *Val* of random variables, disjoint of *Sec*, for communication with oracles. These variables, once given a value through an interaction, can be used in mex-steps.

Formally, we assume given a type *Sec* of formal secret parameters and an interpretation function $\sigma : Sec \rightarrow \mathbb{Z}_q$ that maps formal secret inputs to actual secrets, as well as a type *Val* of random variables, disjoint of *Sec*, and a function $\tau : Val \rightarrow \mathbb{Z}_q$. Further, we assume given a non-repeating list of monomials

$input : list_{mon_{Sec}}$ of length t', and let $m_1, \ldots, m_{t'}$ be the elements of $input$. These monomials constitute the formal inputs of the algorithm; the actual inputs can be defined as $\mathsf{map}\ (Eval_{mon}\ \sigma)\ input : list_{\mathbb{Z}_q}$.

Then, the type of interactive generic algorithms is defined as the record type

$$IGA = \{run : Run;\ ok : \ldots\}$$

where Run is defined inductively by the clauses

$$\frac{}{erun : Run} \qquad \frac{\begin{array}{c} r : Run \\ a : list_{Poly_{\mathbb{Z}_q, Val}} \end{array}}{(step\ r\ a) : Run}$$

$$\frac{\begin{array}{c} r : Run \\ a, b, d : list_{Poly_{\mathbb{Z}_q, Val}} \\ c : Val \end{array}}{(HashQuery\ r\ a\ b\ d\ c) : Run} \qquad \frac{\begin{array}{c} r : Run \\ m : Val \\ c, z : Poly_{\mathbb{Z}_q, Val} \\ \bar{h}, \bar{f} : list_{Poly_{\mathbb{Z}_q, Val}} \end{array}}{(DecQuery\ r\ m\ \bar{h}\ \bar{f}\ c\ z) : Run}$$

and ok is a predicate that guarantees some suitable properties on run.

Further, observe that hash queries implicitly define a hash "function" $H : Poly_{\mathbb{Z}_q, Val} \to Poly_{\mathbb{Z}_q, Val} \to Poly_{\mathbb{Z}_q, Val} \to Val$. Hence ok must contain a conjunct that guarantees that the resulting hash "function" is well-behaved, and in particular that it is indeed a function.

The length of the run is defined as the number of steps taken to type $r : Run$ and is defined by straightforward structural recursion. Further, the output of the run is obtained by computing from the exponents $a_1^i, \ldots, a_{t'}^i$ the polynomial $p_i = \sum_{1 \le j \le t'} a_j^i\ m_j$, then evaluating each polynomial p_i with σ, obtaining in each case an element q_i of $Pol_{\mathbb{Z}_q, Val}$, and then evaluating each polynomial q_i with τ, obtaining in each case an element f_i of $\mathbb{Z}_q$.

5 Conclusion

Using the proof assistant Coq, we have given a formal account of GM and ROM and of some of its applications; in the case of non-interactive generic algorithms, Propositions 1 and 2 generalize existing results to the case of an arbitrary generic algorithm.

Much work remains to be done to provide a more extensive machine-checked account of GM and ROM and its applications. In particular, we have formalized interactive algorithms but have not provided any formal proofs about them. The next natural step to take is to formalize the results about such algorithms. We are planning to prove the security of signed ElGamal encryption, by relying on the results of [27,28]. A more ambitious objective would be to exploit our formalizations to prove the security of realistic protocols, following e.g. [9,30]. An even more far-fetched goal would be to give a machine-checked account of a formalism that integrates the computational and formal views of cryptography.

Acknowledgments. We are grateful to the anonymous referees for their constructive and detailed comments, and to Guillaume Dufay for his help and comments on an earlier version of the paper.

References

1. M. Abadi and B. Blanchet. Secrecy types for asymmetric communication. *Theoretical Computer Science*, 298(3):387–415, April 2003.
2. M. Abadi and R. M. Needham. Prudent engineering practice for cryptographic protocols. *Transactions on Software Engineering*, 22(1):6–15, January 1996.
3. M. Abadi and P. Rogaway. Reconciling two views of cryptography (the computational soundness of formal encryption). *Journal of Cryptology*, 15(2):103–127, 2002.
4. A. Bailey. Representing Algebra in LEGO. Master's thesis, University of Edinburgh, 1993.
5. H. Barendregt and H. Geuvers. Proof assistants using dependent type systems. In A. Robinson and A. Voronkov, editors, *Handbook of Automated Reasoning*, volume II, chapter 18, pages 1149–1238. Elsevier Publishing, 2001.
6. G. Bella, F. Massacci, and L.C. Paulson. Verifying the set registration protocols. *IEEE Journal on Selected Areas in Communications*, 21:77–87, 2003.
7. M. Bellare and P. Rogaway. Random oracles are practical: A paradigm for designing efficient protocols. In *Proceedings of the 1st ACM Conference on Computer and Communications Security*, pages 62–73. ACM Press, November 1993.
8. D. Bolignano. Towards a mechanization of cryptographic protocol verification. In O. Grumberg, editor, *Proceedings of CAV'97*, volume 1254 of *Lecture Notes in Computer Science*, pages 131–142. Springer-Verlag, 1997.
9. D. Brown. Generic groups, collision resistance, and ecdsa, 2002. Available from http://eprint.iacr.org/2002/026/.
10. M. Burrows, M. Abadi, and R. Needham. A logic of authentication. *ACM Transactions on Computer Systems*, 8(1):18–36, February 1990.
11. R. Canetti, O. Goldreich, and S. Halevi. The random oracle methodology, revisited. In *Proceedings of STOC'98*, pages 209–218, New York, 1998. ACM Press.
12. Y. Chevalier, R. Küsters, M. Rusinowitch, and M. Turuani. An NP Decision Procedure for Protocol Insecurity with XOR. In *Proceedings of LICS'03*, pages 261–270. IEEE Computer Society Press, 2003.
13. H. Comon-Lundh and V. Shmatikov. Intruder deductions, constraint solving and insecurity decision in presence of exclusive or. In *Proceedings of LICS'03*, pages 271–280. IEEE Computer Society Press, 2003.
14. Coq Development Team. *The Coq Proof Assistant User's Guide. Version 7.4*, February 2003.
15. A. W. Dent. Adapting the Weaknesses of the Random Oracle Model to the Generic Group Model. In Y. Zheng, editor, *Proceedings of ASIACRYPT'02*, volume 2501 of *Lecture Notes in Computer Science*, pages 100–109. Springer-Verlag, 2002.
16. A. Fiat and A. Shamir. How to Prove Yourself: Practical Solutions to Identification and Signature Problems. In *Proc. CRYPTO'86*, volume 286 of *Lecture Notes in Computer Science*, pages 186–194. Springer-Verlag, 1986.
17. A. Gordon and A. Jeffrey. Authenticity by typing for security protocols. In *Proceedings of CSFW'01*, pages 145–159. IEEE Computer Society Press, 2001.

18. J. Harrison. *Theorem Proving with the Real Numbers.* Distinguished Dissertations. Springer-Verlag, 1998.

19. J. Hurd. *Formal Verification of Probabilistic Algorithms.* PhD thesis, University of Cambridge, 2002.

20. G. Lowe. Breaking and Fixing the Needham-Schroeder Public-Key Protocol Using FDR. In T. Margaria and B. Steffen, editors, *Proceedings of TACAS'96*, volume 1055 of *Lecture Notes in Computer Science*, pages 147–166. Springer-Verlag, 1996.

21. C. Meadows. The nrl protocol analyzer: An overview. *Journal of Logic Programming*, 26(2):113–131, February 1996.

22. C. Meadows. Analysis of the internet key exchange protocol using the NRL protocol analyzer. In *Proceedings of SOSP'99*, pages 216–233. IEEE Computer Society Press, 1999.

23. C. Meadows. Open issues in formal methods for cryptographic protocol analysis. In V.I. Gorodetski, V.A. Skormin, and L.J. Popyack, editors, *Proceedings of MMMACNS*, volume 2052 of *Lecture Notes in Computer Science*. Springer-Verlag, 2001.

24. V. I. Nechaev. Complexity of a determinate algorithm for the discrete logarithm. *Mathematical Notes*, 55(2):165–172, 1994.

25. A. Nedzusiak. σ-fields and probability. *Journal of Formalized Mathematics*, 1, 1989.

26. L.C. Paulson. The inductive approach to verifying cryptographic protocols. *Journal of Computer Security*, 6(1/2):85–128, 1998.

27. C.-P. Schnorr. Security of Blind Discrete Log Signatures against Interactive Attacks. In S. Qing, T. Okamoto, and J. Zhou, editors, *Proceedings of ICICS'01*, volume 2229 of *Lecture Notes in Computer Science*, pages 1–12. Springer-Verlag, 2001.

28. C.-P. Schnorr and M. Jakobsson. Security of Signed ElGamal Encryption. In T. Okamoto, editor, *Proceedings of ASIACRYPT'00*, volume 1976 of *Lecture Notes in Computer Science*, pages 73–89. Springer-Verlag, 2000.

29. V. Shoup. Lower bounds for discrete logarithms and related problems. In W. Fumy, editor, *Proceedings of EUROCRYPT'97*, volume 1233 of *Lecture Notes in Computer Science*, pages 256–266. Springer-Verlag, 1997.

30. N. Smart. The exact security of ecies in the generic group model. In B. Honary, editor, *Cryptography and Coding*, pages 73–84. Springer-Verlag, 2001.

31. J. Stern. Why provable security matters? In E. Biham, editor, *Proceedings of EUROCRYPT'03*, volume 2656 of *Lecture Notes in Computer Science*, pages 449–461. Springer-Verlag, 2003.

Automatic Generation of Classification Theorems for Finite Algebras

Simon Colton[1], Andreas Meier[2]*, Volker Sorge[3]**, and Roy McCasland[4]***

[1] Department of Computing, Imperial College London, UK,
sgc@doc.ic.ac.uk, http://www.doc.ic.ac.uk/~sgc
[2] DFKI GmbH, Saarbrücken, Germany,
ameier@ags.uni-sb.de, http://www.ags.uni-sb.de/~ameier
[3] School of Computer Science, University of Birmingham, UK,
V.Sorge@cs.bham.ac.uk, http://www.cs.bham.ac.uk/~vxs
[4] School of Informatics, University of Edinburgh, UK
rmccasla@inf.ed.ac.uk, http://www.inf.ed.ac.uk/~rmccasla

Abstract. Classifying finite algebraic structures has been a major motivation behind much research in pure mathematics. Automated techniques have aided in this process, but this has largely been at a quantitative level. In contrast, we present a qualitative approach which produces verified theorems, which classify algebras of a particular type and size into isomorphism classes. We describe both a semi-automated and a fully automated bootstrapping approach to building and verifying classification theorems. In the latter case, we have implemented a procedure which takes the axioms of the algebra and produces a decision tree embedding a fully verified classification theorem. This has been achieved by the integration (and improvement) of a number of automated reasoning techniques: we use the Mace model generator, the HR and C4.5 machine learning systems, the Spass theorem prover, and the Gap computer algebra system to reduce the complexity of the problems given to Spass. We demonstrate the power of this approach by classifying loops, groups, monoids and quasigroups of various sizes.

1 Introduction

As witnessed by the classification of finite simple groups – described as one of the major intellectual achievements of the twentieth century [4] – classifying finite algebraic structures has been a major motivation behind much research in pure mathematics. As discussed further in Sec. 2.2, automated techniques have aided this process, but this has largely been at a quantitative level, e.g., to count the number of groups of a particular order. Classification theorems of a more qualitative nature are often more interesting and more informative, sometimes allowing one to use properties of relatively small structures to help

* The author's work supported by EU IHP grant Calculemus HPRN-CT-2000-00102.
** The author's work was supported by a Marie-Curie Grant from the European Union.
*** The author's work was supported by EPSRC MathFIT grant GR/S31099.

D. Basin and M. Rusinowitch (Eds.): IJCAR 2004, LNAI 3097, pp. 400–414, 2004.
© Springer-Verlag Berlin Heidelberg 2004

classify larger structures. For example, Kronecker's classification of finite Abelian groups [6] states that every Abelian group, G, of size n can be expressed as a direct product of cyclic groups, $G = C_{s_1} \times \cdots \times C_{s_m}$, where $n = s_1.s_2 \ldots s_m$ such that each s_{i+1} divides s_i.

In this paper we look at automating the task of generating and fully verifying qualitative classification theorems for algebraic structures of a given size. As a simple example, our system is given the axioms of group theory and told to find a classification theorem for groups of size 6. It returns the following (paraphrased) result: "all groups of size 6 can be classified up to isomorphism as either Abelian or non-Abelian" [an Abelian group, G, is such that $\forall a, b \in G.\ a \circ b = b \circ a$]. The system generates such results, then proves that they provide valid classifications – as specified in Sec. 2.1 – by showing that each concept is a *classifying property*, i.e., true for all members of exactly *one* isomorphism class.

In our first – semi-automated – approach to generating and verifying classification theorems, as discussed in Sec. 3, the Mace model generator [9] was used to generate representatives of each isomorphism class for the given algebra of given size, then the HR [1] and C4.5 [12] machine learning systems were used to induce a set of classifying properties. To guarantee the validity of the classification we construct appropriate verification problems, which we first simplify with the Gap computer algebra system [3] and then prove with the Spass theorem prover [16]. We found various limitations with this approach, and the lessons we learned informed a second approach. As discussed in Sec. 4, we implemented a fully automated bootstrapping procedure that builds a decision tree which can be used to decide the isomorphism class of a given algebra. The system uses Mace to successively construct non-isomorphic algebras and HR to find properties that discriminate between them. The correctness of the decision tree is guaranteed in each step with Spass after using Gap to simplify the problems.

We used both approaches to generate a number of classification theorems for groups, monoids, quasigroups and loops up to size 6, with the results from the second approach presented in Sec 5. The power of this bootstrapping approach is demonstrated by the generation and verification of a classification theorem which covers the 109 loops of size 6. Not only does our approach highlight the utility of employing multiple reasoning systems for difficult tasks such as classification, our technique is neither restricted to the algebraic domain nor the isomorphism equivalence relation. In Sec. 6, we discuss possible applications of our approach to other domains of pure mathematics, along with other future directions for this work, including distributing the bootstrapping algorithm and producing classification theorems which are generative as well as descriptive.

2   Background

2.1   Classification Problems

We define a general classification problem as follows: let $\mathfrak{A}$ be a finite collection of algebraic structures and let $\sim$ be an equivalence relation on $\mathfrak{A}$. Then $\sim$ induces a partition into equivalence classes $[A_1]_\sim, [A_2]_\sim, \ldots, [A_n]_\sim$, where $A_i \in \mathfrak{A}$ for $i = 1, \ldots, n$.

Let P be a property which is invariant with respect to $\sim$. Then P acts as a *discriminant* for any two structures A and B in $\mathfrak{A}$, in the sense that if $P(A)$ and $\neg P(B)$, then $A \not\sim B$. If, in addition, P holds for every element of an equivalence class $[A_i]_\sim$, but does not hold for any element in $\mathfrak{A} \setminus [A_i]_\sim$, then we call P a *classifying property* for $[A_i]_\sim$. A full set of classifying properties – with one property for each equivalence class – comprises a classifying theorem stating that each element of $\mathfrak{A}$ exhibits exactly one of the classifying properties. The classification problem is therefore to find a full set of classifying properties.

At present we consider the isomorphism equivalence relation $\cong$, and the algebraic structures quasigroups, loops, groups and monoids. For these structures, we only need the following four axioms (letting A be a set and $\circ$ be a closed operation on the elements of A):

1. **Associativity:** A is associative with respect to $\circ$.
2. **Divisors:** For every two elements $a, b \in A$ there exist two corresponding divisors $x, y \in A$ such that $a \circ x = b$ and $y \circ a = b$ holds.
3. **Unit element:** There exists a unit element e in A with respect to $\circ$.
4. **Inverse:** Given a unique unit element e, each element x has an inverse x^{-1} such that $x \circ x^{-1} = x^{-1} \circ x = e$.

We call $(A, \circ)$ a *quasigroup* if only axiom 2 holds, a *monoid* if axioms 1 and 3 hold, a *loop* if axioms 2 and 3 hold and a *group* if axioms 1, 3 and 4 hold. Working with relatively simple algebras keeps the necessary axiomatic overhead down, but can lead to large numbers of structures even for small sizes, e.g., there are 109 non-isomorphic loops of size 6.

As a concrete example, consider quasigroups of size 3. There are 12 quasigroups of this size, but only 5 different isomorphism classes. Presented in terms of their Cayley tables, the following three quasigroups of order 3 are pairwise non-isomorphic.

Q_1	a b c		Q_2	a b c		Q_3	a b c
a	b a c		a	a c b		a	a b c
b	a c b		b	c b a		b	b c a
c	c b a		c	b a c		c	c a b

Because if we observe the diagonals of the Cayley tables for Q_1 and Q_2, we find that Q_2 has idempotent elements (i.e., such that $x \circ x = x$), but Q_1 does not. Hence the property $\exists x . x \circ x = x$ is a discriminant for Q_1 and Q_2. However, this property is not enough to distinguish Q_2 from Q_3 (i.e., this property is not a classifying property for $[Q_2]_\cong$), as Q_3 also contains idempotent elements. Since all elements of Q_2 are idempotent, we can strengthen the property to $\forall x . x \circ x = x$, which is sufficient to discriminate Q_2 from both Q_1 and Q_3. As we will see later, this is actually a classifying property for the equivalence class represented by Q_2.

2.2 Previous Work

When constructing classification theorems for algebraic structures, the first question to answer is how many algebras there are for a given size. Automated techniques such as constraint solving and the Davis-Putnam method have been used extensively to determine the number of algebras of a given type and size, and this has answered many open questions. In particular, the Finder system has been used to solve many quasigroup existence problems [14]. Also, representatives of every isotopy and isomorphism class for quasigroups and loops have been generated up to at least order 9 [10]. In addition to classifying structures within an algebraic axiomatisation, automated theorem proving has been used to find new axiomatisations for algebras, thus enabling better intra-classification of algebras. In particular, new axiomatic representations of algebras such as groups have been found [5,8].

As described in [1], integration of automated reasoning systems has always been a major aspect of the HR project. Moreover, we have shown in previous work that the HR system can be used to support proof planning, and that this has some promise for classification tasks [11]. However, to our knowledge, there have been no attempts to produce and verify full classification theorems for particular algebras of a given size from the axioms alone. As described in Sec. 3 and Sec. 4, our approach has been to integrate various reasoning systems, as we believe it is not possible for a single system to solve this problem. In particular, we have employed the following systems:

Spass is a resolution-based automated theorem prover for first-order logic with equality [16]. Spass combines a superposition calculus with specific inference/reduction rules for sorts and a splitting rule for case analysis.

MACE-4 is a model generator that searches for finite models of first-order formulae [9]. For a given domain size, all instances of the formula over the domain are constructed and a decision procedure searches for satisfiable instances of the formula. The distribution package contains several auxiliary programs, including the *isofilter* program, which detects and removes isomorphic models, returning a set of pairwise non-isomorphic models.

HR is a machine learning program which performs automated theory formation by building new concepts from old ones [1] using a set of production rules. It uses the examples of the concepts to empirically form conjectures relating their definitions [2] and employs third party theorem proving and model generation software to prove/disprove the conjectures.

C4.5 is a state of the art decision tree learning system which has been used with much success for many predictive induction tasks [12].

Gap is a computer algebra system with special application to algebraic domains [3]. We used Gap as a toolbox to implement various computer algebra algorithms involving generators and factorisations of algebras.

3   Approach One: Semi-automated

3.1   Stage 1: Generating Non-isomorphic Algebras

The goal of this stage was to produce a covering set of pairwise non-isomorphic algebras of the given type and size, i.e., a single representative of each isomorphism class. To do this, we used Mace in two parts: by constructing all structures with the given properties and cardinality, then applying the isofilter tool to remove isomorphic structures. We also used lists of loops up to size 7 and lists of quasigroups up to size 5 kindly provided by Wendy Myrvold.

3.2 Stage 2: Generating Classifying Concepts

Given a set of pairwise non-isomorphic algebras, $A_1, \ldots, A_n$ such as those generated in stage 1, the problem in stage 2 was to find a set of boolean properties $P_1, \ldots, P_n$ such that P_i is a classifying property for $[A_i]_\cong$, for all i. To do this, we used only HR's concept formation abilities. Starting with some background concepts – in this case, those derived from the axioms of the algebras being considered, e.g., multiplication, inverse, identity, etc., – HR builds new concepts from old ones using a set of production rules. For this application, we used the compose, exists, match and negate production rules. To see how these work, consider starting in group theory with the concept of triples of elements $[a, b, c]$ related via the multiplication concept, i.e., $a \circ b = c$. In its search for concepts, HR might use the compose rule (which conjoins clauses in definitions) to compose this concept with itself, producing the concept of triples of elements related via commutativity: $[a, b, c]$ such that $a \circ b = c \wedge b \circ a = c$. Later, HR might use the negate production rule (which negates clauses in definitions) to produce this concept: $[a, b, c]$ such that $a \circ b = c \wedge b \circ a \neq c$. It then might use the exists production rule (which introduces existential quantification) to produce the concept of non-Abelianess, i.e., groups for which $\exists a, b, c . a \circ b = c \wedge b \circ a \neq c$. Finally, it might use the negate rule again to produce the concept of Abelianess: groups for which $\nexists a, b, c . a \circ b = c \wedge b \circ a \neq c$. In this way, we can see how HR can solve the classification problem for groups of size 6, as mentioned in Sec. 1.

We started by simply using HR to exhaustively generate concepts until it had produced boolean properties $P_1, \ldots, P_n$ as required above. This scheme worked for some simple classification problems, but it did not scale up, and various improvements were made. Firstly, we used a heuristic search whereby the match, exists and negate production rules were used greedily before the compose rule was used. This encourages the generation of the kind of boolean properties we require, and greatly increased the efficiency of HR's search. For instance, using a breadth first search, HR takes over an hour to solve the classification of size 3 quasigroups. However, with the greedy heuristic search, this took only 6 seconds (on a Pentium 4 2GHz machine), producing the following theorem:

Quasigroups of size 3 have exactly one of the following properties:
(i) $\nexists a . a \circ a = a$ (ii) $\forall a . a \circ a = a$ (iii) $\exists a . \nexists b, c . b \circ c = a \wedge a \circ b = c$
(iv) $\exists a . \nexists b, c . b \circ a = c \wedge c \circ b = a$ (v) $\exists a . \nexists b, c . a \circ b = c \wedge c \circ a = b$.

Using this simple scheme, HR was able to solve classification problems for groups of size 6 and 8, loops of size 4 and 5, quasigroups of size 3, qg4-quasigroups of size 5 and qg5-quasigroups of size 7 [note that qg4-quasigroups are quasigroups such that $\forall a, b . (b \circ a) \circ (a \circ b) = a$ and qg5-quasigroups are such that $\forall a, b . (((b \circ a) \circ b) \circ b) = a]$. As an interesting example of the kind of result produced in this way, in a session working with the five isomorphism classes for groups of size

```
p3=true:                        Properties: p1: ∃b. b⁻¹ ≠ b
|  p16=true:class3
|  p16=false:class2             p3: ∃b,c,d. b ∘ c = d ∧ c ∘ b ≠ d
p3=false:                       p10: ∄b. b⁻¹ = b ∧ ∄c. c ∘ c = b
|  p1=false:class0              p16: ∃b. ∄c,d. c ∘ b = d ∧ c ∘ d ≠ b ∧ b ∘ d = c
|  p1=true:
|  |  p10=true:class4
|  |  p10=false:class1
```

Fig. 1. Decision tree produced by HR/C4.5 for groups of size 8

8, we gave HR the additional concept of commutators (elements which can be expressed as $a \circ b \circ a^{-1} \circ b^{-1}$ for some a and b), and HR produced four classifying concepts, which enabled us to form this (paraphrased) classifying theorem:

Groups of order 8 can be classified by their self-inverse elements $(x^{-1} = x)$: they will either have (i) all self inverse elements (ii) an element which squares to give a non-self inverse element (iii) no self-inverse elements which aren't also commutators (iv) a self inverse element which can be expressed as the product of two non-commutative elements or (v) none of these properties.

As a second improvement, we used HR differently: we asked it to form a theory for a given amount of time (up to an hour), then asked it to output all the boolean properties it had produced, regardless of whether they were classifying properties or not. The output was produced in Prolog format, and we used Sicstus Prolog to search for conjunctions of the boolean properties which were true of single algebras. This method helped us scale up further: we were able to find five classifiers for groups of size 8 much more efficiently than with HR alone. More interestingly, we solved the classification problem of quasigroups of size 4 (for which there are 35 isomorphism classes). Using HR for 20 minutes, followed by a Prolog search lasting 5 minutes, the classification theorem produced had 9 classifying properties which were single boolean properties from HR, 25 classifying properties which were conjunctions of two boolean properties and 1 classifying property which was a conjunction of three boolean properties.

As an alternative to using Prolog, we experimented using C4.5: by giving it the boolean properties generated by HR and making the isomorphism classification the one for C4.5 to learn, we were able to produce decision trees such as the one in Fig. 1 (for groups of size 8, presented in the format produced by C4.5). We see that conjoining the nodes of the tree from root to leaf for each leaf provides a set of classifying properties, hence a classification theorem can be derived from the tree. Using C4.5 was problematic, however because (a) due to statistical constraints, we had to multiply our data, e.g., give 10 copies of each algebra, before the algorithm would learn a tree and (b) due to a heuristic search, C4.5 would sometimes learn a tree which was less than 100% accurate, hence could not be used as a theorem. However, we believe that with more experimentation, we will be able to overcome these difficulties.

Using one or other of the above schemes, we found classification theorems for each of the algebra/sizes we tried, with one exception: the classification problem for loops of size 6, with 109 isomorphism classes. For this problem, using the Prolog extension, we found classifying properties for only around half the isomorphism classes, and using the decision tree extension, C4.5 produced a tree, but it was not 100% accurate and didn't cover all the classes.

3.3   Stage 3: Verifying the Classification

None of HR, Mace or C4.5 have been formally verified to work correctly. Hence, it was necessary to use an automated theorem prover to prove that the classification theorems were correct. HR and C4.5 provide classification results in the form of a set of pairs $(A_1, P_1), \ldots, (A_n, P_n)$ of structures A_i and discriminant properties P_i. We call these *classification pairs* since P_i characterises an isomorphism class of algebras with axiomatic properties $\mathcal{P}$ and cardinality c, and A_i is a representant for this isomorphism class. To verify a classification result, we prove the following theorems with Spass:

Discriminant Theorems: Each property P_i is a discriminant, i.e., if P_i holds for one algebra but does not hold for another algebra, then the two algebras are not isomorphic. This also guarantees that P_i is an invariant under isomorphism, since this property is logically equivalent.
Representant Theorems: For each pair (A_i, P_i), A_i satisfies P_i and $\mathcal{P}$.
Non-Isomorphic Theorems: For two pairs (A_i, P_i) and (A_j, P_j) the representants A_i and A_j are not isomorphic. This verifies the results of Mace and guarantees against the construction of too many isomorphism classes.
Isomorphism-Class Theorems: For each pair (A_i, P_i), all algebras of cardinality c that satisfy P_i and $\mathcal{P}$ are isomorphic.
The Covering Theorem: For each algebra of cardinality c that satisfies $\mathcal{P}$, $P_1 \vee \ldots \vee P_n$ holds. This guarantees that all isomorphism classes are found.

Properties P_i are invariants (i.e., discriminants) for arbitrary algebras. Thus, the discriminant theorems do not depend on the properties $\mathcal{P}$ and the cardinality c. To encode the discriminant theorems for proof by Spass, we employ two sort predicates s_1 and s_2 and two binary operations $\circ_1$ and $\circ_2$ to model two algebras, which are closed with respect to s_1 and s_2, respectively. The required discriminant properties are encoded with sorted quantifiers. Both kinds of theorems require the encoding of an arbitrary isomorphism h between the two algebras.

The other theorems depend on the properties $\mathcal{P}$ and P_i as well as the cardinality c (e.g., a property P_i (typically) specifies an isomorphism class only for a particular $\mathcal{P}$ and c). For the covering and isomorphism-class theorems, we have to model arbitrary algebras of cardinality c. In our experiments, we discovered that Spass performed best on these problems when transforming the first-order encoding into a propositional logic encoding without quantifiers (we also tested an encoding with sorts and sorted quantifiers). We first encode the fact that there are c different elements $e_1, \ldots, e_c$. Then, the quantifiers of all used properties are expanded with respect to $e_1, \ldots, e_c$: a universal quantification $\forall x. Q[x]$

results in the conjunction $Q[e_1] \wedge \ldots \wedge Q[e_c]$, whereas an existential quantification $\exists x. Q[x]$ results in the disjunction $Q[e_1] \vee \ldots \vee Q[e_c]$.

For the isomorphism-class theorems, we have to prove the existence of an isomorphism between two arbitrary algebras satisfying P_i and $\mathcal{P}$, which is generally a much harder task than constructing an isomorphism for two concrete structures. As we deal only with finite algebras of given cardinality c, we can enumerate the $c!$ possible bijective functions between two algebras of cardinality c and explicitly axiomatise them as pointwise defined functions. Then, the conclusion of an isomorphism-class theorem is that (at least) one of the bijective functions is a homomorphism, i.e., $homo(h_1) \vee \ldots \vee homo(h_{c!})$. The isomorphism-class theorems turned out to be the most complex problems for Spass. In particular, their complexity heavily depends on the cardinality, c, since there are $c!$ potential bijective functions. In order to reduce the complexity of these theorems, we consider sets of generators and factorisations to decrease the number of potential isomorphism mappings. In this context, a structure $(A, \circ)$ is generated by a set of elements $\{a_1, \ldots, a_m\} \subseteq A$ if every element of A can be expressed as a combination – usually called a factorisation or word – of the a_i under the operation $\circ$. For example, quasigroup Q_3 from Sec. 2.1 is generated by element $b \in A$, as both $c = b \circ b$ and $a = b \circ (b \circ b)$ can be expressed as factorisations in b.

Note that generators and factorisations are invariants of isomorphisms. That is, if a structure $(A, \circ)$ is generated by $\{a_1, \ldots, a_m\} \subseteq A$ and h is an isomorphism mapping $(A, \circ)$ to a structure $(A', \circ')$, then $\{h(a_1), \ldots, h(a_m)\} \subseteq A'$ is a generating set for $(A', \circ')$. Moreover, given a factorisation, in terms of the a_i's, for an element $a \in A$, then one obtains the factorisation for $h(a)$ by simply replacing each of the a_i's with $h(a_i)$, respectively.

To compute generators and factorisations we employed a computer algebra algorithm, which we encoded in the Gap system. For a given structure, this algorithm constructs a minimal set of generators together with a set of factorisations expressing each element of the structure in terms of the generators. The set of generators is constructed by successively combining elements with the longest traces until all elements of the set can be generated. If necessary, the set of generators is then reduced if any of its elements is redundant. Note that this approach works for all types of algebraic structures, regardless of the number of operations, and ensures that the set of generators is minimal in the sense that none of its subsets generates the full set.

If a classification pair (A_i, P_i) characterises an isomorphism class and if our algorithm computes generators $\{a_1, \ldots, a_m\} \subseteq A_i$ and factorisations for A_i, then all algebras with property P_i (and cardinality c and properties $\mathcal{P}$) have a generating set with m elements and the factorisations of the elements of A_i. We prove this as additional theorem which we call the *Gensys-Verification Theorem*. Having proved this, we can express the isomorphism-class theorem for the algebras using the generators and factorisations. This reduces the number of functions which are candidates for isomorphisms. Instead of $c!$, there are only $\frac{c!}{(c-m)!}$ possible mappings, since only the m generators have to be mapped explicitly. This is because each isomorphism is uniquely determined by the images of these generators.

We ran Spass on a Linux PC with four 2GHz Xeon processors and 4GB RAM and proved HR's groups6, loops4, loops5, quasigroups3 and qg4-quasigroups5 results. For the quasigroup4 classification problem, Spass failed due to internal problems while proving the covering theorem. For the qg5-quasigroups7 and the groups8 results, Spass was not able to prove all the gensys-verification theorems within a two day time limit. For the discriminant theorems, Spass typically required less than a second. However, there were a few examples which needed considerably more time (e.g., one proved theorem in the quasigroup4 classification problem was to show that $\forall x. \exists y. (x \circ y) \circ (x \circ y) = y$ is a discriminant; Spass needed 1h 6m to prove this theorem). The performance of Spass on representant and non-isomorphic theorems depended slightly on the cardinality of the examined algebras. Spass needed less than a second to prove the theorems for the smaller algebras, but took several seconds to prove the theorems for the larger algebras.

The complexity of covering theorems depends not only on the cardinality of the examined algebras, but also on the number of isomorphism classes and the classifying properties. Unfortunately, Spass failed on the covering theorem resulting from the 35 quasigroups4 isomorphism classes, but it proved the groups8 covering theorem in about 4 seconds (this theorem is particularly simple since it has the form $Q \vee \neg Q$ where $Q = P_1 \wedge P_2 \wedge P_3 \wedge P_4$), and it took about 4 minutes to prove the loops5 covering theorem. In addition, isomorphism-class and gensys-verification theorems depend heavily on the cardinality of the examined algebras, as we shall discuss further in Sec. 5.

4   Approach Two: Bootstrapping

After experimenting with various schemes in approach 1, we identified some limitations, which enabled us to specify the following requirements for an algorithm in our second approach:

- The process should be entirely automatic and bootstrapping, able to produce verified classification theorems starting from the axioms alone, with no human intervention.
- The process should call HR to find classifying concepts for small numbers of isomorphism classes, as HR struggled to solve larger classification problems.
- The process should not require the production of *every* algebraic structure satisfying the axioms. This is because, when using Mace to generate all structures and reduce them using its isomorphism filter, we found that the intermediate files produced could often be unmanageably large (up to 4GB).
- The process should generate the classification theorem as a decision tree. We found that decision trees often involved fewer concept definitions and enabled easier classification of algebras.

Input: Cardinality c and basic properties $\mathcal{P}$ of the algebraic structures
Output: Decision tree $(r, \mathcal{V}, \mathcal{E})$

1. **Initialise** $\mathcal{V} := \{r\}$, $\mathcal{E} := \emptyset$.
2. Let $\mathcal{S}$ be the set of unprocessed nodes of the tree, initially $\mathcal{S} := \{r\}$.
3. **While** $\mathcal{S} \neq \emptyset$ do
     3.1. Pick $v \in \mathcal{S}$ with $l(v)=P_v$ (i.e. the label of v specifies the properties P_v)
     3.2. **If** there exists a model m that satisfies $\mathcal{P} \cup P_v$ then:
       3.2.1. **If** there exists $m' \not\cong m$ satisfying $\mathcal{P} \cup P_v$ then:
         3.2.1.1. Construct discriminants P_1 and P_2 for m and m', respectively.
         3.2.1.2. **If** $P_1 = \neg P_2$ then create two child vertices v_1, v_2 with edges e_1, e_2 such that $l(e_1) = P_1, l(e_2) = P_2$. $\mathcal{S} := \mathcal{S} \cup \{v_1, v_2\}$.
         3.2.1.3. **Else** create four child vertices v_1, v_2, v_3, v_4 and edges $e_1, \ldots, e_4$ such that $l(e_1) = P_1 \wedge P_2$, $l(e_3) = P_1 \wedge \neg P_2$, $l(e_2) = \neg P_1 \wedge P_2$, and $l(e_4) = \neg P_1 \wedge \neg P_2$. $\mathcal{S} := \mathcal{S} \cup \{v_1, v_2, v_3, v_4\}$.
       3.2.2. **Else** all structures of cardinality c with $\mathcal{P} \cup P_v$ are isomorphic: Mark v as a leaf representing an isomorphism class.
     3.3. **Else** no structures of cardinality c with $\mathcal{P} \cup P_v$ exist: Mark v as empty leaf.
     3.4. $\mathcal{S} := \mathcal{S} \setminus \{v\}$.
4. **Return** $(r, \mathcal{V}, \mathcal{E})$.

Fig. 2. The bootstrapping algorithm used in approach 2.

The algorithm portrayed in Fig. 2 satisfies these criteria. This constructs a decision tree by successively generating non-isomorphic algebraic structures and associated discriminants until the maximal number of isomorphism classes is reached. The result serves as a qualitative classification for the specified algebras up to isomorphism for a given cardinality. The algorithm takes the cardinality c and axiomatic properties $\mathcal{P}$ of the algebraic structures to be considered as input. It returns a decision tree $(r, \mathcal{V}, \mathcal{E})$ for isomorphism classes of the algebraic structure, where $\mathcal{V}$ is a set of vertices, $\mathcal{E}$ a set of edges, and $r \in \mathcal{V}$ is the root of the tree. Each edge $e \in \mathcal{E}$ is labelled with an algebraic property and each vertex $v \in \mathcal{V}$ is labelled with the conjunction of properties on the path from r to v.

In principle the algorithm is complete, i.e., for given cardinality c and axiomatic properties $\mathcal{P}$ it provides a classification theorem after a finite number of steps. In practice, however, the success of the algorithm depends on whether all used systems actually provide the answers they are supposed to deliver (see Sec. 5 for the discussion of the practical limitations of the algorithm).

To illustrate the algorithm, we use the example of the decision tree constructed to classify quasigroups of order 3, given in Fig. 3 (here, the vertices of the tree are enumerated rather than assigned their actual labels, to preserve space.)

Initially, the decision tree consists only of the root node. The single nodes of the tree are expanded by first generating an algebraic structure satisfying the properties specified by the node (step 3.2). For example, for the root node 1 in Fig. 3, an arbitrary quasigroup of order 3 is constructed; for node 3, however, a

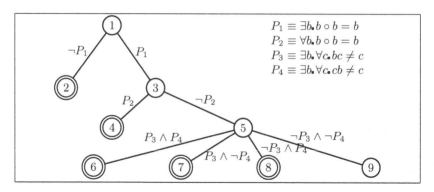

Fig. 3. Decision tree for the classification problem of order 3 quasigroups.

quasigroup that additionally satisfies the property $\exists b. b \circ b = b$ is required. After such a *representant* is generated by Mace, we check with Spass that it actually satisfies the required properties. The first representant generated in our example was quasigroup A_2 given in Fig. 4. If no representant can be generated, we must prove that there exists no algebraic system of cardinality c satisfying both the original axioms $\mathcal{P}$ and the additional properties generated up to this point (step 3.3). This proof is again done by Spass. An example of this case is node 9, i.e., no quasigroup of size 3 can be constructed which satisfies $P_1 \wedge \neg P_2 \wedge \neg P_3 \wedge \neg P_4$.

When Mace does produce a model, we have two cases to consider: there exists a non-isomorphic structure exhibiting the same properties considered so far (step 3.2.1) or the property represented by the node constrains the structures to a single isomorphism class (step 3.2.2). In the latter case, we employ Spass to prove the appropriate isomorphism-class theorem (using Gap as in approach 1 to re-write the theorems in terms of generators and factorisations). For instance, in node 2 we show that all quasigroups of order 3 without idempotent elements belong to one isomorphism class. In the former case, we employ Mace to generate a structure non-isomorphic to the original one. The two structures are then passed to HR to compute two discriminants (step 3.2.1.1). Again, Spass does the necessary verification of the non-isomorphic theorem and the discriminant theorems. In the example, the quasigroup A_4 given in Fig. 4 was constructed as a non-isomorphic counterpart to A_2. Given those two quasigroups, HR came up with discriminants P_1 and $\neg P_1$ as stated in Fig. 3. Depending on the nature of the discriminants, either two or four child nodes are constructed (steps 3.2.1.2 & 3.2.1.3). The case of two child nodes can be observed in the expansion of nodes 1 and 3 in Fig. 3, whereas the expansion of node 5 leads to four children. The

A_2	a b c		A_4	a b c		A_6	a b c		A_7	a b c		A_8	a b c
a	b a c		a	a c b		a	a b c		a	a b c		a	a c b
b	a c b		b	c b a		b	b c a		b	c a b		b	b a c
c	c b a		c	b a c		c	c a b		c	b c a		c	c b a

Fig. 4. Isomorphism class representants for quasigroups of order 3.

advantage of the former case is that when expanding the two child nodes, the models for step 3.2 can be reused and do not have to be produced with Mace, whereas in the second case, Mace has to be called for two of the nodes.

Once the decision tree is fully expanded, the discriminating properties of all the isomorphism classes give the final classification theorem. In our example, this corresponds to a disjunction of the labels of the doubly-circled leaf nodes. Although we omit a formal proof, both the construction of the decision tree as well as the fact that we work in classical logic (i.e. *tertium non datur* holds) guarantee that the final decision tree determines all possible isomorphism classes and a full classification for the algebraic systems specified by the input parameters. Importantly, the correctness of our implementation depends only on the correctness of Spass.

The branching factor of up to four of the tree is influenced by two design decisions for our algorithm: On the one hand, we could have kept the decision tree binary by always branching with respect to one discriminant and its negation. However, this would have meant losing some of the more intuitive discriminants generated by HR, as well as discarding previously computed results and thus increasing the number of calls to HR. Alternatively, we could have given HR more than two non-isomorphic structures to compute discriminants for, thereby reducing the number of calls to HR and increasing the branching factor of the tree. We decided against this, because this increases the risk that HR would not come up with an answer, or produce lengthier, more complex discriminants, which are usually more difficult to verify with Spass.

Both the size of the decision tree and the number of calls to Spass and Mace can become fairly large, even when considering a relatively small number of structures. However, the algorithm offers some potential for parallelism and distribution. We currently exploit this by parallelising steps 3.2.1 and 3.2.2, i.e. generating a non-isomorphic structure with Mace and proving the isomorphism-class theorem with Spass. This actually increases efficiency, since both computations are generally very expensive and take a long time. We gain another small speed up by constructing the proofs for step 3.2.1.1 in parallel with Spass. We highlight more potential for parallelism in Sec. 6.

5   Results from the Bootstrapping Approach

Given the more ad-hoc, semi-automated nature of approach 1 when compared to approach 2, and given that approach 2 was found to be more effective, while we have supplied some illustrative results for approach 1 in Sec. 3, we concentrate here on the testing we undertook for approach 2. We tested the hypothesis that the bootstrapping system can generate and verify full classification theorems for loops, groups, quasigroups and monoids up to size 6. Hence we experimented by using the system to generate classification theorems, and Table 1 summarises the results of these experiments. For each algebra, the table describes the decision tree constructed to classify it, in terms of the number of nodes, the number of identified isomorphism classes, and the maximal depth of the tree. Moreover, it

Table 1. Results of the experiments with the bootstrapping approach.

Algebra	Nodes	IsoClasses	Max-Depth	Isoclass Proof	Gensys-Verification Proof
Monoids3	13	7	5	< 1s	< 1s
Quasigroups3	9	5	4	< 1s	< 1s
Quasigroups4	71	35	9	17s	50s
Loops4	3	2	2	2s	< 1s
Loops5	11	6	5	21s	45s
Loops6	233	109	17	3m4s	668m17s
Groups4	3	2	2	< 1s	< 1s
Groups6	3	2	2	1m40s	6m10s

provides the mean times of Spass runs to prove the necessary isomorphism-class and gensys-verification theorems, i.e., the most complex proof problems.

We see that the proof time taken by Spass significantly increases with the cardinality of the algebra as well as with the depth of the trees (the higher the depth, the more properties associated with the nodes). In particular, these statistics suggest that cardinality 7 is the borderline for the discussed techniques, since we cannot hope to prove, in particular, gensys-verification theorems beyond cardinality 8.[1] However, we believe it is a significant achievement to produce a classification theorem for the 109 isomorphism classes of loops of size 6. Moreover, we currently employ Spass in auto-mode, and it might be possible to push the solvability horizon with settings tailored to our problems.

The time Mace needs to construct a model for a node in a tree depends on both the cardinality of the structures and the depth of the node in the tree. The deeper the node in the tree, the more properties the model has to satisfy. For instance, for the initial nodes of Loops6, Mace needed less than a second to construct a model. For the deep nodes it needed up to 15 seconds. The time Mace needs to construct a non-isomorphic structure is typically considerably longer (up to 4 minutes for Loops6). This is not surprising, as the encoding of these Mace problems comprises many additional symbols for the homomorphisms. For nodes that correspond to isomorphism classes, no non-isomorphic structures exist, and for these cases, Mace had to traverse the entire search space, which could take up to an hour for Loops6.

In the experiments summarised above, the bootstrapping algorithm always computed only two non-isomorphic structures, which it passed to HR, and HR succeeded in finding suitable discriminants in every case. Moreover, in no cases were Mace or HR shown to have performed incorrect calculations. We also experimented for Loops6 with settings that passed 3 and 4 non-isomorphic structures together to HR. In these experiments HR often took considerably longer and sometimes failed to provide discriminants. Moreover, when successful, it created discriminants that were considerably more complex as they involved more quan-

---

[1] HR succeeded to classify groups of order 8. In approach 1, when verifying the classification of HR, Spass was able to prove all 5 isomorphism-class theorems (mean time: around 7 hours) but succeeded to prove only one gensys-verification theorem (in about 19 hours). We interrupted the other runs of Spass after 3 days.

tifiers and sub-formulas. These in turn made the verification problems much more difficult for Spass.

6   Conclusions and Further Work

We have presented a novel approach to constructing classification theorems in pure mathematics, which has produced novel and interesting results of a qualitative nature. The classification theorems produced have often contained classically interesting results such as commutativity and idempotency, and we believe, especially for the larger classification problems, that no such theorems have ever been produced. Our bootstrapping algorithm successfully exploits the strengths of diverse reasoning techniques including deduction, inductive learning, model generation and symbolic manipulation, while avoiding many weaknesses we identified in earlier approaches, which we have also presented. The collaboration of the various systems produces results that clearly cannot be achieved by any single system, and the incorporation of external systems offers improved flexibility, as we can profit from any advances of the individual technologies.

Thus far we have dealt only with isomorphism classes of relatively simple algebraic structures. However, our approach is adaptable to more complicated algebraic domains and indeed Spass, Mace and HR have all been demonstrated to work in algebras over two operators like rings. In addition, we also want to experiment with different types of equivalence relations, which can lead to more insights about the structures under consideration. For instance, quasigroups and loops can also be grouped into isotopy classes, and modules can be distinguished with respect to their uniform dimensions. Moreover, classification is not only interesting in algebraic domains, and our approach could also be applied to other domains such as analysis, differential geometry or number theory.

In addition to scaling up in terms of the complexity of the domains looked at, we also hope to produce classification theorems for larger sizes. The bootstrapping approach provides much potential for parallelisation, some of which we already exploit in our current implementation. In addition, the decision tree offers potential for distribution since new threads can be spawned for already existing vertices by applying the algorithm to the original axioms extended by the properties a vertex is labelled with. Nevertheless, the results of our experiments suggest that cardinality 8 is the borderline of our current approach. A general first-order theorem prover like Spass seems unable to prove isomorphism-class and gensys-verification problems with larger cardinalities. There are two possible solutions to further push the solvability horizon. Firstly, we could try to involve further symbolic computations to simplify the problems at hand. Secondly, special theorem proving techniques for finite algebras could be developed and employed.

In mathematics, classifying theorems are often generative, e.g., Kronecker showed that, by constructing certain direct products of cyclic groups, it is possible to generate every Abelian group. To produce such generative theorems, it may be necessary to more systematically construct discriminants in order to gain comparable properties for structures of diverse cardinality, and to work with more complicated concepts such as maps between algebraic structures and products of algebras, such as the direct product. We believe that as systems such as the one we have presented here become more sophisticated, they may be of use to pure mathematicians. For instance, currently only Abelian quasigroups are classified [13], whereas we have classified all quasigroups, albeit only up to a certain size. We intend to analyse the classification theorems produced by the system in order to identify some concepts which may form a part of a general classification theorem. In particular, we intend to work with algebras associated with Zariski spaces [7], a relatively new domain of pure mathematics which we hope to explore via automated means.

Acknowledgements. We would like to thank Wendy Myrvold and colleagues for providing data on isomorphism classes for loops and quasigroups, Bill McCune for expert advice about Mace, Thomas Hillenbrand for providing us with an improved version of Spass and helping us to encode our proof problems, and Geoff Sutcliffe for helping us to determine that Spass was best suited for our task.

References

1. S Colton. *Automated Theory Formation in Pure Mathematics*. Springer, 2002.
2. S Colton. The HR program for theorem generation. In Voronkov [15].
3. The GAP Group. *GAP – Groups, Algorithms, and Programming, Version 4.3*, 2002. http://www.gap-system.org.
4. J Humphreys. *A Course in Group Theory*. Oxford University Press, 1996.
5. K Kunen. Single Axioms for Groups. *J. of Autom. Reasoning*, 9(3):291–308, 1992.
6. L Kronecker. Auseinandersetzung einiger Eigenschaften der Klassenanzahl idealer komplexer Zahlen. *Monatsbericht der Berliner Akademie*, pages 881–889, 1870.
7. R McCasland, M Moore, and P Smith. An introduction to Zariski spaces over Zariski topologies. *Rocky Mountain Journal of Mathematics*, 28:1357–1369, 1998.
8. W McCune. Single axioms for groups and Abelian groups with various operations. *J. of Autom. Reasoning*, 10(1):1–13, 1993.
9. W McCune. *Mace4 Reference Manual and Guide*. Argonne National Laboratory, 2003. ANL/MCS-TM-264.
10. B McKay, A Meinert, and W Myrvold. Counting small latin squares. European Women in Mathematics Int. Workshop on Groups and Graphs, pages 67–72, 2002.
11. A Meier, V Sorge, and S Colton. Employing theory formation to guide proof planning. In *Proc. of Calculemus-2002, LNAI* 2385, pages 275–289. Springer, 2002.
12. R Quinlan. *C4.5: Programs for Machine Learning*. Morgan Kaufmann, 1993.
13. J. Schwenk. A classification of abelian quasigroups. *Rendiconti di Matematica, Serie VII*, 15:161–172, 1995.
14. J Slaney, M Fujita, and M Stickel. Automated reasoning and exhaustive search: Quasigroup existence problems. *Comp &Math with Applications*, 29:115–132, 1995.
15. A Voronkov, editor. *Proc. of CADE-18, LNAI* 2392. Springer, 2002.
16. C Weidenbach, U Brahm, T Hillenbrand, E Keen, C Theobald, and D Topic. SPASS version 2.0. In Voronkov [15], pages 275–279.

Efficient Algorithms for Computing Modulo Permutation Theories

Jürgen Avenhaus

FB Informatik, Technische Universität Kaiserslautern, Kaiserslautern, Germany,
avenhaus@informatik.uni-kl.de

Abstract. In automated deduction it is sometimes helpful to compute modulo a set E of equations. In this paper we consider the case where E consists of permutation equations only. Here a permutation equation has the form $f(x_1, \ldots, x_n) = f(x_{\pi(1)}, \ldots, x_{\pi(n)})$ where π is a permutation on $\{1, \ldots, n\}$. If E is allowed to be part of the input then even testing E-equality is at least as hard as testing for graph isomorphism. For a fixed set E we present a polynomial time algorithm for testing E-equality. Testing matchability and unifiability is NP-complete. We present relatively efficient algorithms for these problems. These algorithms are based on knowledge from group theory.

1 Introduction

1.1 Overview on the Paper

In automated deduction one is concerned with the following problem: Given a signature sig and a set $\mathcal{A}x$ of axioms, try to prove that a closed formula F follows from $\mathcal{A}x$ or – more precisely – is valid in all models of the specification spec = (sig, $\mathcal{A}x$). For this problem powerful theorem provers have been developed over the past years. The provers work on clauses. See [BG01] for a recent overview on resolution methods. If $\mathcal{A}x$ contains an "appropriate" set E of universally closed equations then it may be of considerable advantage to compute modulo E, i.e. to work on clauses C that represent the whole E-equivalence class of C. The most prominent example is that E is a set of AC-axioms (associativity and commutativity) of some binary function symbols [NR01]. In this paper we assume that E is a set of permutation equations.

The most basic algorithmic problems for computing modulo E are testing for equality, matching and unification modulo E. We will present algorithms for these problems that are based on knowledge about permutation groups.

To give a simple example, assume that E is given by the permutation axioms

$$E = \{ f(\bar{x}) = f(\pi_i(\bar{x})) \mid i = 1, \ldots, n \}$$

where $\bar{x} = (x_1, \ldots, x_k)$ is a vector, the π_i are permutations on $\mathbb{N}_k = \{1, \ldots, k\}$ and $\pi(\bar{x}) = \pi(x_1, \ldots, x_k) = (x_{\pi(1)}, \ldots, x_{\pi(k)})$ for any permutation π. Let $G = \langle \pi_1, \ldots, \pi_n \rangle$ be the permutation group generated by $\pi_1, \ldots, \pi_n$. Then we have

D. Basin and M. Rusinowitch (Eds.): IJCAR 2004, LNAI 3097, pp. 415–429, 2004.

for f-free terms s_i and t_j and $s \equiv f(s_1, \ldots, s_k)$ and $t \equiv f(t_1, \ldots, t_k)$: $s =_E t$ iff $\pi(\bar{s}) = \bar{t}$ for some $\pi \in G$. It turns out that even this problem is at least as hard as testing whether two graphs with k vertices are isomorphic. On the other hand, if the s_i are pairwise distinct, then this problem reduces to test for a given permutation π whether π is in G. This can be done in polynomial time [FHL80]. After a preprocessing of the $\pi_1, \ldots, \pi_n$ this can be done in time $O(k^2)$.

Now assume that the arities of all top symbols of equations in E are bounded by some $k \in \mathbb{N}$. Let $n = |t|$ be the length of the term t. Based on knowledge about permutation groups we present efficient algorithms for the problems equality, matching and unification modulo E. For E fixed, the test for E-equality is polynomial in $n = |s| + |t|$ (see [AP01]).

This paper is not on the theory of unification. (See [BS01] and [JK91] for surveys on that.) The emphasis of the paper is to present methods to enhance the efficiency of known algorithms. These methods are based on knowledge from the theory of permutation groups from the theoretical point of view and on our experience from writing efficient theorem provers such as WALDMEISTER [HL02] from the implementation point of view. From this experience we learned that it is more efficient to use simple algorithms based on simple data structures than to use sophisticated algorithms based on complex data structures (e.g. multisets as in [AP01]) to avoid redundancies in search problems. This does not prevent usage of sharing and memoization techniques to avoid solving the same subproblem repeatedly. Knowledge about permutation groups allows us to enumerate the permutations $\pi \in G$ in an intelligent way (e.g. to detect failure early) and the implementation techniques allow us to reuse already computed information efficiently. For complexity results on related problems see [dITE03].

We agree that for simple problems it is more efficient to use the naive algorithms for testing equality, matchability and unifiability modulo E. But if deep terms and large permutation groups occur our algorithms will outperform the naive ones.

The paper is organized as follows. In section 2 we present what is needed from group theory later on. In section 3 we present our algorithms for computing modulo E. In section 4 we outline some generalizations. Missing proofs can be found in [Ave03].

1.2   Notations

We use standard notations. A (one sorted) signature is a pair $sig = (F, \alpha)$ where F is a set of function symbols and $\alpha : F \to \mathbb{N}$ fixes the arity of any $f \in F$. Let V be a set of variables, then $\text{Term}(F, V)$ is the set of terms over F and V. We write $s \equiv t$ if s and t are syntactically equal. The length of t is denoted by $|t|$ and $\text{depth}(t)$ is its depth. The top symbol of a term t is denoted by $\text{top}(t)$, i.e. $\text{top}(t) = f$ if $t \equiv f(t_1, \ldots, t_n)$ and $\text{top}(t) = x$ if $t \equiv x$. $\text{Var}(t)$ is the set of variables in t and $\text{Sub}(t)$ is the set of all subterms of t. We denote by t/p the subterm of t at position p and write $t[s]_p$ to denote t with the subterm t/p replaced by s. An equation has the form $s = t$ with $s, t \in \text{Term}(F, V)$.

A substitution is $\sigma : V \to \text{Term}(F, V)$ such that the domain $\text{dom}(\sigma) = \{x \mid \sigma(x) \not\equiv x\}$ is finite. Then σ is extended to $\sigma : \text{Term}(F, V) \to \text{Term}(F, V)$

homomorphically as usual. σ is a *match* from s to t iff $\sigma(s) \equiv t$. A *unifier* of s and t is a substitution σ such that $\sigma(s) \equiv \sigma(t)$. As usual we generalize the concept of unifiers to sets E of equations: σ is a unifier for E if $\sigma(s) \equiv \sigma(t)$ for all $s = t$ in E. If there is a unifier of E then there always exists a *most general unifier* $\sigma = \text{mgu}(E)$.

Let E be a set of equations. Then $t_1 \vdash_E t_2$ iff $t_1 \equiv t_1[\sigma(u)]_p$ and $t_2 \equiv t_1[\sigma(v)]_p$ for some $u = v$ in E and some σ. Then $=_E$ is the reflexive, symmetric and transitive closure of $\vdash_E$. Two terms s, t are *E-equal* iff $s =_E t$.

A substitution σ is an *E-match* from s to t iff $\sigma(s) =_E t$. And σ is an *E-unifier* of s and t if $\sigma(s) =_E \sigma(t)$. We write $\sigma_1 =_E \sigma_2$ iff $\sigma_1(x) =_E \sigma_2(x)$ for all $x \in \text{dom}(\sigma_1) \cup \text{dom}(\sigma_2)$.

For doing rewriting or deduction modulo E (see e.g. [BN98], [Ave95], [NR01]) one needs to solve the following problems: For given $s, t \in \text{Term}(F, V)$ decide whether

(1) s is E-equal to t,
(2) s is E-matchable to t,
(3) s is E-unifiable with t.

For problem (3) it may be necessary to compute a $\text{CSU}_E(s, t)$. In the rest of the paper we will study these problems for a set E of permutative equations.

2   Permutative Equivalence on Strings

In this section we consider two problems on strings of a fixed length. These problems (and their solution) are basic to the rest of the paper. We first recall some results on permutation groups to solve these problems.

A *permutation* on $\mathbb{N}_k = \{1, \ldots, k\}$ is a bijective function $\pi : \mathbb{N}_k \to \mathbb{N}_k$. These permutations form a group $\text{Sym}(k)$ by using composition of functions as the group multiplication and the inverse function π^{-1} of π as the group inverse. Then id (the identity mapping) is the unit element of $\text{Sym}(k)$. For $\pi_1, \ldots, \pi_n \in \text{Sym}(k)$ we denote by $G = \langle \pi_1, \ldots, \pi_n \rangle$ the subgroup of $\text{Sym}(k)$ generated by the π_i: G consists of all finite products of the π_i and π_i^{-1}.

Theorem 1 ([FHL80]).
Let $\pi_1, \ldots, \pi_n \in \text{Sym}(k)$ be given and let $G = \langle \pi_1, \ldots, \pi_n \rangle$. One can compute in polynomial time a table $T = (g_{ij})_{1 \le i \le j \le k}$ with $g_{ij} \in G, g_{ii} = \text{id}$ and

$$g_{ij}(m) = m \text{ for } 1 \le m < i \le k \text{ and } g_{ij}(j) = j.$$

If no such $g_{ij} \in G$ exists then the entry at position (i, j) in T is $-$. Now each $\pi \in G$ has a (according to T unique) factorization $\pi = g_{1i_1} \circ \cdots \circ g_{ki_k}$.   □

Example 1. Let $k = 6$ and π_1, π_2 be given as below. This results in T with non-empty entries at positions (i, i) and $(1, j)$ for $j = 2, \ldots, 6$ and $(2, 6)$ only. (See [FHL80] for computing T.) The g_{ij} are given by

$$
\begin{array}{c|c}
 & 1\ 2\ 3\ 4\ 5\ 6 \\
\hline
g_{12} = \pi_1 & 2\ 3\ 4\ 5\ 6\ 1 \\
g_{13} = \pi_1^2 & 3\ 4\ 5\ 6\ 1\ 2 \\
g_{14} = \pi_1^3 & 4\ 5\ 6\ 1\ 2\ 3 \\
g_{15} = \pi_2 & 5\ 4\ 3\ 2\ 1\ 6 \\
g_{16} = \pi_1\pi_2 & 6\ 5\ 4\ 3\ 2\ 1 \\
g_{26} = \pi_1^2\pi_2 & 1\ 6\ 5\ 4\ 3\ 2
\end{array}
$$

$$
\begin{array}{c|c}
i & 1\ 2\ 3\ 4\ 5\ 6 \\
\hline
\pi_1 & 2\ 3\ 4\ 5\ 6\ 1 \\
\pi_2 & 5\ 4\ 3\ 2\ 1\ 6
\end{array}
$$

As a result, each $\pi \in G$ has the canonical form $\pi = g_{1j}g_{2m}$ with $1 \le j \le 6$ and $m \in \{2, 6\}$. So G has exactly $6 \times 2 = 12$ elements. □

We remark that Theorem 1 helps us to factorize $\pi \in G$ in polynomial time. Indeed, we know that $\pi = g_1 \circ g_2 \circ \cdots \circ g_k$ with g_i in row i of table T. This leads to

$$\pi(i) = g_1 \cdots g_k(i) = g_1 \cdots g_i(i) \quad \text{for } i = 1, \ldots, k$$
$$g_i(i) = g_{i-1}^{-1} \cdots g_1^{-1}\pi(i) \qquad \text{for } i = 1, \ldots, k$$

Using table T this allows us to compute g_i for $i = 1, \ldots, k$. For any $\pi \in \mathrm{Sym}(k)$ we have $\pi \in G$ iff this factorization succeeds. So one can decide in polynomial time for any $\pi \in \mathrm{Sym}(k)$ whether $\pi \in G$ holds [FHL80].

Now we come to the permutative equivalence problems on strings. We use the following notation. For any alphabet $A = \{a_1, \ldots, a_r\}$, $u = u_1 \cdots u_k \in A^k$ and $\pi \in \mathrm{Sym}(k)$ we write $\pi(u) = u_{\pi(1)} \cdots u_{\pi(k)}$. So we have $\pi(u)_i = u_{\pi(i)}$.

Definition 1. *The permutative equivalence problem (PEP) for strings is*
*   Input:* $G = \langle \pi_1, \cdots, \pi_n \rangle$ *a subgroup of* $\mathrm{Sym}(k)$
              $u, v \in A^k$.
*   Question: Is there a permutation* $\pi \in G$ *such that* $\pi(u) = v$?

One can easily prove (see [Ave03]).

Theorem 2. *The graph isomorphism problem GI is polynomial time reducible to PEP.* □

It is not known whether GI is NP-complete. But it is widely believed that GI is not solvable in polynomial time. So we do not expect to find a polynomial time algorithm for PEP.

On the other hand, Theorem 1 opens the door to an algorithm that works efficiently in many situations: If $u, v \in A^k$ are given and the table T is already computed then there may exist several $\pi \in \mathrm{Sym}(k)$ such that $\pi(u) = v$. We use T to find such a $\pi \in G$ by backtracking. We outline an algorithm how to do this. Recall that for any $w \in A^k$ and for any $\pi \in Sym(k)$ we have $\pi(w)_i = w_{\pi(i)}$.

Now assume that $\pi(u) = v$ for some $\pi \in G$. Then we have $\pi = g_1 \circ \cdots \circ g_k$ with g_i in row i of T and so $u = g_k^{-1} \cdots g_1^{-1}(v)$. For computing $g_1, \ldots, g_k$ from u and v notice that $u_i = u_{g_{i+1} \cdots g_k(i)} = g_{i+1} \cdots g_k(u)_i = g_i^{-1} \cdots g_1^{-1}(v)_i$, hence

$$u_i = g_i^{-1}(v')_i \qquad where \qquad v' = g_{i-1}^{-1} \cdots g_1^{-1}(v)$$

·   So there is a letter v'_j in v' such that $u_i = v'_j$ and v'_j is mapped to place i by g_i^{-1}, so $g^{-1}(j) = i$. Now we have $g_i(i) = j$ and $i \le j \le k$ and $g_i = g_{ij}$ by

Theorem 1. Computing $g_1, \ldots, g_k$ this way leads to the loop invariant that u and $g_i^{-1} \cdots g_1^{-1}(v)$ have a common prefix of length i.

This leads to a non-deterministic algorithm to test whether there is a $\pi \in G$ such that $\pi(u) = v$:

$v' := v$
for $i = 1$ **to** k **do**
   if there is no g_i in row i of T such that $g_i(i) = j$ and $v'_j = u_i$
      then STOP with NO **else** choose such a g_i and put $v' := g_i^{-1}(v')$ **fi**
od
STOP with YES and $\pi = g_1 \circ \cdots \circ g_k$

Example 2 (Example 1 continued).
   Assume $u = a\,a\,a\,b\,c\,b$ and $v = a\,b\,c\,a\,b\,a$. The first letter of u is a. There are three g_1 in the first row of T such that g_1^{-1} moves an a in v to the first place
(1) $g_1 = id$ gives $g_1^{-1}(v) = a\,b\,c\,a\,b\,a$
(2) $g_1 = g_{14}$ gives $g_1^{-1}(v) = a\,b\,a\,a\,b\,c$
(3) $g_1 = g_{16}$ gives $g_1^{-1}(v) = a\,b\,a\,c\,b\,a$
   In the same way we can compute g_2. The full search tree is as follows

$$v = a\,b\,c\,a\,b\,a$$

```
           id            |g14              g16
  a b c a b a     a b a a b c     a b a c b a
     |g26                             |g26
  a a b a c b                   a a b c a b
```

So there are exactly two strings of the form $a\,a\,u'$ that are permutative equivalent (modulo G) to v, and there is no such string of the form $a\,a\,a\,u''$. □

In the next section we need to consider a slightly more general problem than PEP. To describe that let $Compat(u, v, \pi)$ be a (compatability) predicate on $A^k \times A^k \times \text{Sym}(k)$. We assume that $Compat$ is easy to decide.

Definition 2. *The generalized permutative equivalence problem GPEP is*
    Input: $G = \langle \pi_1, \ldots, \pi_n \rangle$, *a subgroup of* $\text{Sym}(k)$
                  $u, v \in A^k$
                  $Compat(u, v, \pi)$, *a predicate on* $A^k \times A^k \times \text{Sym}(k)$.
    Question: *Is there a* $\pi \in G$ *such that* $Compat(u, v, \pi)$ *and* $\pi(u) = v$?

It is easy to design an algorithm $\mathcal{A}$ for GPEP in the same way we have done it for PEP. We will use this algorithm $\mathcal{A}$ in the next sections.

3   Permutation Axioms

In this section E is a fixed finite set of permutation equations. We will show that for any such fixed E we can decide (1) in deterministic polynomial time (measured in $n = |s| + |t|$) whether $s =_E t$ holds and (2) in non-deterministic polynomial time whether s is E-matchable to t and (3) whether s and t are E-unifiable. Furthermore, we are going to develop fairly efficient algorithms for these problems.

We start with some notations. A *permutation equation* for $f \in F$ is of the form

$$f(x_1, \ldots, x_k) = f(x_{\pi(1)}, \ldots, x_{\pi(k)})$$

for some $\pi \in \text{Sym}(k)$. Let E_f be a set of such permutation equations and let $G_f = \langle \pi \mid f(\bar{x}) = f(\pi(\bar{x})) \text{ is in } E_f \rangle$. If E_f is empty then $G_f = \langle id \rangle$. In this section we assume

$$E = \bigcup_{f \in F} E_f$$

We want to emphasize that for the time bounds the set E is assumed to be fixed. If E is allowed to be part of the input then the graph isomorphism problem is polynomial time reducible to testing E-equality for permutation axioms E. See the end of this section. But this reduction assumes that function symbols of arbitrary arity are allowed. In our setting the size of E and of all G_f are bounded by a constant independent of $n = |s| + |t|$.

We represent a term t by its maximally shared graph, i.e. by $\text{graph}(t) = (V, E)$ where $V = \text{Sub}(t)$ is the set of vertices and $E = \{(s, s_i) \mid s \equiv f(s_1, \ldots, s_k), i = 1, \ldots, k\}$ is the set of edges. The edge (s, s_i) is labeled by i. In node $s \in \text{Sub}(t)$ we also store $|s|$ and $\text{depth}(s)$. Recall that $\text{Sub}(t)$ is the set of all subterms of t.

We now come to some basic algorithms for computing modulo E. In all our algorithms we will assume that the table T from Theorem 1 is already computed (in polynomial time).

Lemma 1. *a)* $f(s_1, \ldots, s_k) =_E f(t_1, \ldots, t_k)$ *iff*
  $s_{\pi(i)} =_E t_i$ *for some* $\pi \in G_f$ *and all* $i = 1, \ldots, k$.
b) If $s =_E t$ *then* $\text{top}(s) = \text{top}(t)$ *and* $|s| = |t|$ *and* $\text{depth}(s) = \text{depth}(t)$.

Proof: By induction on the maximal depth of the terms. ☐

Based on this lemma we give a recursive algorithm for testing E-equality that stores in $Eq(s', t')$ for some $s' \in \text{Sub}(s)$ and $t' \in \text{Sub}(t)$ the result of the test "$s' =_E t'$?". We want to avoid repetitions of this test for fixed s', t'. This is supported by representing s and t by the maximally shared graph for the term $*(s, t)$, where $*$ is a new function symbol. But this alone does not exclude repetitions of this test, especially if $ar(f) > 2$ and G_f is large for some $f \in F$. So we store in $Treated(s', t')$ whether this test is already performed.

Based on Lemma 1, we choose the following compatability predicate for detecting early that $s =_{E, \pi} t$ cannot hold:
  $EqCompat(s, t, \pi)$ iff
  For all $i = 1, \ldots, ar(f)$: $\text{top}(s_{\pi(i)}) = \text{top}(t_i)$, $|s_{\pi(i)}| = |t_i|$ and
  $\text{depth}(s_{\pi(i)}) = \text{depth}(t_i)$.
  So for $s \equiv f(s_1, \ldots, s_k)$, $t \equiv f(t_1, \ldots, t_k)$, if $EqCompat(s, t, \pi)$ does not hold, then $s_{\pi(i)} =_E t_i$ does not hold for some i.

Algorithm 31
Eq-Test

    Input: $s, t \in \text{Term}(F, V)$
    Output: If $s =_E t$ then YES else NO

$Eq\text{-}Test(s, t) =$
for all $(s', t') \in \text{Sub}(s) \times \text{Sub}(t)$ **do** $\text{Treated}(s', t') :=$ false
$TestEq(s, t)$
OUTPUT: $Eq(s, t)$

TestEq
$TestEq(s, t) =$
$Eq(s, t) :=$ false
if $\text{top}(s) = \text{top}(t)$ and $|s| = |t|$ and $depth(s) = depth(t)$
  then
    if $s \in V \cup F$
      then $Eq(s, t) := true$
      else
        let $s \equiv f(s_1, \ldots, s_k)$ and $t \equiv f(t_1, \ldots, t_k)$
        for all $\pi \in G_f$ such that $EqCompat(s, t, \pi)$ **do**
        $ok\pi :=$ true; $i := 1$
        while $i \leq k$ and $ok\pi$ **do**
          if not $\text{Treated}(s_{\pi(i)}, t_i)$ **then** $TestEq(s_{\pi(i)}, t_i)$ **fi**
          if $Eq(s_{\pi(i)}, t_i) =$ false **then** $ok\pi :=$ false **fi**
        od
        if $ok\pi =$ true **then** $\{Eq(s, t) := true; Treated(s, t) := true; exit\}$ **fi**
        od
    fi
fi
$Treated(s, t) :=$ true

Here the enumeration of all $\pi \in G_f$ such that $EqCompat(s, t, \pi)$ holds is supported by algorithm $\mathcal{A}$ from Section 2. This takes also into account for which i the π-loop was interrupted. We demonstrate this by an example.

Example 3. Assume E = $\{f(x_1, x_2, x_3, x_4)$ = $f(x_2, x_3, x_4, x_1), f(x_1, x_2, x_3, x_4)$ = $f(x_2, x_1, x_3, x_4)\}$. Then G_f = $\text{Sym}(4)$ is the full permutation group on $\mathbb{N}_4$.

Question: Is $s \equiv f(s_1, \ldots, s_4) =_E f(t_1, \ldots, t_4) \equiv t$?

The naive algorithm generates all $4! = 24$ permutations $\pi \in G_f$ and tests whether $s_{\pi(i)} =_E t_i$ holds for all $i = 1, \ldots, 4$. This results in $4 \cdot 24 = 96$ subtests of the form "$s_i =_E t_j$?". Note that these tests may be expensive.

Now assume $s_1 \equiv s_2 \equiv \bar{s}$, $t_2 \equiv t_4 \equiv \bar{t}$ and $\bar{s} =_E \bar{t}$, $\bar{s} \neq_E t_1$, $\bar{s} \neq_E t_3$, $s_3 =_E t_1$, $s_3 \neq_E t_3$ and $s_4 \neq_E t_3$. We describe the computation of $TestEq$ for this input by a tree. Here the numbers near the edges at level i denote the possible values of $\pi(i)$. The nodes describe the subtest to be performed and its outcome.

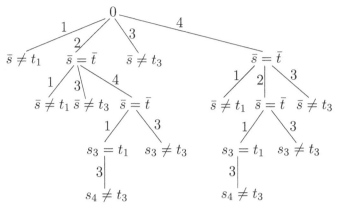

So $TestEq(s,t)$ calls itself with input (s_i, t_j) only six times (as mentioned in the assumption on the input) compared to 96 calls of the naive algorithm. The outcome of the computation is $f(s_1, \ldots, s_4) \neq_E f(t_1, \ldots, t_4)$. □

Theorem 3. *Let E be a fixed set of permutation equations and T its table as in Theorem 1. Algorithm $Eq\text{-}Test=Eq\text{-}Test(E, T, s, t)$ is correct. It runs in time $O(n^2)$, where $n = |s| + |t|$.*

Proof: By Lemma 1 the algorithm $TestEq$ satisfies the loop invariant $Eq(E, T, s, t) = $ true iff $s =_E t$. So the algorithm is correct. To estimate the running time, we notice that $TestEq(s', t')$ is called at most once for each $(s', t') \in \mathrm{Sub}(s) \times \mathrm{Sub}(t)$. So there are at most n^2 calls of $TestEq$ on input(s, t). Each call of $TestEq$ costs constant time since the arity of all $f \in F$ and the size of all G_f are bounded by a constant independent of n. Hence $Eq\text{-}Test(E, T, s, t)$ needs time $O(n^2)$. □

We now comment on some implementation details to enhance efficiency. The basic idea is to propagate more knowledge from recursive calls of $TestEq$. The first idea is to insert the following test before the line "**for** all $\pi \in G_f \ldots$":

If for some i and all j we have $\mathrm{Treated}(s_i, t_j) \wedge \neg Eq(s_i, t_j)$ then $s \neq_E t$.

The second idea is to exploit the fact that $=_E$ is an equivalence relation, i.e. to use its symmetry and transitivity. So we propose to maintain the current part of the E-equality by the well known *Union-Find* algorithm [CLR96]. Then in algorithm $TestEq$ the command $Eq(s', t') := $ true is to be replaced by $Union(s', t')$. For later use we have $s' =_E t'$ iff $Find(s') = Find(t')$. So the vector Eq is not used any more.

In the following we will frequently have to test whether $t' =_E t''$ holds for $t', t'' \in \mathrm{Sub}(s) \cup \mathrm{Sub}(t)$. So it is recommended to preprocess $*(s, t)$ so that for all $t', t'' \in \mathrm{Sub}(*(s, t))$ we have $t' =_E t''$ iff $Find(t') = Find(t'')$. This preprocessing can be done in time $O(n^2)$ by an algorithm similar to $TestEq$: We have to test whether $t' =_E t''$ bottom up in $*(s, t)$ for all t', t'' such that $\mathrm{top}(t') = \mathrm{top}(t'')$ and $|t'| = |t''|$ and $\mathrm{depth}(t')=\mathrm{depth}(t'')$. If we finally put $Eq(t', t'') = $ true iff $Find(t') = Find(t'')$ then we can test E-equality on $\mathrm{Sub}(*(s, t))$ in one step.

By using a lemma similar to Lemma 1 one can design a non-deterministic algorithm for the E-matchability problem that runs in polynomial time, see

[Ave03]. Given the fact that matchability modulo commutativity is already NP-complete [BKN85], we have

Theorem 4. *For fixed E the E-matchability problem may be NP-complete.* $\square$

We do not outline an efficient algorithm for this problem. It can be constructed from the algorithm for the E-unification problem we are going to describe next.

Lemma 2. *a)* $\sigma(f(s_1, \ldots, s_n)) =_E \sigma(f(t_1, \ldots, t_k))$ *iff*
$\sigma(s_{\pi(i)}) =_E \sigma(t_i)$ *for some* $\pi \in G_f$ *and all* $i = 1, \ldots, k$.
b) Assume $s, t \notin V$. *If* $\sigma(s) =_E \sigma(t)$ *then* $\text{top}(s) = \text{top}(t)$.
c) Assume $s \equiv x$. *If* $x \in \text{Var}(t)$ *and* $t \not\equiv x$ *then there is no E-unifier for s and t. If* $x \notin \text{Var}(t)$ *then* $\sigma = \{x \leftarrow t\}$ *is an E-unifier for s and t.*

Proof: a) and b) follow directly from Lemma 1 and c) is easy. $\square$

Again, based on this Lemma one can easily design a non-deterministic polynomial time algorithm for the E-unification problem. So we have

Theorem 5. *For fixed E the E-unification problem may be NP-complete.* $\square$

Sometimes one needs to find, for any equation $s = t$, all "solutions" σ such that $\sigma(s) =_E \sigma(t)$. To do so, we define a search tree with nodes labeled by a constraint C such that C is E-solvable iff one of its sons is E-solvable. We are going to make this precise.

Definition 3. *A **constraint** C is a set of equations. It is **E-solvable** if there is a σ such that $\sigma(s) =_E \sigma(t)$ for all $s = t$ in C. Then σ is an **E-solution** of C and $\text{Sol}(C)$ is the set of all E-solutions of C. C is in **solved form** if it is of the form $C = \{x_i = t_i \mid i = 1, \ldots, m\}$ such that for all $i = 1, \ldots m$ we have (1) $x_i \not\equiv x_j$ for $i \neq j$ and (2) $x_i \notin \text{Var}(t_j)$ for $i \leq j \leq m$.*
*If C is in solved form then it describes a substitution σ_C as usual. σ_C is an **E-unifier** of C. A **complete set of unifiers modulo E** for C (a $CSU_E(C)$) is a set Σ of substitutions such that (1) each $\sigma \in \Sigma$ is an E-unifier for C and (2) for each E-unifier τ for E there is a $\sigma \in \Sigma$ and a τ' such that $\tau =_E \tau' \circ \sigma$.*

Lemma 3. *Let C be a constraint*

a) $C \cup \{f(s_1, \ldots, s_k) = f(t_1, \ldots, t_k)\}$ *is E-solvable iff*
$C \cup \{s_{\pi(i)} = t_i \mid i = 1, \ldots, k\}$ *is E-solvable for some* $\pi \in G_f$.
b) Assume $s, t \notin V$. $C \cup \{s = t\}$ *is not E-solvable if* $\text{top}(s) \neq \text{top}(t)$.
c) Assume $s =_E t$. $C \cup \{s = t\}$ *is E-solvable iff C is E-solvable.*
d) Let $\sigma = \{x \leftarrow t\}$. *If* $x \in \text{Var}(t)$ *and* $t \not\equiv x$ *then* $C \cup \{x = t\}$ *is not E-solvable.*
If $x \notin \text{Var}(t)$ *then* $C \cup \{x = t\}$ *is E-solvable iff* $\sigma(C) \cup \{x = t\}$ *is E-solvable.*
e) $C \cup \{x = t_1, x = t_2\}$ *is E-solvable iff* $C \cup \{x = t_1, t_1 = t_2\}$ *is E-solvable.*

Proof: This follows from Lemma 2. $\square$

If $G_f = \langle id \rangle$ for all $f \in F$ then E-unification reduces to syntactic unification, i.e., to testing whether $\sigma(s) \equiv \sigma(t)$ for some σ. The standard algorithm for syntactic unification is to transform C into C' in solved form by applying the deduction steps indicated by Lemma 3 a) to d). But that may need exponential time. There are well known techniques to design polynomial time algorithms for syntactic unification (see e.g. [BN98]). One idea is to replace deduction step d) of Lemma 3 by deduction step e) and to test C' for cycle-freeness. In our context it is advantageous to adopt this technique.

For computing a $CSU_E(s = t)$, starting with $C = \{s = t\}$ the deduction steps a)-c) and e) transform C into a set $\mathrm{Con}(s,t)$ of constraints each of the form $C' = \{x_i = t_i \mid i = 1, \ldots, m\}$ with $x_i \not\equiv x_j$ for $i \neq j$. We call such a C' *presolved*. Notice that during this transformation only equations $t_1 = t_2$ appear such that $t_1, t_2 \in \mathrm{Sub}(s) \cup \mathrm{Sub}(t)$. So, if we preprocess (s,t) as indicated after Theorem 3 we can test E-equality of these t_1, t_2 in one step.

It is advisable to eliminate unsolvable constraints early in the transformation from $C = \{s = t\}$ into $\mathrm{Con}(s,t)$. Unsolvability of a constraint can be detected by the test on cycle-freeness. We are going to describe this.

We reduce a constraint C to $\mathrm{red}(C)$ by first performing **while** $C = C' \cup \{x = y\}$ with $x \not\equiv y, x \in \mathrm{Var}(C')$ **do** $\sigma := \{x \leftarrow y\}, C := \sigma(C') \cup \{x = y\}$ **od** and then removing all equations of the form $x = x$. We define $\mathrm{graph}(C) = (V_C, E_C)$ by $V_C = \mathrm{Var}(C)$ and $E_C = \{(x,y) \mid x = t$ in $\mathrm{red}(C), y \in \mathrm{Var}(t), y \not\equiv t\}$. We call C *cycle-free* if $\mathrm{graph}(C)$ has no cycle. Testing for cycle-freeness can be done in linear time (see [CLR96]).

Lemma 4. *a) If C is not cycle-free then C is E-unsolvable.*

b) If C is presolved and cycle-free then $\mathrm{Sol}(C) = \mathrm{Sol}(\mathrm{red}(C))$ and $\mathrm{red}(C)$ is in solved form. So C defines the E-solution $\sigma = \sigma_{\mathrm{red}(C)}$ of C.   □

Let us now comment on the design decisions for our algorithm $Unify(s,t)$ to compute a $CSU_E(s,t)$. When transforming the set $C = \{s = t\}$ naively into a $CSU_E(C)$ by applying the inference steps of Lemma 3 then subproblems $C' = \{t' = t''\}$ are solved repeatedly for some fixed $t', t'' \in \mathrm{Sub}(*(s,t))$, though in different contexts. To avoid this we follow the strategy of algorithm $Eq\text{-}Test$ and store the solution of C' in $Con(t', t'')$ as soon as it is computed. To do so we first create the necessary data structure (graph) to support this storing and to support testing for E-equality as mentioned above.

For describing $Unify$ we use the following notations and subalgorithms: C always denotes a constraint and Con denotes a set of constraints.
$UCompat$ is the predicate
     $UCompat(s,t,\pi)$ iff
     For all $i = 1, \ldots, ar(f)$: $s_{\pi(i)} \in V$ or $t_i \in V$ or $\mathrm{top}(s_{\pi(i)}) = \mathrm{top}(t_i)$.
$Preprocess(s,t)$ does the preprocessing of (s,t) and initializes $\mathrm{Treated}(t', t'') :=$ false for all $t', t'' \in \mathrm{Sub}(*(s,t))$.
$Combine(C, Con)$ is a function that produces $Con' = \{C_0 \mid C_0 = UPrune(C \cup C'), C' \in Con, C_0 \neq \bot\}$.
$UPrune(C)$ prunes the unification constraint C.
$PreUnify(Con)$ transforms each $C \in Con$ into a set of presolved constraints.

$Tran(s,t)$ delivers in $Con(s,t)$ a set of constraints simpler than $C = \{s = t\}$ but in general not presolved.

Before presenting the algorithms in detail we shortly describe in words how they work. To describe $Trans(s,t)$ assume $s \equiv f(s_1, ..., s_k)$ and $t \equiv f(t_1, ...t_k)$. (The other cases of the structure of s and t are easy.) Then $Con(s,t)$ is the union of all $Con\pi$ with $\pi \in G_f$. Here $Con\pi$ is the combination of all $C_j \in Con(s_{\pi(j)}, t_j), j = 1, ..., k$. One could compute all $Con(s_{\pi(j)}, t_j)$ first and then put $Con* = \{C \mid C = UPrune(C_1 \cup ... \cup C_k), C_j \in Con(s_{\pi(j)}, t_j)\}$. Then we have $Con\pi = Con*$, but $Con*$ may contain some $C = \bot$. For reasons of efficiency we compute instead $Coni = \{C \mid C = UPrune(C_1 \cup ... \cup C_i), C_j \in Con(s_{\pi(j)}, t_j), C \neq \bot\}$ for $i = 1, ..., k$. This will eliminate early all $C \in Con*$ with $C = \bot$. Specifically, if $Coni$ becomes empty then immediately $Con\pi$ becomes empty. Finally, for $i = k$ we have $Coni = Con\pi$.

If, in general, $Sol(Con)$ denotes the union of all $Sol(C)$ with $C \in Con$ then $Trans(s,t)$ satisfies the loop invariant $Sol(\{s = t\}) = Sol(Con(s,t))$. Some $C \in Con(s,t)$ again may contain equations $s = t$ with $top(s) = top(t) \in F$. These equations are processed by $PreUnify(Con)$. In the loop of this algorithm each $C = C' \cup \{s = t\}$ in Con with $top(s) = top(t) \in F$ is transformed into $Con_C = Combine(C', Con(s,t))$. Since $Sol(C) = Sol(Con_C)$ we have the loop invariant for $PreUnify$ that $Sol(Con)$ is fixed.

Algorithm 32
Unify

    Input: $s, t \in \text{Term}(F, V)$
    Output: Con, describing a $CSU_E(s,t)$

$Unify(s,t) =$
$Preprocess(s,t)$
$C := \{s = t\}; \; UPrune(C);$
if $C = \bot$ **then** $Con := \varnothing$ **else** $\{Con := \{C\}; \; PreUnify(Con)\}$ **fi**
OUTPUT $: Con$

    UPrune
$UPrune(C) =$
while $t = x$ in C with $t \notin V$ **do** $C := (C - \{t = x\}) \cup \{x = t\}$ **od**
while $C = C' \cup \{x = t_1, x = t_2\}$ **do** $C := C' \cup \{x = t_1, t_1 = t_2\}$ **od**
if $(s = t$ in C with $s, t \notin V$ and $top(s) \neq top(t))$ or (C is not cycle-free)
    then $C := \bot$
    else while $s = t$ in C with $s =_E t$ **do** $C := C - \{s = t\}$ **od fi**

    PreUnify
$PreUnify(Con) =$
while $\exists C \in Con, s = t$ in $C, top(s) = top(t) \in F$ **do**
   $Con := Con - \{C\}$
   $C := C - \{s = t\}$
   if not $Treated(s,t)$ **then** $Trans(s,t)$ **fi**
   $Con := Con \cup Combine(C, Con(s,t))$
od

Trans

$Trans(s,t) =$
$Con := \varnothing$
if $s \in V$ or $t \in V$ **then** $Con := \{UPrune(\{s = t\})\}$ **fi**
if $top(s) = top(t) \in F$ **then**
    let $s \equiv f(s_1, \ldots, s_k), t \equiv f(t_1, \ldots, t_k)$
    for all $\pi \in G_f$ such that $UCompat(s,t,\pi)$ **do**
        $Con\pi := \{\varnothing\}; \ i := 1$
        while $i \leq k$ and $Con\pi \neq \varnothing$ **do**
          $Coni := \varnothing$
          if not $\text{Treated}(s_{\pi(i)}, t_i)$ **then** $Trans(s_{\pi(i)}, t_i)$ **fi**
          for all $C \in Con(s_{\pi(i)}, t_i)$ **do** $Coni := Coni \cup Combine(C, Con\pi)$ **od**
          $Con\pi := Coni; \ i := i + 1$
        od
        $Con := Con \cup Con\pi$
    od fi
$Con(s,t) := Con$; $\text{Treated}(s,t) := \text{true}$

We assume here that E is fixed and T is already computed. Again, the enumeration of all $\pi \in G_f$ such that $UCompat(s,t,\pi)$ holds is supported by algorithm $\mathcal{A}$ from section 2 and this also takes into account for which i the π-loop was interrupted. See Example 3 to demonstrate that.

Theorem 6. *Algorithm* $Unify(s,t)$ *stops and produces a* $CSU_E(s,t)$. $\square$

We want to justify that algorithm $Unify$ is correct and terminates. Let $Sol(Con)$ be the union of all $Sol(C)$ with $C \in Con$. Then $Sol(\{s = t\}) = Sol(Con(s,t))$. The while-loop in $PreUnify$ satisfies the loop invariant: $Sol(Con)$ is fixed. All $C \in Con$ in the output of $PreUnify$ are presolved and cycle-free. So the output of $Unify(s,t)$ describes a $CSU_E(s,t)$. To prove the termination of $Unify$, we show that the while-loop in $PreUnify$ decreases Con in a well-founded order: For a constraint C let $set(C) = \{s, t \mid s = t$ in $C\}$ be the set of all terms in C and let $mult(C)$ be the corresponding multiset. For any ordered set $(M, >)$ we denote by $(\mathcal{M}(M), >_{mult})$ the multiset extension of $(M, >)$, see [BN98]. Let $>_{st}$ be the subterm order. We define the well-founded order $\succ$ on constraints by
$$C \succ C' \text{ iff } mult(C) >_{st,mult} mult(C').$$
This defines the well-founded order $\succ_{mult}$ on sets Con of constraints. Now let C' be the result of $UPrune(C)$. We have $set(C') \subseteq set(C)$ and so $\{s = t\} \succ C$ implies $\{s = t\} \succ C'$. This gives $\{s = t\} \succ C_0$ for all $C_0 \in Con(s,t)$ produced by $Trans(s,t)$ and hence $\{\{s = t\}\} \succ_{mult} Con(s,t)$. Now it is easy to see that the while-loop in algorithm $PreUnify$ decreases Con in the order $\succ_{mult}$. This proves that algorithm $Unify$ terminates.

One can check that $Unify(s,t)$ runs in polynomial time provided that the number of constraints C generated is bounded by a polynomial in $n = |s| + |t|$. (Notice that $Trans$ is called at most n^2 times.)

Example 4. Let E be given by $E = \{f(x,y) = f(y,x), g(x,y) = g(y,x)\}$ and

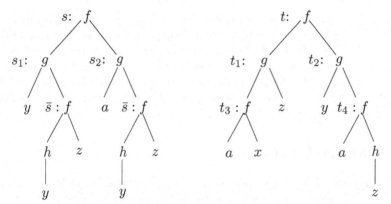

Algorithm $Trans(s,t)$ basically computes the sets of constraints $Con(s',t')$ left-to-right bottom-up. This gives (constraints already in solved form)

$Con(\bar{s},t_3) : \{\{h(y) = a, z = x\}, \{h(y) = x, z = a\}\} \rightsquigarrow \{\{z = a, x = h(y)\}\}$
$Con(\bar{s},t_4) : \{\{h(y) = a, z = h(z)\}, \{h(y) = h(z), z = a\}\} \rightsquigarrow \{\{y = a, z = a\}\}$
$Con(s_1,t_1) : \{\{z = a, y = a, x = h(a)\}\}$
$Con(s_1,t_2) : \{\{z = a, y = a\}\}$
$Con(s_2,t_1) : \{\{z = a, x = h(y)\}\}$
$Con(s_2,t_2) : \{\{z = a, y = a\}\}$
$Con(s,t) : \{\{z = a, y = a, x = h(a)\}\}$

$Unify(s,t)$ produces only one constraint, it is solved. So there is exactly one E-unifier of s and t: $\sigma = \{x \leftarrow h(a), y \leftarrow a, z \leftarrow a\}$.

Now let $s_0 \equiv f(x',x')$ and $t_0 \equiv f(s,t)$. We call $Unify(s_0,t_0)$. Then $Pre\text{-}Unify(\{\{s_0 = t_0\}\})$ calls $Trans(s_0,t_0)$ and $Trans(s_0,t_0)$ produces $Con(s_0,t_0) = \{C\}$ with $C = \{x' = t, s = t\}$. Then $PreUnify$ calls $Trans(s,t)$ which produces $Con(s,t) = \{C'\}$ with $C' = \{x = h(a), y = a, z = a\}$. Combining $\{x' = t\}$ with $Con(s,t)$ leads to $Con = \{C''\}$ with $C'' = \{x = h(a), y = a, z = a, x' = t\}$ which is in solved form. Now $PreUnify$ stops with output $Con = \{C''\}$. So there is exactly one (up to E-equality) E-unifier $\sigma_0 = \sigma_{C''}$ for s_0 and t_0. $\square$

Up to now we have considered the problems of E-equality, E-matchability and E-unification only for a fixed set E of permutation equations. We now come to the generalized version of these problems where E is allowed to be part the input.

Theorem 7. *The E-unifiability for permutation equations E is NP-complete.*

Proof: Algorithm $Unify$ from above is easily transformed into a NP-algorithm: Replace in the algorithm $Trans$ the line "for all $\pi \in G_f$ ..." by "choose $\pi \in G_f$" Now E-unifiability is NP-complete since so is E-matchability. $\square$

Theorem 8. *The E-equality for permutation equations E is in NP, but it is at least as hard as the graph isomorphism problem. The second claim already holds for those E that contain two equations.*

Proof: This follows from the fact that the graph isomorphism problem GI is polynomial time reducible to the permutation equivalence problem PEP for strings. $\square$

In our setting, to any function symbol f only one permutation group G_f is associated. This assumption is reasonable for computing modulo E. In [dlTE03] the NP-completeness of several problems for permutative equations (see below) is shown. Translated to our context, these results imply that the E-equality problem is NP-complete if several permutation groups can be associated to some function symbols. We did not try hard to transform the proof of [dlTE03] to our more special setting.

4   Permutative Equations

For the correctness of the algorithms in section 3 we have used basically two facts about E: (1) Applying an equation to a term preserves the term structure. (2) Applying equations commutes. We use this observation to extend our results.

Let $c[\bar{x}] \equiv c[x_1, \dots, x_k]$ denote a term such that $top(c) \in F$ and each x_i appears exactly once. This is also called a *context*. An equation $c[\bar{x}] = c[\pi(\bar{x})]$ with $\pi \in \text{Sym}(k)$ is called a *permutative equation* in [AP01] and in that paper algorithms are presented to compute modulo permutative theories E. For example, $f(x, a, g(y, z)) = f(y, a, g(z, x))$ is a permutative equation. Unfortunately, in general it is not so easy to represent the whole E-equivalence class $[t]_E$ by a unique term t and to compute with these representatives efficiently. But if E satisfies some conditions, algorithms similar to those presented in section 3 apply. Let us shortly comment on that.

Let I be a set of contexts, for each $c \equiv c[x_1, \dots, x_k] \in I$ let $G_c = \langle \pi_1^c, \dots, \pi_n^c \rangle$ be a group of permutations and let $E = \bigcup_{c \in I} E_c$, where $E_c = \{c[\bar{x}] = c[\pi_i^c(\bar{x})] \mid i = 1, \dots, n\}$. Finally, let $E^* = \{c[\bar{x}] = c[\pi(\bar{x})] \mid c \in I, \pi \in G_c\}$.

First assume that no $c, c' \in I$ overlap, i.e., c is not unifiable with any non-variable subterm of c'. Then E^* is ground confluent, i.e. confluent on $\text{Term}(F \cup K)$ where K is any set of constants. And, for any $t \in \text{Term}(F \cup K)$ there is at most one $c \in I$ such that c is matchable on t. This implies that we can prove $s =_E t$ "level by level", i.e., we have for $s \equiv f(s_1, \dots, s_n)$ and $t \equiv f(t_1, \dots, t_n)$

$s =_E t$ iff

   $s \equiv c[s'_1, \dots, s'_k]$, $t \equiv c[t'_1, \dots, t'_k]$ and $s'_{\pi(i)} =_E t'_i$ for some $c \in I$ and $\pi \in G_c$

   or else $s_i =_E t_i$ for all i.

Using this result we can design an efficient algorithm for deciding E-equality. For computing a $\text{CSU}_E(s = t)$ we have to modify Lemma 2 accordingly: Let I' consist of I and all $c_f \equiv f(x_1, \dots, x_n), f \in F$ and let $G_{c_f} = <id>$. Then we have for $s \equiv f(s_1, \dots, s_n)$ and $t \equiv f(t_1, \dots, t_n)$

   $\sigma(s) =_E \sigma(t)$ iff

   there are $c \in I', \pi \in G_c$ and τ such that $c[\bar{x}] \equiv f(c_1, \dots, c_n)$, $c[\pi(\bar{x})] \equiv f(c'_1, \dots, c'_n)$, τ is an E-solution of $C = \{s_i = c_i, \ t_i = c'_i \mid i = 1, \dots, n\}$ and $\sigma(x) =_E \tau(x)$ for all $x \in Var(s = t)$.

This way we can design an efficient algorithm for computing a $\text{CSU}_E(s = t)$.

Next assume that there are only overlaps of the form that c' is a subterm of c, $c, c' \in I$. Then we can extend G_c so that $c[\bar{x}] =_{E_c} c[\pi(\bar{x})]$ iff $c[\bar{x}] =_{E_c \cup E_{c'}} c[\pi(\bar{x})]$. This "saturation" of E stops with E_0, say. Then E_0^* is ground confluent, so we can prove $s =_E t$ level by level, and our techniques apply.

If there are more general overlaps then one can first try to transform E into E_0 such that E_0 is ground confluent. See [AHL03] how to do that.

In [AP01] we have extended our methods from Section 3 so that one can compute more efficiently modulo permutative equations in the style of [AP01].

References

[AHL03] Jürgen Avenhaus, Thomas Hillenbrand, and Bernd Löchner. On using ground joinable equations in equational theorem proving. *Journal of Symbolic Computation*, (36):217–233, 2003.

[AP01] Jürgen Avenhaus and David Plaisted. General algorithms for permutations in equational inference. *Journal of Automated Reasoning*, (26):223–268, 2001.

[Ave95] Jürgen Avenhaus. *Reduktionssysteme (in German)*. Springer, 1995.

[Ave03] Jürgen Avenhaus. Computing in theories defined by permutative equations. Technical report, Uni. Kaiserslautern, 2003. http://www-avenhaus. informatik.uni-kl.de/ag-avenhaus/berichte/2003/permutative-equations/.

[BG01] Leo Bachmair and Harald Ganzinger. *Resolution theorem proving, in [RV01]*, volume 1, pages 19–99. Elsevier, 2001.

[BKN85] Dan Benanav, Deepak Kapur, and Paliath Narendran. Complexity of matching problems. In J.P. Jouannaud, editor, *Int. Conf. on Term Rewriting and Applications*, number 202 in LNCS, pages 417–429. Springer, 1985.

[BN98] Franz Baader and Tobias Nipkow. *Term rewriting and all that*. Cambridge University Press, 1998.

[BS01] Franz Baader and Wayne Snyder. *Unification theory, in [RV01]*, volume 1, pages 445–533. Elsevier, 2001.

[CLR96] Thomas Cormen, Charles Leiserson, and Ronald Rivest. *Introduction to algorithms*. MIT Press, 1996.

[dlTE03] Thierry Boy de la Tour and Mnacho Echenim. NP-completeness results for deductive problems on stratified terms. In M. Y. Vardi and A. Voronkov, editors, *10th Int. Conf. on Logic for Programming, Artificial Intelligence, and Reasoning*, number 2850 in LNCS, pages 317–331. Springer, 2003.

[FHL80] M. Furst, J. Hopcroft, and E. Luks. Polynomial time algorithms for permutation groups. In *Proc. 21st IEEE Symp. on Foundations of Computer Science*, pages 36–41, 1980.

[HL02] Thomas Hillenbrand and Bernd Löchner. The next WALDMEISTER loop. In A. Voronkov, editor, *Int. Conf. on Automated Deduction*, number 2392 in LNAI, pages 486–500. Springer, 2002.

[JK91] Jean-Pierre Jouannaud and Claude Kirchner. Solving equations in abstract algebras: A rule-based survey of unification. In J.J. Lassez, G. Plotkin, editors, *Computational Logic: Essays in Honor of Alan Robinson*. MIT Press, 1991.

[NR01] Robert Nieuwenhuis and Albert Rubio. *Paramodulation-based theorem proving, in [RV01]*, volume 1, pages 371–443. Elsevier, 2001.

[RV01] Alan Robinson and Andrei Voronkov, editors. *Handbook of Automated Reasoning*. Elsevier, 2001.

Overlapping Leaf Permutative Equations

Thierry Boy de la Tour and Mnacho Echenim

LEIBNIZ laboratory, IMAG - CNRS
INPG, 46 avenue Félix Viallet F-38031 Grenoble Cedex
Thierry.Boy-de-la-Tour@imag.fr, Mnacho.Echenim@imag.fr

Abstract. Leaf permutative equations often appear in equational reasoning, and lead to dumb repetitions through rather trivial though profuse variants of clauses. These variants can be compacted by performing inferences *modulo* a theory E of leaf permutative equations, using group-theoretic constructions, as proposed in [1]. However, this requires some tasks that happen to be **NP**-complete (see [7]), unless restrictions are imposed on E. A natural restriction is orthogonality, which allows the use of powerful group-theoretic algorithms to solve some of the required tasks. If sufficient, this restriction is however not necessary. We therefore investigate what kind of overlapping can be allowed in E while retaining the complexity obtained under orthogonality.

1 Introduction

An equation is *leaf permutative* (in short: is a lp-equation) if it joins two similar linear terms, i.e. if one side is obtained from the other by some permutation of their variables. Commutativity is the simplest non-trivial example, and is notoriously difficult to handle in equational theorem proving. The problem is even more difficult when we consider several lp-equations, and their possible interactions. Take for instance:

$$f(x, y, z) = f(y, x, z),$$
$$f(x, y, z) = f(x, z, y).$$

We see that z is a fixpoint of the first equation, and x a fixpoint of the second. By transitivity, we get $f(y, x, z) = f(x, z, y)$, which has no fixpoint. It is easy to see that any permutation can be obtained from these two equations, i.e. given any terms t_1, t_2 and t_3 and any permutation σ of $\{1, 2, 3\}$, we may deduce

$$f(t_1, t_2, t_3) = f(t_{\sigma(1)}, t_{\sigma(2)}, t_{\sigma(3)}).$$

We thus obtain a theory that we may call *3-commutativity*, in which any term $f(t_1, t_2, t_3)$ is provably equal to at least 6 different terms (including $f(t_1, t_2, t_3)$). Similarly, two lp-equations yield a theory of n-commutativity, in which any term $f(t_1, \ldots, t_n)$ is provably equal to at least $n!$ different terms.

This example shows how profuse a few simple lp-equations can be. The idea of [1] is therefore to lift deduction to so-called stratified terms, representing some

D. Basin and M. Rusinowitch (Eds.): IJCAR 2004, LNAI 3097, pp. 430–444, 2004.
© Springer-Verlag Berlin Heidelberg 2004

sets of permuted terms. Even though commutative unification is **NP**-complete (see [2]), the potential reduction of the search space is worth considering. Moreover, the relationship of lp-equations with *permutation groups* provides a nice mathematical structure, bearing promises of efficient algorithms.

Unfortunately, even the simplest tasks involved in the processing of stratified terms happen to be rather complex, and even **NP**-complete (see [7]). The proof of this result relies on the use of overlapping equations in order to express group constraints. However, we have proved in [5] that most of the **NP**-completeness can be avoided on pseudo-orthogonal theories (with only a restricted form of overlapping). Our aim is to generalise this result by allowing as much overlapping as possible and still being able to use efficient group-theoretic algorithms.

In the next section we introduce the notion of *stratified term trees*, built on signatures shaped after the lp-equations from E. We adopt a formalism[1] that leads to the definition of *stratified sets* as group orbits. In section 3 we analyse a limited form of completeness, namely *group completeness*, introducing *stable* signatures. Section 4 is devoted to completeness of stratified sets under a further restriction to *unify-stable* signatures. This directly yields our complexity results in section 5. We finally give an example of a unify-stable theory.

2   Stratified Term Trees

In order to foster the group theoretic structure of some lp-equation $t = t'$, we want to extract an explicit permutation from it. An obvious candidate, since t and t' are variants, is the variable renaming σ such that $t\sigma$ is identical to t'. However, σ is a permutation of variables, whose particular names are not relevant. We replace these variables by (a simple form of) de Bruijn's indices, thus avoiding explicit α-conversions. Hence we build the *context* c from t (or t') by replacing the variables by $\circ$, and also replace the variables of t by the indices $1, 2, \ldots, n$ from left to right. The *same* replacement is then applied to t', which yields an equation of the form:

$$c[1, 2, \ldots, n] = c[1^\sigma, 2^\sigma, \ldots, n^\sigma]$$

where σ is a permutation of $\{1, 2, \ldots, n\}$, i.e. $\sigma \in \mathrm{Sym}(n)$. We use the group-theoretic notation i^σ for $\sigma(i)$. The lp-equation $t = t'$ can then be noted $[\![c, \sigma]\!]$.

As an example, the two axioms for 3-commutativity of f are $[\![f(\circ, \circ, \circ), (1\ 2)]\!]$ and $[\![f(\circ, \circ, \circ), (2\ 3)]\!]$. The reason that makes these two axioms sufficient to generate all possible permutations is that $\{(1\ 2), (2\ 3)\}$ is a *generating set* of the group $\mathrm{Sym}(3)$. The essential idea behind [1] is to realize that, given equations $[\![c, \sigma_1]\!], \ldots, [\![c, \sigma_m]\!]$, we can deduce any equation $[\![c, \sigma]\!]$ for $\sigma \in G$, where G is the group generated by $\{\sigma_1, \ldots, \sigma_m\}$. It is therefore convenient to represent the set of these equations by $[\![c, G]\!]$.

Given a set E of lp-equations, the difficulty is now to account for deduction modulo E in a group-theoretic setting. We note $t =_E t'$ if $E \models t = t'$; by Birkhoff's completeness theorem, $=_E$ reduces to finite applications of E's axioms,

---

[1] a little bit loose compared to the strict formalisation in [5].

and we would like to restrict these applications so that they can be modelled by a group of permutations of subterms.

The problem is that such permutations are inherently context-dependent, and may not always be composed. Consider for example the theory $E = \{[\![f(a, \circ, \circ), \text{Sym}(2)]\!], [\![f(\circ, \circ, c), \text{Sym}(2)]\!]\}$. We have

$$f(a, b, c) =_E f(a, c, b) =_E f(b, a, c).$$

Hence to $f(a, b, c)$ we can apply the permutations $(b\ c)$ and $(a\ b)$, but not $(a\ b)(b\ c) = (a\ c\ b)$ since

$$f(a, b, c)^{(a\ c\ b)} = f(c, a, b) \neq_E f(a, b, c).$$

A restricted form of equational reasoning is therefore necessary, and [1] proposed *stratification*, a way of applying equations $[\![c, G]\!]$ at fixed non-overlapping positions. This is done by labelling occurrences of function symbols by $[\![c, G]\!]$, thus obtaining *stratified terms*. By representing stratified terms as *term graphs* (see [4]), or more precisely as term trees, we have been able in [6] to express the restricted reasoning on stratified terms as the action of a permutation group on the graph's vertices.

Definition 1. *We are given a signature Σ and a set E of lp-equations over Σ. We then consider a signature Σ' of symbols $\langle f, c, G \rangle$ of arity m, where $f \in \Sigma$ has arity m, c is a Σ-context with n holes whose head symbol is f, and G is a subgroup of $\text{Sym}(n)$, such that $E \models [\![c, G]\!]$.*

A stratified term tree T is a term tree (V, s, a) such that: V is a set of vertices; the symbols are taken in $\Sigma \cup \Sigma'$, i.e. s is a function from V to $\Sigma \cup \Sigma'$; the number of arguments is correct, i.e. $\forall v \in V, a(v)$ is a list of vertices whose length is the arity of $s(v)$; and moreover, for all $v \in V$ such that $s(v) = \langle f, c, G \rangle$, the subterm of T starting at v, noted $T|v$, after replacing its head symbol by f, is an instance of c.

Note that since c is a Σ-context, the term $T|v$ can contain symbols from Σ' only at its root and below c. This is how stratification is enforced: two different equations $[\![c, G]\!]$ and $[\![c', G']\!]$, when applied to a stratified term tree, cannot modify each other's context occurrences, and can therefore be applied *independently*.

Stratified term trees are related to Σ-terms by the simple operation of *unmarking*: it simply consists in translating T by replacing each occurrence of $\langle f, c, G \rangle$ by f. The resulting term is noted $\text{um}(T)$.

Example 1. We consider the theory E axiomatised by

$$g(x, y) = g(y, x),$$
$$f(g(x, y), z) = f(g(x, z), y).$$

Let $g' = \langle g, g(\circ, \circ), \text{Sym}(2) \rangle$ and $f' = \langle f, f(g(\circ, \circ), \circ), \text{Sym}(3) \rangle$, then we can use the signature $\Sigma' = \{f', g'\}$ where f' has arity 3, and g' has arity 2. We now define a stratified term tree T on $V = \{1, 2, 3, 4, 5, 6, 7\}$ by defining the functions s and a (here . is string formation, and ε is the empty string):

v	1	2	3	4	5	6	7
$s(v)$	f'	g	b	d	g'	b	d
$a(v)$	2.5	3.4	ε	ε	6.7	ε	ε

The *root* of T is root$(T) = 1$. Of course, if we take $s(2) = g'$ then T is no longer a stratified term tree. It is possible to represent T in string notation by $f'(g(b,d), g'(b,d))$. We have um$(T) = f(g(b,d), g(b,d))$.

The restricted form of equational reasoning applied to stratified term trees is rather simple: only if $s(v) = \langle f, c, G \rangle$ may we apply the equations $[\![c, G]\!]$ to $T|v$. This amounts to applying permutations of vertices, and therefore to translating the elements of G into elements of Sym(V). This requires some more formalism.

We first know that the term tree T' obtained from $T|v$ by replacing its head symbol by f is an instance of c. By this we mean that there is a 1-1 function h from the set Var$(c) = \{1, \ldots, n\}$ (assuming c has n holes) to the set of vertices of T', such that T' is of the form $c[T|h(1), \ldots, T|h(n)]$. In Example 1, $T.1$ is an instance of $f(g(\circ, \circ), \circ)$ with $h(1) = 3$, $h(2) = 4$, and $h(3) = 5$.

Each element $\sigma \in G$ is a permutation of Var(C), hence corresponds to a permutation of $h(\text{Var}(c))$, the *conjugate* σ^h defined by:

$$\forall i \in \text{Var}(c), \quad [h(i)]^{\sigma^h} = h(i^\sigma).$$

The set of conjugates $G^h = \{\sigma^h \,|\, \sigma \in G\}$ is a subgroup of Sym(V), and we note it $\mathcal{G}_T(v)$ (the function h is unique). In Example 1, we have $\mathcal{G}_T(1) = \text{Sym}(\{3, 4, 5\})$, and $\mathcal{G}_T(5) = \text{Sym}(\{6, 7\})$. It is easy to see that the *product* of these groups is also a subgroup of Sym(V), noted $\mathcal{G}_T$.

Definition 2. *For any stratified term tree $T = (V, s, a)$ and any permutation $\pi \in \mathcal{G}_T$, we define $T^\pi = (V, s, a^\pi)$, where for all $v \in V$ the list $a^\pi(v)$ is obtained from the list $a(v) = v_1. \cdots .v_n$ as: $a^\pi(v) = v_1^\pi. \cdots .v_n^\pi$.*

We have proved in [5] that this defines an action of $\mathcal{G}_T$ on a set of stratified term trees, and we may then consider the orbit of T by $\mathcal{G}_T$, i.e. the set $T^{\mathcal{G}_T} = \{T^\pi \,|\, \pi \in \mathcal{G}_T\}$. This is the stratified set $[T]_s$ *from [1]. The* unmarked stratified set *of T is* S$[T] = \text{um}([T]_s)$.

It is easy to see that the term um(T^π), for any $\pi \in \mathcal{G}_T(v)$ is obtained from the term um(T) by applying an equation $[\![c, \sigma]\!]$ at the position corresponding to v, where $\sigma \in G$ (and c, G as above).

Example 2. Following Example 1, we have $\mathcal{G}_T = \text{Sym}(\{3, 4, 5\})\text{Sym}(\{6, 7\})$. Let $\pi = (3\ 4\ 5)(6\ 7)$, we compute T^π by applying π to a, yielding:

v	1	2	3	4	5	6	7
$s(v)$	f'	g	b	d	g'	b	d
$a^\pi(v)$	2.3	4.5	ε	ε	7.6	ε	ε

In string notation T^π is written $f'(g(d, g'(d, b)), b)$, and um$(T^\pi) = f(g(d, g(d, b)), b)$, which is equal to um$(T)$ modulo E. We have applied simultaneously the equations $[\![f(g(\circ, \circ), \circ), (1\ 2\ 3)]\!]$ and $[\![g(\circ, \circ), (1\ 2)]\!]$ at non-overlapping positions.

$S[T]$ is therefore included in the congruence class of $t = \mathrm{um}(T)$ modulo E. However, this congruence class may not be covered by a single stratified set $S[T]$. And it may be difficult to compute a list of stratified term trees $T_1, \ldots, T_n$ such that the corresponding sets $S[T_1], \ldots, S[T_n]$ cover the congruence class of t modulo E, at least if we require to minimise n.

3 Group Completeness

In Definition 1 we have imposed a condition of correction on Σ' (relative to E), i.e. that for any $\langle f, c, G \rangle \in \Sigma'$ we should have $E \models [\![c, G]\!]$. This is a rather weak condition, which would be fulfilled by taking $\Sigma' = \emptyset$. If we want to reach some kind of completeness of stratified sets, we clearly have to impose stricter conditions, at least that Σ' should "contain" the axioms of E. But the elements of Σ' are also supposed to gather all permutations pertaining to a context, which leads to the following definition.

Definition 3. *The signature Σ' covers E if for any equation $[\![c, \sigma]\!] \in E$ there is a symbol $\langle f, c, G \rangle \in \Sigma'$ such that $\sigma \in G$. Σ' is group complete w.r.t. E if Σ' covers E and for any symbol $\langle f, c, G \rangle \in \Sigma'$, if for some permutation σ we have $E \models [\![c, \sigma]\!]$, then $\sigma \in G$.*

Group completeness ensures for instance that if E contains two equations $[\![c, \sigma]\!], [\![c, \sigma']\!]$ on the same context, then there is a single symbol $\langle f, c, G \rangle$ in Σ' on this context, and of course both σ and σ' must be in G.

3.1 Substitutions and Automorphisms

In the general case of possibly overlapping contexts, we need to investigate how an lp-equation $[\![c', \sigma']\!]$ can be deduced from E, hence how some $[\![c, \sigma]\!] \in E$ can be applied to c'. We therefore need a notion of subsumption between contexts.

Definition 4. *A substitution is a partial function μ from $\mathbb{N}$ to the set of Σ-contexts. Given two contexts c and c', we write $\mu : c \sqsubseteq c'$ if c' is exactly the context $c[\mu(1), \ldots, \mu(n)]$, which is noted $\mu(c)$.*
    We say that c subsumes c', or that c' is an instance of c, noted $c \sqsubseteq c'$, if there is a substitution μ defined on $\mathrm{Var}(c)$ such that $\mu(c) = c'$ (i.e. such that $\mu : c \sqsubseteq c'$). c and c' are unifiable if there are substitutions μ, μ' such that $\mu(c) = \mu'(c')$.

Example 3. Let $c = f(\circ, \circ, \circ)$, and $c' = f(\circ, g(\circ, \circ), g(\circ, \circ))$. Then $c \sqsubseteq c'$ since by taking $\mu(1) = \circ$ and $\mu(2) = \mu(3) = g(\circ, \circ)$, we have $\mu(c) = c'$.

Similarly to term subsumption, it is easy to see that $\sqsubseteq$ is an order on Σ-contexts (contexts can be variants only if they are identical). This order is a lower semi-lattice, and any pair of contexts $\{c, c'\}$ that is bounded from above (i.e. unifiable) has a least upper bound $c \sqcup c'$. For instance,

$$f(g(\circ, \circ), \circ) \sqcup f(\circ, a) = f(g(\circ, \circ), a).$$

We may apply an lp-equation $[\![c,\sigma]\!]$ to a term t if there is a position p in t such that $c \sqsubseteq t|p$. As usual, a *position* p is a string of integers indicating the path from the root of a term t to a subterm $t|p$, such that $t|\varepsilon = t$, $t|(p.p') = (t|p)|p'$ and $f(t_1,\ldots,t_n)|i = t_i$.

Definition 5. *We write* $\mu, p : c \trianglelefteq c'$ *for* $\mu : c \sqsubseteq c'|p$. *We also write* $p : c \trianglelefteq c'$ *if there is a substitution* μ *such that* $\mu, p : c \trianglelefteq c'$, *and we write* $c \trianglelefteq c'$ *if there is a position* p *such that* $p : c \trianglelefteq c'$. *The relation* $\trianglelefteq$ *is the* encompassment ordering. *Note that* $\varepsilon : c \trianglelefteq c'$ *is equivalent to* $c \sqsubseteq c'$.

If $\mu, p : c \trianglelefteq t$, *the* result *of applying the lp-equation* $[\![c,\sigma]\!]$ *at position* p *in the term* t *is the term* t' *obtained from* t *by grafting the term* $\mu \circ \sigma(c)$ *at position* p. *We then write* $[\![c,\sigma]\!] \vdash_p t = t'$.

Example 4. Consider the term $t = f(1, g(2,3), g(4,5))$ associated to the context c' of Example 3. There is a substitution μ' such that $\mu', \varepsilon : c \trianglelefteq t$, hence we may apply the equation $[\![c,(1\ 2)]\!]$ at the root ε of t. This yields the term

$$t' = [\mu' \circ (1\ 2)](c) = c[\mu'(2), \mu'(1), \mu'(3)] = f(g(2,3), 1, g(4,5)).$$

Note that the equation $t = t'$ is not leaf permutative, and that $\mu(1) \neq \mu(2)$.

Definition 6. *The* automorphism group *of a substitution* μ *on* $\mathrm{Var}(c)$ *is*

$$\mathrm{E}(\mu) = \{\sigma \in \mathrm{Sym}(\mathrm{Var}(c)) \mid \forall x \in \mathrm{Var}(c),\ \mu(x^\sigma) = \mu(x)\}.$$

If $\mu, p : c \trianglelefteq c'$ *and* $\sigma \in \mathrm{E}(\mu)$, *we write* $\mu, p : c \trianglelefteq_\sigma c'$. *The meaning of* $p : c \trianglelefteq_\sigma c'$ *and* $c \trianglelefteq_\sigma c'$ *are as above. Moreover,* $c \sqsubseteq_\sigma c'$ *stands for* $\varepsilon : c \trianglelefteq_\sigma c'$, *and we write* $c \sqsubseteq_G c'$ *if for all* $\sigma \in G$, *we have* $c \sqsubseteq_\sigma c'$.

Example 5. Following Examples 3 and 4, we have $\mathrm{E}(\mu) = \mathrm{Sym}(\{2,3\})$. If we apply the equation $[\![c,\sigma]\!]$, where $\sigma = (2\ 3)$ at the root of t we get the term

$$t' = c[\mu'(1), \mu'(3), \mu'(2)] = f(1, g(4,5), g(2,3)).$$

Of course the equation $t = t'$ is leaf-permutative; it is the equation $[\![c',\sigma']\!]$ with $\sigma' = (2\ 4)(3\ 5)$.

We now show how to compute σ', provided that $\sigma \in \mathrm{E}(\mu)$. The problem is that the indices of the holes in c' corresponding to the holes in the contexts $\mu(x)$ for $x \in \mathrm{Var}(c)$ are not the elements of $\mathrm{Var}(\mu(x))$: these have to be shifted according to the number of holes in the contexts $\mu(y)$ for $y < x$, and to the number s of holes in c' occurring to the left of $c'|p$ (in our example, $s = 0$ since $p = \varepsilon$). The amount of shifting is given by the function I_μ^s, defined inductively by:

$$\begin{cases} \mathrm{I}_\mu^s(1) = s, \\ \mathrm{I}_\mu^s(x+1) = \mathrm{I}_\mu^s(x) + |\mathrm{Var}(\mu(x))|. \end{cases}$$

In Example 5, we have

$$I_\mu^0(1) = 0, \ I_\mu^0(2) = |\text{Var}(\mu(1))| = 1, \ I_\mu^0(3) = 1 + |\text{Var}(\mu(2))| = 3, \ I_\mu^0(4) = 5.$$

The indices of the holes from $\mu(x)$ in c' are the integers y in $]I_\mu^s(x), I_\mu^s(x+1)]$. For any $\sigma \in E(\mu)$ we have

$$I_\mu^s(x+1) - I_\mu^s(x) = |\text{Var}(\mu(x))| = |\text{Var}(\mu(x^\sigma))| = I_\mu^s(x^\sigma + 1) - I_\mu^s(x^\sigma),$$

and therefore, for any integer y in $]I_\mu^s(x), I_\mu^s(x+1)]$, we have

$$I_\mu^s(x^\sigma) \ < \ y + I_\mu^s(x^\sigma) - I_\mu^s(x) \ \leq \ I_\mu^s(x+1) + I_\mu^s(x^\sigma) - I_\mu^s(x) \ = \ I_\mu^s(x^\sigma + 1).$$

We may then formally define the permutation σ' on $\text{Var}(c')$ corresponding to σ:

$$\begin{cases} \forall x \in \text{Var}(c), \forall y \in]I_\mu^s(x), I_\mu^s(x+1)], \ \ y^{\sigma'} = y + I_\mu^s(x^\sigma) - I_\mu^s(x), \\ \forall y \in \text{Var}(c')\backslash]s, s'], \ \ y^{\sigma'} = y, \end{cases}$$

where $s' = I_\mu^s(1 + |\text{Var}(c)|)$. We have just proved that $y^{\sigma'}$ belongs to the interval $]I_\mu^s(x^\sigma), I_\mu^s(x^\sigma + 1)]$, i.e. refers to a hole from $\mu(x^\sigma) = \mu \circ \sigma(x)$ in c', when y refers to a hole from $\mu(x)$ in c'. It is then easy to see that $[\![c, \sigma]\!] \vdash_p [\![c', \sigma']\!]$.

In Example 5, we correctly obtain, for:

$$\begin{aligned}
x = 1, \ y \in]0, 1], \ y^{\sigma'} \ &= \ y + I_\mu^0(1^{(2\ 3)}) - I_\mu^0(1) \ = \ y, \\
x = 2, \ y \in]1, 3], \ y^{\sigma'} \ &= \ y + I_\mu^0(2^{(2\ 3)}) - I_\mu^0(2) \ = \ y + 2, \\
x = 3, \ y \in]3, 5], \ y^{\sigma'} \ &= \ y + I_\mu^0(3^{(2\ 3)}) - I_\mu^0(3) \ = \ y - 2.
\end{aligned}$$

The definition of σ' obviously depends on σ, but also on μ and s, which depend solely on c, c' and p. We note σ' as $\Phi_{[c,c',p]}(\sigma)$, and we therefore have:

Lemma 1. *If* $\mu, p : c \trianglelefteq_\sigma c'$ *then* $[\![c, \sigma]\!] \vdash_p [\![c', \Phi_{[c,c',p]}(\sigma)]\!]$.

$\Phi_{[c,c',p]}$ is obviously a function, and moreover:

Lemma 2. *If* $\mu, p \ : \ c \ \trianglelefteq \ c'$ *then* $\Phi_{[c,c',p]}$ *is a morphism from* $E(\mu)$ *to* $\text{Sym}(\text{Var}(c'))$.

Proof. Let $\sigma, \sigma' \in E(\mu)$, the integer s as above, $x \in \text{Var}(c)$ and y any integer in $]I_\mu^s(x), I_\mu^s(x+1)]$. We let

$$y' = y^{\Phi_{[c,c',p]}(\sigma)} = y + I_\mu^s(x^\sigma) - I_\mu^s(x),$$

then we have $y' \in]I_\mu^s(x^\sigma), I_\mu^s(x^\sigma + 1)]$, so that

$$\begin{aligned}
y^{(\Phi_{[c,c',p]}(\sigma))(\Phi_{[c,c',p]}(\sigma'))} &= (y')^{\Phi_{[c,c',p]}(\sigma')} \\
&= y' + I_\mu^s(x^{\sigma\sigma'}) - I_\mu^s(x^\sigma) \\
&= y + I_\mu^s(x^{\sigma\sigma'}) - I_\mu^s(x) \\
&= y^{\Phi_{[c,c',p]}(\sigma\sigma')}.
\end{aligned}$$

This of course proves that $\Phi_{[c,c',p]}(\sigma\sigma') = [\Phi_{[c,c',p]}(\sigma)][\Phi_{[c,c',p]}(\sigma')]$.

We thus obtain a necessary condition of group completeness, by testing every pair of symbols in Σ' according to the following

Theorem 1. *If Σ' is group complete, then for all $\langle f, c, G \rangle, \langle f', c', G' \rangle \in \Sigma'$ such that $\mu, p : c \trianglelefteq c'$, we have that $\Phi_{[c,c',p]}(G \cap E(\mu))$ is a subgroup of G'.*

Proof. For all $\sigma \in G \cap E(\mu)$, by correction we have $E \models [\![c, \sigma]\!]$, and since $\mu, p : c \trianglelefteq_\sigma c'$ we have by Lemma 1 that $E \models [\![c', \Phi_{[c,c',p]}(\sigma)]\!]$, and hence by group completeness $\Phi_{[c,c',p]}(\sigma) \in G'$. The set $\Phi_{[c,c',p]}(G \cap E(\mu))$ is therefore included in G', and is a group according to Lemma 2.

This is not a sufficient condition of group completeness, since we may deduce a new lp-equation in a nontrivial way, by destroying its context and then reconstructing it, as we now illustrate.

Example 6. Consider the following three contexts and groups

$$c = f(\circ, \circ), \qquad c_1 = f(g(\circ, \circ), \circ), \qquad c_2 = f(\circ, g(\circ, \circ)),$$
$$G = \mathrm{Sym}(\{1, 2\}), \quad G_1 = \mathrm{Sym}(\{1, 3\}), \quad G_2 = \mathrm{Sym}(\{2, 3\}),$$

and the set E of lp-equations and signature Σ',

$$E = \{[\![c, G]\!], [\![c_1, G_1]\!], [\![c_2, G_2]\!]\}, \quad \Sigma' = \{\langle f, c, G \rangle, \langle f, c_1, G_1 \rangle, \langle f, c_2, G_2 \rangle\}.$$

We see that Σ' trivially satisfies the condition in Theorem 1, since we have $\mu_1 : c \sqsubseteq c_1$ and $\mu_2 : c \sqsubseteq c_2$ where the automorphism groups $E(\mu_i)$ are reduced to the trivial group I, hence so are the groups $G \cap E(\mu_i)$, and their image by $\Phi_{[c,c_i,\varepsilon]}$. The signature Σ' is however not group complete, since:

$$f(g(1, 2), 3) = f(3, g(1, 2)) \quad \text{by } [\![c, (1\ 2)]\!]$$
$$= f(3, g(2, 1)) \quad \text{by } [\![c_2, (2\ 3)]\!]$$
$$= f(g(2, 1), 3) \quad \text{by } [\![c, (1\ 2)]\!],$$

so that $E \models [\![c_1, (1\ 2)]\!]$ and yet $(1\ 2) \notin G_1$.

3.2   Stable Contexts and Signatures

In order to obtain a simple criterion for group completeness we now rule out such destructions and reconstructions with the following definition.

Definition 7. *A context c with n holes is E-stable if all terms in the congruence class of $c[1, \ldots, n]$ modulo E are instances of c. A signature Σ' is E-stable if $\forall \langle f, c, G \rangle \in \Sigma'$, c is E-stable.*

Since E is a set of lp-equations, it is easy to see that the E-class of the term $c[1, \ldots, n]$ for an E-stable context c only contains terms of the form $c[1^\sigma, \ldots, n^\sigma]$, with $\sigma \in \mathrm{Sym}(n)$. It is not difficult to test the E-stability of a signature:

Lemma 3. *A signature Σ' that covers E is E-stable if and only if for all $\langle f, c, G \rangle, \langle f', c', G' \rangle \in \Sigma'$, if $\mu, p : c \trianglelefteq c'$ then G is a subgroup of $E(\mu)$.*

Proof. If Σ' is E-stable, let $\langle f, c, G \rangle, \langle f', c', G' \rangle \in \Sigma'$ such that $\mu, p : c \trianglelefteq c'$. Then for all $\sigma \in G$, the equation $[\![c, \sigma]\!]$ can be applied to $c'|p$, and since c' is E-stable (and $E \models [\![c, \sigma]\!]$ by correction of Σ'), we have $\sigma \in \mathrm{E}(\mu)$.

Conversely, we show that $\forall \langle f', c', G' \rangle \in \Sigma'$, the context c' is E-stable. In order to apply an equation $[\![c, \sigma]\!] \in E$ to a term $t = c'[1^\pi, \ldots, n^\pi]$, where $\pi \in \mathrm{Sym}(n)$, we need $\mu, p : c \trianglelefteq c'$. Since Σ' covers E, there is a $\langle f, c, G \rangle \in \Sigma'$ such that $\sigma \in G$, hence by hypothesis we have $\sigma \in \mathrm{E}(\mu)$. Applying $[\![c, \sigma]\!]$ at $c'|p$ therefore turns t into $c'[1^{\pi'}, \ldots, n^{\pi'}]$, where $\pi' = \pi^{-1} \Phi_{[c,c',p]}(\sigma)\pi$. A trivial induction therefore shows that all terms in the E-class of $c[1, \ldots, n]$ are instances of c.

With this restriction to E-stable signatures, we are able to characterise group completeness, by considering the following group.

Definition 8. *For any context c' we define the group $\Gamma_{\Sigma'}(c')$ as generated by*

$$\cup \{ \Phi_{[c,c',p]}(G) \mid \langle f, c, G \rangle \in \Sigma' \ \wedge \ p : c \trianglelefteq c' \}.$$

Note that if $\langle f', c', G' \rangle \in \Sigma'$, since $\Phi_{[c',c',\varepsilon]}$ is the identity, then G' is always a subgroup of $\Gamma_{\Sigma'}(c')$.

Theorem 2. *If Σ' covers E and is E-stable, then Σ' is group complete iff $\forall \langle f', c', G' \rangle \in \Sigma'$, $\Gamma_{\Sigma'}(c') = G'$.*

Proof. If Σ' is group complete then by Theorem 1 we have that $\forall \langle f, c, G \rangle, \langle f', c', G' \rangle \in \Sigma'$, if $\mu, p : c \trianglelefteq c'$ then $\Phi_{[c,c',p]}(G \cap \mathrm{E}(\mu))$ is a subgroup of G'. But since Σ' is E-stable, by Lemma 3 we know that $G \subseteq \mathrm{E}(\mu)$, i.e. that $G \cap \mathrm{E}(\mu) = G$. Hence all generators of $\Gamma_{\Sigma'}(c')$ are in G', so that $\Gamma_{\Sigma'}(c') = G'$.

Conversely, let $\langle f', c', G' \rangle \in \Sigma'$ where c' has n holes, and $\sigma' \in \mathrm{Sym}(n)$ such that $E \models [\![c', \sigma']\!]$, since c' is E-stable then there is a finite sequence of permutations $\pi_1, \ldots, \pi_m$, with $\pi_1 = \mathrm{id}$ and $\pi_m = \sigma'$, and a finite sequence of equations $[\![c_1, \sigma_1]\!], \ldots, [\![c_{m-1}, \sigma_{m-1}]\!] \in E$ such that

$$\forall i < m, \ [\![c_i, \sigma_i]\!] \vdash_{p_i} c'[1^{\pi_i}, \ldots, n^{\pi_i}] = c'[1^{\pi_{i+1}}, \ldots, n^{\pi_{i+1}}].$$

We therefore have $\mu_i, p_i : c_i \trianglelefteq_{\sigma_i} c'$, and we can write $\pi_{i+1} = \pi_i^{-1} \Phi_{[c_i,c',p_i]}(\sigma_i)\pi_i$. Since Σ' covers E there is a $\langle f_i, c_i, G_i \rangle \in \Sigma'$ such that $\sigma_i \in G_i$, hence by hypothesis $\Phi_{[c_i,c',p_i]}(\sigma_i) \in G'$. It is obvious that $\pi_1 \in G'$, hence by induction we obtain $\pi_m = \sigma' \in G'$, which proves group completeness.

Of course, if Σ' is E-stable and we have $\langle f, c, G \rangle, \langle f', c', G' \rangle \in \Sigma'$ such that $\mu, p : c \trianglelefteq c'$, then $E \models [\![c', \Phi_{[c,c',p]}(G)]\!]$ according to Lemma 1, since G is a subgroup of $\mathrm{E}(\mu)$. Therefore $E \models [\![c', \Gamma_{\Sigma'}(c')]\!]$, and it is correct w.r.t. E to replace each symbol $\langle f', c', G' \rangle \in \Sigma'$ by $\langle f', c', \Gamma_{\Sigma'}(c') \rangle$, which clearly yields a group complete signature (provided Σ' covers E) according to Theorem 2. This process is called *group completion*.

It should be noted that the lp-equations thus obtained could also be obtained (as many others) by Knuth-Bendix completion. Group completion is a restriction of this general form of completion, which can be performed in polynomial time.

4   Completeness of Stratified Sets

We now investigate how to enforce the best possible kind of completeness of unmarked stratified sets. As mentioned above, the class of a term t modulo E can always be covered by stratified sets $S[T_1] \cup \ldots \cup S[T_n]$, but efficiency requires that their number n be as small as possible. The optimum being $n = 1$ we will try, starting from any term t, to construct a stratified term tree T and obtain a *complete stratified set* $S[T]$, equal to the E-class of t. We then say that T is E-complete. Our aim will be to obtain reasonably efficient procedures to check E-completeness, and to compute E-complete stratified term trees.

Unfortunately, the group completeness of signatures does not ensure the completeness of stratified sets, because it only concerns the groups attached to *existing* contexts. For example, if we take the group complete signature $\Sigma' = \{\langle f(\circ, \circ), \mathrm{Sym}(\{1,2\})\rangle, \langle f(g(\circ, \circ), \circ), \mathrm{Sym}(\{1,2\})\rangle\}$, then we will never be able to construct a stratified term tree T (on a correct signature) such that $S[T] = \{f(g(a,b),c), f(g(b,a),c), f(c,g(a,b)), f(c,g(b,a))\}$. Note however that Σ' is not stable.

4.1   Unify-Stable Signatures

When trying to build an E-complete stratified term tree T from a term t, a major problem we encounter is the necessity to choose between different contexts. In some degenerate cases, it may be possible to choose one of them over the others, but generally, selecting one context when building a stratified term tree means leaving out E-consequences that could have been obtained with the other one. Take for instance

$$\Sigma' = \{\langle f(g(\circ), g(\circ), \circ, \circ), \mathrm{Sym}(\{3,4\})\rangle, \langle f(\circ, \circ, h(\circ), h(\circ)), \mathrm{Sym}(\{1,2\})\rangle\}$$

and $t = f(g(a), g(b), h(c), h(d))$; we may build T with any of these two contexts, but they would both leave some consequences of t out of $S[T]$.

When having to choose between two unifiable contexts c and c', we could therefore try to use a context $\bar{c}$ that is an instance of both c and c', in case the stratified term tree constructed with it contains all of the E-consequences of c and c'. The best choice is of course $\bar{c} = c \sqcup c'$. In the previous example, we can use the new context $f(g(\circ), g(\circ), h(\circ), h(\circ))$, associated to the group obtained by group completion. Then any stratified term tree constructed with it will contain all of the E-consequences of the two previous contexts.

As for group completion, the lp-equations obtained on the context $c \sqcup c'$ could be obtained by Knuth-Bendix completion, but we will design a more efficient (restricted) form of completion.

We have to make sure that the group associated to $c \sqcup c'$ will contain all of the consequences of c and c'. In particular, no permutation associated to c or c' must destroy $c \sqcup c'$, otherwise, for example, we could have a term t that is both an instance of c and c', but after applying some equation $[\![c, \sigma]\!]$, we would obtain a term t' that is still an instance of c, but not one of c'. The generalisation of these requirements leads to the definition of a *unify-stable* signature.

Definition 9. Σ' is unify-stable *if for all* $\langle f, c, G \rangle, \langle f', c', G' \rangle \in \Sigma'$,

1. *if c and c' are unifiable then $c \sqsubseteq_G c \sqcup c'$,*
2. *if there is a position $p \neq \varepsilon$ such that c and $c'|p$ are unifiable, then $c \sqsubseteq_G c'|p$.*

We now prove that unify-stability is actually stronger than stability.

Theorem 3. *If Σ' is unify-stable and covers E then it is E-stable.*

Proof. By Lemma 3, we only need to prove that $\forall \langle f, c, G \rangle, \langle f', c', G' \rangle \in \Sigma'$, if $\mu, p : c \trianglelefteq c'$ then G is a subgroup of $E(\mu)$, i.e. $c \sqsubseteq_G c'|p$. If $p = \varepsilon$ we have $c \sqsubseteq c'$, hence c and c' are unifiable, hence by Definition 9 item 1, we get $c \sqsubseteq_G c \sqcup c' = c' = c'|p$. If $p \neq \varepsilon$ we have $c \sqsubseteq c'|p$ and by item 2 we get $c \sqsubseteq_G c'|p$.

This means that group completion as defined above can be applied on unify-stable signatures. We now show that the first item of Definition 9 allows to unify more than two contexts. We first need a result of stability of $\sqsubseteq_G$ by unification.

Lemma 4. *For any pair $\{c_1, c_2\}$ of unifiable contexts, and c, G such that $c \sqsubseteq_G c_1$ and $c \sqsubseteq_G c_2$, we have $c \sqsubseteq_G c_1 \sqcup c_2$.*

Proof. Let μ, μ_1, μ_2 be the substitutions such that $\mu_i(c) = c_i$ and $\mu(c) = c_1 \sqcup c_2$. For all $x \in \mathrm{Var}(c)$, it is clear that $\mu_i(x) \sqsubseteq \mu(x)$ (since $c_i \sqsubseteq c_1 \sqcup c_2$), hence that $\mu_1(x) \sqcup \mu_2(x) \sqsubseteq \mu(x)$. But since $c_1 \sqcup c_2$ is obtained by unification of c_1 and c_2, considered as linear terms with no common variables, then unification can be performed independently on the subterms of $c_1 = \mu_1(c)$ and $c_2 = \mu_2(c)$, i.e. on $\mu_1(x)$ and $\mu_2(x)$. Therefore, $\mu(x) = \mu_1(x) \sqcup \mu_2(x)$.

For all $y \in x^G$, we have $\mu_i(y) = \mu_i(x)$ by hypothesis, and therefore

$$\mu(y) \;=\; \mu_1(y) \sqcup \mu_2(y) \;=\; \mu_1(x) \sqcup \mu_2(x) \;=\; \mu(x).$$

We now show that the stability of contexts through pairwise unification (which can be tested in polynomial time) implies their stability under any number of unifications.

Lemma 5. *For any finite set C of unifiable contexts, and c, G such that $\forall c' \in C$, $c \sqsubseteq_G c \sqcup c'$, we have $c \sqsubseteq_G c \sqcup (\sqcup C)$.*

Proof. By induction; suppose it is true for C, and let $C' = \{c'\} \uplus C$, we have $c \sqsubseteq_G c \sqcup c'$ and $c \sqsubseteq_G c \sqcup (\sqcup C)$. By Lemma 4 we get $c \sqsubseteq_G c \sqcup c' \sqcup c \sqcup (\sqcup C) = c \sqcup (\sqcup C')$.

Theorem 4. *If Σ' is unify-stable and $\forall i \in \{1, \ldots, n\}$, $\langle f', c_i, G_i \rangle \in \Sigma'$ such that $c' = \sqcup \{c_1, \ldots, c_n\}$ exists, then $E \models [\![c', \Gamma_{\Sigma'}(c')]\!]$.*

Proof. Let $\langle f, c, G \rangle \in \Sigma'$ such that $\mu, p : c \trianglelefteq c'$. If $p = \varepsilon$, i.e. we have $c \sqsubseteq c'$, hence $\forall i$, c and c_i are unifiable, so by unify-stability of Σ' we get $c \sqsubseteq_G c \sqcup c_i$. We may then apply Lemma 5 to $\{c_1, \ldots, c_n\}$, and get $c \sqsubseteq_G c \sqcup c' = c'$. By Lemma 1 we easily obtain $E \models [\![c', \Phi_{[c,c',\varepsilon]}(G)]\!]$ from $E \models [\![c, G]\!]$.

We now suppose $p \neq \varepsilon$, i.e. we have $c \sqsubseteq c'|p$. For all c_i that have a position p, c unifies with $c_i|p$, and if C is the set of these $c_i|p$, then $c'|p = \sqcup C$. By unify-stability of Σ' we have $c \sqsubseteq_G c_i|p$, hence $c_i|p = c \sqcup c_i|p$, and we can apply Lemma 5, which yields $c \sqsubseteq_G c \sqcup (\sqcup C) = c'|p$. As above we get $E \models [\![c', \Phi_{[c,c',p]}(G)]\!]$, and conclude $E \models [\![c', \Gamma_{\Sigma'}(c')]\!]$.

This means that it is correct w.r.t. E to extend Σ' with $\langle f', c', \Gamma_{\Sigma'}(c') \rangle$.

lift_rec(v) =
  let $f = s(v)$ **in**
  let $C = \{c \mid \langle f, c, G \rangle \in \Sigma' \wedge c \sqsubseteq t|v\}$ **in**
  if $C = \emptyset$ **then for** u **in** $a(v)$ **do** lift_rec(u) **done**
  else
    let $c = \sqcup C$ **in**
    let $\mu : \text{Var}(c) \to V$ such that $\mu(c) = t|v$ **in**
    $s(v) := \langle f, c, \Gamma_{\Sigma'}(c) \rangle$;
    for i **in** $\text{Var}(c)$ **do** lift_rec($\mu(i)$) **done**

lift(t) = **let** $(V, s, a) = t$ **in** lift_rec(root(t))

Fig. 1. *Algorithm* lift

4.2 E-completeness

We now show how to build E-complete stratified term trees on extensions of unify-stable signatures. See the lifting procedure in Figure 1. It first (implicitly) transforms the term t into a term tree (V, s, a), which will then be transformed into a stratified term tree by replacing symbols $f \in \Sigma$ by symbols $\langle f, c, G \rangle$ at suitable vertices. In order to prove E-completeness we need a short lemma.

Lemma 6. *If* $c \sqsubseteq_\sigma c' \sqsubseteq t$ *and* $[\![c, \sigma]\!] \vdash_\varepsilon t = t'$ *then* $c' \sqsubseteq t'$.

Proof. Let $\sigma' = \Phi_{[c,c',\varepsilon]}(\sigma)$, we have by Lemma 1 that $[\![c, \sigma]\!] \vdash_\varepsilon [\![c', \sigma']\!]$, and since $c' \sqsubseteq t$ we have $[\![c', \sigma']\!] \vdash_\varepsilon t = t'$, so obviously $c' \sqsubseteq t'$.

We will use a notion from [5]. We say that two stratified term trees T and T' are *equivalent*, noted $T \bowtie T'$, if there exists a $\pi \in \mathcal{G}_T$ such that T^π is[2] T'.

Theorem 5. *If* Σ' *covers* E *and is unify-stable, then* $t =_E t' \Rightarrow$ lift(t) $\bowtie$ lift(t').

Proof. Suppose there is an equation $[\![e, \sigma]\!] \in E$ such that $[\![e, \sigma]\!] \vdash_p t_1 = t_2$, and let $T_i = $ lift(t_i). Obviously t_1 and t_2 are identical except below p, and since the procedure lift_rec is top-down, the computations of lift(t_i) are parallel until the subterms $t_i|p$ are reached. Let $(V_i, s_i, a_i) = t_i$, and $v_i \in V_i$ be the vertex corresponding to position p in t_i, then either there are calls on both lift_rec(v_i), or on none.

    If both calls lift_rec(v_i) do happen, we let $f = s_1(v_1) = s_2(v_2)$, and $C_i = \{c \mid \langle f, c, G \rangle \in \Sigma' \wedge c \sqsubseteq t_i|v_i\}$. We prove that $C_1 = C_2$: let $c \in C_1$, since we have both $c \sqsubseteq t_1|v_1$ and $e \sqsubseteq t_1|v_1$, then c and e are unifiable, and of course there is a symbol in Σ' with the context e. Since Σ' is unify-stable then $e \sqsubseteq_\sigma c \sqcup e \sqsubseteq t_1|v_1$, hence by Lemma 6 we have $c \sqcup e \sqsubseteq t_2|v_2$, and therefore $c \in C_2$. By symmetry

---

[2] they are isomorph term trees, or simply they represent the same term in string notation. For sake of simplicity we avoid explicit mention to isomorphisms, although a more orthodox version of the following proof would explicitly build a term tree isomorphism.

we have $C_2 \subseteq C_1$ (since $[\![e, \sigma^{-1}]\!] \vdash_p t_2 = t_1$), so that $C_1 = C_2$, which proves that T_1 and T_2 have the same symbol $\langle f, c, \Gamma_{\Sigma'}(c) \rangle$ at v_1 and v_2, where $c = \sqcup C_i$. We have also proved that Lemma 5 can be applied, yielding $e \sqsubseteq_\sigma e \sqcup (\sqcup C_i) = c$, and then the permutation $\Phi_{[e,c,\varepsilon]}(\sigma) \in \Gamma_{\Sigma'}(c)$ turns T_1 into T_2, so that $T_1 \bowtie T_2$.

We now suppose there is no call to lift_rec(v_i), which means that a symbol $\langle f, c, \Gamma_{\Sigma'}(c) \rangle$ has been given to a vertex u_i above v_i, the same for both terms as we have just proved, and v_i occurs "inside" the context c, i.e. there is a position $p' \neq \varepsilon$ in c such that v_i is reached from u_i by following the path p' in an obvious way. Therefore $c|p' \sqsubseteq t_i|p$, so that $c|p'$ and e are unifiable, and since Σ' is unify-stable we get $e \sqsubseteq_\sigma c|p'$. The permutation $\Phi_{[e,c,p']}(\sigma) \in \Gamma_{\Sigma'}(c)$ turns T_1 into T_2, so that $T_1 \bowtie T_2$.

Finally, if $t =_E t'$ then we have a sequence of equations $[\![e_i, \sigma_i]\!] \in E$ and terms t_i such that $[\![e_i, \sigma_i]\!] \vdash_{p_i} t_i = t_{i+1}$, with $t_1 = t$ and $t_n = t'$. We therefore have lift(t_i) $\bowtie$ lift(t_{i+1}), and since $\bowtie$ is transitive (see [5]) we get lift(t) $\bowtie$ lift(t').

Corollary 1. *If Σ' covers E and is unify-stable, then* lift(t) *is E-complete.*

Proof. Let $T = $ lift(t), it is easy to see that $t = $ um(T), so that $t \in S[T]$. For any term t' such that $t =_E t'$ by Theorem 5 we have $T \bowtie$ lift(t'), i.e. there exists a $\pi \in \mathcal{G}_T$ such that T^π is lift(t'), so that um(T^π) $= t'$, which proves that $t' \in S[T]$.

Note that in the lifting procedure, the computation of each new context c and of the group attached to c are relative to Σ', hence independent of the new symbols (equations) previously introduced in the process. This of course contrasts with Knuth-Bendix completion, and ensures polynomial time computation.

5   Deduction Problems

We now prove that, when restricted to stratified term trees built in that way, then the equality, the inclusion and the intersection problems on unmarked stratified sets are equivalent to the equivalence problem on the stratified term trees. The same holds for the membership problem.

We therefore assume that from E we can build a unify-stable signature Σ' that covers E (we say that E is unify-stable), and consider the set $\mathfrak{U} = $ lift($\mathfrak{T}$), where $\mathfrak{T}$ is the set of Σ-terms. Variables are treated as constant symbols.

Theorem 6. *For all $T, T' \in \mathfrak{U}$ we have:*

$$S[T] = S[T'] \Leftrightarrow S[T] \subseteq S[T'] \Leftrightarrow S[T] \cap S[T'] \neq \emptyset \Leftrightarrow T \bowtie T'.$$

Proof. If $S[T] = S[T']$ then obviously $S[T] \subseteq S[T']$. If $S[T] \subseteq S[T']$ then obviously $S[T] \cap S[T'] = S[T] \neq \emptyset$.

Now suppose that $S[T] \cap S[T'] \neq \emptyset$, and let $t \in S[T] \cap S[T']$. We have um(T) $=_E t =_E$ um(T'), hence by Theorem 5 we get lift(um(T)) $\bowtie$ lift(t). But lift(um(T)) is T, so $T \bowtie$ lift(t), and similarly $T' \bowtie$ lift(t), so that $T \bowtie T'$.

Finally, if $T \bowtie T'$, any member of $[T]_s$ is a T^π, with $\pi \in \mathcal{G}_T$. But $T' \bowtie T^\pi$, so there is a $\pi' \in \mathcal{G}_{T'}$ such that $T'^{\pi'}$ is T^π, hence such that um($T'^{\pi'}$) $= $ um(T^π). This proves that $S[T] \subseteq S[T']$, and by symmetry that $S[T] = S[T']$.

This trivially shows that the intersecting and inclusion problems are polynomial time equivalent to the problem $\text{EQUIV}_\mathfrak{U}$ of testing $T \bowtie T'$ for $T, T' \in \mathfrak{U}$.

Theorem 7. *The membership problem* $t \in S[T]$ *on* $\mathfrak{T} \times \mathfrak{U}$ *is polynomial time equivalent to* $\text{EQUIV}_\mathfrak{U}$.

Proof. We first show that $\forall t \in \mathfrak{T}, \forall T \in \mathfrak{U}$, $t \in S[T] \Leftrightarrow \text{lift}(t) \bowtie T$. Let $T' = \text{lift}(t)$, if $t \in S[T]$ then $t \in S[T] \cap S[T']$, and by Theorem 6 we get $\text{lift}(t) \bowtie T$. Conversely from $\text{lift}(t) \bowtie T$ we get $S[T] = S[T']$, hence $t \in S[T]$.

Since the function lift can be computed in polynomial time this result provides a polynomial time transformation from $\text{EQUIV}_\mathfrak{U}$ to membership. Conversely, for all $T, T' \in \mathfrak{U}$, we have $T = \text{lift}(\text{um}(T)) \bowtie T' \Leftrightarrow \text{um}(T) \in S[T']$ by the same result, and this provides a polynomial time transformation from membership to $\text{EQUIV}_\mathfrak{U}$, since um is polynomial.

In [6] can be found a polynomial Turing reduction from the equivalence problem to a problem in Luks class (see [3]), which therefore holds for its restriction $\text{EQUIV}_\mathfrak{U}$. The reverse reduction has been proved in [5], and the proof uses a signature Σ' with only one symbol on a flat context $f(\circ, \dots, \circ)$, which is trivially unify-stable, hence the reduction holds to $\text{EQUIV}_\mathfrak{U}$. This proves that $\text{EQUIV}_\mathfrak{U}$ is in Luks class, and can be solved by efficient (though non polynomial) group-theoretic algorithms (see [9,8]). Incidentally, the problem of determining whether $t =_E t'$ holds or not is equivalent to $t \in S[\text{lift}(t')]$, hence also equivalent to the problem $\text{EQUIV}_\mathfrak{U}$. This means that:

Corollary 2. *The word problem for unify-stable theories is in Luks class.*

6   Example and Conclusion

We consider the set of lp-equations E given by the following contexts and associated groups:

$$c_1 = f(g(h(\circ, \circ), h(\circ, \circ)), \circ, \circ), \quad c_2 = f(\circ, h(\circ, \circ), \circ), \quad c_3 = g(\circ, \circ),$$
$$G_1 = \{(), (1\ 2)(3\ 4)\}, \quad\quad\quad G_2 = \text{Sym}(\{2, 3\}), \quad G_3 = \text{Sym}(\{1, 2\}).$$

The signature $\Sigma' = \{\langle f, c_1, G_1 \rangle, \langle f, c_2, G_2 \rangle, \langle g, c_3, G_3 \rangle\}$ obviously covers E, and is E-stable: we have $\mu, 1 : c_3 \trianglelefteq c_1$ with $\mu(1) = \mu(2) = h(\circ, \circ)$, hence $E(\mu) = G_3$. We obtain a group complete signature by replacing in Σ' the group G_1 by $G_1' = \Gamma_{\Sigma'}(c_1)$, which is generated by $G_1 \cup \Phi_{[c_3, c_1, 1]}(G_3)$. We have $\Phi_{[c_3, c_1, 1]}((1\ 2)) = (1\ 3)(2\ 4)$, hence $G_1' = \{(), (1\ 2)(3\ 4), (1\ 3)(2\ 4), (1\ 4)(2\ 3)\}$.

Now that Σ' is group-complete, we may check whether it is unify-stable:

- c_1 and c_2 are unifiable, let $c = c_1 \sqcup c_2 = f(g(h(\circ, \circ), h(\circ, \circ)), h(\circ, \circ), \circ)$, and μ_i such that $\mu_i(c_i) = c_1 \sqcup c_2$. We have $E(\mu_1) = \text{Sym}(\{1, 2, 3, 4, 6, 7\})$, which contains G_1' and $E(\mu_2) = G_2$, hence $c_1 \sqsubseteq_{G_1'} c$ and $c_2 \sqsubseteq_{G_2} c$.
- $\mu, 1 : c_3 \trianglelefteq c_1$ with $\mu(1) = \mu(2) = h(\circ, \circ)$, so $c_3 \sqsubseteq_{G_3} c_1 | 1$.

Σ' is therefore unify-stable, and we can apply the lift algorithm to any term and obtain an E-complete stratified term tree.

For example, let $t = f(g(h(a,b), h(c,d)), h(a,b), g(a,b))$. We start by representing it by a term tree, as in Example 1, and call the procedure lift on this term tree. So, lift_rec is first called on the root of the term tree. t is an instance of both c_1 and c_2, so the label of the root becomes $\langle f, c, G \rangle$ where $G = \Gamma_{\Sigma'}(c)$ is generated by $\Phi_{[c_1,c,\varepsilon]}(G_1') = G_1'$, $\Phi_{[c_2,c,\varepsilon]}(G_2) = \mathrm{Sym}(\{5,6\})$, and $\Phi_{[c_3,c,1]}(G_3)$ contained in G_1'. Hence $G = G_1'\mathrm{Sym}(\{5,6\})$ has 8 elements.

Then 7 other calls to lift_rec are made, corresponding to the 7 holes in c, and the only one that modifies a label is the call on $t|3$ (the subterm $g(a,b)$), which labels this vertex with $\langle g, c_3, G_3 \rangle$. We finally obtain the stratified term tree $T = \langle f, c, G \rangle (g(h(a,b), h(c,d)), h(a,b), \langle g, c_3, G_3 \rangle (a,b))$, and $\mathcal{G}_T$ is isomorphic to $G \times G_3$. $S[T]$ contains the 16 terms in the E-class of t.

As a conclusion, we have to mention that the restriction to unify-stable theories may not be a necessary condition of E-completeness of stratified sets. We believe however that it would be very difficult to generalise this result to a significantly larger class of theories, provided that we can test membership in this class in polynomial time.

Another possibility is to drop E-completeness of single stratified sets, and try to obtain a lifting procedure that would yield a limited number of stratified term trees covering all E-consequences. But it would then be more difficult to reduce the deduction problems to the equivalence of stratified term trees.

References

1. J. Avenhaus and D. Plaisted. General algorithms for permutations in equational inference. *Journal of Automated Reasoning*, 26:223–268, April 2001.
2. F. Baader and T. Nipkow. *Term Rewriting and All That*. Cambridge University Press, 1998.
3. László Babai. Automorphism groups, isomorphism, reconstruction. In R. L. Graham, M. Grotschel, and L. Lovasz, editors, *Handbook of Combinatorics*, volume 2. Elsevier and The MIT Press, 1995.
4. H. P. Barendregt, M. C. J. D. van Eekelen, J. R. W. Glauert, J. R. Kennaway, M. J. Plasmeijer, and M. R. Sleep. Term graph rewriting. In J. W. de Bakker, A. J. Nijman, and P. C. Treleaven, editors, *PARLE'87*, LNCS 259, pages 141–158. Springer Verlag, june 1987.
5. T. Boy de la Tour and M. Echenim. On the complexity of deduction modulo leaf permutative equations. To appear in Journal of Automated Reasoning.
6. T. Boy de la Tour and M. Echenim. On leaf permutative theories and occurrence permutation groups. In I. Dahn and L. Vigneron, editors, *FTP'2003*, volume 86 of *Electronic Notes in Theoretical Computer Science*. Elsevier, 2003.
7. T. Boy de la Tour and M. Echenim. **NP**-completeness results for deductive problems on stratified terms. In M. Vardi and A. Voronkov, editors, *10th International Conference LPAR*, LNAI 2850, pages 315–329. Springer Verlag, 2003.
8. G. Butler. *Fundamental algorithms for permutation groups*. Lecture Notes in Computer Science 559. Springer Verlag, 1991.
9. C. Hoffmann. *Group-theoretic algorithms and graph isomorphism*. Lecture Notes in Computer Science 136. Springer Verlag, 1981.

TaMeD : A Tableau Method for Deduction Modulo

LIP6*

Abstract. Deduction modulo is a formalism introduced to separate cleanly computations and deductions by reasoning modulo a congruence on propositions. A sequent calculus modulo has been defined by Dowek, Hardin and Kirchner as well as a resolution-based proof search method called Extended Narrowing And Resolution (ENAR), in which the congruences are handled through rewrite rules on terms and atomic propositions.
We define a tableau-based proof search method, called *Tableau Method for Deduction modulo* (TaMeD), for theorem proving modulo. We then give a syntactic proof of the completeness of the method with respect to provability in the sequent calculus modulo. Moreover, we follow in our proofs the same steps as the ENAR method in such a way that it allows to try and compare the characteristics of both methods.

Introduction

As noted in [10] and [11] automated theorem proving methods such as resolution or tableau might lead to ineffective procedure. If we try to prove the following example

$$(a + b) + ((c + d) + e) = a + ((b + c) + (d + e))$$

with the associativity and identity axioms, a program with a poor strategy could run endlessly without finding the right solution whereas we would like it to apply a deterministic and terminating strategy to check that the two terms are indeed the same modulo associativity. This problem would be more efficiently solved by *computation* (i.e blind execution) instead of *deduction* (non-deterministic search), using a rewrite rule to express the associativity axiom. In addition to terms, rewrite rules can as well apply to propositions of the considered language.

Some tableau-based methods allow to add equational theories to the standard first-order tableau method described by Smullyan in [16]. Many propositions have been made in this field — descriptions can be found in [1], [2], [4], [8] and [12] — and it permits to get complete methods combining deduction and term-rewriting steps. However, rewriting on propositions is usually not handled in the tableau methods described in these papers, although this is of practical use.

---

* Laboratoire d'Informatique de Paris 6, 8 rue du Capitaine Scott, 75015 Paris, France
`richard.bonichon@lip6.fr`

D. Basin and M. Rusinowitch (Eds.): IJCAR 2004, LNAI 3097, pp. 445–459, 2004.

For example, take the following axiom, which is part of the theory of integral domains

$$\forall x \, \forall y \, (x * y = 0 \Leftrightarrow (x = 0 \vee y = 0))$$

It yields the corresponding rewrite rule: $x * y = 0 \rightarrow x = 0 \vee y = 0$

In this rule, an atomic proposition is turned into a disjunction and it is hard to see how it could be replaced by a rule rewriting terms. Having the rewriting rule above, we can prove the proposition:

$$\exists z \, (a * a = z \Rightarrow a = z)$$

but the closed tableau can not be derived from the disjunctive normal form of its negation which yields the following propositions

$$a * a = z \qquad \neg(a = z)$$

since the traditional branch closure rule does not see that z can be instantiated by 0. If we had the **Extended Narrowing** rule of Fig. 2, we would be suggested in this case the instantiation $z := 0$ so that the tableau could be closed.

This paper begins with a short presentation of the sequent calculus modulo and its principal properties in Sect. 1. Then, the proof search method TaMeD is introduced in Sect. 2, before stating the main theorem we intend to prove in Sect. 3 and the plan of the proof. We then proceed to show in Sect. 4 the soundness and completeness of an intermediate method called IC-TaMeD . After that we can establish the soundness and completeness of the TaMeD method in Sect. 5. Section 6 shows a small example. Eventually, a brief comparison with the ENAR method ([10]) as well as hints at further research are given in the conclusion (Sect. 7).

Note that, as the proofs are generally long, they are mostly omitted. However, the full-length proofs are available in [5][1].

1   The Sequent Calculus Modulo

The notions of *terms, atomic propositions, propositions, formulas, sentences* are defined as usual in first-order logic, i.e. as in [12], [13] or [15]. The notions of *term rewrite rule, proposition rewrite rule, equational axiom, class rewrite system* are those of [10] and the other notations as well (e.g. $\overset{\mathcal{R}}{\rightarrow}, \overset{\mathcal{RE}}{\rightarrow}$), unless otherwise stated. The standard substituion avoiding capture of the term t for the variable x in a proposition P is written $P[x := t]$.

1.1   The Sequent Calculus Modulo

The sequent calculus modulo is an extension of the sequent calculus defined for first-order logic. This calculus is defined in Fig. 1. One might notice that the axiom rule requires not just unifiability of the left proposition and the right one but they have to be identical modulo the congruence. If $=_{\mathcal{RE}}$ is taken to be the identity, this sequent calculus becomes the usual one.

---

[1] under http://www-spi.lip6.fr/~bonichon/papers/tamed.ps.gz

Proposition 1. *If* $=_{\mathcal{RE}}$ *is a decidable congruence, then proof checking for the sequent calculus modulo is decidable. This is in particular the case when the rewrite relation* $\longrightarrow_{\mathcal{RE}}$ *is confluent and (weakly) terminating.*

Proposition 2. *If* $P =_{\mathcal{RE}} Q$ *then* $\Gamma \vdash_{\mathcal{RE}} P, \Delta$ *if and only if* $\Gamma \vdash_{\mathcal{RE}} Q, \Delta$ *and* $\Gamma, P \vdash_{\mathcal{RE}} \Delta$ *if and only if* $\Gamma, Q \vdash_{\mathcal{RE}} \Delta$ *and the proofs have the same size.*

Proposition 3. *If a closed sequent* $\Gamma \vdash_{\mathcal{RE}} \Delta$ *has a proof, then it also has a proof where all the sequents are closed.*

1.2   Equivalence Between $\vdash$ and $\vdash_{\mathcal{RE}}$

We mention in this subsection a really important property regarding the equivalence between the classical sequent calculus and the sequent calculus modulo, as proved in [10]. Indeed, it states the soundness and completeness of the sequent calculus modulo with respect to first-order logic.

Definition 1 (Compatibility). *A set of axioms* $\mathcal{K}$ *and a class rewrite system* $\mathcal{RE}$ *are said to be* compatible *if:*

- $P =_{\mathcal{RE}} Q$ *implies* $\mathcal{K} \vdash P \Leftrightarrow Q$.
- *for every proposition* $P \in \mathcal{K}$, *we have* $\vdash_{\mathcal{RE}} P$.

Proposition 4 (Equivalence). *If the set of axioms* $\mathcal{K}$ *and the class rewrite system* $\mathcal{RE}$ *are compatible then we have:*

$$\mathcal{K}, \Gamma \vdash \Delta \text{ if and only if } \Gamma \vdash_{\mathcal{RE}} \Delta$$

From this proposition, we now know that the same theorems can be deduced from the two formalisms. Of course, corresponding proofs of the same theorem may be of different size depending on the formalism used. Actually, proof are generally smaller in the sequent calculus modulo.

2   The TaMeD Method

We are now going to extend the standard tableau method for first-order classical logic as defined in [12] and [16] to another method where congruences are built-in. *In the rest of this paper, the relation* $\longrightarrow^{*}_{\mathcal{RE}}$ *is assumed to be confluent.*

2.1   Labels

The usual first step of a tableau based proof search method is to transform the propositions to be proved into a set of branches. This step generally involves skolemization. In fact, several skolemized forms are possible for a proposition. Take for example the closed formula $\forall x \; \exists y \; P(0, y)$ where the variable x does not occur: the Skolem constant f could as well be nullary as unary, yielding respectively $P(0, f)$ or $P(0, f(x))$. The latter is chosen in this paper (as in [10])

$$\frac{}{P \vdash_{\mathcal{RE}} Q}\text{axiom} \ \ \text{if } P =_{\mathcal{RE}} Q$$

$$\frac{\Gamma, P \vdash_{\mathcal{RE}} \Delta \quad \Gamma \vdash_{\mathcal{RE}} Q, \Delta}{\Gamma \vdash_{\mathcal{RE}} \Delta}\text{cut} \ \ \text{if } P =_{\mathcal{RE}} Q$$

$$\frac{\Gamma, Q_1, Q_2 \vdash_{\mathcal{RE}} \Delta}{\Gamma, P \vdash_{\mathcal{RE}} \Delta}\text{contr-l} \ \ \text{if } P =_{\mathcal{RE}} Q_1 =_{\mathcal{RE}} Q_2$$

$$\frac{\Gamma \vdash_{\mathcal{RE}} Q_1, Q_2, \Delta}{\Gamma \vdash_{\mathcal{RE}} P, \Delta}\text{contr-r} \ \ \text{if } P =_{\mathcal{RE}} Q_1 =_{\mathcal{RE}} Q_2$$

$$\frac{\Gamma \vdash_{\mathcal{RE}} \Delta}{\Gamma, P \vdash_{\mathcal{RE}} \Delta}\text{weak-l}$$

$$\frac{\Gamma \vdash_{\mathcal{RE}} \Delta}{\Gamma \vdash_{\mathcal{RE}} P, \Delta}\text{weak-r}$$

$$\frac{\Gamma, P, Q \vdash_{\mathcal{RE}} \Delta}{\Gamma, R \vdash_{\mathcal{RE}} \Delta}\wedge\text{-l} \ \ \text{if } R =_{\mathcal{RE}} (P \wedge Q)$$

$$\frac{\Gamma \vdash_{\mathcal{RE}} P, \Delta \quad \Gamma \vdash_{\mathcal{RE}} Q, \Delta}{\Gamma \vdash_{\mathcal{RE}} R, \Delta}\wedge\text{-r} \ \ \text{if } R =_{\mathcal{RE}} (P \wedge Q)$$

$$\frac{\Gamma, P \vdash_{\mathcal{RE}} \Delta \quad \Gamma, Q \vdash_{\mathcal{RE}} \Delta}{\Gamma, R \vdash_{\mathcal{RE}} \Delta}\vee\text{-l} \ \ \text{if } R =_{\mathcal{RE}} (P \vee Q)$$

$$\frac{\Gamma \vdash_{\mathcal{RE}} P, Q, \Delta}{\Gamma \vdash_{\mathcal{RE}} R, \Delta}\vee\text{-r} \ \ \text{if } R =_{\mathcal{RE}} (P \vee Q)$$

$$\frac{\Gamma \vdash_{\mathcal{RE}} P, \Delta \quad \Gamma, Q \vdash_{\mathcal{RE}} \Delta}{\Gamma, R \vdash_{\mathcal{RE}} \Delta}\Rightarrow\text{-l} \ \ \text{if } R =_{\mathcal{RE}} (P \Rightarrow Q)$$

$$\frac{\Gamma, P \vdash_{\mathcal{RE}} Q, \Delta}{\Gamma \vdash_{\mathcal{RE}} R, \Delta}\Rightarrow\text{-r} \ \ \text{if } R =_{\mathcal{RE}} (P \Rightarrow Q)$$

$$\frac{\Gamma \vdash_{\mathcal{RE}} P, \Delta}{\Gamma, R \vdash_{\mathcal{RE}} \Delta}\neg\text{-l} \ \ \text{if } R =_{\mathcal{RE}} \neg P$$

$$\frac{\Gamma, P \vdash_{\mathcal{RE}} \Delta}{\Gamma \vdash_{\mathcal{RE}} R, \Delta}\neg\text{-r} \ \ \text{if } R =_{\mathcal{RE}} \neg P$$

$$\frac{}{\Gamma, P \vdash_{\mathcal{RE}} \Delta}\bot\text{-l} \ \ \text{if } P =_{\mathcal{RE}} \bot$$

$$\frac{\Gamma, \{t/x\}Q \vdash_{\mathcal{RE}} \Delta}{\Gamma, P \vdash_{\mathcal{RE}} \Delta}(Q, x, t) \ \forall\text{-l} \ \ \text{if } P =_{\mathcal{RE}} \forall x \ Q \text{ and } t \text{ arbitrary term}$$

$$\frac{\Gamma \vdash_{\mathcal{RE}} \{c/x\}Q, \Delta}{\Gamma \vdash_{\mathcal{RE}} P, \Delta}(Q, x, c) \ \forall\text{-r} \ \ \text{if } P =_{\mathcal{RE}} \forall x \ Q \text{ and } c \text{ fresh constant}$$

$$\frac{\Gamma, \{c/x\}Q \vdash_{\mathcal{RE}} \Delta}{\Gamma, P \vdash_{\mathcal{RE}} \Delta}(Q, x, c) \ \exists\text{-l} \ \ \text{if } P =_{\mathcal{RE}} \exists x \ Q \text{ and } c \text{ fresh constant}$$

$$\frac{\Gamma \vdash_{\mathcal{RE}} \{t/x\}Q, \Delta}{\Gamma \vdash_{\mathcal{RE}} P, \Delta}(Q, x, t) \ \exists\text{-r} \ \ \text{if } P =_{\mathcal{RE}} \exists x \ Q \text{ and } c \text{ fresh constant}$$

Fig. 1. The sequent calculus modulo

because of the following fact: if we had an equation $x * 0 = 0$ leading to an $\mathcal{E}$-equivalence between $\forall x \ \exists y \ P(0, y)$ and $\forall x \ \exists y \ P(x * 0, y)$, the Skolem symbols would have the same arity in both cases. This choice is implemented by memorizing the universal quantifier scope of each subformula during the tableau form computation by associating a *label*.

Definition 2 (Labeled proposition). *A labeled proposition is a pair P^l formed by a proposition P and a finite set l of variables containing all the free variables (and possibly more) of P called its* label.

Definition 3 (Substitution in a labeled proposition). *When we apply a substitution Θ to a labeled proposition, each variable x of the label is replaced by the free variables of Θx. Two labeled proposition P^l and $Q^{l'}$ are $\mathcal{E}$-equivalent if $P =_{\mathcal{RE}} Q$ and $l = l'$. The labeled proposition P^l $\mathcal{R}$-rewrites to $Q^{l'}$ if P $\mathcal{R}$-rewrites to Q and $l = l'$*

2.2   Constrained Tableau

The notations we will use throughout the paper to represent transformations of a tableau are inspired by the ones that can be found either in [10] for clausal form transformations and in [7] and [8] for tableau methods. Hence the more classical presentation of [12] and [16] will not be followed, mostly because it does not give at first glance a global view of the tableau transformation in the expansion rules. The names $\alpha-, \beta-, \gamma-, \delta-$formulas are those commonly used in the above references.

Definition 4 (Tableau-related vocabulary).

- *A branch is a finite set $\{Q_1, \ldots, Q_n\}$ of propositions.*
- *A tableau is a finite set $\{\mathcal{B}_1 \mid \ldots \mid \mathcal{B}_p\}$ of branches.*
- *A fully expanded branch ($\mathcal{F}exp$-branch) of a tableau is a finite set $\{P_1, \ldots, P_n\}$ of propositions such that every P_i is a literal, i.e. either an atomic proposition or the negation of an atomic proposition.*
- *A branch is said to be closed if it both contains a formula P and its negation $\neg P$.*
- *A tableau is then said closed if each of its branch is closed.*
- *The closed tableau is denoted $\odot$.*

Let $\mathcal{B}$ be a set of labeled propositions and P^l a labeled proposition, we write $\mathcal{B}, P^l$ for the set $\mathcal{B} \cup \{P^l\}$. Let $\mathcal{T}$ be a set of sets of labeled propositions and $\mathcal{B}$ a set of labeled propositions, we write $\mathcal{T} \mid \mathcal{B}$ for the set $\mathcal{T} \cup \{\mathcal{B}\}$.

We now give a presentation of the tableau expansion calculus. In this calculus, we use an integer to forbid infinite computations. Iterative deepening is performed if it is chosen too small at the beginning.

Definition 5 (Tableau form transformation: Tab). *To put a set of non-labeled sentences in tableau form, we first label them with the empty set. The universally quantified propositions are also labeled with a given integer n_x denoting the allowed number of γ-expansions per γ-formula, where x is the universally quantified variable bound in the γ-formula. We consider the following transformations on sets of sets of labeled propositions.*

β-expansion steps

- $\mathcal{T} \mid \mathcal{B}, (P \vee Q)^l \xrightarrow{tab} \mathcal{T} \mid \mathcal{B}, P^l \mid \mathcal{B}, Q^l$

- $\mathcal{T} \mid \mathcal{B}, \neg(P \wedge Q)^l \xrightarrow{tab} \mathcal{T} \mid \mathcal{B}, \neg P^l \mid \mathcal{B}, \neg Q^l$

- $\mathcal{T} \mid \mathcal{B}, (P \Rightarrow Q)^l \xrightarrow{tab} \mathcal{T} \mid \mathcal{B}, \neg P^l \mid \mathcal{B}, Q^l$

α-expansion steps

- $\mathcal{T} \mid \mathcal{B}, (P \wedge Q)^l \xrightarrow{tab} \mathcal{T} \mid \mathcal{B}, P^l, Q^l$

- $\mathcal{T} \mid \mathcal{B}, \neg(P \vee Q)^l \xrightarrow{tab} \mathcal{T} \mid \mathcal{B}, \neg P^l, \neg Q^l$

- $\mathcal{T} \mid \mathcal{B}, \neg(P \Rightarrow Q)^l \xrightarrow{tab} \mathcal{T} \mid \mathcal{B}, P^l, \neg Q^l$

- $\mathcal{T} \mid \mathcal{B}, \neg\neg P^l \xrightarrow{tab} \mathcal{T} \mid \mathcal{B}, P^l$

γ-expansion steps

- $\mathcal{T} \mid \mathcal{B}, (\forall x P)^l_{n_x} \xrightarrow{tab} \mathcal{T} \mid \mathcal{B}, (\forall x P)^l_{n_x-1}, P^{l,x}$ where x is a fresh variable and $n_x > 1$

- $\mathcal{T} \mid \mathcal{B}, (\forall x P)^l_1 \xrightarrow{tab} \mathcal{T} \mid \mathcal{B}, P^{l,x}$ where x is a fresh variable

- $\mathcal{T} \mid \mathcal{B}, \neg(\exists x P)^l_{n_x} \xrightarrow{tab} \mathcal{T} \mid \mathcal{B}, \neg(\exists x P)^l_{n_x-1}, \neg P^{l,x}$ where x is a fresh variable and $n_x > 1$

- $\mathcal{T} \mid \mathcal{B}, \neg(\exists x P)^l_1 \xrightarrow{tab} \mathcal{T} \mid \mathcal{B}, \neg P^{l,x}$ where x is a fresh variable

δ-expansion steps

- $\mathcal{T} \mid \mathcal{B}, (\exists x P)^{y_1,\dots,y_n} \xrightarrow{tab} \mathcal{T} \mid \mathcal{B}, (P\{x := f(y_1,\dots,y_n)\})^{y_1,\dots,y_n}$ where f is a fresh Skolem symbol

- $\mathcal{T} \mid \mathcal{B}, \neg(\forall x P)^{y_1,\dots,y_n}) \xrightarrow{tab} \mathcal{T} \mid \mathcal{B}, (\neg P[x := f(y_1,\dots,y_n)])^{y_1,\dots,y_n}$ where f is a fresh Skolem symbol

Other expansion steps:

- $\mathcal{T} \mid \mathcal{B}, \bot^l \xrightarrow{tab} \mathcal{T}$

- $\mathcal{T} \mid \mathcal{B}, \neg\bot^l \xrightarrow{tab} \mathcal{T} \mid \mathcal{B}$

Proposition 5 (Termination). *The Tab transformation terminates for a given number of γ-expansions allowed per γ-proposition.*

Definition 6. *For some equational theory $\mathcal{E}$, an equation modulo $\mathcal{E}$ is a pair of terms or of atomic propositions denoted $t =_{\mathcal{E}}^? t'$. A substitution σ is a $\mathcal{E}$-solution of $t =_{\mathcal{E}}^? t'$ when $\sigma t =_{\mathcal{E}} \sigma t'$. It is a $\mathcal{E}$-solution of an equation system C when it is a solution of all the equations in C.*

Definition 7 (Constrained tableau). *A constrained tableau is a pair $\mathcal{T}[C]$ such that $\mathcal{T}$ is a tableau and C is a set of equations called constraints.*

2.3   The TaMeD Method

Definition 8. *Let $\mathcal{RE}$ be a class rewrite system and $\mathcal{T}[C]$ a constrained tableau, we write*

$$\mathcal{T}[C] \mapsto \mathcal{T}'[C']$$

*if the constrained tableau $\mathcal{T}'[C']$ can be deduced from the constrained tableau $\mathcal{T}[C]$ using finitely many applications of the **Extended Narrowing** and **Extended Branch Closure** rules described in Fig. 2. This means there is a derivation of the tableau $\mathcal{T}'[C']$ under the assumptions $\mathcal{T}[C]$, i.e. a sequence $\mathcal{T}_1[C_1], \ldots, \mathcal{T}_n[C_n]$ such that either $n = 0$ and $\mathcal{T}'[C'] = \mathcal{T}[C]$ or $n > 0, \mathcal{T}_0[C_0] = \mathcal{T}[C], \mathcal{T}_n[C_n] = \mathcal{T}'[C']$ and each $\mathcal{T}_i[C_i]$ is produced by the application of a rule in TaMeD to $\mathcal{T}_{i-1}[C_{i-1}]$.*

The first rule, **Extended Branch Closure** is a simple extension of the usual branch closure rule for first-order equational tableau, where the $\mathcal{E}$-unification constraints are not solved but stored in the constraints part. This can be compared to a certain extent with the constrained tableau method described by Degtyarev and Voronkov in [7] and [8]. Although propositions are labeled with variables, these play no role when applying the **Extended Branch Closure** rule. In particular, they are removed from the constraints part of the tableau. Moreover, there appears a major difference between resolution and tableau methods: the **Extended Branch Closure** is only binary, whereas the **Extended Resolution** rule for ENAR needs to be applied to all relevant propositions.

The **Extended Narrowing** rule is much the same as the one proposed for resolution in [10] and the narrowing is only applied to atomic propositions and not directly to terms. As atomic propositions may be rewritten to non-atomic ones, it must be ensured that they are transformed back into tableau form.

When $\mathcal{R}$ is empty the **Extended Narrowing** rule is never used and we get a method for equational tableau. When both $\mathcal{R}$ and $\mathcal{E}$ are empty, then we get back a first-order free variable tableau method.

3   Main Theorem

The main result of this paper states the soundness and completeness of the TaMeD method with respect to the sequent calculus modulo.

Extended Branch Closure

$$\Gamma_1, P, \neg Q \mid \Gamma_2 \mid \ldots \mid \Gamma_n \ [C]$$
$$\overline{\Gamma_2 \mid \ldots \mid \Gamma_n \ [C \cup \{P \overset{?}{=}_\mathcal{E} Q\}]}$$

Extended Narrowing

$$\Gamma_1, U \mid \Gamma_2 \mid \ldots \mid \Gamma_n \ [C]$$
$$\overline{\mathcal{B}_1 \mid \ldots \mid \mathcal{B}_p \mid \Gamma_2 \mid \ldots \mid \Gamma_n \ [C \cup \{U_{|\omega} \overset{?}{=}_\mathcal{E} l\}]}$$

if $l \to r \in \mathcal{R}$, $U_{|\omega}$ is an atomic proposition and $\mathcal{B}_1 \mid \ldots \mid \mathcal{B}_p = Tab(\Gamma_1, U[r]_\omega)$

Fig. 2. TaMeD rules

Theorem 1 (Main theorem). *Let* $\mathcal{RE}$ *be a class rewrite system such that* $\longrightarrow_{\mathcal{RE}}$ *is confluent. For every* $\mathcal{B}$ *and* Γ *sets of closed formulas, if* C *is a* $\mathcal{E}$-*unifiable set of constraints, then we have the following implications:*

$$Tab(\mathcal{B} \wedge \neg \Gamma)[\emptyset] \underset{\mathcal{RE}}{\mapsto} \odot[C] \quad \Rightarrow \quad \mathcal{B} \vdash_{\mathcal{RE}} \Gamma$$

where $\neg \Gamma = \{\neg P | P \in \Gamma\}$.
If the sequent $\mathcal{B} \vdash_{\mathcal{RE}} \Gamma$ *has a cut-free proof then there exists a derivation*

$$Tab(\mathcal{B} \wedge \neg \Gamma)[\emptyset] \underset{\mathcal{RE}}{\mapsto} \odot[C] \text{ for some } \mathcal{E}\text{-unifiable set } C$$

We can deduce the following corollary from the same hypothesis when the cut-elimination property holds

$$Tab(\mathcal{B} \wedge \neg \Gamma)[\emptyset] \underset{\mathcal{RE}}{\mapsto} \odot[C] \quad \Leftrightarrow \quad \mathcal{B} \vdash_{\mathcal{RE}} \Gamma$$

4   Soundness and Completeness of the IC-TaMeD Method

In order to prove the theorem stated in Sec. 3, we first define an intermediate calculus simply called *Intermediate Calculus for TaMeD* (IC-TaMeD). The allowed derivations in this calculus are described in Fig. 3 .

Definition 9 (IC-TaMeD derivation). *Let* $\mathcal{RE}$ *be a class rewrite system and* $\mathcal{T}$ *a tableau, we note:*

$$\mathcal{T} \hookrightarrow \mathcal{T}'$$

if the tableau $\mathcal{T}'$ *can be obtained from* $\mathcal{T}$ *using finitely many applications of the IC-TaMeD rules described in Fig. 3. This means there is a sequence* $\mathcal{T}_1, \ldots, \mathcal{T}_n$ *such that either* $n = 0$ *and* $\mathcal{T} = \mathcal{T}'$ *or* $n > 0, \mathcal{T} = \mathcal{T}_0, \mathcal{T}' = \mathcal{T}_n$ *and each* $\mathcal{T}_i$ *is produced by the application of a rule of IC-TaMeD to the tableau* $\mathcal{T}_{i-1}$.

$$\frac{\Gamma_1 \mid \dots \mid \Gamma_n}{(\Gamma_1 \mid \dots \mid \Gamma_n)[x \mapsto t]} \text{ Instantiation}$$

$$\frac{\Gamma_1, P \mid \Gamma_2 \mid \dots \mid \Gamma_n}{\Gamma_1, P' \mid \Gamma_2 \mid \dots \mid \Gamma_n} \text{ Conversion if } P =_{\mathcal{RE}} P'$$

$$\frac{\Gamma_1, P \mid \Gamma_2 \mid \dots \mid \Gamma_n}{\mathcal{B}_1 \mid \dots \mid \mathcal{B}_p \mid \Gamma_2 \mid \dots \mid \Gamma_n} \text{ Reduction}$$

$$\text{if } P \xrightarrow{\mathcal{R}} Q \text{ and } \mathcal{B}_1 \mid \dots \mid \mathcal{B}_p = Tab(\Gamma_1, Q)$$

$$\frac{\Gamma_1, P^{l_1}, \neg P^{l_2} \mid \Gamma_2 \mid \dots \mid \Gamma_n}{\Gamma_2 \mid \dots \mid \Gamma_n} \text{ Identical Branch Closure}$$

Fig. 3. IC-TaMeD

In the **Instantiation** rule, the instantiated variable is replaced in the label by the free variables of the substituted term. The difference with ENAR is in the fact that the instantiation must be global because of the tableau rigid free variables. In the **Conversion** rule, because of Definition 3 of $\mathcal{E}$-equivalent labeled propositions, the labels are kept by the transformed propositions, thus forbidding in particular to introduce free variables in P' that were not present in the labels of P. In the **Reduction** rule, labels are extended by the tableau form transformation algorithm. In the **Identical Branch Closure**, eliminated propositions need not have the same label.

The proof of Theorem 1 is detailed in this section and the next ones. The plan of this proof is the following: first the soundness and completeness of the IC-TaMeD method are proved, then the soundness and completeness of the TaMeD method with respect to IC-TaMeD . It is schematized as follows:

$$\mathcal{K}, \mathcal{B}, \neg \Gamma \vdash \overset{\underline{Lem.}4}{\Longleftrightarrow} \mathcal{B}, \neg \Gamma \vdash_{\mathcal{RE}} \overset{\underset{Prop.7}{Prop.6}}{\Longleftrightarrow} \mathcal{B}, \neg \Gamma \hookrightarrow \odot \overset{\underset{Prop.9}{Prop.8}}{\Longleftrightarrow} \mathcal{B}, \neg \Gamma [\emptyset] \mapsto \odot [C]$$

Finally, we can deduce from all this that the set of branches $Tab(\Gamma \wedge \neg \Delta)$ can be refuted if and only if the sequent $\mathcal{K}, \Gamma \vdash \Delta$ is provable in the sequent calculus.

4.1 Tableau Form Algorithm Soundness

Before proving the theorem corresponding to this subsection, we define some more specific notations. The notations for *terms, propositions, branches, tableaux* are the same as in [5].

Definition 10 (Free variables of a set of propositions). *Let* $\mathcal{B} = \{P_1, \dots P_n\}$ *be a set of propositions (atomic or not). The free variables of* $\mathcal{B}$ *are defined as the union of the sets of free variables of* $P_1, \dots, P_n$.

Definition 11 ($\overline{\forall}$ notation). *Let $\Gamma_1, \ldots, \Gamma_n$ be sets of propositions. Let $x_1, \ldots, x_n$ be the free variables of the $\Gamma_i s$. We will use the following notation:*

$$\overline{\forall}(\Gamma_1 \vee \ldots \vee \Gamma_n) = \forall x_1, \ldots, \forall x_n (\Gamma_1 \vee \ldots \vee \Gamma_n)$$

The tableau form algorithm from Definition 5 has the following necessary property.

Lemma 1 (Tableau form algorithm soundness). *Let $\Gamma_1, \ldots, \Gamma_m$ and $\mathcal{B}_1, \ldots, \mathcal{B}_n$ be sets of labeled propositions. If, by using the Tab computation of Def. 5, we have:*

$$\{\Gamma_1 \mid \ldots \mid \Gamma_m\} \xrightarrow{tab} \{\mathcal{B}_1 \mid \ldots \mid \mathcal{B}_n\}$$

Then

$$\overline{\forall}(\Gamma_1 \vee \ldots \vee \Gamma_m) \vdash \iff \overline{\forall}(\mathcal{B}_1 \vee \ldots \vee \mathcal{B}_n) \vdash$$

4.2   IC-TaMeD Soundness

We will now concentrate on the specific IC-TaMeD soundness proof. The next lemma is the main intermediate lemma used to prove the correctness of the IC-TaMeD method.

Lemma 2. *Let $(\Gamma_1 \mid \ldots \mid \Gamma_n)$ be a set of $\mathcal{F}exp$-branches . If*

$$(\Gamma_1 \mid \ldots \mid \Gamma_n) \hookrightarrow \odot$$

then

$$\overline{\forall}(\Gamma_1 \vee \ldots \vee \Gamma_n) \vdash_{\mathcal{RE}}$$

Proposition 6 (IC-TaMeD soundness). *Let $P_1, \ldots, P_n, Q_1, \ldots, Q_m$ be closed formulas. If*

$$Tab(P_1 \wedge \ldots \wedge P_n \wedge \neg Q_1 \wedge \ldots \wedge \neg Q_m) \hookrightarrow \odot$$

then

$$P_1, \ldots, P_n \vdash_{\mathcal{RE}} Q_1, \ldots, Q_m$$

Proof. First use Lemma 2 on $Tab(P_1 \wedge \ldots \wedge P_n \wedge \neg Q_1 \wedge \ldots \wedge \neg Q_m)$, and then conclude with Lemma 1 and Prop. 4.

4.3   IC-TaMeD Completeness

The following definition and two lemmas deal with the problems linked to the interfering use of the rewriting steps and skolemization steps in IC-TaMeD .

Definition 12 (Function symbol transformation). *Let t be a term (resp. a proposition), f a function symbol of arity n and u a term whose free variables are among $x_1, \dots, x_n$. The individual transformation of symbol f into u is denoted by $(x_1, \dots, x_n)u/f$. Its application on a term (resp. a proposition) t, denoted by $\{(x_1, \dots, x_n)u/f\}t$, is obtained by replacing in t any subterm of the form $f(v_1, \dots, v_n)$, where $v_1, \dots, v_n$ are arbitrary terms by the term $u[x_1 := v_1, \dots, x_n := v_n]$.*

For a finite set of indexes I, we define the result of the application of a transformation of function symbols $\rho = \{(x_1^i, \dots, x_n^i)u^i/f^i\}_{i \in I}$ to a term (resp. a proposition) t as the simultaneous application of the individual symbol transformations on t.

Labels are not affected by such transformations.

Lemma 3. *Let $\mathcal{T}$ be a set of sets of labeled propositions and ρ a transformation of function symbols. We assume that the Skolem symbols introduced when putting $\mathcal{T}$ in tableau form are fresh, i.e. not transformed by ρ. Then we have $Tab(\rho\mathcal{T}) = \rho Tab(\mathcal{T})$ up to some renaming.*

Lemma 4. *Let $\mathcal{T}$ be a set of branches and ρ a transformation of function symbols not appearing in $\mathcal{RE}$. If $\mathcal{T} \hookrightarrow \odot$ then $\rho\mathcal{T} \hookrightarrow \odot$ and the derivations have the same length.*

Let us now state some lemmas used to prove the completeness of IC-TaMeD.

Lemma 5. *Let t be a closed term, $\mathcal{B}$ a set of labeled propositions, then*

$$Tab(\mathcal{B}[x := t]) \hookrightarrow \odot \Rightarrow Tab(\mathcal{B}) \hookrightarrow \odot$$

Lemma 6. *Let $\mathcal{B} = \{P_1, \dots, P_n\}$ and $\Gamma = \{Q_1, \dots, Q_n\}$ be two sets of labeled propositions such that, for every i, $P_i \longrightarrow_{\mathcal{RE}}^* Q_i$. If $\mathcal{T}Tab(\Gamma) \hookrightarrow \odot$, then $Tab(\mathcal{B}) \hookrightarrow \odot$*

Lemma 7. *Let R, S be closed terms and Γ, Δ sets of closed terms. If*

$$Tab(R, \Gamma, \neg\Delta) \hookrightarrow \odot \quad and \quad Tab(S, \Gamma, \neg\Delta) \hookrightarrow \odot$$

Then we can build a derivation of:

$$Tab(R \vee S, \Gamma, \neg\Delta) \hookrightarrow \odot$$

The following lemma allows the restriction of the use of the congruence to reductions in proofs.

Lemma 8. *If the relation $\longrightarrow_{\mathcal{RE}}$ is confluent, we have:*

- *If P and $Q \wedge R$ (resp. $Q \vee R$) are sentences such that $P =_{\mathcal{RE}} Q \wedge R$ (resp. $P =_{\mathcal{RE}} Q \vee R$) , then there exists a sentence $Q' \wedge R'$ (resp. $Q' \vee R'$) such that $P \longrightarrow_{\mathcal{RE}}^* Q' \wedge R'$ (resp. $P \longrightarrow_{\mathcal{RE}}^* Q' \vee R'$), $Q =_{\mathcal{RE}} Q'$ and $P =_{\mathcal{RE}} P'$.*

- If P and $\neg Q$ are sentences such that $P =_{\mathcal{RE}} \neg Q$, then there exists a sentence $\neg Q'$ such that $P \longrightarrow^*_{\mathcal{RE}} \neg Q'$ et $Q =_{\mathcal{RE}} Q'$.
- If P is a sentence such that $P =_{\mathcal{RE}} \bot$, then $P \longrightarrow^*_{\mathcal{RE}} \bot$.
- If P and $\forall x\ Q$ (resp. $\exists x\ Q$) are sentences such that $P =_{\mathcal{RE}} \forall x\ Q$ (resp. $P =_{\mathcal{RE}} \exists x\ Q$), then there exists a sentence $\forall x\ Q'$ (resp. $\exists x\ Q'$)such that $P \longrightarrow^*_{\mathcal{RE}} \forall x\ Q'$ (resp. $P \longrightarrow^*_{\mathcal{RE}} \exists x\ Q'$) and $Q =_{\mathcal{RE}} Q'$.

Proposition 7 (IC-TaMeD completeness). *Let $\longrightarrow^*_{\mathcal{RE}}$ be a confluent relation and $P_1,\dots,P_m, Q_1,\dots,Q_n$ be closed terms. If the sequent:*

$$P_1,\dots,P_m \vdash_{\mathcal{RE}} Q_1,\dots,Q_n$$

has a cut-free proof then:

$$Tab(P_1,\dots,P_m,\neg Q_1,\dots,\neg Q_n) \hookrightarrow \odot$$

Proof. Using Prop. 2 and Lemmas 5, 6, 7, 8 , we reason by induction on the size of a closed cut-free proof of $P_1,\dots,P_m \vdash_{\mathcal{RE}} Q_1,\dots,Q_n$.

5   Soundness and Completeness of the TaMeD Method

The soundness and completeness results obtained for IC-TaMeD will be lifted to TaMeD in the following subsections. Interactions between tableau forms and substitutions are handled by special lemmas in both cases.

5.1   TaMeD Soundness

Lemma 9. *Let $\mathcal{B}$ be a set of labeled propositions and Θ a closed substitution such that the variables bound in $\mathcal{B}$ are not in the domain of Θ. Then, there exists a transformation ρ of the function symbols introduced by putting $\mathcal{B}$ in tableau form such that $Tab(\Theta\mathcal{B}) = \rho\Theta Tab(\mathcal{B})$.*

Lemma 10. *Let $\mathcal{T}[C]$ be constrained tableau such that*

$$\mathcal{T}[C] \mapsto \odot$$

and Θ be a closed substitution, unifier of C, mapping all variables of C to a closed term. Then $\Theta\mathcal{T} \hookrightarrow \odot$.

Proposition 8 (TaMeD soundness). *Let $\mathcal{T}$ be a set of non-constrained branches such that*

$$\mathcal{T} \mapsto \odot[C]$$

where C is a unifiable set of constraints. Then $\mathcal{T} \hookrightarrow \odot$.

Proof. The set of constraints C is unifiable and thus it has a unifier Θ mapping all variables of C. By Lemma 10 $\Theta\mathcal{T} \hookrightarrow \odot$. All the branches of $\Theta\mathcal{T}$ can be derived from those of $\mathcal{T}$ itself with the **Instantiation** rule. Hence $\mathcal{T} \hookrightarrow \odot$.

Since IC-TaMeD is sound, the previous proposition entails immediately TaMeD soundness.

5.2   TaMeD Completeness

The completeness result will now be lifted from IC-TaMeD to TaMeD much in the same way. Let us first introduce a lemma similar to Lemma 9.

Lemma 11. *Let $\mathcal{B}$ and Γ be two sets of labeled propositions. Let Θ be a substitution such that no variable bound in $\mathcal{B}$ is in the domain of Θ. Suppose that $\Theta\mathcal{B} =_{\mathcal{E}} \Gamma$. Then there is a transformation ρ of function symbols introduced by putting Γ in tableau form such that $\Theta Tab(\mathcal{B}) =_{\mathcal{E}} \rho Tab(\Gamma)$*

Proposition 9 (TaMeD completeness). *Let $\mathcal{U}$ be a set of constrained branches and Θ be a $\mathcal{E}$-unifier of the constraints of $\mathcal{U}$, and $\mathcal{T}$ a set non-constrained branches such that*

$$\Theta\mathcal{U} =_{\mathcal{E}} \mathcal{T} \text{ and } \mathcal{T} \hookrightarrow \odot$$

then

$$\mathcal{U} \mapsto \odot[C]$$

where C is a unifiable set of constraints .

Proof. By induction on the structure of the IC-TaMeD proof of $\mathcal{T}$, using Lemmas 4 and 11.

6   Example

We will finally use TaMeD with the short example suggested in the introduction to show how it works. We want to prove $\exists z\ (a * a = z \Rightarrow a = z)$ with the following rewrite rule: $x * y = 0 \rightarrow x = 0 \lor y = 0$.

By the tableau form transformation on $\neg\exists z\ (a * a = z \Rightarrow a = z)$ with $n_x = 1$ we get the lone branch $\{(a * a = Z)^Z, \neg(a = Z)^Z\}$.

We cannot close it yet. However we can have the following derivation in TaMeD (where the labels are omitted for the sake of more readablility):

$$
\frac{\{(a * a = Z), \neg(a = Z)\}[\emptyset]}{Tab\{(a = 0 \lor a = 0), \neg(a = 0)\}[a * a = 0 \overset{?}{\underset{\mathcal{E}}{=}} x * y = 0]} \; Narrowing
$$

$$=$$

$$
\frac{(a = 0), \neg(a = 0)|(a = 0), \neg(a = 0)[a * a = 0 \overset{?}{\underset{\mathcal{E}}{=}} x * y = 0]}{\odot[a * a = 0 \overset{?}{\underset{\mathcal{E}}{=}} x * y = 0, a = 0 \overset{?}{\underset{\mathcal{E}}{=}} a = 0]} \; Branch\ Closure
$$

In this derivation, the narrowing is done with the free variable Z instantiated by 0 and the branch closure rule is applied on each branch, yielding twice the same constraint. As the constraints are clearly solvable, we know we have found a TaMeD derivation proving $\exists z\ (a * a = z \Rightarrow a = z)$.

7 Conclusion

After introducing the sequent calculus modulo of [10], we have presented a tableau-based proof search method for deduction modulo. Furthermore we have shown that TaMeD is indeed sound and complete with respect to this sequent calculus modulo.

Integrating rewrite rules for computational operations in sequent rules dedicated to deduction allow powerful and expressive theories to be expressed as deduction modulo. Small congruences can be defined such as in the propositional rewriting example given in the introduction, but Dowek, Hardin and Kirchner have developed in [9] a theory called $HOL_{\lambda\sigma}$, which is an intentionally equivalent first-order presentation of higher-order logic. Actually, $HOL_{\lambda\sigma}$ can be expressed as sets of rewrite rules and integrated in deduction modulo and thus could also give us almost for free a framework for a higher-order tableau method, adding to the one of [14].

We have also tried to underline the differences with the ENAR method, which principally come from the fact that free variables introduced by our *Tab*-calculus can not be locally dealt with and therefore must be treated rigidly. This difference appears not only in the calculus rules but also and above all in the proofs (see [5]). Rigid free variables might be a drawback when compared to resolution but, on the other side, the tableau closure rule of TaMeD allows binary closure, which ENAR's Extended Resolution rule does not.

All in all, the calculus presented here is is a simple extension of the standard method which cleanly separates the *Tab*-form algorithm and the TaMeD calculus. Yet, one could want to introduce some restrictions or extensions. What easily comes to mind would be to allow non-literal branch closure, which would ensure faster closure in many cases (and not change the soundness and completeness of the method). Furthermore, the skolemization process defined through the use labels could be discussed (and refined) in order to improve the method.

Strong assumptions have been made in our proofs, for example on the confluence property of the considered $\mathcal{RE}$ rewrite system or omitted, such as the method we would use to unify the constraints. Actually, the problems linked to unification have not been presented because it is not the purpose of this paper. But we can assume the rigidity of free variables introduced by the tableau method will once again be a major point (see [3] on rigid $\mathcal{E}$-unification and [7] on how to deal with it), since simultaneous rigid $\mathcal{E}$-unification has been proved undecidable in [6].

First of all, further research could be made to get a model-based completeness for TaMeD ([17] gives this proof for ENAR). Moreover, on the practical side and since at least one ENAR implementation has already been made, it would be interesting to implement TaMeD in order to confront the two methods at work. Finally, a generalization of TaMeD could be considered, along the same lines of the similar proposed extension of ENAR at the end of [10].

References

[1] B. Beckert. Konzeption und Implementierung von Gleichheit für einen tableau-basierten Theorembeweiser. IKBS report 208, Science Center, Institute for Knowledge Based Systems, IBM Germany, 1991.

[2] B. Beckert. Adding equality to semantic tableaux. In K. Broda, M. D'Agostino, R. Goré, R. Johnson, and S. Reeves, editors, *Proceedings, 3rd Workshop on Theorem Proving with Analytic Tableaux and Related Methods, Abingdon*, pages 29–41, Imperial College, London, TR-94/5, 1994.

[3] B. Beckert. Rigid e-unification. In W. Bibel and P.H. Schmidt, editors, *Automated Deduction: A Basis for Applications. Volume I, Foundations: Calculi and Methods.* Kluwer Academic Publishers, Dordrecht, 1998.

[4] B. Beckert and R. Hähnle. An improved method for adding equality to free variable semantic tableaux. In D. Kapur, editor, *Proceedings, 11th International Conference on Automated Deduction (CADE), Saratoga Springs, NY*, LNCS 607, pages 507–521. Springer, 1992.

[5] R. Bonichon. TaMeD: A tableau method for deduction modulo. To be published as a LIP6 report, 2004.

[6] A. Degtyarev and A. Voronkov. The undecidability of simultaneous rigid e-unification. *Theoretical Computer Science*, 166(1-2):291–300, 1996.

[7] A. Degtyarev and A. Voronkov. What you always wanted to know about rigid e-unification. *J. Autom. Reason.*, 20(1-2):47–80, 1998.

[8] A. Degtyarev and A. Voronkov. Equality reasoning in sequent-based calculi. In A. Robinson and A. Voronkov, editors, *Handbook of Automated Reasoning*, chapter 10. Elsevier Science Publishers B.V., 2001.

[9] G. Dowek, T. Hardin, and C. Kirchner. HOL-$\lambda\sigma$: An intentional first-order expression of higher-order logic. In *RTA*, pages 317–331, 1999.

[10] G. Dowek, T. Hardin, and C. Kirchner. Theorem proving modulo. *Journal of Automated Reasoning*, (31):33–72, 2003.

[11] G. Dowek, T. Hardin, and C. Kirchner. Theorem proving modulo, revised version. Rapport de recherce 4861, INRIA, 2003.

[12] M. Fitting. *First Order Logic and Automated Theorem Proving*. Springer-Verlag, 2nd edition, 1996.

[13] J. Goubault-Larrecq and I. Mackie. *Proof Theory and Automated Deduction.* Applied Logic Series. Kluwer Academic Publishers, 1997.

[14] M. Kohlhase. Higher-order automated theorem proving. In W. Bibel and P.H. Schmidt, editors, *Automated Deduction: A Basis for Applications. Volume I, Foundations: Calculi and Methods.* Kluwer Academic Publishers, Dordrecht, 1998.

[15] R. Letz. First-order tableau calculi. In M. D'Agostino, D.M. Gabbay, R. Hähnle, and J. Posegga, editors, *Handbook of Tableau Methods*. Kluwer Academic Publishers, Dordrecht, 2000.

[16] R. Smullyan. *First Order Logic.* Springer, 1968.

[17] J. Stuber. A model-based completeness proof of extended narrowing and resolution. *Lecture Notes in Computer Science*, 2083:195+, 2001.

Lambda Logic

Michael Beeson*

San Jose State University, San José CA 95192, USA,
beeson@cs.sjsu.edu
www.cs.sjsu.edu/faculty/beeson

Abstract. Lambda logic is the union of first order logic and lambda calculus. We prove basic metatheorems for both total and partial versions of lambda logic. We use lambda logic to state and prove a soundness theorem allowing the use of second order unification in resolution, demodulation, and paramodulation in a first-order context.

1 Introduction

The twentieth century saw the flowering of first order logic, and the invention and development of the lambda calculus. When the lambda calculus was first developed, its creator (Alonzo Church) intended it as a foundational system, i.e., one in which mathematics could be developed and with the aid of which the foundations of mathematics could be studied. His first theories were inconsistent, just as the first set theories had been at the opening of the twentieth century. Modifications of these inconsistent theories never achieved the fame that modifications of the inconsistent set theories did. Instead, first order logic came to be the tool of choice for formalizing mathematics, and lambda calculus is now considered as one of several tools for analyzing the notion of algorithm.

The point of view underlying lambda logic is that lambda calculus is a good tool for representing the notion of function, not only the notion of computable function. First-order logic can treat functions by introducing function symbols for particular functions, but then there is no way to construct other functions by abstraction or recursion. Of course, one can consider set theory as a special case of first order logic, and define functions in set theory as univalent functions, but this requires building up a lot of formal machinery, and has other disadvantages as well. It is natural to consider combining lambda calculus with logic. That was done long ago in the case of typed logics; for example Gödel's theory T had what amounted to the ability to define functions by lambda-abstraction. But typed lambda calculus lacks the full power of the untyped (ordinary) lambda calculus, as there is no fixed-point theorem to support arbitrary recursive definitions.

In this paper, we combine ordinary first order lambda calculus with ordinary first order logic to obtain systems we collectively refer to as lambda logic. We are not the first to define or study similar systems.[1] The applicative theories

---

* Research supported by NSF grant number CCR-0204362.

[1] For example, John McCarthy told me that he lectured on such systems years ago, but never published anything.

D. Basin and M. Rusinowitch (Eds.): IJCAR 2004, LNAI 3097, pp. 460–474, 2004.
© Springer-Verlag Berlin Heidelberg 2004

proposed by Feferman in [6] are similar in concept. They are, however, different in some technical details that are important for the theorems proved here. Lambda logic is also related to the systems of illative combinatory logic studied in [2], but these are stronger than lambda logic. As far as we know, the systems defined in this paper have not been studied before.

Both ordinary and lambda logic can be modified to allow "undefined terms". In the context of ordinary logic this has been studied in [8,4,3]. In the context of applicative theories, [3] defined and studied "partial combinatory algebras"; but application in the λ-calculus is always total. Moggi [7] was apparently the first to publish a definition of partial lambda calculus; see [7] for a thorough discussion of different versions of partial lambda calculus and partial combinatory logic.

The system Otter-λ [5] uses second-order unification in an untyped context to enhance the capabilities of the automated theorem prover Otter. Inference rules such as resolution and paramodulation are allowed to use second-order unification instead of only first order unification. Higher order unification has in the past been used with typed systems. Lambda logic provides a theoretical foundation for its application in an untyped setting, as we show with a soundness theorem in this paper. Lambda logic answers the question, "In what formal system can the proofs produced by Otter-λ be naturally represented?".[2]

2   Syntax

We will not repeat the basics of first order logic or the basics of lambda calculus, which can be found in [9] and [1], respectively. We start with any of the usual formulations of first order logic with equality. This includes a stock of variables, constants, and function symbols, from which one can build up terms and formulas as usual. In addition, there is a distinguished unary function symbol Ap. As usual in lambda calculus, we optionally abbreviate $Ap(x, y)$ to $x(y)$ or even xy, with left association understood in expressions like xyz, which means $(xy)z$. But we do this less often than is customary in lambda calculus, since we also have $f(x)$ where f is a function symbol; and this is not the same as $Ap(f, x)$. Of course, theoretically there is only one syntactically correct way of reading the abbreviated term.

Terms are created by the following term formation rules: variables and constants are terms; if t is a term and x is a variable, then $\lambda x. t$ is a term; if $t_1, \ldots, t_n$ are terms and f is an n-ary function symbol then $f(t_1, \ldots, t_n)$ is a term.

The notion of a variable being free in a term is defined as usual: quantifiers and lambda both bind variables. The notion of substitution is defined as in [1]. The notation is $t[x := s]$ for the result of substituting s for the free variable x in t. Note that this is literal substitution, i.e. $t[x := s]$ does not imply the renaming

---

[2] For further information about Otter-λ, including example input and output files, the ability to run the system over the web, a precise description of its algorithm for λ-unification, see the Otter-λ web site, accessible from the author's home page www.cs.sjsu.edu/faculty/beeson. Two proofs which were too long to include in this paper can also be found there.

of bound variables of t to avoid capture of free variables of s. We also define $t[x ::= s]$, which does imply an algorithmic renaming of bound variables of t to avoid capture of free variables of s.

Alpha-conversion means renaming a bound variable in such a way that there is no capture of formerly free variables by the renamed variable. The induced equivalence relation on terms is called alpha-equivalence. Beta-reduction is defined as usual. As usual we distinguish two particular lambda terms $\mathbf{T}$ and $\mathbf{F}$.

Example: in first order logic we can formulate the theory of groups, using a constant e for the identity, and function symbols for the group operation and inverse. The use of infix notation $x \cdot y$ can either be regarded as official or as an informal abbreviation for $m(x, y)$, just as it can in first order logic. If we formulate the same theory in lambda logic, we use a unary predicate G for the group, and relativize the group axioms to that predicate, just as we would do in first order logic if we needed to study a group and a subgroup. Then in lambda logic we can define the commutator $c := \lambda x, y.((i(x) \cdot i(y)) \cdot x) \cdot y$, and derive the following:

$$G(x) \wedge G(y) \to c(x, y) = ((i(x) \cdot i(y)) \cdot x) \cdot y$$

The hypothesis $G(x) \wedge G(y)$ is needed because we relativized the group axioms to G. We want to formulate the theory of groups, not the theory of models of the lambda calculus that can be turned into groups. Alternately, we can replace $G(x)$ by $Ap(G, x) = \mathbf{T}$, and $x \cdot y$ by $Ap(Ap(\cdot, x), y)$, so that we can discuss subgroups, homomorphisms, etc. quite naturally.

3   Axioms

Lambda logic can be formulated using any of the usual approaches to predicate calculus. We distinguish the sequent-calculus formulation, the Hilbert-style formulation, and the resolution formulation. For definiteness we choose the Hilbert-style formulation as the definition (say as formulated in [9], p. 20), for the standard version. There will then be two further variants for different logics of partial terms, but these will be discussed in another section.

(*Prop*) propositional axioms

(*Q*) standard quantifier axioms

(α) $t = s$ if t and s are alpha-equivalent. The alternative would be to include the axiom only in case t alpha-converts to s.

(β) $Ap(\lambda x.t, s) \cong t[x ::= s]$

(ξ) (*weak extensionality*) $\forall x(Ap(t, x) \cong Ap(s, x)) \to \lambda x.Ap(t, x) \cong \lambda x.Ap(s, x)$

(*non-triviality*) $\mathbf{T} \neq \mathbf{F}$, where $\mathbf{T} = \lambda x \lambda y.x$ and $\mathbf{F} = \lambda x \lambda y.y$

The following additional formulae are not part of lambda logic, but are of interest.

(η) (*extensionality*) $\lambda x.Ap(t, x) \cong t$.

(*AC*) (*Axiom of Choice*) $\forall x \exists y P(x, y) \to \exists f \forall x P(x, Ap(f, x))$.

4   Semantics

There is a standard definition of λ-model that is used in the semantics of the lambda calculus (see [1], p. 86, with details on p. 93). There is also a well-known notion of a model of a first order theory. In this section our goal is to define the concept M *is a model of the theory T in lambda logic* in such a way that it will imply that, neglecting λ, M is a first order model of T, and also, neglecting the function symbols other than Ap, M is a λ-model.

The cited definition of λ-model involves the notion of terms, which we shall call M-terms, built up from Ap and constants c_a for each element a of the model. It requires the existence, for each term t of this kind, and each variable x, of another M-term $\lambda^*(x, t)$ such that M will satisfy

$$Ap(\lambda^*(x, t), x) = t.$$

Note that this does not yet make sense, as we must first define the notion of "the interpretation of a term t in M". We cannot simply refer to [1] for the definition, since we need to extend this definition to the situation in which we have a theory T with more function symbols than just Ap, although the required generalization is not difficult.

We first define a λ-structure. As usual in first order logic we sometimes use "M" to denote the carrier set of the structure M; and we use f_M or $\bar{f}$ for the function in the structure M that serves as the interpretation of the function symbol f, but we sometimes omit the bar if confusion is unlikely. $(M, \lambda*)$ is a λ-structure for T if (1) M is a structure with a signature containing all the function symbols and constants occurring in T, and another binary function Ap_M to serve as the interpretation of Ap, and (2) there is an operation λ^* on pairs (x, t), where t is an M-term and x is a variable, producing an element $\lambda^*(x, t)$ of M.

If (M, λ^*) is a λ-structure for T, then by a *valuation* we mean a map g from the set of variables to (the carrier set of) M. If g is a valuation, and $v = v_1, \ldots, v_n$ is a list (vector) of variables, then by $g(v)$ we mean $g(v_1), \ldots, g(v_n)$. If t is a term, then by $t[v := g(v)]$ we mean the M-term resulting from replacing each variable v_i by the constant $c_{g(v_i)}$ for the element $g(v_i)$ of M. If g is a valuation, then we can then extend g to the set of terms by defining

$$g[f(t_1, \ldots, t_n)] = \bar{f}(g(t_1), \ldots, g(t_n))$$
$$g[\lambda x.t] = \lambda^*(x, t[v := g(v)])$$

Now we have made sense of the notion "the interpretation of term t under valuation g". We define the notion $M \models \phi$ as it is usually defined for first order logic. We are now in a position to define λ-model. This definition coincides with that in [1] in case T has no other function symbols than Ap.

Definition 1 (λ-model). (M, λ^*) *is a λ-model of a theory T in lambda logic if* (M, λ^*) *satisfies the axioms α, β, and ξ, and M satisfies the axioms of T.*

5 Basic Metatheorems

Define the relation $t \equiv s$ on terms of T to mean that t and s have a common reduct (using β and α reductions). The Church-Rosser theorem for λ calculus ([1], p. 62) implies that this is an equivalence relation, when the language includes only Ap and λ. The following theorem says that adding additional function symbols does not destroy the Church-Rosser property.

Theorem 1. *The Church-Rosser theorem is valid for lambda logic.*

Proof. For each function symbol f we introduce a constant $\bar{f}$. We can then eliminate f in favor of $\bar{f}$ as follows. For each term t we define the term t° as follows:

$$x^\circ = x \quad \text{for variables } x$$
$$c^\circ = c \quad \text{for variables } c$$
$$f(t)^\circ = Ap(\bar{f}, t^\circ)$$
$$f(t, r)^\circ = Ap(Ap(\bar{f}, t^\circ), r^\circ))$$
$$Ap(t, r)^\circ = Ap(t^\circ, r^\circ)$$
$$(\lambda x.\, t)^\circ = \lambda x.\, t^\circ$$

and similarly for functions of more than two arguments. Since there are no reduction rules involving the new constants, t reduces to q if and only if t° reduces to q°. Moreover, if t° reduces to v, then v has the form u° for some u. (Both assertions are proved by induction on the length of the reduction.) Suppose t reduces to q and also to r. Then t° reduces to q° and to r°. By the Church-Rosser theorem, q° and r° have a common reduct v, and v is u° for some u, so q and r both reduce to u. That completes the proof.

The following theorem is to lambda logic as Gödel's completeness theorem is to first order logic. As in first order logic, a theory T in lambda logic is called *consistent* if it does not derive any contradiction. In this section, we will prove the lambda completeness theorem: any consistent theory has a λ-model. First, we need some preliminaries.

Theorem 2 (Lambda Completeness Theorem). *Let T be a consistent theory in lambda logic. Then T has a λ-model.*

Remark. There is a known "first order equivalent" of the notion of λ-model, namely *Scott domain*. See [1](section 5.4). However, we could not use Scott domains to reduce the lambda completeness theorem to Gödel's first order completeness theorem, because there is no syntactic interpretation of the theory of Scott domains in lambda logic.

Proof of the completeness theorem. The usual proof (Henkin's method) of the completeness theorem, as set out for example in [9] pp. 43-48, can be imitated for lambda logic. If T does not contain infinitely many constant symbols, we begin by adding them; this does not destroy the consistency of T since in any proof of contradiction, we could replace the new constants by variables not occurring elswhere in the proof. We construct the "canonical structure" M for a theory T. The elements of M are equivalence classes of closed terms of T under the equivalence relation of provable equality: $t \sim r$ iff $T \vdash t = r$. Let $[t]$ be the equivalence class of t. We define the interpretations of constants, function symbols, and predicate symbols in M as follows:

$$c^M = [c]$$
$$f^M([t_1], \ldots, [t_n]) = [f(t_1, \ldots, t_n)]$$
$$P^M([t_1], \ldots, [t_n]) = [P(t_1, \ldots, t_n)]$$

In this definition, the right sides depend only on the equivalence classes $[t_i]$ (as shown in [9], p. 44).

Exactly as in [9] one then verifies that M is a first order model of T. To turn M into a λ-model, we must define $\lambda^*(x, [t])$, where x is a variable and t is an M-term, i.e. a closed term with parameters from M. The "parameters from M" are constants $c_{[q]}$ for closed terms q of T. If t is an M-term t, let $[t]^\circ$ be the closed term of T obtained from t by replacing each constant $c_{[q]}$ by a closed term q in the equivalence class $[q]$. Then $[t]^\circ$ is well-defined. Define $\lambda^*(x, [t]) = [\lambda x. [t]^\circ]$. By axiom (ξ), this is a well-defined operation on equivalence classes: if T proves $t = s$ then T proves $[t]^\circ = [s]^\circ$ and hence $\lambda x. [t]^\circ = \lambda x. [s]^\circ$.

We verify that the axioms of lambda logic hold in M. First, the (β) axiom: $Ap(\lambda x. t, r) = t[x := r]$. It suffices to consider the case when t has only x free. The interpretation of the left side in M is the equivalence class of $Ap(\lambda x. t, r)$. The interpretation of the right side is the class of $t[x := r]$. Since these two terms are provably equal, their equivalence classes are the same, verifying axiom (β). Now for axiom (ξ). Suppose t and s are closed terms and $Ap(t, x) = Ap(s, x)$ is valid in M. Then for each closed term r, we have tr provably equal to sr. Since T contains infinitely many constant symbols, we can select a constant c that does not occur in t or s, so $tc = sc$ is provable. Replacing the constant c by a variable in the proof, $tx = sx$ is provable. Hence by axiom (ξ), $\lambda x. t = \lambda x. s$ is probable, and hence that equation holds in M. In verifying axiom (ξ), it suffices to consider the case when s and t are closed terms. The axiom (α) holds in M since it simply asserts the equality of pairs of provably equal terms. The axiom $\mathbf{T} \neq \mathbf{F}$ holds since T does not prove $\mathbf{T} = \mathbf{F}$, because T is consistent. That completes the proof.

Theorem 3 (Axiomatization of first-order theorems). *Let T be a first order theory, and let A be a first order sentence. Then T proves A in lambda logic if and only if for some positive integer n, T plus "there exist n distinct things" proves A in first order logic.*

Proof. First suppose A is provable from T plus "there exist n distinct things". We show A is provable in lambda logic, by induction on the length of the proof of A. Since lambda logic includes first order logic, the induction step is trivial. For the basis case we must show that lambda logic proves "there exist n distinct things" for each positive integer n. The classical constructions of numerals in lambda calculus produce infinitely many distinct things. However, it must be checked that their distinctness is provable in lambda logic. Defining numerals $\ulcorner n \urcorner$ as on p. 130 of [1] we verify by induction on n that for all $m < n$, $\ulcorner m \urcorner \neq \ulcorner n \urcorner$ is provable in lambda logic. If $m < n + 1$ then either $m < n$, in which case we are done by the induction hypothesis, or $m = n$. So what has to be proved is that for each n, lambda logic proves $\ulcorner n \urcorner \neq \ulcorner n + 1 \urcorner$. This in turn is verifiable by induction on n.

Conversely, suppose that A is not provable in T plus "there exist n distinct things" for any n. Then by the completeness theorem for first order logic, there is an infinite model M of $\neg A$; indeed we may assume that M has infinitely many elements not denoted by closed terms of T. Then M can be expanded to a lambda model $\hat{M}$ satisfying the same first order formulas, by defining arbitrarily the required operation λ^* on M-terms, and then inductively defining relations $E(x, y)$ and Ap_M to serve as the interpretations of equality and Ap in $\hat{M}$. A detailed proof is available at the author's web site.

6   Skolemization

We now consider the process of Skolemization. This is important for automated deduction, since it is used to prepare a problem for submission to a theorem prover that requires clausal form. This can be extended from ordinary logic to lambda logic, but unlike in ordinary logic, the axiom of choice is required:

Theorem 4 (Skolemization). *Let T be a theory in lambda logic. Then we can enlarge T to a new theory S by adding new function symbols (Skolem symbols) such that (1) the axioms of S are quantifier-free and (2) in lambda logic $+$ AC, S and T prove the same theorems in the language of T.*

Proof. The proof is the same as for ordinary logic–we eliminate one alternating quantifier at a time. Consider $\forall x \exists y P(x, y)$. We add a new function symbol f and the (Skolem) axiom

$$\forall x \exists y P(x, y) \rightarrow \forall x P(x, f(x)).$$

In the presence of this Skolem axiom, if $\forall x \exists y P(x, y)$ is an axiom of T it can be eliminated in favor of $\forall x P(x, f(x))$, which has one quantifier fewer. It remains to prove that adding such a Skolem axiom to $T + AC$ does not add any new theorems in the language of T. By the lambda completeness theorem, it suffices to show that any λ-model of $T + AC$ can be expanded to a λ-model of the Skolem axiom. Let (M, λ^*) be a λ-model of $T + AC$. Assume M satisfies $\forall x \exists y P(x, y)$. Since M satisfies AC, there is some u in M such that M satisfies $\forall x P(x, Ap(c_u, x))$. (Here c_u is a constant denoting the element u of M; technically λ^* is defined on M-terms.) Interpret the Skolem function symbol f as the function $x \mapsto Ap(u, x)$. Similarly, if r is any M-term possibly involving f, let r' be defined as follows:

$$r' = r \quad \text{if } r \text{ does not contain } f$$
$$f(t)' = Ap(c_u, t')$$

Now we extend λ^* to a function λ' defined on terms involving the new symbol f as well as the other symbols of T and constants for elements of M, by defining

$$\lambda'(x,t) = \lambda^*(x,t')$$

The right side is defined since t' does not contain f. Note that when t does not contain f, we have $t' = t$ and hence $\lambda^*(x,t) = \lambda'(x,t)$.

Consider the λ-structure (M', λ'), where M' is M augmented with the given interpretation of f. We claim that this λ-structure is actually a λ-model of $T + AC$ that satisfies the Skolem axiom. Formulas which do not contain the Skolem symbol f are satisfied in (M, λ^*) if and only they are satisfied in (M', λ'), since on such terms we have $\lambda^*(x,t) = \lambda'(x,t)$ as noted above. Therefore (M', λ') is a λ-model of $T + AC$, and we have already showed that it satisfies the Skolem axiom. That completes the proof.

7   The Logic of Partial Terms

In the group theory example near the end of the introduction, it is natural to ask whether $x \cdot y$ needs to be defined if x or y does not satisfy $G(x)$. In first order logic, $\cdot$ is a function symbol and hence in any model of our theory in the usual sense of first order logic, $\cdot$ will be interpreted as a function defined for all values of x and y in the model. The usual way of handling this is to say that the values of $x \cdot y$ for x and y not satisfying $G(x)$ or $G(y)$ are defined but irrelevant. For example, in first order field theory, $1/0$ is defined, but no axiom says anything about its value. As this example shows, the problem of "undefined terms" is already of interest in first order logic, and two different (but related) logics of undefined terms have been developed. We explain here one way to do this, known as the *Logic of Partial Terms* (LPT). See [4] or [3], pp. 97-99.

LPT has a term-formation operator $\downarrow$, and the rule that if t is a term, then $t \downarrow$ is an atomic formula. One might, for example, formulate field theory with the axiom $y \neq 0 \rightarrow x/y \downarrow$ (using infix notation for the quotient term). Thereby one would avoid the (sometimes) inconvenient fiction that $1/0$ is some real number, but it doesn't matter which one because we can't prove anything about it anyway; many computerized mathematical systems make use of this fiction. Taking this approach, one must then modify the quantifier axioms. The two modified axioms are as follows:

$$\forall x \, A \wedge t \downarrow \rightarrow A[x := t]$$
$$A[x := t] \wedge t \downarrow \rightarrow \exists x \, A$$

Thus from "all men are mortal", we are not able to infer "the king of France is mortal" until we show that there *is* a king of France. The other two quantifier axioms, and the propositional axioms, of first order logic are not modified. We also add the axioms $x \downarrow$ for every variable x, and $c \downarrow$ for each constant c.

In LPT, we do not assert anything involving undefined terms, not even that the king of France is equal to the king of France. The word "strict" is applied here to indicate that subterms of defined terms are always defined. LPT has the following "strictness axioms", for every atomic formula R and function symbol f. In these axioms, the x_i are variables and the t_i are terms.

$$R(t_1, \ldots, t_n) \to t_1 \downarrow \wedge \ldots \wedge t_n \downarrow$$

$$f(t_1, \ldots, t_n) \downarrow \to t_1 \downarrow \wedge \ldots \wedge t_n \downarrow$$

$$t_1 \downarrow \wedge \ldots \wedge t_n \downarrow \wedge f(x_1, \ldots, x_n) \downarrow \to f(t_1, \ldots, t_n) \downarrow$$

Remark. In LPT, while terms can be undefined, formulas have truth values just as in ordinary logic, so one never writes $R(t) \downarrow$ for a relation symbol R. That is not legal syntax.

We write $t \cong r$ to abbreviate $t \downarrow \vee r \downarrow \to t = r$. That is, For example, a special case of this schema is

$$t = r \to t \downarrow \wedge r \downarrow$$

It follows that $t \cong r$ really means "t and r are both defined and equal, or both undefined."

The equality axioms of LPT are as follows (in addition to the one just mentioned):

$$x = x$$

$$x = y \to y = x$$

$$t \cong r \wedge \phi[x := t] \to \phi[x := r]$$

8   Partial Lambda Calculus

In lambda calculus, the issue of undefined terms arises perhaps even more naturally than in first order logic, as it is natural to consider partial recursive functions, which are sometimes undefined.[3]

Partial lambda calculus is a system similar to lambda calculus, but in which Ap is not necessarily total. There can then be "undefined terms." Since lambda calculus is a system for deducing (only) equations, the system has to be modified. We now permit two forms of statements to be deduced: $t \cong r$ and $t \downarrow$. The axioms (α), (β), and (ξ) are modified by changing $=$ to $\cong$, and the rules for deducing $t \downarrow$ are specified as follows: First, we can always infer (without premise) $x \downarrow$ when x is a variable. Second, we can apply the inference rules

$$\frac{t \cong s \quad t \downarrow}{s \downarrow} \qquad \frac{t \cong s \quad s \downarrow}{t \downarrow}$$

$$\frac{t \downarrow \quad r \downarrow}{t[x := r] \downarrow} \qquad \lambda x.\, t \downarrow$$

---

[3] There is, of course, an alternative to λ-calculus known as *combinatory logic*. Application in combinatory logic is also total, but in [3], the notion of a *partial combinatory algebra* is introduced and studied, following Feferman, who in [6] first introduced partial applicative theories. See [7] for some relationships between partial combinatory logic and partial lambda calculus.

Note that we do not have strictness of Ap. As an example, we have $(\lambda y.\, a)b) \cong a$, whether or not $b \downarrow$. We could have formulated a rule "strict (β)" that would require deducing $r \downarrow$ before concluding $Ap(\lambda x.\, t, r) \cong t[x := r]$, but not requiring strictness corresponds better to the way functional programming languages evaluate conditional statements. Note also that $\lambda x.\, t$ is defined, whether or not t is defined.

9 Partial Lambda Logic

Partial lambda logic results if we make similar modifications to lambda logic instead of to first order logic or lambda calculus. In particular we modify the rules of logic and the equality axioms as in LPT, add the strictness axiom, and modify the axioms (α), (β), and (ξ) by replacing $=$ with $\cong$. In LPT, $\cong$ is an abbreviation, not an official symbol; in partial lambda calculus it is an official symbol; in partial lambda logic we could make either choice, but for definiteness we choose to make it an official symbol, so that partial lambda logic literally extends both LPT and partial lambda calculus. The first three rules of inference listed above for partial lambda calculus are superfluous in the presence of LPT. The fourth one is replaced in partial lambda logic by the following axiom:

$$t \downarrow \rightarrow \lambda x.\, t \downarrow .$$

We do not apply the strictness axiom of *LPT* to the function symbol Ap. Instead we rely only on (β) for deducing theorems of the form $Ap(t, r) \downarrow$. There is one additional axiom, as in partial lambda calculus:

$$\lambda x.\, t \downarrow \qquad \text{for each term } t.$$

We review the semantics of LPT as given in [4,3]. A model consists of a set and relations on that set to interpret the predicate symbols; the function symbols are interpreted by partial functions instead of total functions. Given such a *partial structure* one defines simultaneously, by induction on the complexity of terms t, the two notions $M \models t \downarrow$ and t_M, the element of M that is denoted by t.

We now discuss the semantics of partial lambda logic. The definition of partial λ-model is similar to that of λ-model, except that now Ap and the other function symbols can be interpreted by partial functions instead of total functions. The function λ^* in the definition of λ-model (which takes a variable and an M-term as arguments) is required to be total, so that the axiom $\lambda x.\, t \downarrow$ will be satisfied.

Definition 2 (Partial lambda model). (M, λ^*) *is a partial λ-model of a theory T in partial lambda logic if* (M, λ^*) *satisfies the axioms* (α), (ξ), *and* (β), *and M satisfies all the axioms of T and LPT, except that Ap need not be strict.*

The following theorem generalizes the completeness theorem for LPT to partial lambda logic.[4]

---

[4] In [4] there is a completeness theorem for LPT, generalizing Gödel's completeness theorem. The strictness axiom is important in the proof.

Theorem 5 (Lambda Completeness Theorem for LPT). *Let T be a consistent theory in partial lambda logic. Then T has a partial lambda model.*

The proof is available at the author's web site. The theorem on Skolemization also generalizes to partial lambda logic, with the same proof.

10   Soundness of Second Order Unification

We define σ to be a *second order unifier* of t and s if $t\sigma = s\sigma$ is provable in lambda logic. The usual rules of inference used in first-order logic, such as resolution and paramodulation, are extended by allowing second order unifiers. We give a proof of the soundness of such extended rules of inference. This proof does not refer to any specific algorithm for finding second order unifiers, but it does apply to such algorithms: all one has to do is verify that the algorithm in question does indeed produce second order unifiers. This theorem thus provides a theoretical basis for the practical use of such algorithms, e.g. in our theorem prover Otter-λ.

We define (second order) paramodulation to be the following rule: if $\alpha = \beta$ (or $\beta = \alpha$) has been deduced, and $P[x := \gamma]$ has been deduced, and for some substitution σ we have $\alpha\sigma = \gamma\sigma$, and the free variables of $\beta\sigma$ either occur in γ or are not bound in P, then we can deduce $P[x := \beta\sigma]$. This differs from the first order version of paramodulation only in the extra condition about the free variables of $\beta\sigma$.

The rules of inference that we prove sound are paramodulation (as just defined), demodulation (including β-reduction), binary resolution (using second order unification), and factoring. Specifically binary resolution allows the inference from $P|Q$ and $-S|R$ to $Q\sigma|R\sigma$ provided that $P\sigma = S\sigma$ is provable in λ-logic. Here of course Q and R stand for lists of literals, and the literals P and S can occur in any position in the clause, not just the first position as shown. Factoring allows us to unify different literals in the same clause: specifically we can infer $P\sigma|R\sigma$ from $P|Q|R$ when $P\sigma = R\sigma$ is provable in λ-logic.

An algorithm for untyped λ-unification is used in our theorem prover Otter-λ. This algorithm is not the same as second-order or higher-order unification algorithms, which are used in a typed setting. There is a natural question about its soundness, considering that (a) the algorithm is not a subset of typed λ-unification, in that it can solve problems whose solution is not typeable, and (b) it may surprise non-experts that the axiom of choice is provable (the theorem shows there will be no more such surprises), and (c) the paramodulation rule has to be carefully formulated to work correctly with λ-bound variables. Lambda logic is the correct tool to analyze this algorithm, as shown by the generality of the soundness theorem. To verify the soundness of Otter-λ, we need only verify that the rules of inference implemented are special cases of the rules mentioned above; in particular the extra condition on paramodulation has been enforced, and the unifiers produced are second order unifiers in the sense defined above.

We identify a clause with the formula of λ-logic which is the disjunction of the literals of the clause. If Γ is a set of clauses, then Γ can also be considered as a set of formulas in λ-logic.

Theorem 6 (Soundness of second order unification).

(i) Suppose
there is a proof of clause C from a set of clauses Γ using binary resolution,
factoring, demodulation (including β-reduction), and paramodulation, and the
clause $x = x$, allowing second order unification in place of first order unification.
Then there is a proof of C from Γ in lambda logic.

(ii) Let P be a formula of lambda logic. Suppose there is a refutation of the
clausal form of the negation of P as in part (i). Then P is a theorem of lambda
logic plus the axiom of choice AC.

Proof. (We treat demodulation as an inference rule.) Ad (i): We proceed by induction on the lengths of proofs, In the base case, if we have a proof of length 0, the clause C must already be present Γ, in which case certainly $\Gamma \vdash C$ in lambda logic.

For the induction step, we first suppose the last inference is by paramodulation. Then one of the parents of the inference is an equation $\alpha = \beta$ (or $\beta = \alpha$) where the other parent ϕ has the form $\psi[x := \gamma]$ where for some substitution σ we have $\gamma\sigma = \alpha\sigma$, and the free variables of $\beta\sigma$ either occur in γ or are not bound in ψ, according to the definition of paramodulation. Then the newly deduced formula is $\psi[x := \beta\sigma]$. We have to show that this formula is derivable in lambda logic from the parents ϕ and $\alpha = \beta$. Apply the substitution σ to the derived formula ϕ. We get

$$\phi\sigma = \psi[x := \gamma]\sigma$$
$$= (\psi\sigma)[x := \gamma\sigma]$$
$$= (\psi\sigma)[x := \alpha\sigma]$$

Now using the other deduced equation $\alpha = \beta$, we can deduce $\alpha\sigma = \beta\sigma$ and hence we can, according to the rules of lambda logic, substitute $\beta\sigma$ for $\alpha\sigma$, provided the free variables in $\beta\sigma$ either occur already in $\alpha\sigma$ or are not bound in $\psi\sigma$. Since $\gamma = \alpha\sigma$, this is exactly the condition on the variables that makes the application of paramodulation legal. That completes the induction step in the case of a paramodulation inference.

If the last inference is by factoring, it has the form of applying a substitution σ to a previous clause. This can be done in lambda logic. (In the factoring rule, as in lambda logic, substitution must be defined so as to not permit capture of free variables by a binding context.)

If the last inference is by binary resolution, then the parent clauses have the form $P|R$ and $-Q|S$ (using R and S to stand for the remaining literals in the clauses), and substitution σ unifies P and Q. The newly deduced clause is then $R\sigma|S\sigma$. Since the unification steps are sound, $P\sigma = Q\sigma$ is provable in λ-logic. By induction hypothesis, λ-logic proves both $P|R$ and $-Q|S$ from assumptions Γ, and since the substitution rule is valid in λ-logic, it proves $P\sigma|R\sigma$ and $-Q\sigma|S\sigma$. But since $P\sigma = Q\sigma$ is provable, λ-logic proves $R\sigma|S\sigma$ from Γ. This completes the induction step, and hence the proof of (i).

If the last inference is by demodulation, the corresponding inference can be made in lambda logic using an equality axiom or (β).

Ad (ii): We take Γ to be the clausal form of the negation of P, and C to be the empty clause. The theorem follows from (i) together with the following assertion: if P is refutable in λ-logic plus AC, then the clausal form Γ of the negation of P is contradictory in λ-logic plus AC. In first order logic, this is true without AC–see the remark following the proof. However, when AC is allowed, then every formula A is equivalent to its clausal form, in the following precise sense. To bring a formula A to clausal form, we first bring it to prenex form, with the matrix in conjunctive normal form, and then use Skolem functions to eliminate the existential quantifiers. We then arrive at a formula $\forall x_1, \ldots, x_n A^o$, where A^o contains new Skolem function symbols $f_1, \ldots, f_n$. Within lambda logic, using AC, we can perform these same manipulations, but instead of introducing new Skolem function symbols, we instead show the A is equivalent to $\exists g_1, \ldots, f_m A'$, where A' is the same as A^o except that it has $Ap(g_i(x))$ where A^o has $f_i(x)$. (Here x stands for $x_1, \ldots, x_k$ for some k depending on i.) In particular, if P is refutable, then P^o is refutable. Replacing each Skolem function term $f_i(x)$ by $Ap(g_i, x)$, where g_i is a variable, and using the law for $\exists$ in first order logic, we see that $\exists g_1, \ldots, g_m P'$ is refutable. That completes the proof.

Remark. The proof was entirely syntactical. No discussion, or even definition, of models of lambda logic was needed; but it may help to consider why AC is needed in lambda logic but not in first order logic. In first order logic, every model of $\forall x \exists y A(x, y)$ can be expanded to a model of $\forall x A(x, f(x))$ by interpreting the new function symbol f appropriately. But if we are dealing with λ-models, this is no longer the case. The function required may not be represented by a λ-term in the model. For example, it could happen that all possible interpretations of the Skolem function are non-recursive, but all operations representable in the model are recursive.

Corollary 1. *If the system Otter-λ [5] refutes the clausal form of the negation of P, using the clause $x = x$ and some other axioms Γ, but with the option to generate* **cases** *terms in unification off, then P is provable from the conjunction of Γ in lambda logic plus A C.*

Proof. The second order unification algorithm implemented in Otter-λ produces second order unifiers; the restriction on the application of paramodulation is also implemented; the rules of hyperresolution, etc. are all theoretically reducible to binary resolution (they are only used to make the proof search more efficient). Otter-λ also uses demodulation β-reduction to post-process (simplify) deduced conclusions. These correspond to additional infererences in lambda logic using the equality axioms and the β axiom. Note: we have not formally verified any specific computer program; this is an informal proof.

Some of the interesting applications of second order unification involve undefined terms, and therefore we need a soundness theorem relative to partial lambda logic. When we use LPT in the resolution-based theorem prover Otter-λ, we replace the axiom $x = x$ by the clause $-E(x)|x = x$, where we think of $E(t)$ as representing $t \downarrow$. We also add the clause $x \neq x|E(x)$, thus expressing that $t \downarrow$ is equivalent to $t = t$. The soundness theorem takes the following form:

Theorem 7 (Soundness of second order unification for LPT).
(i) Suppose there is a proof of clause C from a set of clauses Γ using binary resolution, factoring, demodulation (including β-reduction), and paramodulation, the clauses $x = x| - E(x)$ and $-E(x)|x = x$, and clauses expressing the strictness axioms, allowing second order unification in place of first order unification. Then there is a proof of C from Γ in partial lambda logic.

(ii) Let P be a formula of partial lambda logic. Suppose there is a refutation of the clausal form of the negation of P as in part (i). Then P is a theorem of partial lambda logic plus AC.

Proof. The proof is similar to the proof for lambda logic, except for the treatment of β-reduction steps. When β-reduction is applied, we only know that $\cong$ holds in partial lambda logic. But since in partial lambda logic, we have the substitutivity axioms for $\cong$ as well as for $=$, it is still the case in partial lambda logic that if $P[x := Ap(\lambda z.t, r)]$ is used to deduce $P[x := t[z := r]]$, and if (by induction hypothesis) the former is derivable in partial lambda logic plus AC, then so is the latter. For example if $Ap(\lambda z.a, \Omega) = a$ is derivable, then $a = a$ is derivable. Each of these formulas is in fact equivalent to $a \downarrow$.

Corollary 2. *If the system Otter-λ [5] refutes the clausal form of the negation of P, using the clauses $x = x| - E(x)$ and $-E(x)|x = x$, and some other axioms Γ, but with the option to generate **cases** terms in unification off, then P is provable from the conjunction of Γ in partial lambda logic plus AC.*

References

1. Barendregt, H., *The Lambda Calculus: Its Syntax and Semantics*, Studies in Logic and the Foundations of Mathematics **103**, Elsevier Science Ltd. Revised edition (October 1984).
2. Barendregt, H., Bunder, M., and Dekkers, W., Completeness of two systems of illative combinatory logic for first order propositional and predicate calculus *Archive für Mathematische Logik* **37**, 327–341, 1998.
3. Beeson, M., *Foundations of Constructive Mathematics*, Springer-Verlag, Berlin Heidelberg New York (1985).
4. Beeson, M., Proving programs and programming proofs, in: Barcan, Marcus, Dorn, and Weingartner (eds.), *Logic, Methodology, and Philosophy of Science VII, proceedings of the International Congress, Salzburg, 1983*, pp. 51-81, North-Holland, Amsterdam (1986).
5. Beeson, M., Otter Two System Description, submitted to IJCAR 2004.
6. S. Feferman, Constructive theories of functions and classes, pp. 159-224 in: M. Boffa, D. van Dalen, and K. McAloon (eds.), *Logic Colloquium '78: Proceedings of the Logic Colloquium at Mons*, North-Holland, Amsterdam (1979).

7. E. Moggi. The Partial Lambda-Calculus. PhD thesis, University of Edinburgh, 1988. http://citeseer.nj.nec.com/moggi88partial.html
8. Scott, D., Identity and existence in intuitionistic logic, in: Fourman, M. P., Mulvey, C. J., and Scott, D. S. (eds.), *Applications of Sheaves*, Lecture Notes in Mathematics **753** 660-696, Springer–Verlag, Berlin Heidelberg New York (1979).
9. Shoenfield, J. R., *Mathematical Logic*, Addison-Wesley, Reading, Mass. (1967).

Formalizing Undefinedness Arising in Calculus

William M. Farmer

McMaster University
Hamilton, Ontario, Canada
wmfarmer@mcmaster.ca

Abstract. Undefined terms are commonplace in mathematics, particularly in calculus. The *traditional approach to undefinedness* in mathematical practice is to treat undefined terms as legitimate, nondenoting terms that can be components of meaningful statements. The traditional approach enables statements about partial functions and undefined terms to be stated very concisely. Unfortunately, the traditional approach cannot be easily employed in a standard logic in which all functions are total and all terms are defined, but it can be directly formalized in a standard logic if the logic is modified slightly to admit undefined terms and statements about definedness. This paper demonstrates this by defining a version of simple type theory called Simple Type Theory with Undefinedness (STTWU) and then formalizing in STTWU examples of undefinedness arising in calculus. The examples are taken from M. Spivak's well-known textbook *Calculus*.

1 Introduction

A mathematical term is *undefined* if it has no prescribed meaning or if it denotes a value that does not exist. Undefined terms are commonplace in mathematics, particularly in calculus. As a result, any approach to formalizing mathematics must include a method for handling undefinedness.

There are two principal sources of undefinedness. The first source are terms that denote an application of a function. A function f usually has both a *domain of definition* D_f consisting of the values at which it is defined and a *domain of application* D_f^* consisting of the values to which it may be applied. (The domain of definition of a function is usually called simply the *domain* of the function.) These two domains are not always the same. A *function application* is a term $f(a)$ that denotes the application of a function f to an argument $a \in D_f^*$. $f(a)$ is *undefined* if $a \notin D_f$. We will say that a function is *partial* if $D_f \neq D_f^*$ and *total* if $D_f = D_f^*$.

As an example, consider the square root function $\sqrt{} : \mathbf{R} \to \mathbf{R}$, where $\mathbf{R}$ denotes the set of real numbers. $\sqrt{}$ is a partial function; it is defined only on the nonnegative real numbers, but it can be applied to negative real numbers as well. That is, $D_{\sqrt{}} = \{x \in \mathbf{R} \mid 0 \leq x\}$ and $D_{\sqrt{}}^* = \mathbf{R}$. Hence, a statement like

$$\forall\, x : \mathbf{R} \,.\, 0 \leq x \Rightarrow (\sqrt{x})^2 = x$$

makes perfectly good sense even though $\sqrt{x}$ is undefined when $x < 0$.

D. Basin and M. Rusinowitch (Eds.): IJCAR 2004, LNAI 3097, pp. 475–489, 2004.

The second source of undefinedness are terms that are intended to uniquely describe a value. A *definite description* is a term t of the form "the x that has the property P". t is *undefined* if there is no unique x (i.e., none or more than one) that has property P. Definite descriptions are quite common in mathematics but often occur in a disguised form. For example, "the limit of $\sin\frac{1}{x}$ as x approaches 0" is a definite description—which is undefined since the limit does not exist.

There is a *traditional approach to undefinedness* that is widely practiced in mathematics and even taught to some extent to students in high school. This approach treats undefined terms as legitimate, nondenoting terms that can be components of meaningful statements. The traditional approach is based on three principles:

1. Atomic terms (i.e., variables and constants) are always defined—they always denote something.
2. Compound terms may be undefined. A function application $f(a)$ is undefined if f is undefined, a is undefined, or $a \notin D_f$. A definite description "the x that has property P" is undefined if there is no x that has property P or there is more than one x that has property P.
3. Formulas are always true or false, and hence, are always defined. To ensure the definedness of formulas, a function application $p(a)$ formed by applying a predicate p to an argument a is *false* if p is undefined, a is undefined, or $a \notin D_p$.

Formalizing the traditional approach to undefinedness is problematic. The traditional approach works smoothly in informal mathematics, but it cannot be easily employed in mathematics formalized in standard logics like first-order logic or simple type theory. This is due to the fact that in a standard logic all functions are total and all terms denote some value. As a result, in a standard logic partial functions must be represented by total functions and undefined terms must be given a value.

The mathematics formalizer has basically three choices for how to formalize undefinedness. The first choice is to formalize partial functions and undefined terms in a standard logic. There are various methods by which this can be done (see [8,21]). Each of these methods, however, has the disadvantage that it is a significant departure from the traditional approach. This means in practice that a concise informal mathematical statement S involving partial functions or undefinedness is often represented by a verbose formal mathematical statement S' in which unstated, but implicit, definedness assumptions within S are explicitly represented.

The second choice is to select or develop a three-valued logic in which the traditional approach can be formalized. Such a logic would admit both undefined terms and undefined formulas. The logic needs to be three valued since formulas can be undefined as well as true and false. For example, M. Kerber and M. Kohlhase propose in [19] a three-valued version of many-sorted first-order logic that is intended to support the traditional approach to undefinedness. A three-valued logic of this kind provides a very flexible basis for formalizing and reasoning about undefinedness. However, it is nevertheless a significant departure from the traditional approach: it is very unusual in mathematical practice to use truth values beyond true and false. Moreover, the presence of a third truth value makes the logic more complicated to use and implement.

It is possible to *directly* formalize the traditional approach in a standard logic if the logic is modified slightly to admit undefined terms and statements about definedness but not undefined formulas (see [11]). The resulting logic remains two-valued and can be viewed as a more convenient version of the original standard logic. Let us call a logic of this kind a *logic with undefinedness*. As far as we know, the first example of a logic with undefinedness is a version of first-order logic presented by R. Schock in [23]. Other logics with undefinedness have been derived from first-order logic [3,4,5,6,15,18,20,22], simple type theory [8,9,10], and set theory [12,15].

The third choice is to select or develop an appropriate logic with undefinedness. Using a logic with undefinedness to formalize the traditional approach has two advantages and one disadvantage. The first advantage is that the use of the traditional approach in informal mathematics can be preserved in the formalized mathematics. This helps keep the formalized mathematics close to the informal mathematics in form and meaning. The second advantage is that assumptions about the definedness of terms and functions often do not have to be made explicit. This helps keep the formalized mathematics concise. The disadvantage, of course, is that one is committed to using a nonstandard logic that is based on (slightly) different principles and requires different techniques for implementation.

This disadvantage is not as bad as one might think. By virtue of directly formalizing the traditional method, a logic with undefinedness is closer to traditional mathematical practice than the corresponding standard logic in which all functions are total and all terms are defined. Hence, logics with undefinedness are more practice-oriented than standard logics. Moreover, the logic LUTINS [10], a logic with undefinedness derived from simple type theory, is the basis for the IMPS theorem proving system [16,17]. IMPS has been used to prove hundreds of theorems in traditional mathematics, especially in mathematical analysis. Most of these theorems involve undefinedness in some manner. The techniques used to implement LUTINS in IMPS have been thoroughly tested and can be applied to the implementation of other logics with undefinedness.

Even though LUTINS has been successfully implemented in IMPS and has proven to be highly effective for mechanizing traditional mathematics, there is still a great deal of scepticism and misunderstanding concerning logics with undefinedness. The goal of this paper is to try to dispel some of this scepticism and misunderstanding by illustrating how undefinedness arising in calculus can be directly formalized in a very simple higher-order logic called Simple Type Theory with Undefinedness (STTwU). We will show that the conciseness that comes from the use of the traditional approach can be fully preserved in a logic like STTwU. For example,

$$f(x) = \sqrt{x^2 - 1},$$

a common-style definition of a partial function in which the domain of the function is implicitly defined, can be formalized precisely in STTwU as

$$\forall x \,.\, f(x) \simeq \sqrt{x^2 - 1}$$

(where $a \simeq b$ means a and b are both defined and equal or both undefined).

All of our examples from calculus come from M. Spivak's well-known textbook *Calculus* [24]. Spivak's book is a masterpiece; it is elegantly rigorous, replete with interesting examples and exercises, and exceptionally careful about important issues such as undefinedness that are not always given proper attention in other calculus textbooks. More than just a book on calculus, *Calculus* is also an uncompromising introduction to mathematical practice and mathematical thinking. It is an excellent place to see how undefinedness is handled in standard mathematical practice.

The paper is organized as follows. Section 2 presents the syntax and semantics of STTwU, a version of simple type theory that directly formalizes the traditional approach to undefinedness. Section 2 also give a proof system for STTwU. Section 3 describes a theory of STTwU that formalizes the 13 properties of the real numbers on which Spivak's *Calculus* is based. How partial functions are defined in *Calculus* and how their definitions are formalized in STTwU is the subject of section 4. Section 5 deals with the important notion in calculus of a limit of a function at a point, which is a rich source of undefinedness. Finally, a conclusion and a recommendation are given in section 6.

2   Simple Type Theory with Undefinedness

In this section we present a version of simple type theory called *Simple Type Theory with Undefinedness* (STTwU) that formalizes the traditional approach to undefinedness. STTwU is a variant of Church's type theory [7] with a standard syntax but a nonstandard semantics. STTwU is very similar to **PF** [8], **PF*** [9], and LUTINS [10], the logic of the IMPS. **PF** and **PF*** are simple versions of LUTINS that are primarily intended for study unlike LUTINS. STTwU is much simpler than these three logics but much less practical than **PF*** and LUTINS. The definition of STTwU will show that the traditional approach can be formalized in Church's type theory by just a small modification of its semantics.

2.1   Syntax

The syntax of STTwU is exactly the same syntax as the syntax of STT, a very simple variant of Church's type system presented in [13]. STTwU has two kinds of syntactic objects. "Expressions" denote values including the truth values T (true) and F (false); they play the role of both terms and formulas. "Types" denote nonempty sets of values; they are used to restrict the scope of variables, control the formation of expressions, and classify expressions by value.

A *type* of STTwU is defined by the formation rules given below. **type**$[\alpha]$ asserts that α is a type.

T1 $\dfrac{}{\textbf{type}[\iota]}$ (**Type of individuals**)

T2 $\dfrac{}{\textbf{type}[*]}$ (**Type of truth values**)

T3 $\dfrac{\textbf{type}[\alpha],\ \textbf{type}[\beta]}{\textbf{type}[(\alpha \rightarrow \beta)]}$ (**Function type**)

Let $\mathcal{T}$ denote the set of types of STTwU.
The *logical symbols* of STTwU are:

1. *Function application*: @.
2. *Function abstraction*: λ.
3. *Equality*: =.
4. *Definite description*: I (capital iota).
5. An infinite set $\mathcal{V}$ of symbols called *variables*.

A *language* of STTwU is a pair $L = (\mathcal{C}, \tau)$ where:

1. $\mathcal{C}$ is a set of symbols called *constants*.
2. $\mathcal{V}$ and $\mathcal{C}$ are disjoint.
3. $\tau : \mathcal{C} \rightarrow \mathcal{T}$ is a total function.

That is, a language is a set of symbols with assigned types (what is also called a "signature").

An *expression* E of *type* α of an STTwU language $L = (\mathcal{C}, \tau)$ is defined by the formation rules given below. **expr**$_L[E, \alpha]$ asserts that E is an expression of type α of L.

E1 $\dfrac{x \in \mathcal{V},\ \textbf{type}[\alpha]}{\textbf{expr}_L[(x : \alpha), \alpha]}$ (**Variable**)

E2 $\dfrac{c \in \mathcal{C}}{\textbf{expr}_L[c, \tau(c)]}$ (**Constant**)

E3 $\dfrac{\textbf{expr}_L[A, \alpha],\ \textbf{expr}_L[F, (\alpha \rightarrow \beta)]}{\textbf{expr}_L[(F\ @\ A), \beta]}$ (**Function application**)

E4 $\dfrac{x \in \mathcal{V},\ \textbf{type}[\alpha],\ \textbf{expr}_L[B, \beta]}{\textbf{expr}_L[(\lambda x : \alpha\ .\ B), (\alpha \rightarrow \beta)]}$ (**Function abstraction**)

E5 $\dfrac{\textbf{expr}_L[E_1, \alpha],\ \textbf{expr}_L[E_2, \alpha]}{\textbf{expr}_L[(E_1 = E_2), *]}$ (**Equality**)

E6 $\dfrac{x \in \mathcal{V},\ \textbf{type}[\alpha],\ \textbf{expr}_L[A, *]}{\textbf{expr}_L[(I\,x : \alpha\ .\ A), \alpha]}$ (**Definite description**)

We will see shortly that the value of a definite description $(\mathrm{I}\,x : \alpha \,.\, A)$ is the unique value x of type α satisfying A if it exists and is "undefined" otherwise.

"Free variable", "closed expression", and similar notions are defined in the obvious way. An expression of L is a *formula* if it is of type $*$, a *sentence* if it is a closed formula, and a *predicate* if it is of type $(\alpha \to *)$ for any $\alpha \in \mathcal{T}$. Let $A_\alpha, B_\alpha, C_\alpha, \ldots$ denote expressions of type α. Parentheses and the types of variables may be dropped when meaning is not lost. An expression of the form $(F \,@\, A)$ will usually be written in the more compact and standard form $F(A)$.

2.2   Semantics

The semantics of STTwU is the same as the semantics of STT except that:

1. A model contains partial and total functions instead of just total functions.
2. The value of an "undefined" function application is F if it is a formula and is undefined if it is not a formula.
3. The value of a function abstraction is a function that is possibly partial.
4. The value of an equality is F if the value of either of its arguments is undefined.
5. The value of an "undefined" definite description is F if it is a formula and is undefined if it is not a formula.

A *model*[1] for a language $L = (\mathcal{C}, \tau)$ of STTwU is a pair $M = (\mathcal{D}, I)$ where:

1. $\mathcal{D} = \{D_\alpha : \alpha \in \mathcal{T}\}$ is a set of nonempty domains (sets).
2. $D_* = \{\mathrm{T}, \mathrm{F}\}$, the domain of truth values.
3. For $\alpha, \beta \in \mathcal{T}$, $D_{\alpha \to \beta}$ is the set of all *total* functions from D_α to D_β if $\beta = *$ and is the set of all *partial and total* functions from D_α to D_β if $\beta \neq *$.[2]
4. I maps each $c \in \mathcal{C}$ to a member of $D_{\tau(c)}$.

Fix a model $M = (\mathcal{D}, I)$ for a language $L = (\mathcal{C}, \tau)$ of STTwU. A *variable assignment* into M is a function that maps each variable expression $(x : \alpha)$ to a member of D_α. Given a variable assignment φ into M, an expression $(x : \alpha)$, and $d \in D_\alpha$, let $\varphi[(x : \alpha) \mapsto d]$ be the variable assignment φ' into M such that $\varphi'((x : \alpha)) = d$ and $\varphi'(v) = \varphi(v)$ for all $v \neq (x : \alpha)$.

The *valuation function* for M is the partial binary function V^M that satisfies the following conditions for all variable assignments φ into M and all expressions E of L:

---

[1] This is the definition of a *standard model* for STTwU. There is also the notion of a *general model* for STTwU in which functions domains are not fully "inhabited" (see [14]).

[2] The condition that a domain $D_{\alpha \to *}$ contains only total functions is needed to ensure that the law of extensionality holds for predicates. This condition is weaker than the condition used in the semantics for LUTINS and its simple versions, **PF** and **PF***. In these logics, a domain D_γ contains only total functions iff γ has the form $(\alpha_1 \to (\alpha_2 \to \cdots (\alpha_n \to *) \cdots))$ where $n \geq 1$. The weaker condition, which is due to A. Stump [25], yields a semantics that is somewhat simpler.

1. Let E_α be of the form $(x : \alpha)$. Then $V_\varphi^M(E_\alpha) = \varphi((x : \alpha))$.
2. Let $E \in \mathcal{C}$. Then $V_\varphi^M(E) = I(E)$.
3. Let E_β be of the form $(F @ A)$. If $V_\varphi^M(F)$ is defined, $V_\varphi^M(A)$ is defined, and $V_\varphi^M(A)$ is in the domain of $V_\varphi^M(F)$, then $V_\varphi^M(E_\alpha) = V_\varphi^M(F)(V_\varphi^M(A))$, the result of applying the function $V_\varphi^M(F)$ to the argument $V_\varphi^M(A)$. Otherwise, $V_\varphi^M(E_\beta) = \text{F}$ if $\beta = *$ and $V_\varphi^M(E_\beta)$ is undefined if $\beta \neq *$.
4. Let $E_{\alpha \to \beta}$ be of the form $(\lambda x : \alpha \, . \, B)$. Then $V_\varphi^M(E_{\alpha \to \beta})$ is the (partial or total) function $f : D_\alpha \to D_\beta$ such that, for each $d \in D_\alpha$, $f(d) = V_{\varphi[(x:\alpha) \mapsto d]}^M(B)$ if $V_{\varphi[(x:\alpha) \mapsto d]}^M(B)$ is defined and $f(d)$ is undefined if $V_{\varphi[(x:\alpha) \mapsto d]}^M(B)$ is undefined.
5. Let E_* be of the form $(E_1 = E_2)$. If $V_\varphi^M(E_1)$ is defined, $V_\varphi^M(E_2)$ is defined, and $V_\varphi^M(E_1) = V_\varphi^M(E_2)$, then $V_\varphi^M(E_*) = \text{T}$; otherwise $V_\varphi^M(E_*) = \text{F}$.
6. Let E_α be of the form $(\text{I} x : \alpha \, . \, A)$. If there is a unique $d \in D_\alpha$ such that $V_{\varphi[(x:\alpha) \mapsto d]}^M(A) = \text{T}$, then $V_\varphi^M(E_\alpha) = d$. Otherwise, $V_\varphi^M(E_\alpha) = \text{F}$ if $\alpha = *$ and $V_\varphi^M(E_\alpha)$ is undefined if $\alpha \neq *$.

Let E be an expression of type α of L. When $V_\varphi^M(E)$ is defined, $V_\varphi^M(E)$ is called the *value* of E in M with respect to φ and $V_\varphi^M(E) \in D_\alpha$. Whenever E is a formula, $V_\varphi^M(E)$ is defined. A formula A is *valid* in M, written $M \models A$, if $V_\varphi^M(A) = \text{T}$ for all variable assignments φ into M. A *theory* of STTwU is a pair $T = (L, \Gamma)$ where L is a language of STTwU and Γ is a set of sentences of L called the *axioms* of T. A *model* of T is a model M for L such that $M \models B$ for all $B \in \Gamma$.

2.3 Definitions and Abbreviations

We define in STTwU the standard propositional connectives and quantifiers as well as some special operators concerning definedness.

T	means	$(\lambda x : * \, . \, x) = (\lambda x : * \, . \, x)$
F	means	$(\lambda x : * \, . \, \text{T}) = (\lambda x : * \, . \, x)$
$\neg A_*$	means	$A_* = \text{F}$
$(A_\alpha \neq B_\alpha)$	means	$\neg(A_\alpha = B_\alpha)$
$(A_* \wedge B_*)$	means	$(\lambda f : (* \to (* \to *)) \, . \, f(\text{T})(\text{T})) =$ $(\lambda f : (* \to (* \to *)) \, . \, f(A_*)(B_*))$
$(A_* \vee B_*)$	means	$\neg(\neg A_* \wedge \neg B_*)$
$(A_* \Rightarrow B_*)$	means	$\neg A_* \vee B_*$
$(A_* \Leftrightarrow B_*)$	means	$A_* = B_*$
$\Pi_{(\alpha \to *) \to *}$	means	$\lambda p : (\alpha \to *) \, . \, p = (\lambda x : \alpha \, . \, \text{T})$
$(\forall x : \alpha \, . \, A_*)$	means	$\Pi_{(\alpha \to *) \to *}(\lambda x : \alpha \, . \, A_*)$
$(\exists x : \alpha \, . \, A_*)$	means	$\neg(\forall x : \alpha \, . \, \neg A_*)$
$(A_\alpha \downarrow)$	means	$\exists x : \alpha \, . \, x = A_\alpha$
		where x does not occur in A_α.
$(A_\alpha \uparrow)$	means	$\neg(A_\alpha \downarrow)$

$(A_\alpha \simeq B_\alpha)$ means $(A_\alpha\!\downarrow \lor B_\alpha\!\downarrow) \Rightarrow A_\alpha = B_\alpha$

$\perp_\alpha$ means $\mathrm{I}\,x : \alpha \,.\, x \neq x$

$\mathsf{if}(A_*, B_\alpha, C_\alpha)$ means $\mathrm{I}\,x : \alpha \,.\, (A_* \Rightarrow x = B_\alpha) \land (\neg A_* \Rightarrow x = C_\alpha)$

            where x does not occur in A_*, B_α, or C_α

Notice that we are using the syntactic conventions mentioned in section 2.1. For example, the meaning of T is officially the expression

$$((\lambda x : * \,.\, (x : *)) = (\lambda x : * \,.\, (x : *))).$$

$(A_\alpha\!\downarrow)$ says that A_α is defined, $(A_\alpha\!\uparrow)$ says that A_α is undefined, and $A_\alpha \simeq B_\alpha$ says that A_α and B_α are *quasi-equal*, i.e., that A_α and B_α are either both defined and equal or both undefined. $\perp_\alpha$ is a canonical undefined expression of type α. if is an if-then-else expression constructor such that $\mathsf{if}(A_*, B_\alpha, C_\alpha)$ denotes B_α if A_* holds and denotes C_α if $\neg A_*$ holds.

We will write a formula of the form $\Box\, x_1 : \alpha \,.\, \cdots \,\Box\, x_n : \alpha \,.\, A$ as simply $\Box\, x_1, \ldots, x_n : \alpha \,.\, A$ where $\Box$ is $\forall$ or $\exists$. If we fix the type of a variable x, say to α, then an expression of the form $\Box\, x : \alpha \,.\, E$ may be written as simply $\Box\, x \,.\, E$ where $\Box$ is λ, I, $\forall$, or $\exists$. If desired, all types can be removed from an expression by fixing the types of the variables occurring in the expression.

2.4   Proof System

We present now a Hilbert-style proof system for STTwU called $\mathbf{A_u}$ that is sound and complete with respect to the general models semantics for STTwU [14]. It is a modification of the proof system $\mathbf{A}$ for STT given in [13] which is based on P. Andrews' proof system [1,2] for Church's type theory.

Define $B_\beta[(x : \alpha) \mapsto A_\alpha]$ to be the result of simultaneously replacing each free occurrence of $(x : \alpha)$ in B_β by an occurrence of A_α. Let $(\exists!\, x : \alpha \,.\, A)$ mean

$$\exists\, x : \alpha \,.\, (A \land (\forall y : \alpha \,.\, A[(x : \alpha) \mapsto (y : \alpha)] \Rightarrow y = x))$$

where y does not occur in A. This formula asserts there exists a unique value x of type α that satisfies A.

For a language $L = (\mathcal{C}, \tau)$, the proof system $\mathbf{A_u}$ consists of the following sixteen axiom schemas and two rules of inference:

A1 (Truth Values)

$\forall f : (* \rightarrow *) \,.\, (f(\mathsf{T}) \land f(\mathsf{F})) \Leftrightarrow (\forall x : * \,.\, f(x)).$

A2 (Leibniz' Law)

$\forall x, y : \alpha \,.\, (x = y) \Rightarrow (\forall p : (\alpha \rightarrow *) \,.\, p(x) \Leftrightarrow p(y)).$

A3 (Extensionality)

$\forall f, g : (\alpha \rightarrow \beta) \,.\, (f = g) \Leftrightarrow (\forall x : \alpha \,.\, f(x) \simeq g(x)).$

A4 (Beta-Reduction)

$$A_\alpha \downarrow \,\Rightarrow (\lambda x : \alpha \,.\, B_\beta)(A_\alpha) \simeq B_\beta[(x : \alpha) \mapsto A_\alpha]$$

provided A_α is free for $(x : \alpha)$ in B_β.

A5 (Equality and Quasi-Quality)

$$A_\alpha \downarrow \,\Rightarrow (B_\alpha \downarrow \,\Rightarrow (A_\alpha \simeq B_\alpha) \simeq (A_\alpha = B_\alpha)).$$

A6 (Expressions of Type $*$ are Defined)

$$A_* \downarrow.$$

A7 (Variables are Defined)

$$(x : \alpha) \downarrow \quad \text{where } x \in \mathcal{V} \text{ and } \alpha \in \mathcal{T}.$$

A8 (Constants are Defined)

$$c \downarrow \quad \text{where } c \in \mathcal{C}.$$

A9 (Function Abstractions are Defined)

$$(\lambda x : \alpha \,.\, B_\beta) \downarrow$$

A10 (Improper Function Application)

$$(F_{\alpha \to \beta} \uparrow \, \vee A_\alpha \uparrow) \Rightarrow F_{\alpha \to \beta}(A_\alpha) \uparrow \quad \text{where } \beta \neq *.$$

A11 (Improper Predicate Application)

$$(F_{\alpha \to *} \uparrow \, \vee A_\alpha \uparrow) \Rightarrow \neg F_{\alpha \to *}(A_\alpha).$$

A12 (Improper Equality)

$$(A_\alpha \uparrow \, \vee B_\alpha \uparrow) \Rightarrow \neg(A_\alpha = B_\alpha).$$

A13 (Proper Definite Description of Type $\alpha \neq *$)

$$(\exists ! x : \alpha \,.\, A_*) \Rightarrow ((\mathrm{I}\, x : \alpha \,.\, A_*) \downarrow \wedge A_*[(x : \alpha) \mapsto (\mathrm{I}\, x : \alpha \,.\, A_*)])$$

where $\alpha \neq *$ and provided $(\mathrm{I}\, x : \alpha \,.\, A_*)$ is free for $(x : \alpha)$ in A_*.

A14 (Improper Definite Description of Type $\alpha \neq *$)

$$\neg(\exists ! x : \alpha \,.\, A_*) \Rightarrow (\mathrm{I}\, x : \alpha \,.\, A_*) \uparrow \quad \text{where } \alpha \neq *.$$

A15 (Proper Definite Description of Type $*$)

$$(\exists ! x : * \,.\, A_*) \Rightarrow A_*[(x : *) \mapsto (\mathrm{I}\, x : * \,.\, A_*)]$$

provided $(\mathrm{I}\, x : * \,.\, A_*)$ is free for $(x : *)$ in A_*.

A16 (Improper Definite Description of Type $*$)

$$\neg(\exists ! x : * \,.\, A_*) \Rightarrow \neg(\mathrm{I}\, x : * \,.\, A_*).$$

R1 (Modus Ponens) From A_* and $A_* \Rightarrow B_*$ infer B_*.

R2 (Quasi-Equality Substitution) From $A_\alpha \simeq B_\alpha$ and C_* infer the result of replacing one occurrence of A_α in C_* by an occurrence of B_α, provided that the occurrence of A_α in C_* is not immediately preceded by λ.

$\mathbf{A_u}$ is sound and complete with respect to the general models semantics for STTwU [14]. The completeness proof is closely related to the completeness proofs for **PF** [8] and **PF*** [9]. All the standard laws of predicate logic hold in STTwU except some of those involving equality and instantiation. However, the laws of equality and instantiation do hold if certain "definedness requirements" are satisfied. See [8,9,14] for details.

3   A Theory of the Real Numbers

Spivak's development of calculus in his textbook *Calculus* [24] begins with a presentation of 13 basic properties of the real numbers [24, pp. 9, 113]. These properties are essentially the axioms of the theory of a complete ordered field—which has exactly one model up to isomorphism, namely, the standard model of the real numbers.

We will begin our exploration of undefinedness in Spivak's *Calculus* by formulating the 13 properties as a theory in STTwU. Let $\mathsf{COF} = (L, \Gamma)$ be the theory of STTwU such that:

− $L = (\{+, 0, -, \cdot, 1, {}^{-1}, \mathsf{pos}, <, \leq, \mathsf{ub}, \mathsf{lub}\}, \tau)$ where τ is defined by:

Constant c	Type $\tau(c)$
0,1	ι
$-, {}^{-1}$	$\iota \to \iota$
pos	$\iota \to *$
$+, \cdot$	$\iota \to (\iota \to \iota)$
$<, \leq$	$\iota \to (\iota \to *)$
ub, lub	$(\iota \to *) \to (\iota \to *)$

Note: The type ι is being used to represent the set of real numbers.
− Γ is the set of the 19 formulas given below. We assume that the variables a, b, c are of type ι and the variable s is of type $(\iota \to *)$.

P1 $\forall\, a, b, c\ .\ a + (b + c) = (a + b) + c.$
P2 $\forall\, a\ .\ a + 0 = a \wedge 0 + a = a.$
P3 $\forall\, a\ .\ a + -a = 0 \wedge -a + a = 0.$
P4 $\forall\, a, b\ .\ a + b = b + a.$
P5 $\forall\, a, b, c\ .\ a \cdot (b \cdot c) = (a \cdot b) \cdot c.$
P6a $\forall\, a\ .\ a \cdot 1 = a \wedge 1 \cdot a = a.$
P6b $0 \neq 1.$

P7a $\forall a \,.\, a \neq 0 \Rightarrow (a \cdot a^{-1} = 1 \wedge a^{-1} \cdot a = 1)$.

P7b $0^{-1}\uparrow$.

P8 $\forall a,b \,.\, a \cdot b = b \cdot a$.

P9 $\forall a,b,c \,.\, a \cdot (b + c) = (a \cdot b) + (a \cdot c)$.

P10 $\forall a \,.\, (a = 0 \wedge \neg\mathsf{pos}(a) \wedge \neg\mathsf{pos}(-a)) \vee$
$(a \neq 0 \wedge \mathsf{pos}(a) \wedge \neg\mathsf{pos}(-a)) \vee$
$(a \neq 0 \wedge \neg\mathsf{pos}(a) \wedge \mathsf{pos}(-a))$.

P11 $\forall a,b \,.\, (\mathsf{pos}(a) \wedge \mathsf{pos}(b)) \Rightarrow \mathsf{pos}(a + b)$.

P12 $\forall a,b \,.\, (\mathsf{pos}(a) \wedge \mathsf{pos}(b)) \Rightarrow \mathsf{pos}(a \cdot b)$.

D1 $\forall a,b \,.\, a < b \Leftrightarrow \mathsf{pos}(b - a)$.

D2 $\forall a,b \,.\, a \leq b \Leftrightarrow (a < b \vee a = b)$.

D3 $\forall s,a \,.\, \mathsf{ub}(s)(a) \Leftrightarrow (\forall b \,.\, s(b) \Rightarrow b \leq a)$.

D4 $\forall s,a \,.\, \mathsf{lub}(s)(a) \Leftrightarrow (\mathsf{ub}(s)(a) \wedge (\forall b \,.\, \mathsf{ub}(s)(b) \Rightarrow a \leq b))$.

P13 $\forall s \,.\, ((\exists a \,.\, s(a)) \wedge (\exists a \,.\, \mathsf{ub}(s)(a))) \Rightarrow \exists a \,.\, \mathsf{lub}(s)(a)$.

Notes:

1. Axioms **P1–P13** correspond to Spivak's 13 properties P1–P13. Properties P6 and P7 are both expressed by pairs of axioms. Axioms **D1–D4** are definitions.
2. We write the additive and multiplicative inverses of a as $-a$ and a^{-1} instead of as $-(a)$ and $^{-1}(a)$, respectively.
3. $+$ and $*$ are formalized by constants of type $(\iota \to (\iota \to \iota))$ representing curried functions. However, we write the application of $+$ and $*$ using infix notation, e.g., we write $a + b$ instead of $+(a)(b)$. $<$ and $\leq$ are handled in a similar way. In our examples below, we will also write $a - b$ instead of $a + -b$ and $\frac{a}{b}$ instead of $a \cdot b^{-1}$.
4. pos is a predicate that represents the set of positive real numbers.
5. $\mathsf{ub}(s)(a)$ and $\mathsf{lub}(s)(a)$ say that a is an upper bound of s and a is the least upper bound of s, respectively.
6. Axiom **P13** expresses the *completeness principle* of the real numbers, i.e., that every nonempty set of real numbers that has an upper bound has a least upper bound.

COF is an extremely direct formalization of Spivak's 13 properties. (The reader is invited to check this for herself.) The only significant departure is Axiom **P7b**. The undefinedness of the multiplicative inverse of 0 is not stated in property P7. However, Spivak says in the text that "0^{-1} is meaningless" and implicitly that 0^{-1} is always undefined [24, p. 6].

COF is categorical, i.e., it has exactly one model $(\mathcal{D}, I)$ up to isomorphism where $D_\iota = \mathbf{R}$, the set of real numbers, and I assigns $+$, 0, $-$, $\cdot$, 1, $^{-1}$, pos, $<$, $\leq$, ub, lub their usual meanings.

4   Partial Functions

In this section and the next section we will examine six examples of statements from Spivak's *Calculus* involving undefinedness. The examples in this section are definitions of partial functions, and the examples in the next section involve limits that can be undefined. For each example, we will display how Spivak expresses it in *Calculus* followed by how it can be formalized in STTwU. We will assume that in the STTwU formalizations the variables f, g are of type $(\iota \to \iota)$ and the variables $a, l, m, x, \delta, \epsilon$ are of type ι. The listed page numbers refer to the first edition [24] of *Calculus*.

Spivak devotes two chapters in *Calculus* to functions. He emphasizes that functions are often partial and discusses several examples of partial functions. He handles partial functions according to the traditional approach. In fact, he says "the symbol $f(x)$ makes sense only for x in the domain of f; for other x the symbol $f(x)$ is not defined" [p. 38].

Example 1. This is a definition of a partial function in which the function's domain is given explicitly.

Spivak: $k(x) = \frac{1}{x} + \frac{1}{x-1}$, $x \neq 0, 1$ [p. 39].

STTwU: $\forall x \,.\, \text{if}(x \neq 0 \wedge x \neq 1, \ k(x) = \frac{1}{x} + \frac{1}{x-1}, \ k(x)\!\uparrow)$.

The formalization of the definition in STTwU is very explicit but certainly more verbose than Spivak's definition. A second formalization in STTwU as an equation is

$$k = \lambda x \,.\, \text{if}(x \neq 0 \wedge x \neq 1, \ \tfrac{1}{x} + \tfrac{1}{x-1}, \ \perp_\iota).$$

Although the second formalization directly identifies what is being defined and is somewhat more succinct, we prefer the first formalization because it more faithfully captures the form and meaning of Spivak's definition.

Example 2. This is a shortened version of the previous example in which the function's domain is implicit and, as Spivak says, "is understood to consist of all [real] numbers for which the definition makes any sense at all" [p. 39].

Spivak: $k(x) = \frac{1}{x} + \frac{1}{x-1}$ [p. 39].

STTwU: $\forall x \,.\, k(x) \simeq \frac{1}{x} + \frac{1}{x-1}$.

Notice that in the formalization quasi-equality $\simeq$ must be used instead of ordinary equality $=$. Spivak does not distinguish in *Calculus* between these two kinds of equality, but he is certainly aware of the important distinction between them. See his comment after problem 28 of Chapter 5 on p. 92.

This example illustrates that a partial function can be precisely described in STTwU, as in informal mathematics, without mentioning the domain of the function. The definition is thus more succinct than it would be if the domain were made explicit. The implicit domain can always be determined later *if necessary*.

Example 3. This example defines the quotient of two functions.

Spivak: $\left(\frac{f}{g}\right)(x) = \frac{f(x)}{g(x)}$ [p. 41].

STTwU: $\forall\, f, g, x : \mathsf{fun\_div}(f)(g)(x) \simeq \frac{f(x)}{g(x)}$.

fun_div is a constant of type $((\iota \to \iota) \to ((\iota \to \iota) \to (\iota \to \iota)))$ that denotes the quotient of two functions. Notice that the domain of $\mathsf{fun\_div}(f)(g)$ is not explicitly defined, but it can be computed to be $D_f \cap D_g \cap \{x : \mathbf{R} \mid g(x) \neq 0\}$.

5 Limits

The notion of a limit of a function at a point in its domain of application is perhaps the most important concept in calculus. Spivak devotes an entire chapter to it. Since a limit does not always exist, this concept is a rich source of undefinedness.

Example 4. This is Spivak's definition of a limit—which is actually formed from two definitions: (1) what is means for a function to approach a limit near a point [p. 78] and (2) what the notation $\lim_{x \to a} f(x)$ denotes [p. 81][3].

Spivak: $\lim_{x \to a} f(x)$ denotes the real number l such that, for every $\epsilon > 0$, there is some $\delta > 0$ such that, for all x, if $0 < |x - a| < \delta$, then $|f(x) - l| < \epsilon$ [pp. 78, 81].

STTwU: $\forall\, f, a \,.\, \mathsf{lim}(f)(a) \simeq$
   $(I l \,.$
      $(\forall\, \epsilon \,.\, 0 < \epsilon \Rightarrow$
        $(\exists\, \delta \,.\, 0 < \delta \,\wedge$
          $(\forall\, x \,.\, (0 < \mathsf{abs}(x - a) \wedge \mathsf{abs}(x - a) < \delta) \Rightarrow$
          $\mathsf{abs}(f(x) - l) < \epsilon))))$

abs denotes the absolute value function. Both the informal and formal definitions of a limit at a point are defined as the value of a definite description provide the value exists. Other limit concepts, such as the limit of a sequence, are defined by similar definition descriptions.

Example 5. This is a theorem about the limit of the quotient of the identity function and another function g.

Spivak: If $\lim_{x \to a} g(x) = m$ and $m \neq 0$, then $\lim_{x \to a} \left(\frac{1}{g}\right)(x) = \frac{1}{m}$ [Theorem 2(3), p. 84].

STTwU: $\forall\, g, a, m \,.\, (\mathsf{lim}(g)(a) = m \wedge m \neq 0) \Rightarrow \mathsf{lim}(\mathsf{fun\_div}(\mathsf{id})(g))(a) = \frac{1}{m}$.

---

[3] Spivak remarks that a "more logical symbol" like $\lim_a f$ is "so infuriatingly rigid that almost no one has seriously tried to use it" [p. 81].

id is $\lambda x \, . \, x$, the identity function. Notice that there is no explicit mention of the definedness of the limit of g at x in either Spivak's theorem or in its formalization in STTwU. This is because the hypothesis that the limit $\lim(g)(a)$ equals m automatically implies that $\lim(g)(a)$ is defined by the third principle of the traditional approach.

Example 6. This is the definition of the notion of a function being continuous at a point.

Spivak: The function f is continuous at a if $\lim_{x \to a} f(x) = f(a)$ [p. 93].

STTwU: $\forall f, a \, . \, \mathrm{cont}(f)(a) \Leftrightarrow \lim(f)(a) = f(a)$.

It is crucial to use equality $=$ instead of quasi-equality $\simeq$ because f is continuous at a only if $\lim(f)(a)$ and $f(a)$ are both defined and equal to each other.

6   Conclusion

In this paper we have presented STTwU, a version of simple type theory that directly formalizes the traditional approach to undefinedness. We have also formalized in STTwU several examples from Spivak's *Calculus* involving undefinedness. The formalizations are exceedingly faithful in form and meaning to how the examples are expressed by Spivak. Moreover, the conciseness that comes from Spivak's use of the traditional approach is fully preserved in the formalizations in STTwU.

It is our recommendation that logics with undefinedness, such as STTwU, be considered as the logical basic for mechanized mathematics systems. They are much closer to mathematical practice with respect to undefinedness than standard logics, and as IMPS has demonstrated, they can be effectively implemented.

Acknowledgments. The author would like to thank Jacques Carette for many valuable discussions on formalizing undefinedness, partial functions, and countless other aspects of mathematics.

References

1. P. B. Andrews. A reduction of the axioms for the theory of propositional types. *Fundamenta Mathematicae*, 52:345–350, 1963.
2. P. B. Andrews. *An Introduction to Mathematical Logic and Type Theory: To Truth through Proof, Second Edition*. Kluwer, 2002.
3. M. Beeson. Formalizing constructive mathematics: Why and how? In F. Richman, editor, *Constructive Mathematics: Proceedings, New Mexico, 1980*, volume 873 of *Lecture Notes in Mathematics*, pages 146–190. Springer-Verlag, 1981.
4. M. J. Beeson. *Foundations of Constructive Mathematics*. Springer-Verlag, Berlin, 1985.

5. T. Burge. *Truth and Some Referential Devices*. PhD thesis, Princeton University, 1971.

6. T. Burge. Truth and singular terms. In K. Lambert, editor, *Philosophical Applications of Free Logic*, pages 189–204. Oxford University Press, 1991.

7. A. Church. A formulation of the simple theory of types. *Journal of Symbolic Logic*, 5:56–68, 1940.

8. W. M. Farmer. A partial functions version of Church's simple theory of types. *Journal of Symbolic Logic*, 55:1269–91, 1990.

9. W. M. Farmer. A simple type theory with partial functions and subtypes. *Annals of Pure and Applied Logic*, 64:211–240, 1993.

10. W. M. Farmer. Theory interpretation in simple type theory. In J. Heering et al., editor, *Higher-Order Algebra, Logic, and Term Rewriting*, volume 816 of *Lecture Notes in Computer Science*, pages 96–123. Springer-Verlag, 1994.

11. W. M. Farmer. Reasoning about partial functions with the aid of a computer. *Erkenntnis*, 43:279–294, 1995.

12. W. M. Farmer. STMM: A set theory for mechanized mathematics. *Journal of Automated Reasoning*, 26:269–289, 2001.

13. W. M. Farmer. The seven virtues of simple type theory. SQRL Report No. 18, McMaster University, 2003.

14. W. M. Farmer. A sound and complete proof system for STTWU. Technical Report No. CAS-04-01-WF, McMaster University, 2004.

15. W. M. Farmer and J. D. Guttman. A set theory with support for partial functions. *Studia Logica*, 66:59–78, 2000.

16. W. M. Farmer, J. D. Guttman, and F. J. Thayer. IMPS: An Interactive Mathematical Proof System. *Journal of Automated Reasoning*, 11:213–248, 1993.

17. W. M. Farmer, J. D. Guttman, and F. J. Thayer Fábrega. IMPS: An updated system description. In M. McRobbie and J. Slaney, editors, *Automated Deduction—CADE-13*, volume 1104 of *Lecture Notes in Computer Science*, pages 298–302. Springer-Verlag, 1996.

18. S. Feferman. Polymorphic typed lambda-calculi in a type-free axiomatic framework. *Contemporary Mathematics*, 106:101–136, 1990.

19. M. Kerber and M. Kohlhase. A mechanization of strong Kleene logic for partial functions. In A. Bundy, editor, *Automated Deduction—CADE-12*, volume 814 of *Lecture Notes in Computer Science*, pages 371–385. Springer-Verlag, 1994.

20. L. G. Monk. PDLM: A Proof Development Language for Mathematics. Technical Report M86-37, The MITRE Corporation, Bedford, Massachusetts, 1986.

21. O. Müller and K. Slind. Treating partiality in a logic of total functions. *The Computer Journal*, 40:640–652, 1997.

22. D. L. Parnas. Predicate logic for software engineering. *IEEE Transactions on Software Engineering*, 19:856–861, 1993.

23. R. Schock. *Logics without Existence Assumptions*. Almqvist & Wiksells, Stockholm, Sweden, 1968.

24. M. Spivak. *Calculus*. W. A. Benjamin, 1967.

25. A. Stump. Subset types and partial functions. In F. Baader, editor, *Automated Deduction—CADE-19*, volume 2741 of *Lecture Notes in Computer Science*, pages 151–165. Springer-Verlag, 2003.

The CADE ATP System Competition

Geoff Sutcliffe[1] and Christian Suttner[2]

[1] Department of Computer Science, University of Miami
geoff@cs.miami.edu
[2] Cirrus Management
christian@suttner.info

1 Introduction

The CADE ATP System Competition (CASC) is an annual evaluation of fully automatic, first-order Automated Theorem Proving (ATP) systems. In addition to the primary aim of evaluating the relative capabilities of ATP systems, CASC aims to stimulate ATP research in general, to stimulate ATP research towards autonomous systems, to motivate implementation and fixing of systems, to provide an inspiring environment for personal interaction between ATP researchers, and to expose ATP systems both within and beyond the ATP community. Fulfillment of these objectives provides stimulus and insight for the development of more powerful ATP systems, leading to increased and more effective usage.

CASC-J2 was held on 6th July 2004, as part of the 2nd International Joint Conference on Automated Reasoning, in Cork, Ireland. It was the ninth competition in the CASC series. CASC-J2 was organized by Geoff Sutcliffe and Christian Suttner, and overseen by a panel consisting of Alan Bundy, Uli Furbach, and Jeff Pelletier. The competition machines were supplied by the University of Manchester. The CASC-J2 WWW site provides access to details of the competition design, competition resources, and the systems that were entered:
http://www.cs.miami.edu/~tptp/CASC/J2/

2 The Design of CASC-J2

CASC-J2 was (like all CASCs) divided into divisions according to problem and system characteristics. There were competition divisions in which systems were explicitly ranked, and a demonstration division in which systems could demonstrate their abilities without being formally ranked.

The CASC-J2 competition divisions were:

- The MIX division: Mixed CNF really-non-propositional theorems.
- The FOF division: Mixed FOF non-propositional theorems.
- The SAT division: Mixed CNF really-non-propositional non-theorems.
- The EPR division: CNF effectively propositional theorems and non-theorems.
- The UEQ division: Unit equality CNF really-non-propositional theorems.

D. Basin and M. Rusinowitch (Eds.): IJCAR 2004, LNAI 3097, pp. 490–491, 2004.
© Springer-Verlag Berlin Heidelberg 2004

The MIX, FOF, and SAT divisions each had two classes: an assurance class, ranked according to the number of problems solved (just a "yes" output), and a proof/model class, ranked according to the number of problems solved with an acceptable proof/model output. For analysis purposes, the divisions were additionally categorized according to the syntactic characteristics of the problems.

The problems for CASC-J2 were selected from the next unreleased version of the TPTP problem library, so that new problems had not previously been seen by the entrants. The problems were in a difficulty range such that it was expected that no problems would be solved by all systems and no problems would be unsolved. There was a reasonable distribution over different problem types, and the selection of problems was biased towards up to 50% new problems. A CPU time limit was imposed on each system run, and a wall clock time limit was imposed to limit very high memory usage that causes swapping.

The systems that were entered into CASC-J2 were required to be reasonably general purpose, so that their performance on TPTP problems could be expected to extend usefully to new unseen problems. The systems were required to have simple installation procedures and robust execution characteristics. Only their output to stdout was considered in their evaluation, and they were required to output distinguished strings to indicate what they had established about each given problem. The systems were all tested for soundness before the competition.

The main change in CASC-J2 since CASC-19 was the addition of the proof class in the FOF division, which includes a requirement that proofs by CNF refutation should adequately document the FOF to CNF transformation.

3 Conclusion

Experimental system evaluation is a crucial research tool, and competitions provide stimulus and insight that can lay the basis for the development of future ATP systems. A key to sustaining the value of CASC in the future is continued growth of the TPTP. Developers and users are strongly encouraged to contribute new problems to the TPTP, particularly problems from emerging commercial applications of ATP.

Author Index

Lecture Notes in Artificial Intelligence (LNAI)